Caterpillars

Caterpillars

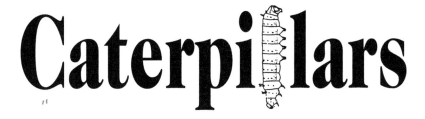

Ecological and Evolutionary Constraints on Foraging

Edited by

Nancy E. Stamp
and
Timothy M. Casey

Chapman & Hall
New York • London

First published in 1993 by
Chapman and Hall
an imprint of
Routledge, Chapman & Hall, Inc.
29 West 35th Street
New York, NY 10001-2291

Published in Great Britain by
Chapman and Hall
2-6 Boundary Row
London SE1 8HN

Printed in the United States of America on acid free paper.

Library of Congress Cataloging in Publication Data available.

British Library Cataloguing in Publication Data also available.

ISBN 0-412-02681-3 (HB)

Contributors

Pedro Barbosa
Department of Entomology
University of Maryland
College Park, MD 20742, USA

Matthew Baylis
Tsetse Research Laboratory
Department of Veterinary Sciences
Langford House, Langford
Bristol BS18 7DU, UK

Elizabeth A. Bernays
Department of Entomology
University of Arizona
Tucson, AZ 85721, USA

M. Deane Bowers
University of Colorado Museum &
Department of EPO Biology
University of Colorado
Boulder, CO 80309-0334, USA

Timothy M. Casey
Department of Entomology
Cook College
Rutgers University
New Brunswick, NJ 08903, USA

Hans Damman
Department of Biology
Carleton University
Ottawa, Ontario
K1S 5B6, Canada

David Dussourd
Department of Biology
University of Central Arkansas
Conway, AR 72032, USA

Terrence D. Fitzgerald
Department of Biology
State University of New York College at
Cortland
Cortland, NY 13045, USA

Erkki Haukioja
Laboratory of Ecological Zoology
Department of Biology
University of Turku
SF-20500 Turku 50
Finland

Bernd Heinrich
Department of Zoology
University of Vermont
Burlington, VT 05405, USA

Daniel H. Janzen
Department of Biology
University of Pennsylvania
Philadelphia, PA 19104, USA
and
Instituto Nacional de Biodiversidad
(INBio)
3100 Santo Domingo
Heredia, Costa Rica

Olga Kukal
Department of Biology
University of Victoria
P.O. Box 1700
Victoria, British Columbia
V8W 2Y2, Canada

Clytia B. Montllor
ISK Mountain View Research Center, Inc.
1195 W. Fremont Ave.
Sunnyvale, CA 94087, USA

Naomi Pierce
Museum of Comparative Zoology
Harvard University
Cambridge, MA 02138, USA

Duncan Reavey
Museum of Comparative Zoology
Harvard University
Cambridge, MA 02138, USA
and
Department of Biology
University of York
York Y01 5DD, UK

Frank Slansky, Jr.
Entomology and Nematology Department
University of Florida
Gainesville, FL 32611, USA

Nancy E. Stamp
Department of Biological Sciences
Binghamton University
State University of New York
Binghamton, NY 13902-6000, USA

Ronald M. Weseloh
Connecticut Agricultural Experiment Station
University of Connecticut
New Haven, CT 06504, USA

Richard T. Wilkens
Department of Biological Sciences
Binghamton University
State University of New York
Binghamton, NY 13902-6000, USA

Contents

II Ecological and Evolutionary Consequences: Caterpillar Life-Styles

III Environmental Variation in Time and Space

Preface

Caterpillars are an important part of every terrestrial ecosystem; there are species that feed on lichens, herbs, grasses, shrubs, and trees. In most of the earth's forests, caterpillars eat more leaves than all other herbivores combined. Some of the most harmful insect pests are caterpillars.

Caterpillars are often used in basic ecological and evolutionary studies. Caterpillars provide models for studying insect–plant interactions, predator–prey interactions, host–parasite interactions, and insect physiology. The theories of plant defense against herbivores largely reflect research on caterpillar–plant interactions because caterpillars are such excellent representative organisms for the study of feeding behavior, tactics, and constraints. Our understanding of how birds learn to forage for insects comes mainly from studies with caterpillars as prey, or with the adult stages (moths and butterflies) that may or may not be toxic, depending on what the larvae ate and how they processed their food. Furthermore, research with caterpillars has contributed greatly to our picture of how insects deal with the abiotic environment.

However, despite numerous studies worldwide, we have a limited understanding of the constraints on foraging patterns of caterpillars. Recent articles, including those of a special feature section on "Insect Host Range" in the journal Ecology (1988), express frustration in the development of an adequate overview of the ecology of foraging patterns in insects and hostplant resistance. Those authors argue for greater consideration of the multiplicity of factors that constrain insect herbivores.

A consensus seems to be developing: research on foraging strategies of insect herbivores and caterpillars in particular must take more of a multiple factor approach. To develop a more unified theory of foraging patterns, we need to take into account how two or more constraints, such as food quality and natural enemies combined or interaction effects of nutrients and allelochemicals, influence caterpillars. Although studies often do examine more than one constraint, the complexities of investigating two or more constraints *within an experiment* frequently discourage that approach. Yet recent studies show that the patterns generated when two or more factors are considered in one experiment were often not predicted based on the results from examining those factors each in isolation. Therefore, if we are to understand real-world processes, we need to investigate this kind of complexity. The excitement generated by such research is fostering more work along these lines, but we would benefit from an explicit framework. This volume is an attempt to provide such a framework by evaluating the constraints individually and in combination ecologically and evolutionarily.

We thank the reviewers of the chapters for their careful consideration and insightful critiques: M. D. Bowers, C. M. Bristow, H. Damman, S. Faeth, L. Fink, R. Karban, C. Montllor, M. Raveret Richter, D. W. Tallamy, P. Turchin, G. P. Waldbauer, R. Weseloh, R. T. Wilkens, and a number of anonymous reviewers. G. Payne at Chapman and Hall was especially helpful and patient with us. NES wishes to express her appreciation to M. D. Bowers and A. Tan-Wilson for their support and encouragement throughout this project. Finally, we also thank the contributors for their considerable efforts and the opportunity to work with them on this volume.

Introduction

Caterpillars have served appropriately and well as a model insect herbivore for study of plant–herbivore interactions. Caterpillars often provide the centerpiece for study of comparatively simple systems because of their relatively large size and limited mobility and thus ease with which they can be monitored in the field and tested in the laboratory, under a variety of conditions. Consequently, our current understanding of plant–insect herbivore interactions rests heavily on lepidopteran studies. Of 20 seminal papers contributing to the development of the theory of plant–herbivore interactions since 1950, 70% relied at least to some and often to a large extent (40% of them) on biology of Lepidoptera.

As this volume illustrates, insect herbivores deal with diluted nutrients in a matrix of allelochemicals (= the hostplant) in an environment that is almost always thermally suboptimal and occupied by predators, parasites, and other herbivores (= potential competitors). Such constraints, in some ways in concert and in other ways in opposition, presumably have shaped caterpillar foraging patterns. The result has been a dichotomous list of traits: specialist vs. generalist feeding preferences, cryptic vs. warningly colored appearance, palatable vs. unpalatable to enemies, solitary vs. gregarious behavior, and ant-mutualist vs. non-ant-mutualist. In turn, the set of traits for any particular caterpillar species reflects the impact it can have on hostplants and interactions with its enemies in both natural and managed systems.

More specifically, this volume contributes to further development of the theory of plant–insect herbivore interactions, in particular what shapes host plant range of herbivores, by focusing on the following issues. Caterpillar response to temperature ranges from thermal conformation, which is probably advantageous in avoiding enemies, to behavioral thermoregulation, which allows faster growth but increases the risk of predation (Casey, Chapter 1). Caterpillars may modify their feeding rate to regulate intake of nutrients, which may affect their exposure to allelochemicals and predators, and vice versa, plant allelochemicals and predators may influence feeding rate of caterpillars. Slansky (Chapter 2) indicates how important such modifications in feeding rate are to understanding the impact of insects and plants on each other. Because even an individual leaf is a heterogeneous resource that includes tissues high in allelochemicals, caterpillars benefit by behaviorally circumventing these plant defenses, and caterpillars that lack such behaviors are constrained in choice of host plant species (Dussourd, Chapter 3). Direct competition seems to be a relatively minor force in structuring insect herbivore communities. However, indirect, resource- or enemy-mediated interactions are common and frequently include positive, as well as negative, effects (Damman, Chapter 4). Montllor and Bernays (Chapter 5), Weseloh (Chapter 6), and Heinrich (Chapter 7) indicate how the different foraging abilities and behaviors of invertebrate predators, parasitoids, and birds could shape foraging patterns of caterpillars. Heinrich argues that in general the appearance and behavior of caterpillars reflect bird predation rather than attack by invertebrate enemies. Montllor and Bernays provide further evidence that invertebrate predators have contributed to narrowing of host plant range, and Weseloh marshalls more evidence that parasitoids have the opposite effect. Of course, the behavior of a caterpillar toward its host plant and enemies will reflect its size. Because size changes dramatically during larval development, so does response by the caterpillar to its environment (Reavey, Chapter 8). Particular sets of caterpillar traits (i.e., cryptic, aposematic, gregarious, and ant-tended life-styles) appear useful for particular sets of constraints. But there are costs as well as benefits to these life-styles. Crypsis may reduce predation at the cost of a slow growth rate, but the outcome may be similar to that of an aposematic life-style (Stamp and Wilkens, Chapter 9). Caterpillars that advertize their unpalatability may obtain their thermal optima more often and high quality food at less risk to predators, but not without some costs (Bowers, Chapter 10). Group living seems to reflect thermoregulatory and defensive advantages coupled with modified foraging behavior to accommodate living in groups (Fitzgerald, Chapter 11). Mutualism with ants constrains caterpillars to living near ants and providing them with food, but it also allows a wider use of host plant species (Baylis and Pierce, Chapter 12).

Caterpillars must contend with the seasonality of their environment, which includes seasonal and herbivory-induced changes in food quality, and alterations of wet and dry or warm and cold periods and thus availability of food (Haukioja, Chapter 13; Janzen, Chapter 14; Stamp, Chapter 15; Kukal, Chapter 16). In addition, both temperature and predator pressure change through the caterpillar season. Response to these constraints yields different foraging and life history patterns within and among latitudinal regions. These constraints on caterpillar foraging have important consequences for management of forest insect pests. Caterpillars face the same constraints in agroecosystems that they face in natural systems. But the levels of these constraints in agroecosystems differ markedly from those in natural systems and thus there are different consequences (Barbosa, Chapter 17). Together these views indicate how far we have come in developing an understanding of plant–herbivore interactions, and that abiotic and biotic factors must be considered simultaneously.

This volume indicates what the next major plant–herbivore synthesis will have to include if we hope to account for the degree of host specificity of insect herbivores. In addition to plant–herbivore interactions per se, the synthesis must address from an ecological and evolutionary perspective herbivore–predator (also –parasitoid and –pathogen) interactions, plant–predator (–parasitoid and –pathogen) interactions, and herbivore–herbivore interactions, including direct and indirect effects for each. Furthermore, these interactions must take into account abiotic conditions because the effect of food quality and enemies is modified by temperature, through physiological and behavioral responses of the herbivores.

PART I

Constraints on Foraging Patterns of Caterpillars

In the following section, abiotic and biotic factors as potential constraints on the foraging patterns of caterpillars are examined. The particular topics in the first section were chosen for their relative importance *and* lack of adequate coverage elsewhere in the literature. One major topic, host plant chemistry, is dealt with in many of the chapters but overall not as explicitly as it usually is in volumes addressing plant-herbivore interactions. Page limitations prevented us from including yet additional coverage of this important topic. But in contrast to the other topics we have included here, the effects of hostplant chemistry on insect herbivores and especially relative to caterpillars are addressed in numerous excellent reviews and in many volumes (e.g., Rosenthal and Janzen 1979; Bell and Carde 1983; Denno and McClure 1983; Futuyma 1983; Hedin 1983; Brattsten and Ahmad 1986; Miller and Miller 1986; Barbosa and Letourneau 1988; Heinrichs 1988; Spencer 1988; Isman and Roitberg 1991; Rosenthal and Berenbaum 1992; Tallamy and Raupp 1991). We refer the reader to these for further discussion of host plant chemistry as a constraint on foraging patterns.

Our objective in the first section is to discuss abiotic and biotic factors as constraints on caterpillars. Among and in some cases within caterpillar species, the continuum of feeding patterns ranges from opportunistic feeding at appropriate temperatures to tightly controlled foraging independent of temperature (Casey, Chapter 1). These responses to thermal conditions have consequences for larval acquisition of food and evasion of enemies. Leaves are extremely heterogeneous as an environment and as food for caterpillars. Caterpillars may or may not circumvent local plant defenses and thus experience different consequences of such plant defenses (Dussourd, Chapter 3). Food quality affects larval growth rate and final size. In general, growing faster is advantageous because it reduces exposure to mortality factors. Bigger larvae pupate at a larger size and consequently adults are more fecund than smaller individuals. If food quality is poor, caterpillars may compensate by adjusting their feeding rate in a way that lessens variation in growth performance (Slansky, Chapter 2). Although competition is less common in herbivore communities than among organisms at other trophic levels, indirect resource- or enemy-mediated interactions are important in herbivore communities for their positive as well as negative effects (Damman, Chapter 4). As visually oriented and quick-learning predators, insectivorous birds very likely are a potent force in shaping foraging patterns of caterpillars (Heinrich, Chapter 7). Predatory wasps may act as a constraint on foraging patterns of caterpillars in ways similar to that of insectivorous birds, and other invertebrate predators, even of quite different size and behavior, may also be strong selective agents on caterpillars (Montllor and Bernays, Chapter 5). Invertebrate predators may have contributed to narrowing of host plant range by insect herbivores. Parasitoids are more specialized on microhabitat and/or host species than predators. Selective pressure by parasitoids presumably accounts for some broadening of microhabitat and host plant ranges by caterpillar species (Weseloh, Chapter 6). The effects of abiotic conditions on a caterpillar, how a caterpillar uses a host plant, and how it evades enemies are determined in part by its size, which changes dramatically over its development (Reavey, Chapter 8).

For a caterpillar, meeting the challenge of one constraint (ecologically or evolutionarily) may facilitate dealing with another constraint yet complicate or even preclude successfully countering another constraint. For instance, acquiring high quality food, such as new foliage, during the daytime may place a caterpillar in direct sunlight because new leaves are often at stem tips *and* therefore place the caterpillar in a more favorable thermal environment, but as a consequence the caterpillar may be quite conspicuous to predators. Addressing separately the potential constraints on caterpillar foraging should facilitate identifying and testing the possible ramifications for various sets of these constraints (e.g., particular circumstances for a caterpillar species).

Literature Cited

Barbosa, P., and Letourneau, D. (eds.) 1988. Novel Aspects of Insect-Plant Interactions. Wiley, New York.

Bell, W. J., and Carde, R. T. (eds.) 1983. Chemical Ecology of Insects. Sinauer, Sunderland, MA.

Brattsten, L. B., and Ahmad, S. (eds.) 1986. Molecular Aspects of Insect-Plant Associations. Plenum, New York.

Denno, R. F., and McClure, M. S. (eds.) 1983. Variable plants and herbivores in natural and managed systems. Academic, New York.

Futuyma, D. J. 1983. Evolutionary interactions among herbivorous insects and plants, pp. 207–231. In D. J. Futuyma and M. Slatkin (eds.), Coevolution. Sinauer, Sunderland, MA.

Hedin, P. A. (ed.) 1983. Plant Resistance to Insects. ACS Symposium Series 208. American Chemical Society, Washington, D.C.

Heinrichs, E. A. (ed.) 1988. Plant Stress—Insect Interactions. Wiley, New York.

Isman, M. B., and Roitberg, B. (eds.) 1991. Chemical Ecology of Insects: An Evolutionary Approach. Chapman and Hall, New York.

Miller, T. A., and Miller, J. (eds.) 1986. Insect-Plant Interactions. Springer-Verlag, New York.

Rosenthal, G. A. and Berenbaum, M. R. (eds.) 1992. Herbivores: Their Interactions with Secondary Plant Metabolites. Academic Press, New York.

Rosenthal, G. A., and Janzen, D. H. (eds.) 1979. Herbivores: their Interaction with Secondary Plant Metabolites. Academic Press, New York.

Spencer, K. C. (ed.) 1988. Chemical Mediation of Coevolution. Academic Press, San Diego.

Tallamy, D. W., and Raupp, M. J. (eds.) 1991. Phytochemical Induction by Herbivores. Wiley, New York.

1

Effects of Temperature on Foraging of Caterpillars

Timothy M. Casey

Introduction

Temperature has a pervasive effect on physiological processes and therefore on almost all aspects of organismal performance. Caterpillars are ectothermic; that is they produce insufficient metabolic heat to elevate their body temperature and must rely on external heat sources and sinks if they are to control their body temperature. If they do not thermoregulate, their body temperature tracks the surrounding temperature. There are several ways to adapt to this situation. Although physiological adaptations can prevent the animal from being totally at the mercy of the environment, there are energetic consequences to such a strategy (Hochachka and Somero 1973; Heinrich 1977).

In this chapter I examine physiological responses of caterpillars to temperature and where possible their relation to foraging. I also examine behavioral response to temperature associated with caterpillars' physiological responses, thereby attempting an ecological as well as a physiological assessment. The behavioral responses of caterpillars encompass a broad continuum from more or less continuous thermoregulators to complete thermoconformers. The thermal environment should differ (often dramatically) for species along this continuum. I examine the question of whether there is a corresponding range of physiological responses favorable to different species that exhibit different thermal behaviors.

Physiological Responses to Temperature

It is well known that growth of caterpillars is strongly temperature dependent (Scriber and Slansky 1981). However, the degree of temperature dependence varies from species to species and the optimal temperature for growth is also variable (Taylor 1981). Before examining the response of particular species, it is perhaps useful to present an overview of an ectotherm's physiological response to temperature. As discussed by Weiser (1973), our understanding of the physiological response of ectotherms has changed as we gain further insight into the processes involved. The classical view of metabolic processes of ecotherms is that physiological rate functions have a Q_{10} of about 2 over the normal temperature range. Initially, ectotherms were considered to be at the mercy of their environment, being stuck on a physiological response curve on which rate functions doubled for every 10°C (equivalent to the effects of temperature on biochemical reaction rates—the van't Hoff effect). This opinion, characterized as the "Krogh normal curve" (Krogh, 1914; Weiser, 1973), was supplanted in the 1950s by H. Precht and C. L. Prosser, who examined the ability of ectotherms to develop temperature independent regions on the temperature–performance curve. These "plateaus" of thermal independence on a relationship that was otherwise strongly temperature dependent (i.e., Krogh normal) suggested a homeostatic response comparable to that seen in endotherms for controlling body temperature. Moreover, and more importantly, these plateaus routinely occurred in the region of body temperature (and in most cases, air temperature) that the animal experienced in the field. The fact that such plateaus occur in a variety of animals, including intertidal invertebrates, a variety of insects (in several stages of development), and in several vertebrate taxa (Cossins and Bowler 1987), suggests that this is a very general adaptation by ectotherms to environmental temperature.

More recently, the thermal response of these animals has been interpreted as an adaptation to a specific series of temperatures experienced under normal field conditions during the activity phase with different regions of the curve serving different functions. That is, one part of the curve (the plateau) provides relatively temperature independent response over a range of temperatures actually experienced in the field. The other half of the curve at low temperatures (high Q_{10}s) could function to reduce maintenance costs for an animal at night while it is inactive. Thus, thermal response curves may indicate a multistable system rather than a homeostatic one (Weiser 1973).

The metabolic response to temperature of three caterpillar species is shown in Figure 1.1. The response of *Manduca sexta* shows a typical metabolic thermal compensation with the plateau (thermal insensitivity with Q_{10} just over 1 from T_as of 25–35) occurring over the normal range of daytime temperatures experienced by the caterpillars in the field (Casey 1977). At lower (15–25°C) and higher temperatures (35–40°C) oxygen consumption has a Q_{10} of about 2.0. The gypsy moth *Lymantria dispar* exhibits a similar pattern (Fig. 1.1). In contrast, the oxygen consumption of the tent caterpillar, *Malacosoma americanum,* shows a continuous increase (Q_{10} of about 2) in response to temperature (Casey and Knapp 1987).

With such variability in physiological responses to temperature it is not surprising that performance of caterpillars at fluctuating temperatures is different from that of caterpillars at constant temperature having equivalent heat units (Taylor

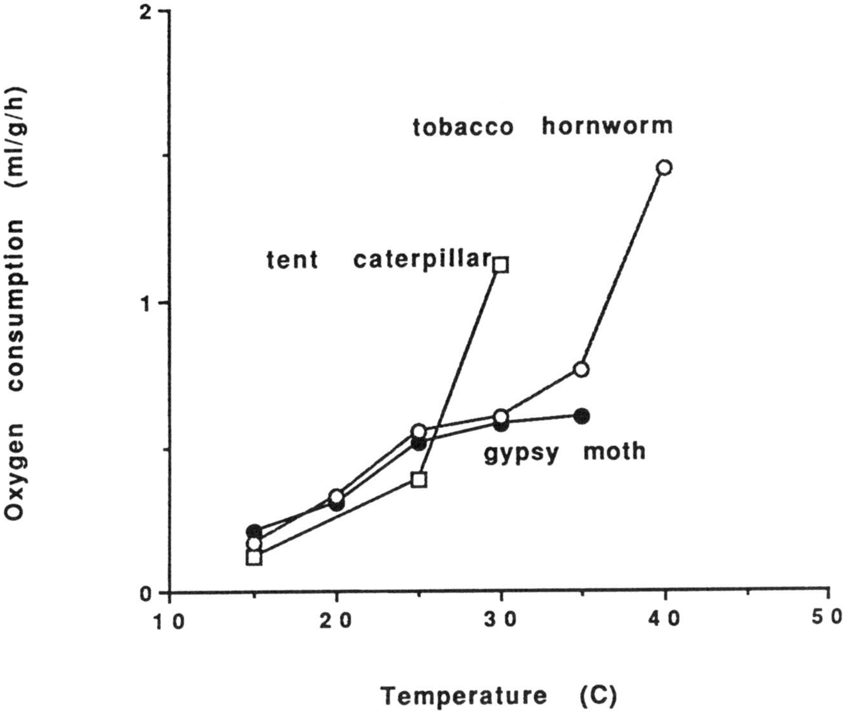

Figure 1.1 Metabolic responses to temperature for three species of caterpillars. Data for *M. sexta* from Casey (1977), data for *M. americanum* and *L. dispar* from Casey and Knapp (1987).

and Shields 1990). However, the ways in which a given species will respond to a given thermal regime should reflect its physiological responses to temperature. It is difficult to generalize the response of different species of caterpillars to such conditions unless equivalent data are available for constant temperature (such as those given by Stamp, Chapter 15).

When discussing the thermal optimum temperature care must be exercised because not all physiological processes have the same optimum temperature. For example, in *Manduca sexta,* biting rates increase continuously with temperature, unlike the change shown in energy metabolism (Fig. 1.1), heart rate, or cuticular water loss (Casey 1977).

In general, a physiological rate function doubles for every 10°C increase in environmental temperature (Hochachka and Somero 1973). However, the physiological response curve of an animal is characteristic, with a high thermal sensitivity (large changes in performance for small changes in temperature) at the lower range of temperatures and leveling off at the upper region (the optimal temperature) followed by a rather sharp decline at still higher temperatures.

Selection can operate on the position of the curve. By shifting the position of the curve, left or right, the optimal temperature for performance can shift to lower or higher temperatures respectively. In addition, selection can operate on the shape of the curve. By shifting the shape of the curve the thermal sensitivity of the performance can be modified (Hochachka and Somero 1973).

Thermoconformation

Although thermoregulation is usually assumed to be adaptive, many ectotherms are forced to allow their body temperatures (T_b) to follow the surrounding temperatures passively because the costs associated with thermoregulation (e.g., exposure to predators/parasites, etc.) often outweigh the benefits involved (Huey and Slatkin 1976; Heinrich 1979; Schultz 1983). Most caterpillars are slow moving prey that appear to be incapable of outrunning predators and parasitoids once they are detected. Consequently, they seem to rely on an elaborate game of hide and seek strategies (Heinrich and Collins 1983). Foremost of these is cryptically blending into background, nocturnal feeding bouts, and seeking sheltered locations away from feeding sites. Behaviors that enhance seclusion are not compatible with behavioral thermoregulation, which requires the animals to be in view of potential predators to benefit from radiant heating (Heinrich 1979; Knapp and Casey 1986; Stamp and Bowers 1988). Consequently, thermal conformation is a common pattern among caterpillars.

Classifying caterpillars according to their thermal strategies provides a useful point of departure but it may obscure the considerable range of adaptations within each of the two thermal categories. By placing a species in a box ("thermoconformer" or "thermoregulator") we may be inadvertently ascribing attributes to that species that are inappropriate. What is probable, given the selective pressures from the plant side (Rhoades 1985), and the predator/parasitoid side (Heinrich 1979; Schultz 1983), is that thermoconformers are very often "locked in" to their behavioral strategy. Nevertheless, the degree to which a thermoconforming species is sluggish or opportunistic in its feeding strategy probably varies widely. Thermoconformers may display a range of adaptive physiological options, which is potentially as large a continuum as that between the thermoconformers and the thermoregulators outlined below.

Adaptations of Caterpillars Facilitating Thermoregulation

Because of their small size and high surface-to-volume ratios, caterpillars are usually only a few degrees warmer than the surrounding air temperatures. Heat transfer between the caterpillar and its environment occurs primarily by radiation and convection (see Casey 1981 for further discussion). Control of body temperature by physiological means has not been demonstrated and seems unlikely due to the homogeneous body composition, continuous body wall, lack of compartmentalization between tagma, and lack of variable insulation or control of blood flow to different body regions. Physiological control of heat loss by evaporative cooling has not been documented in caterpillars but in view of the demonstrated effectiveness of evaporative cooling in sawfly larvae (Seymour 1974), it seems reasonable to postulate that the capacity for such physiological control exists in caterpillars.

Behavior

Caterpillars must rely on their behavior to elevate body temperatures above the surrounding temperature. Orientation and postural adjustments allow caterpillars to maximize the body surface area exposed to solar radiation. Basking behavior has often been reported in caterpillars (Sherman and Watt 1973; Casey 1976; Capinera et al. 1980; Rawlins and Lederhouse 1981; Porter 1982; Knapp and Casey 1986; Fields and McNeil 1988; Stamp and Bowers 1990a). Microhabitat selection can also significantly increase body temperature. *Hyles lineata* (Sphingidae) caterpillars bask on the desert floor during cold periods enhancing heat gain by conduction and reradiation from the warm substrate (Casey 1976). Since the body is in the boundary layer of the ground, heat loss by forced convection is reduced. Body temperatures of these caterpillars are several degrees warmer than those in vegetation. As temperatures warm up during the day, negative orientation (long axis of the body oriented parallel to the sun), movement off the ground to high in the vegetation, and shade seeking occur. These behaviors minimize radiant heat uptake and help stabilize the body temperature (Casey 1976; see Capinera et al. 1980 for another example).

Setae

Variation in morphology also affects the body temperatures of caterpillars. A variety of caterpillars are "hairy." Setae provide several useful functions for caterpillars whether they are thermoconformers or thermoregulators. The distribution of the setae on many caterpillars is such that they increase the effective cross-sectional area of the caterpillar. The setae provide selective insulation and reduce convective heat exchange without affecting radiative heat gain, resulting in about 2°C increase in body temperature compared with caterpillars of equal size without setae. This effect can be abolished by orientation and postural adjustment (Casey and Hegel 1981).

Qualitatively similar patterns of setae enhancing body temperatures have been reported for other caterpillars (Kevan et al. 1982; Fields and McNeil 1988; Kukal et al. 1988) where there is clearly adaptive value to the insulation. Shaved arctic wooly bear caterpillars (*Gynaephora rossii*) maintained a body temperature excess of only 6.9°C compared to 10°C for intact animals (Kevan et al. 1982). *Gynae-*

phora groenlandica caterpillars maintain body temperature excesses of up to 20°C, suggesting that the setae may be even more effective in this species.

Setae have other functions for caterpillars as well. While they serve a thermoregulatory function in some caterpillars, they occur on a wide variety of species which are not known to display thermoregulatory behavior (Heinrich and Collins 1983). It is likely that they developed as a deterrence to predators. Nevertheless, their presence on thermoregulating species enhances temperature excess.

Coloration

The surface color of caterpillars may have important consequences for their thermal balance. Since about 50% of the energy in solar radiation is in the visible region of the spectrum, different surface color could potentially affect the quantity of radiant energy absorbed. Darker coloration enhances radiant heating. It is significant that many basking caterpillars are darkly colored (Casey 1976; Porter 1982; Knapp and Casey 1986; Fields and McNeil 1988; Stamp and Bowers 1990a). However, as in the case of setae, coloration is not necessarily a thermoregulatory characteristic. Cryptically colored animals can bask (Sherman and Watt 1973) and a wide variety of dark caterpillars adapt "stealthy" strategies.

Coloration may be changed in response to changing environmental conditions. In *Ctenucha virginica* (Arctiidae) the color of the setae change markedly on a seasonal basis. During the spring and autumn larvae are black and yellow, while in the summer they are predominantly yellow. The black and yellow caterpillars achieved significantly higher body temperatures than the yellow ones. Thus color enhanced T_b during cool periods and reduced the temperature excess during the warmer season (Fields and McNeil 1988).

Aggregation

Basking caterpillars often form aggregations (Wellington 1950; Porter 1982; Knapp and Casey 1986; Stamp and Bowers 1990a) allowing them to maximize radiant uptake while reducing convective heat exchange by reducing the relative body surface exposed. The equilibrium temperature achieved by a basking caterpillar under given microclimatic conditions is a function of its size (Stevenson 1985). Aggregation increases the effective body size and the equilibrium temperature that can be achieved.

Tent Making

Eastern tent caterpillars *Malacosoma americanum* are highly gregarious, and during larval development they spin a tent that provides protection from predators and parasitoids during inactive periods (Fitzgerald 1980). Emerging from their egg cases in early April, they are among the earliest of spring caterpillars. Although such early emergence is advantageous because many predators and parasitoids will not be abundant until later in the season, the mean air temperature during April is close to the developmental threshold for growth (Knapp and Casey 1986). Despite their small size, even second instar larvae, weighing only tens of milligrams, are capable of maintaining body temperature in excess of 20°C above air temperature (Knapp and Casey 1986). By mid-May the increased size of the caterpillars, warmer air temperatures, and thermal properties of the tent put them in serious danger of overheating.

Joos et al. (1988) examined heat exchange to evaluate the thermal significance of aggregation behavior and the tent under field conditions. In this study we used operative temperature (T_e) models (Bakken and Gates 1975) of the tent caterpillar, consisting of freeze-dried specimens with thermocouples implanted in the body cavity. To evaluate the role of aggregation on body temperature, we made "clumps" of individual caterpillars (Fig. 1.2B). To examine the temperature distribution throughout the tent, we forced a colony of caterpillars to build their tent on a wooden frame with thermocouples placed on each of the six arms at precise intervals. Placing the entire apparatus out into the field, we put operative temperature models at various locations on the tent and on a small branch as a control. We also measured solar radiation and wind velocity during the experiments.

As a consequence of their behavior and the architecture of the tent, tent caterpillars have a wide range of operative temperatures at various locations on the tent (Fig. 1.2A). A T_e model placed on a branch in full sunshine nearby the tent achieved a temperature only a few degrees C above air temperature while a similar model placed on the tent surface in full sunshine can be as much as 30°C above air temperature (Fig. 1.2A). Clumping together further increases body temperatures. Compared with individual models, clumps of 5–6 caterpillars exhibited T_es 6 to 7°C greater. The elevated temperatures are due to several factors. The tent provides a large boundary layer that should reduce convective heat loss. Aggregation further decreases the surface-to-volume ratio and convection, while still allowing radiant heat uptake. The tent itself warms several degrees above air temperature, further enhancing the T_b of the caterpillars. In addition to allowing the caterpillars to heat up far above the levels they could achieve without the tent, the tent provides a range of temperatures that the caterpillars can utilize by microhabitat selection. During the middle of the day the body temperature of caterpillars can differ by as much as 20°C, depending on their location on or in the tent (Fig. 1.2A). This highly heterogeneous environment allows the caterpillars to regulate their T_b quite precisely (Fig. 1.2C). It is interesting to note that the caterpillars feed at routine intervals, unaffected by the microclimate. Their synchronized behavior ensures that a large tent is constructed due to communal spinning of silk. The caterpillars routinely feed before dawn when environmental temperatures are lowest, and at dusk. Behavioral thermoregulation is not apparent during feeding so it is not associated with the rate of feeding as such. The behavior maximizes digestion, thereby strongly affecting growth rates and insuring that

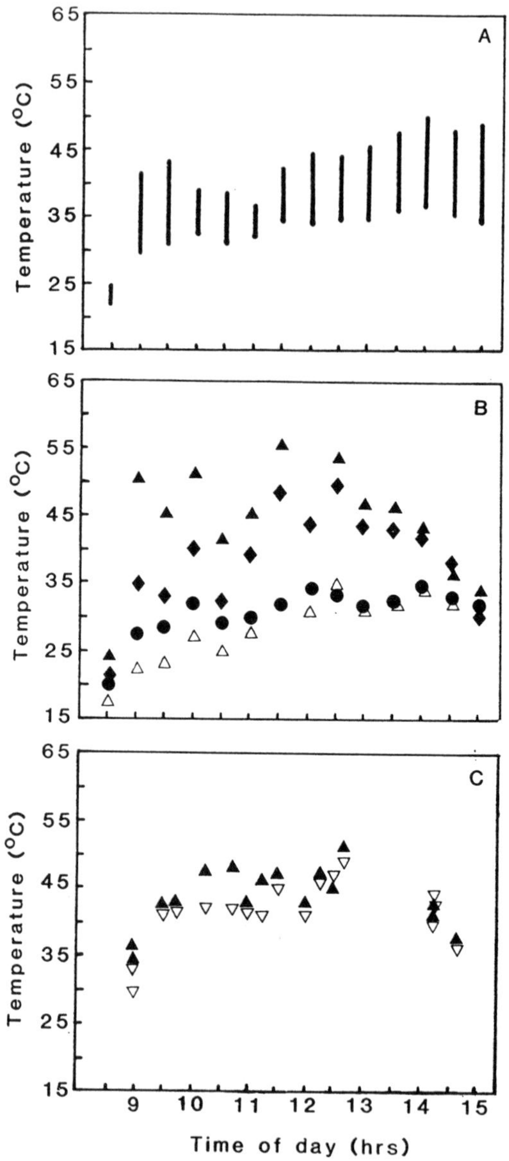

Figure 1.2. Thermal relations of Eastern tent caterpillars in the field. (A) The range of air temperatures in different locations inside the tent. (B) Operative temperatures (T_e) of models placed in various locations [closed circles = underside of tent (in shade); diamonds = single model on dorsal surface in sunlight; closed triangles = "clump model" on the same dorsal surface consisting of five freeze-dried caterpillars attached to resemble a natural caterpillar aggregation; open triangles = a single caterpillar model placed on a branch away from the Tent.] (C) Body temperatures of living caterpillars under the same environmental conditions utilizing the thermal heterogeneity of the tent microclimate (after Joos *et al.* 1988).

the caterpillars have an empty gut when they leave the tent to feed (Casey et al. 1988).

Precision of Thermoregulation

In contrast to the tent caterpillar and several other species mentioned above, many caterpillars are capable of producing only modest increases in their body temperature. As Rawlins and Lederhouse (1981) point out in their study, Monarch caterpillars elevate their body temperature by behavior but they do not regulate it. Monarch butterfly larvae increase their T_b by 3–8°C by basking and this behavior significantly reduces development time particularly at low ambient temperatures (Rawlins and Lederhouse 1981). Thermoregulation might be more obvious if the species were examined over longer periods of time or in different habitat where the range of thermal conditions varied. For example, *Colias* caterpillars are so small that even with behavioral thermoregulation they are at best a few degrees above the air temperature (Sherman and Watt 1973). However, the range of air temperatures that they experience often exceeds their thermal optimum causing them to seek shade and curtail feeding. In many studies reported field conditions were sufficiently cool that the caterpillars were simply attempting to maximize T_b.

The advantage for basking to produce elevated T_b is that even with small increments of temperature excess there are large reductions in development time (Sherman and Watt 1973; Rawlins and Lederhouse 1981; Scriber and Lederhouse 1983). Thus, small changes in body temperature can have important consequences on larval survival and therefore on individual fitness. This pattern suggests that "hotter is better." But this interpretation is clearly unacceptable as a general rule for caterpillars because only perhaps a dozen species clearly thermoregulate (Stamp and Bowers 1990a).

Foraging Patterns in Relation to Temperature

As with thermal behavior, it is difficult to generalize the effects of temperature on the foraging behavior of caterpillars because it is intimately related to the natural history of the species, which varies enormously across the taxa. For example, in the tent caterpillar, foraging schedules are tightly regimented, highly synchronized, and clearly temperature independent. The caterpillars forage during the coldest time of the day (just before sunrise) when thermoregulation is impossible and during the warmest part of the day (Fitzgerald et al. 1988). Consequently, these caterpillars may forage at body temperatures as low as 5°C and as high as 35°C (Joos et al. 1988) *on the same day*. The elaborate social development of this species (Fitzgerald, Chapter 11) is undoubtedly responsible for this rather unusual foraging.

Activity patterns of caterpillars are entrained by photoperiod (Beck 1968) but temperature can modify this. There are many examples of temperature having a direct effect on the foraging of caterpillars. In the Mojave Desert, nighttime temperatures during April and May are so cool (4°C or less) that all activity of *Hyles lineata* is curtailed (Casey 1976). Similarly, at high ambient temperatures this species adopts heat avoidance postures, seeks cool microhabitats, and curtails feeding. This behavioral response is clearly opportunistic since on cool days during the same period the caterpillars forage continuously. A similar pattern of foraging occurs in *Manduca sexta* (Sphingidae). However, this species occurs in the Mojave desert much later in the summer (July–September) when the environment is much hotter and drier. These caterpillars feed continuously during the nighttime period which is considerably warmer than that encountered by *H. lineata* in early May. In contrast, during the daytime, ambient temperatures are much higher and feeding by *M. sexta* is routinely curtailed during midday. These two sphingids adopt different thermoregulatory behaviors with *H. lineata* exhibiting obvious thermoregulatory behavior and *M. sexta* operating as a thermal conformer typical to its cryptic coloration pattern of predator avoidance. In the laboratory at moderate air temperatures, feeding by *M. sexta* larvae is essentially continuous throughout the 24-hour cycle (Casey 1976).

Ctenucha caterpillars feeding behavior is strongly correlated with temperature. In the field these caterpillars exhibit two feeding bouts per day on warm days corresponding to morning and afternoon. They ceased feeding activity at body temperatures of 25°C or more (Fields and McNeil 1988). To confirm that temperature was driving the feeding, these authors designed an ingenious laboratory experiment. They produced a double temperature cycle over a 24-hour period. In response to this artificial cycle the caterpillars had four feeding periods. If caterpillars were placed at 21°C constant temperature their feeding was essentially continuous throughout the entire photophase (Fields and McNeil 1988).

In addition to temperature, predator pressure may shape activity patterns (Young 1972), causing feeding at suboptimal times and temperatures. The feeding behavior of late instar gypsy moths in uncrowded populations appears to be predator driven (Campbell 1981). Late instars feed in the evening and in the early morning before sunrise. Activity is curtailed throughout the day and the larvae spend the day resting at the base of trees and in the leaf litter (Knapp and Casey 1986).

Relations between Physiological Processes and Behavior

The advantage to being a thermoconformer or a thermoregulator may be based on whether or not the costs (i.e., exposure to predators and or parasitoids) outweigh the benefits (increased rates of food intake and growth, etc.) of thermoregulation. Implicit in these arguments is that, other things being equal, ectotherms are better off at a high, constant body temperature than a variable one. However, thermoregulation is only one way to maintain the biochemical machinery at a relatively constant preferred level of operation.

If a caterpillar is basking, one might predict that having a high thermal sensitivity for growth in the region of body temperatures achieved is advantageous. For example, if body temperature excess is 10°C above T_a, with a Q_{10} of 2 the growth rate at T_b would be twice that if $T_b = T_a$. If, however, the growth rate had a Q_{10} of 1, there would be no advantage to thermoregulation in that temperature range.

Thus, low thermal sensitivity and broad thermal optima at low (ambient) temperatures should favor the thermal conformer, while the reverse would be favorable to a thermoregulator.

Huey and Kingsolver (1989) recently presented an analysis of the evolution of thermal sensitivity of ectotherm performance. They describe two different hypotheses to explain variation in an ectotherm's physiological response to temperature. The first hypothesis is that "hotter is better" (Fig. 1.3a). In this situation the performance curve is shifted to the right and the thermal optimum is greater at the higher temperature. An alternative hypothesis suggests "a jack of all temperatures is a master of none" (Fig. 1.3b). In this scheme an animal must sacrifice catalytic power in exchange for a broader (less thermally sensitive) optimum. The biochemical underpinnings of caterpillars that are responsible for these differences have not been examined, but some reasonable explanations are possible given the large database available for enzyme response in ectotherms (Heinrich 1977, 1981).

Assuming that the thermoconformer represents the primitive condition, we might expect that the optimum temperature for growth is in the region of body temperature (i.e., air temperature) normally encountered. One way of enhancing physiological rates at a particular temperature is to produce more enzyme. By packing the cell with more of the same enzyme, the maximum rate of the function can be increased at the same optimum temperature (Fig. 1.4). This is the "quantitative strategy" (Hochachka and Somero 1973). It would allow an animal to increase its physiological maximum without affecting the thermal characteristics of the performance curve (Fig. 1.4).

Alternatively, temperature independent rates could be developed by producing isozymes that function at different temperatures (Hochachka and Somero 1973). This must come at a cost to the animal in overall catalytic power because cells can hold only so much enzyme. At one temperature with one isozyme operating at near maximal activity the other is barely functional (Fig. 1.4). Although this may not be an acceptable tradeoff for maximal activity in such activities as flight, which requires relatively narrow thermal optima at high temperatures, it may be acceptable for thermal conformers that generally are not highly active. Moreover, it is the resting level of metabolism that is exhibiting thermal compensation. Continuous selection by predators and parasitoids for seclusion could favor such a strategy.

The performance of thermoregulators suggests very different enzyme packaging. Adapting enzymes to a given thermal regime involves a compromise between catalytic potential and structural stability. If an enzyme functions effectively at low temperature it must be held together by a few weak bonds to retain flexibility. However, such a structural arrangement will be significantly impaired at higher body temperatures. If an enzyme is adapted to high temperatures, it must be held together by numerous bonds making it relatively inflexible and poorly functional at low temperatures (Heinrich 1981). Thus, "is hotter better"? Yes, if the animal

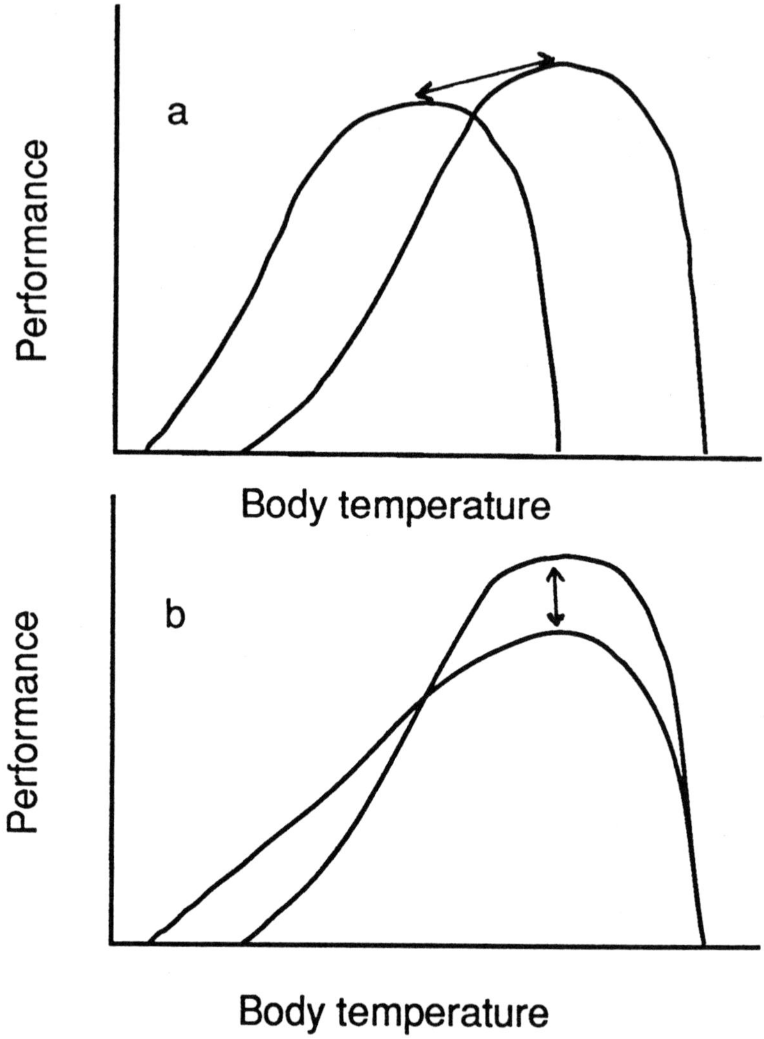

Figure 1.3. Models for evolution of performance curves (modified from Huey and Kingsolver 1989). (a) The "hotter is better" model. (b) The "jack of all temperatures is master of none" model. In the former, maximum performance increases as the optimum temperature increases. In the latter, development of lower thermal sensitivity results in a reduction in the maximum performance.

can achieve the optimal body temperature through behavior. Heinrich (1977) suggests that to maximize activity, enzymes should be adapted to the highest body temperature experienced by the animal. As long as a thermoconformer is operating at relatively low temperatures and cannot control its body temperature there may be some selection pressure for thermal independence and thermal

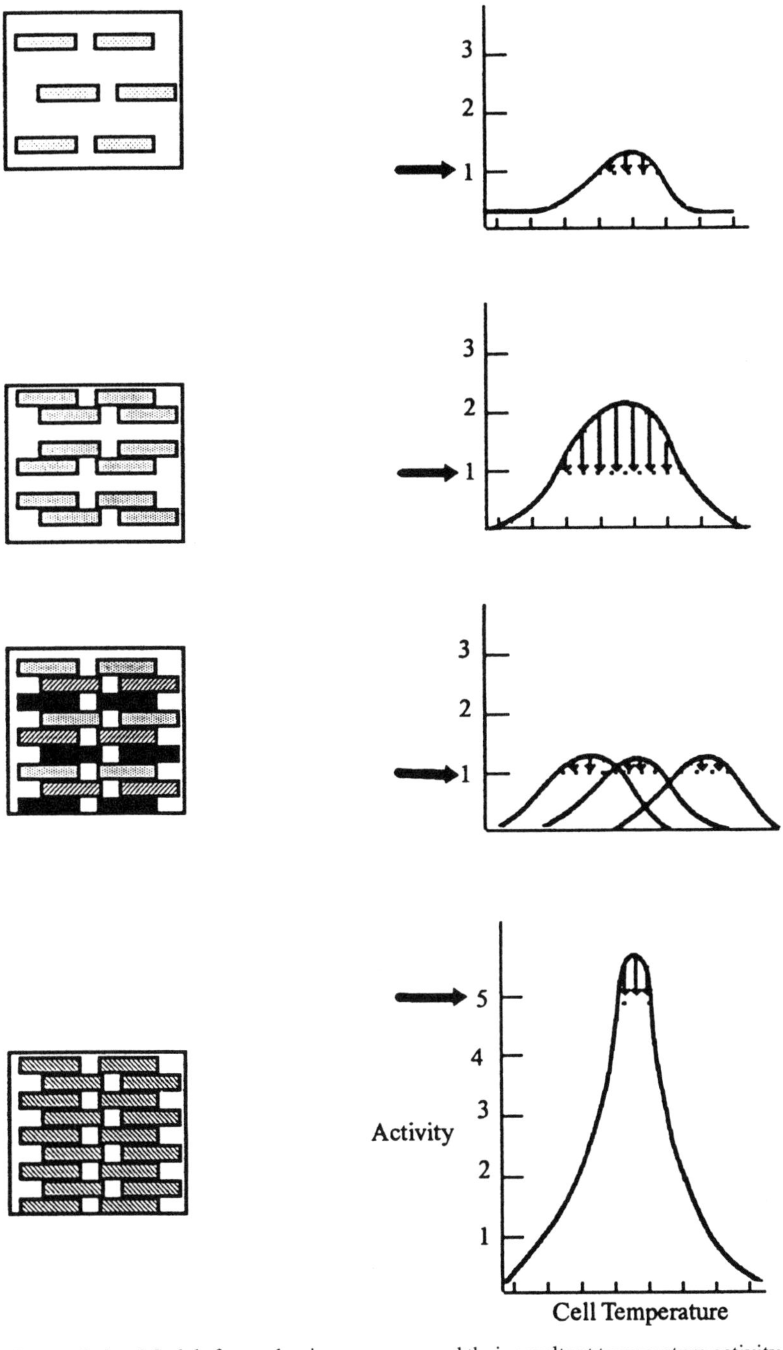

Figure 1.4. Models for packaging enzymes and their resultant temperature activity curves (adapted from Heinrich 1981). As different amounts of types (indicated by different colors) of enzymes are added to the cell, the magnitude of the performance and the shape of the performance curve can be altered.

optima in the range of air temperatures (equals body temperatures) routinely experienced by the caterpillars. If for some reason there is selection on caterpillars to thermoregulate (see below), they will experience considerably higher body temperatures. If selection for enzyme function is based on experienced body temperatures, a shift in enzyme structure and function could occur, enhancing catalytic power, but at the expense of function at low body temperatures.

To what extent are the physiological responses of caterpillars to temperature related to their ecology and behavior patterns? In this section I present case studies of thermoregulators and thermoconformers to see if any generalizations are possible.

Thermoregulators

Eastern Tent Caterpillar Malacosoma americanum (Lasiocampidae)

As discussed above, tent caterpillars exhibit external traits suitable for thermoregulation (dark coloration, setae). They are highly conspicuous in the field, and their elaborate behavioral thermoregulation is enhanced by aggregation and altered microclimate associated with the tent. They maintain high body temperatures when exposed to solar radiation (up to 40°C) and as much as 30°C above T_a. They occur in the habitat in early April when daytime temperatures are cool and nighttime temperatures are very cold (4°C). Their foraging is independent of temperature occurring at low and high T_a (Fitzgerald et al. 1988). These caterpillars exhibit high Q_{10}s for metabolism over the range of environmental temperatures normally encountered and high Q_{10}s for growth over the same T_a range. Optimum temperature for growth is significantly higher than the environmental temperatures they experience (Casey and Knapp 1987). There is presently no indication whether the thermal optimum temperature range is broad or narrow.

Arctic Wooly Bear Caterpillars Gynaephora groenlandica (Lymantriidae)

Like the tent caterpillars, wooly bears exhibit dark coloration and dense, long setae. They occur on the Arctic tundra where low summer environmental temperatures (4–10°C) are routinely encountered. Behavioral thermoregulation appears to be essential for growth (Kukal et al. 1988). They have a high Q_{10} of energy metabolism over the range of environmental temperatures experienced. Growth rates are similar at 15 and 30°C, suggesting an optimum temperature somewhere in between and well above the ambient temperatures they experience (Kukal, Chapter 16).

Thermoconformers

Gypsy Moth Lymantria dispar (Lymantriidae)

Late instar caterpillars hide during daylight hours at the base of tree trunks and in leaf litter. They occur in New Jersey in May–June, when air temperatures are relatively warm (25°C) during the day, and moderate at night (15°C). The caterpillars avoid exposure to solar radiation and exhibit no obvious thermoregulation. Feeding behavior of the later instars is only during the nighttime period, presumably in response to predation pressure (Campbell 1981). Metabolic thermal independence is apparent for field collected animals over a range of body temperatures normally experienced in the field during the day. Growth rates are maximal and thermally independent over a similar range of body temperatures. Optimum temperature for growth occurs over a range of normally experienced air temperatures (Casey and Knapp 1987).

Tobacco Hornworm Manduca sexta (Sphingidae)

This case is interesting because there are a lot of data for this species, much of it being contradictory. These caterpillars routinely experience T_as of 35°C or more in the desert population I studied. But they are also prevalent over a much wider range of cooler habitats. Their behavior in the field is that of a thermoconformer (Casey 1976). They feed from the underside of leaves, usually avoid solar radiation and are cryptically colored. Activity patterns are closely coupled to temperature over a broad range. Biting rate (i.e., rate of food intake) is strongly temperature dependent, showing no reduction in thermal sensitivity (even in the temperature range where feeding has voluntarily been curtailed during the day 29–35°C). Feeding behavior is opportunistic over a wide range of temperatures. Metabolic rate shows little thermal dependence (25–35°C) in either lab reared (Reynolds and Nottingham 1985) or field collected animals (Casey 1976). There is no indication of acclimation of metabolic rate. Growth rate is temperature independent and maximal at T_bs of 25 to 35°C (Reynolds and Nottingham 1985). In lab-reared populations there is a correlation between thermal independence of energy metabolism and growth (Reynolds and Nottingham 1985). In contrast, growth is more thermally sensitive in desert collected animals with an optimum temperature of about 30°C (Casey 1976).

Data presented in this section suggest that there may be a correlation between whether or not a caterpillar thermoregulates and the shape and position of the performance curve. Thermoregulators experience thermal optima above the air temperatures they normally encounter while thermal conformers have lower optima, corresponding to ambient temperatures which they routinely encounter in the field. There is some indication that there is a broader thermal optima in the thermoconformers associated with temperature compensated metabolism but more data are needed to evaluate this idea. The comparison of the gypsy moth and the tent caterpillar (Fig. 1.5) could represent the models, "jack of all temperatures is master of none" and "hotter is better" as presented by Huey and Kingsolver (1989). Unfortunately the study by Knapp and Casey (1986) was not conducted at high enough temperatures to completely flesh out the performance curves of these species. Note though that the tent caterpillar's thermal sensitivity is high at

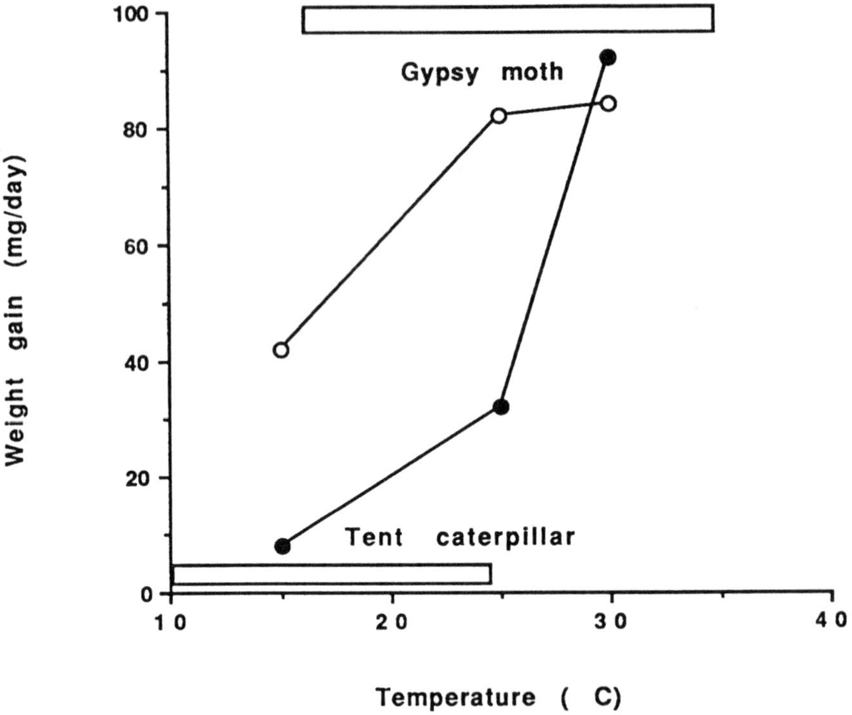

Figure 1.5. Comparison of energy metabolism and growth of gypsy moth caterpillars (thermoconformers) and eastern tent caterpillars (thermoregulators). Ambient temperature ranges experienced by gypsy moth are shown at the top of the graph. Those for tent caterpillars are shown at the bottom of the graph (data from Knapp and Casey 1986). Note that the thermal conformer grows more rapidly at low temperature and exhibits lower thermal sensitivity in the range of air temperatures that is routinely encounters in the field.

the daytime temperatures experienced and that the reverse is true in the case of the gypsy moth. The high thermal sensitivity at low temperatures in the case of the tent caterpillar may be a consequence of enzymes being adapted to high temperature (see above).

Plasticity of Thermal Response within Species

There are several spectacular cases where the foraging patterns of species change markedly in response to changing environmental conditions. Although these behavioral and activity changes are not associated with temperature per se, they significantly affect both foraging and growth and may have consequences for both thermoregulation and physiological thermal response.

One of the best studied systems of phenotypic plasticity is the African armyworm *Spodoptera exempta*. This species exhibits a density-dependent polyphenism somewhat analogous to that of the migratory locusts. Faure (1943) characterized caterpillars from populations having different densities as solitary and gregarious phases. The solitary phase larvae, which occur in low density populations, are green and highly cryptic in their behavior, feeding in the low grasses. At high densities, the gregarious phase larvae are black and feed voraciously with a preference for sunlit vegetation (Gatehouse 1986). In the laboratory the gregarious caterpillars grow faster than the solitary phase animals at the same temperature (Khassimudin 1981; Simmonds and Blaney 1986). This effect is undoubtedly enhanced in the field due to their basking behavior and black pigmentation (Rose and Dewhurst 1979).

The different growth rates by the phases at constant temperature suggest that the thermal sensitivity of important physiological processes has been altered and this is coupled with a change from a cryptic thermoconforming strategy to that of a thermoregulator. The mechanism responsible for this transformation has not been examined. It is possible that the caterpillars reduce the duration of the larval stage by spending more time feeding. However, this would assume that they are processing food more rapidly. If this were the case presumably either solitary animals were not processing it as rapidly as they could (i.e., behaving "sluggishly") or that the gregarious caterpillars had an enhanced physiological capacity. Support for the latter is given by Simmonds and Blaney (1986) who report that crowded armyworm caterpillars had a larval duration of about 16 days compared to 24 days for the solitary forms under the same conditions of temperature and radiant heating, despite the fact that they spend *less* rather than more time feeding.

While the armyworm is probably the most dramatic case of phase polymorphism at different population densities, the phenomenon has been described repeatedly throughout the Lepidoptera occurring in the families Tortricidae, Plusidae, Saturniidae, Notodontidae, Geometridae, Sphingidae, and Noctuidae (Rhoades 1985). The dark larvae tend to grow more rapidly than their pale counterparts in low-density populations. In general, the solitary phases are relatively inactive while the dark phases are highly active. Rhoades (1985) characterizes these phases as "stealthy" and "opportunistic," respectively. Stealthy strategies (thermoconformers) minimize the damage to the plant and maximize efficiency of conversion of ingested food to biomass. This strategy requires a low metabolic rate promoted by low body temperatures and pale cryptic phenotype. Opportunists use a hit and run strategy on host plants, which requires a high metabolic rate and high levels of physical activity (Rhoades 1985). The ability to utilize a food source quickly and migrate out of the area is often used to characterize the highly active and aggressive stages of these species (Casey 1976; Rawlins and Lederhouse 1981; Gatehouse 1986; Simmonds and Blaney 1986). Rhoades (1985) suggests that these attack strategies of herbivores may be of overriding importance in determining the various nutritional indices such as growth rate, efficiencies of conversion, and food digestibility.

In this type of life history, a caterpillar species can be transformed in a single generation from a behavioral thermoconformer to a behavioral thermoregulator. The changes in activity levels, behavior, and growth rates suggest that the physiological processes have also changed. Moreover, it is also apparent that the continuum between the thermoregulator and the thermoconformer is routinely crossed by a number of species. Examining the thermal response of biochemical and physiological parameters of these species should give us important clues not only to the physiological basis of their thermal strategies but also, perhaps, insight toward generalizing the response for strict thermoregulators and conformers as well.

The gypsy moth, *Lymantria dispar,* also shows large changes in activity patterns and behavior in response to crowding (Campbell 1981; Leonard 1981). Lance et al. (1987) reported that crowded gypsy moths showed dramatic increases in growth (1–3 weeks faster) compared to uncrowded populations. The caterpillars had body temperatures 2–6°C warmer than those in a nearby low-density population. Although the authors observed no obvious thermoregulatory behavior, they attribute this change to an increase in the microclimatic temperature associated with the more open canopy and the higher degree of diurnal activity. Evidence for crowding causing enhanced growth rates of caterpillars is equivocal. Leonard (1981) reported that crowded gypsy moth caterpillars grew more rapidly. Lance et al. (1987) reevaluated the data and concluded that crowding does not cause accelerated growth. However, given the thermal insensitivity of growth of uncrowded gypsy moth caterpillars (Knapp and Casey 1986), one would not expect there to be such an impressive increase. However, a shift in the performance curve by qualitative or quantitative changes in enzyme activity would be clearly highly adaptive. Given the major shifts in activity patterns, and feeding behavior, it is not unreasonable to postulate that shifts also occur in thermal sensitivities of physiological processes as originally postulated by Leonard (1981) and seen in other opportunistic species.

Evolution of Thermoregulatory Behavior

The behavioral response to temperature (thermoregulation or thermoconformation) employed by different caterpillars is a compromise shaped by physical and biotic components of the environment (see Stamp, Chapter 15). It is usually assumed that it is adaptive to grow as rapidly as possible since this will reduce the duration of the larval stage thereby decreasing effects of predation and/or competition. The major benefit of behavioral thermoregulation is clearly associated with enhanced growth. However, to thermoregulate, the animals utilize solar radiation and consequently must be conspicuous, thereby exposing themselves to predators and parasitoids. Predator and parasitoid pressure have forced many species into a cryptic life-style (Heinrich, Chapter 7; Stamp and Wilkens, Chapter

9). What circumstances can account for a caterpillar's strategy to change from thermal conformation to thermoregulation?

Predator Deterrence

Species that have developed aposematic coloration as a consequence of sequestering harmful chemicals should benefit by basking since their behavioral patterns rely on their being conspicuous (Brower et al. 1967). For example, Monarch butterfly larvae increase their T_b by 3–8°C by basking and this behavior significantly reduces development time particularly at low ambient temperatures (Rawlins and Lederhouse 1981).

Caterpillars that have developed hair or spines are often avoided by vertebrate predators (Heinrich, Chapter 7). As setae also enhance the effectiveness of thermoregulation (though they were obviously not developed for that purpose), it would also seem that such caterpillars would benefit from basking.

Extreme Environments

Thermoregulatory behavior could be expected to develop if the environmental temperature is sufficiently extreme that growth is severely curtailed. For example, the hatching of tent caterpillar egg masses is tied to bud break, which usually occurs in early April. At that time of year the average shade air temperature is only above the caterpillars developmental threshold temperature for about 10 hours and is never more than 5°C higher. Without access to solar radiation the caterpillars would be incapable of growth (Knapp and Casey 1986). Even early instars clumped in the center of the small transparent tent are capable of experiencing temperature excesses of as much as 18°C and body temperatures occasionally in excess of 40°C. If the tent provides protection from predators, it also functions to reduce convective heat exchange thereby increasing the body temperatures of the caterpillars (Wellington 1950; Knapp and Casey 1986). Given the high body temperatures experienced by the caterpillars even in the early instars (Fig. 1.2C), Heinrich's (1977) hypothesis that the performance curve is adapted to the highest temperatures experienced is supported.

Euphydryas aurinia (Nymphalidae) seem to have developed a similar thermal strategy. Like the tent caterpillars, they begin feeding in early spring and are exposed to low air temperatures. They are black and form feeding aggregations on highly nutritious new leaves. Their thermoregulatory behavior causes elevated body temperatures (Porter 1982). The success or failure of these caterpillars can be associated with the climatic conditions. If the spring is clear and cold with lots of sunshine, the caterpillars can complete development before predators and parasitoids accumulate (Porter 1983). Alternatively, if the season is rainy and cloudy, and thus reducing growth rates, the population can be decimated by parasitoids. Buckmoth caterpillars (*Hemileuca lucinia*) provide another similar example (see Stamp, Chapter 15).

The thermal situation is even more critical for Arctic caterpillars. Experiencing continuously low temperature (0 or less to 6°C), these caterpillars can take as many as 14 years to complete development (Downes, 1965). Even in the face of high parasite load these caterpillars bask continually in the midnight sun (Kukal and Kevan 1987) achieving body temperatures considerably above ambient.

Outbreaks

During outbreaks of a particular species in a habitat, caterpillars may be so numerous that predator populations appear to be saturated. Under such conditions intraspecific competition is probably more important than predation and there should be an advantage for rapid development because the resource is liable to become decimated. Large outbreaks of sphingid caterpillars (Casey 1976), armyworms (Gatehouse 1986), and gypsy moths (Leonard 1981) routinely occur, and in each of these species aggressive, continuous, diurnal activity and heavy resource depletion occur. In the sphingids and armyworms basking behavior is clearly apparent.

The outbreak pattern would appear to be more unusual than the other two situations discussed above because it represents a change from a thermoconformer mode at low densities to a thermoregulatory one at high density. This scenario points out the continuum of thermal behavior and, at least in the case of the armyworm, exhibits the physiological plasticity which also accompanies the change in behavior.

In all of the cases cited above, movement from a cryptic or sedentary life-style to that of a more active thermoregulator will undoubtedly expose the caterpillars to higher heat loads. Under these circumstances Heinrich's hypothesis that the caterpillars adapt to the highest body temperatures they normally encounter could cause a shift in the performance curve to the right (i.e., to higher temperatures).

Different thermal solutions are bound to occur for different species, for the same species at different times within a habitat, and for the same species in different habitats (see Stamp, Chapter 15). The degree of flexibility in moving to and from a thermoregulatory or thermoconformer behavior pattern is not clear. Certainly, for species with elaborate behavior and social organization such as tent caterpillars, it is unlikely that facultative shifting between the two patterns will occur. On the other hand, the variety of species exhibiting either pattern in different phases suggests that the transition is relatively common. Experiments examining the interaction of temperature with other biotic factors should be important to examine flexibility of different species (Stamp and Bowers, 1988, 1990b).

Conclusions and Future Directions

Caterpillars differ in their behavioral responses to temperature along a broad continuum from strict thermoconformation to more or less continuous thermoregulation. Physiological responses to temperature have been examined for a wide variety of species but often these data have not been placed in context with the behavior and natural history. Caterpillars differ in thermal sensitivity and thermal optima of important physiological processes. The demonstration that the shape and position of the performance curve for important physiological variables (growth, metabolism and feeding rates) vary for different species and thus this variation may be related to the thermal strategy (conformer or thermoregulator) represents an issue that needs to be addressed. It is tempting to suggest that thermal conformers are sluggish (like the solitary phases of the armyworm). Low activity and low metabolism could favor the development of thermal independent rate processes through isozyme production. However, it is likely that there is a continuum of activity within thermoconformers and generalizations at this stage are not warranted. Physiologically speaking, caterpillars are not necessarily all alike and physiological differences coupled with appropriate thermal strategy may have important adaptive consequences. Coupling physiological performance with the continuum of thermal behavior patterns will provide important information about caterpillar adaptations and sets the stage nicely for more detailed examination of seasonal and geographical (both latitude and altitude) influences on caterpillar life history.

Literature Cited

Bakken, G. S., and Gates, D. M. 1975. A heat transfer analysis of animals: some implications for field ecology, physiology and evolution, pp. 225–290. In D. M. Gates and R. B. Schmerl (eds.), Perspectives of Biophysical Ecology, Springer-Verlag, Berlin.

Beck, S. D. 1968. Insect Photoperiodism. Academic Press, New York.

Brower, L. P., Brower, J. V. Z., and Corvino, J. M. 1967. Plant poisons in a terrestrial food chain. Proc. Natl. Acad. Sci. (U.S.A.) 57:893–898.

Campbell, R. W. 1981. Population dynamics, pp. 65–214. In C. C. Doane and M. L. McManus (eds.), The Gypsy Moth: Research toward Integrated Pest Management. United States Department of Agriculture, Washington, D.C.

Capinera, J. L., Wiener, L. F., and Anamosa, P. R. 1980. Behavioral thermoregulation by late-instar range caterpillar larvae *Hemileuca oliviae* Cockerell (Lepidoptera: Saturniidae). J. Kansas Entomol. Soc. 53:631–638.

Casey, T. M. 1976. Activity patterns, body temperature and thermal ecology of two desert caterpillars (Lepidoptera: Sphingidae). Ecology 56:485–497.

Casey, T. M. 1977. Physiological responses to temperature of caterpillars of desert populations of *Manduca sexta*. Comp. Biochem. Physiol. 57A:679–682.

Casey, T. M. 1981. Behavioural mechanisms of thermoregulation, pp. 79–113. In B. Heinrich (ed.), Insect Thermoregulation. Wiley, New York.

Casey, T. M. and Hegel, J. R. 1981. Caterpillar setae: Insulation for an ectotherm. Science 214:1131–1133.

Casey, T. M., and R. Knapp 1987. Caterpillar thermal adaptation: Behavioral differences reflect metabolic thermal sensitivities. Comp. Biochem. Physiol. GSA:679–682.

Casey, T. M., Joos, B., Fitzgerald, T. D., Yurlina, M. E., and Young, P. A. 1988. Synchronized group foraging, thermoregulation and growth of Eastern tent caterpillars. Physiol. Zool. 61:372–377.

Cossins, A. R., and Bowler, K. 1987. Temperature Biology of Animals. Chapman and Hall, New York.

Downes, J. A. 1965. Adaptations of insects in the Arctic. Annu. Rev. Entomol. 106:257–274.

Edwards, D. K. 1964. Activity rhythms in lepidopterous defoliators. II. *Halisidota argentata* Pack (Arctiidae) and *Nephystia phastasmaria* Stkr. (Geometridae). Can J. Zool. 42:939–958.

Faure, J. C. 1943. Phase variation in the armyworm *Laphygma exempta* (Walk.). Sci. Bull. Dep. Agric. S. Afr. 18:69–78.

Fields, P. G., and McNeil, J. N. 1988. The importance of seasonal variation in hair coloration for thermoregulation of *Ctenucha virginica* larvae (Lepidoptera: Arctiidae). Physiol. Zool. 13:165–175.

Fitzgerald, T. D. 1980. An analysis of daily foraging patterns of laboratory colonies of the eastern tent caterpillar, *Malacosoma americanum* (Lepidoptera: Lasiocampidae), recorded photoelectrically. Can. Entomol. 112:731–738.

Fitzgerald, T. D., Casey, T. M., and Joos, B. 1988. Daily foraging schedule of the Eastern tent caterpillar *Malacosoma americanum*. Oecologia (Berlin) 76:574–578.

Gatehouse, A. G. 1986. Migration in the African armyworm *Spodoptera exempta:* genetic determination of migratory capacity and a new synthesis, pp. 128–146 In W. Danthanarayana, (ed.), Insect-Flight Dispersal and Migration. Springer-Verlag, Berlin.

Heinrich, B. 1977. Why have some animals evolved to regulate a high body temperature? Am. Nat 111:623–640.

Heinrich, B. 1979. Foraging strategies of caterpillars: Leaf damage and possible avoidance strategies. Oecologia (Berlin) 42:325–337.

Heinrich, B. 1981. Ecological and evolutionary perspectives, pp. 235–302. In B. Heinrich (ed.), Insect Thermoregulation. Wiley, New York.

Heinrich, B., and Collins, S. L. 1983. Caterpillar leaf damage, and the game of hide-and-seek with birds. Ecology 64:592–602.

Hochachka, P. W. and Somero, G. N. 1973. Strategies of Biochemical Adaptation. W. B. Saunders, Philadelphia.

Huey, R. B., and Kingsolver, J. G. 1989. Evolution of thermal sensitivity of physiological performance. TREE 4(5):131–135.

Huey, R. B., and Slatkin, M. 1976. Cost and benefit of lizard thermoregulation. Quart. Rev. Biol. 51:363–384.

Joos, B., Casey, T. M., Fitzgerald, T. D., and Buttemer, W. A. 1988. Roles of the tent in behavioral thermoregulation of Eastern tent caterpillars. Ecology 69:2004–2011.

Kevan, P. G., Jensen, T. W., and Shorthouse, J. D. 1982. Body temperatures and behavioral thermoregulation of high arctic wooly-bear caterpillars and pupae (*Gynaephora rossi,* Lymantridae: Lepidoptera) and the importance of sunshine. Arctic Alpine Res. 14:125–136.

Khasimuddin, S. 1981. Phase variation and off-season survival of the African armyworm, *Spodoptera exempta* (Walker) (Lepidoptera: Noctuidae). Insect Sci. Appl. 1:357–360.

Knapp, R., and Casey, T. M. 1986. Thermal ecology, behavior and growth of gypsy moth and eastern tent caterpillars. Ecology 67:598–608.

Krogh, A. 1914. The quantitative relation between temperature and standard metabolism in animals. Int. Z. Physik.-chem. Biol. 1: 491–508.

Kukal, O., and Kevan, P. G. 1987. The influence of parasitism on the life history of a high arctic insect, *Gynaephora groenlandica* (Wocke) (Lepidoptera: Lymantriidae). Can J. Zool. 65:156–163.

Kukal, O., Heinrich, B., and Duman J. G. 1988. Behavioral thermoregulation in the freeze-tolerant arctic caterpillar *Gynaephora groenlandica*. J. Exp. Biol. 138:181–193.

Lance, D. R., Elkinton, J. S., and Schwalbe, C. P. 1987. Microhabitat and temperature effects explain accelerated development during outbreaks of the gypsy moth (Lepidoptera: Lymantria) Environ. Entomol **16:** 202–205.

Leonard, D. E. 1981. Bioecology of gypsy moths. In C. E. Doane and M. L. McManus (eds.), The Gypsy Moth: Research toward Integrated Pest Management. United States Department of Agriculture, Washington, D.C.

Porter, K. 1982. Basking behaviour in larvae of the butterfly *Euphydras aurinia*. Oikos 38:308–312.

Porter, K. 1983. Multivoltinism in *Apanteles bignellii* and the influence of weather on synchronisation with its host *Euphydrayas aurinia*. Entomol. Exp. Appl. 34:155–162.

Rawlins, J. E. and Lederhouse, R. C. 1981. Developmental influences of thermal behavior on monarch caterpillars (*Danaus plexippus*): An adaptation for migration (Lepidoptera: Nymphalidae: Danaidae). J. Kansas Entomol. Soc. 54:387–408.

Reynolds, S. E., and Nottingham, S. F. 1985). Effects of temperature and efficiency of food utilization in fifth-instar caterpillars of the tobacco hornworm, *Manduca sexta*. J. Insect Physiol. 31:129–134.

Rhoades, D. F. 1985. Offensive and defensive interactions between herbivores and plants and their relevance in herbivore population dynamics and ecological theory. Am. Nat. 125:205–238.

Rose, D. W. J., and C. F. Dewhurst 1979. The significance of low density populations of the African armyworm, *Spodoptera exempta* (Walker). Phil. Trans. R. Soc. London B. Biol. Sci. 287:393–402.

Scriber, J. M., and Lederhouse, R. C. 1983. Temperature as a factor of the development and feeding ecology of tiger swallowtail caterpillars, *Papilio glaucus* (Lepidoptera). Oikos 40:95–102.

Scriber, J. M., and Slansky, F., Jr. 1981. The nutritional ecology of immature insects. Annu. Rev. Entomol. 83:25–40.

Schultz, J. C. 1983. Habitat selection and foraging tactics of caterpillars in heterogeneous trees, pp. 61–90. In R. F. Denno and M. S. McClure (eds.), Variable Plants and Herbivores in Natural and Managed Systems. Academic Press, New York.

Seymour, R. S. 1974. Convective and evaporative cooling in sawfly larvae. J. Insect Physiol. 20:2447–2457.

Sherman, P. W., and Watt, W. B. 1973. The thermal ecology of some *Colias* butterfly larvae. J. Comp. Physiol. 83:25–40.

Simmonds, M. J. S. and Blaney, W. M. 1986. Effects of rearing density on development and feeding behavior in larvae of *Spodoptera exempta*. J. Insect Physiol. 32:1043–1053.

Stamp, N. E., and Bowers, M. D. 1988. Direct and indirect effects of predatory wasps (*Polistes* sp.: Vespidae) on gregarious caterpillars (*Hemileuca lucina:* Saturniidae). Oecologia 75:619–624.

Stamp, N. E., and Bowers, M. D. 1990a. Body temperature, behavior and growth of early spring caterpillars (*Hemileuca lucina:* Saturniidae). J. Lepid. Soc. 44:143–155.

Stamp, N. E., and Bowers, M. D. 1990b. Variation in food quality and temperature constrain foraging of gregarious caterpillars. Ecology 71:1031–1039.

Stevenson, R. D. 1985. Body size and limits to the daily range of body temperatures in terrestrial ectotherms. Am. Nat. 125:102–117.

Taylor, F. 1981. Ecology and evolution of physiological time in insects. Am. Nat. 117:1–23.

Taylor, P. S., and Shields, E. J. 1990. Development of the armyworm (Lepidoptera: Noctuidae) under fluctuating daily temperature regimes. Environ. Entomol. 19:1422–1431.

Wellington, W. G. 1950. Effect of radiation on the temperature of insectan habitats. Sci. Agr. 30:209–234.

Weiser, W. 1973. Temperature relations of ectotherms: A speculative review, pp. 1–24. W. Weiser (ed.), Effects of Temperature on Ectothermic Organisms. Springer-Verlag, New York.

Young, A. M. 1972. Adaptive strategies of feeding and predator avoidance in the larvae of the neotropical butterfly, *Morpho peleides limpeda* (Lepidoptera: Morphidae), J. N.Y. Entomol. Soc. 80:66–82.

2

Nutritional Ecology: The Fundamental Quest for Nutrients

Frank Slansky Jr.

The consumption of food provides the energy and nutrients (including water) necessary to carry out the remainder of an insect's (or other animal's) life activities: growth and development, storage of metabolic reserves, movement, defense, and eventual reproduction (Townsend and Calow 1981; Slansky and Rodriguez 1987a). Feeding is thus one of the most fundamental behaviors; its widespread ramifications provide the underlying framework for the adaptive strategies of consumers and its broad consequences for consumer fitness have likely lead to its evolution as a highly regulated behavior. In their review of insect foraging strategies, Hassell and Southwood (1978, p. 75) clearly elucidate the crux of insect (and other animal) feeding behavior:

> The evolutionary fitness of an animal depends significantly upon an optimal diet in both quantity and quality. Foraging strategies are therefore rigorously shaped by natural selection and should be considered in terms of the degree to which they maximize the net nutrient gain from feeding, and to which they minimize the risks to survival.

The rate of feeding is a key component of an insect's foraging strategy (Fig. 2.1); it in large part determines the quantity of food consumed and the net nutrient gain from feeding. In addition, the quality of an insect's food can affect its feeding rate and therefore the quantity eaten, and feeding subjects insects to risks to survival. It is therefore obviously important to understand the factors affecting feeding and the extent to which insects can regulate their food consumption.

Hassell and Southwood (1978) did not address feeding rate as a major component of an insect's foraging strategy. However, the central focus of insect nutritional ecology is to determine the adaptive tradeoffs involved in the consumption, utilization, and allocation of food (Slansky and Rodriguez 1987a) and we are beginning to elucidate the adaptations exhibited by insects that allow them to

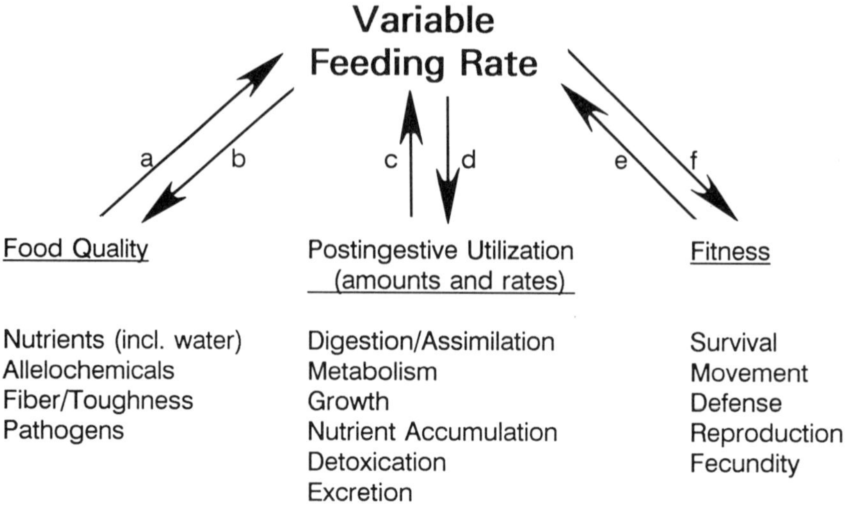

Figure 2.1. Interactions between feeding rate and food quality, postingestive utilization, and fitness. (a) Food quality can affect feeding rate, such as when a compensatory increase in feeding occurs in response to nutrient dilution. (b) Feeding may affect food quality, such as through a feeding damage-induced increase in allelochemical concentration. (c) Postingestive food utilization can affect feeding rate, such as due to gut stretch-receptor and hemolymph-composition feedback involved in the regulation of feeding. (d) Feeding rate can affect postingestive food utilization, such as when insufficient food intake results in slowed growth. (e) Fitness components may affect feeding rate, such as when feeding increases in reproductive females. (f) Feeding affects fitness, such as due to greater exposure to natural enemies during feeding.

cope with the costs and risks of feeding. Nonetheless, the role of these constraints in the evolution of insect feeding behavior, and especially their relationships to feeding rate and its regulation, even among the well-studied phytophagous caterpillars, remain little understood. In this chapter I address the feeding behavior of phytophagous insects (especially caterpillars) and, in particular, feeding rate, from an "adaptive response" perspective. I emphasize the need to go beyond the feeding rate values typically calculated in quantitative food utilization studies to understand fully the adaptive significance of feeding behavior. Many questions are raised that will hopefully be the subject of future research in this area.

Components of Food Consumption

For a caterpillar or other herbivore, feeding involves much more than merely filling its gut with the nearest plant tissue. As aptly stated by Lawton and McNeill (1979, p. 223):

> plant-feeding insects live in a world dominated on the one hand by their natural enemies and on the other by a sea of food that, at best, is often nutritionally inadequate and, at worst, is simply poisonous.

The nutritive quality of plants is probably seldom if ever optimal in terms of meeting an herbivore's nutritional requirements, and it varies both within and between various plant tissues. Many factors contribute to this variation, as well as to variation in allelochemicals and other plant characteristics (e.g., indigestible fiber and toughness) relevant to feeding by phytophagous caterpillars, including tissue and plant age, diurnal cycles, abiotic factors (especially temperature, water and light), soil fertility, impact of damage from pathogens, insects and other herbivores, and species differences (Harvey 1974; Haukioja et al. 1978a,b; McKey 1979; Tingey and Singh 1980; Denno and McClure 1983; Coley et al. 1985; Rhoades 1985; Slansky and Scriber 1985; Mattson and Haack 1987; Mattson and Scriber 1987; Tabashnik and Slansky 1987; Heinrichs 1988; Fajer et al. 1989; Karban and Myers 1989; Haukioja 1991).

In addition to being subjected to food of variable nutritional quality, caterpillars incur various costs and face assorted risks while feeding, including the presence of potentially toxic allelochemicals in plants, natural enemies, and abiotic factors that constrain their ability to obtain adequate nutrition in a timely manner (Fig. 2.1). Further, a caterpillar's nutritional needs may differ depending on age, sex, previous diet and other factors, and the consumption of food affects and is affected by its postingestive utilization (Slansky and Scriber 1985; Fig. 2.1). Thus, feeding by caterpillars involves, in large part, their attempt to obtain appropriate nutrients at an adequate rate in the face of qualitatively and quantitatively variable supplies and changing requirements, while experiencing various costs and risks. As a consequence, their feeding behavior exhibits various degrees of flexibility, which is likely related to the extent of variation in food quality they typically experience, to life-style differences, and to other factors, as discussed in this chapter.

Before a plant can be consumed, it must first be chosen as a foodplant. This choice of foodplant and of the particular plant part fed on can be made by the ovipositing adult female and/or by the immature insect, especially based on visual and chemical (nutritional and allelochemical) cues (Ahmad 1983; Fitzgerald and Peterson 1983; Miller and Strickler 1984; Hsiao 1985; Finch 1986; Frazier 1986; Singer 1986; Tallamy 1986; Schoonhoven 1987; Waldbauer and Friedman 1991). At one extreme, a caterpillar may not consume a plant leaf because of the lack of feeding stimulants (these can be nutrients and allelochemicals) or presence of repellents and deterrents (usually allelochemicals). These are generally well-documented responses (see above references) and mostly beyond the scope of this chapter.

If a food is consumed, it may be eaten in different amounts and at different rates. As for food choice, stimulants and deterrents may influence food consumption. In addition, various factors (e.g., suboptimal temperatures and ingestion of a sublethal dose of toxic allelochemical) may act to reduce food consumption through deleterious effects on metabolism and growth. Furthermore, caterpillars may alter their feeding in an apparent adaptive response to changes in the quantitative intake of nutrients brought about by variation in food composition. In the remainder of this chapter I focus on the rate and pattern of feeding once a foodplant has been chosen. These aspects of feeding behavior comprise a major component of the foraging strategies of caterpillars as they attempt to respond to variable food composition to obtain adequate nutrition while presumably limiting the costs and risks associated with feeding. The amount of food consumed by a caterpillar depends on its rate of feeding and the length of time during which feeding occurs. In this section I address feeding rates and rhythms; in a later section I discuss the duration of development of immature insects during which feeding occurs.

Feeding Rate and Its Components

Feeding rate is a broad term encompassing several different components. As commonly used in the literature on quantitative food utilization by insects, feeding rate is the amount of food consumed per instar or caterpillar over its life, generally expressed on a "per day" basis (consumption rate, CR) or on a "per day per unit body mass" basis (relative consumption rate, RCR). The former measures the overall rate at which food is being consumed, and thus is useful, for example, in assessing the impact of insect feeding on the plant being consumed, whereas the latter is especially useful in comparing different instars, species, or treatment individuals that differ considerably in body mass. In the literature on insects, RCR is usually calculated in terms of the dry mass (generally termed dry weight, dw) of the food consumed (RCRdw) per unit dry body mass (occasionally fresh body mass) (see Kogan 1986; Farrar et al. 1989). As described below, the conventional use of RCRdw (rather than also calculating the relative consumption rate based on the food's fresh weight, RCRfw), although essential to understanding food consumption and utilization processes, has hampered our more general understanding of feeding activity and its costs and risks.

At a finer scale, the "overall" feeding rates described above (whether absolute or body mass relative) are comprised of two key components: the proportion of time spent feeding (determined by the number and duration of feeding bouts or meals) and the speed of "biting and chewing" (i.e., instantaneous feeding rate; Jones et al. 1981; Bowdan 1988a). While usually studied from a physiological perspective in terms of understanding the regulation of feeding (Simpson and Abisgold 1985; Bowdan 1988a,b; Timmins et al. 1988), there may be important ecological relevance to these fine-scale components, such as when considering the consequences of an increase in overall feeding rate. A caterpillar may feed at a faster rate by increasing either or both of these components, but different risks may be involved. For example, an increase in the number of feeding bouts may result in greater exposure to natural enemies whereas a faster instantaneous consumption rate, while not increasing feeding time, may increase the rate at which a potentially toxic allelochemical must be detoxified. Unfortunately, only a few of the studies quantifying changes in the consumption and utilization of food by insects have included these fine-scale measures of feeding behavior (see Mechanisms of Compensatory Feeding, below).

Dry Weight and Fresh Weight Consumption Rates

A certain amount of dietary water is an important requirement of terrestrial caterpillars, and foliage water content seems to be a major factor influencing the evolution of maximum relative growth rates (RGRdw= mg dw gained mg^{-1} dw d^{-1}) of different species, as well as the success of individual caterpillars of a particular species in achieving their maximum rate (Slansky and Scriber 1985; Martin and Van't Hof 1988; Van't Hof and Martin 1989). Caterpillars typically contain 80–90% water (expressed as a percentage of their fresh weight, fw) whereas their food often is drier than this (see below) and the ambient humidity is frequently low enough to cause them to lose water (Wharton 1985; Willmer 1986). Thus, caterpillars (along with other terrestrial insects) exhibit various biochemical, physiological, and behavioral adaptations that allow them to concentrate water, the bulk of which they typically obtain from their food (Reynolds et al. 1985; Martin and Van't Hof 1989), and effectively conserve it (Wharton 1985). Some of these adaptations associated with feeding include living within plants (i.e., leaf miners, stem borers, gall formers, etc.), creating protective structures (e.g., leaf tiers, leaf rollers, and bagworms), group living (including both web-building and aggregation behavior) and particular temporal feeding patterns (Slansky and Scriber 1985; Willmer 1986). For example, although the performance of caterpillars of the European corn borer (*Ostrinia nubilalis*) can be affected by dietary moisture, individuals feeding within maize stalks are buffered from substantial external changes in moisture because the plants maintain a relatively stable internal water level (Ellsworth et al. 1989). One response apparently generally lacking in lepidopteran caterpillars, however, is increased feeding to obtain additional water when dietary moisture levels decline, in contrast to the common increase in food consumption exhibited in response to reduced food nutrient level (i.e., especially protein), often caused by an increase in dietary moisture (see Compensatory Feeding, below).

Despite the importance of dietary water to caterpillar performance, quantitative studies of the consumption and utilization of food by insects typically concentrate on dry mass budgets (Waldbauer 1968; Slansky 1985; Slansky and Scriber 1985; Kogan 1986). This is done in large part because of the difficulty in measuring water budgets accurately and because the main nutrient components of the food (i.e., protein, carbohydrate and lipid), and thus its energy content, comprise a major portion of the dry mass. Such studies have in general contributed greatly to our understanding of insect nutritional ecology, but consumption rate values based on dry weight provide only partial information about the ecological aspects of caterpillar feeding. Our emphasis on quantitative food consumption and utilization data based on dry mass has tended to divert attention away from gathering the full complement of data necessary to understand the ecological relevance of feeding behavior. One of the main situations in which dry weight consumption rate, as the sole measure of food consumption, is of limited use and can even be misleading involves comparisons among foods that differ in their proportion of water and dry mass. This is a common occurrence not only when comparing insect species in different feeding guilds (e.g., forb- vs. tree-foliage chewers, and leaf-chewers vs. phloem fluid-suckers), but also within a feeding guild (e.g., tree-foliage feeding caterpillars).

In addition to its beneficial physiological importance, water can serve as a diluent of dietary nutrients. The water content of plant tissues varies greatly; for example, plant foliage water may range from less than 50 to over 90% fw across species, and within a species it may change as much as ca. 10 percentage points diurnally and ca. 20 percentage points as the leaves mature (e.g., Scriber 1977; Haukioja et al. 1978a,b; Fig. 2.2). Rainfall, soil moisture, and other environmental factors also have a large impact on foliage water levels, with consequent effects on insect performance (Mattson and Haack 1987; Holtzer et al. 1988). This variation in foliage water content requires substantially different consumption rates based on fresh weight to obtain the same dry weight consumption rate value (see below). In these situations, consumption rate based on fresh weight should be calculated along with that based on dry weight, because the former gives a more realistic measure of the amount of food actually consumed and processed, and thus would presumably better reflect the costs and risks associated with feeding. Unfortunately, this is seldom done in studies of insect nutritional ecology. The examples that follow demonstrate the importance of measuring both RCRdw and RCRfw when attempting to understand caterpillar feeding behavior (see also When Is It Compensatory Feeding? below).

In a broad, interspecific comparison of food consumption and utilization by phytophagous caterpillars, Scriber and Slansky (1981; later updated in Slansky and Scriber 1985) found a general increase in RCRdw as foliage water (% fw) and nitrogen (% dw) increased (Fig. 2.3). Maximum RCRdw values of about 4–5 mg dw mg^{-1} dw d^{-1} occurred on "high water" foliage (i.e., 80–90% water, typically leaves of forbs), whereas the maximum values on "low water" foliage (i.e., 50–60% water, often tree leaves) were about 2–3 mg dw mg^{-1} dw d^{-1}, implying that caterpillars on the high water foliage can consume food approximately 1.5 to 2 times faster than those feeding on the low water foliage. While this interpretation is of course correct for RCRdw, it is misleading in terms of these caterpillars' actual feeding behavior. Because the higher RCRdw values occur on leaves with greater water content, caterpillars consuming these leaves have substantially larger RCRfw values, ranging from about 4 to 5 times greater than those consuming the drier leaves (Fig. 2.4). Thus, assuming the same

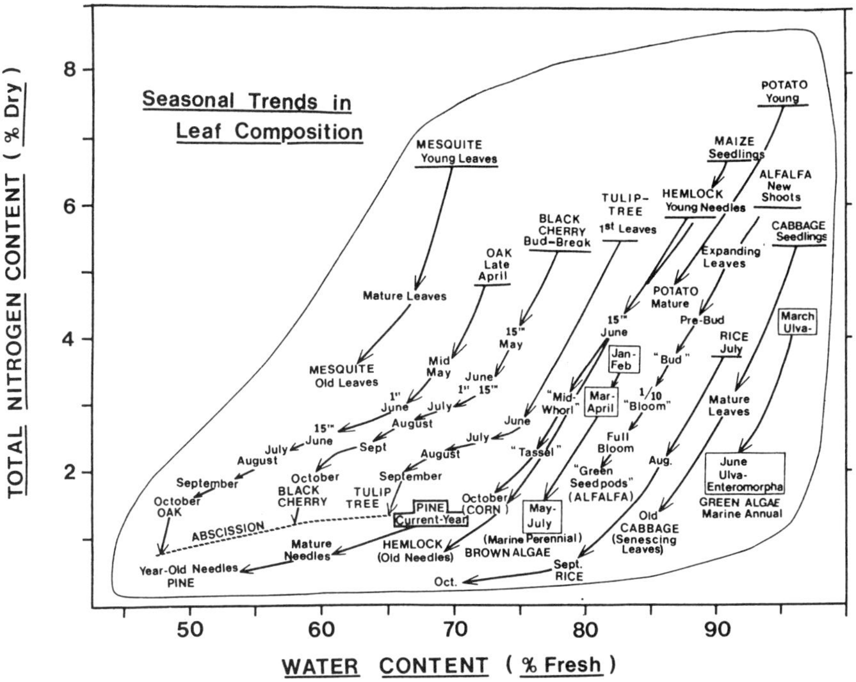

Figure 2.2. Seasonal trends in foliage water (% fw) and nitrogen (% dw) contents of selected plant types, including annual and perennial forbs, grasses, deciduous and coniferous trees, and marine aquatic species. (Reprinted from Slansky and Scriber 1985, by permission.)

instantaneous feeding rate (the data are too limited to determine if this is a valid assumption, however), to achieve their higher RCRdw values, caterpillars on the higher water leaves must spend not 1.5- to 2-fold more time feeding, as it would appear from the RCRdw values, but from 4- to 5-fold more time feeding and must process physiologically that much more food.

A comparison of two species of caterpillars feeding on oaks (*Quercus*) further illustrates the importance of calculating both RCRdw and RCRfw values. Lawson et al. (1984) measured dry mass budgets for caterpillars of the geometrid *Alsophila pometaria* consuming early-season foliage of several oak species and of the saturniid *Anisota senatoria* consuming late-season foliage of the same species. The mean RCRdw value for the former species, across all oak species (ca. 3.03 mg dw mg^{-1} d^{-1}), was 1.3-fold greater than that of the latter species (ca. 2.35). However, if the authors also had calculated RCRfw values, it would have been evident that the early-foliage feeder was consuming food at a substantially faster rate, associated with the higher water content of the early-season foliage. The RCRfw for *A. pometaria* was over twice as great as that of *A. senatoria* (i.e.,

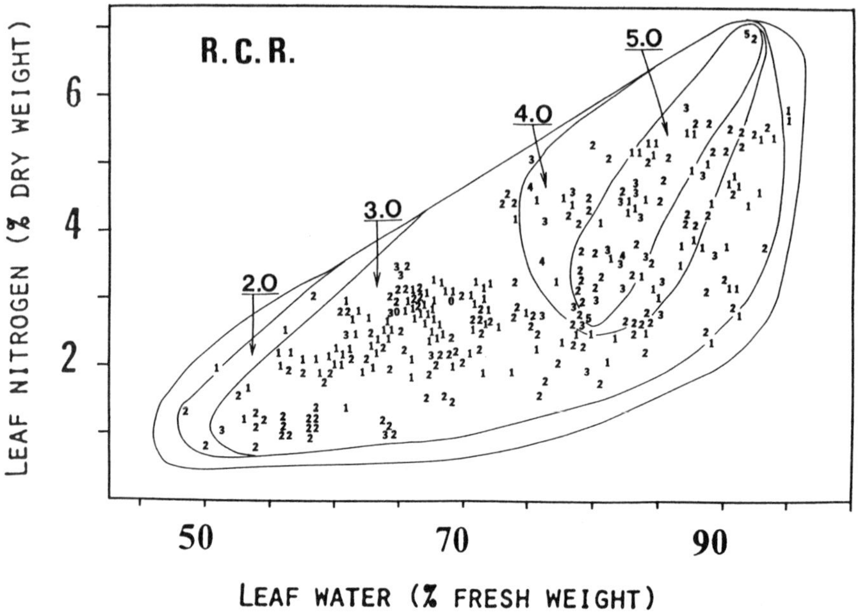

Figure 2.3. Relative consumption rate based on dry weight of the penultimate instar larvae of phytophagous insects (25 species of Lepidoptera and four species of Hymenoptera), as related to the water (% fw) and nitrogen (% dw) contents of the foliage consumed. Lines enclose all RCR values equal or greater than those indicated for that zone, but they do not exclude lower values. Number code: 1 = 0.5–1.49, 2 = 1.5–2.49, ..., 5 = 4.5–5.49 mg dw mg^{-1} dw d^{-1}. (From Slansky and Scriber 1985, by permission.)

mean RCRfw= 10.1 vs 4.8 mg fw mg^{-1} d^{-1}, respectively, calculated from the author's data).

From these examples, it is evident that RCRfw values must be higher on leaves with greater water content for caterpillars consuming these leaves merely to obtain the same RCRdw value as caterpillars consuming drier leaves (see also Fig. 2.4). Because water and nitrogenous compounds (especially protein and amino acids) are both generally important to caterpillar growth (Scriber 1984; Slansky and Scriber 1985) and often covary (e.g., both being relatively high in young leaves and declining with leaf maturation; Fig. 2.2), it is also important to consider the changes in nutrient level (% dw) within the context of changes in leaf water content (% fw) (i.e., to examine the change in nutrient level on a fresh weight basis). For example, compared with a caterpillar feeding on a leaf with 50% water and 2% dw N (representative of a mature tree leaf), one consuming a leaf with 90% water (e.g., young foliage of a forb), in order to obtain the same relative consumption rate for nitrogen (RCRN), must either feed on a leaf with 10% dw N (at the same RCRfw) or have

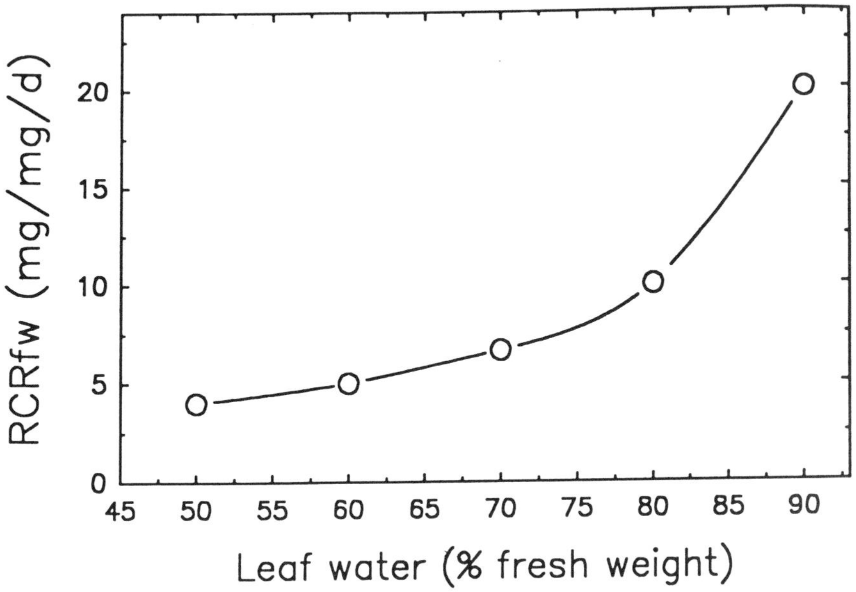

Figure 2.4. Relative consumption rate based on the fresh weight of the food (RCRfw) required to obtain the same RCRdw (in this case, RCRdw = 2.0 mg dw mg^{-1} dw d^{-1}) as a function of leaf water content (% fw).

a 5-fold greater RCRfw (at the same % dw N in the leaf). High water leaves often have a greater nutrient level (% dw) than leaves with lower water, but most plant foliage generally has less than 5–6% dw N (Slansky and Scriber 1985). Thus, it is evident that a caterpillar consuming high water content leaves must typically exhibit substantially greater RCRfw values than one ingesting drier leaves to obtain the same or greater RCRN value. However, an increase in food utilization efficiencies on high water food (Slansky and Scriber 1985) can partially offset the magnitude of the required increase in food consumption.

The study by Lawson et al. (1984) mentioned above also provides a representative example of this point. Although the early-season oak foliage had a significantly higher N content (% dw) compared with the late-season foliage, when calculated on a fresh weight basis, the N content was 1.3-fold greater in late-season foliage. Thus, caterpillars consuming the higher-water, early-season foliage, despite its higher % dw N level, would require a greater RCRfw merely to obtain the same RCRN. That the *A. pometaria* caterpillars achieved substantially greater RCRN values on the early-season foliage (averaging 1.7-fold higher) than the *A. senatoria* caterpillars consuming the late-season foliage further attests to the former's higher RCRfw values (see above). Clearly, both RCRfw and RCRdw values are essential for understanding insect feeding behavior.

Feeding Rhythms

In addition to the rate of food consumption, another component of feeding that undoubtedly has ecological relevance is its daily pattern, in terms of both when and how frequently it occurs during the day/night cycle. For example, three caterpillars with the same instantaneous feeding rate could also have similar overall feeding rates and yet the first could spend 10 minutes out of each of 12 daylight hours eating, the second might feed only twice (i.e., for about 1 hour at dawn and at dusk), and the third could feed nocturnally. Such diverse feeding patterns would likely result in differences between the three species in exposure to natural enemies (Evans 1990; see below) or unfavorable temperature and humidity conditions (Edwards 1964; Willmer 1986); indeed, these factors may provide the selection pressures leading to the evolution of particular feeding rhythms. In addition, there might be a relationship between insect feeding rhythms and plant allelochemical defenses (see below). Baseline consumption rates and ability to increase food intake, such as in response to a reduction in food nutrient level, of species with various feeding rhythms might be constrained to different degrees, with associated differential effects on growth. Unfortunately, there are few studies that have examined these issues, and especially whether feeding rhythm might constrain food consumption and thus growth rate.

Daily feeding rhythms of caterpillars span the hypothetical range mentioned above. For example, late instar caterpillars of two sphingids, the tobacco hornworm (*Manduca sexta*) and the white-lined sphinx moth (*Hyles lineata*), feed more or less continually (i.e., 10–30 minute meals alternating with 20–30 minute nonfeeding periods) during both day and night (provided that the temperature is suitable; Casey 1976; Reynolds et al. 1986), as do caterpillars of the cabbage butterfly (*Pieris rapae*) (Slansky 1974a) and several other species (Alexander 1961; Heinrich 1979). Other species, such as preultimate instar tent caterpillars (*Malacosoma americanum*) exhibit three to four daily feeding bouts of about 3 hours each (including travel time between the tent and food), separated by 5–7 hour nonfeeding periods, although the caterpillars were most active at dawn and dusk (Fitzgerald 1980; Fitzgerald et al. 1988). Dawn-dusk feeding, with general daytime inactivity, is also exhibited by caterpillars of *Morpho peleides* (Young 1972). Last instar caterpillars of *M. americanum* feed only during the dark, and many other species are also nocturnal feeders (Herrebout et al. 1963; Edwards 1964, 1965; Heinrich 1979; Lance et al. 1986).

Regardless of their daily rhythm, most species appear to spend only a small portion of their time actually consuming food. For example, *P. rapae* caterpillars, which typically have high RCRfw values (i.e., >20 mg fw mg^{-1} dw d^{-1}; calculated from Slansky and Feeny 1977), cycle between an average feeding period of about 4 minutes and an average "resting" period of about 22 minutes, such that they spend only about 13–15% of their time feeding (data for mid-fifth instar caterpillars feeding on excised leaves in petri dishes; Slansky 1974a). This conclusion was supported by direct observations of their activity pattern on whole plants in the field; caterpillars were feeding during about 10–15% of the observations (Mauricio and Bowers 1990). Based on observations of feeding time for these and other leaf-chewing insect species (Young 1972; Heinrich 1979; Fitzgerald 1980; Lance et al. 1986; Bergelson and Lawton 1988; Dethier 1988; Fitzgerald et al. 1988; Chapman and Beerling 1990; Mauricio and Bowers 1990), it appears that feeding typically occupies <20% of their time, and often much less, although last instar *M. sexta* caterpillars may spend about 70% of their last two days feeding [Reynolds 1990; see also Heinrich (1979), Rausher (1981), and Schmidt et al. (1988) for other cases of feeding apparently exceeding >50% of a caterpillar's time].

Short-duration feeding rhythms (e.g., less than 1 hour feeding separated by about 30 minutes to 2 hours of "resting") likely reflect the time required to digest and absorb an optimal portion of the ingested food (Reynolds 1990). In contrast, the longer term dawn–dusk and nocturnal-only feeding rhythms are probably not set by the time required to digest ingested food but more likely function to reduce exposure to harsh abiotic conditions and/or to day-active natural enemies (Edwards 1964; Lance et al. 1986; Fitzgerald et al. 1988). For example, while about 10 hours separate the dawn and dusk feeding periods of late instar *M. americanum* caterpillars, the ingested food is probably processed within about 3 hours after feeding (Fitzgerald et al. 1988). Caterpillars that are cryptic, especially on twigs or bark, are often inactive during the day, hiding away from their feeding site (Herrebout et al. 1963; Heinrich 1979; Heinrich and Collins 1983). In contrast, aposematically colored species may feed in highly exposed positions during the day. Species whose caterpillars are cryptic on leaves may feed during the day, but they also exhibit behaviors that can be interpreted as antienemy, including feeding from the underside of leaves, eating leaves so as not to produce irregular damage, chewing off partially eaten leaves (the latter two behaviors presumably remove leaf damage as a cue to their presence), and moving away from the feeding site (Heinrich 1979; Heinrich and Collins 1983; Stamp 1984; Mauricio and Bowers 1990; but see Bergelson and Lawton 1988). Some of these behaviors might be involved in preventing the induction of defensive allelochemicals by the plant or in avoiding allelochemicals produced at or translocated to the site of feeding damage (Rhoades 1985; Tallamy 1986; Karban and Myers 1989; Haukioja 1991). Whether the evolution of a particular feeding rhythm or feeding rate is influenced by the rate of detoxication/excretion of potentially deleterious allelochemicals appears to have not been examined (Slansky 1992), although insects consuming plants containing phototoxic allelochemicals may avoid feeding during the day, remaining in dark places (e.g., in a leaf roll or burrowed in the soil; Fields et al. 1990).

Few studies have linked daily foraging rhythms with quantitative food consumption and utilization, but there are some examples demonstrating that caterpillars can alter their feeding rhythm in response to changing conditions. For example, late instar caterpillars of the gypsy moth (*Lymantria dispar*) generally feed nocturnally, but during outbreaks they also feed during the day (Lance et al. 1986). This alteration of their feeding rhythm appears to be caused by at least one of the changes in the quality of the damaged leaves, namely an increase in tannin content. In the laboratory, the increase in daytime feeding did not result in greater overall time spent feeding on either the highly damaged foliage or the tannic acid-containing artificial diet compared with caterpillars feeding on foliage with little damage or on the tannic acid-free diet, respectively, nor was their developmental time shortened. In contrast, in the field the increase in daytime feeding subjects caterpillars in an outbreak population to warmer temperatures, and they develop 1–3 weeks faster than those in a low density population (Lance et al. 1986). For species whose caterpillars feed during the day, the typical increase in temperature toward midday may generally limit feeding to periods of the morning and afternoon. For example, the grass-feeding caterpillars of the arctiid moth *Ctenucha virginica* seek shade at high midday temperatures, which precludes feeding during this time (Fields and McNeil 1988; see also Casey, Chapter 1). On overcast days when the temperature is moderated, feeding may occur throughout the day. Edwards (1964, 1965) found both temperature- and age-dependent changes in the feeding rhythms of various lepidopteran caterpillars feeding on Douglas fir.

The developmental consequences of feeding rhythms have seldom been addressed. One of the few studies relevant to this issue is that of Herrebout et al. (1963) on caterpillars of various moths feeding on Scots pine. They found that the two species which probably feed both day and night complete larval development in 6 weeks or less, whereas two of the species that are primarily nocturnal feeders "spend many months" in the larval stage. These data suggest that feeding rhythms, whatever the selective forces molding their evolution, likely have substantial ramifications for caterpillar food consumption, growth, and associated life cycle components (e.g., number of generations per year and diapause stage; Slansky 1974b), but clearly much more data are required to assess this.

Intraspecific Differences in Feeding

Parental Effects: Genetic and Nongenetic

Variation in feeding and other performance criteria occurs among individuals of the same age and sex reared under the same conditions, reflecting in large part the genetic component to caterpillar performance (Table 2.1). Individuals with divergent genotypes (either within a population or from different populations) can vary in their consumption and utilization of food, reflecting in part differential adaptation to foodplant species (Rausher 1982; Scriber 1983; Futuyma et al. 1984; Tabashnik and Slansky 1987; but not always, e.g., see last reference and

Table 2.1. Coefficient of Variation (100 × SD/Mean) for Measures of Consumption and Growth of Select Species of Lepidopteran Caterpillars and a Larval Sawfly

Species	Experimental conditions	Sex	Coefficient of variation: Consumption	Coefficient of variation: Growth	Reference
Anticarsia gemmatalis	Control diet	M	11.0[a]	14.9[b]	Slansky and Wheeler (unpublished)
		F	21.0[a]	16.1[b]	
Cyclophragma leucosticta	*Eucalyptus*	M	—	4.4[b]	Mackey (1978)
		F	—	9.1[b]	
Dineura viridorsata[c]	*Betula*: a.m. vs p.m. leaves	M	—	12.6–19.4[d, e]	Haukioja et al. (1978b)
		F	—	9.3–10.6[d, e]	
Junonia coenia	Low CO_2	M	—	10.7[e]	Fajer et al. (1989)
		F	—	11.3[e]	
	High CO_2	M	—	9.8[e]	Fajer et al. (1989)
		F	—	14.0[e]	
Malacosoma americanum	*Prunus*	M	10.5[f]	7.2[b]	Futuyma and Wasserman (1981)
		F	6.4[f]	14.8[b]	
Malacosoma disstria	*Prunus*	M	27.8[f]	45.1[b]	Futuyma and Wasserman (1981)
		F	20.8[f]	20.6[b]	
Oporinia autumnata	*Betula*: a.m. vs p.m. leaves	M	—	8.8–10.2[d, e]	Haukioja et al (1978b)
		F	—	8.1–11.3[d, e]	
Papilio polyxenes	NY pop. on carrot	M	24.0[f]	16.1[e]	Lederhouse et al. (1982)
		F	17.6[f]	15.4[e]	
	Wisc pop. on parsnip	M	20.8[f]	14.5[e]	Lederhouse et al. (1982)
		F	20.0[f]	18.8[e]	
Pieris rapae	Collards	M	11.6[g]	14.4[h]	Slansky (1974a)
		F	17.2[g]	4.7[h]	
Spodoptera eridania	Control diet	M	—	9.1[i]	Gunderson et al. (1985)
		F	—	16.3[i]	
Spodoptera frugiperda	Cellulose-diluted diet	M	12.6[a]	20.8[h]	Slansky and Wheeler (unpublished)
		F	10.7[a]	33.9[h]	

[a]RCRfw
[b]Pupal dw.
[c]Hymenoptera: Tenthredinidae.
[d]Depending on season.
[e]Pupal fw.
[f]RCRdw.
[g]CRdw.
[h]Dry mass gain.
[i]Adult fw.

Williams et al. 1983a). Differences in the performance of caterpillars may also result from presumably nongenetic effects transmitted via the parents, especially due to the amount (and possibly quality) of nutrients available to them during embryonic development. Variable nutrient allocation to eggs, reflected in egg size variation, can result from food quality differences among female caterpillars of the parental generation, and can also occur between a female's early- and later-laid eggs, with subsequent effects on caterpillar performance (i.e., dispersal, developmental rate, weight gain and survival; Capinera and Barbosa 1976, 1977; Capinera et al. 1977; Capinera 1979; Wagner and Leonard 1979; but not always, e.g., Harvey 1977). The differences in performance listed above strongly suggest that food consumption is also affected, although this was not examined in these studies. However, there do appear to be differences in general activity and feeding behavior among individuals of certain gregarious species, although the causes are uncertain (see discussion in Papaj and Rausher 1983).

Age Changes

Relative consumption rate may change with age both within and between instars, presumably reflecting physiological changes during development (Lafont et al. 1975; Simpson 1983; Reynolds et al. 1985; Sehnal 1985; Slansky and Scriber 1985). Later instars tend to have lower RCRdw values than earlier instars (Slansky and Scriber 1985), but only a few species have been studied and these data must be interpreted with caution, both because of the technical difficulty of assessing accurately food consumption by early instars and because food quality may change between early and late instars. The latter can occur because of natural changes in leaf quality, such as associated with leaf age, and/or because of differences in feeding behavior between early and late instars, such as early instars "skeletonizing" leaves (i.e., avoiding leaf veins) whereas later instars may consume most of the leaf including veins (e.g., Wagner and Leonard 1979; Salinas 1984). Further, because these data refer to "overall" RCRdw values, they combine potential differences in instantaneous feeding rate and time spent feeding, and thus do not allow a determination of how the differences in RCRdw are derived. For example, in *Manduca sexta*, one of the few caterpillars for which the fine structure of feeding has been studied, the increase in daily food consumption during the last instar results from more time spent feeding, especially due to longer meals; bite frequency also increases (Reynolds et al. 1986; Bowden 1988a). Whatever the cause, real differences in feeding rate with age may have important ecological implications, such as by subjecting early instars to greater relative rates of ingestion of potentially toxic allelochemicals that must be detoxified; this may contribute to the often seen greater sensitivity of early instars to dietary allelochemicals, although lower relative detoxication enzyme activity may also be involved (Gould 1984; Ahmad et al. 1986; Hedin et al. 1988).

In addition to food consumption, other feeding-associated behaviors may change with age, such as feeding rhythm (see above) and from gregarious to solitary living (Breden and Wade 1987; Cornell et al. 1987; Blackwell 1988; Vulinec 1990). Dark body coloration of instars occurring in the typically cooler spring and fall, in contrast with lighter color during the warm summer, contributes to a caterpillar's thermoregulatory abilities and thus likely facilitates feeding (Fields and McNeil 1988), which is prevented by extreme body temperatures (Capinera et al 1980; Fields and McNeil 1988; see also Casey, Chapter 1). Age-associated changes in body coloration (e.g., from cryptic to aposematic coloration or vice versa; Young, 1972; Edmunds 1990) and other components of defense from natural enemies (Cornell et al. 1987) also occur; these likely alter the risks associated with feeding and may be accompanied by changes in feeding behavior (see Reavey, Chapter 8).

Table 2.2. Percentage Difference in Food Consumption and Body Weight (mg dw, Unless Indicated Otherwise) between Female [+, Greater, s, similar (difference ≤ 10%)] and Male Immature Lepidoptera and Orthoptera (updated from Slansky and Scriber 1985)

Species	Food consumption (%)	Body weight (%)	References
Lepidoptera			
Anticarsia gemmatalis	s	+14[a]	Slansky and Wheeler (unpublished data)
Argyrotaenia velutinana	+48	+57	Chou et al. (1973)
Bombyx mori	s	+57	Nakano and Monsi (1968)
	+18	+21	Horie et al. (1976)
Malacosoma americanum	+56	+39	Futuyma and Wasserman (1981)
Malacosoma disstria	+31	+41	Futuyma and Wasserman (1981)
Papilio polyxenes	+33	+14	Lederhouse et al. (1982)
Pieris rapae	s	s	Slansky (1974a)
Samea multiplicalis			Taylor (1989)
High nitrogen food	+21[b]	+35	
Low nitrogen food	s[b]	+24	
Spodoptera frugiperda	s-+11	s-+11[a, c]	Wheeler and Slansky (unpublished data)
Orthoptera			
Chorthippus montanus	+66	+87	Chlodny (1969)
Chorthippus dorsatus	+29	+121	Chlodny (1969)
Oxya velox	+58	+22	Delvi and Panaidan (1971)
Schistocerca gregaria	+27[d]	+19[d]	Carefoot (1977)

[a]Male heavier than female.

[b]Calculated from the author's data by multiplying the RCRdw times the mean weight (calculated by dividing the sum of the initial and final weights by 2) times the duration of the instar.

[c]Depending on experiment.

[d]kcal.

Sexual Differences

There can be differences in food consumption between the sexes, even in the immature stage (Table 2.2; see also Raubenheimer and Simpson 1990), because during this period the often different-sized sexes are being "built" and nutrient storage for future reproductive needs (which are typically greater in females) frequently occurs. For some of the studies listed in Table 2.2 in which food consumption was measured but consumption rate was not calculated, greater consumption by the larger sex occurred during the same time period as the smaller sex, or the former's percentage increase in food consumption was substantially greater than its increase in developmental time. These differences indicate that the larger sex was feeding at least at a greater absolute rate, and possibly at a greater RCR ("possibly" is used here because, for example, a greater absolute CR, if directly proportional to the larger individual's greater body mass, would yield a RCR value similar to that of the smaller individual). For those cases in which there was a greater percentage increase in body weight than consumption, food utilization efficiencies were likely greater for that sex (these were not always measured in the studies cited; see Slansky and Scriber 1985). Thus, different reproductive strategies between the sexes as adults are frequently reflected in differences in food consumption and other performance criteria as immatures, but the ecological consequences of this have seldom been addressed [e.g., are there sex differences during the immature stage in (1) exposure to natural enemies, and hence survival probability, associated with their differences in food consumption, (2) ability to compensate for reduced nutrients, and (3) the impact of altered developmental time on fitness?]

Further, because insect species differ in their reproductive strategies, such as in the contribution of immature and adult nutrition to reproduction, differences between the sexes as immatures, in terms of feeding and other performance characteristics, are likely to vary interspecifically. Adult nutrition in Lepidoptera ranges from none (i.e., reproductive allocation is totally dependent on nutrients accumulated by the caterpillars) to various combinations of intake of carbohydrates, lipids, amino acids, salts, etc.; in some cases the male provides nutrition to the female during mating (e.g., Murphy et al. 1983; Boggs 1987; Pivnick and McNeil 1987; Marshall and McNeil 1989). For example, about 95% of the nitrogen (N) allocated to eggs by adult females of *Dryas julia* was accumulated by the caterpillar; these females consume only nectar (generally, a low N food source) and receive relatively little N from mating (Boggs 1981). In contrast, adult females of *Heliconius cydno*, which obtain additional N from pollen feeding and from mating, allocate 75% of the caterpillar's N to eggs. Associated with this diversity in adult nutrition, there are probably interspecific and intersexual differences in caterpillar feeding rate, developmental time, body size and other performance criteria, such as their ability to accumulate adequate nutrition in response to variable food quality and other factors. Unfortunately, such issues have seldom been studied (see Boggs 1981, 1986; and Prolonged Development, below).

Environmental Effects on Feeding

Abiotic environmental factors such as suboptimal temperatures, as well as biotic factors such as toxic allelochemicals in plants, can alter food consumption and other performance criteria through affecting metabolic processes (Slansky and Scriber 1985; Slansky 1992). In these situations, the changes in caterpillar performance are often passive consequences of the environmental change. In other cases, however, caterpillars exhibit active responses to environmental changes (whether abiotic or biotic), altering their behavior in an apparent adaptive manner (Slansky 1982a,b; Slansky and Rodriguez 1987a). These resources can compensate, to various degrees, for the environmental change, such as behavioral thermoregulation in response to suboptimal ambient temperatures (see Casey, Chapter 1) and increased feeding in response to a declining dietary nutrient level (see Compensatory Feeding, below).

In addition, various environmental factors serve as cues to insects, allowing them to predict the future environment, and can result in the induction of physiological and behavioral changes. These inductive changes include alterations in feeding behavior, such as increased feeding associated with greater lipid storage in preparation for diapause, induced by short photoperiods (Slansky and Scriber 1985; Tauber et al. 1986). Food quality may also serve as an inducer, such as for caterpillars of the geometrid moth *Nemoria arizonaria*, in which individuals feeding on oak catkins develop into a morph that mimics catkins, whereas individuals consuming oak foliage (which occurs during the moth's second generation in the summer when catkins are no longer present) develop into twig-mimics (Greene 1989). The differences between morphs include not just coloration but body texture, hiding behavior, and head and jaw morphology. It was unclear whether the relatively larger mandibles and head capsules of the leaf-feeding twig morph are induced in response to the dietary cue (which was not determined; see Faeth and Hammon 1989) or result from the physical exercise of feeding on the tougher leaves (the latter effect occurs in other caterpillars; Bernays 1986a), but in either case, the difference would likely be adaptive, improving the ability of the foliage-feeding caterpillars to consume the tough oak leaves while not diverting resources to unnecessarily large mouthparts and associated musculature in the caterpillars consuming the more tender catkins.

Environmental factors (including food quality) acting during early instars may carry over to affect the performance of later instars (Scriber 1982; Taylor 1989), and abiotic and biotic environmental factors may interact to influence feeding and growth. For example, gregarious caterpillars of the saturniid *Hemileuca lucina* feeding on the young, sun-exposed branch tip leaves of their foodplant (*Spiraea latifolia*) exhibit their highest relative growth rate (RGRdw), but they may also be attacked and killed by *Polistes* wasps (Stamp and Bowers 1988). Caterpillars surviving wasp attacks tend to move toward the shaded center of the plant where they feed on mature leaves; as a consequence of the cooler temperature and the poorer quality food, their RGRdw can be reduced by about half (Stamp and Bowers 1990). Whether they exhibited compensatory feeding on the mature leaves was not investigated.

Interguild and Interspecific Differences in Feeding

As might be expected from the variety of substances used as food by various insect species, and the latter's wide size range, the amount and rate of food consumption vary greatly among species. In terms of RCRdw, foliage-feeding caterpillars tend to have the highest values (Slansky and Scriber 1985), although fluid-feeders (e.g., phloem and xylem sap-suckers, with their food consisting of >98% water) likely have the highest RCRfw values. In addition to interguild differences, there is wide variation within each feeding guild, especially due to environmental effects (see Environmental Effects on Feeding, above) and to interspecific life-style differences. Space limitations prevent a detailed review of the interguild and interspecific differences in feeding and other performance criteria. For additional information on the nutritional ecology of insects the reader is directed to other chapters in this volume, and to the following: forb foliage-chewers (Tabashnik and Slansky 1987; Bernays and Janzen 1988), grass foliage-chewers (Bernays and Barbehenn 1987; Capinera 1987), tree foliage-chewers (Mattson and Scriber 1987; Bernays and Janzen 1988), "internal" plant feeders (West 1985; Abrahamson and Weis 1987; Hemminga and van Soelen 1988; Cornell 1989; Craig et al. 1989), insects in other feeding guilds (see select chapters in Slansky and Rodriguez 1987b), specialists versus generalists (Fox and Morrow 1981; Scriber 1983; Futuyma and Moreno 1988), solitary versus gregarious species (Stamp 1980; Breden and Wade 1987; Stamp and Bowers 1990; Vulinec 1990; Fitzgerald, Chapter 11), species with holometabolous versus hemimetabolous development (Bernays 1988b; Berenbaum and Isman 1989), and various compilations (e.g., Denno and Dingle 1981; Denno and McClure 1983; Bell and Cardé 1984; Vane-Wright and Ackery 1984; Miller and Miller 1986; Barbosa and Schultz 1987; Barbosa and Letourneau 1988; Heinrichs 1988; Spencer 1988; Abrahamson 1989; Evans and Schmidt 1990). Despite this wealth of information, our understanding of the adaptive significance of feeding rate variation among insect species remains extremely limited.

Compensatory Feeding

The amount and rate of food consumption are strongly influenced by food quality, both in terms of passive consequences (e.g., reduced feeding resulting from decreased growth due to deleterious metabolic activity of an ingested toxic allelochemical) and active responses (e.g., refusal to eat after detecting a feeding deterrent, and increased feeding in response to a decline in dietary nutrient level). In this section, I focus primarily on the active feeding responses of insects, although it is sometimes difficult to distinguish these from passive effects, especially when attempting to separate phagodeterrent from toxic activity of allelochemicals (Berenbaum 1986; Slansky 1992) and when assessing whether compensatory feeding has occurred based solely on measurement of dry mass consumption (see When Is It Compensatory Feeding?, below). I first discuss the importance of measuring food consumption based on fresh weight as well as dry weight when investigating whether compensatory feeding for a decline in nutrient level has occurred, and then describe the compensatory feeding abilities of phytophagous insects (primarily caterpillars and orthopterans, which are the best studied groups). Next I address some of the mechanisms underlying compensatory feeding, especially from an ecological perspective, and finally I describe the benefits of compensatory feeding. The potential costs and risks to feeding that may constrain an insect's ability to compensate for variation in food quality, and that more broadly have likely influenced the evolution of feeding behavior, are discussed in the following major section of this chapter. Consumption of different proportions of various foods to obtain a particular nutrient mix (i.e., dietary self-selection) is a form of feeding compensation for inadequate food quality; this subject has been reviewed recently (Waldbauer and Friedman 1991) and will not be addressed in detail here.

When Is It Compensatory Feeding?

An increase in feeding associated with a decline in dietary nutrient level is generally viewed as an adaptive response in which an insect attempts to compensate for a reduction in nutrient intake by consuming more food (see Benefits of Compensatory Feeding, below). If the foods being compared do not differ in their proportions of water and dry mass, then an increase in the amount or rate of intake of dry mass (which is usually how food consumption is expressed in studies of insect quantitative nutrition) would indicate that increased feeding has occurred. However, differences in feeding are often assessed for foods differing not only in the nutrient content of their dry mass, but also in their proportion of water. In these instances, measurement of the intake of food dry mass, by itself, is insufficient to determine whether an insect has actually exhibited a compensatory increase in feeding, but rather must be accompanied by measurement of fresh mass intake, as the following examples illustrate.

Fall armyworm (*Spodoptera frugiperda*) caterpillars fed an artificial diet increasingly diluted with either water and/or indigestible cellulose exhibit a substantial compensatory increase in RCRfw (Wheeler and Slansky 1991; Fig. 2.5A). On the diets diluted with cellulose, RCRdw (which includes the dry mass of nutrients and the added cellulose) shows a corresponding increase compared with that on the undiluted diet, whereas it declines on the diets diluted with water (Fig. 2.5B) despite the increase in RCRfw (Fig. 2.5A). The body mass-relative consumption rate of nutrients (RCRnu, based on the intake of dry mass minus the added cellulose) declines with dilution regardless of diluent (Fig. 2.5C). If only RCRdw were calculated, as is frequently done in such studies, one might be led to conclude that the caterpillars showed compensatory feeding only on the cellulose-diluted diets; the declining RCRnu values would suggest that the caterpillars did not compensate at all for diet dilution. However, the RCRfw values clearly indicate the large compensatory increase in feeding exhibited by these caterpillars.

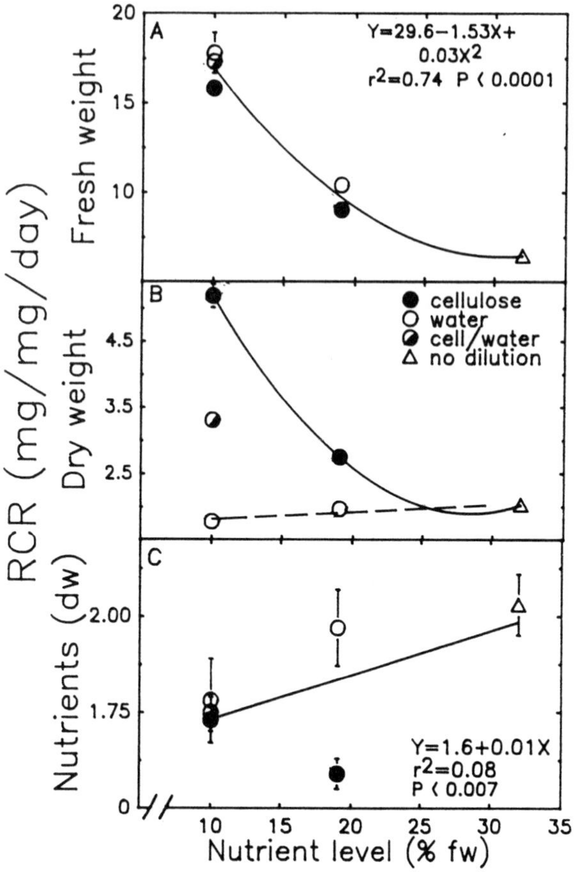

Figure 2.5. Changes in relative consumption rate based on (A) fresh weight, (B) dry weight, and (C) total nutrients (cellulose-free dw) of late instar *Spodoptera frugiperda* caterpillars reared on an artificial diet with different nutrient levels (% fw) obtained through dilution with cellulose and/or water. Regression parameters are shown in (A) and (C). For the solid line in (B), $Y=9.8-(0.6X)+(0.01X^2)$, $r^2=0.92$, $P<0.0001$; the dashed line in (B) connects the RCRdw on the water-diluted diets with that on the undiluted diet although the regression equation is not significant. (Modified from Wheeler and Slansky 1991, by permission.)

From the above, it is evident that if foods differ in their water content, consumption values based on fresh weight are essential in determining whether compensatory feeding has occurred in response to a decline in nutrient concentration (% dw). For example, given the same RCRfw, a caterpillar consuming a leaf with 50% water will have a 5-fold greater RCRdw than one feeding on a leaf with 90% water. If the nitrogen content of the high-water leaf was 5% dw, the low-water leaf could have a nitrogen level as low as 1% dw and caterpillars consuming the two leaves would have the same RCRN (again, given the same RCRfw). If only RCRdw were calculated in this example, the results would suggest a classic case of compensatory feeding, with caterpillars on the low-nitrogen leaves (1% dw) presumably having increased their RCRdw 5-fold as a compensatory response to the reduced leaf nitrogen. However, no feeding compensation would have occurred because caterpillars in each treatment had the same RCRfw. Clearly, both RCRdw and RCRfw values (and ideally the RCRs for energy and select nutrients as well) are required when considering caterpillar feeding rate from an ecological perspective, and especially when assessing compensatory feeding on foods differing in both their water (% fw) and nutrient (% dw) levels.

Compensatory Feeding Scope

The most clearcut examples allowing an evaluation of the compensatory feeding scope of insects involve consumption of artificial diets in which the nutrient level has been reduced with known diluents, relative to a control diet. Typically, this is accomplished using either water or an indigestible material (e.g., cellulose; see Fig. 2.5), but there are different implications depending on which diluent is used. With water as the diluent, nutrient level as a percentage of diet fresh weight declines, but there is no change in the nutrient content on a percentage dw basis. In contrast, when indigestible material is added, it generally replaces some portion of the nutrients (either a particular major nutrient such as protein, or the overall mixture of nutrients) and thus the nutrient level declines both on a percentage dw and a percentage fw basis. When the indigestible material replaces nutrients, the water content of the diet (percentage fw) is not altered; however, if it is added to the usual nutrient mix per standard amount of water, then diet water level (percentage fw) would decrease. Using high levels of water to dilute an artificial diet may subject the insect to stress from excessive water intake, independent of nutrient dilution (Timmins et al. 1988; Slansky and Wheeler 1989; Wheeler and Slansky 1991); thus when relatively low levels of nutrients are required, an indigestible material may be the preferred diluent because the water content (percentage fw) of the diet could be kept at the control level. Of course, in trying to understand the feeding behavior of insects consuming plant leaves, artificial diets should be formulated to reflect the types of changes in fiber, water, etc., occurring in the leaves of their foodplants.

Table 2.3. Variation in Food Consumption (Fresh Weight, Unless Indicated) Exhibited by Select Insects Feeding on Diluted Artificial Diets[a]

		X-fold range in		
Species	Diluent	Dietary nutrients (% fw)	Food consumption (fw)	Reference
Lepidoptera				
Agrotis ipsilon	Water	1.7	4.2[b]	Schmidt and Reese (1988)
Anticarsia gemmatalis	Water	3.0	1.8[b]	Slansky and Wheeler (1989)
	Water and/or cellulose	3.2	2.7[c]	Slansky and Wheeler (1991)
Celerio euphorbiae	Water	2.0	1.3[b]	House (1969)
Euphydryas chalcedona	Cellulose	4.1[e]	2.2[c, d]	Lincoln et al. (1982)
		2.6[f]	2.5[c, d]	
Malacosoma americanum	Water and/or cellulose	3.4	2.0[c]	Slansky and Wheeler (unpublished data)
Manduca sexta	Water	5.5	2.6[g, h]	Timmins et al. (1988)
	Cellulose	3.6	2.2[g, i]	
	Water	1.9	1.9[c, j]	Martin and Van't Hof (1988); Van't Hof and Martin (1989)
	Water	1.8	1.4–1.9[c, k]	Stamp (1990)

[a]The overall nutrient level (% fw) or a major nutrient class (protein or carbohydrate) was reduced through dilution with water and/or an indigestible material. In most cases, fresh weight consumption had to be calculated from the authors' data by dividing dry weight consumption by the % dw of the diet.

[b]Total consumption (fw).

[c]RCRfw.

[d]Response to protein dilution.

[e]Resin-free diet.

[f]Resin-containing diet.

[g]Absolute CRfw.

Results of such diet dilution studies demonstrate that the compensatory feeding abilities of a variety of phytophagous insects are quite substantial, with the magnitude of the increase in food consumption (on a fw basis) often equalling or at least approaching the magnitude of the dilution of nutrients (percentage fw; Table 2.3). For example, caterpillars of two species of noctuids, one a forb-foliage feeder (velvetbean caterpillar, *Anticarsia gemmatalis*) and the other a forb/grass feeder (fall armyworm, *Spodoptera frugiperda*), increased their RCRfw almost 3-fold as the nutrient level was diluted through addition of water or cellulose from 32 to 10% fw (Fig. 2.6). There were some differences in

Table 2.3. *(continued)*

Species	Diluent	*X*-fold range in: Dietary nutrients (% fw)	*X*-fold range in: Food consumption (fw)	Reference
Orgyia leucostigma	Water	1.9	1.8–1.9[c, j]	Van't Hof and Martin (1989)
Spodoptera frugiperda	Water and/or cellulose	3.2	2.8[c]	Wheeler and Slansky (1991)
Spodoptera eridania	Silica and/or cellulose	1.3	1.2[c]	Peterson et al. (1988)
	Agar	2.1	1.5[c, d]	Karowe and Martin (1989)
Orthoptera				
Locusta migratoria	Cellulose	2.0[l]	1.7[m]	Dadd (1960)
	Cellulose	2.0[l]	1.5[d, m] 1.0[m, n] 1.4[m, o]	Simpson and Abisgold (1985)
Melanoplus sanguinipes	Cellulose	32.3	6.9[b, p]	McGinnis and Kasting (1967)
Schistocerca gregaria	Cellulose	3.3[l]	3.1[m]	Dadd (1960)

[h]Data for 94% water diet excluded.

[i]Data for 90% cellulose diet excluded.

[j]Depending on diet.

[k]Depending on temperature and presence/absence of rutin.

[l]Cellulose added to the diet spanned this range as a percentage of dry weight; these diets were presented as basically dry material, with drinking water provided separately (see Simpson and Absigold 1985).

[m]Total consumption (dw).

[n]Response to carbohydrate dilution.

[o]Response to both protein and carbohydrate dilution.

[p]Data for 100% cellulose diet excluded.

response depending upon whether the diluent was water or fiber, possibly due to different effects of these substances on the processes regulating feeding (see Mechanisms of Compensatory Feeding, below). In comparison, tree-leaf feeding tent caterpillars (*Malacosoma americanum*) showed a lesser response over a similar range of nutrient dilution, although the maximum increase in RCRfw (ca. 2-fold) was nonetheless substantial (Fig. 2.6; Slansky and Wheeler unpubl. data). In contrast to the former two species, which often consume high-water foliage, performance of the latter species declined considerable on the high water diet, perhaps because this water level is much higher than it naturally experiences. As

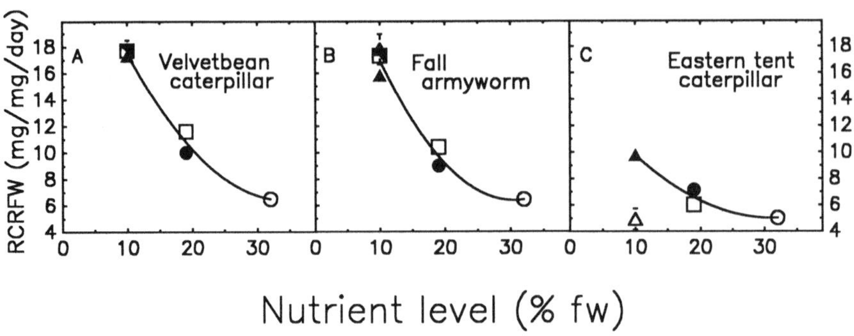

Figure 2.6. Relative consumption rate based on fresh weight (RCRfw) as a function of diet nutrient level (% fw) for late instar caterpillars of three lepidopteran species: (A) velvetbean caterpillar, *Anticarsia gemmatalis;* (B) fall armyworm, *Spodoptera frugiperda;* and (C) eastern tent caterpillar, *Malacosoma americanum*). Open circle, undiluted diet; open square and open triangle, water dilution; solid circle and solid triangle, cellulose dilution; half-filled square, water and cellulose dilution. Not tested: high-water diluted diet (open triangle) in (A) and water-cellulose diluted diet (half-filled square) in (C). Data from (A) Slansky and Wheeler (1991), (B) Wheeler and Slansky (1991), and (C) Slansky and Wheeler (unpublished data).

demonstrated in the previous section, it is important to express food consumption on a fresh weight basis when changes in dietary water are involved because the consumption of dry mass may decline (suggesting a failure to compensate) even though the insects are consuming more fresh mass of food.

As discussed above, foliage nutrient level (both % fw and % dw) can differ considerably within and between plant species. If the increased feeding responses summarized in Table 2.3 are not artifacts of using artificial diets, then insects should exhibit such responses when consuming plant leaves, which they appear to do (Table 2.4). However, unlike artificial diets in which precise changes in nutrient level can be made, plants exhibit many qualitative and quantitative changes in nutrients and allelochemicals with leaf maturation, with chemical fertilization, between species, etc., that might affect food consumption independent of a compensatory response to decreased nutrient level. In an effort to minimize the effect of unusual allelochemistry, I attempted to limit the examples in Table 2.4 to those in which the plants are known to or likely to be foodplants, although differences in allelochemicals that might contribute to the variation in feeding (especially reducing consumption below the typically exhibited values) cannot be ruled out.

For several of the studies in which plant water and nitrogen contents were given, the change in food consumption (fw) could be interpreted as a compensatory response (Table 2.4). In some cases, consumption increased as leaf dry mass, expressed as percentage fw, decreased (i.e., as leaf water content increased) whereas in others it increased as leaf nitrogen content (percentage fw) declined.

Again, I emphasize the importance of assessing food consumption on a fresh weight as well as dry weight basis in order to understand an insect's feeding behavior. For example, *M. americanum* caterpillars feeding on spring, summer or fall *Prunus serotina* leaves exhibited similar RCRdw values (Schroeder 1986b), and it might be concluded that they fed at similar rates. However, RCRfw values indicate that they actually consumed spring leaves (RCRfw= 5.6 mg mg^{-1} d^{-1}) about 1.2 times faster than summer leaves (RCRfw= 4.7) and 1.5 times faster than fall leaves (RCRfw= 3.8; values calculated from the author's data), apparently compensating for the early season dilution of leaf dry mass such that RCRdw values were more or less stabilized, rather than compensating for the seasonal decline in leaf N (percentage dw). Thus, caterpillars consuming spring leaves had a greater N intake and also experienced a significantly higher relative growth rate (RGRdw) than those consuming summer leaves, and similarly for the latter compared with those fed fall leaves. Using an artificial diet, Slansky and Wheeler (unpublished data) found that *M. americanum* also compensated for a similar magnitude of decline in diet dw (from a diet with 34% dw to one with 20% dw, a 1.7-fold decline; the decrease was 1.6-fold in Schroeder's study, although the actual % dw values were different, from 42 to 27%). However, when fed the diluted artificial diet, their compensatory increase in RCRfw was only 1.2-fold compared with 1.5-fold when fed leaves, such that RCRdw declined on the diluted diet, unlike the stabilization that occurred when feeding on leaves.

In other studies, measures of caterpillar performance (e.g., pupal weight) were independent of foodplant nutrient level, suggesting that feeding compensation may have occurred, such as for *Tyria jacobaeae* caterpillars consuming *Senecio* plants varying in their percentage dw N content (van der Meijden et al. 1984). However, because either food consumption or water content of the foliage was not measured in these studies, it is not possible to determine if compensatory feeding actually occurred. Adult folivorous insects also may exhibit compensatory feeding in response to a decline in foliage nutrient level, such as seen in the curculionid *Strophosomus melanogrammus* consuming birch (*Betula*) leaves with less than 1.5% dw N (Parry et al. 1990). Thus, there seems to be enough evidence to at least strongly suggest that insects exhibit compensatory feeding in response to changes in nutrient levels (either total dry mass, nitrogen, or some unmeasured factor correlated with these) in their typical foodplants, although obviously other factors (e.g., allelochemicals) may also affect food intake.

There appears to be little research investigating the extent to which compensatory feeding ability may vary with caterpillar age, sex, feeding guild, or lifestyle. In one of the few studies in which the sexes of the caterpillars were distinguished, there were no consistent differences in the magnitude of the increase in RCRfw values of female and male *C. fumiferana* when fed diets with foliage of two different ages from different spruce species (Thomas 1989). Although only a few comparisons have been made, there do not appear to be any consistent trends between specialists or generalists in their apparent compensatory

Table 2.4. Variation in Food Consumption (Fresh Weight, Unless Indicated) Exhibited by Select Lepidopteran Caterpillars Feeding on Different Plant Species, Varieties, or Treatments, Suggesting Various Capacities for Compensatory Alteration in Feeding[a]

		X-fold range in				
		Leaf				
Species	Food	dw (% fw)	N (% fw)	Food consumption (fw)	Compensatory feeding	References
Alsophila pometaria	*Quercus* spp.	1.2	1.2	1.2[b]	Yes for dw	Lawson et al. (1984)
Anisota senatoria	*Quercus* spp.	1.2	1.5	1.5[b]	Yes for dw,N	Lawson et al. (1984)
Anticarsia gemmatalis	*Glycine*[c]	?	?	1.4[d]	?	Moscardi et al. (1981a)
	Legume spp.	1.5	1.3	1.6[b]	No	Slansky (1989)
	Legume spp.	3.3	?	2.0[b]	Yes for dw	Waters and Barfield (1989)
Choristoneura fumiferana	Spruce spp.[e]	—	1.2–1.4[e]	1.1–1.4[b]	Yes for N	Thomas (1989)
Datana ministra	*Tilia*[f]	1.1	1.6	1.2[b]	Yes for dw,N	Schroeder (1986a)
Hyalophora cecropia	*Prunus*[c]	1.6	Same	1.8[b]	Yes for dw	Schroeder (1986b)
Malacosoma americanum	*Prunus*[c]	1.6	Same	1.5[b]	Yes for dw	Schroeder (1986b)

[a]Compensatory feeding is here defined as an increase in fresh weight (fw) food consumption either in association with a decrease in the proportion of dry weight (dw) or nitrogen (N) of the food, each expressed as percentage fw. In most cases, the values presented in this table were calculated from the various studies' dry mass data and water content of the foods, and the conclusions regarding compensatory feeding are mine, based on visual inspect of the resulting values.

[b]RCRfw.

[c]Various leaf ages.

[d]Total consumption (fw).

[e]Lyophilized foliage diets, consisting primarily of foliage with a gelling agent, were used; all diets contained 78.3% fw water.

[f]Control leaves vs leaves coated with protein.

[g]Based on three experiments with different species and varieties, and one species chemically fertilized with NH_4NO_3, excluding data from three chemically unique species on which consumption was exceptionally low (if these were included, the range in consumption would be 1.9–3.0).

Table 2.4. (continued)

		X-fold range in				
		Leaf				
Species	Food	dw (% fw)	N (% fw)	Food consumption (fw)	Compensatory feeding	References
Pieris rapae	Crucifer var and spp.[g]	1.6–1.7	2.0–2.6	1.3–1.6[h]	Yes for N	Slansky and Feeny (1977)
Plathypena scabra	*Glycine*[i]	?	?	2.0–2.6[j, k]	?	Hammond et al. (1979)
Samea multiplicatis	*Salvinia*[l]	?	2.6[m]	1.5–2.0[n]	Yes for N	Talor (1989)
Spodoptera eridania	Legume spp. and var	2.2	1.6	3.4[o]	?	Scriber (1979a)
Spodoptera frugiperda	*Arachis* var and spp	?	?	1.8[j]	?	Lynch et al. (1981)
	Bermuda-grass clones	1.3	1.4	1.5[b]	No	Quisenberry and Wilson (1985)
	Susceptible grasses	1.8	?	2.9[b]	Yes for dw	Chang et al. (1987)

[h]Absolute CRfw.

[i]Greenhouse vs field grown plants.

[j]Leaf area (cm^2).

[k]Depending on temperature and sex.

[l]Unfertilized vs fertilized plants.

[m]Nitrogen is expressed as percentage dw, because leaf water content was not given. *Salvinia* is an aquatic plant and thus its water content is likely to be similarly high for both the control and fertilized plants.

[n]Depending on sex (males exhibit greater response than females).

[o]RCRdw. Plants ranged from about 80–90% water, but the value for each variety or species was not given and thus RCRdw values could not be converted to RCRfw values, nor could it be determined whether the variation in feeding was related to dw or N variation.

abilities in response to changes in food quality (Futuyma and Saks 1981; Schroeder 1986b). Nymphs of *Locusta migratoria* respond through compensatory feeding for nutrient deficiency much more rapidly than do caterpillars of *Spodoptera littoralis*, suggesting inherent differences in compensatory feeding ability between insects in these two orders (Simpson et al. 1988, 1990).

The hypothesis that organisms operating at slower rates (these generally have relatively larger body size) may have greater compensatory scope than those with higher rates (and typically smaller body size), which are presumably operating closer to their physiological maximum performance values, has received some experimental support, such as from snails (Rollo and Hawryluk 1988). Two studies for caterpillars, one comparing two tree-foliage feeders (the early season, faster growing, smaller sized *M. americanum* with the later season, slower growing, larger sized *H. cecropia*; Schroeder 1986b) and the other comparing a faster growing (although larger in body size) forb-foliage feeder (*M. sexta*) with a slower growing tree-leaf feeder (*Orgyia leucostigma*; Van't Hof and Martin 1989) appear relevant to this hypothesis (although they were not explicitly designed to test it). In the former study, the slower growing *H. cecropia* exhibited a somewhat greater increase in RCRfw compared with *M. americanum* (1.8 versus 1.5-fold) when feeding on spring foliage (with higher water content) compared with fall leaves of *P. serotina*, whereas in the latter study both species showed similar increases in RCRfw on water-diluted artificial diets. In each study, however, associated with various changes in food utilization efficiencies, the RGRdw of the faster growing species was affected to a greater degree by the changes in food composition, appearing to support the hypothesis that the slower growing species are better buffered against changes in food quality because of their greater compensatory abilities (which may reflect feeding compensation and/or alteration in utilization efficiencies).

These comparisons, however, may not appropriately test the "compensatory scope" hypothesis. Because the two slower growing species probably feed on foods of lower water content (mature compared with young, spring tree leaves, and tree leaves versus forb leaves, respectively), and are likely adapted to this type of food, in these experiments the change in food composition that they were being subjected to was food of higher water content (spring leaves or diluted artificial diet, respectively). The higher water foods in these experiments reflected the typical foods for the spring tree-foliage feeder and the forb feeder, and therefore the reduced food quality these two species were being subjected to was food of lower water content (fall leaves and the drier artificial diet, respectively). Thus, although the species in each pair were being subjected to the same treatments, relative to their typical foods they were experiencing different changes in food composition, and their compensatory responses would likely differ and not be directly comparable. To better test the hypothesis, different species should be subjected to the same type and magnitude of change relative to the food composition they typically experience. Clearly, more research is required to test whether slow- and fast-growing caterpillars differ in their ability to compensate for reduced food quality.

Not all phytophagous species appear to exhibit compensatory feeding, as evidenced by their food consumption not being inversely correlated with leaf nutrient content (percentage dw), and/or by their growth performance being positively correlated with plant nutrient content (percentage dw) [e.g., caterpillars of the geometrid *Operophtera brumata* on foliage of various host trees differing in N and available protein (Wint 1983), and the chrysomelid *Paropsis atomaria* consuming leaves of several *Eucalyptus* species differing in their percentage dw N (Fox and Macauley 1977; Morrow and Fox 1980)]. However, these studies measured food consumption based only on dry mass and they do not present data allowing calculation of fresh weight consumption; thus it is not possible to determine if compensatory feeding based on leaf fresh weight occurred. As described previously, an increase in food consumption based on fresh mass, such as in response to high-water food, might not prevent a decline in dry mass consumption or growth performance; therefore, dry mass values alone are insufficient to determine whether compensatory feeding has occurred, unless the water content of the different foods is similar. In some situations, changes in allelochemical content associated with changes in nutrient level may interfere with a caterpillar's ability to compensate [e.g., Mihaliak et al. 1987; see also Costs and Risks to Feeding, (4) Coping with potentially toxic allelochemicals, below]. Whether adults can compensate through increased feeding for reduced larval performance has apparently seldom been investigated for leaf-chewing insects or insects in other feeding guilds (see Slansky and Rodriguez 1987b).

In general, there appears to be a lack of compensatory feeding for low leaf water in phytophagous caterpillars (e.g., Scriber 1977; Schmidt and Reese 1988; Timmins et al. 1988; Reynolds and Bellward 1989; Slansky and Wheeler 1989, 1991; Van't Hof and Martin 1989). Again, it is important to distinguish fresh weight and dry weight consumption in making this evaluation. For example, dry weight consumption by cecropia moth (*H. cecropia*) caterpillars increased almost 1.5-fold as the water level of their food (*Prunus serotina* leaves) declined from about 70 to 49% fw (Scriber 1977). However, this was not a compensatory feeding response to the declining dietary water, because consumption based on fresh weight actually declined (i.e., from about 3.3 to 2.8 mg fw total consumption and from 2.8 to 1.6 RCRfw; calculated from the author's data). Dry weight consumption increased only because the large decline in water content of the leaves increased the amount of dry mass per unit fresh mass consumed.

One explanation for the lack of compensatory feeding for low dietary water involves the physiological mechanisms underlying feeding regulation, especially feedback from the levels of particular nutrients in the hemolymph (see Machanisms of Compensatory Feeding, below). In addition, low water food can reduce nitrogen utilization, other food utilization efficiencies, and growth of caterpillars (Scriber 1977, 1979b; Martin and Van't Hof, 1988; Schmidt and Reese 1988).

Further, because a low water diet implies that the dietary nutrients are highly concentrated on a percentage fw basis, increased consumption would likely lead to excess nutrient intake, at the same time the low water diet is limiting growth and thus reducing the demand for nutrients. Absorption of excess nutrients would increase the caterpillar's metabolic demands of catabolism and excretion (Maddrell 1981; Horie and Watanabe 1983; Martin and Van't Hof 1988), creating additional physiological stress. Finally, increased consumption of food with a suboptimal water level may contribute to desiccation, especially because the caterpillar may need to add water to the ingested food (Reynolds and Bellward 1989; Timmins et al. 1988), probably to facilitate digestion and absorption (Turunen 1985). However, the highly effective water-resorbing capacity of the caterpillar rectum, which allows excretion of fecal pellets considerably drier than the ingested food and/or gut contents (Sriber 1977; Maddrell 1981; Martin and Van't Hof 1988; Reynolds and Bellward 1989; Van't Hof and Martin 1989), reduces fecal water loss.

Caterpillars similarly may not have evolved to compensate through increased feeding for reduced protein quality, probably because greater intake would only increase the metabolic costs of catabolizing and excreting the imbalanced amino acids. For example, caterpillars of *Spodoptera eridania* consuming diets with zein (a poor quality protein), although ingesting similar amounts of nitrogen as larvae consuming the diets with casein (a high quality protein), excreted over three times more uric acid and had respiration rates about twice as high as the latter caterpillars (Karowe and Martin 1989; see also Broadway and Duffey 1986a).

Mechanisms of Compensatory Feeding

The physiological mechanisms underlying compensatory feeding in insects are beginning to be elucidated. In brief, of key importance is nutritional feedback, especially associated with the osmolality and amino acid content of the hemolymph (which are strongly influenced by the nutrient level of ingested food), and volumetric feedback, which involves the bulk of ingested food and its rate of passage through the gut. These and other mechanisms (e.g., learning and neurotransmitter imbalances) likely vary in importance among species of insects and with the type of variation in food composition (Simpson and Simpson 1990; Simpson et al. 1990; Waldbauer and Friedman 1991). The nutritional feedback mechanism may contribute to the failure of caterpillars to increase feeding in response to a low water diet (see Compensatory Feeding Scope, above). Although there is homeostatic regulation of hemolymph composition (Willmer 1986), a caterpillar becoming dehydrated due to a low intake of dietary water is likely to experience an increase in the relative concentration of hemolymph nutrients, which should feed back to reduce rather than increase feeding.

Compensatory feeding may be accompanied by changes in food utilization efficiencies, some of which might be considered as additional compensatory responses. There is often an increase in the efficiency of nutrient absorption on reduced-nutrient diets (e.g., Fox and Macauley 1977; Slansky and Feeny 1977; Schroeder and Malmer 1980; Martin and Van't Hof 1988; Slansky and Wheeler 1989; Wheeler and Slansky 1991; but not always, e.g., Schroeder and Malmer 1980; Taylor 1989). Although such changes may help mitigate the impact of reduced nutrient intake, they may not be physiologically active, compensatory responses. For example, an increase in the efficiency of absorption of food from the gut when consuming a food with higher water content may be a passive consequence of the increased dietary water facilitating absorption, rather than, say, due to increased production of digestive enzymes. Although seldom studied in lepidopteran caterpillars (e.g., Broadway and Duffey 1986a), the production of digestive enzymes in many insects is regulated by a secretagogue mechanism (e.g., greater production of proteases in response to a greater amount of protein in the gut; Chapman 1985) and it is therefore unlikely that insects would produce greater amounts of digestive enzymes in response to a decline in nutrient intake on the reduced-nutrient diet (Simpson and Simpson 1990). However, if an insect produces the same amount of digestive enzymes regardless of the amount of nutrients in the gut (see Broadway and Duffey 1986a), then digestion may increase as the rate of nutrient ingestion declines, associated with a greater enzyme/substrate ratio. Finally, the apparent increase in nutrient utilization efficiency on a reduced-nutrient diet may not result from any physiological effect (whether active or passive) that actually increases the efficiency, but rather it may be a relative increase due to a decline in the nutrient utilization efficiency of caterpillars consuming the more concentrated control diet, as they excrete excess nutrients (Schroeder 1986a; Martin and Van't Hof 1988; Karowe and Martin 1989; Raubenheimer 1992). More research is obviously required before we can ascertain whether any of the changes in food utilization efficiencies associated with consuming different foods result from physiologically regulated, compensatory responses.

Fine Structure of Feeding

The manner in which food consumption is increased to bring about compensatory feeding is probably of ecological relevance. Greater consumption can result from feeding at a faster rate, which may involve more rapid instantaneous consumption, longer meals and/or more frequent meals, and/or from prolonging larval developmental time, which may include extra instars. It is expected that there will be different ecological consequences depending on how food consumption is increased, but the few studies that have assessed how the fine structure of feeding behavior changes (see below) were carried out from a physiological rather than ecological perspective. For example, although there is some evidence that caterpillars and other insects respond in an adaptive manner to the presence of natural enemies, I am aware of only one study (Loader and Damman 1991) that has evaluated if increasing the proportion of time feeding increases a caterpillar's exposure to natural enemies (and hence decreases its probability of survival). Further, apparently, there are no published studies that have determined if, given an increase in feeding, exposure and associated survival probability differ depending on whether feeding occurs more often but for a shorter overall time period compared with feeding less often but over a longer developmental period (see Costs and Risks to Feeding, below).

One of the few caterpillar species for which changes in the fine structure of feeding have been studied is the sphingid *Manduca sexta*. When caterpillars were fed either tobacco leaves (85% water) or an artificial diet (77% water), those consuming the former exhibited compensatory feeding, increasing their fresh food consumption by spending more time feeding, accomplished through longer, more frequent meals (Reynolds et al. 1986); fresh food consumption was increased in a similar manner when caterpillars were fed diluted artificial diets (Timmins et al. 1988). Similarly, Bowdan (1988b) found that *M. sexta* caterpillars increased their food consumption by increasing meal duration, both when given access to food after a period of starvation, compared with unstarved individuals, and when fed tomato leaves versus an artificial diet, but in contrast to the previously cited studies, instantaneous feeding rate (i.e., bite frequency) also increased whereas meal frequency did not change. Similar responses occurred in adults of the coccinellid *Epilachna varivestis* following starvation (Jones et al. 1981). When nymphs of *Locusta migratoria* were fed a diet with reduced protein content brought about by addition of cellulose, they increased food consumption by feeding more frequently (Simpson and Abisgold 1985; see also Raubenheimer and Simpson 1990). Clearly, considerably more data are required before we can begin to formulate ecological hypotheses regarding changes in the fine structure of feeding associated with compensatory increases in food consumption.

Prolonged Development

Prolonging larval development is another means to increase overall food consumption, although RCR values will likely decline; slowed growth and lengthened development frequently occur for caterpillars feeding on suboptimal foods. However, because the disadvantages to fitness can be substantial (see below), slowed growth and extended development are probably more commonly consequences of the inability to adequately compensate for reduced food quality (or other environmental stresses through increased feeding (or other means), rather than compensatory mechanisms themselves. Nonetheless, there is likely an adaptive component to the magnitude of prolonged development, both because its ecological consequences may differ among species, and because its underlying cause (i.e., the need to obtain a species-specific minimum body size or mass prior to pupation) is likely an evolved trait, as discussed below.

One deleterious aspect of prolonged development is that it can increase a caterpillar's exposure to natural enemies by lengthening the overall time in the caterpillar stage; this may be especially critical if the normal timing of development is such that the larval stage is usually completed before there is an increase in certain natural enemy populations, and if the natural enemies undergo a learning process that improves their ability to locate the caterpillars (Heinrich and Collins 1983; Gould and Jeanne 1984; Fitzgerald et al. 1988; Vet et al. 1990). Slowed growth resulting in prolonged development may also maintain the caterpillar longer within the vulnerable size range of various, especially invertebrate, natural enemies, which frequently attack only certain size categories of prey or host (Enders 1975; Slansky 1986; Hagen 1987; Isenhour et al. 1989; but see Clancy and Price 1987; Riechert and Harp 1987).

Asynchrony with foodplant phenology can also be a consequence of prolonged development. If growth is being slowed by an ongoing deterioration in food quality (e.g., due to leaf aging or leaf damage), prolonged development is likely to exacerbate the deleterious impact on the caterpillar. The life cycle of many phytophagous insect species is synchronized for feeding on a particular stage of their foodplants (e.g., the early season flush of nutrient-rich leaves), and their performance is reduced when feeding at other times, especially associated with altered food quality (Slansky and Scriber 1985; Schroeder 1986b; Tallamy 1986; Aide and Londoño 1989; Potter and Redmond 1989; Straw 1989; Thomas 1989). For example, a 1 week delay in establishing first instar caterpillars of the geometrid *Oporinia autumnata* on newly expanding birch leaves reduced their survival to pupation about 30 to 80%, depending on latitude (Haukioja 1980). Caterpillar performance can be reduced when feeding on plants or leaves damaged by herbivores and pathogens (Haukioja and Hanhimäki 1985; Faeth 1986; Karban and Myers 1989; Haukioja 1991), and natural enemies may use damaged leaves or odors emanating therefrom, as cues for locating their host or prey (Heinrich and Collins 1983; Faeth 1986; Nordland et al. 1988). Thus, a further disadvantage to prolonged development is the greater likelihood of feeding on increasingly damaged plants. Early leaf abscission may select against prolonged development, especially in species restricted to developing on or in a single leaf, such as leaf miners (Faeth et al. 1981; but see Stiling and Simberloff 1989).

Asynchrony with mates could be another consequence of prolonged development. For example, adult males of many lepidopterans and other insects typically emerge before females (Lederhouse et al. 1982; Thornhill and Alcock 1983), and slow-developing males will probably have reduced mating success. The need to complete a certain period of development before being able to withstand or emigrate from an upcoming harsh environment, such as entering diapause prior to a freezing temperature or achieving a particular body size prior to overwintering (Chew 1975; Scriber 1983; Williams and Bowers 1987; Nylin 1988; Wolda 1988; Wickman et al. 1990), and producing a migratory adult before lethally high summer temperatures (Malcolm et al. 1987), would also select against prolonged development as a compensatory response. These same consequences may also select against the use of foodplants on which caterpillar development is slowed sufficiently to alter deleteriously the individual insect's phenology.

Given the above, it would appear that insects generally should be strongly selected to maintain a rather specific duration of development. This could occur if caterpillars had highly effective mechanisms to compensate for variable food quality and other environmental stresses, such that they could achieve their genetically determined potential body size or mass without prolonged development. However, from the previous discussion it is evident that growth performance varies despite compensatory responses by the insects (see also Benefits of Compensatory Feeding, below). Alternatively, caterpillars could pupate at body sizes below their genetic potential in order to avoid prolonging their duration of development. However, various degrees of prolonged development do occur, in large part because probably all insects must achieve some minimum body size or mass before pupation will occur (Nijhout 1975). Thus, if growth is slowed, development will be prolonged until this minimum is reached or death ensues.

If we assume that the values of minimum weight for pupation (as a percentage of the maximal weight) reflect life-style differences, then various predictions can be made. For example, species having a high probability of experiencing depleted or deteriorated larval food supplies, and species in which the timing of pupation is more important to fitness than pupal size or weight (e.g., those with the need to achieve a cold-hardy diapausing pupa prior to the first freeze) might evolve a low minimum weight percentage for pupation (e.g., Banno 1990). In other words, for these species prolonged development to achieve their maximal body size or weight may not be a viable strategy because it leads to death without reproducing. Under such circumstances, a substandard sized adult might experience a reduced reproductive contribution compared with a full-sized adult, but it would nonetheless have greater fitness than if it attempted to achieve its maximum body mass through prolonged development. A low minimum weight percentage might also be seen in species in which the adult feeds and can partially compensate for poor larval performance. Species with low minimum weight values will tend to produce adults exhibiting a broad range of size or weight. In contrast, there might be selection for higher minimum weight values in species with more consistent larval food supplies, in those in which temporal asynchrony is less critical to fitness, and in insects in which body weight may be more important to survival and reproductive output than developmental time, such as those requiring a high level of metabolic reserves to withstand diapause (Fields and McNeil 1988) and species with nonfeeding adults. In such species, prolonged development might be a more viable strategy than for the previously described species, and adult body size or weight will tend to show less variation than for the species with low minimum weight values (for further discussion see, e.g., Taylor 1980; Schroeder 1986b; Stearns and Koella 1986; Gebhardt and Stearns 1988; Nylin et al. 1989).

Insect species do seem to differ in their minimum size or weight for pupation, with the smallest individuals ranging from about 10 to 60% of the largest (Slansky and Scriber 1985). These data must be evaluated with caution because this issue has seldom been addressed systematically (e.g., Anderson and Nilssen 1983), and the different studies often measured different parameters (e.g., dry mass, fresh mass, body length, etc.), with the variability sometimes measured in natural populations and sometimes generated experimentally. Nonetheless, there appear to be some trends in the relationship between minimum body size and life-style that support some of the above-stated predictions, especially among wood-feeding insects and dung- and carrion-feeders (Haack and Slansky 1987; Hanski 1987). Even though lepidopteran species span the range in minimum size for pupation indicated above, the significance of these values to life-style differences has not been rigorously investigated (see Angelo and Slansky 1984; Banno 1990). Further, we probably lack sufficient information for most insect species (and especially species with phytophagous larvae) to adequately assess the precise fitness tradeoffs between prolonged development and reduced body size or mass. The work of Smith et al. (1987) examining tradeoffs between fecundity, juvenile survival, and developmental rate in *Pieris rapae* is a step in the right direction (see also general discussion by Gatto et al. 1989), but more research on this topic is required before we can begin to understand the ecological significance of prolonged development and its role as a compensatory mechanism.

Benefits of Compensatory Feeding

The growth and survival probability of caterpillars consuming food with a suboptimal level of one or more nutrients is likely to be reduced unless they increase food intake and/or exhibit other compensatory responses. Fitness-associated performance criteria of insects (i.e., fecundity, dispersal ability, mating success, life span, susceptibility to natural enemies, etc.) are often positively correlated with body size (e.g., Yamada and Umeya 1972; Hespenheide 1973; Sanders and Lucuik 1975; Blum and Blum 1979; Barbosa et al. 1981; Hinton 1981; Moscardi et al. 1981b; Isenhour et al. 1989; Wickman and Karlsson 1989; Banno 1990), and probably also with the level of nutrient storage within a particular body size (Slansky and Scriber 1985; Wickman and Karlsson 1989). Maintaining a particular developmental period is also important to insect fitness (see Prolonged Development, above). Thus, the finding that, for many of the species listed in Tables 2.3 and 2.4, the substantial increases in feeding in response to reduced food nutrient level are accompanied by considerably less variation in measures of growth performance, suggests that the ability to increase feeding is truly a compensatory, adaptive response (see also Slansky 1982b).

This conclusion is most obvious when growth performance is maintained at a level more or less similar to that of the control insects. However, even if growth performance declines despite the compensatory feeding, the response may be considered adaptive in that it prevents a greater reduction in growth or nutrient storage, or limits the length of prolonged development. For example, without their increase in RCRfw as diet nutrient level was reduced from 35 to 14% fw through dilution with water, caterpillars of *A. gemmatalis* would have experienced a decline in their RCRdw of about 54%, rather than the 15% reduction that actually occurred, and their growth also would have declined substantially more than it did (Fig. 2.7). Thus, while reductions in food consumption (based on dry mass) and in growth performance of insects on certain foods might result from a lack of compensatory feeding, it cannot be assumed that this is the case; clearly, food consumption based on fresh mass must be quantified to determine whether compensatory feeding has occurred.

That caterpillars fail to achieve their genetic potential in body size, mass, nutrient storage, and developmental time, despite compensatory changes in the consumption and utilization of food, and other compensatory responses (e.g., behavioral thermoregulation), indicates that these responses are not always effective at fully mitigating the impact of a variable environment on their performance. In regard to increased feeding in response to a decline in dietary nutrient level, various costs and risks associated with feeding likely constrain a caterpillar's ability to respond adequately. These costs and risks to feeding are addressed in the following section.

Costs and Risks to Feeding

The high end of the range of RCRdw values for leaf-chewing insects is about 3–6 mg dw mg^{-1} d^{-1} (Slansky and Scriber 1985), which would translate into presumably maximal RCRfw values attainable by leaf-chewing insects of 30–60 mg fw mg^{-1} d^{-1} on high water (90% fw) foliage. However, the values exhibited by many of the species whose food consumption has been quantified are substantially lower than this, even when consuming their usual foodplants under more or less optimal abiotic environmental conditions. Feeding rates that are less than maximal may be the norm because insects, like other animals, have presumably evolved in a manner that maximizes the benefit/cost fitness ratio of their behavior, and associated with various tradeoffs, this often appears to be accomplished by exhibiting less than maximal performance rates (Hassell and Southwood 1978; Calow 1982; Altmann 1984; Stephens and Krebs 1986; Reynolds 1990). The capacity for compensatory increases in feeding rate under certain circumstances supports this hypothesis; presumably, caterpillars do not typically feed at their maximal rate because there are costs and risks to feeding faster that are endured only when failure to respond would on average result in a greater reduction in fitness than would responding.

Despite their so-called ravenous appetites and characterization as "feeding machines," food intake by caterpillars does not appear to be limited by time per se (i.e., they do not feed continuously; see Feeding Rhythms, above). The

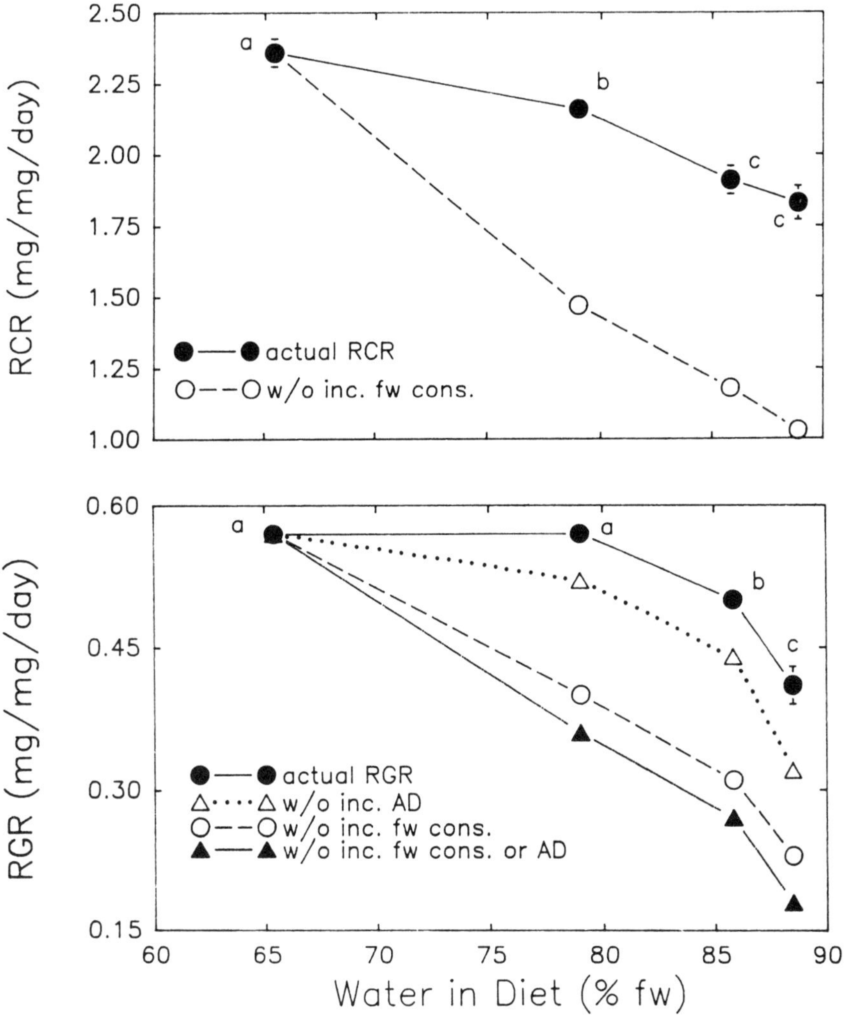

Figure 2.7. (Upper panel) Actual relative consumption rate (RCR, based on dry weight) values of *A. gemmatalis* caterpillars fed an artificial diet in which the nutrients were diluted with increasing amounts of water, and expected RCR values if the caterpillars had not increased their fresh weight food consumption as the diet was diluted. (Lower panel) Actual relative growth rate (RGR, based on dry weight) values of *A. gemmatalis* caterpillars fed an artificial diet in which the nutrients were diluted with increasing amounts of water, and expected RGR values if approximate digestibility (AD), fresh weight food consumption or both had not increased as the diet was diluted. Means followed by the same letter are not significantly different. (From Slansky and Wheeler 1989, by permission.)

question thus becomes, what limits the amount of time a caterpillar spends feeding, and thus its overall consumption rate? Given a set duration of development, one factor limiting the time available for feeding is the proportion of a caterpillar's time spent molting, which may approach and even exceed 50% (Ayres and MacLean 1987; Stamp 1990). However, caterpillars do not appear to feed continuously even during the nonmolting time available for feeding. A number of "generic" costs and risks can be postulated that may well make feeding at a maximal rate a suboptimal endeavor under average conditions; these are described below. However, before we can understand the constraints on feeding rate, we need to know how these costs vary as a function of feeding rate and what benefits are derived from the costs; that is, the energetic, nutritional and other costs not only need to be quantified but also must be translated into their impact on individual fitness. No leaf-chewing insect species (and probably no other insect species) has been studied adequately enough to derive a benefit/cost analysis that would allow us to determine the advantage of one feeding rate over another; studies of the feeding behavior of certain aquatic predatory insects probably come closest to this goal (Sih 1987; Sih et al. 1990; see also Abrams 1989).

Potential costs and risks to feeding include the following:

1. Reduced food processing time, and thus reduced absorption rate. For phytophagous caterpillars, the rate of passage of food through the gut (i.e., the time from ingestion of a meal to the appearance of the first fecal pellets from the meal) probably commonly ranges from about 1 to 6 hours (Slansky 1974; Santos et al. 1983; Reynolds 1990), with the larval gut being more or less continually full. Absorption of nutrients from the midgut may, however, peak relatively rapidly (i.e., within 30 to 60 minutes; Reynolds 1990). It has been postulated that the particular value of the average feeding rate for a species, and the presumably correlated rate of passage of food through the gut, have evolved to provide a particular value for the efficiency of digestion and absorption (termed "approximate digestibility," AD, in the literature on quantitative food utilization by insects) that yields the maximal rate of digestion and absorption of nutrients (Slansky 1974; Reynolds et al. 1985; Reynolds 1990; see also Taghon 1981).

While this seems logical based on the kinetics of digestion and absorption, if this model were correct, then a faster than average rate of passage of food through the gut, expected to occur when a caterpillar exhibits compensatory feeding, should lower AD. In some cases, AD and feeding rate appear inversely correlated (e.g., Slansky and Feeny 1977; Schroeder 1986a; Karowe and Martin 1989), but in many cases, AD did not decline with increased feeding rate, and in some cases it actually increased (Schroeder 1986b; Peterson et al. 1988; Slansky and Wheeler 1989; 1991; Wheeler and Slansky 1991). One explanation for the lack of decline in AD with increased feeding rate is that the latter may not be closely correlated with retention time in the gut (see Simpson 1983; Timmins et al. 1988); also, the rates of digestion and absorption may be increased in some manner (e.g., through greater digestive enzyme production), thereby mitigating an increase in the rate of throughput. However, because of the lack of data, it is at present unclear if food processing time limits feeding rate; if it does, this would most likely occur in species with frequent feeding bouts (i.e., < ca. 1–2 hours between bouts). That the time between feeding bouts in some species appears to be longer than the time required to digest and absorb the ingested nutrients (see Feeding Rhythms, above) suggests that other factors limit feeding rate in these species. A time lag in the physiological response to a reduced-nutrient diet may, however, limit the magnitude of a compensatory feeding response (Slansky and Wheeler 1991); locusts appear to respond faster to nutritional feedback than caterpillars (Simpson et al. 1989, 1990).

2. Energy costs of consuming and processing food. The respiration rate of caterpillars increases with feeding, digestion and absorption, and excretion of excess or imbalanced nutrients (McEvoy 1984; Karowe and Martin 1989; Kukal and Dawson 1989; Van't Hof and Martin 1989). However, the relationship of these energy costs to feeding rate and their impact on caterpillar performance remain to be demonstrated. The emerging view that caterpillar growth is probably most commonly nutrient (e.g., nitrogen) or water limited, rather than energy limited, downplays the impact of energy costs of feeding and food processing. Further, in studies of quantitative food utilization by insects, increased energy costs are expected to be reflected in a lowering of the net growth efficiency [also termed efficiency of conversion of digested (and absorbed) food to biomass, ECDdw], which reflects the percentage of digested and absorbed food converted to biomass, with the remainder being used in metabolic activities. In many cases, however, a compensatory increase in feeding (RCRfw) is not accompanied by a decline in ECDdw (e.g., Schroeder 1986b; Martin and Van't Hof 1988; Karowe and Martin 1989; Slansky and Wheeler 1989; Van't Hof and Martin 1989; Stamp 1990). For example, an almost 2-fold higher RCRfw of *Orgyia leucostigma* caterpillars on an 82% versus 65% water diet was accompanied by a significantly higher ECDdw (Van't Hof and Martin 1989), suggesting that the energy costs of feeding did not divert energy from growth.

However, there are at least three problems with expecting that changes in ECDdw values will reflect differences in energy costs. First the energy content of food, feces, and insect biomass can vary. For example, a caterpillar experiencing increased energy costs may accumulate less high-energy lipid, but it may gain the same dry mass as one experiencing lower energy costs. Assuming the two caterpillars absorbed the same amount of their ingested food, the ECDdw of each would be the same despite the greater energy costs of the former. Thus, the ECD based on energy (ECDen) is the more appropriate measure for metabolic costs, but unfortunately most food utilization studies of insects measure only dry mass budgets (see Slansky 1985). In a study of compensatory feeding by *A. gemmatalis* caterpillars fed nutrient-diluted diets, Slansky and Wheeler (1991) found that RCRfw was more highly correlated (negatively) with ECDen than with ECDdw (although both equations were statistically significant), in large part because the biomass (mg dw) gained by the caterpillars and their lipid content (% dw) both declined as feeding rate increased with diet dilution.

A second problem in using changes in ECD (even when based on energy) to indicate differential energy costs is that the ECD reflects the "overall" metabolic cost. Many metabolic activities (feeding, tissue syntheses, uric acid production, ion pumping, etc.) combine to make up the overall metabolic cost, and a change in any one component (e.g., energy used in feeding) may be balanced or masked by other changes in metabolism and thus not be reflected by a change in ECD (see Martin and Van't Hof 1988). For example, both ECDdw and ECDen of *H. cecropia* caterpillars declined dramatically when fed desiccated leaves of *P. serotina*, even though their RCRfw was 43% less than larvae fed moister leaves; any energy savings due to the lower feeding rate were probably overwhelmed by the greater energy costs associated with producing metabolic water and possibly with resorbing a high proportion of water from the gut to produce very dry feces (Scriber 1977). In addition, the water limitation may have prevented the larvae from converting a higher proportion of their energy into tissue growth (see Martin and Van't Hof 1988). The ECD of *Manduca sexta* larvae declined when fed various allelochemicals, not because of an increase in metabolic rate (e.g., associated with detoxifying the allelochemicals), but because of the prolonged development and consequent increase in the proportion of assimilated food used in overall respiration caused by the allelochemicals (Appel and Martin 1992). A third problem is that food utilization budgets are based on several estimations that limit their accuracy (See Schmidt and Reese 1986), and thus ECD values may be too imprecise to detect changes in insect metabolism. Finally, even if presumed energy costs are detected, we have a poor understanding of their impact on fitness. For example, the lipid content of caterpillars may decline, perhaps because of greater energy costs, but the effects on survival to the adult stage and on adult fecundity have seldom been assessed (Angelo and Slansky 1984; Karowe and Martin 1989; Mason et al. 1990; Slansky and Wheeler 1991; Wheeler and Slansky 1991). Thus, the impact of the energy costs of consuming and processing food on limiting feeding rate remains uncertain. If feeding can have biologically significant energy costs, this is most likely to occur for those species in which energy stored by the caterpillar is critical to pupal survival and/or adult performance, especially when the insect is feeding at a high rate on a food with reduced energy content (such as occurs when nutrients are highly diluted with an indigestible substance; Wheeler and Slansky 1991) and when substantial excretory demands occur (such as when consuming foods that are highly concentrated or imbalanced in their nutrient content; Schroeder 1986a; Karowe and Martin 1989; Van't Hof and Martin 1989).

3. Nutrient costs of consuming and processing food. The most evident potential nutrient costs to feeding by caterpillars are the production of digestive enzymes and peritrophic membrane, but their relationship to increased feeding rate and

any subsequent drain on a caterpillar's nutrient budgets apparently have not been assessed. Synchronizing the production of digestive enzymes with the amount of nutrients in the gut (i.e., secretagogue stimulation of digestive enzymes; Chapman 1985; Broadway and Duffey 1986a; Simpson and Simpson 1990) and recycling them (Terra 1990) likely limit nutrient loses via excreted digestive enzymes. An insect may produce more digestive enzymes in response to their inhibition by certain allelochemicals, which could thus create a nutrient drain (Broadway and Duffey 1986b); reduced nutrient absorption, whatever the cause, might interfere with the action of detoxication enzymes and thus an insect's ability to handle absorbed allelochemicals (Lindroth et al. 1990). The probable cost in water of increased feeding on low-water foods (especially losses via the feces leading to desiccation) was discussed previously (see Compensatory Feeding Scope, above). If increased feeding causes greater physical exercise of the mouthpart musculature, this could lead to greater allocation of nutrients to building larger muscles and an associated larger head capsule (Bernays 1986a). For example, when *Pseudaletia unipuncta* caterpillars were fed tough grass foliage, their head mass doubled to about 16% of total dry body mass, compared with artificial diet-reared individuals; the benefit of this was an increased ability to consume the tough grass. Because few studies have addressed the nutrient costs of feeding, their role in limiting feeding rate is not clear. If nutrient costs can have a biologically significant effect, this would most likely occur in a situation such as a caterpillar producing a high amount of enzymes to digest a low-nutrient food, but additional research is required before such an impact can be confirmed.

4. Coping with potentially toxic allelochemicals. There is no doubt that allelochemicals have a major influence on insect feeding behavior and associated lifestyle characteristics. The choice of a species as a foodplant frequently involves allelochemicals acting as attractants, stimulants, repellents, and deterrents (Ahmad 1983; Hsiao 1985; Frazier 1986; Bernays and Chapman 1987; Schoonhoven 1987). Further, insects may avoid particular individuals or certain leaves of their foodplants, sometimes even the most nutrient-rich ones, because these have higher concentrations of potentially deleterious allelochemicals (e.g., van der Meijden et al. 1984; Williams et al. 1983b; Johnson et al. 1985). Insects may manipulate their food, such as by trenching and by chewing through petioles, thereby preventing an increase in allelochemical concentration at the feeding site that might otherwise affect them negatively (Tallamy 1986; Dussourd, Chapter 3), and they may move relatively long distances between feeding sites on a plant, possibly in part to avoid locally induced toxins (Rhoades 1985). Caterpillars may construct and live within leaf rolls or burrow in the soil during the day, allowing them to avoid the deleterious impact of photoactive allelochemicals (Berenbaum 1987; Fields et al. 1990). The extent to which such phenomena limit a caterpillar's feeding time and thus its overall consumption rate has seldom been investigated, but these and other such adaptations clearly indicate that allelochemicals are important forces in the evolution of insect biochemistry, physiology, and behavior

(see Spencer 1988). Indeed, some insect species derive beneficial uses from ingested allelochemicals that are toxic to other species, including their use as antienemy agents, pheromone precursors, and nutrients (Bowers 1990; Slansky 1992).

Caterpillars of different species can tolerate the ingestion of various potentially toxic compounds, either because they exhibit target-site insensitivity or because they have evolved to detoxify, excrete, or otherwise neutralize the compounds, thereby preventing them from reaching pharmacologically active levels (Brattsten and Ahmad 1986). In addition, for those dietary allelochemicals that are tolerated, caterpillars can often cope with a wide concentration range, even doses higher than those occurring naturally (e.g., Fox and Macauley 1977; Blau et al. 1978; Johnson et al. 1985; Manuwoto and Scriber 1985), probably because either their target-site insensitivity is more or less independent of concentration effects or their detoxication/excretion rates are increased correspondingly (Brattsten and Ahmad 1986). The induction of detoxication enzymes in response to absorption of allelochemicals suggests that there are costs to maintaining high levels of these enzymes which are incurred only when necessary (Krieger et al. 1971). Recent evidence indicates that these costs are likely not energetic or nutritional; induction of detoxication enzyme levels up to 7 to 9 times greater than baseline values did not significantly impact the consumption and utilization of food by *Helicoverpa (Heliothis) zea* caterpillars, and the amount of protein allocated to the induced enzymes was estimated to be much less than 1% of the dry mass of the caterpillars (Neal 1987; see Slansky 1992 for further discussion of this topic).

Although insects exhibit diverse means of coping with allelochemicals, their deleterious impact on insect performance is well documented (Holyoke and Reese 1987; Reese and Holyoke 1987), indicating either that insects have not evolved to detect and reject food containing certain of these compounds or that the quest for nutrients overrides the repulsion of the allelochemical. This latter situation is expected to occur especially when the insect has no choice of foods and thus must either feed on the allelochemical-containing food or starve. Despite the important impact of allelochemical intake on insect performance, however, at least two key questions have seldom been addressed: (1) Is a species' baseline feeding rate on its usual foodplants set so as to minimize the costs of or to not exceed the capacity of its coping systems (i.e., detoxication, excretion, etc.)? and (2) Can these coping systems be overwhelmed (despite their capacity for induced activity) because a compensatory increase in feeding rate in response to reduced-nutrient food results in the too-rapid ingestion of an allelochemical? Many species exhibit reduced feeding in response to certain allelochemicals in the food, but it is difficult to distinguish whether this results from an active response by the insect or is merely a passive consequence of the deleterious action of the ingested allelochemical slowing growth and thus consumption, and few studies have done so (see discussions in Berenbaum 1986; Slansky 1992). Indeed, in many studies assessing allelochemical effects on insects, food consumption is not even quantified (Slansky 1992). Further, to determine the adaptive nature of these responses, whether the benefit of reduced allelochemical ingestion outweighs the cost of reduced nutrient intake must be assessed. In several cases allelochemicals acting as feeding deterrents are not toxic to the responding insect (Bernays and Chapman 1987; Cottee et al. 1988; Bernays 1990).

That caterpillars can tolerate higher relative consumption rates of allelochemicals (RCRal) than typically experienced suggests that feeding rates, at least over a certain range, are not limited by the need to limit RCRal. Similarly, substantial compensatory increases in feeding rate in response to foods with reduced levels of nutrients have led to little apparent deleterious impact due to the accompanying increased RCRal values (Slansky and Feeny 1977; Lincoln et al. 1982: Johnson et al. 1985; Lincoln and Couvet 1989; Raubenheimer and Simpson 1990), further suggesting that caterpillars can adequately cope with a broad range in RCRal values for the allelochemicals they typically encounter. However, such results do not necessarily establish whether the limit to a compensatory increase in feeding might be set by the maximum tolerated ingested dose of the allelochemicals. That is, the results of these studies could indicate that the compensatory increase in feeding is limited by other factors below the values of RCRal that would be deleterious, or that the species have indeed evolved to limit their feeding rate to values that prevent an active dose of the allelochemical from being ingested. The studies cited above were not designed to test these hypotheses. For example, to adequately test the latter hypothesis, the experimental design must first include a determination of the pharmacologically active dose of the allelochemical (in terms of RCRal) and then a manipulation of the insects' diet such that they would ingest this active dose (or exceed it) if they exhibited their typical compensatory increase in RCRfw in response to a reduction in food nutrient level. If the insects have the ability to adjust their RCRfw to prevent the ingestion of a deleterious dose of allelochemical, then they should moderate their increase in RCRfw to limit RCRal. Presumably, in this situation the negative impact on fitness of the reduced nutrient intake (due to the moderated RCRfw as the nutrients were increasingly diluted) would be less than the impact of the increased RCRal if RCRfw were not moderated.

Such an experimental protocol has been used with caterpillars of *A. gemmatalis* consuming an artificial diet containing the pharmacologically active methylxanthine alkaloid, caffeine (Slansky and Wheeler 1992). As the nutrient content of the diet was increasingly diluted, while keeping the caffeine concentration (on a percentage fw basis) constant, the RCRfw of caterpillars increased to the point at which an active (and in this case lethal) dose of caffeine was ingested. If the caterpillars had not exhibited such a large compensatory increase in RCRfw, they would not have ingested a toxic dose. Thus, in this case the caterpillars adjusted their feeding rate in response to changes in dietary nutrient level and not to RCRal. Because caffeine does not occur in the food of this species, it is perhaps not surprising that they have not evolved to respond adaptively to it. However, larvae of *Spodoptera frugiperda* also failed to moderate their increased-consumption response to diluted nutrients when the diet contained an "evolutionarily familiar" allelochemical (the coumarin esculetin). Consequently, their relative growth rate (RGR) decreased with the increase in RCRal (Slansky, unpublished data).

In a few other studies, greater RCRal values resulting from increased consumption of nutrient-diluted diets have also had deleterious impacts on caterpillar performance. For example, Stamp (1990) found reduced biomass gain of *M. sexta* at 20° and 30°C (but not at 25°C) associated with an approximately 1.7-fold increase in consumption of a reduced-nutrient diet containing rutin, compared with caterpillars consuming the rutin-free diet. Although older caterpillars of *Spodoptera eridania* were not affected deleteriously by their 1.6-fold increase in RCRal of the alkaloid sparteine on a low protein diet, mortality was greater for younger larvae on this diet compared with those fed the high protein diet (presumably the young caterpillars also exhibited compensatory feeding but this was not quantified) (Johnson and Bentley 1988). For a fifth instar *Locusta migratoria*, the presence of tannic acid limited their compensatory increase in feeding on a low protein diet compared with the tannin-free diet (Raubenheimer 1992), suggesting that feeding was being moderated to limit RCRal, but whether this was indeed an adaptive response was not assessed. From the above discussion it should be obvious that much more research is necessary before we can determine whether there is an adaptive link between feeding rate and a species' ability to cope with ingested allelochemicals.

5. Increased exposure to natural enemies. The impact of natural enemies on caterpillars is broadly evident both in their diverse antienemy adaptations and in their high mortality from natural enemies despite these adaptations (Rausher 1981; Gould and Jeanne 1984; Smiley 1985; Bernays 1989; Evans and Schmidt 1990; Fritz and Nobel 1990; also see appropriate chapters in this volume). It is likely that several feeding-associated factors influence a caterpillar's mortality risk, especially exposure during movement to a feeding site or during feeding, and creation of leaf damage that serves as a cue to enemies (Hassell and Southwood 1978; Heinrich 1979; Price et al. 1980; Heinrich and Collins 1983; Schultz 1983; Barbosa 1988; Nordlund et al. 1988; see also Gross and Price 1988; Bernays 1989), although the increase in mortality may be slight (Bergelson and Lawton 19;88). In a recent study, Loader and Damman (1991) experimentally confirmed previous suggestions (e.g., Slansky and Feeny 1977) that increased feeding (in this case, an approximately 2-fold increase on low versus high-nitrogen collards by *Pieris rapae* larvae) can lead to greater mortality from natural enemies. Avoidance of natural enemies may be an important component of food choice (Atsatt 1981) and of seasonal period of feeding (Kukal and Kevan 1987), and may force caterpillars to feed under suboptimal conditions (Stamp and Bowers 1990; see Environmental Effects on Feeding, above). Reduced exposure to day-active natural enemies is probably a key selective advantage of certain feeding

rhythms (i.e., dawn/dusk and nocturnal feeding; see Feeding Rhythms, above). Thus, the time available for feeding may be limited because of antienemy adaptations, but how this affects the baseline feeding rate and compensatory feeding ability of caterpillars remains to be systematically addressed. The rate of feeding, as well as the composition of the ingested food, may also influence a caterpillar's exposure to pathogens (Gallardo et al. 1990; Meisner et al. 1990).

Conclusions

As is evident from the previous discussion and the other chapters in this volume, we have gained considerable knowledge about the variety of life-styles and the diverse biochemical, physiological, and behavioral adaptations evolved by insects, especially phytophagous caterpillars, that allow them to fulfill their fundamental quest for nutrients while coping with less than ideal abiotic and biotic environmental conditions. Equally evident are the many basic questions about how insects obtain and allocate resources for which we yet have no answers, such as: How and why do insects differ in their ability to compensate for altered food quality? Do insects regulate their rate of food consumption to achieve an optimal balance in their ingestion of beneficial nutrients and potentially deleterious allelochemicals? To what extent does the evolution of adaptations for coping with natural enemies and other deleterious factors, as well as the short-term responses of individual insects to these factors, constrain food choice and feeding rates of insects, and thus presumably their growth rates, voltinism patterns, and other components of resource allocation? Although the list of unanswered questions is lengthy, it is encouraging that we have developed the broad framework of understanding that allows us to identify these and other key questions and to design and implement the appropriate experiments and comparisons to answer them. Knowledge of the consumption and utilization of food by insects is also of utmost importance to the management of insect pests. For example, effective development and use of host plant resistance tactics to manipulate insect pest performance requires an understanding of the relationships between the attributes of crop plants, their consumption and utilization by insect pests, and subsequent pest performance (Smith 1989; Slansky 1990). Clearly, the burgeoning of research in the basic and applied contexts of insect nutritional ecology that has occurred during the past 20 years will continue, and during the next few years many of these important questions should be answered.

Acknowledgments

I thank Greg Wheeler and Enwood Nevis for helping track down copies of published studies for this chapter, Dottie Clements for assistance with the graphics, Kris Faircloth and Sue Ann Cotton for helping with word processing, and

Nancy Stamp and G. P. Waldbauer for their helpful comments on an early draft of this manuscript. This material is based on work supported in part by the National Science Foundation under Award No. BSR-8918254, Hatch project FLA-ENY-02362, and state project ENY-03012. Florida Agricultural Experiment Station Journal Series No. R-01555.

Literature Cited

Abrahamson, W. G. (ed.) 1989. Plant-Animal Interactions. McGraw-Hill, New York.

Abrahamson, W. G., and Weis, A. E. 1987. Nutritional ecology of arthropod gall makers, pp. 235–258. In F. Slansky Jr. and J. G. Rodriguez (eds.), Nutritional Ecology of Insects, Mites, Spiders, and Related Invertebrates. Wiley, New York.

Abrams, P. A. 1989. Decreasing functional responses as a result of adaptive consumer behavior. Evol. Ecol. 3:95–114.

Ahmad, S. (ed.) 1983. Herbivorous Insects. Host-Seeking Behavior and Mechanisms. Academic, New York.

Ahmad, S., Brattsten, L. B., Mullen, C. A., and Yu, S. J. 1986. Enzymes involved in the metabolism of plant allelochemicals, pp. 73–151. In L. B. Brattsten and S. Ahmad (eds.), Molecular Aspects of Insect-Plant Associations. Plenum, New York.

Aide, T. M., and Londoño, E. C. 1989. The effects of rapid leaf expansion on the growth and survivorship of a lepidopteran herbivore. Oikos 55:66–70.

Alexander, A. J. 1961. A study of the biology and behavior of the caterpillars, pupae and emerging butterflies of the subfamily Heliconiinae in Trinidad, West Indies. Part I. Some aspects of larval behavior. J. N.Y. Zool. Soc. 146:1–24.

Altmann, S. A. 1984. What is the dual of the energy-maximization problem? Am. Nat. 123:433–451.

Anderson, J., and Nilssen, A. C. 1983. Intrapopulation size variation of free-living and tree-boring Coleoptera, Can. Entomol. 115:1453–1464.

Angelo, M. J. and Slansky, F., Jr. 1984. Body building by insects: Trade-offs in resource allocation with particular reference to migratory species. Florida Entomol. 67:22–41.

Appel, H. M., and Martin, M. M. 1992. Significance of metabolic load in the evolution of host specificity of *Manduca sexta*. Ecology 73:216–228.

Atsatt, P. R. 1981. Lycaenid butterflies and ants: Selection for enemy-free space. Am. Nat. 188:638–654.

Ayres, M. P., and MacLean, S. F., Jr. 1987. Molt as a component of insect development: *Galerucella sagittariae* (Chrysomelidae) and *Epirrita autumnata* (Geometridae). Oikos 48:273–279.

Banno, H. 1990. Plasticity of size and relative fecundity in the aphidophagous lycaenid butterfly, *Taraka Hamada*. Ecol. Entomol. 15:111–113.

Barbosa, P. 1988. Natural enemies and herbivore-plant interactions: influence of plant allelochemicals and host specificity, pp. 201–229. In P. Barbosa and D. K. Letourneau (eds.), Novel Aspects of Insect-Plant Interactions. Wiley, New York.

Barbosa, P., and Letourneau, D. K. (eds.) 1988. Novel Aspects of Insect-Plant Interactions. Wiley, New York.

Barbosa, P., and Schultz, J. C. (eds.) 1987. Insect Outbreaks. Academic, San Diego.

Barbosa, P., Cranshaw, W., and Greenblatt, J. 1981. Influence of food quantity and quality on polymorphic dispersal behaviors in the gypsy moth, *Lymantria dispar*. Can. J. Zool. 59:293–296.

Bell, W. J., and Cardé, R. T. (eds.) 1984. Chemical Ecology of Insects. Sinauer, Sunderland, M.A.

Berenbaum, M. 1986. Postingestive effects of phytochemicals on insects: On Paracelsus and plant products, pp. 121–153. In J. R. Miller and T. A. Miller (eds.), Insect-Plant Interactions. Springer-Verlag. New York.

Berenbaum, M. 1987. Charge of the light brigade; insect adaptations to phototoxins, pp. 206–216. In J. R. Heitz and K. R. Downum (eds.), Light Activated Pesticides. ACS Symp. Ser. 339, Washington, D.C.

Berenbaum, M. R., and Isman, M. B. 1989. Herbivory in holometabolous and hemimetabolous insects: Contrasts between Orthoptera and Lepidoptera. Experientia 45:229–236.

Bergelson, J. M., and Lawton, J. H. 1988. Does foliage damage influence predation on the insect herbivores of birch? Ecology 69:434–445.

Bernays, E. A. 1986a. Diet-induced head allometry among foliage-chewing insects and its importance for graminivores. Science 231:495–497.

Bernays, E. A. 1986b. Evolutionary contrasts in insects: Nutritional advantage of holometabolous development. Physiol. Entomol. 11:377–382.

Bernays, E. A. 1989. Host range in phytophagous insects: The potential role of generalist predators. Evol. Ecol. 3:299–311.

Bernays, E. A. 1990. Plant secondary compounds deterrent but not toxic to the grass specialist acridid *Locusta migratoria*: Implications for the evolution of graminivory. Entomol. Exp. Appl. 54:53–56.

Bernays, E. A. and Barbehenn, R. 1987. Nutritional ecology of grass foliage-chewing insects, pp. 147–175. In Slansky, Jr. and J. G. Rodriguez (eds.), Nutritional Ecology of Insects, Mites, Spiders, and Related Invertebrates. Wiley, New York.

Bernays, E. A., and Chapman, R. 1987. The evolution of deterrent responses in plant-feeding insects, pp. 159–173. In R. F. Chapman, E. A. Bernays, and J. G. Stoffolano, Jr. (eds.), Perspectives in Chemoreception and Behavior. Springer-Verlag, New York.

Bernays, E. A., and Janzen, D. H. 1988. Saturniid and sphingid caterpillars: Two ways to eat leaves. Ecology 69:1153–1160.

Blackwell, A. 1988. The disruption of aggregation behavior in *Pieris brassicae* larvae by chlordimeform. Entomol. Exp. Appl. 48:141–147.

Blau, P. A., Feeny, P., Contardo, L. and Robson, D. S. 1978. Allylglucosinolate and herbivorous caterpillars: A contrast in toxicity and tolerance. Science 200:1296–1298.

Blum, M. S., and Blum, N. A. (eds.) 1979. Sexual Selection and Reproductive Competition in Insects. Academic, New York.

Boggs, C. L. 1981. Nutritional and life-history determinants of resource allocation in holometabolous insects. Am. Nat. 117:692–709.

Boggs, C. L. 1986. Reproductive strategies of female butterflies: variation in and constraints on fecundity. Ecol. Entomol. 11:7–15.

Boggs, C. L. 1987. Ecology of nectar and pollen feeding in Lepidoptera, pp. 369–391. In F. Slansky, Jr. and J. G. Rodriguez (eds.), Nutritional Ecology of Insects, Mites, Spiders, and Related Invertebrates. Wiley, New York.

Bowdan, E. 1988a. Microstructure of feeding by tobacco hornworm caterpillars, *Manduca sexta*. Entomol. Exp. Appl. 47:127–136.

Bowdan, E. 1988b. The effect of deprivation on the microstructure of feeding by the tobacco hornworm caterpillar. J. Insect Behav. 1:31–50.

Bowers, M. D. 1990. Recycling plant natural products for insect defense, pp. 353–386. In D. L. Evans and J. O. Schmidt (eds.), Insect Defenses. Adaptive Mechanisms and Strategies of Prey and Predators. SUNY Press, Albany, New York.

Brattsten, L. B.., and Ahmad, S. 1986. Molecular Aspects of Insect-Plant Associations. Plenum, New York.

Breden, F., and Wade, M. J. 1987. An experimental study of the effect of group size on larval growth and survivorship in the imported willow leaf beetle, *Plagiodera vesicolora* (Coleoptera: Chrysomelidae). Environ. Entomol. 16:1082–1086.

Broadway, R. M., and Duffey, S. S. 1986a. The effect of dietary protein on the growth and digestive physiology of larval *Heliothis zea* and *Spodoptera exigua*. J. Insect Physiol. 12:673–680.

Broadway, R. m., and Duffey, S. S. 1986b. Plant proteinase inhibitors: mechanisms of action and effect on the growth and digestive physiology of larval *Heliothis zea* and *Spodoptera exigua*. J. Insect Physiol. 32:827–833.

Calow, P. 1982. Homeostasis and fitness. Am. Nat. 120:416–419.

Capinera, J. L. 1979. Qualitative variation in plants and insects: effect of propagule size on ecological plasticity. Am. Nat. 114:350–361.

Capinera, J. L. (ed.) 1987. Integrated Pest Management on Rangeland. A Shortgrass Prairie Perspective. Westview, Boulder.

Capinera, J. L., and Barbosa, P. 1976. Dispersal of first-instar gypsy moth larvae in relation to population quality. Oecologia 26:53–60.

Capinera, J. L., and Barbosa, P. 1977. Influence of natural diets and larval density on gypsy moth, *Lymantria dispar* (Lepidoptera: Orgyiidae), egg mass characteristics. Can. Entomol. 109:1313–1318.

Capinera, J. L., Barbosa, P., and Hagedorn, H. H. 1977. Yolk and yolk depletion of gypsy moth eggs: Implications for population quality. Ann. Entomol. Soc. Am. 70:40–42.

Capinera, J. L., Wiener, L. F., and Anamosa, P. R. 1980. Behavioral thermoregulation by late-instar range caterpillar larvae *Hemileuca oliviae* Cockerell (Lepidoptera: Saturniidae). J. Kans. Entomol. Soc. 53:631–638.

Carefoot, T. H. 1977. Energy partitioning in the desert locust, *Schistocerca gregaria* (Forsk.). Acrida 6:85–107.

Casey, T. M. 1976. Activity patterns, body temperature and thermal ecology in two desert caterpillars (Lepidoptera: Sphingidae). Ecology 57:485–497.

Chang, N. T., Lynch, R. E., Slansky, F., Jr., Wiseman, B. R., and Habeck, D. H. 1987. Qualitative utilization of selected grasses by fall armyworm larvae. Entomol. Exp. Appl. 45:29–35.

Chapman, R. F. 1985. Coordination of digestion, pp. 213–240. In G. A. Kerkut and L. I. Gilbert (eds.), Comprehensive Insect Physiology, Biochemistry and Pharmacology. Vol. 4. Pergamon, Oxford.

Chapman, R. F., and Beerling, E. A. M. 1990. The pattern of feeding of first instar nymphs of *Schistocerca americana*. Physiol. Entomol. 15:1–12.

Chew, F. S. 1975. Coevolution of pierid butterflies and their cruciferous foodplants. Oecologia 20:117–127.

Chlodny, J. 1969. The energetics of larval development of two species of grasshopper from the genus *Chorthippus* Fieb. Ekol. Polska. 17A:391–407.

Chou, Y. M., Rock, G. C., and Hodgson, E. 1973. Consumption and utilization of chemically defined diets by *Argyrotaenia velutinana* and *Heliothis virescens*. Ann. Entomol. Soc. Am. 66:627–632.

Clancy, K. M., and Price, P. W. 1987. Rapid herbivore growth enhances enemy attack: Sublethal plant defenses remain a paradox. Ecology 68:733–737.

Coley, P. D., Bryant, J. P., and Chapin, F. S., III. 1985. Resource availability and plant antiherbivore defense. Science. 230:895–899.

Cornell, H. V. 1989. Endophage-ectophage ratios and plant defense. Evol. Ecol. 3:64–76.

Cornell, J. C., Stamp, N. E., and Bowers, M. D. 1987. Developmental changes in aggregation, defense and escape behavior of buckmoth caterpillars, *Hemileuca lucina* (Saturniidae). Behav. Ecol. Sociobiol. 20:383–388.

Cottee, P. K., Bernays, E. A., and Mordue, A. J. 1988. Comparisons of deterrency and toxicity of selected secondary plant compounds to an oligophagous and a polyphagous acridid. Entomol. Exp. Appl. 46:241–247.

Craig, T. P., Itami, J. K., and Price, P. W. 1989. A strong relationship between oviposition preference and larval performance in a shoot-galling sawfly. Ecology. 70:1691–1699.

Dadd, R. H. 1960. Observations on the palatability and utilization of food by locusts, with particular reference to the interpretation of performance in growth trials using synthetic diets. Entomol. Exp. Appl. 3:283–304.

Delvi, M. R., and Pandian, T. J. 1971. Ecophysiological studies on the utilization of food in the paddy field grasshopper, *Oxya velox*. Oecologia 8:267–275.

Denno, R. F., and Dingle, H. (eds.) 1981. Insect Life History Patterns. Habitat and Geographic Variation. Springer-Verlag, New York.

Denno, R. F., and McClure, M. S. (eds.) 1983. Variable Plants and Herbivores in Natural and Managed Systems. Academic, New York.

Dethier, V. G. 1988. The feeding behavior of a polyphagous caterpillar (*Diacrisia virginica*) in its natural habitat. Can. J. Zool. 66:1280–1288.

Edmunds, M. 1990. The evolution of cryptic coloration pp. 3–21. In D. L. Evans and J. O. Schmidt (eds.), Insect Defenses. Adaptive Mechanisms and Strategies of Prey and Predators. SUNY Press, Albany, N Y.

Edwards, D. K. 1964. Activity rhythms of lepidopterous defoliators II. *Halisidota argentata* Pack. (Arctiidae), and *Nepytia phantasmaria* Stkr. (Geometridae). Can. J. Zool. 42:939–958.

Edwards, D. K. 1965. Activity rhythms of lepidopterous defoliators III. The Douglas-fir tussock moth, *Orgyia pseudotsugata* (McDunnough) (Liparidae). Can. J. Zool. 43:673–681.

Ellsworth, P. C., Patterson, R. P., Bradley, J. R., Jr., Kennedy, G. G., and Stinner, R. E. 1989. Developmental consequences of water and temperature in the European corn borer-maize interaction. Entomol. Exp. Appl. 53:287–296.

Enders, F. 1975. The influence of hunting manner on prey size, particularly in spiders with long attack distances (Araneidae, Linyphiidae, and Salticidae). Am. Nat. 109:737–763.

Evans, D. L. 1990. Phenology as a defense: a time to die, a time to live, pp. 191–202. In D. L. Evans and J. O. Schmidt (eds.), Insect Defenses, Adaptive Mechanisms and Strategies of Prey and Predators. SUNY Press, Albany, N Y.

Evans, D. L. and Schmidt, J. O. (eds.) 1990. Insect Defenses. Adaptive Mechanisms and Strategies of Prey and Predators. SUNY Press, Albany, N Y.

Faeth, S. H. 1986. Indirect interactions between temporally separated herbivores mediated by the host plant. Ecology. 67:479–494.

Faeth, S. H., and Hammon, K. E. 1989. Caterpillars and polymorphisms. Science 246:1639.

Faeth, S. H., Connor, E. F., and Simberloff, D. 1981. Early leaf abscission: A neglected source of mortality for folivores. Am. Nat. 117:409–415.

Fajer, E. D., Bowers, M. D., and Bazzaz, F. A. 1989. The effects of enriched carbon dioxide atmospheres on plant-insect herbivore interactions. Science 243:1198–1200.

Farrar, R. R., Jr., Barbour, J. D., and Kennedy, G. G. 1989. Quantifying food consumption and growth in insects. Ann. Entomol. Soc. Am. 82:593–598.

Fields, P. G., and McNeil, J. N. 1988. Characteristics of the larval diapause in *Ctenucha virginica* (Lepidoptera: Arctiidae). J. Insect Physiol. 34:111–115.

Fields, P. G., Arnason, J. T., and Philogène, J. R. 1990. Behavioral and physical adaptations of three insects that feed on the phototoxic plant *Hypericum perforatum*. Can. J. Zool. 68:339–346.

Finch, S. 1986. Assessing host-plant finding by insects, pp. 23–63. In J. R. Miller and T. A. Miller (eds.), Insect-Plant Interactions. Springer-Verlag. New York.

Fitzgerald, T. D. 1980. An analysis of daily foraging patterns of laboratory colonies of the eastern tent caterpillar, *Malacosoma americanum* (Lepidoptera: Lasiocampidae), recorded photoelectronically. Can. Entomol. 112:731–738.

Fitzgerald, T. D., and Peterson, S. C. 1983. Elective recruitment by the eastern tent caterpillar (*Malacosoma americanum*). Anim. Behav. 31:417–423.

Fitzgerald, T. D., Casey, T., and Joos, B. 1988. Daily foraging schedule of field colonies of the eastern tent caterpillar *Malacosoma americanum*. Oecologia 76:574–578.

Fox, L. R., and Macauley, B. J. 1977. Insect grazing on *Eucalyptus* in response to variation in leaf tannins and nitrogen. Oecologia 29:145–162.

Fox, L. R., and Morrow, P.A. 1981. Specialization: species property or local phenomenon? Science 211:887–893.

Frazier, J. L. 1986. The perception of plant allelochemicals that inhibit feeding, pp. 1–24. In L. B. Brattsten and S. Ahmad (eds.), Molecular Aspects of Insect-Plant Associations. Plenum, New York.

Fritz, R. S., and Nobel, J. 1990. Host plant variation in mortality of the leaf-binding sawfly on the arroyo willow. Ecol. Entomol. 15:25–35.

Futuyma, D. J., and Moreno, G. 1988. The evolution of ecological specialization. Annu. Rev. Ecol. Syst. 19:207–233.

Futuyma, D. J., and Saks, M. 1981. The effect of variation in host plant on the growth of an oligophagous insect, *Malacosoma americanum* and its polyphagous relative, *Malacosoma disstria*. Entomol. Exp. Appl. 30:163–168.

Futuyma, D. J., and Wasserman, S. S. 1981. Food plant specialization and feeding efficiency in the tent caterpillars *Malacosoma disstria* and *M. americanum*. Entomol. Exp. Appl. 30:106–110.

Futuyma, D. J., Cort, R. P., and Noordwijk, I. V. 1984. Adaptation to host plants in the fall cankerworm (*Alsophila pometaria*) and its bearing on the evolution of host affiliation in phytophagous insects. Am. Nat. 123:287–295.

Gallardo, F., Boethel, D. J., Fuxa, J. R., and Richter, A. 1990. Susceptibility of *Heliothis zea* (Boddie) larvae to *Nomuraea rileyi* (Farlow) Samson. Effects of α-tomatine at the third trophic level. J. Chem. Ecol. 16:1751–1759.

Gatto, M., Matessi, C., and Slobodkin, L. B. 1989. Physiological profiles and demographic rates in relation to food quantity and predictability: An optimization approach. Evol. Entomol. 3:1–30.

Gebhardt, M. D., and Stearns, S. C. 1988. Reaction norms for developmental time and weight at eclosion in *Drosophila mercatorum*. J. Evol. Biol. 1:335–354.

Gould, F. 1984. Mixed function oxidases and herbivore polyphagy: the devil's advocate position. Ecol. Entomol. 9:29–34.

Gould, W. P., and Jeanne, R. L. 1984. *Polistes* wasps (Hymenoptera: Vespidae) as control agents for lepidopterous cabbage pests. Environ. Entomol. 15:150–156.

Greene, E. 1989. A diet-induced developmental polymorphism in a caterpillar. Science 243:643–646.

Gross, P., and Price, P. W. 1988. Plant influences on parasitism of two leafminers: a test of enemy-free space. Ecology 69:1506–1516

Gunderson, C. A., Samuelian, J. H., Evans, C. K., and Brattsten, L. B. 1985. Effects of the mint monoterpene pulegone on *Spodoptera eridania* (Lepidoptera: Noctuidae). Environ. Entomol. 14:859–863.

Haack, R. A., and Slansky, F., Jr. 1987. Nutritional ecology of wood-feeding Coleoptera, Lepidoptera and Hymenoptera, pp. 449–486. In F. Slansky, Jr. and J. G. Rodriguez (eds.), Nutritional Ecology of Insects, Mites, Spiders, and Related Invertebrates. Wiley, New York.

Hagen, K. S. 1987. Nutritional ecology of terrestrial insect predators, pp. 533–577. In F. Slansky, Jr. and J. G. Rodriguez (eds.), Nutritional Ecology of Insects, Mites, Spiders, and Related Invertebrates. Wiley, New York.

Hammond, R. B., Pedigo, L. P., and Poston, F. L. 1979. Green cloverworm leaf consumption on greenhouse and field soybean leaves and development of a leaf-consumption model. J. Econ. Entomol. 72:714–717.

Hanski, I. 1987. Nutritional ecology of dung- and carrion-feeding insects, pp. 837–884. In F. Slansky, Jr. and J. G. Rodriguez (eds.), Nutritional Ecology of Insects, Mites, Spiders, and Related Invertebrates. Wiley, New York.

Harvey, G. T. 1974. Nutritional studies of eastern spruce budworm (Lepidoptera: Tortricidae) I. Soluble sugars. Can. Entomol, 106:353–365.

Harvey, G. T. 1977. Mean weight and rearing performance of successive egg clusters of eastern spruce budworm. (Lepidoptera: Tortricidae). Can. Entol. 109:587–496.

Hassell, M. P., and Southwood, T. R. E. 1978. Foraging strategies of insects. Annu. Rev. Ecol. Syst. 9:75–98.

Haukioja, E. 1980. On the role of plant defenses in the fluctuation of herbivore populations. Oikos 35:202–213.

Haukioja, E. 1991. Induction of defenses in trees. Annu. Rev. Entomol. 36:25–42.

Haukioja, E., and Hanhimäki, S. 1985. Rapid wound-induced resistance in white birch (*Betula pubescens*) foliage to the geometrid *Epirrita autumnata*: A comparison of trees and moths within and outside the outbreak range of the moth. Oecologia 65:223–228.

Haukioja, E., Niemelä, P., Iso-Iivari, L., Ojala, H., and Aro, E. 1978a. Birch leaves as a resource for herbivores. I. Variation in the suitability of leaves. Rep. Kevo Subarctic Res. Sta. 14:5–12.

Haukioja, E., Niemelä, P., Iso-Iivari, L., 1978b. Birch leaves as a resource for herbivores. II. Diurnal variation in the usability of leaves for *Oporinia autumnata* and *Dineura viridirsata*. Rep. Kevo Subarctic Res. Sta. 14:21–24.

Hedin, P. A., Parrott, W. L., Jenkins, J. N., Mulrooney, J. E., and Menn, J. J. 1988. Elucidating mechanisms of tobacco budworm resistance to allelochemicals by dietary tests with insecticide synergists. Pestic. Biochem. Physiol. 32:55–61.

Heinrich, B. 1979. Foraging strategies of caterpillars. Leaf damage and possible predator avoidance strategies. Oecologia 42:325–337.

Heinrich, B., and Collins, S. L. 1983. Caterpillar leaf damage and the game of hide and seek with birds. Ecology 64:592–602.

Heinrichs, E. A. (ed.) 1988. Plant Stress-Insect Interactions. Wiley, New York.

Hemminga, M. A., and van Soelen, J. 1988. Estaurine gradients and the growth and development of *Agapanthia villosoviridescens*, (Coleoptera), a stem-borer of the salt marsh halophyte *Aster tripolium*. Oecologia 77:307–312.

Herrebout, W. M., Kuyten, P. J., and DeRuiter, L. 1963. Observations on colour patterns and behaviour of caterpillars feeding on Scots pine. Archiv. Neerl. Zool. 15:315–357.

Hespenheide, H. A. 1973. Ecological inferences from morphological data. Annu. Rev. Ecol. Syst. 4:213–229.

Hinton, H. E. 1981. Biology of Insect Eggs, Vol. 1. Pergamon, Oxford.

Holtzer, T. O., Archer, T. L., and Norman, J. M. 1988. Host plant suitability in relation to water stress, pp. 111–137. In E. A. Heinrichs (ed.), Plant Stress-Insect Interactions. Wiley, New York.

Holyoke, C. W., and Reese, J. C. 1987. Acute insect toxicants from plants, pp. 67–118. In E. D. Morgan and N. B. Mandava (eds.), Handbook of Natural Pesticides. Vol. III: Insect Growth Regulators, part B. CRC, Boca Raton, FL.

Horie, Y., and Watanabe, K. 1983. Effect of various kinds of dietary protein and supplementation with limiting amino acids on growth, haemolymph components and uric excretion in the silkworm *Bombyx mori*. J. Insect Physiol. 29:187–189.

Horie, Y., Inokuchi, T., Watanabe, K., Nakasone, S., and Yanagawa, H. 1976. Quantitative study on food utilization by the silkworm, *Bombyx mori*, through its life cycle. I. Economy of dry matter, energy, and carbon. Bull. Seric. Exp. Sta. 26:411–442.

House, H. L. 1969. Effects of different proportions of nutrients on insects. Entomol. Exp. Appl. 12:651–669.

Hsiao, T. H. 1985. Feeding behavior, pp. 471–512. In G. A. Kerkut and L. I. Gilbert (eds.), Comprehensive Insect Physiology, Biochemistry and Pharmacology, Vol. 9. Pergamon, Oxford.

Isenhour, D. J., Wiseman, B. R., and Layton, R. C. 1989. Enhanced predation by *Orius insidiosus* (Hemiptera: Anthocoridae) on larvae of *Heliothis zea* and *Spodoptera frugiperda* (Lepidoptera: Noctuidae) caused by prey feeding on resistant corn genotypes. Environ. Entomol. 18:418–422.

Johnson, N. D., and Bentley, B. L. 1988. Effects of dietary protein and lupine alkaloids on growth and survivorship of *Spodoptera eridania*. J. Chem. Ecol. 14:1391–1403.

Johnson, N. D., Brian, S. A., and Ehrlich, P. R. 1985. The role of leaf resin in the interaction between *Eriodictyon californicum* (Hydrophyllaceae) and its herbivore, *Trirhabda diducta* (Chrysomelidae). Oecologia 66:106–110.

Jones, C. G., Hoggard, M. P., and Blum, M. S. 1981. Pattern and process in insect feeding behaviour: A quantitative analysis of the Mexican bean beetle, *Epilachna varivestis*. Entomol. Exp. Appl. 30:254–264.

Karban, R., and Myers, J. H. 1989. Induced plant responses to herbivory. Annu. Rev. Ecol. Syst. 20:331–348.

Karowe, D. N., and Martin, M. M. 1989. The effects of quantity and quality of diet nitrogen on the growth, efficiency of utilization, nitrogen budget, and metabolic rate of fifth-instar *Spodoptera eridania* larvae (Lepidoptera: Noctuidae). J. Insect Physiol. 35:699–708.

Kogan, M. 1986. Bioassays for measuring quality of insect food, pp. 155–189. In J. R. Miller and T. A. Miller (eds.), Insect-plant Interactions. Springer-Verlag. New York.

Krieger, R. I., Feeny, P. P., and Wilkinson, C. F. 1971. Detoxication enzymes in the guts of caterpillars: An evolutionary answer to plant defenses. Science 172:579–581.

Kukal, O., and Dawson, T. E. 1989. Temperature and food quality influences on feeding behavior, assimilation efficiency and growth rate of arctic woolly-bear caterpillars. Oecologia 79:526–532.

Kukal, O., and Kevan, P. G. 1987. The influence of parasitism on the life history of a high arctic insect, *Gynaephora groenlandica* (Wöcke) (Lepidoptera: Lymantriidae). Can. J. Zool. 65:156–163.

Lafont, R., Mauchamp, B., Boulay, G., and Tarroux, P. 1975. Developmental studies in *Pieris brassicae*. I. Growth of various tissues during the last larval instar. Comp. Biochem. Physiol. Ser. B 51:430–444.

Lance, D. R., Elkinton, J. S., and Schwalbe, C. P. 1986. Feeding rhythms of gypsy moth larvae: Effect of food quality during outbreaks. Ecology 67:1650–1654.

Lawson, D. L., Merritt, R. W., Martin, M. M., Martin, J. S., and Kukor, J. J. 1984. The nutritional ecology of larvae of *Alsophila pometaria* and *Anisota senatoria* feeding on early- and late-season oak foliage. Entomol. Exp. Appl. 35:105–115.

Lawton, J. H., and McNeill, S. 1979. Between the devil and the deep blue sea: on the problem of being a herbivore, pp. 223–244. In R. M. Anderson, B. D. Turner, and L. R. Taylor (eds.), Population Dynamics. 20th Symp. British Ecol. Soc. Blackwell, Oxford.

Lederhouse, R. C., Finke, M. D., and Scriber, J. M. 1982. The contributions of larval growth and pupal duration to protandry in the black swallowtail butterfly, *Papilio polyxenes*. Oecologia 53:296–300.

Lincoln, D. E., and Couvet, D. 1989. The effect of carbon supply on allocation to allelochemicals and caterpillar consumption of peppermint. Oecologia 69:112–114.

Lincoln, D. E., Newton, T. S., Ehrlich, P. R., and Williams, K. S. 1982. Coevolution of the checkerspot butterfly *Euphydryas chalcedona* and its larval food plant *Diplacus aurantiacus*: Larval response to protein and leaf resin. Oecologia 52:216–223.

Lindroth, R. L., Anson, B. D., and Weisbrod, A. V. 1990. Effects of protein and juglone on gypsy moths: growth performance and detoxification enzyme activity. J. Chem. Ecol. 16:2533–2547.

Loader, C., and Damman, H. 1991. Nitrogen content of food plants and vulnerability of *Pieris rapae* to natural enemies. Ecology 72:1586–1590.

Lynch, R. E., Branch, W. D., and Garner, J. W. 1981. Resistance of *Arachis* species to the fall armyworm, *Spodoptera frugiperda*. Peanut Sci. 8:106–109.

Mackey, A. P. 1978. Growth and bioenergetics of the moth *Cyclophragma leucosticta* Grünberg. Oecologia 32:367–376.

Maddrell, S. H. P. 1981. The functional design of the insect excretory system. J. Exp. Biol. 90:1–15.

Malcolm, S. B., Cockrell, B. J., and Brower, L. P. 1987. Monarch butterfly voltinism: Effects of temperature constraints at different latitudes. Oikos 49:77–82.

Manuwoto, S., and Scriber, J. M. 1985. Consumption and utilization of experimentally altered corn by southern armyworm: iron, nitrogen, and cyclic hydroxamates. J. Chem. Ecol. 11:1469–1483.

Marshall, L. D., and McNeil, J. N. 1989. Spermatophore mass as an estimate of male nutrient investment: A closer look on *Pseudaletia unipuncta* (Haworth) (Lepidoptera: Noctuidae). Funct. Ecol. 3:605–612.

Martin, M. M., and Van't Hof, H. M. 1988. The cause of reduced growth of *Manduca sexta* larvae on a low-water diet: Increased metabolic processing costs or nutrient limitation? J. Insect Physiol. 34:515–525.

Mason, L. J., Johnson, S. J., and Woodring, J. P. 1990. Influence of age and season on whole-body lipid content of *Plathypena scabra* (Lepidoptera: Noctuidae). Environ. Entomol. 19:1259–1262.

Mattson, W. J., and Haack, R. A. 1987. The role of drought stress in provoking outbreaks of phytophagous insects, pp. 365–407. In P. Barbosa and J. C. Schultz (eds.), Insect Outbreaks. Academic Press, San Diego.

Mattson, W. J., and Scriber, J. M. 1987. Nutritional ecology of insect folivores of woody plants: Nitrogen, water, fiber, and mineral considerations, pp. 105–146. In F. Slansky, Jr. and J. G. Rodriguez (eds.), Nutritional Ecology of Insects, Mites, Spiders, and Related Invertebrates. Wiley, New York.

Mauricio, R., and Bowers, M. D. 1990. Do caterpillars disperse their damage?: Larval foraging behaviour of two specialist herbivores, *Euphydryas phaeton* (Nymphalidae) and *Pieris rapae* (Pieridae). Ecol. Entomol. 15:153–161.

McEvoy, P. B. 1984. Increase in respiratory rate during feeding in larvae of the cinnabar moth *Tyria jacobaeae*. Physiol. Entomol. 9:191–195.

McGinnis, A. J., and Kasting, R. 1967. Dietary cellulose: Effect on food consumption and growth of a grasshopper. Can. J. Zool. 45:365–367.

McKey, D. 1979. The distribution of secondary compounds within plants, pp. 55–133. In G. A. Rosenthal and D. H. Janzen (eds.), Herbivores: Their interaction with Secondary Plant Metabolites. Academic Press, New York.

Meisner, J., Hadar, D., Wysoki, M., and Harpaz, I. 1990. Phagostimulants enhancing the efficacy of *Bacillus thuringiensis* formulations against the giant looper, *Boarmia (Ascotis) selenaria*, in avocado. Phytoparasitica 18:107–115.

Mihaliak, C. A., Couvet, D., and Lincoln, D. E. 1987. Inhibition of feeding by a generalist insect due to increased volatile leaf terpenes under nitrate-limiting conditions. J. Chem. Ecol. 13:2059–2067.

Miller, J. R., and Miller, T. A. (eds.) 1986. Insect-Plant Interactions. Springer-Verlag, New York.

Miller, J. R., and Strickler, K. L. 1984. Finding and accepting host plants, pp. 127–157. In W. J. Bell and R. T. Cardé (eds.), Chemical Ecology of Insects. Sinauer, Sunderland, MA.

Morrow, P.A., and Fox, L. R. 1980. Effects of variation in *Eucalyptus* essential oil yield on insect growth and grazing damage. Oecologia 45:209–219.

Moscardi, F., Barfield, C. S., and Allen, G. E. 1981a. Consumption and development of velvetbean caterpillar as influenced by soybean phenology. Environ. Entomol. 10:880–884.

Moscardi, F., Barfield, C. S., and Allen, G. E. 1981b. Impact of soybean phenology on velvetbean caterpillar (Lepidoptera: Noctuidae): Oviposition, egg hatch and adult longevity. Can. Entomol. 113:113–119.

Murphy, D. D., Launer, A. E., and Ehrlich, P. R. 1983. The role of adult feeding in egg production and population dynamics of the checkerspot butterfly *Euphydryas editha*. Oecologia 56:257–263.

Nakano, K., and Monsi, M. 1968. An experimental approach to some quantitative aspects of grazing by silkworms, (*Bombyx mori*). Jpn. J. Ecol. 18:212–229.

Neal, J. J. 1987. Metabolic costs of mixed-function oxidase induction in *Heliothis zea*. Entomol. Exp. Appl. 43:175–179.

Nijhout, H. F. 1975. A threshold size for metamorphosis in the tobacco hornworm, *Manduca sexta* (L). Biol. Bull. 149:214–225.

Nordland, D. A., Lewis, W. J., and Altieri, M. A. 1988. Influences of plant-produced allelochemicals on the host/prey selection behavior of entomophagous insects, pp. 65–90. In P. Barbosa and D. K. Letourneau (eds.), Novel Aspects of Insect-Plant Interactions. Wiley, New York.

Nylin, S. 1988. Host plant specialization and seasonality in a polyphagous butterfly, *Polygonia c-album* (Nymphalidae). Oikos 53:381–386.

Nylin, S., Wickman, P.-O., and Wiklund, C. 1989. Seasonal plasticity in growth and development of the speckled wood butterfly, *Pararge aegeria* (Satyrinae). Biol. J. Linn. Soc. 38:155–171.

Papaj, D. R., and Rausher, M. D. 1983. Individual variation in host location by phytophagous insects, pp. 77–124. In S. Ahmad (ed.), Herbivorous Insects. Host-Seeking Behavior and Mechanisms. Academic Press, New York.

Parry, W. H., Kelly, J. M., and McKillop, A. R. 1990. The role of nitrogen in the feeding strategy of *Strophosomus melanogrammus* (Forster) (Col. Curculionidae) in a mixed woodland habitat. J. Appl. Entomol. 109:367–376.

Peterson, S. S., Scriber, J. M., and Coors, J. G. 1988. Silica, cellulose and their interactive effects on the feeding performance of the southern armyworm, *Spodoptera eridania* (Cramer) (Lepidoptera: Noctuidae). J. Kans. Entomol. Soc. 61:169–177.

Pivnick, K. A., and McNeill, J. N. 1987. Puddling in butterflies: Sodium affects reproductive success in *Thymelicus lineola*. Physiol. Entomol. 82:3–26.

Potter, D. A., and Redmond, C. T. 1989. Early spring defoliation, secondary leaf flush, and leafminer outbreaks on American holly. Oecologia 81:192–197.

Price, P. W., Bouton, C. E., Gross, P., McPheron, B. A., Thompson, J. N., and Weis, A. E. 1980. Interactions among three trophic levels: influence of plants on interactions between insect herbivores and natural enemies. Annu. Rev. Ecol. Syst. 11:41–65.

Quisenberry, S. S., and Wilson, H. K. 1985. Consumption and utilization of Bermuda grass by fall armyworm (Lepidoptera: Noctuidae) larvae. J. Econ. Entomol. 78:820–824.

Raubenheimer, D. 1992. Tannic acid, protein and digestible carbohydrate: Dietary imbalance and nutritional compensation in the African migratory locust. Ecology (in press).

Raubenheimer, D., and Simpson, S. J. 1990. The effect of simultaneous variation in protein, digestible carbohydrate and tannic acid on the feeding behaviour of larval *Locusta migratoria* (L.) and *Schistocerca gregaria* (Forskal). I. Short-term studies. Physiol. Entomol. 15:219–233.

Rausher, M. D. 1981. Host plant selection by *Battus philenor* butterflies: The roles of predation, nutrition, and plant chemistry. Ecol. Monogr. 51:1–20.

Rausher, M. D. 1982. Population differentiation in *Euphydryas editha* butterflies: Larval adaptation to different hosts. Evolution 36:581–590.

Reese, J. C., and Holyoke, C. W., Jr. 1987. Allelochemicals affecting insect growth and development, pp. 21–66. In E. D. Morgan and N. B. Mandava (eds.), Handbook of Natural Pesticides. Vol. III. Insect Growth Regulators, part B. CRC, Boca Raton, FL.

Reynolds, S. E. 1990. Feeding in caterpillars: maximising or optimising food acquisition?, pp. 106–118. In J. Mellinger (ed.), Animal Nutrition and Transport Processes. 1. Nutrition in Wild and Domestic Animals. Karger, Basel.

Reynolds, S. E., and Bellward, K. 1989. Water balance in *Manduca sexta* caterpillars: Water recycling from the rectum. J. Exp. Biol. 141:33–45.

Reynolds, S. E., Nottingham, S. F., and Stephens, A. E. 1985. Food and water economy and its relation to growth in fifth-instar larvae of the tobacco hornworm, *Manduca sexta*. J. Insect Physiol. 31:119–127.

Reynolds, S. E., Yeomans, M. R., and Timmins, W. A. 1986. The feeding behaviour of caterpillars (*Manduca sexta*) on tobacco and on artificial diet. Physiol. Entomol. 11:39–51.

Rhoades, D. F. 1985. Offensive-defensive interactions between herbivores and plants: Their relevance in herbivore population dynamics and ecological theory. Am. Nat. 125:205–238.

Riechert, S. E., and Harp, J. M. 1987. Nutritional ecology of spiders, pp. 645–672. In F. Slansky, Jr. and J. G. Rodriguez (eds.), Nutritional Ecology of Insects, Mites, Spiders, and Related Invertebrates. Wiley, New York.

Rollo, C. D., and Hawryluk, M. D. 1988. Compensatory scope and resource allocation in two species of aquatic snails. Ecology 69:146–156.

Salinas, P. J. 1984. Studies on the behaviour of the larvae of *Plutella xylostella* (Linnaeus) (Lepidoptera: Plutellidae), a world pest of cruciferous crops. Normal and "spacing" behaviour. Turrialba 34:77–84.

Sanders, C. J., and Lucuik, G. S. 1975. Effects of photoperiod and size on flight activity and oviposition in the eastern spruce budworm (Lepidoptera: Tortricidae). Can. Entomol. 107:1289–1299.

Santos, C. D., Ferreira, C., and Terra, W. R. 1983. Consumption of food and spatial organization of digestion in the cassava hornworm, *Erinnyis ello*. J. Insect Physiol. 29:707–714.

Schmidt, D. J., and Reese, J. C. 1986. Sources of error in nutritional index studies of insects on artificial diet. J. Insect Physiol. 32:193–198.

Schmidt, D. J., and Reese, J. C. 1988. The effects of physiological stress on black cutworm (*Agrotis ipsilon*) larval growth and food utilization. J. Insect Physiol. 34:5–10.

Schmidt, K. M., Benedict, J. H., and Walmsley, M. H. 1988. Behavioral responses (time budgets) of bollworm (Lepidoptera: Noctuidae) larvae for three cotton cultivars. Environ. Entomol. 17:350–353.

Schoonhoven, L. M. 1987. What makes caterpillars eat? The sensory code underlying feeding behavior, pp. 69–97. In R. F. Chapman, E. A. Bernays, and J. G. Stoffolano, Jr. (eds.), Perspectives in Chemoreception and Behavior. Springer-Verlag, New York.

Schroeder, L. A. 1986a. Protein limitation of a tree leaf feeding lepidopteran. Entomol. Exp. Appl. 41:115–120.

Schroeder, L. A. 1986b. Changes in tree leaf quality and growth performance of lepidopteran larvae. Ecology 67:1628–1636.

Schroeder, L. A., and Malmer, M. 1980. Dry matter, energy and nitrogen conversion by Lepidoptera and Hymenoptera larvae fed leaves of black cherry. Oecologia 45:63–71.

Schultz, J. C. 1983. Habitat selection and foraging tactics of caterpillars in heterogeneous trees, pp. 61–90. In R. F. Denno and M. S. McClure (eds.), Variable Plants and Herbivores in Natural and Managed Systems. Academic, New York.

Scriber, J. M. 1977. Limiting effects of low leaf-water content on the nitrogen utilization, energy budget, and larval growth of *Hyalophora cecropia* (Lepidoptera: Saturniidae). Oecologia 28: 269–287.

Scriber, J. M. 1979a. Post-ingestive utilization of plant biomass and nitrogen by Lepidoptera: Legume feeding by the southern armyworm. J. N.Y. Entomol. Soc. 87:141–153.

Scriber, J. M. 1979b. Effects of leaf-water supplementation upon post-ingestive nutritional indices of forb-, shrub-, vine- and tree-feeding Lepidoptera. Entomol. Exp. Appl. 25:240–252.

Scriber, J. M. 1982. The behavior and nutritional physiology of southern armyworm larvae as a function of plant species consumed in earlier instars. Entomol. Exp. Appl. 31:359–369.

Scriber, J. M. 1983. Evolution of feeding specialization, physiological efficiency, and host races in selected Papilionidae and Saturniidae, pp. 373–412. In R. F. Denno and M. S. McClure (eds.), Variable Plants and Herbivores in Natural and Managed Systems. Academic Press, New York.

Scriber, J. M. 1984. Host-plant suitability, pp. 159–202. In W. J. Bell and R. T. Cardé (eds.), Chemical Ecology of Insects. Sinauer, Sunderland, MA.

Scriber, J. M., and Slansky, F., Jr. 1981. The nutritional ecology of immature insects. Annu. Rev. Entomol. 26:183–211.

Sehnal, F. 1985. Growth and life cycles, pp. 1–86. In G. A. Kerkut and L. I. Gilbert (eds.), Comprehensive Insect Physiology, Biochemistry and Pharmacology, Vol. 2. Pergamon, Oxford.

Sih, A. 1987. Nutritional ecology of aquatic insect predators, pp. 579–607. In F. Slansky, Jr. and J. G. Rodriguez (eds.), Nutritional Ecology of Insects, Mites, Spiders, and Related Invertebrates. Wiley, New York.

Sih, A., Krupa, J., and Travers, S. 1990. An experimental study on the effects of predation risk and feeding regime on the mating behavior of the water strider. Am. Nat. 135:284–290.

Simpson, S. J. 1983. Changes during the fifth instar of *Locusta migratoria* in the rate of crop emptying and their relationship to feeding and food utilisation. Entomol. Exp. Appl. 33:225–243.

Simpson, S.J., and Abisgold, J. D. 1985. Compensation by locusts for changes in dietary nutrients: behavioural mechanisms. Physiol. Entomol. 10:443–452.

Simpson, S. J., and Simpson, C. L. 1990. The mechanisms of nutritional compensation by phytophagous insects, pp. 111–160. In E. A. Bernays (ed.), Insect-Plant Interactions, Vol. II. CRC, Boca Raton, FL.

Simpson, S. J., Simmonds, S. J., and Blaney, W. M. 1988. A comparison of dietary selection behaviour in larval *Locusta migratoria* and *Spodoptera littoralis*. Physiol. Entomol. 13:225–238.

Simpson, S. J., Simmonds, M. S. J., Blaney, W. M., and Jones, J. P. 1990. Compensatory dietary selection occurs in larval *Locusta migratoria* but not *Spodoptera littoralis* after a single deficient meal during ad libitum feeding. Physiol. Entomol. 15:235–242.

Singer, M. C. 1986. The definition and measurement of oviposition preference in plant-feeding insects, pp. 65–94. In J. R. Miller and T. A. Miller (eds.), Insect-Plant Interactions. Springer-Verlag, New York.

Slansky, F., Jr. 1974a. Energetic and nutritional interactions between larvae of the imported cabbage butterfly, *Pieris rapae* L., and cruciferous food plants. Ph.D. dissertation, Cornell, University, Ithaca, New York.

Slansky, F., Jr. 1974b. Relationship of larval foodplants and voltinism patterns in temperate butterflies. Psyche 81:243–253.

Slansky, F., Jr. 1982a. Insect nutrition: An adaptationist's perspective. Florida Entomol. 65:45–71.

Slansky, F., Jr. 1982b. Toward a nutritional ecology of insects, pp. 253–259. In Proc. 5th Intern. Symp. Insect-Plant Relationships, Wageningen, The Netherlands, Pudoc, Wageningen.

Slansky, F., Jr. 1985. Food utilization by insects: Interpretation of observed differences between dry weight and energy efficiencies. Entomol. Exp. Appl. 39:47–60.

Slansky, F., Jr. 1986. Nutritional ecology of endoparasitic insects and their hosts: An overview. J. Insect Physiol. 32:255–261.

Slansky, F., Jr. 1989. Early season weedy legumes: Potential larval food plants for the migratory velvetbean caterpillar (Lepidoptera: Noctuidae). J. Econ. Entomol. 82:819–824.

Slansky, F., Jr. 1990. Insect nutritional ecology as a basis for studying host plant resistance Fla. Entomol. 73:359–378.

Slansky, F., Jr. 1992. Allelochemical–nutrient interactions in herbivore nutritional ecology, pp. 135–174. In G. A. Rosenthal and M. R. Berenbaum (eds.), Herbivores: Their Interactions with Secondary Plant Metabolites, Vol. II, 2nd ed. Academic Press, New York.

Slansky, F., Jr., and Feeny, P. 1977. Stabilization of the rate of nitrogen accumulation by larvae of the cabbage butterfly on wild and cultivated food plants. Ecol. Monogr. 47:209–228.

Slansky, F., Jr., and Rodriguez, J. G. 1987a. Nutritional ecology of insects, mites, spiders, and related invertebrates: An overview, pp. 1–69. In F. Slansky, Jr. and J. G. Rodriguez (eds.), Nutritional Ecology of Insects, Mites, Spiders, and Related Invertebrates. Wiley, New York.

Slansky, F., Jr., and Rodriguez, J. G. (eds.) 1987b. Nutritional ecology of insects, mites, spiders, and related invertebrates. Wiley, New York.

Slansky, F., Jr., and Scriber, J. M. 1985. Food consumption and utilization, pp. 87–163. In G. A. Kerkut and L. I. Gilbert (eds.), Comprehensive Insect Physiology, Biochemistry and Pharmacology, Vol. 4. Pergamon, Oxford.

Slansky, F., Jr., and Wheeler, G. S. 1989. Compensatory increases in food consumption and utilization efficiencies by velvetbean caterpillars mitigate impact of diluted diets on growth. Entomol. Exp. Appl. 51:175–187.

Slansky, F., Jr., and Wheeler, G. S. 1991. Food consumption and utilization responses to dietary dilution with cellulose and water by velvetbean caterpillars, *Anticarsia gemmatalis*. Physiol. Entomol. 16:99–116.

Slansky, F., Jr., and Wheeler, G. S. 1992. Caterpillars' compensatory feeding response to diluted nutrients leads to toxic allelochemical dose. Entomol. Exp. Appl. (in press).

Smiley, J. T. 1985. *Heliconius* caterpillar mortality during establishment on plants with and without attending ants. Ecology 66:845–849.

Smith, C. M. 1989. Plant Resistance to Insects. A Fundamental Approach. Wiley, New York.

Smiley, R. H., Silby, R. M., and Moller, H. 1987. Control of size and fecundity in *Pieris rapae*: Towards a theory of butterfly life cycles. J. Appl. Ecol. 56:341–350.

Spencer, K. C. (ed.) 1988. Chemical mediation of coevolution. Academic Press, San Diego.

Stamp, N. E. 1980. Egg deposition patterns in butterflies: Why do some species cluster their eggs rather than deposit them singly? Am. Nat. 115:367–380.

Stamp, N. E. 1984. Foraging behavior of tawny emperor caterpillars (Nymphalidae: *Asterocampa clyton*). J. Lepidop. Soc. 38:186–191.

Stamp, N. E. 1990. Growth versus molting time of caterpillars as a function of temperature, nutrient concentration and the phenolic rutin. Oecologia 82:107–113.

Stamp, N. E., and Bowers, M. D. 1988. Direct and indirect effects of predatory wasps

(*Polistes* sp: Vespidae) on gregarious caterpillars (*Hemileuca lucina*: Saturniidae). Oecologia 75:619–624.

Stamp, N. E., and Bowers, M. D. 1990. Variation in food quality and temperature constrain foraging of gregarious caterpillars. Ecology 71:1031–1039.

Stearns, S. C., and Koella, J. C. 1986. The evolution of phenotypic plasticity in life-history traits: Predictions of reaction norms for age and size at maturity. Evolution 40:893–913.

Stephens, D. W., and Krebs, J. R. 1986. Foraging theory. Princeton University Press, Princeton, NJ.

Stiling, P., and Simberloff, D. 1989. Leaf abscission: Induced defense against pests or response to damage? Oikos 55:43–49.

Straw, N. A. 1989. The timing of oviposition and larval growth by two tephritid fly species in relation to host-plant development. Ecol. Entomol. 14:443–454.

Tabashnik, B. E., and Slansky, F., Jr. 1987. Nutritional ecology of forb foliage-chewing insects, pp. 71–103. In F. Slansky, Jr. and J. G. Rodriguez (eds.), Nutritional Ecology of Insects, Mites, Spiders, and Related Invertebrates. Wiley, New York.

Taghon, G. L. 1981. Beyond selection: Optimal ingestion rate as a function of food value. Am. Nat. 118:202–214.

Tallamy, D. W. 1986. Behavioral adaptations in insects to plant allelochemicals, pp. 273–300. In L. B. Brattsten and S. Ahmad (eds.), Molecular Aspects of Insect-Plant Associations. Plenum, New York.

Tauber, M. J., Tauber, C. A., and Masaki, S. 1986. Seasonal adaptations of insects. Oxford University Press, New York.

Taylor, F. 1980. Timing in the life histories of insects. Theoret. Pop. Biol. 18:112–124.

Taylor, M. F. J. 1989. Compensation for variable dietary nitrogen by larvae of the salvinia moth. Funct. Ecol. 3:407–416.

Terra, W. R. 1990. Evolution of digestive systems of insects. Annu. Rev. Entomol. 35:181–200.

Thomas, A. W. 1989. Food consumption and utilization by 6th-instar larvae of spruce budworm, *Choristoneura fumiferana*: A comparison of three *Picea* (spruce) species. Entomol. Exp. Appl. 52:205–214.

Thornhill, R., and Alcock, J. 1983. The Evolution of Insect Mating Systems. Harvard University Press, Cambridge, MA.

Timmins, W. A., Bellward, K., Stamp, A. J., and Reynolds, S. E. 1988. Food intake, conversion efficiency, and feeding behaviour of tobacco hornworm caterpillars given artificial diet of varying nutrient and water content. Physiol. Entomol. 13:303–314.

Tingey, W. M., and Singh, S. R. 1980. Environmental factors influencing the magnitude and expression of resistance, pp. 87–113. In F. G. Maxwell and P. R. Jennings (eds.), Breeding Plants Resistant to Insects. Wiley, New York.

Townsend, C. R., and Calow, P. (eds.) 1981. Physiological Ecology. Sinauer, Sunderland, MA.

Turunen, S. 1985. Absorption, pp. 241–277. In G. A. Kerkut and L. I. Gilbert (eds.),

Comprehensive Insect Physiology, Biochemistry and Pharmacology, Vol. 4. Pergamon, Oxford.

van der Meijden, E., van Bemmelen, M., Kooi, R., and Post, B. J. 1984. Nutritional quality and chemical defence in the ragwort-cinnabar moth interaction. J. Anim. Ecol. 53:443–453.

Vane-Wright, R. I., and Ackery, P. R. (eds.) 1984. The biology of butterflies. Princeton University Press, Princeton.

Van't Hof, H. M., and Martin, M. M. 1989. Performance of the tree-feeder *Orgyia leucostigma* (Lepidoptera: Liparidae) on artificial diets of different water content: A comparison with the forb-feeder *Manduca sexta* (Lepidoptera: Sphingidae). J. Insect Physiol. 35:635–641.

Vet, L. E. M., Lewis, W. J., Papaj, D. R., and van Lenteren, J. C. 1990. A variable-response model for parasitoid foraging behavior. J. Insect Behav. 3:471–490.

Vulinec, K. 1990. Collective security: aggregation by insects as a defense, pp. 251–288. In D. L. Evans and J. O. Schmidt (eds.), Insect Defenses. Adaptive Mechanisms and Strategies of Prey and Predators, SUNY Press, Albany, NY.

Wagner, T. L., and Leonard, D. E. 1979. The effects of parental and progeny diet on development, weight gain, and survival of pre-diapause larvae on the satin moth, *Leucoma salicis* (Lepidoptera: Lymantriidae). Can. Entomol. 111:721–729.

Waldbauer, G. P. 1968. The consumption and utilization of food by insects. Adv. Insect Physiol. 5:229–288.

Waldbauer, G. P., and Friedman, S. 1991. Self-selection of optimal diets by insects. Annu. Rev. Entomol. 36:43–63.

Waters, D. J., and Barfield, C. S. 1989. Larval development and consumption by *Anticarsia gemmatalis* (Lepidoptera: Noctuidae) fed various legume species. Environ. Entomol. 18:1006–1010.

West, C. 1985. Factors underlying the late season appearance of the lepidopterous leaf-mining guild on oak. Ecol. Entomol. 10:111–120.

Wharton, G. W. 1985. Water balance of insects, pp. 565–601. In G. A. Kerkut and L. I. Gilbert (eds.), Comprehensive Insect Physiology, Biochemistry and Pharmacology, Vol. 4. Pergamon, Oxford.

Wheeler, G. S., and Slansky, F., Jr. 1991. Compensatory response of the fall armyworm (*Spodoptera frugiperda*) when fed water- and cellulose-diluted diets. Physiol. Entomol. 16:361–374.

Wickman, P.-O., and Karlsson, B. 1989. Abdomen size, body size and the reproductive effort of insects. Oikos 56:209–214.

Wickman, P.-O., Wiklund, C., and Karlsson, B. 1990. Comparative phenology of four satyrine butterflies inhabiting dry grasslands in Sweden. Holarc. Ecol. 13:238–246.

Williams, E. H., and Bowers, M. D. 1987. Factors affecting host-plant use by the montane butterfly *Euphydryas gillettii* (Nymphalidae). Am. Midl. Nat. 118:153–161.

Williams, K. S., Lincoln, D. E., and Ehrlich, P. R. 1983a. The coevolution of *Euphydryas chalcedona* butterflies and their larval host plants. II. Maternal and host plant effects on larval growth, development, and food-use efficiency. Oecologia 56:330–335.

Williams, K. S., Lincoln, D. E., and Ehrlich, P. R. 1983b. The coevolution of *Euphydryas chalcedona* butterflies and their larval host plants. I. Larval feeding behavior and host plant chemistry. Oecologia. 56:323–329.

Willmer, P. 1986. Microclimatic effects on insects at the plant surface, pp. 65–80. In B. Juniper and R. Southwood (eds.), Insects and the Plant Surface. Arnold, London.

Wint, G. R. W. 1983. The effect of foliar nutrients upon the growth and feeding of a lepidopteran larva, pp. 301–320. In J. A. Lee, S. McNeill, and J. H. Rorison (eds.), Nitrogen as an Ecological Factor. Blackwell, Oxford.

Wolda, H. 1988. Insect seasonality: Why? Annu. Rev. Ecol. Syst. 19:1–18.

Yamada, H., and Umeya, K. 1972. Seasonal changes in wing length and fecundity of the diamond-back moth, *Plutella xylostella* (L.). Jpn. J. Entomol. Zool. 16:180–186.

Young, A. M. 1972. Adaptive strategies of feeding and predator-avoidance in the larvae of the Neotropical butterfly, *Morpho peleides limpida* (Lepidoptera: Morphidae). J. N.Y. Entomol. Soc. 80:66–82.

3

Foraging with Finesse: Caterpillar Adaptations for Circumventing Plant Defenses

David E. Dussourd

Introduction

The suggestion, initially advanced by Stahl (1888), that plant allelochemicals play a central role in the host selection of herbivores has now been confirmed by a century of investigation (reviewed by Rosenthal and Janzen 1979; Denno and McClure 1983). A pivotal early discovery was provided by Verschaffelt (1910), who documented the importance of secondary compounds in plant recognition. Working with *Pieris* caterpillars, he found that the application of host juices to a nonhost (also to neutral substrates) elicited feeding, as did the addition of a pure host chemical, sinigrin. Verschaffelt's techniques have since been replicated with hundreds of plant extracts leading to the discovery of innumerable chemicals of ecological and evolutionary significance.

In a typical protocol for testing caterpillar response to allelochemicals, plant tissues are extracted in solvent, concentrated, and applied to innocuous disks (Lewis and van Emden 1986; Doss and Shanks 1986). Caterpillar feeding on control and treated disks is then monitored. Active extracts are fractionated and tested further, leading eventually to the isolation and identification of the individual chemicals responsible for simulating or deterring feeding. The technique is powerful, but not without limitations. Chemicals are tested in isolation, removed from the botanical matrix in which they are synthesized, sequestered, and deployed for defense. This matrix mediates interactions with the herbivore, potentially increasing or decreasing the activity of metabolites (Berenbaum and Zangerl 1988; Bloem et al. 1989). In addition, the herbivore, through behavioral adaptation, may further modify the efficacy of defenses.

In this chapter, I consider how allelochemicals are packaged in leaves, and how their varied presentation affects host choice by caterpillars. Foraging behaviors, particularly those that alter the food resource, will be examined in detail. Such behaviors offer clues to constraints imposed on caterpillars, constraints that potentially include host defenses. In essence, the behaviors present an opportunity for identifying what factors are truly significant to a caterpillar—indeed of sufficient importance to select for behavioral countermeasures.

I combine an analysis of plant anatomy and caterpillar behavior in order to test the proposition that herbivores respond not just to phytochemicals, but also to the structures in which the compounds reside. Canal systems will be emphasized, while other types of repositories and physical defenses will be only briefly reviewed. Although I recognize that a defense function for many structures and associated chemicals is still tentative, I will bypass this topic (also evidence for alternative roles, e.g., Tallamy and Krischik 1989) since it has been discussed extensively elsewhere (e.g., Juniper and Southwood 1986; Langenheim 1990; Farrell et al. 1991).

The chapter is organized into five sections. The first one discusses some of the major ways that plants deploy defenses, considering their advantages, disadvantages, and particularly vulnerabilities to caterpillar attack. In the second section, I summarize four avenues available to caterpillars for feeding on leaves despite the presence of defenses. Next, two adaptations, vein-cutting and trenching, are discussed in detail. Evidence that both behaviors function specifically to deactivate defensive canal systems is summarized, together with observations supporting alternative roles. In the fourth section, I present experimental evidence that caterpillars lacking behavioral counteradaptations are indeed constrained in host choice. Finally, the last section explores the effects of secretory canals on other aspects of caterpillar biology, with consideration given to promising areas of future research.

Deployment of Plant Defenses

The distribution of allelochemicals differs enormously in different plants: mint leaves are coated with surface glands loaded with aromatic terpenes (Fahn 1979; Kelsey et al. 1984); maple leaves are impregnated with protein-binding tannins (enclosed in vacuoles) and toughtened with lignified veins (Schultz et al. 1982; Hagen and Chabot 1986), whereas milkweed leaves contain a ramifying network of latex canals pressurized with a lethal brew of toxic cardenolides in a quick-setting glue (Blaser 1945; Nelson et al. 1981; Dussourd and Eisner 1987). Individual chemicals are often deployed in diverse ways, such as the monoterpene, α-pinene, found in both the glandular trichomes of *Artemisia* (Kelsey et al. 1984) and the resin ducts of conifers (Cates and Alexander 1982) (also in the gland secretions of termites, ants, bugs, and caterpillars, Blum 1981). Some and probably most plants have multiple layers of defense. Leaves of the "Evil Woman," *Cnidoscolus urens*, for example, are protected by both urticating hairs and latex canals (Dillon et al. 1983).

Clearly, plants package defensive substances in a variety of ways. In the following section, I explore their relative merits and limitations. The emphasis will not be on the particular chemicals involved, but on their location and method of deployment. Previous reviews have analyzed temporal and spatial patterns in the distribution of allelochemials (McKey 1979; Denno and McClure 1983), such as between young and mature leaves, and related that distribution to the perceived value of the plant part. I will consider a finer scale of immediate consequence to a foraging caterpillar—within an individual leaf.

Surface Structures

Leaves of many species are covered with a dense vestiture of trichomes, dotted with urticating hairs, or coated with resinous secretions. The anatomy, chemistry, and varied functions of these surface features have been treated in several excellent reviews (Levin 1973; Stipanovic 1983; Rodriguez et al. 1984; Juniper and Southwood 1986), which document the extraordinary variability of these ubiquitous plant structures. This variation permits multiple modes of action against herbivores. Glandular trichomes effectively deploy both toxins and adhesives. Nonglandular structures act by preventing attachment, blocking access to the leaf surface, or impaling herbivores. Direct evidence of detrimental effects on caterpillars has now been repeatedly obtained (Levin 1973; Stipanovic 1983; Kelsey et al. 1984; Southwood 1986). The hook-like trichomes of *Passiflora,* for example, have been shown to puncture and fatally entrap *Heliconius* caterpillars (Gilbert 1971).

An obvious advantage of surface features is that insects encounter the defense without first rupturing the leaf surface (Levin 1973). Damage is thus minimized and the plant spared exposure to pathogen attack. In sequestering defensive chemicals away from the photosynthetic tissues and in releasing the substances externally, surface structures also avoid problems with autotoxicity.

A major disadvantage is that only the leaf surface is protected. The interior remains vulnerable to miners, which may in fact be protected from natural enemies by the plant's external defensive investment. We might expect plant species well endowed with surface defenses to be particularly prone to mining caterpillars. A notable counterexample is supplied by Gross and Price (1988), who document that a glechiid miner in *Solanum* leaves with stellate trichomes is particularly vulnerable to parasitoids because the miner is unable to escape the mine when attacked.

All plant defenses, of course, can have ramifying effects on the third trophic level. The influence of surface features, however, is likely to be particularly severe since a predator or parasitoid need only walk on the plant to be affected. Although leaf structures, such as veins and hairs, may offer a foothold and thus enhance predator/parasitoid movement, negative effects also occur (Southwood 1986). The glandular trichomes of tobacco, for example, entrap egg parasitoids

(Rabb and Bradley 1968) and inhibit movement of predacious coccinellids (Belcher and Thurston 1982). Gravid Lepidoptera females may similarly be aided or deterred in oviposition (Levin 1973; Southwood 1986).

Unlike internal defenses, surface structures are directly exposed to the elements. Glandular secretions may be removed by wind and rain, or volatized in direct sunlight, creating the necessity of continuous replenishment. Physical impediments, such as nonglandular trichomes, while largely immune to the weather, suffer a clear vulnerability to mandibulate herbivores, including caterpillars, which may bite off the inedible hairs before feeding (Hulley 1988). Larger structures, such as thorns, spines, and urticating hairs, would appear to be effective only against large herbivores, and of little use against insects which might simply feed around them. This need not always be the case. The midrib of *Lactuca serriola* (prickly lettuce), for example, is protected by a single column of prickles. For plusiine noctuids to cut a trench across the leaf, they must first remove the prickles blocking the transection (Dussourd) 1986).

Finally, like all defenses, surface features suffer a vulnerability to herbivores small enough to feed between adjacent structures. Unless the leaf is totally coated (as with resins in creosote bush, Thompson et al. 1979), small caterpillars and particularly haustellate herbivores may squeeze their mouthparts into undefended spaces. However, the insect must still reside on the plant, a risky business on species such as *Phaseolus* and *Solanum* where herbivore entrapment on hooked or glandular trichomes has been well documented (Stipanovic 1973; Juniper and Southwood 1986).

Internal Repositories

Internal leaf defenses include an array of structures, diverse in form and function. Common features potentially deterrent to caterpillars include thickened cell walls, specialized cellular receptacles, intercellular pockets, and ramifying canals. I will somewhat artificially separate these disparate forms into three categories: dispersed, patchy, and canalicular. Dispersed defenses occur throughout the leaf and are essentially homogeneous, at least with regard to feeding caterpillars. Examples might include secondary metabolites (such as some tannins and alkaloids) that are stored in subcellular compartments (McKey 1979; Luckner et al. 1980), and structural elements (such as lignins and silica grains), that are often incorporated into cell walls (Swain 1979; Metcalfe and Chalk 1983). Although no substance is truly uniform in distribution, storage sites may be so small and evenly dispersed as to be indistinguishable to caterpillars.

Patchy repositories differ in their greater size and in their disjunct distribution within the leaf. Common examples include intercellular cavities and cells termed idioblasts that are specialized as receptacles for substances as diverse as tannins, oils, and raphides (Foster 1956; Esau 1965; Fahn 1979; Metcalfe and Chalk 1983). Larger discontinuities in the leaf are produced by the venational network.

The veins not only interrupt the uniformity of the palisade layer, but also the distribution of structural and defensive chemicals. The leaves of sugar maple (*Acer saccharum*), for example, are toughened by bundle sheaths that form a lignified fibrous encasement around vascular traces (Hagen and Chabot 1986). Crystalline deposits occur below but not above the bundle sheaths. Such irregularities present herbivores with different constraints and opportunities than those offered by more homogeneous features.

The distinction between dispersed and patchy defenses of course depends on caterpillar size. For large saturniid larvae, the contents of idioblasts and cavities may appear to be evenly distributed, whereas small larvae on the same leaf may perceive and respond to three-dimensional patterns.

Cellular repositories are commonly elongate in shape, in some cases forming complex systems of canals that ramify throughout the plant. These canalicular structures (my third category) vary tremendously in anatomy and in the chemistry of contained fluids. Many, however, share the conspicuous trait of oozing copious secretions when damaged, a function of their large size and pressurized contents. Familiar examples include the oleoresin exudate of conifers, the noxious secretion of poison ivy, and the milky lattices of figs and euphorbs (Fahn 1979; Metcalfe and Chalk 1983). Even phloem can act in similar fashion; cucurbits, for example, release a viscous, quick-gelling sap from leaf veins when damaged (Alosi et al. 1988; Dussourd and Denno 1991).

In the following sections, I explore the ramifications of sequestering secondary substances within dispersed, patchy, or canalicular structures, emphasizing their defensive properties and potential effects on foraging caterpillars.

Dispersed Defenses

Distributing defenses evenly throughout the leaf offers a clear advantage—the entire structure is protected. A caterpillar's only feasible response may be to avoid the leaf altogether or to ingest (and biochemically detoxify) its entire defensive endowment. This benefit to the plant is countered by a distinct disadvantage: the defense is diluted by surrounding tissues. Caterpillars may never encounter a threshold concentration necessary for deterrence. Adhesives, for example, would be of little use unless concentrated. Another disadvantage is the potential risk of autotoxicity, which may limit what chemicals can be safely dispersed throughout the leaf. Even secretions sequestered in intracellular repositories could prove hazardous due to the high surface area of small compartments.

Patchy Defenses

Sequestering substances within specialized cellular or intercellular receptacles minimizes dangers of autotoxicity, at least as long as the leaf remains intact. With large repositories, adhesives as well as toxins can be effectively deployed.

An insect attacking a leaf would sporadically encounter high concentrations of the defense, perhaps enough to deter further feeding. However, the small volume of most idioblasts and cavities (relative to canals) must surely limit their effectiveness. In addition, plants employing patchy defenses suffer a serious vulnerability to herbivores small enough to feed between the patches. With haustellate insects, this distance need not be large at all.

Interspecific differences in the distribution, size, and abundance of storage sites have been documented (e.g., Welch 1920; Langenheim et al. 1982; Curtis and Lersten 1986). The significance of this variation remains largely unknown, although the topic (also comparisons between cavities and canals) appears to be eminently suited to experimental manipulation and mathematical modeling.

Canalicular Defenses

Canal systems solve many of the problems of dispersed and patchy defenses while presenting new vulnerabilities. When ruptured, canals release secretion precisely at the site of attack, immediately on injury. The volume of secretion emitted is large relative to the amount initially present at this location since secretion flows through the canals down a pressure gradient toward the rupture. In *Cryptostegia grandiflora* (Asclepiadaceae), severance of the stem elicits latex flow from a distance of at least 70 cm within the plant (Buttery and Boatman 1976). Plants damaged after a shower sometimes emit a stream of latex over a meter in length that lasts 2–3 seconds (Bangham 1934). Eventually latex outflow from wounds ceases due to clotting agents that plug ruptures and stem the loss of additional secretion. Pressure is then restored within the canals (Buttery and Boatman 1976; d'Auzac et al. 1989).

Several characteristics of canal systems appear to be ideally suited as a defense. The canals commonly ramify throughout leaves, thus providing thorough coverage against attack. Pinpricks in intact leaves almost invariably elicit a visible release of secretion (Dussourd and Denno 1991). Since large volumes of fluid are released, adhesives as well as toxins can be deployed in sufficient quantities to entrap or deter caterpillars. Secretions may also act by coating receptors, thereby preventing the insect from recognizing acceptable food. Exudation from canals flushes and seals wounds, presumably thereby discouraging pathogen attack (Farrell et al. 1991; Van Parijs et al. 1991). Finally, because the canals typically follow veins, the essential transport system of the leaf would appear to be well protected.

The widespread use of canal secretions as fish and arrow poisons (Shaw et al. 1963; Rizk 1987), and occasional use as birdlimes (Biffen 1898), attests to their toxicity and stickiness. Effects on caterpillars can be extraordinary. When a microliter of *Asclepias curassavica* latex is daubed on the mouthparts of the generalist, *Trichoplusia ni,* the caterpillars respond by vigorously wiping their mouthparts, thus removing most secretion (Dussourd unpublished observations).

The tiny amount remaining, however, is often sufficient to cause final-instar larvae to convulse with spasms a few minutes following contact. Within a half hour, some of the larvae hang immobile from the leaf with a large drop of regurgitate engulfing their head. Eventually, they fall from the leaf and over a period of 1 to 2 days shrivel up to a fraction of their former size. Then miraculously, they recover and if offered suitable food successfully complete development.

The toxic agent in *A. curassavica* is almost surely cardenolides, which are known to be concentrated in the latex (Seiber et al. 1982). Another cardenolide, digitoxin, causes similar short-term immobility in a coccinellid beetle, *Coleomegilla maculata,* which falls over on its back, but recovers within a few hours (Dussourd 1986). Whereas the effects of *A. curassavica* latex on *T. ni* may be only temporary, the furanocoumarin-containing secretion of *Pastinaca sativa* can be lethal to the larvae (A. R. Zangerl personal communication).

Secretory canals thus provide a suitable depot for secondary metabolites and a means of transporting the compounds to sites under attack. The greater the length and width of canals (and pressure of secretion within canals), the greater the advantages they offer. More fluid is delivered to damaged tissues posing a greater threat to herbivores. The extent of canals in some plants is astonishing. Milkweeds (*Asclepias* spp.), for example, contain only 16 latex cells (Wilson 1986). These nonarticulated laticifers extend in length as the plant grows, branching repeatedly to form a complex system throughout the plant. Other species likewise produce only small numbers of these enormously long, multinucleate cells (i.e., *Euphorbia marginata*—12, Mahlberg and Sabharwal 1968; *Jatropha dioica*—5 to 7, Cass 1985; *Nerium oleander*—usually 28, Mahlberg 1961). In other plants, latex cells are compound in origin (articulated laticifers), beginning as a chain of cells, that often unite to form continuous tubes through the disintegration of connecting cell walls. Adjacent tubes may then join to form an elaborate net-like structure (Esau 1965, Fahn 1979).

Nonarticulated and articulated laticifers differ in their vulnerability to attack. Damage to either structure causes the release and concentration of secretion at a distinct point that can be subsequently avoided by the herbivore. But in species with nonarticulated laticifers, secretion pressures are rapidly restored proximal, but not distal, to a rupture (Dussourd and Denno 1991). The distal section containing several branches is then highly vulnerable to attack. Not only are the branches isolated from the main canal system, but they are also partially drained of fluid and any remaining secretion is diluted with water (the release of pressure associated with a rupture causes an influx of water, Buttery and Boatman 1976). Systematic attacks on canals may thereby severely compromise the effectiveness of latex as a defense. Not surprisingly, insects take full advantage of this feature either by cutting veins (as discussed later) or by feeding on damaged tissues, a common habit in generalist caterpillars (Dussourd unpublished observations) as well as grasshoppers (Lewis 1979).

Thus, while the advantages of canal systems increase with increasing size, so do their liabilities. Ruptures may not only cause a greater loss of secretion, but also render a larger area distal to the cut vulnerable to attack. With canals just a few millimeters in length, only small sections of the plant would be left undefended. But with canals that encompass entire branches, whole leaves can be readily isolated. Nonarticulated laticifers in particular are vulnerable because the canals do not interconnect. Articulated laticifers forming networks are less susceptible since areas distal to a cut can be resupplied with secretion through adjacent strands of the net. All strands would have to be severed to isolate distal canals. However, even these plants have a weakness: all canals converge in the petiole. Damage to the petiole may thus isolate the entire leaf. Clasping leaves with their multiple connections to the stem (as in many Cichorieae) and large leaves with their own sizable reservoirs of secretion (perhaps as in *Carica papaya*) offer potential solutions.

Plants that sequester secretions within patchy repositories or canals share a vulnerability to insects small enough to feed (or oviposit, Joel 1980) between the structures. Patchy and canalicular defenses may also be relatively ineffective against large herbivores that eat entire leaves, thereby encountering secretions only in dilute form (see Becerra and Venable 1990 for possible exception). The effectiveness of both defenses, and particularly of canals, may thus be maximal against intermediate-sized herbivores (such as most caterpillars), which are too large to feed between canals and too small to be unaffected by the secretory response.

Secretions sequestered within canals pose a potential risk of autotoxicity if the canals are damaged. Ruptures produced by herbivores may cause minimal injury since the secretions can be voided to the leaf surface. Internal leakage may be a more serious problem. In lettuce, necrotic zones (termed tipburn) form when laticifers rupture during conditions conducive to latex production (Tibbitts and Read 1976). Such problems are no doubt exacerbated by the high pressures necessary for forcing viscous secretions through narrow canals to sites of attack.

Two additional liabilities of canal systems deserve mention. Many canal secretions are effective only if released into air. Latex and other water-based exudates would presumably have little value for submerged aquatic plants since the fluids would be rapidly diluted and washed away. Furthermore, secretions that require oxygen to gel, such as phloem sap (Alosi et al. 1988), may be completely ineffective against haustellate insects that contact the fluids inside the leaf. Certainly, aphids attack cucurbit phloem successfully (Blackman and Eastop 1985) without any reported difficulty from stylet occlusion or entrapment due to gelling sap.

Finally, canals that contain substances of value to animals may become a liability rather than an asset. Humans have for centuries damaged plants to collect such valuable products as oleoresin, rubber and opium (Schery 1952; Hillis 1987). The milky latex of *Brosimum galactodendron,* appropriately named the cow tree, is even harvested for human consumption (Haberlandt 1914). Humans are not the only ones exploiting these fluid reservoirs. Buffy-headed marmosets gouge holes in the bark of acacias and consume the nutritious gums that excude (Farrari 1991; also see Heinrich 1991). Rothschild and Ford (1970) proposed that monarch caterpillars similarly damage milkweed stems and petioles to drink exuding latex and thereby obtaining cardenolides necessary for defense against predators.

Caterpillar Counteradaptations

Insects respond to spatial and temporal variation in plant defenses by selectively attacking species, individuals, or tissues low in noxious traits (Schultz 1983; Tallamy 1986). Yet even relatively undefended plants are not without protective barriers to feeding. Herbivores have developed a variety of ploys for circumventing defenses (Rhoades 1983, 1985; Tallamy 1986): four general categories—avoidance, biochemical/physiological adaptation, behavioral deactivation, and appropriation—are discussed below.

Avoidance

Insects with the capability of recognizing leaf defenses, and with mouthparts smaller than the distance between adjacent repositories, can avoid the defense altogether through selective feeding. Any nonhomogeneous defense is potentially vulnerable. First instar arctiid and geometrid caterpillars on maple, for example, avoid sclerenchymous bundle sheaths by "window-feeding"; they remove discrete patches of mesophyll and overlying epidermis while avoiding the veins (Hagen and Chabot 1986). Likewise, first instar larvae of *Trichoplusia ni* (Noctuidae) feed between parsnip veins containing furanocoumarins (Zangerl 1990), and young *Pterogon proserpina* (Sphingidae) caterpillars avoid the raphides in grape leaves (Merz 1959 cited by Ehrlich and Raven 1964). This option, although immeasurably important for haustellate insects with narrow mouthparts, is of only limited value for mandibulate caterpillars. Early instar larvae may be able to bypass defenses, but the caterpillars must grow to complete development, necessitating the adoption of other strategies (Reavey, Chapter 8).

Biochemical/Physiological Adaptation

Chemicals damaging to cellular processes are widespread in nature. Animals rely heavily on biochemical mechanisms at all levels (from digestive tracts to subcellular vesicles) for neutralizing toxicants. Mixed-function oxidases in the midgut and fatbody of caterpillars have received the most attention due to their importance in degrading secondary compounds and pesticides (Brattsten et al. 1977). As documented in several recent reviews (Dowd et al. 1983; Brattsten and Ahmad 1986; Brattsten 1988), these enzyme systems are extensive and accompanied by a plethora of other biochemical and physiological adaptations that include target-site insensitivity, enhanced excretion, sequestration, and high gut pH.

Such mechanisms are clearly of value no matter where a defensive substance is stored in the leaf. They are not, however, without limitations. High concentrations of toxicants may overwhelm enzyme batteries, at least until inducible responses occur. Plants may thwart digestive processes by producing compounds resistant to degradation (McKey 1979). Enzymes can themselves be neutralized. Proteinase inhibitors, for example, effectively deactivate digestive enzymes and reduce caterpillar growth (Broadway et al. 1986). Some secondary substances are "bioactivated" by insect enzymes to produce metabolites more reactive and thus potentially more toxic than the parent molecule (Brattsten 1988). Finally, biochemical and physiological mechanisms provide little protection against defenses that engage the insect externally, such as sticky exudates and nonglandular trichomes.

Behavioral Deactivation

As tabulated in Table 3.1, caterpillars exhibit a variety of complex behaviors before, during and after feeding, at least some of which function specifically to remove, inhibit, or deactivate plant defenses. Most of the behaviors involve mandibular modification of host plants. Some caterpillars on leaves coated with trichomes, for example, bite off the trichomes before feeding on the underlying leaf (Table 3.1A) Plusiine noctuids (*Anagrapha falcifera, Autographa precationis,* and *Trichoplusia ni*) similarly remove the midrib prickles of *Lactuca serriola* either by eating the prickles outright (Fig. 3.1A) or by chewing them off at the base, an arduous task on older leaves with toughened spines. The prickles neither interfere with feeding nor pose a risk of entrapment. They do, however, block the caterpillars from transecting the leaf with a trench (Fig. 3.1B), thus necessitating their removal.

Many other insects cut trenches or partially sever leaf veins, petioles or stems before feeding (Table 3.1B, Fig. 3.1C–H). A variety of functions have been proposed for these widespread behaviors, a detailed account of which is provided in the following section.

Other species transect petioles *after* feeding, discarding the leaf remnants in a seemingly wasteful process (Table 3.1C). Petiole clipping has been recorded only on trees and vines, where it presumably functions to eliminate cues attractive to birds (Heinrich and Collins 1983) or parasitoids (Weinstein 1990). An alternative explanation was provided by Edwards and Wanjura (1989), who proposed that caterpillars and sawflies on *Eucalyptus* cut petioles to block the transmission of induced defenses from damaged leaves to adjacent foliage. This suggestion was recently refuted by Weinstein (1990), who prevented sawflies from severing *Eucalyptus* petioles and found no concomitant reduction in growth.

Table 3.1. Behavioral Modification of Host Plants by Caterpillars.[a]

Caterpillar species	Plant genus (Family)	Behavior	Proposed function	Reference[b]
A. Circumvent Trichomes or Spines				
Nymphalidae (Ithomiinae)				
Mechanitis isthmia	*Solanum* (Solanaceae)	Spin silk scaffolding over trichomes	Prevents entrapment on trichomes	1, 2
Sphingidae				
Erinnyis ello	*Cnidoscolus* (Euphorbiaceae)	Eat urticating hairs on stem/ petiole	Removes obstacle to reaching leaf	3
Noctuidae				
Anagrapha falcifera, also *Autographa precationis*, *Trichoplusia ni*	*Lactuca* (Asteraceae)	Bite off or eat midrib spines	Allows caterpillar to cut trench	4, 5
Pardasena sp. nr. *diversipennis*	*Solanum* (Solanaceae)	Bite off stellate trichomes	Removes inedible trichomes	6
B. Damage Leaf Veins or Petioles before Feeding				
Gelechiidae				
Telphusa longifasciella	*Rhus* (Anacardiaceae)	Cut leaf veins	Deactivates resin canals	7
Papilionidae				
Euphoeades troilus	Lauraceae	Chew channel, fold over flap	Creates hiding place	8
Papilio polyxenes	*Daucus*, *Petroselinum* (Apiaceae)	Bite petiolules, presumably before feeding	Deactivates resin canals	4
Nymphalidae				
Nymphalinae				
Aglais milberti	*Urtica* (Urticaceae)	Notch base of leaf, tie leaf with silk	Provides shelter	8
Colobura dirce	*Cecropia* (Moraceae)	Droop leaves by cutting veins	—	9
Polygonia comma	*Boehmeria* (Urticaceae)	Cut veins, tie leaf edges with silk	Provides shelter	8
	Ulmus (Ulmaceae)	Cut channels across leaf, tie leaf with silk	Provides shelter	8
Tigridia acesta	*Cecropia*, *Paruma* (Moraceae)	Droop leaves into tent by cutting veins	—	9

Table 3.1. (continued)

Caterpillar species	Plant genus (Family)	Behavior	Proposed function	Reference[b]
Nymphalidae *(continued)*				
Vanessa atalanta	*Boehmeria* (Urticaceae)	Cut midrib and veins, tie leaf with silk	Shelter from enemies/weather	8, 10, 11
Heliconiinae				
Dione spp. *Dryas julia*	*Passiflora* (Passifloraceae)	Cut channels across leaf	Territorial marker	12, 13
Dryadula phaetusa	*Passiflora* (Passifloraceae)	Cut channels across leaf	Territorial marker	12
Heliconius spp.	*Passiflora* (Passifloraceae)	Chew midrib furrows before or during feeding	Territorial marker	12
Philaethria spp.	*Passiflora* (Passifloraceae)	Cut channels across leaf	Territorial marker	12, 13
Danainae				
Danaus spp.	Asclepiadaceae	Some early instars cut circular trenches, later instars damage veins, midrib, or petiole	Deactivates latex canals	14–16, but also see 17, 18
Euploea core, also *Idea leuconoe nigriana*	*Parsonsia spiralis* (Apocynaceae)	Early instars cut circular trench and regurgitate froth on outer perimeter	Froth ring protects against ants	19
Ideopsis similis	*Tylophora* (Asclepiadaceae)	Damange petiole	—	20
Lycorea cleobaea	Asclepiadaceae, also Caricaceae, Moraceae	Early instars cut circular trenches, late instars may cut veins	Prevents mobilization of defense Stops latex flow	9 21
Parantica sita	Asclepiadaceae	Damage midrib	—	20
Ithomiinae				
Aeria eurimedea	*Prestonia* (Apocynaceae)	Cut veins in older leaves, sometimes also lateral channels	Stops latex flow Protects against rain/predators	9 22
Mechanitis isthmia	*Solanum* (Solanaceae)	Cut midrib	—	2
Melinaea ethra	*Markea* (Solanaceae)	Early instars trench, late instars cut veins	Prevents mobilization of defense	9
Pyralidae				
Palpita flegia	*Thevetia* (Apocynaceae)	Damage midrib or petiole, tie leaves with silk	Deactivates latex canals	7

Table 3.1. (continued)

Caterpillar species	Plant genus (Family)	Behavior	Proposed function	Reference[b]
Pyralidae *(continued)*				
Saucrobotys futilalis	*Apocynum* (Apocynaceae)	Sometimes cut midrib	Deactivates latex canals	7
Saturniidae				
Sphingicampa albolineata	*Acacia* (Fabaceae)	Damage petiole	Facilitates feeding	23
Sphingidae				
Erinnyis alope	*Carica* (Caricaceae)	Cut trenches across leaf tips	Deactivates latex canals	7
Erinnyis ello	Euphorbiaceae, also Caricaceae, Moraceae	Early instars cut trenches, late instars constrict/cut petioles	Deactivates latex canals	3, 5
Manduca sexta	*Nicotiana* (Solanaceae)	Cut midrib	Facilitates feeding by bending leaf	24, 25
Pseudoclanis postica	*Ficus* (Moraceae)	Early instars cut channels	Sabotages latex defense	26
Notodontidae				
Heterocampa sp.	*Acer* (Aceraceae)	Partially sever petiole	Prevents induction of defenses	27
Schizura ipomoeae	*Aesculus* (Hippocastanaceae)	Cut midrib, gridle petiole	—	5
Arctiidae				
Cycnia inopinatus	*Asclepias* (Asclepiadaceae)	Cut midrib (also flower stalks)	Deactivates latex canals	5
Cycnia oregonensis, C.tenera	*Apocynum* (Apocynaceae)	Cut midrib	Deactivates latex canals	7
Empyreuma affinis	*Nerium* (Apocynaceae)	Cut midrib or petiole	Deactivates latex canals	7
Eucereon carolina	*Asclepias* (Asclepiadaceae)	Cut midrib	Deactivates latex canals	7
Euchaetes egle	*Apocynum* (Apocynaceae) *Asclepias* (Asclepiadaceae)	Cut midrib or petiole	Deactivates latex canals	7
Syntomeida epilais	*Nerium* (Apocynaceae)	Cut midrib or petiole	Deactivates latex canals	7
Utetheisa ornatrix	*Crotalaria* (Fabaceae)	Partially sever petiole	—	7
Hypsidae				
Aganais speciosa	*Ficus* (Moraceae)	Cut leaf veins and midrib	Sabotages latex defense	16, 26
Noctuidae				
Amphipyra tragopoginis	*Lactuca, Taraxacum* (Asteraceae)	Cut trench across leaf	Deactivates latex canals	7
Anagrapha falcifera	Apiaceae, Asteraceae (Cichorieae)	Cut trench across leaf	Deactivates canal systems	4, 7
Autographa biloba, A. californica	*Lactuca* (Asteraceae)	Cut trench across leaf	Deactivates latex canals	5, 7

Table 3.1. (continued)

Caterpillar species	Plant genus (Family)	Behavior	Proposed function	Reference[b]
Noctuidae *(continued)*				
Autographa precationis	Apiaceae, Asteraceae (Cichorieae)	Cut trench across leaf	Deactivates canal systems	4, 7
Chrysodeixis acuta	*Ficus* (Moraceae)	Cut trench across leaf	Sabotages latex defense	26
Paectes oculatrix	*Rhus* (Anacardiaceae)	Cut leaf veins	Deactivates resin canals	7
Pseudoplusia includens	*Cucumis* (Cucurbitaceae)	Cut trench across leaf	Eliminates phloem exudate	7
Spargaloma sexpunctata	*Apocynum* (Apocynaceae)	Cut leaf midrib	Deactivates latex system	7
Trichoplusia ni	Apiaceae, Asteraceae, Cucurbitaceae	Cut trench across leaf	Deactivates canal systems	4, 7, 28
C. Sever Leaf Petiole after Feeding				
Papilionidae				
Papilio glaucus	Lauraceae, Magnoliaceae	Sever petiole, discard leaf remnants	Eliminates evidence of presence	29
Limacodidae				
Doratifera quadrigutata	*Eucalyptus* (Myrtaceae)	Sever petiole, discard leaf remnants	Prevents induction of defenses	30, but see 31
Anthelidae				
Anthela varia	*Eucalyptus* (Myrtaceae)	Sever petiole, discard remnants	Prevents induction of defenses	30, but see 31
Saturniidae				
Callosamia promethea	*Fraxinus* (Oleaceae)	Sever petiole, discard remnants	Avoids attracting predacious birds	32
Antheraea polyphemus	*Acer* (Aceraceae)	Sever petiole, discard remnants	Avoids attracting predacious birds	32
Sphingidae				
Ceratomia undulosa, *Sphinx* spp.	*Fraxinus* (Oleaceae)	Sever petiole, discard remnants	Avoids attracting predacious birds	32
Pachysphinx modesta	*Populus*, *Salix* (Salicaceae)	Sever petiole, discard remnants	Avoids attracting predacious birds	32
Paonias spp.	*Salix* (Salicaceae), *Prunus* (Rosaceae)	Sever petiole, discard remnants	Avoids attracting predacious birds	32
Smerinthus jamaicensis	*Populus* (Salicaceae)	Sever petiole, discard remnants	Avoids attracting predacious birds	32
Sphecodina abbottii	*Vitis* (Vitaceae)	Sever petiole, discard remnants	Avoids attracting predacious birds	32
Notodontidae				
Cerura cinerea	*Populus* (Salicaceae)	Sever petiole, discard remnants	Avoids attracting predacious birds	32
Heterocampa spp.	*Acer* (Aceraceae)	Sever petiole, discard remnants	Avoids attracting predacious birds	32
Notodonta stragula	*Salix* (Salicaceae)	Sever petiole, discard remnants	Avoids attracting predacious birds	32
Schizura unicornis	*Betula* (Corylaceae)	Sever petiole, discard remnants	Avoids attracting predacious birds	32

Table 3.1. (continued)

Caterpillar species	Plant genus (Family)	Behavior	Proposed function	Reference[b]
Noctuidae				
Acronicta spp.	*Acer* (Aceraceae), *Populus* (Salicaceae)	Sever petiole, discard remnants	Avoids attracting predacious birds	25, 32
Catocala spp.	*Populus* (Salicaceae), *Tilia* (Tiliaceae)	Sever petiole, discard remnants	Avoids attracting predacious birds	32, 33
Zale sp.	*Populus* (Salicaceae)	Sever petiole, discard remnants	Avoids attracting predacious birds	32
D. Girdle or Transect Stem				
Cosmopterigidae				
Periploca nigra	*Juniperus* (Cupressaceae)	Girdle stems	—	34
Tortricidae				
Archips purpurana	*Polygonum* (Polygonaceae)	Cut petioles or stems, tie leaves with silk	—	4
Noctuidae				
Agrotis spp., also *Feltia jaculifera*, *Nephelodes minians*	herbs	Cut off stems at base	—	35

[a]Some, but not necessarily all, behaviors serve to deactivate host defenses.

[b](1) Rathcke and Poole 1975; (2) Young and Moffett 1979; (3) Dillon et al. 1983; (4) Dussourd 1986; (5) Dussourd unpublished observations; (6) Hulley 1988; (7) Dussourd and Denno 1991; (8) Scudder 1889; (9) DeVries 1987; (10) Edwards 1882; (11) Edwards 1883; (12) Alexander 1961; (13) Brown 1981; (14) Brewer 1977; (15) Dussourd and Eisner 1987; (16) Compton 1987; (17) Rothschild and Ford 1970; (18) Rothschild 1977; (19) P.J. DeVries personal communication; (20) Fukuda et al. 1985; (21) Janzen 1985; (22) Young 1978; (23) Tuskes 1985; (24) Heinrich 1971; (25) Heinrich 1980; (26) Compton 1989; (27) Carroll and Hoffman 1980; (28) McKinney 1944; (29) Lederhouse 1990; (30) Edwards and Wanjura 1989; (31) Weinstein 1990; (32) Heinrich and Collins 1983; (33) Heinrich 1979; (34) Johnson and Lyon 1976; (35) Metcalf et al. 1962.

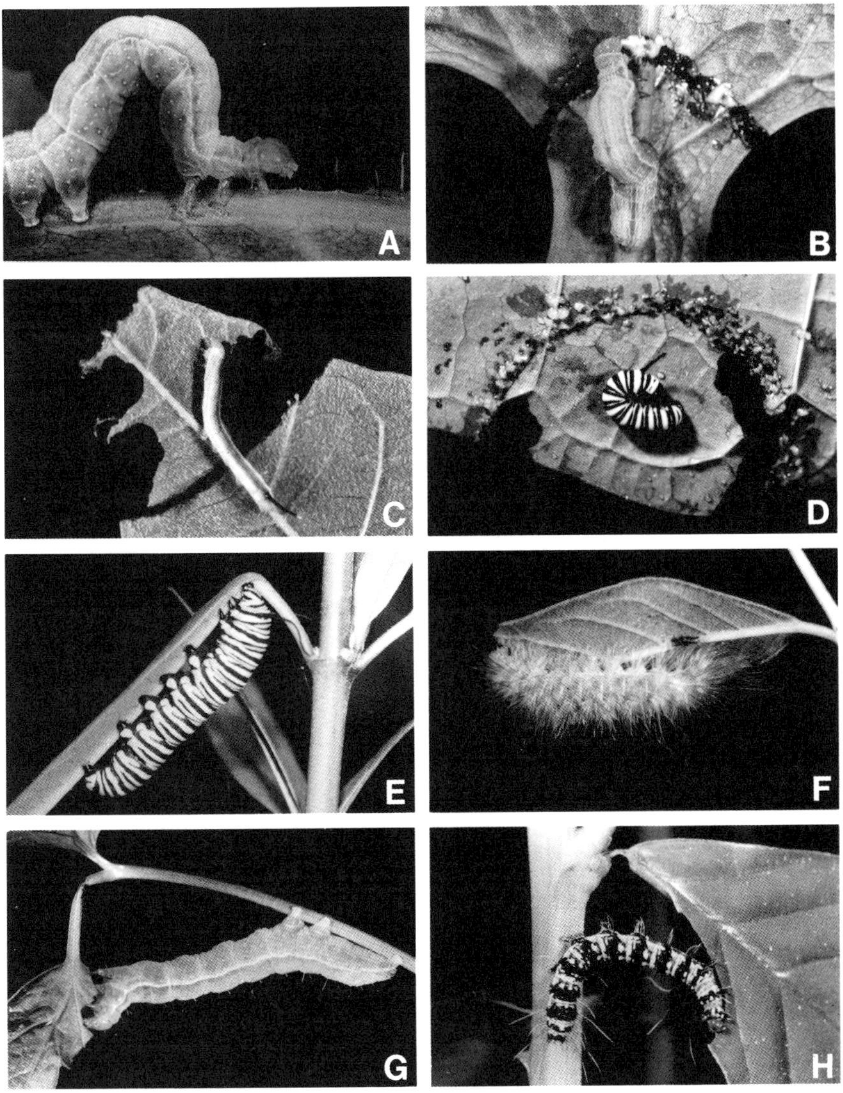

Figure 3.1. (A) Final instar *Trichoplusia ni* (Noctuidae) consuming spines along the midrib of prickly lettuce, *Lactuca serriola* (Asteraceae), before cutting a trench (B) across the narrow portion of the lobed leaf. (C) Second instar *Erinnyis ello* (Sphingidae) feeding on *Euphorbia pulcherrima* (Euphorbiaceae) after isolating the leaf tip by transecting the blade and partially severing the midrib. (D) *Lycorea* caterpillar (Danainae) within a trench it produced along the leaf margin of *Carica papaya* (Caricaceae). (E) Monarch caterpillar, *Danaus plexippus* (Danainae), drooping an *Asclepieas curassavica* leaf (Asclepiadaceae) by pinching and cutting the petiole, prior to consuming the leaf. (F) *Cycnia tenera* (Arctiidae) feeding after severing the midrib of an *Apocynum cannabinum* leaf. (Apocynaceae). (G) *Anagrapha falcifera* (Noctuidae) feeding on a leaflet of Italian parsley, *Petroselinum crispum* (Apiaceae), distal to a single cut in the petiolule. (H) *Utetheisa ornatrix* (Arctiidae) consuming a leaf of *Crotalaria spectabilis* (Fabaceae) after first cutting into the petiole.

Caterpillars not only modify plants by cutting, but also by tying leaves together with silk, thus creating an enclosure in which to rest and feed (Fitzgerald et al. 1991). Although protection from enemies and inclement weather seems a likely explanation, this need not always be the case. Caterpillars on umbellifers, for example, are shielded from UV light by the leaves, thereby gaining protection from phototoxic compounds in their diet (Berenbaum 1978).

Silk can also be used to absorb plant secretions. A boring caterpillar, *Rhyacionia buoliana,* collects resins oozing into its tunnels in *Pinus* buds and spreads them out of harms way on a silk lining constructed in the tunnel (Harris 1960). The saliva of the insect emulsifies the resin, perhaps thereby preventing the entrapment suffered by insects entombed in amber (Langenheim 1990).

Aggregated *Mechanitis* larvae use silk in a third way. They create a silk scaffolding over trichomes from which they can feed safely on the unprotected leaf edges. Multiple larvae may be necessary to produce sufficient silk and also to break off the trichomes (Rathcke and Poole 1975; Young and Moffett 1979). Aggregation may likewise benefit other caterpillar species by facilitating feeding on tough leaves (Hagen and Chabot 1986) or by overwhelming canalicular defenses (Dussourd and Denno 1991).

In summary, behavioral modifications of host plants are diverse and widely employed. Some undoubtedly function to deactivate plant defenses, although experimental evidence so far is limited. Behavioral strategies offer a clear advantage over biochemical processes since host defenses are typically deactivated before ingestion. Any benefits, however, are achieved at a price that often includes a substantial investment of time and energy, as well as exposure to enemies (Schultz 1983; Tallamy 1986; Becerra and Venable 1990). In addition, behavioral adaptations may be effective against only a limited subset of plant defenses.

Appropriation

Whereas many insects have evolved means of avoiding or deactivating plant defenses, others employ the defenses for their own protection. This benefit may be acquired simply by feeding on a plant. Leaf miners, for example, may be protected from parasitoids by a bilayer of grandular trichomes on the leaf surface. Likewise, caterpillars on toxic plants may feed freely with little risk of an inadvertent death in the jaws of a passing bovine or other large herbivore (Rothschild and Reichstein 1976; Southwood 1986). Caterpillars also use their meal as a defense by regurgitating gut contents when disturbed. Deterrent substances, such as cyanogenic glycosides in the host plants of *Malacosoma americanum,* are effectively deployed in this manner (Peterson et al. 1987).

In many cases though, there is clear unequivocal evidence that insects seek out and sequester secondary substances, a process termed pharmacophagy (Boppre 1984). The larvae or adults of several hundred Lepidoptera species, for example, collect pyrrolizidine alkaloids from plants, in some cases damaging leaves apparently for this purpose alone (Boppre 1990). Collected alkaloids are utilized not only for protection of larvae and adults against predators (Eisner 1980, Brown 1987), but also as a courtship pheromone that communicates a male's possession of the alkaloidal resource (Conner et al. 1990). Females mate preferentially with males disseminating alkaloid perfume (Pliske and Eisner 1969; Conner et al. 1981), thereby ensuring their acquisition of a nuptial gift of alkaloids (Brown 1987; Dussourd et al. 1988, 1989). Male-derived alkaloids are incorporated into the eggs where they serve as a defense against predators (Dussourd et al. 1988).

The complex interlacing of plant chemistry and insect ecology exemplified by Lepidoptera and pyrrolizidine alkaloids is not unique (e.g. Nishida and Fukami 1990) and probably common. Certainly, caterpillars selectively sequester a variety of secondary compounds, and in many cases retain them during the adult and egg stages (Brower 1984; Bowers 1990). Caterpillar responses to plant allelochemicals may thus affect all stages of the life cycle.

Allelochemicals of course are not the only plant defense that can be appropriated by caterpillars. Mutualistic ants that protect plants in exchange for bribes of nectar, protein bodies, and/or shelter would appear to be particularly susceptible to recruitment. A notable example is provided by the riodinid caterpillar, *Thisbe irenea,* which produces a secretion rich in amino acids that is highly attractive to ants—the same ants that visit the extrafloral nectaries of the host plant (DeVries and Baker 1989). Evidently, the riodinid gains a double benefit. Not only do the ants protect the larvae from enemies, they also defend the riodinid's food source. In addition, the caterpillars drink nectar from the extrafloral nectaries, thus adding insult to herbivory (De Vries and Baker 1989).

Vein-Cutting/Trenching: Counterploy to Canal-Borne Defenses?

Before feeding, many insects engage in a stereotypic sequence of behaviors in which they selectively sever one or more leaf veins, or cut a trench or channel across the leaf blade (Table 3.1B, Fig. 3.1B–H). Feeding then takes place beyond the cuts, in the area supplied by the severed veins. Several explanations have been proposed for these behaviors, which appear to have evolved repeatedly within the Lepidoptera. Below I review the evidence for each of the hypotheses, and then consider in detail the function of trenching by one insect, the cabbage looper (*Trichoplusa ni*). A similar analysis with regards to *Asclepias* folivores was presented in Dussourd (1986).

1. Manipulate leaf position to facilitate feeding. Heinrich (1971) proposed that *Manduca sexta* larvae sever tobacco midribs to create a hinge for bending flimsy leaves to their mouthparts. Heavy caterpillars that have difficulty reaching leaf edges may facilitate feeding in this manner. However, since many insects crawl out beyond vein cuts and initiate

feeding at the leaf edge (eg. Fig. 3.1D, E, F), other explanations must be sought for their behavior.

2. Create shelter. Damage to leaf veins, and in particular to the petiole, sometimes causes the leaf to droop (Fig. 3.1E, G, H), creating a hiding place where a caterpillar can feed somewhat sheltered from inclement weather, sunlight, predators and parasitoids (Edwards 1883; Scudder 1889; Brewer 1977; Young 1978). Certainly the vein-cutting habits of tropical bats (Timm and Clauson 1990) serve to form protected roosting sites. Caterpillars too are adept at constructing shelters, often produced by bending leaves with silk (Fitzgerald and Peterson 1988; Fitzgerald et al. 1991). Vein cuts or trenches, such as those of *Aglais milberti, Polygonia comma* and *Vanessa atalanta* (Nymphalidae), may facilitate the process. However, the cuts of many caterpillar species (Fig. 3-1D, F) do not appreciably alter leaf position. Even when leaves are drooped, the net effect may be increased vulnerability because the caterpillars are exposed during the vein-cutting/trenching procedure and the bent leaves may attract visual predators (Heinrich and Collins 1983).

3. Alter water content or the nutritional/allelochemical composition of the leaf. Severance of leaf veins may induce wilting with concomitant changes in leaf quality. Water stress is known to produce a variety of biochemical changes (Levitt 1972; Mattson and Haack 1987), some potentially beneficial to foraging insects. Notably, some grasshoppers prefer wilted, senescent, or damaged tissue, including leaves distal to stems girdled by a cerambycid beetle (Lewis 1979; Bernays et al. 1977; Bernays and Lewis 1986). Forcella (1982) found that girdling by another cerambycid increased nitrogen levels in the stem, presumably by trapping nutrients released by senescing leaves in autumn. Voles likewise improve the quality of plant resources through behavioral modification. They gnaw off conifer branches and allow them to age for two or more days before feeding, thereby decreasing titers of phenolics and condensed tannins in their prospective meal (Roy and Bergeron 1990). Vein cuts might similarly benefit caterpillars, although any changes would have to occur rapidly since most species begin feeding immediately after transecting veins. However, some larvae feed for lengthy periods beyond cuts (cutworms, for example, snip off entire plants at the base) and might profit from changes induced by wilting. Days or weeks might not be necessary for improvements to occur. In cassava leaves, for example, cyanide release diminishes rapidly within an hour following leaf removal (Bernays et al. 1977).

4. Increase uptake of allelochemicals. Rothschild (1977, also Rothschild and Ford 1970) observed monarchs (*Danaus plexippus*) in Trinidad eliciting latex emission from *Asclepias curassavica* by damaging stems.

The caterpillars reportedly consumed the latex like a cat drinking milk, perhaps to acquire protective cardenolides concentrated in the latex (Seiber et al. 1982). My observations of monarchs on *A. curassavica* planted near Beltsville, Maryland do not confirm this appetite for latex, although the monarchs did routinely transect leaf veins or petioles before feeding (Fig. 3.1E).

The effect of vein cuts on cardenolide sequestration has not been determined for monarchs or any other milkweed herbivore. However, the following observations suggest that the behaviors are designed to minimize, rather than maximize, ingestion of latex. Monarchs and queen (*D. gilippus*) larvae often pinch milkweed veins repeatedly with their mandibles during the cutting procedure, apparently rupturing canals internally and thus minimizing the external release of latex. Some milkweed folivores (e.g., *Labidomera clivicollis* and *Tetraopes tetrophthalmus*) make multiple cuts along veins; such cuts are always produced from the base of the leaf toward the periphery. Since each incision diminishes latex flow to tissues beyond that point, the insects encounter less latex with each cut (Dussourd and Denno 1991). To maximize latex consumption, milkweed insects should make successive cuts in the opposite direction, or feed on the leaf without severing veins.

5. Utilize exudates as defensive barrier. Secretions released by caterpillar bites might create a protective moat deterrent to competitors or enemies. Indeed, red-cockaded woodpeckers repeatedly damage pine trunks around their nest hole to elicit the outflow of resin which effectively repels climbing snakes (Rudolph et al. 1990). De Vries (personal communication) likewise found that circular trenches cut by early instar *Euploea core* and *Idea leuconoe* larvae deter foraging ants, although protection was apparently due to a froth emitted by the larvae rather than plant secretions. Nevertheless, copious exudates (for example from the circular trenches of *Lycorea,* Fig. 3.1D) could potentially constitute a formidable barrier to predators and parasitoids.

6. Block inducible defenses. By severing vascular traces, insects might prevent plants from mobilizing defensive chemicals through the veins, or from transmitting damage signals to adjacent foliage. Thus, beetles in the genus *Epilachna* reportedly trench cucurbitaceous plants to prevent an influx of bitter cucurbitacins to their feeding site within the trench (Carroll and Hoffman 1980; Tallamy 1985). This view has been supported by chemical analysis of damaged and undamaged leaves, and by behavioral assays, not only with *Epilachna,* but also with Diabroticite beetles that are stimulated to feed by cucurbitacins (Carroll and Hoffman 1980; Tallamy 1985, 1986). More recently, however, Tallamy and McCloud (1991) suggest instead that *Epilachna* beetles cut trenches to

reduce their exposure to sticky sap. They note that beetles fail to trench wilted leaves with reduced sap, and that the insects require several hours for grooming following a facial dip in sap. Dussourd and Denno (1991) reported that both *Epilachna borealis* and *Trichoplusia ni* (Noctuidae) trench nonbitter cucumber plants, which lack cucurbitacins, but release similar amounts of sap as bitter plants. Furthermore, *T. ni* larvae can be induced to trench by applying nonbitter sap to their mouthparts (Dussourd unpublished data), implicating sap as the behavioral stimulant. The release of phloem sap from damaged veins resembles exudation of latex in many respects (Buttery and Boatman 1976), and indeed *T. ni* larvae cut similar trenches in cucurbits and latex-bearing composites. Thus, although inducible defenses (or damage signals) mobilized through the vascular system (i.e., Baldwin 1989) would appear to be a likely target of vein cutters and trenchers, evidence for this role is presently inconclusive.

7. Prevent outflow of secretion from pressurized canals. Caterpillars that sever leaf veins or cut trenches occur predominantly on plants with either latex canals (Apocynaceae, Asclepiadaceae, Asteraceae—Liguliflorae, Caricaceae, Euphorbiaceae, Moraceae, Uricaceae), oil/resin canals (Apiaceae, Anacardiaceae), or exuding phloem (Cucurbitaceae). Of the species listed in Table 3.1, about three-quarters (32 of 45 genera) feed on plants within these categories. In contrast, all of the species that sever petioles *after* feeding (18 genera in Table 3.1) occur on plants that lack secretory canals. Metcalfe and Chalk (1983) and Farrell et al. (1991) were consulted to determine the distribution of canals. Some *Eucalyptus* (series Corymbosae—Welch 1921; Carr and Carr 1969) produce oil canals in the leaves, but species with petiole-clipping herbivores (Edwards and Wanjura 1989; Weinstein 1990) lack these structures, although oil glands are common.

As described previously, secretory canals can be effectively deactivated by vein severance. In all cases examined (Brewer 1977; Dillon et al. 1983; Dussourd and Eisner 1987; Compton 1987, 1989; Dussourd 1990), caterpillar cuts substantially diminish the distal outflow of secretion. This need not be the case since canals occur in different arrangements. The midrib cuts of *Cycnia tenera* (Fig. 3.1F) and other *Apocynum* herbivores, for example, would be ineffective on plants such as *Lactuca serriola* with net-like canal systems. On *Lactuca,* caterpillars must cut trenches to depressurize the articulated laticifers, which indeed they do (Table 3.1, Fig. 3.1B). Indeed, there is a direct correspondence between insect behavior and canal architecture (Dussourd and Denno 1991), suggesting that insects on plants bearing secretory canals sever veins and cut trenches specifically to deactivate the canals.

The insects encounter some secretion during the vein-cutting/trenching procedure. However, the amount is typically trivial in comparison with the substantial quantities they would experience during feeding on leaves with intact veins (Dussourd and Denno 1991). Furthermore, the fresh exudate oozing from vein cuts is liquid and thus more easily swallowed or wiped off than secretions that have begun to congeal. Some caterpillars (e.g., *Danaus gilippus* feeding on milkweed flowers, Dussourd and Denno 1991) avoid exudation altogether by pinching the pedicels of florets, thus causing the internal rupture of laticifers. Other insects, notably beetles and katydids, disrupt the canals with a rapid succession of bites, removing their mandibles quickly before exuding latex accumulates. Loopers such as *Trichoplusia ni* do encounter secretions during their lengthy trenching procedure, but pause repeatedly to remove exudate from their mandibles by grooming.

In summary, it appears that insects may benefit in a variety of ways by severing leaf veins. Experimental evidence for most proposed functions is noticeably lacking, as are experiments designed to discriminate between competing hypotheses. What is clearly needed are detailed investigations of single organisms to document function in individual cases. Below, I present such an analysis with the cabbage looper, *Trichoplusia ni*.

Like other plusiine noctuids (Table 3.1), cabbage loopers transect leaves with trenches before beginning to feed (Fig. 3.1B). The larvae repeatedly nibble back and forth across the leaf, deepening and broadening the trench with each pass. The entire operation requires from a few minutes to over an hour depending on plant species. The section beyond the trench is typically consumed in its entirety; larvae then often produce a second trench below the first and consume the intervening section as well.

Trichoplusia ni has many merits as a study organism not the least of which are its abundance and significance as an agricultural pest (Sutherland 1966; Flint 1987). The caterpillars reportedly attack plants in 28 families (Eichlin and Cunningham 1978). Although some of these records (e.g., Asclepiadaceae) are almost surely incorrect, host plants certainly include species with and without secretory canals. If loopers trench to deactivate canal systems (hypothesis 7) or to employ secretions for defense (4 and 5), we might expect the caterpillars to cut trenches only in plants that exude secretions when damaged. Conversely, if any of the other hypotheses are correct, trenches should be cut in all species, or at least without respect to the presence or absence of secretory canals. For example, if the larvae cut trenches to facilitate feeding or to create shelters, perhaps only leaves of certain shapes would be trenched.

In a comparative study, Dussourd (unpublished data) found that *T. ni* larvae cut trenches only in plants that release secretions when damaged (Apiaceae, Asteraceae—Liguliflorae, Cucurbitaceae). Representatives of seven other groups that lack secretions (Asteraceae—Tubiflorae, Brassicaceae, Chenopodiaceae, Fabaceae, Lamiaceae, Malvaceae, and Plantaginaceae) were eaten without any preliminary modification of the leaves. Leaf shape varied considerably in both cases. For example, *T. ni* trenched the lobed leaves of *Lactuca serriola* (Fig. 3.1B), compound leaves of *Daucus carota*, and round leaves of *Cucumis sativus*. The uniform response to these species is all the more remarkable given the variable secretions they emit: latex in *Lactuca*, essential oils in *Daucus*, and phloem sap in *Cucumis*.

The facultative trenching response of *T. ni* offers a convenient assay for identifying the behavioral stimulant. Larvae reared on *Plantago lanceolata* normally do not cut trenches at any point during development. Yet, if a single dose (1 μl) of *Lactuca serriola* latex or *Cucumis sativus* (nonbitter) sap is applied to their mouthparts, many respond by cutting a trench in *Plantago* (Dussourd unpublished data).

The association between trenching and plant exudates eliminates all hypotheses except 4, 5, and 7. It seems unlikely that cabbage loopers cut trenches to drink exudates, and thus acquire protective chemicals (hypothesis 4), since the larvae are cryptic green in all instars and acceptable to diverse predators (Bernays 1988; Bernays and Cornelius 1989) and parasitoids (Flint 1987). Furthermore, to maximize ingestion of secretion, caterpillars should eat leaves without trenching, or feed proximal, not distal, to the trench.

Detailed observation of caterpillar behavior provides a means of discriminating between the two remaining hypotheses (5 and 7). Cabbage loopers normally cut trenches while resting along the midrib and facing toward the leaf tip. Feeding invariably begins distal to the trench on the isolated section. However, lab-reared larvae transferred to *Lactuca serriola* sometimes attempt to cut trenches while facing toward the base. In this situation, they often cut unusually extensive trenches (sometimes greater than 1 cm thick) before either abandoning the leaf or turning around and feeding at the tip. Their inability to initiate feeding on the leaf base is inconsistent with hypothesis 5 (using secretions as a protective moat). A standard trench would presumably suffice as a moat irrespective of the direction of the larva. However, looper behavior is readily explained by the only remaining hypothesis (7—deactivate secretory canals). Trenches create an asymmetry in *Lactuca* leaves with high latex pressure on one side and low pressure on the other. Feeding takes place only on the low pressure side. The thick trenches fail to eliminate exudation from the base of the leaf because this section remains connected with laticifers in the stem.

Plucking the leaf eliminates these connections and reduces latex pressure throughout the leaf. Larvae on plucked leaves (with the base in water to maintain turgidity) often cut shallow trenches while facing in the "wrong" direction, but in this case feed without hesitation beyond the trench (on the base of the leaf) and continue feeding on both sides. Clearly, it is the presence or absence of pressurized latex canals, not location or direction, that dictates caterpillar feeding.

I conclude from this analysis that *T. ni* larvae cut trenches specifically to deactivate canal systems in host plants and thereby prevent the outflow of secretion during feeding. This explanation is the only hypothesis supported by experimental evidence, not only for *T. ni,* but also for most of the other caterpillars listed in Table 3.1B that cut veins or trenches. Although their behaviors can thus be interpreted as a counterploy to canal-borne secretions, this role does not appear to be universal. The Passifloraceae, for example, supports a sizable fauna of vein cutters and trenchers (Table 3.1B), yet apparently lacks secretory canals (Metcalfe and Chalk 1983). The well-studied defensive attributes of this family (Spencer 1988) bear further consideration in light of these widespread, but not universal, behaviors in *Passiflora* herbivores.

Canalicular Constraints: Do Secretory Canals Deter Foraging Caterpillars?

Many, and probably most, caterpillar species lack behavioral adaptations suitable for deactivating secretory canals. Are these species thereby constrained from feeding on plants protected by canals, or can the caterpillars compensate through other mechanisms (i.e., avoidance or biochemical detoxification). To address this question, I compared the ability of yellow-striped armyworms (*Spodoptera ornithogalli*) and cabbage loopers (*Trichoplusia ni*) to develop on the laticiferous *Lactuca serriola* (=*scariola,* Asteraceae). Both caterpillar species have polyphagous feeding habits (Eichlin and Cunningham 1978; Tietz 1972), but only cabbage loopers cut trenches. For both species, I compared the survivorship of each larval instar to the following instar on plucked and live leaves (N = 10 larvae/instar/treatment). Caterpillars were reared to the desired instar on plucked leaves, then the freshly molted larvae were either sleeved on a newly mature leaf on a potted plant (1 caterpillar per plant), or enclosed in a petri dish with a comparable leaf plucked from another plant. Plucked leaves were cut in three pieces to completely depressurize the latex canals. All plants were 2–3 months old; none had begun to bolt. During the test, caterpillars were maintained at 22°C (18:6 light/dark cycle) within environmental chambers and checked every 8 hours for signs of trenching and molting. New leaves were provided as needed (at least every two days for plucked leaves), and plants were replaced every 3 days.

Spodoptera ornithogalli larvae of all instars survived well on plucked leaves (Fig. 3.2A), showing rapid growth and development (Dussourd unpublished data). However, on live leaves, only a few early instar larvae successfully molted to the next instar. Dead first and second instar larvae were commonly found with their mandibles stuck together with dried latex, sometimes still glued to the leaf where they had attempted to feed. Their demise provided compelling evidence of the defensive capabilities of the latex system, and particularly of the effectiveness of canal-borne adhesives. Older larvae riddled leaves with dozens of bite marks, many covered with the dried drops of exuded latex, before eventually dying after days of starvation. Evidently, the latex deterred feeding since larvae on plucked leaves fed readily, creating large holes in the leaves.

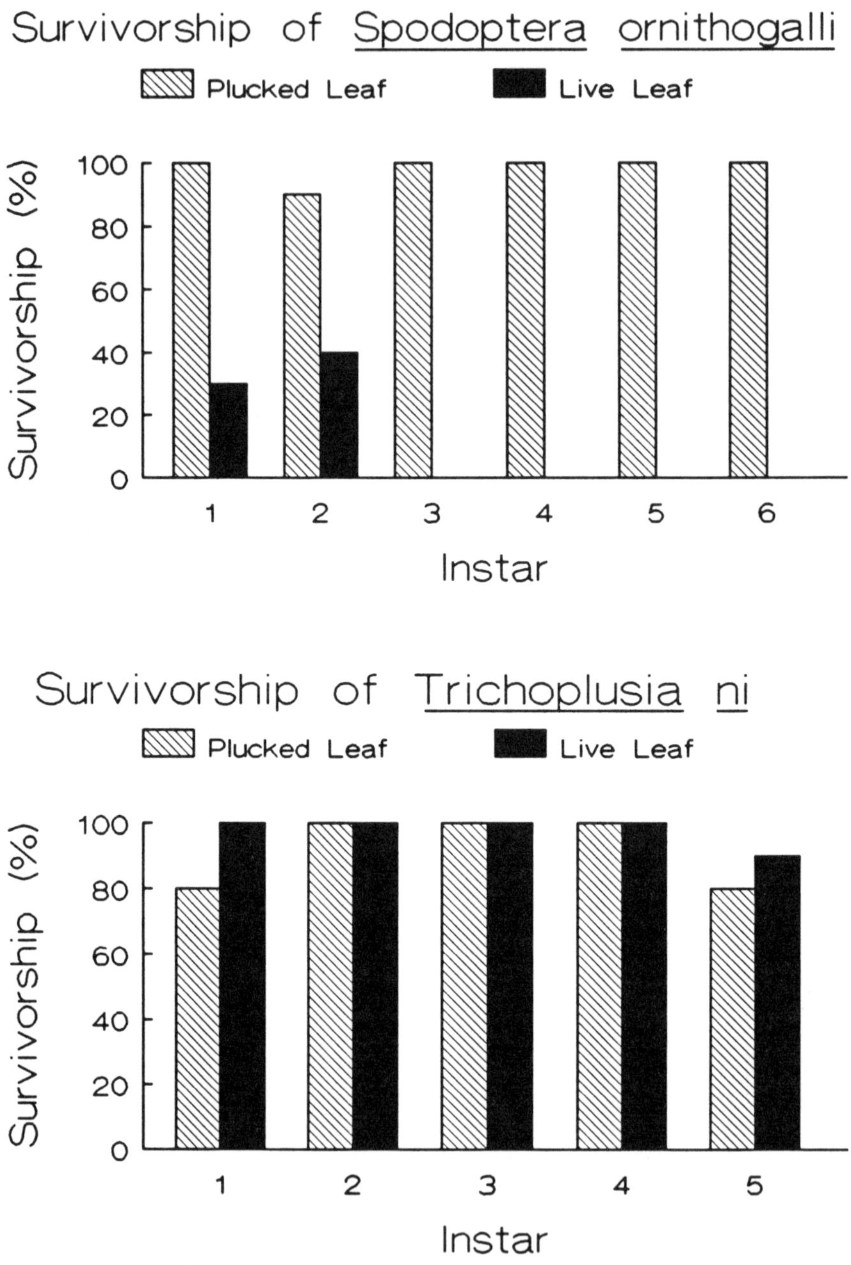

Figure 3.2. Percent of *Spodoptera ornithogalli* (top) and *Trichoplusia ni* (bottom) larvae that survived to the following instar on prickly lettuce, *Lactuca serriola*. (n = 10 larvae per category). Unlike *T. ni* larvae, *S. ornithogalli* caterpillars lack trenching behaviors and perform poorly on live leaves.

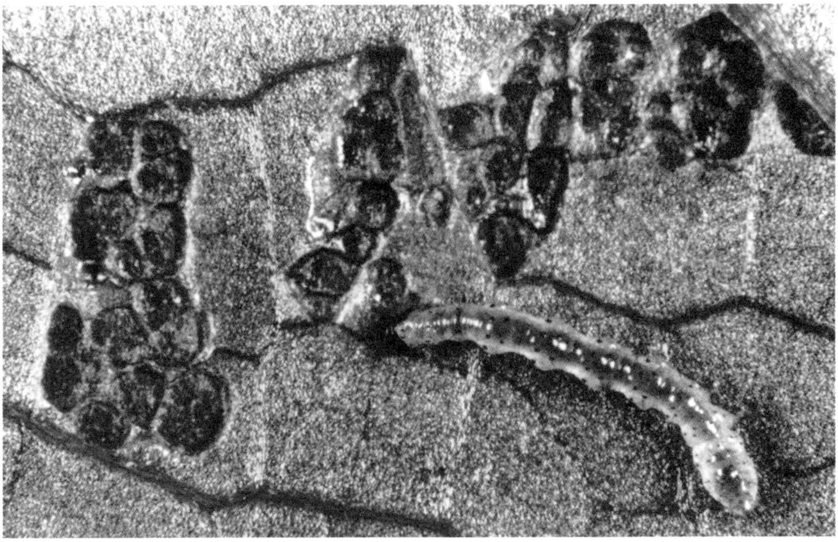

Figure 3.3. First instar *Trichoplusia ni* feeding between leaf veins of *Lactuca serriola*, thereby avoiding laticifers concentrated in the veins.

Cabbage loopers, in contrast, survived well on both plucked and live leaves (Fig. 3.2B). The tiny first instar larvae fed between leaf veins (Fig. 3.3) and in this manner largely avoided latex canals concentrated in the veins. On occasion, a canal was ruptured and exuding latex rapidly filled the pit produced by their feeding. Larvae responded by rearing up on their prolegs, thereby avoiding entrapment; after grooming, they resumed feeding elsewhere. Caterpillars in the second instar began cutting trenches; third to fifth instar larvae were never observed feeding except beyond trenches.

The above observations illustrate the importance of larval size relative to the distribution of plant defenses. The latex canals were relatively ineffective against tiny cabbage loopers. Larger larvae could no longer feed between canals, and instead adopted the alternative strategy of cutting trenches.

The differential success of *T. ni* and *S. ornithogalli* in attacking *Lactuca serriola* supports the assertion that behaviorally adapted herbivores are able to exploit defended plants that are unacceptable to nonadapted herbivores. Differences in survivorship, however, could simply reflect peculiarities of *Trichoplusia, Spodoptera,* or *Lactuca*. The results therefore require confirmation with additional plant and herbivore species. Dussourd (unpublished data) compared the performance of *T. ni* and *S. ornithogalli* larvae in the field on plucked vs. live leaves of 13 plant species. Larvae were reared to the final instar on plucked leaves, then tested over a 24-hour period on the same species on which they were reared. The *Spodoptera* larvae performed comparably on plucked and live leaves of all species lacking secretions, including representatives of Asteraceae (Tubiflorae), Brassicaceae, Chenopodiaceae, Fabaceae, Lamiaceae, Malvaceae, and Plantaginaceae. But on plants that release secretions when damaged (Apiaceae, Asteraceae—Liguliflorae, and Cucurbitaceae), growth on live leaves was sharply reduced (with one exception—*Daucus carota*) and armyworms often lost weight. Cabbage loopers, in contrast, gained weight on all plants tested. However, even their performance was somewhat poorer on the live leaves (relative to plucked leaves) of plants with secretions, suggesting that adapted herbivores are also negatively affected by canals.

Dussourd (unpublished data) also tested 9 additional species of generalist noctuids and arctiids on *Lactuca serriola* using the same procedure. In all cases, larval growth was high on plucked leaves. However, on live leaves, only species adept at cutting trenches (2 plusiine noctuids) gained weight; the other 7 species (2 noctuids, 5 arctiids) all lost weight.

I conclude from the above experiments (also Dussourd and Eisner 1987) that plant secretions stored within canal systems constrain foraging by nonadapted caterpillars. Adapted species are able to overcome the canalicular constraint by cutting trenches. The necessity of modifying plants before feeding may impose costs on these caterpillars that potentially include an investment of time and energy in cutting the trench (and perhaps in learning to cut trenches), and increased risk of predation and parasitism. Nevertheless, possession of this behavioral trait permits the exploitation of plants unavailable to the nonadapted herbivore.

General Considerations

The effects of canalicular defenses on herbivores no doubt extend well beyond the host choices of individual caterpillars in ecological time. Lepidoptera have encountered secretory canals throughout their evolution. Fossilized nonarticulated laticifers still replete with rubber have been reported from the Eocene (at least 50 million years old, Mahlberg et al. 1984), while resin canals extend back at least 225 million years to the Permian (Langenheim 1990). In this final section, I briefly consider possible evolutionary effects of secretory canals on host shifts, plant/insect diversity, and community structure. Finally, I close by noting the occurrence of behavioral counterploys to plant defenses among caterpillar pests and among other insects besides Lepidoptera.

Host Shifts

Insect herbivores have been widely acclaimed for their skill in discerning the taxonomic relations of host plants. Individual species commonly feed only on plants within a single family, and related species often attack related plants. However, many exceptions exist. The presence of shared allelochemicals in unrelated plants has frequently been cited as an explanation for disparate host records (Strong et al. 1984). The danaines, for example, feed not only on the closely related Apocynaceae and Asclepiadaceae, but also on the Moraceae. All three groups produce alkaloids, pyridines, and cardenolides (Ehrlich and Raven 1964, Ackery and Vane-Wright 1984). However, as noted by Ehrlich and Raven (1964), all three families (also the Caricaceae on which a few danaines feed) share an additional trait—latex canals. The ability of danaines to sever leaf veins and petioles and to cut trenches adapts them (preaptation, Gould and Vrba 1982) for colonizing plants bearing a variety of canal anatomies, including both nonarticulated (Apocynaceae, Asclepiadaceae, and Moraceae) and articulated (Caricaceae) laticifers.

Another group, the sphingid tribe Dilophonotini, likewise attacks diverse laticiferous plants (Hodges 1971; Harris 1972). A single species, *Erinnyis ello,* reportedly feeds on latex-bearing plants in seven families (Winder 1976). Its actual range is probably smaller due, in part, to the inability of young caterpillars to trench when faced with copious exudates from plants such as *Ficus benjamina* (Dussourd unpublished observations). Even newly emerged larvae of this species are too large to feed between veins. Nevertheless, the caterpillars clearly attack multiple laticiferous species, an ability no doubt enhanced by their possession of two general purpose behaviors: petiole constriction and trenching (Dillon et al. 1983; Dussourd unpublished data).

Given the effectiveness of behavioral counteradaptations to canalicular defenses, it is somewhat surprising that so many of the species in Table 3.1 are specialists. Various ecological or historical factors could account for their restricted ranges. The diverse chemistries of plant secretions may also be significant. Although vein cuts and trenches depressurize distal canals, they do not totally purge the contents. Exudation (and risk of entrapment) is eliminated, but some allelochemicals undoubtedly remain. In addition, the caterpillars must tolerate direct exposure to secretions during the vein-cutting/trenching operation. As illustrated with the *Trichoplusia ni–Asclepias curassavica* example previously discussed, such exposure may be extremely detrimental to caterpillars lacking the metabolic machinery necessary for tolerating toxins. Thus, while behavioral adaptations may facilitate host shifts to unrelated canal-bearing plants, the presence of toxins in secretions may require some biochemical accommodation as well.

Plant–Insect Diversity

Ehrlich and Raven (1964) proposed that the radiation of plants and phytophagous insects has proceeded in a stepwise coevolutionary sequence. Attack from herbivores would select for plants possessing unusual traits conferring protection. Their escape from herbivory might permit diversification into new "adaptive zones." Any herbivore capable of attacking this underexploited plant resource might in turn diversify, thereby providing selection pressure for the evolution of additional defensive adaptations in the plants. Thus, under this scenario, antagonistic interactions between plants and insects would lead to the evolution of novel defenses in plants and effective countermeasures in herbivores. Clearly, this scheme need not apply only to the evolution of allelochemicals (i.e., Berenbaum 1983), but also to the packaging of these chemicals. Plants presenting defensive substances in novel ways might escape herbivores as successfully as species evolving new chemicals. Farrell et al. (1991) tested for possible effects of latex and resin canals on plant diversification. Canal-bearing clades were found to have significantly greater diversities than sister groups that lacked the canals. Although their conclusion remains tentative due to the lack of sufficient systematic and anatomical information for many canal-bearing clades, the result clearly implies that the evolution of secretory canals has promoted plant diversification, presumably due to protection conferred by the canals against herbivores and pathogens. Whether insect groups evolving behavioral counteradaptations, such as vein cutting, have likewise radiated relative to behaviorally impoverished kin remains to be determined.

Taxonomic patterns are certainly present in the available, albeit limited, data set (Table 3.1). For example, all plusiine noctuids that have been examined (*Anagrapha, Autographa, Chrysodeixis, Pseudoplusia*) cut trenches in canal-bearing plants (Compton 1989; Dissourd and Denno 1991), whereas generalist arctiids (*Apantesis phalerata, Estigmene acrea, Pyrrharctia isabella, Spilosoma congrua, S. virginica*) lack the behavior altogether (Dussourd unpublished data). Behavioral adaptations are well represented in danaine caterpillars (Table 3.1B), but not observed in *Spodoptera* (based on detailed study of *S. ornithogalli* and limited observations of *S. frugiperda* and *S. eridania* from lab colonies).

The evolution of behaviorally adapted herbivores would presumably select for countermeasures in their hosts. One wonders, for example, if net-like canals (such as articulated laticifers) have evolved in response to herbivory by vein-cutting insects.

Community Structure

As discussed previously, plants differ enormously in their vulnerability to attack due in part to differences in the deployment of defenses. The size distribution and guild structure of herbivore assemblages should reflect these differences. One might predict, for example, that leaf-miners would be particularly abundant on plants with patchy defenses (e.g., oaks, Feeny 1970). In contrast, plants with impenetrable canals might be deficient in leaf-miners, but rich in sap-tappers and in external feeders that cut veins or trenches. A partial test is provided by E. A. Bernays (personal communication) who examined the diversity of British leaf-miners on the Asteraceae, and found a lower number on the laticiferous tribe Cichorieae. Although this result certainly suggests that secretory canals affect herbivore communities, sister group comparisons with multiple lineages (as promoted by Mitter et al. 1988) will be necessary to control for possible effects of plant phylogeny.

Crop Resistance

The results discussed in this chapter (also Langenheim 1990; Farrell et al. 1991) implicate canal-borne secretions as major plant defenses. As such, they offer a valuable source of resistance for economically important plants in agriculture (e.g., lettuce, carrots, squash), horticulture (e.g., poinsettia, oleander, figs), and forestry (e.g., conifers). Behavioral counterploys in herbivores, such as *Erinnyis ello* (a major pest of cassava, *Manihot esculenta,* and the rubber tree, *Hevea brasiliensis*) and plusiine noctuids (pests of truck crops), limit the usefulness of such defenses. Although secretory canals and behavioral counteradaptations (i.e., mass attack of bark beetles) have been studied extensively in forestry (Berryman 1972; Mitton and Sturgeon 1982; Mattson et al. 1988), their potential importance in agriculture has received little attention. Lettuce offers a case in point. During thousands of years of domestication (Ryder and Whitaker 1976), latex titers in lettuce have been reduced dramatically relative to the likely ancestor, *Lactuca serriola* (Dussourd and Denno 1991). Predictably, lettuce is now vulnerable not only to behaviorally adapted caterpillars (e.g. *Trichoplusia ni*), but also to generalist caterpillars that lack trenching behavior and are unable to feed on *L. serriola* (Dussourd unpublished data).

Behavioral Adaptation in Other Insects

Caterpillars are by no means the only insect herbivores that manipulate plant tissues before feeding (see Rhoades 1983, 1985; Tallamy 1986 for reviews). Vein-cutting and trenching behaviors occur in multiple groups including beetles (Cerambycidae, Chrysomelidae, Coccinellidae), katydids (Tettigoniidae), and sawflies (Tenthredinidae) (Becerra and Venable 1990; Dussourd and Denno 1991 and references cited). Girding behaviors are also commonly employed, not only by mandibulate herbivores (Lindsey 1959, but also by haustellate insects (Mitchell and Newsom 1984). Unlike Lepidoptera, adults of many groups have biting mouthparts capable of modifying plants to create suitable oviposition sites (e.g., Stride and Warwick 1962). Such behaviors are particularly well developed among the weevils, which demonstrate prodigious strength and skill in fashioning enclosures for their offspring (Hinton 1981; Sakurai 1988). Many herbivorous insects attack plants in aggregations, perhaps thereby facilitating feeding on tough tissues (Ghent 1960) or overwhelming defensive responses (Mitton and Sturgeon 1982). In addition, some species inoculate plant tissue with mutualistic fungi or yeast to counterdefensive responses, to improve plant quality, or to serve as food (Batra 1979; Mitton and Sturgeon 1982; Sakurai 1985).

Behaviors for modifying plants are clearly widespread among insect herbivores. With few exceptions (notably bark beetles/conifers), these foraging behaviors have not been investigated experimentally. Such an analysis offers considerable potential, not only for illuminating the marvels of insect behavior or elucidating mechanisms of plant resistance, but also for providing insights into what constraints are truly significant to an herbivore.

Summary

As noted by Hagen and Chabot (1986), a plant leaf presents a highly heterogeneous environment when viewed on the spatial scale of most insects. Secondary compounds are not distributed homogeneously, but are typically concentrated in specialized depots that vary within and between species in size, shape, and distribution. These varied repositories differ in their relative merits and in their vulnerability to caterpillar attack. Canal systems, for example, augment the efficacy of sequestered secretions by delivering the fluids to sites of damage immediately on injury. Potential herbivores confront a secretory barrier of toxins and adhesives that deter further contact. Such a defense, however, is vulnerable to behavioral deactivation. Caterpillars from diverse groups sever leaf veins or cut trenches prior to feeding, thereby creating a localized section of reduced defense. The cuts partially drain canals, dilute any remaining secretion (due to movement of water into the depressurized canals), and prevent an influx of secretion from the main canal system. Other caterpillars lack appropriate behaviors and are thereby constrained from feeding on canal-bearing plants, even though plucked leaves with depressurized canals are often highly acceptable. These results support the assertion that plant palatability is determined not just by the presence or absence of allelochemicals, but also by the localization of chemicals and method of deployment. Specialized behavioral adaptations in caterpillars (noted in over 30 genera) severely compromise the effectiveness of many defenses and allow larvae to feed on a relatively unprotected food resource.

Acknowledgments

Special thanks to R. F. Denno for guidance, encouragement, and support during my postdoctoral stay at the University of Maryland, to J. C. Schultz and J. G. Franclement for stimulating my interest in Lepidoptera, to J. Ballarino, B. Benrey, H. G. Doebel, B. D. Farrell, E. Lawson, E. S. McCloud, C. Mitter, K. L. Olmstead, G. K. Roderick, D. W. Tallamy, and D. F. Wiemer for helpful discussion, to N. E. Stamp, D. W. Tallamy, and an anonymous reviewer for constructive criticism on the manuscript, to J. G. Franclement and R. W. Poole for identifications, and to E. A. Bernays, P. J. DeVries, and A. R. Zangerl for sharing unpublished information. Thanks also to USDA (Competitive Research

Grant 88-37153-3534) and the Maryland Agricultural Experiment Station for their financial support.

Literature Cited

Ackery, P. R., and Vane-Wright, R. I. 1984. Milkweed Butterflies: Their Cladistics and Biology. Cornell University Press, Ithaca, NY.

Alexander, A. J. 1961. A study of the biology and behavior of the caterpillars, pupae and emerging butterflies of the subfamily Heliconiinae in Trinidad, West Indies. Part 1. Some aspects of larval behavior. Zoologica 46:1–24.

Alosi, M. C., Melroy, D. L., and Park, R. B. 1988. The regulation of gelation by phloem exudate from *Cucurbita* fruit by dilution, glutathione, and glutathione reductase. Plant Physiol. 86:1089–1094.

Baldwin, I. T. 1989. Mechanism of damage-induced alkaloid production in wild tobacco. J. Chem. Ecol. 15:1661–1680.

Bangham, W. N. 1934. Internal pressure in latex system. Science 80:290.

Batra, L. R. 1979. Insect-Fungus Symbiosis: Nutrition, Mutualism, and Commensalism. Wiley, New York.

Becerra, J. X., and Venable, D. L. 1990. Rapid-terpene bath and "squirt-gun" defense in *Bursera schlechtendalii* and the counterploy of chrysomelid beetles. Biotropica 22:320–323.

Belcher, D. W., and Thurston, R. 1982. Inhibition of movement of larvae of the convergent lady beetle by leaf trichomes of tobacco. Environ. Entomol. 11:91–94.

Berenbaum, M. 1978. Toxicity of a furanocoumarin to armyworms: A case of biosynthetic escape from insect herbivores. Science 201:532–534.

Berenbaum, M. 1983. Coumarins and caterpillars: A case for coevolution. Evolution 37:163–179.

Berenbaum, M. R., and Zangerl, A. R. 1988. Stalemates in the coevolutionary arms race: Syntheses, synergisms, and sundry other sins, pp. 113–132. In K. C. Spencer (ed.), Chemical Mediation of Coevolution. Academic Press, New York.

Bernays, E. A. 1988. Host specificity in phytophagous insects: selection pressure from generalist predators. Entomol. Exp. Appl. 49:131–140.

Bernays, E. A., and Cornelius, M. L. 1989. Generalist caterpillar prey are more palatable than specialists for the generalist predator *Iridomyrmex humilis*. Oecologia 79:427–430.

Bernays, E. A., and Lewis, A. C. 1986. The effect of wilting on palatability of plants to *Schistocerca gregaria*, the desert locust. Oecologia 70:132–135.

Bernays, E. A., Chapman, R. F., Leather, E. M., McCaffery, A. R., and Modder, W. W. D. 1977. The relationship of *Zonocerus variegatus* (L.)(Acridoidea: Pyrgomorphidae) with cassava (*Manihot esculenta*). Bull. Entomol. Res. 67:391–404.

Berryman, A. A. 1972. Resistance of conifers to invasion by bark beetle-fungus associations. BioScience 22:598–602.

Biffen, R. H. 1898. The coagulation of latex. Ann. Bot. 12:165–171.

Blackman, R. L., and Eastop, V. F. 1985. Aphids on the World's Crops: An Identification Guide. Wiley, New York.

Blaser, H. W. 1945. Anatomy of *Cryptostegia grandiflora* with special reference to the latex system. Am. J. Bot. 32:135–141.

Bloem, K. A., Kelley, K. C., and Duffey, S. S. 1989. Differential effect of tomatine and its alleviation by cholesterol on larval growth and efficiency of food utilization in *Heliothis zea* and *Spodoptera exigua*. J. Chem. Ecol. 15:387–398.

Blum, M. S. 1981. Chemical Defenses of Arthropods. Academic Press, New York.

Boppre, M. 1984. Redefining "pharmacophagy." J. Chem. Ecol. 10:1151–1154.

Boppre, M. 1990. Lepidoptera and pyrrolizidine alkaloids: Exemplification of complexity in chemical ecology. J. Chem. Ecol. 16:165–185.

Bowers, M. D. 1990. Recycling plant natural products for insect defense, pp. 353–386. In D. L. Evans and J. O. Schmidt (eds.), Insect Defenses: Adaptive Mechanisms and Strategies of Prey and Predators. State University of New York Press, Albany, NY.

Brattsten, L. B. 1988. Enzymic adaptations in leaf-feeding insects to host-plant allelochemicals. J. Chem. Ecol. 14:1919–1939.

Brattsten, L. B., and Ahmad, S. 1986. Molecular Aspects of Insect-Plant Associations. Plenum, New York.

Brattsten, L. B., Wilkinson, C. F., and Eisner, T. 1977. Herbivore-plant interactions: Mixed-function oxidases and secondary plant substances. Science 196:1349–1352.

Brewer, J. 1977. Short lived phenomena. News Lepid. Soc. 4:7.

Broadway, R. M., Duffey, S. S., Pearce, G., and Ryan, C. A. 1986. Plant proteinase inhibitors: A defense against herbivorous insects? Entomol. Exp. Appl. 41:33–38.

Brower, L. P. 1984. Chemical defense in butterflies, pp. 109–134. In R. I. Vane-Wright and P. R. Ackery (eds.), The Biology of Butterflies. Academic Press, New York.

Brown, K. S., Jr. 1981. The biology of *Heliconius* and related genera. Annu. Rev. Entomol. 26:427–456.

Brown, K. S., Jr. 1987. Chemistry at the Solanaceae/Ithomiinae interface. Ann. Missouri Bot. Gard. 74:359–397.

Buttery, B. R., and Boatman, S. G. 1976. Water deficits and flow of latex, pp. 233–289. In T. T. Kozlowski (ed.), Water Deficits and Plant Growth, Vol. IV. Academic Press, New York.

Carr, S. G. M., and Carr, D. J. 1969. Oil glands and ducts in *Eucalyptus* L'Herit. I. The phloem and the pith. Aust. J. Bot. 17:471–513.

Carroll, C. R., and Hoffman, C. A. 1980. Chemical feeding deterrent mobilized in response to insect herbivory and counteradaptation by *Epilachna tredecimnotata*. Science 209:414–416.

Cass, D. D. 1985. Origin and development of the non-articulated laticifers of *Jatropha dioica*. Phytomorphology 35:133–140.

Cates, R. G., and Alexander, H. 1982. Host resistance and susceptibility, pp. 212–263.

In J. B. Mitton and K. B. Sturgeon (eds.), Bark Beetles in North American Conifers. University of Texas Press, Austin.

Compton, S. G. 1987. *Aganais speciosa* and *Danaus chrysippus* (Lepidoptera) sabotage the latex defenses of their host plants. Ecol. Entomol. 12:115–118.

Compton, S. G. 1989. Sabotage of latex defenses by caterpillars feeding on fig trees. South African J. Sci. 85:605–606.

Conner, W. E., Eisner, T., Vander Meer, R. K., Guerrero, A., and Meinwald, J. 1981. Precopulatory sexual interaction in an arctiid moth (*Utetheisa ornatrix*): Role of a pheromone derived from dietary alkaloids. Behav. Ecol. Sociobiol. 9:227–235.

Conner, W. E., Roach, B., Benedict, E., Meinwald, J., and Eisner, T. 1990. Courtship pheromone production and body size as correlates of larval diet in males of the arctiid moth, *Utetheisa ornatrix*. J. Chem. Ecol. 16:543–552.

Curtis, J. D., and Lersten, N. R. 1986. Development of bicellular foliar secretory cavities in white snakeroot, *Eupatorium rugosum* (Asteraceae). Am. J. Bot. 73:79–86.

d'Auzak, J., Jacob, J., and Chrestin, H. 1989. Physiology of Rubber Tree Latex. CRC Press, Boca Raton, FL.

Denno, R. F., and McClure, M. S., eds. 1983. Variable Plants and Herbivores in Natural and Managed Systems. Academic Press, New York.

DeVries, P. J. 1987. The Butterflies of Costa Rica and Their Natural History. Princeton University Press, Princeton, NJ.

DeVries, P. J., and Baker, I. 1989. Butterfly exploitation of an ant-plant mutualism: Adding insult to herbivory. J. N.Y. Entomol. Soc. 97:332–340.

Dillon, P. M., Lowrie, S., and McKey, D. 1983. Disarming the "Evil Woman": Petiole constriction by a sphingid larva circumvents mechanical defenses of its host plant, *Cnidoscolus urens* (Euphorbiaceae). Biotropica 15:112–116.

Doss, R. P., and Shanks, C. H. 1986. Use of membrane filters as a substrate in insect feeding bioassays. Bull. Entomol. Soc. Am. 32:248–249.

Dowd, P. F., Smith, C. M., and Sparks, T. C. 1983. Detoxification of plant toxins by insects. Insect Biochem. 13:453–468.

Dussourd, D. E. 1986. Adaptations of Insect Herbivores to Plant Defenses. Ph.D. Dissertation. Cornell University, Ithaca, NY.

Dussourd, D. E. 1990. The vein drain; or, how insects outsmart plants. Nat. Hist. 90:44–49.

Dussourd, D. E., and Denno, R. F. 1991. Deactivation of plant defense: correspondence between insect behavior and secretory canal architecture. Ecology 72:1383–1396.

Dussourd, D. E., and Eisner, T. 1987. Vein-cutting behavior: Insect counterploy to the latex defense of plants. Science 237:398–901.

Dussourd, D. E., Ubik, K., Harvis, C., Resch, J., Meinwald, J., and Eisner, T. 1988. Biparental defensive endowment of eggs with acquired plant alkaloid in the moth *Utetheisa ornatrix*. Proc. Natl. Acad. Sci. U.S.A. 85:5992–5996.

Dussourd, D. E., Harvis, C. A., Meinwald, J., and Eisner, T. 1989. Paternal allocation of sequestered plant pyrrolizidine alkaloid to eggs in the danaine butterfly, *Danaus gilippus*. Experientia 45:896–898.

Edwards, W. H. 1882. Description of the preparatory stages of *Pyrameis atalanta*. Can. Entomol. 14:229–234.

Edwards, W. H. 1883. Description of the preparatory stages of *Pyrameis atalanta*. Can. Entomol. 15:14–20.

Edwards, P. B., and Wanjura, W. J. 1989. Eucalypt-feeding insects bite off more than they can chew: Sabotage of induced defenses? Oikos 54:246–248.

Ehrlich, P. R., and Raven, P. H. 1964. Butterflies and plants: A study in coevolution. Evolution 18:586–608.

Eichlin, T. D., and Cunningham, H. B. 1978. The Plusiinae (Lepidoptera: Noctuidae) of America north of Mexico, emphasizing genitalia and larval morphology. U.S. Dept. Agric. Tech. Bull. 1567:1–122.

Eisner, T. 1980. Chemistry, defense, and survival: Case studies and selected topics, pp. 847–878. In M. Locke and D. S. Smith (eds.), Insect Biology in the Future. Academic Press, New York.

Esau, K. 1965. Plant Anatomy. Wiley, New York.

Fahn, A. 1979. Secretory Tissues in Plants. Academic Press, New York.

Farrell, B. D., Dussourd, D. E., and Mitter, C. 1991. Escalation of plant defense: Do latex/resin canals spur plant diversification? Am. Nat. 138:891–900.

Feeny, P. 1970. Seasonal changes in oak leaf tannins and nutrients as a cause of spring feeding by winter moth caterpillars. Ecology 51:565–581.

Ferrari, S. F. 1991. Diet for a small primate. Nat. Hist. 91:39–43.

Fitzgerald, T. D., and Peterson, S. C. 1988. Cooperative foraging and communication in caterpillars. BioScience 38:20–25.

Fitzgerald, T. D., Clark, K. L., Vanderpool, R., and Phillips, C. 1991. Leaf-shelter-building caterpillars harness forces generated by axial retraction of stretched and wetted silk. J. Insect Behav. 4:21–32.

Flint, M. L. 1987. Integrated pest management for cole crops and lettuce. University of California Division of Agriculture and Natural Resources Publication 3307.

Forcella, F. 1982. Why twig-girdling beetles girdle twigs. Naturwissenschaften 69:398–400.

Foster, A. S. 1956. Plant idioblasts: Remarkable examples of cell specialization. Protoplasma 46:184–193.

Fukuda, H., Hama, E., Kuzuya, T., Takahashi, A., Takahashi, M., Tanaka, B., Tanaka, H., Wakabayashi, M., and Watanabe, Y. 1985. The Life Histories of Butterflies of Japan, Vol. 1. Hoikusha, Osaka, Japan.

Ghent, A. W. 1960. A study of the group-feeding behaviour of larvae of the jack pine sawfly, *Neodiprion pratti banksianae* Roh. Behaviour 16:110–148.

Gilbert, L. E. 1971. Butterfly-plant coevolution: Has *Passiflora adenopoda* won the selectional race with heliconiine butterflies? Science 172:585–586.

Gould, S. J., and Vrba, E. S. 1982. Exaptation—a missing term in the science of form. Paleobiology 8:4–15.

Gross, P., and Price, P. W. 1988. Plant influences on parasitism of two leafminers: A test of enemy-free space. Ecology 69:1506–1516.

Haberlandt, G. 1914. Physiological Plant Anatomy. Macmillan, London.

Hagen, R. H., and Chabot, J. F. 1986. Leaf anatomy of maples (*Acer*) and host use by Lepidoptera larvae. Oikos 47:335–345.

Harris, P. 1960. Production of pine resin and its effect on survival of *Rhyacionia buoliana* (Schiff.) (Lepidoptera: Olethreutidae). Can. J. Zool. 38:121–130.

Harris, P. 1972. Food-plant groups of the Semanophorinae (Lepidoptera: Sphingidae): A possible taxonomic tool. Can. Entomol. 104:71–80.

Heinrich, B. 1971. The effect of leaf geometry on the feeding behaviour of the caterpillar of *Manduca sexta* (Sphingidae). Anim. Behav. 19:119–124.

Heinrich, B. 1979. Foraging strategies of caterpillars. Oecologia 42:325–337.

Heinrich, B. 1980. Artful diners. Nat. Hist. 89:42–51.

Heinrich, B. 1991. Nutcracker sweets. Nat. Hist. 91:4–8.

Heinrich, B., and Collins, S. L. 1983. Caterpillar leaf damage, and the game of hide-and-seek with birds. Ecology 64:592–602.

Hillis, W. E. 1987. Heartwood and Tree Exudates. Springer-Verlag, New York.

Hinton, H. E. 1981. Biology of Insect Eggs. Pergamon Press, New York.

Hodges, R. W. 1971. The Moths of America North of Mexico. Fasc. 21. Sphingoidea. E. W. Classey, London.

Hulley, P. E. 1988. Caterpillar attacks plant mechanical defense by mowing trichomes before feeding. Ecol. Entomol. 13:239–241.

Janzen, D. H. 1985. Plant defenses against animals in the Amazonian rainforest, pp. 207–217. In G. T. Prance and T. E. Lovejoy (eds.), Amazonia. Pergamon, New York.

Joel, D. M. 1980. Resin ducts in the mango fruit: a defence system. J. Exp. Bot. 31:1707–1718.

Johnson, W. T., and Lyon, H. H. 1976. Insects That Feed on Trees and Shrubs. Cornell University Press, Ithaca, NY.

Juniper, B., and Southwood, R., eds. 1986. Insects and the Plant Surface. Edward Arnold, Baltimore.

Kelsey, R. G., Reynolds, G. W., and Rodriguez, E. 1984. The chemistry of biologically active constituents secreted and stored in plant glandular trichomes, pp. 187–241. In E. Rodriguez, P. L. Healey, and I. Mehta (eds.), Biology and Chemistry of Plant Trichomes. Plenum, New York.

Langenheim, J. H. 1990. Plant resins. Am. Sci. 78:16–24.

Langenheim, J. H., Lincoln, D. E., Stubblebine, W. H., and Gabrielli, A. C. 1982. Evolutionary implications of leaf resin pocket patterns in the tropical tree *Hymenaea* (Caesalpinioideae: Leguminosae). Am. J. Bot. 69: 595–607.

Lederhouse, R. C. 1990. Avoiding the hunt: Primary defenses of lepidopteran caterpillars, pp. 175–189. In D. L. Evans and J. O. Schmidt (eds.), Insect Defenses: Adaptive Mechanisms and Strategies of Prey and Predators. State University of New York Press, Albany, NY.

Levin, D. A. 1973. The role of trichomes in plant defense. Quart. Rev. Biol. 48:3–15.

Levitt, J. 1972. Responses of Plants to Environmental Stresses. Academic Press, New York.

Lewis, A. C. 1979. Feeding preference for diseased and wilted sunflower in the grasshopper, *Melanoplus differentialis*. Entomol. Exp. Appl. 26:202–207.

Lewis, A. C., and van Emden, H. F. 1986. Assays for insect feeding, pp. 95–119. In J. R. Miller and T. A. Miller (eds.), Insect-Plant Interactions. Springer-Verlag, New York.

Linsley, E. G. 1959. Ecology of Cerambycidae. Annu. Rev. Entomol. 4:99–138.

Luckner, M., Diettrich, B., and Lerbs, W. 1980. Cellular compartmentation and channelling of secondary metabolism in microorganisms and higher plants, pp. 103–142. In L. Reinhold, J. B. Harborne, and T. Swain (eds.), Progress in Phytochemistry, Vol. 6. Pergamon, New York.

Malhberg, P. G. 1961. Embryogeny and histogenesis in *Nerium oleander* II. Origin and development of the non-articulated laticifer. Am. J. Bot. 48:90–99.

Mahlberg, P. G., and Sabharwal, P. S. 1968. Origin and early development of nonarticulated laticifers in embryos of *Euphorbia marginata*. Am. J. Bot. 55:375–381.

Mahlberg, P. G., Field, D. W., and Frye, J. S. 1984. Fossil laticifers from Eocene brown coal deposits of the Geiseltal. Am. J. Bot. 71:1192–1200.

Mattson, W. J., and Haack, R. A. 1987. The role of drought stress in provoking outbreaks of phytophagous insects, pp. 365–407. In Barbosa and J. C. Schultz (eds.), Insect Outbreaks. Academic Press, New York.

Mattson, W. J., Levieux, J., and Bernard-Dagan, C. 1988. Mechanisms of Woody Plant Defenses against Insects: Search for Pattern. Springer-Verlag, New York.

McKey, D. 1979. The distribution of secondary compounds within plants, pp. 55–133. In G. A. Rosenthal and D. H. Janzen (eds.), Herbivores: Their Interaction with Secondary Plant Metabolites. Academic Press, New York.

McKinney, K. B. 1944. The cabbage looper as a pest of lettuce in the southwest. U.S. Dept. Agric. Tech. Bull. 846:1–30.

Merz, E. 1959. Aflanzen und raupen. Über einge prinzipien der Futterwahl be: grossschmetterlingsraupen. Biol. Zentr. 78:152–188.

Metcalf, C. L., Flint, W. P., and Metcalf, R. L. 1962. Destructive and Useful Insects: Their Habits and Control. McGraw-Hill, New York.

Metcalfe, C. R., and Chalk, L. 1983. Anatomy of the Dicotyledons, Vol. II. Clarendon Press, Oxford, England.

Mitchell, P. L., and Newsom, L. D. 1984. Histological and behavioral studies of threecornered alfalfa hopper (Homoptera: Membracidae) feeding on soybean. Ann. Entomol. Soc. Am. 77:174–181.

Mitter, C., Farrell, B., and Wiegmann, B. 1988. The phylogenetic study of adaptive zones: Has phytophagy promoted insect diversification? Am. Nat. 132:107–128.

Mitton, J. B., and Sturgeon, K. B. 1982. Bark beetles in North American conifers. University Texas Press, Austin.

Nelson, C. J., Seiber, J. N., and Brower, L. P. 1981. Seasonal and intraplant variation of cardenolide content in the California milkweed, *Asclepias eriocarpa,* and implications for plant defense. J. Chem. Ecol. 7:981–1010.

Nishida, R., and Fukami, H. 1990. Sequestration of distasteful compounds by some pharmacophagous insects. J. Chem. Ecol. 16:151–164.

Peterson, S. C., Johnson, N. D., and LeGuyader, J. L. 1987. Defensive regurgitation of allelochemicals derived from host cyanogenesis by eastern tent caterpillars. Ecology 68:1268–1272.

Pliske, T. E., and Eisner, T. 1969. Sex pheromone of the queen butterfly: Biology. Science 164:1170–1172.

Rabb, R. L., and Bradley, J. R. 1968. The influence of host plants on parasitism of eggs of the tobacco hornworm. J. Econ. Entomol. 61:1249–1252.

Rathcke, B. J., and Poole, R. W. 1975. Coevolutionary race continues: Butterfly larval adaptation to plant trichomes. Science 187:175–176.

Rees, S. B., and Harborne, J. B. 1985. The role of sesquiterpene lactones and phenolics in the chemical defense of the chicory plant. Phytochemistry 24:2225–2231.

Rhoades, D. F. 1983. Herbivore population dynamics and plant chemistry, pp. 155–220. In R. F. Denno and M. S. McClure (eds.), Variable Plants and Herbivores in Natural and Managed Systems. Academic Press, New York.

Rhoades, D. F. 1985. Offensive-defensive interactions between herbivores and plants: Their relevance in herbivore population dynamics and ecological theory. Am. Nat. 125:205–238.

Rizk, A. M. 1987. The chemical constituents and economic plants of the Euphorbiaceae. Bot. J. Linn. Soc. 94:293–326.

Rodriguez, E., Healey, P. L., and Mehta, I. (eds.) 1984. Biology and Chemistry of Plant Trichomes. Plenum Press, New York.

Rosenthal, G. A., and Janzen, D. H. (eds.) 1979. Herbivores: Their Interaction with Secondary Plant Metabolites. Academic Press, New York.

Rothschild, M. 1977. The cat-like caterpillar. News Lepid. Soc., No. 6, 9.

Rothschild, M., and B. Ford. 1970. Heart poisons and the monarch. Nat. Hist. 79:36–37.

Rothschild, M., and Reichstein, T. 1976. Some problems associated with the storage of cardiac glycosides by insects. Nova Acta Leopold. Suppl. 7:507–550.

Rothschild, M., Von Euw, J., Reichstein, T., Smith, D. A. S., and Pierre, J. 1975. Cardenolide storage in *Danaus chrysippus* (L.) with additional notes on *D. plexippus* (L.). Proc. R. Soc. London B 190:1–31.

Roy, J. and Bergeron, J. 1990. Branch-cutting behavior by the vole (*Microtus pennsylvani-*

cus): A mechanism to decrease toxicity of secondary metabolites in conifers. J. Chem. Ecol. 16:735–741.

Rudolph, D. C., Kyle, H., and Conner, R. N. 1990. Red-cockaded woodpeckers vs rat snakes: The effectiveness of the resin barrier. Wilson Bull. 102:14–22.

Ryder, E. J., and Whitaker, T. W. 1976. Lettuce, pp. 39–41. In N. W. Simmonds (ed.), Evolution of Crop Plants. Longman, New York.

Sakurai, K. 1985. An attelabid weevil (*Euops splendida*) cultivates fungi. J. Ethnol. 3:151–156.

Sakurai, K. 1988. Leaf size recognition and evaluation of some attelabid weevils (2) *Apoderus balteatus*. Behaviour 106:301–317.

Schery, R. W. 1952. Plants for Man. Prentice-Hall, New York.

Schultz, J. C. 1983. Habitat selection and foraging tactics of caterpillars in heterogeneous trees, pp. 61–90. In R. F. Denno and M. S. McClure (eds.), Variable Plants and Herbivores in Natural and Managed Systems. Academic Press, New York.

Schultz, J. C., Nothnagle, P. J., and Baldwin, I. T. 1982. Seasonal and individual variation in leaf quality of two northern hardwoods tree species. Am. J. Bot. 69:753–759.

Scudder, S. H. 1889. Butterflies of the Eastern United States and Canada. W. H. Wheeler, Cambridge, Mass.

Seiber, J. N., Nelson, C. J., and Lee, S. M. 1982. Cardenolides in the latex and leaves of seven *Asclepias* species and *Calotropis procera*. Phytochemistry 21:2343–2348.

Shaw, E. M., Woolley, P. L., and Rae, F. A. 1963. Bushman arrow poisons. Cimbebasia 7:1–41.

Southwood, R. 1986. Plant surfaces and insects—an overview, pp. 1–22. In B. Juniper and R. Southwood (eds.), Insects and the Plant Surface. Edward Arnold, Baltimore.

Spencer, K. C. 1988. Chemical mediation of coevolution in the *Passiflora-Heliconius* interaction, pp. 167–240. In K. C. Spencer (ed.), Chemical Mediation of Coevolution. Academic Press, New York.

Stahl, E. 1988. Pfanzen und schnecken. Jena. Z. Med. Naturwiss. 22:557–684.

Stipanovic, R. D. 1983. Function and chemistry of plant trichomes and glands in insect resistance, pp. 69–100. In P. A. Hedin (ed.), Plant Resistance to Insects. American Chemical Society, Washington, D.C.

Stride, G. O., and Warwick, E. P. 1962. Ovipositional girdling in a North American cerambycid beetle, *Mecas saturnina*. Anim. Behav. 10:112–117.

Strong, D. R., Lawton, J. H., and Southwood, R. 1984. Insects on Plants: Community Patterns and Mechanisms. Harvard University Press, Cambridge, MA.

Sutherland, D. W. S. 1966. Biological investigations of *Trichoplusia ni* (Hubner) and other Lepidoptera damaging cruciferous crops on Long Island, New York. Cornell Agric. Expt. St. Mem. 399:1–98.

Swain, T. 1979. Tannins and lignins, pp. 657–682. In G. A. Rosenthal and D. H. Janzen (eds.), Herbivores: Their Interaction with Secondary Plant Metabolites. Academic Press, New York.

Tallamy, D. W. 1985. Squash beetle feeding behavior: An adaptation against induced cucurbit defenses. Ecology 66:1574–1579.

Tallamy, D. W. 1986. Behavioral adaptations in insects to plant allelochemicals, pp. 373–300. In L. B. Brattsten and S. Ahmad (eds.), Molecular Aspects of Insect-Plant Associations. Plenum, New York.

Tallamy, D. W., and Krischik, V. A. 1989. Variation and function of cucurbitacins in *Cucurbita:* An examination of current hypotheses. Am. Nat. 133:766–786.

Tallamy, D. W., and McCloud, E. S. 1991. Squash beetles, cucumber beetles and inducible cucurbit responses. pp. 155–181. In D. W. Tallamy and M. J. Raupp (eds.), Phytochemical Induction by Herbivores. Wiley, New York.

Thomson, W. W., Platt-Aloia, K., and Koller, D. 1979. Ultrastructure and development of the trichomes of *Larrea* (creosote bush). Bot. Gaz. 140:249–260.

Tibbitts, T. W., and Read, M. 1976. Rate of metabolite accumulation into latex of lettuce and proposed association with tipburn injury. J. Am. Soc. Hort. Sci. 101:406–409.

Tietz, H. M. 1972. An Index to the Described Life Histories, Early Stages, and Hosts of the Macrolepidoptera of the Continental United States and Canada. A. C. Allyn, Sarasota, FL.

Timm, R. M., and Clauson, B. L. 1990. A roof over their feet. Nat. Hist. 90:54–59.

Tuskes, P. M. 1985. The biology and immature stages of *Sphingicampa albolineata* and *S. montana* in Arizona (Saturniidae). J. Lepid. Soc. 39:85–94.

Van Parijs, J., Broekaert, W. F., Goldstein, I. J., and Peumans, W. J. 1991. Hevein: an antifungal protein from rubber-tree (*Hevea brasiliensis*) latex. Planta 183:258–264.

Verschaffelt, E. 1910. The cause determining the selection of food in some herbivorous insects. Proc. K. Ned. Akad. Wet. 13:536–542.

Weinstein, P. 1990. Leaf petiole chewing and the sabotage of induced defenses. Oikos 58:231–233.

Welch, M. B. 1920. *Eucalyptus* oil glands. J. Proc. Roy. Soc. N.S.W. 54:208–217.

Welch, M. B. 1921. The occurrence of oil ducts in certain eucalypts and angophoras. Proc. Linn. Soc. N. S. W. 46:475–486.

Wilson, K. J. 1986. Immunological-cytochemical localization of cell products in plant tissue culture, pp. 212–230. In H. F. Linskens and J. F. Jackson (eds.), Immunology in Plant Sciences. Springer-Verlag, New York.

Winder, J. A. 1976. Ecology and control of *Erinnyis ello* and *E. alope,* important insect pests in the New World. Pest Articles News Summ. 22:449–466.

Young, A. M. 1978. The biology of the butterfly *Aeria eurimedea agna* (Nymphalidae: Ithomiinae: Oberiini) in Costa Rica. J. Kansas Entomol. Soc. 51:1–10.

Young, A. M., and Moffett, M. W. 1979. Behavioral regulatory mechanisms in populations of the butterfly *Mechanitis isthmia* in Costa Rica: Adaptations to host plants in secondary and agricultural habitats. Deutsche Entomol. Ztschr. 26:21–38.

Zangerl, A. R. 1990. Furanocoumarin induction in wild parsnip: Evidence for an induced defense against herbivores. Ecology 71:1926–1932.

4

Patterns of Interaction among Herbivore Species

Hans Damman

Introduction

Competitive, predator/prey, and mutualistic interactions occupy a central position in the attempts of ecologists to understand the structure of communities. Competitive interactions, in particular, have been viewed as important in influencing patterns of species packing and resource use (e.g., May 1984), and in driving many of the evolutionary interactions that shape the biology of individual species (e.g., Ehrlich and Raven 1964).

Beginning with a theoretical paper by Hairston et al. (1960), and continuing with more recent reviews of experimental studies (Lawton and Strong 1981; Connell 1983; Schoener 1983; Strong et al. 1984), the importance of within-trophic-level interactions, such as competition, in structuring herbivore communities has been questioned. Hairston et al. (1960) argued that because herbivores rarely denuded plants, plant tissue could not represent a limiting resource for which herbivores must compete. In contrast to the resource-limited producer, carnivore and detritivore trophic levels, Hairston et al. concluded that herbivores must be limited by predators, parasites, and pathogens. The studies drawn together by Lawton and Strong (1981), Connell (1983), Schoener (1983), and Strong et al. (1984) seem to corroborate the pattern proposed by Hairston et al. (1960), and, taken together, suggest that within-trophic-level interactions, like direct competition, are relatively less important than between-trophic-level interactions. As a result, the focus of studies of herbivorous insects has shifted away from interactions between herbivores and toward the interactions between herbivores and their food plants or their natural enemies.

However, Faeth (1987, 1988) indicated that it may be premature to discount the importance of within-trophic-level interactions in herbivore communities. By considering only direct competition between species, many of the interactions

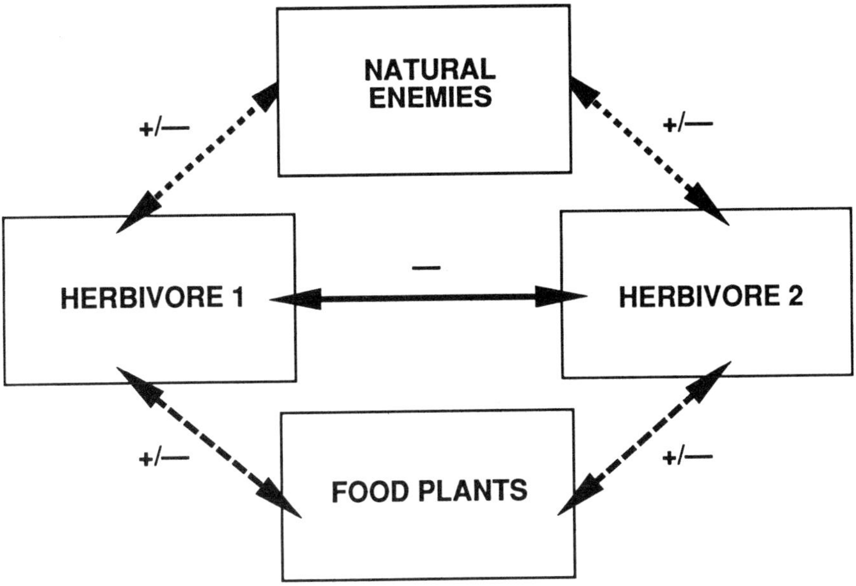

Figure 4.1. Pathways of direct (solid arrow), resource-mediated (long dashed arrows), and enemy-mediated (short dashed arrows) interactions among herbivores. (Modified from Price *et al.* 1986).

between herbivorous insects, both negative and positive, may have been overlooked. A fundamental assumption of the argument against the impact of within-trophic-level interactions on herbivore communities, namely that herbivores rarely deplete their food, may not be entirely true. Finally, only a few studies involving experimental manipulations of populations of herbivorous insects were included in the reviews of Lawton and Strong (1981), Connell (1983), Schoener (1983), and Strong et al. (1984).

Direct competition for food, in which an individual exploits or denies access to food that would otherwise be available to other herbivores, forms the focus of most investigations of within-trophic-level interactions (Fig. 4.1, solid arrow). However, two other types of within-trophic-level interactions, resource- and enemy-mediated interactions, may be as important to herbivorous insects. In resource-mediated interactions, feeding by one herbivore species causes changes in a plant that in turn influence the plant's quality as food for other herbivore species (Fig. 4.1, long dashed arrows). In enemy-mediated interactions (Price et al. 1986, equivalent to competition for enemy-free space, sensu Atsatt 1981b), herbivores interact indirectly by influencing the population dynamics of foraging behavior of shared natural enemies (Fig. 4.1, short dashed arrows). Herbivorous insects may be particularly prone to resource- and enemy-mediated interactions: (1) Unlike predators that kill their prey outright or plants and detritivores that

Table 4.1. Percentage of Leaves on a Plant Suffering at Least Some Herbivore Damage

Plant species	Percentage of leaves damaged	Reference
Calystegia sepium	100	Edwards and Wratten (1983)
Solanum dulcamara	97	Edwards and Wratten (1983)
Betula pubescens	55	Edwards and Wratten (1983)
Corylus avellana	62	Edwards and Wratten (1983)
Sambucus nigra	90	Edwards and Wratten (1983)
Prunus avium	85	Edwards and Wratten (1983)
Salix capraea	82	Edwards and Wratten (1983)
Hammamelis vernalis	98	Edwards and Wratten (1983)
Spartina alterniflora	65	Stiling et al. (1984)
Heliconia imbricata	100	Strong (1984)
Quercus emoryi	≥30	Faeth (1985)
Quercus emoryi	19–25	Faeth (1986)
Juglans arizonica	70–90	Renaud (1987) in Faeth (1987)

derive nutrients from inert substrates, herbivorous insects feed on plants that typically survive the feeding damage and can alter their nutritional and chemical composition in response to attack (Schultz 1988; Karban and Myers 1989). (2) Predators, parasites, and pathogens consistently inflict heavy mortality on the feeding stages of herbivorous insects (Lawton and McNeill 1979; Dempster 1983; Feeny et al. 1985). Moreover, resource- and enemy-mediated interactions may be easily overlooked when investigating within-trophic-level interactions because the effects of changes in resource quality and in foraging patterns of enemies commonly extend to points spatially and temporally distant from the site of feeding of the herbivore causing the changes (e.g., Karban and Myers 1989).

At the heart of the conclusion of Hairston et al. (1961) that food does not limit herbivores, lies the observation that herbivores eat only a small fraction of the available leaf tissue. Defoliation of forests during periodic outbreaks can approach 100%, but such conditions are not chronic and involve relatively few insect species (Faeth 1987; Barbosa and Wagner 1989). Usually, herbivores eat less than 10% of the annual leaf area produced by trees (Nielsen 1978; Springett 1978; Strong 1984; Southwood et al. 1986; Faeth 1987). Because the changes induced by herbivores can extend well beyond the leaves actually suffering damage, total leaf area consumed may severely underestimate the percentage of foliage affected by herbivory. Edwards and Wratten (1983) and Faeth (1987) both indicate that the percentage of leaves receiving at least some damage may provide a more meaningful estimate of the impact of herbivory. Viewed from this perspective, most foliage on most plants may be affected by the activities of herbivores (Table 4.1).

In light of the potential scope of interactions among herbivores, the reviews of between-species competition do not provide a secure foundation for the contention that within-trophic-level interactions have little effect on herbivore communities. While Connell (1983) and Strong et al. (1984) recognize that resource- and enemy-mediated interactions may play an important role, they nonetheless focus on direct competition in evaluating the impact of within-trophic-level interactions on community organization. Connell (1983, 72 studies) and Schoener (1983, 164 studies) reviewed evidence for direct competition in the field extensively, but between them consider only five experimental studies that involve insect herbivores. Strong et al. (1984) list 41 studies assessing between-species competition in herbivorous insects, but at least 17 of these, primarily those showing no evidence of competition, are based on patterns of distribution rather than experimental manipulations. Hastings (1987) notes that such observational studies remain open to alternative interpretations. A number of recent reviews that consider the importance of resource- (Faeth 1987, 1988; Karban and Myers 1989) and enemy-mediated interactions (Price et al. 1986; Holt and Kotler 1987) give no indication of how common these indirect interactions are as compared to either direct interactions or the absence of interactions between herbivores.

In this review, I attempt to reevaluate the importance of within-trophic-level interactions between herbivorous insects in determining their distribution over the available food plants and in shaping their foraging behaviors. I begin by assessing the relative frequency of within-trophic-level interactions by compiling published studies testing for direct, resource-mediated, and enemy-mediated interactions among herbivorous insects. I then compare the data in my compilation to those available in the reviews of Connell (1983), Schoener (1983), and Strong et al. (1984). Focusing on patterns in life-history characteristics for the herbivores included in my survey, I consider how different feeding behaviors predispose insects to the various mechanisms of within-trophic-level interaction. Finally, I use the simple herbivore fauna associated with pawpaws (*Asimina* spp.) to pull together some of the patterns that emerge from the literature review.

The Importance of Interactions among Herbivorous Insects

The maturing of the study of resource-mediated interactions and a growing recognition of the role of enemy-mediated interactions during the past decade have made available a large literature documenting the indirect as well as the direct interactions between herbivores. Including resource- and enemy-mediated interactions may alter our perception of the relative importance of the selective force that structure herbivore communities. Specifically, I will address two questions: (1) When resource- and enemy-mediated interactions are considered along with competition, do herbivorous insects still show a low incidence of within-trophic-level interactions relative to members of other trophic levels? (2) How important are resource- and enemy-mediated interactions to herbivores when compared to direct interactions?

Methods

In compiling the studies included in this review, I concentrated almost exclusively on those that used experimental manipulations to detect interactions. Manipulations generally involved monitoring the change in population size or behavior of an insect or guild of insects following the exposure to or exclusion of another insect or guild attacking a shared food plant. In several studies of resource-mediated interactions, simulated damage was used to mimic damage done by leaf-chewing insects. I excluded studies that inferred the presence of interactions solely from the distributional patterns or abundances of insects, because such patterns can often be explained just as convincingly by host selection behaviors acting independently of any within-trophic-level interactions (Hastings 1987). I included a few observational studies that provided additional information that effectively eliminated the possibility that host selection behavior could be confounded with between-species interactions.

Ferson et al. (1986) found that the rules by which studies are included in literature surveys can have a strong effect on the conclusions. A survey such as this will inherently be biased toward studies that detect interactions over those that provide no evidence of an interaction both because positive results are more likely to be published and because the presence of interaction is more likely to attract the attention of ecologists than is the lack of interaction (Connell 1983). However, as long as studies involving interactions among organisms other than herbivorous insects are similarly biased, the results obtained here can be compared productively to those reported elsewhere.

The studies included in this survey fell into two distinct categories: "Controlled" studies, in which insect densities or severity of damage were manipulated to levels not related to those encountered in the field, and "Field" studies, in which insect populations were manipulated in the field or were manipulated to levels related directly to those encountered in the field. Controlled experiments were generally run in isolation from other variables that might obscure interactions in the field, and are therefore probably a better indicator of the potential for interaction rather than of the actual incidence of interaction in a natural setting. At the same time, Controlled experiments may more effectively reveal the mechanisms that may mediate the interactions between herbivorous insects. Usually, I analyzed the results of Controlled and Field studies separately. For each insect species involved in a study I recorded a series of natural history characteristics (Table 4.2) that might relate to the tendency to be involved in within-trophic-level interactions.

The positive and negative interactions between species were classified as "Strong" if all experiments in a study consistently provided statistical evidence for the interaction. I considered the overall intensity of interaction as "Weak" when interactions were deleted in only some of the experiments in a study. Weak interactions could be analyzed either by ignoring them, by lumping them with

Table 4.2. Natural History Characteristics Recorded for Herbivorous Insects in Published Studies[a] on the Interactions between Herbivores

Characteristic	Categories
Feeding behavior	Leaf chewing, leaf tying, leaf mining, sap sucking (vascular tissues + cell contents), gall makers, stem borers, seed predators (attacking seeds while on the plant), fruit borers, root feeders
Feeding position	External (exposed to natural enemies, not bound to any sort of shelter), internal (living inside plant tissue or surrounded by a shelter)
Gregariousness	Solitary (larvae feed independently of one another), gregarious (larvae feed in groups during at least the early instars)
Specialization	Specialists (restricted to one plant family), generalists (feeding on more than one plant family)
Growth form of food plant	Herbaceous (feed on grasses or dicots of which the above-ground tissues die back between growing seasons), woody (dicots with above-ground tissues that persist from growing season to growing season)

[a]Data were drawn from studies by Addicott (1978, 1979), Akimoto (1981), Antolin and Addicott (1988), Atsatt (1981a), Auerbach and Simberloff (1984), Baldwin (1988b), Bergelson and Lawton (1988), Blakely and Dingle (1978), Bultman and Faeth (1985), Carne (1965), Carroll et al. (1979), Compton and Robertson (1988), Crawley and Pattrasudhi (1988), Cushman and Addicott (1989), Damman (1989), Danell and Huss-Danell (1985), DeBach et al. (1978), Dixon and Barlow (1979), Doutt and Nakata (1973), Dussourd and Eisner (1987), Edson (1985), Edwards and Wratten (1985), Evans (1989), Faeth (1985, 1986, 1987, 1988, 1990), Fellows and Heed (1972), Finch and Jones (1989), Fitt (1984), Fowler and Lawton (1985), Fowler and MacGarvin (1986), Fritz (1983), Fritz and Price (1990), Fritz et al. (1986), Gange and Brown (1989), Gardner (1938), Gibson (1976, 1980), Gibson and Visser (1982), Hajek and Dahlsten (1986), Hanhimäki (1989), Harrison (1964), Harrison and Karban (1986), Hartley (1988), Hartley and Lawton (1987), Hassell (1968), Hawkins (1988), Heie (1972), Heinrich and Collins (1983), Horvitz and Schemske (1984), Hsaio and Holdaway (1966), Hunter (1987, 1990), Hunter and Willmer (1989), Jones et al. (1988), Jutsum et al. (1981), Karban (1986, 1988, 1989), Kareiva (1982), Keiser et al. (1974), Kidd et al. (1985), Klijnstra (1985), Laine and Niemelä (1980), Landsberg (1990), Lewis (1979), Mattson (1986), Mattson et al. (1989), McClure (1980, 1981), McClure and Price (1975), McLain (1981), Messina (1981), Mopper et al. (1990), Moran and Whitham (1990), Neuvonen et al. (1988), Nickerson et al. (1977), Niemelä et al. (1984), Pullin (1987), Raupp and Denno (1984), Raupp et al. (1986), Room (1972), Shapiro (1981), Shearer (1976), Skinner and Whittaker (1981), Smiley et al. (1988), Stamp (1984, 1987), Starý (1978), Stiling (1980), Stiling et al. (1987), Stiling and Strong (1983, 1984), Strong (1982a,b), Valentine et al. (1983), Varley (1947), Waloff (1968), Webb and Moran (1978), West (1985), Williams and Myers (1984), Wratten et al. (1984), and Zwölfer (1979). A complete listing of the data on which the review is based is available from the author.

strong interactions, or by equating them with evidence for no interaction. Unless stated otherwise, I chose to exclude these ambiguous results from analysis.

I analyzed all data using statistics appropriate to categorical information. When the frequency in every cell of a contingency table exceeded five, I used a *G*-statistic (Sokal and Rohlf 1981). In many of the contingency tables, however, frequencies in many of the cells were below five. In these cases, I used either a

Table 4.3. Frequency and Type of Interaction among Herbivorous Insects in Controlled and Field Experiments

	Number of studies reporting interactions that were						
Type of experiment[a]	Negative	Weakly negative[b]	Not detectable	Negative and positive[c]	Weakly positive[b]	Positive	Total
Controlled	23	3	0	2	2	3	33
Field	35	6	7	3	1	14	66
Total	58	9	7	5	3	17	99

[a]See text for explanation of type of experiment.

[b]Interactions were classified as weakly negative or positive if experiments sometimes showed significant interactions and other times showed no detectable interaction. These ambiguous interactions were not included in the statistical analyses of the data.

[c]This category includes studies that clearly and consistently showed interactions, but in which some of the species clearly responded positively and other clearly responded negatively.

Fisher's exact test for 2×2 tables or a Monte Carlo simulation procedure for R×C tables (Fish6 v. 1.001 and Monte Carlo R×C v. 1.0, B. Engels, Genetics Department, University of Wisconsin, 1988).

Do Herbivorous Insects Interact Less Frequently Than Other Trophic Groups?

The vast majority (91%) of the 99 experimental studies included in this review showed evidence of interactions between herbivores (Table 4.3). Moreover, in 25% of the studies, at least one species benefited from the interaction. Sixty-seven percent of the studies detected negative interactions. Field studies detected positive interactions more frequently than did Controlled experiments (Monte Carlo simulation: $P<0.0001$), probably because Controlled experiments generally did not extend beyond potentially interacting pairs of herbivores and their food plant.

A comparison of the Field data for herbivores compiled here to that for field studies compiled by Cornell (1983), Schoener (1983), and Strong et al. (1984) revealed that direct competition was indeed relatively rare between herbivorous insects, but that within-trophic-level interactions were as common as at other trophic levels. When positive interactions were classified as evidence of no competition, as was explicitly true in Connell's analysis and presumably true in Schoener's review, it appeared that direct competition affected herbivores less frequently than it did plants, carnivores, and detritivores (Table 4.4). A greater proportion of Field studies detected competition among herbivorous insects (41 of 66 studies, including weak evidence of interaction) than reported by Strong et al. (1984; 17 of 40 studies, which also includes weak evidence of interaction), even when I combined studies reporting positive interactions with those reporting no competition [$G(1) = 3.88$, $P < 0.05$].

Table 4.4. The Relative Frequency of between-Species Competition between Herbivores as Compared to Organisms at Other Trophic Levels

	Herbivorous insects[a]	Connell (1983)[b]	Schoener (1983)[a, b, c]
When positive interactions are equated with a lack of competition			
Competing	56	110	95
Not competing	71	90	38
		$G=3.70$	$G=19.9$
		$0.05<P<0.1$	$P<0.0001$
When interactions include both positive and negative (competitive) interactions			
Interacting	80	123	
Not interacting	47	77	
		$G=0.074$	
		$P>0.75$	

[a]Species involved in weakly competitive interactions were not included in the analysis.

[b]*G*-statistics calculated for herbivorous insects versus Connell (1983) or Schoener (1983) data.

[c]Data from Table 4 in Schoener (1983), for experiments involving terrestrial organisms and no exclosures.

When resource- and enemy-mediated interactions are considered together with direct interactions, the evidence suggests that within-trophic-level interactions are important for herbivorous insects as they seem to be for organisms at other trophic levels. Because Connell (1983) noted which of the interactions he encountered were positive, it was possible to compare the frequency of interactions (negative or positive) to the frequency of no interaction in my data for herbivorous insects to that in Connell's data for organisms at all trophic levels. Viewed in this way, the frequency of interaction was independent of trophic level (Table 4.4). In addition, positive interactions, which traditionally have received less attention than competitive interactions, make up a large proportion of the interactions between herbivores (25%, Table 4.3). Connell (1983) noted that 32% of 22 studies that he compiled for herbivores involved at least one species that benefited from association with the others. Sih et al. (1985), in a review of predator/prey interactions, reported positive interactions among 21% of the herbivores that they surveyed. These values approximate those reported for plants, but both cases exceeded those reported for omnivores and carnivores.

The tendency to be involved in positive, neutral, or negative interactions depends on the foraging pattern of the herbivores (Table 4.5). Leaf chewers in the Field studies that I surveyed tended to be involved in neutral interactions (Fig. 4.2A, C). Gall makers and leaf miners showed the opposite pattern, being involved in neutral interactions relatively infrequently (Fig. 4.2A). Sap-sucking insects generally interacted negatively with the other herbivores (Fig. 4.2A,C). The more sedentary feeders, therefore, seemed to interact more frequently with other herbivores than did more mobile insects, perhaps because they could not

Table 4.5. Effect of Various Life Histories on Whether an Herbivore Was Affected Positively, Negatively, or Neutrally by Other Insects and Whether It Affected Other Insects Positively, Negatively, or Neutrally

Life history characteristic	Experiment type	Effect on insect by other herbivores[a]	Effect of insect on other herbivores[a]
Feeding behavior	Field	$0.05<P<0.1$	$P<0.0001$
	Controlled	$P<0.01$	$P<0.01$
Feeding position	Field	$P>0.1$	$P>0.1$
	Controlled	$P<0.05$	$P<0.01$
Gregariousness	Field	$P>0.5$	$P<0.05$
	Controlled	$P>0.25$	$P>0.1$
Specialization	Field	$P>0.1$	$P<0.25$
	Controlled	$P>0.5$	$P<0.25$
Growth form of food plant[b]	Field	$P<0.0001$	
	Controlled	$P<0.0001$	

[a]Probability values calculated using a Monte Carlo simulation run for ≥10,000 iterations.

[b]Growth form of food plant was done per study rather than per species. The probability value gives the probability that the frequencies of positive, negative, and neutral interactions are equal for insects on herbaceous and woody plants.

avoid other insects or perhaps because leaf miners, gall makers, and phloem feeders are all thought to have the ability to manipulate plant metabolism (Pritchard and James 1984; Whittaker 1984; Faeth 1987). Positive and neutral interactions occurred relatively more frequently on woody plants than on herbaceous plants than did negative interactions (Fig. 4.3), perhaps because there was more room to escape negative interactions on trees.

The Mechanisms Mediating Interactions among Herbivorous Insects

Interactions within trophic levels have emphasized direct competition. This emphasis may have been maintained by short-term experiments that are unlikely to detect the more slowly unfolding resource- and enemy-mediated interactions (Bender et al. 1984). Faeth (1987) indicates that the links connecting herbivores involved in resource-mediated interactions can be very subtle. Similarly, the connections between herbivores via shared natural enemies may also be difficult to discern, and therefore to incorporate into experiments designed to detect between-species interactions.

To address the broad question of the relative importance of direct and indirect interactions to herbivores I tabulated the mechanisms suggested to be at work by the authors of the experimental studies that I surveyed. The results from Field and Controlled studies combined indicate that direct, resource-mediated, and enemy-mediated interactions all contribute significantly to interactions between herbivorous insects (Fig. 4.4). The contributions of each to competitive interac-

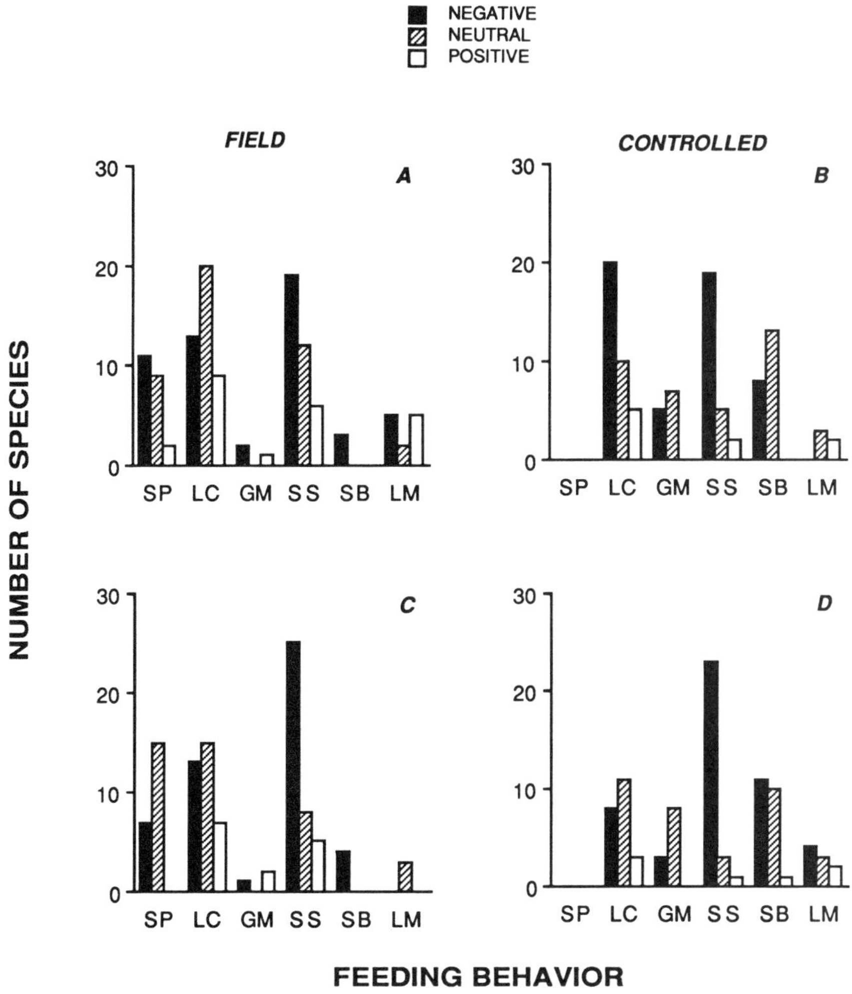

Figure 4.2. The tendency for insect species with different feeding behaviors to be affected by other herbivores (A and B) or to affect other herbivores (C and D) through negative, neutral, or positive interactions as recorded in published experiments. A and C depict data from Field experiments; B and D depict data from Controlled experiments. Refer to Table 4.5 for statistics. SP, seed predators; LC, leaf chewers; GM, gall makers; SS, sap suckers; SB, shoot borers.

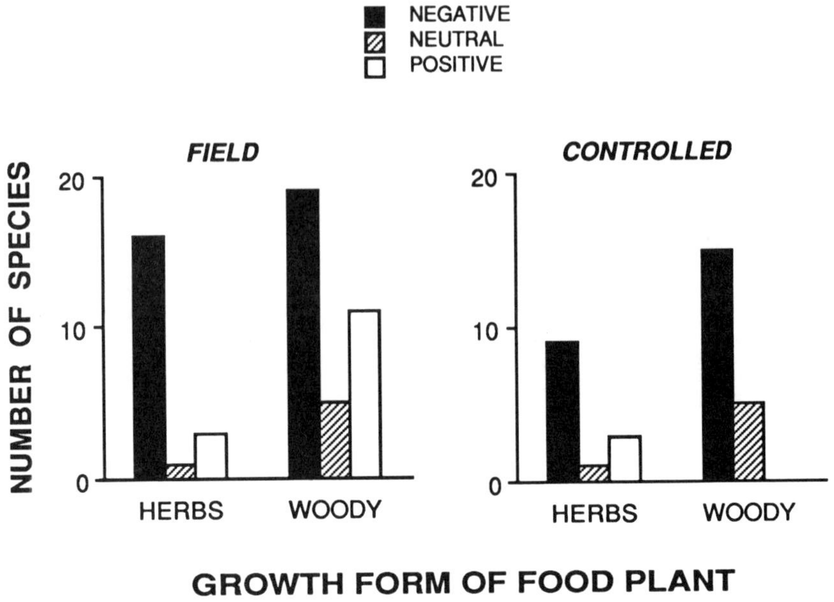

Figure 4.3. The relationship between the frequency of negative, neutral, or positive interactions and the growth form of the plant bearing the insects as recorded in published experiments. Refer to Table 4.5 for statistics.

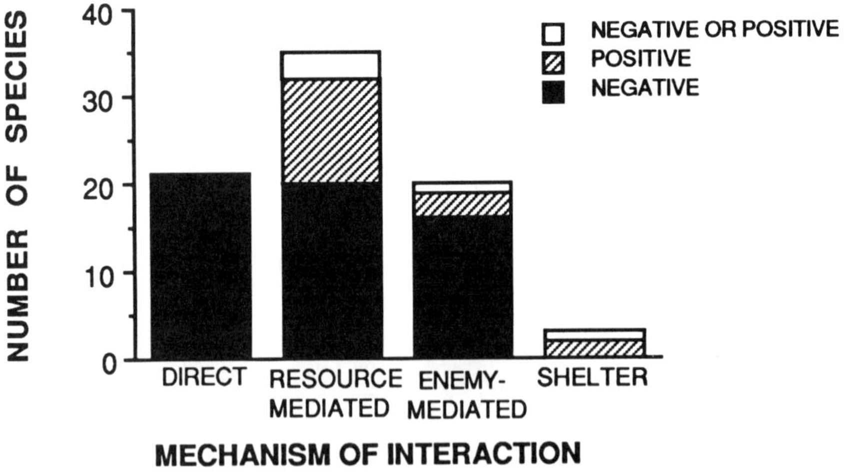

Figure 4.4. The relative importance of the mechanisms implicated in 79 Field and Controlled studies showing clear evidence of interactions between hervirous insects. The mechanisms differ significantly in their tendency to generate negative, positive, and both negative and positive interactions (Monte Carlo simulation: $P<0.001$).

tions are roughly equal, but enemy-mediated and, especially, resource-mediated interactions include a greater proportion of positive interactions than do direct interactions. A fourth type of interaction, the provision of shelters by one species for another herbivore, has been reported infrequently, but invariably benefits the species acquiring the shelter. The indirect, resource-, and enemy-mediated interactions represent extremely important mechanisms of interaction between herbivores not only because they are the most frequent forms of interaction, but also because they commonly generate positive interactions between insects.

Direct Interactions

Direct interactions among herbivorous insects take the form of both exploitative and interference competition. Exploitative competition, where the species involved compete directly for shared resources represented roughly half (13 of 21) of the direct interactions.

Interference competition, in which one herbivore retains exclusive access to a resource by excluding others, affected foraging by both larval and adult stages. For caterpillars, territorial behavior sometimes takes the form of direct aggression between individuals, in which the larger or more aggressive species invariably defeated the small or less aggressive species (Rathcke 1976; Stiling and Strong 1983). Web-spinning by larvae of the moth *Platyptilia williamsii* denied spittlebugs access to their preferred feeding sites (Karban 1986, 1989) and space-filling by aphid colonies on balsam fir leaders effectively prevented feeding by spruce budworm larvae (Mattson et al. 1989). Raupp et al. (1986) reported that larvae of the beetle *Plagiodera versicolora* produced a volatile chemical that deterred feeding on willow by caterpillars of *Nymphalis antiopa*. Adults of many insects avoid laying eggs on plants already occupied by conspecific eggs or larvae (Rothschild and Schoonhoven 1977; Prokopy 1981; Messina and Renwick 1985; Pilson and Rausher 1988; but see Singer and Mandracchia 1982). Several recent studies indicate that ovipositing females also can recognize and avoid plants occupied by other species (Shapiro 1981; West 1985; Klijnstra 1985; Jones et al. 1988; Finch and Jones 1989). Except for direct aggression between larvae, interference competition probably will not reduce population densities of competitors as much as it will alter the distribution pattern of the feeding stages over their food plant. Interestingly, the instances of direct aggression between herbivores both involve stem-borers that cannot escape competitors by feeding elsewhere.

In all cases, direct interactions between species require that the species involved forage on the same plant at the same time. This would occur most commonly among species that feed in similar ways and that can attain high densities (e.g., Denno et al. 1982). An often repeated finding of studies on insect communities, however, is that the density of plant-feeding insects is sufficiently low and their feeding preferences are sufficiently divergent that direct encounters are likely to

Table 4.6. Effect of Various Life Histories on Whether an Herbivore Was Affected by Other Insects Through Direct, Resource-Mediated, or Enemy-Mediated Interactions and Whether It Affected Other Insects through Direct, Resource-Mediated, or Enemy-Mediated Interactions.

<table>
<tr><th>Life history characteristic</th><th>Experiment type</th><th>Effect on insect by other herbivores[a]</th><th>Effect of insect on other herbivores[a]</th></tr>
<tr><td>Feeding behavior</td><td>Field</td><td>$P<0.01$</td><td>$P<0.0001$</td></tr>
<tr><td></td><td>Controlled</td><td>$P<0.0001$</td><td>$P<0.0001$</td></tr>
<tr><td>Feeding position</td><td>Field</td><td>$P>0.5$</td><td>$0.05<P<0.1$</td></tr>
<tr><td></td><td>Controlled</td><td>$P>0.25$</td><td>$P>0.25$</td></tr>
<tr><td>Gregariousness</td><td>Field</td><td>$P>0.5$</td><td>$P<0.05$</td></tr>
<tr><td></td><td>Controlled</td><td>$P<0.01$</td><td>$P>0.25$</td></tr>
<tr><td>Specialization</td><td>Field</td><td>$0.05<P<0.1$</td><td>$P<0.9$</td></tr>
<tr><td></td><td>Controlled</td><td>$0.05<P<0.1$</td><td>$0.05<P<0.1$</td></tr>
<tr><td>Growth form</td><td>Field</td><td colspan="2">$P<0.01$</td></tr>
<tr><td>of food plant[b]</td><td>Controlled</td><td colspan="2">$P>0.1$</td></tr>
</table>

[a]Probability values calculated using a Monte Carlo simulation run for $\geq$10,000 iterations.

[b]Growth form of food plant was done per study rather than per species. The probability value gives the probability that the frequencies of positive, negative, and neutral interactions are equal for insects on herbaceous and woody plants.

be the exception rather than the rule (e.g., Rothschild 1971; Denno et al. 1982; Lawton 1982, 1984; Strong 1984; Bultman and Faeth 1985).

The mechanisms mediating interactions between herbivores were consistently related to the feeding behavior of the insects (Table 4.6). Sap-sucking homopterans, seed predators, stem borers, and fruit borers were involved in direct interactions disproportionately frequently (Fig. 4.5). For stem borers, evidence of interaction disappeared when the insects were studied in the field where likelihood of encounter was low (compare Stiling and Strong 1983, 1984). Seed predators may be particularly prone to direct interactions because they frequently occur at high densities (e.g., Varley 1947; Zwölfer 1979) and because they feed within flower heads, which represent isolated resource patches that are particularly prone to depletion (Prokopy 1981). In contrast, free-living leaf chewers are affected by direct interactions only rarely.

Resource-Mediated Interactions

Resource-mediated interactions among herbivores are both common (Fig. 4.4) and complex. As defined here, resource-mediated interactions involve induced responses (sensu Karban and Myers 1989, 27 studies) and regrowth following defoliation (8 studies). Both types of resource-mediated interactions represent changes in food quality caused by herbivore feeding. Induced responses can involve changes in the absolute or relative concentrations in plant tissues of a number of secondary chemicals and nutrients (e.g., Haukioja and Niemelä 1977,

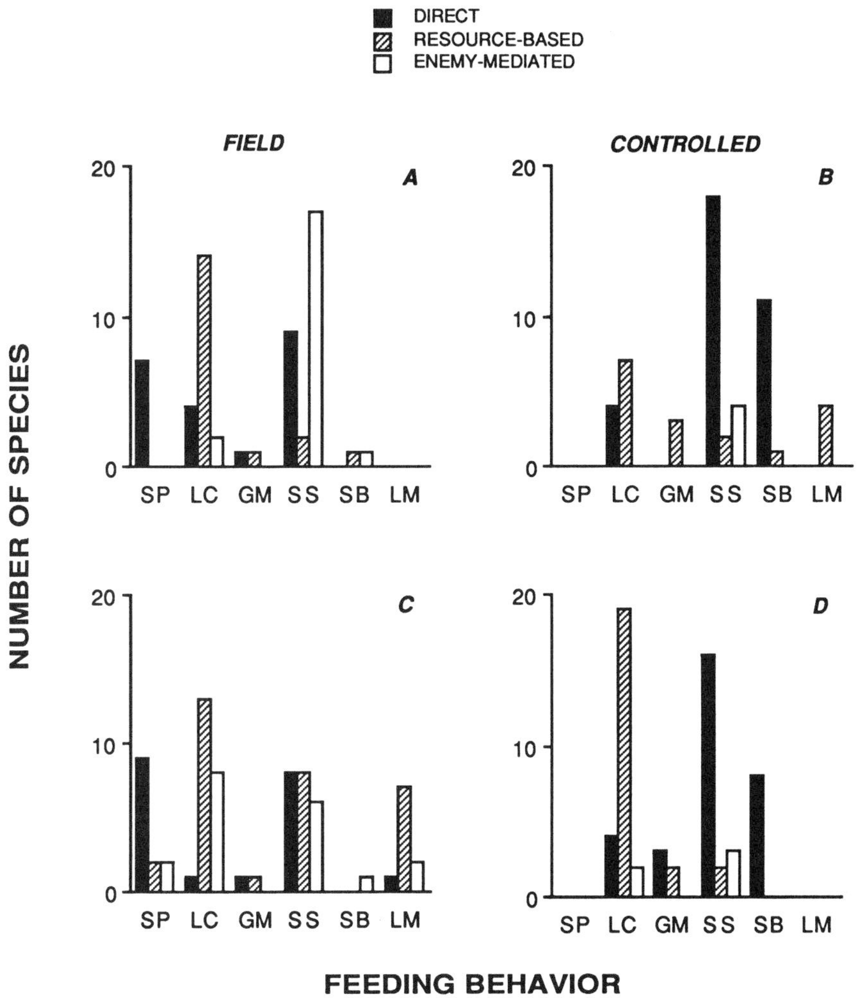

Figure 4.5. The tendency for insect species with different feeding behaviors to be affected by other herbivores (A and B) or to affect other herbivores (C and D) through direct, resource-mediated, or enemy-mediated interactions as recorded in published experiments. A and C depict data from Field experiments; B and D depict data from Controlled experiments. Refer to Table 4.6 for statistics. SP, seed predators; LC, leaf chewers; GM, gall makers; SS, sap suckers; SB, shoot borers.

1979; Bryant 1981; Schultz and Baldwin 1982; Ryan 1983; Pullin 1987; Schultz 1988). The changes may be an active response to tissue damage, as in the increase in alkaloid content of tobacco (Baldwin 1988a), or the cucurbitacin content of cucurbits following herbivory (Tallamy 1985), or simply a passive change resulting from an imbalance in the source/sink relationships in plants following tissue removal (Ericsson et al. 1985; Karban and Myers 1989).

In contrast, induction of new growth following damage does not change the quality of plant tissues per se, but rather alters the timing of production of nutritious, young foliage (Schultz 1988). If the new foliage is produced at a time when most plant foliage is mature, the plants bearing the new growth usually become more attractive relative to surrounding plants (Opler 1974; Rockwood 1974; Auerbach and Simberloff 1984; Potter and Redmond 1989; du Toit et al. 1990). Sometimes, however, new growth may be higher not only in nutrients, but also higher in noxious secondary chemicals, making the plant less attractive than its neighbors (Bryant 1981). If damage disrupts the timing of production of new foliage, herbivores with life cycles closely synchronized with the plant may emerge too soon or too late to obtain suitable foliage (Came 1965; Faeth 1987).

The importance of resource-mediated interactions to herbivorous insects attacking a particular plant depends on the ability of the plant to respond to damage. Many plant species produce new growth following severe defoliation (Heichel and Turner 1976; Stamp 1984; Damman 1989). Not all plants, however, undergo a change in nutritional or chemical content in response to attack (e.g., Webb and Moran 1978; Crawley and Nachapong 1984; Fowler and Lawton 1985; Myers and Williams 1987; Damman 1989), and are therefore not as likely to be involved in resource-mediated interactions. Karban and Myers (1989) suggest that such differences may in part stem from the environment in which a plant grows. Plants adapted to nutrient-poor soils may not change in nutritional quality if they happen to be growing under relatively nutrient-rich conditions. On the other hand, Tuomi et al. (1984) propose that induced changes may persist for longer periods on nutrient-poor soils, increasing the likelihood of resource-mediated interactions. Coley et al. (1985) predicted that plants growing in resource-poor microhabitats will protect their valuable, long-lived foliage with constitutive defenses, while those growing in resource-rich microhabitats can afford to dispose of leaves after a short photosynthetic life and will therefore rely on induced defenses. Thus the environment in which food plants grow may influence the potential for resource-mediated interactions. It remains unclear, however, whether resource-poor or resource-rich environments should favor resource-mediated interactions, though woody plants seem more likely to be involved in resource-mediated interactions than herbaceous plants (Fig. 4.6).

While most studies of induced changes in plant quality have focused on the impact of changes in leaf quality on the herbivore causing the change (e.g., Haukioja 1980; Rhoades 1985; Schultz 1988), it is now becoming clear that many of the responses of plants to herbivores can affect organisms other than the inducer

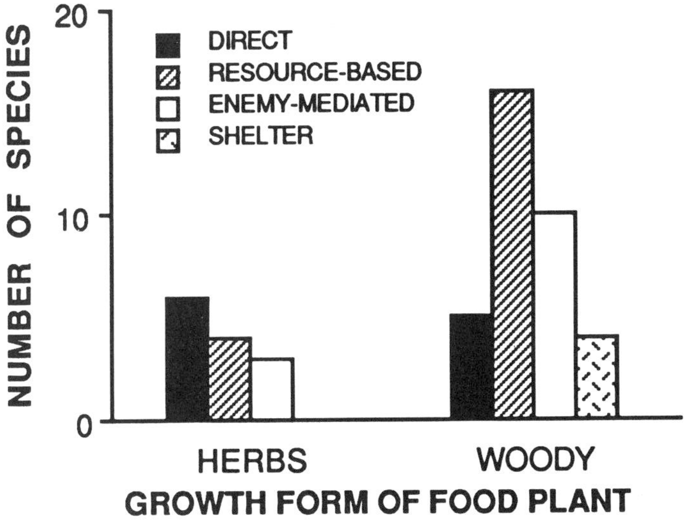

Figure 4.6. The relationship between the frequency of direct, resource-mediated, enemy-mediated, or shelter-mediated interactions and the growth form of the plant bearing the insects as recorded in Field experiments. Refer to Table 4.6 for statistics.

(Karban and Myers 1989). For example, Karban et al. (1987) and Karban (1988) found that damage caused by the two-spotted spider mite induced resistance of cotton to organisms as disparate as the verticillium wilt fungus and beet armyworms. Danell and Huss-Danell (1985) reported that damage to birch by moose browsing influenced the quality of foliage for herbivorous insects. Similarly, some studies involving mechanical clipping of leaves to simulate damage from leaf-chewing caterpillars found that response to the simulated damage closely paralleled that of natural damage (Green and Ryan 1972; Neuvonen et al. 1987). Hartley and Lawton (1987), however, discovered that a number of birch-feeding caterpillars responded differently to changes in leaf quality induced by mechanical clipping and by similar levels of damage caused by a leaf-miner.

The timing and intensity of damage also determine the response that herbivory elicits from a plant. Several studies have shown that severe defoliation early in the growing season more reliably stimulates a second flush of growth than does light defoliation or defoliation late in the growing season (Heichel and Turner 1976; Hodson 1981; Damman 1989). Increasing intensity of damage may also lead to more dramatic induced changes in foliage quality (Haukioja and Neuvonen 1987). The pattern of leaf removal can affect the response of the plants dramatically (Baldwin 1990). Lowman (1982) found that gradual removal of leaves, as would be characteristic of the feeding damage of many leaf-chewing insects (but see Heinrich 1979), stimulated less new growth than did clipping of entire leaves.

Baldwin (1988b) demonstrated, however, that tobacco plants mobilized alkaloids only in response to gradual leaf removal and not to clipping of entire leaves. Sap-sucking, gall-making, and leaf-mining insects are thought to affect plants out of proportion to their abundance because they can interfere with plant metabolism (Pritchard and James 1984; Whittaker 1984; Abrahamson and McCrea 1986; Faeth 1987).

Rhoades (1985) suggested that herbivore foraging behaviors may reflect the patterns of damage most likely to induce a defensive response from the plant. He proposed that "stealthy" herbivores feed solitarily and at low densities to reduce the chances that plant quality will deploy its defenses. "Opportunistic" herbivores, in contrast, attack plants gregariously and inflict severe damage, relying on rapid growth to allow them to complete the feeding stage before the plants have time to respond to their assault. Interactions between herbivores fitting these stereotypes should result in a strong negative effect on the slowly developing stealthy herbivores trapped on plants responding to feeding by opportunists, while the opportunistic herbivores should not suffer any effects. However, the gregarious insects included in my literature review showed no greater tendency to affect other species via resource-mediated interactions than did solitary insects (Fisher's exact test: $P > 0.5$), nor did feeding alone reduce the likelihood that an insect would be affected by resource-mediated interactions (Fisher's exact test: $P > 0.25$). Watt (1965) found that gregarious insects were most likely to outbreak, which could lead them to affect other insects on their host plants on a time scale to which the short-term experiments reviewed here are not sensitive.

One of the distinguishing features of resource-mediated interactions is that the changes in resource quality may be manifested at some time and distance from the original damage. The response time of plants to damage ranges from within hours of the start of herbivory (e.g., Green and Ryan 1972; Carroll and Hoffmann 1980; Berglelson, et al. 1986) to slow enough that the herbivores inducing the response have completed development (e.g., Pritchard and James 1984; Neuvonen et al. 1987; Damman 1989). Persistence of the changes in resource quality ranges from a few weeks (e.g., Röttger and Klingauf 1976) to several years (Neuvonen et al. 1987; Schultz 1988). The spatial scale of resource-mediated changes in quality can encompass a single leaf (e.g., Bergelson et al. 1986), leaves within a few meters of the damage (e.g., Carroll and Hoffmann 1980), or entire plants (e.g., Rhoades 1983).

The impact of damage-induced changes in plant quality on an herbivore will depend largely on the scale of the induced changes relative to the mobility of the caterpillars (Schultz 1983). Larvae capable of moving long distances relative to the area affected by the change (e.g., Dethier 1988, 1989) may be able to avoid induced changes with relative ease, reducing the potential for resource-mediated interactions. Edwards and Wratten (1983, 1985) and Gibberd et al. (1988) proposed that the small amounts of damage suffered by a high proportion of the leaves of many plants (Table 4.1) may stem from movement stimulated by localized changes in leaf quality induced by feeding. On the other hand, leaf miners and gall makers trapped within their feeding site may be vulnerable to even small-scale changes in plant quality (Whitham 1987). This notion led West (1985) to suggest that a major selective force favoring late-season feeding by leaf-mining insects is the lower chance of being trapped in a leaf of which the quality is changing in response to feeding damage by leaf chewers. Data for leaf miners suggested that they are disproportionately prone to resource-mediated interactions (Fig. 4.5). There was no general tendency for internal feeders (e.g., leaf miners and gall makers) to be more prone to resource-mediated interactions than external feeders. In fact, the potentially mobile external leaf chewers appear prone to resource-mediated interactions (Fig. 4.5).

Insects that do not escape induced responses by their mobility can, by virtue of their divergent metabolisms and feeding behaviors, differ in their response to a particular change in tissue quality. Thus Danell and Huss-Danell (1985), Hartley and Lawton (1987), Hunter (1987), and Neuvonen et al. (1988) found that in response to a single type of damage to plant tissue some of the insects responded positively, while others responded negatively or not at all. Faeth (1987) noted that young-leaf specialists foraging on oaks late in the growing season responded positively to young growth, while old-leaf specialists seemed unaffected by the new growth. Late-season sawflies, which typically feed on low quality mature foliage, were much less affected by the response of birch attacked by the geometrid *Epirritia autumnata* in the spring than were a suite of mid-season sawflies, which required higher quality foliage, even though the induced changes persist over several growing seasons. While many species actively move away from damaged foliage, Rose and Lindquist (1982) noted that the pyralid, *Tetralopha expandans,* actively sought out damaged leaves.

Enemy-Mediated Interactions

Enemy-mediated interactions represent the third major interaction between herbivorous insects (Fig. 4.4). Although predators, parasites, and pathogens have long been recognized as major determinants of the success of insect herbivores on plants (Lawton and McNeill 1979; Price et al. 1980), only recently has their potential importance in mediating within-trophic-level interactions been appreciated (Holt 1984; Jeffries and Lawton 1984; Lawton 1986; Price et al. 1986; Holt and Kotler 1987). Two identifiable types of enemy-mediated interactions appeared in my review: 8 of the 20 enemy-mediated interactions involved the impact of mutually shared natural enemies on herbivore populations and were invariably negative interactions (or apparent competition, sensu Holt 1984); the remainder involved the interaction between ant-tended lycaenid caterpillars or homopterans and other herbivores, and included some positive interactions.

Whenever two herbivores share natural enemies, the potential for enemy-mediated interactions exists. Probably most predators, and many parasites and pathogens, of herbivorous insects attack more than one species (Atsatt and O'Dowd 1976; Jervis 1980; Hawkins and Goeden 1984; Cappuccino 1987; Waring and Price 1989; Hawkins et al. 1990). Often parasites require different herbivores as hosts for summer and overwintering generations (Hsaio and Holdoway 1966; Cole 1967; Doutt and Nakata 1973; Perrin 1975; Plakides 1978; Starý 1978). If the shared natural enemies act in either a density-dependent or inversely density-dependent fashion, as is commonly true of both vertebrate and invertebrate enemies (Tullock 1971; Dempster 1983; Lessells 1985; Cappuccino 1987; Stiling 1987; Whitham 1987), then a change in the population size of one herbivore should influence mortality suffered by the other. Holt and Kolter (1987) concluded that enemies that responded in a density-dependent manner will lead to apparent competition, because the high density of one herbivore will increase disproportionately the rate of attack on the other. Natural enemies that respond in an inversely density-dependent manner can generate positive interactions as the herbivores that aggregate will suffer lower risks of mortality (Holt and Kotler 1987).

The different feeding behavior of herbivores can strongly affect the potential for enemy-mediated interactions. Hawkins et al. (1990) noted that internally feeding herbivores were more likely to be attacked by generalist parasites than externally feeding herbivores. Similarly, Askew (1980) found that a higher proportion of tree-feeding than herb-feeding herbivores were attacked by generalist parasites. Thus feeding position and host-plant choice can both influence the potential for enemy-mediated interaction. The insects involved in the studies examined here showed no sign of a higher frequency of enemy-mediated interactions among internal feeders, but did show a tendency for a higher frequency of enemy-mediated interactions among tree feeders (Fig. 4.6).

When natural enemies use damage to the food plant as a cue in locating prey or hosts (Bess 1936; Hassell 1968; Vinson 1974; Greenberg and Gradwohl 1980; Odell and Godwin 1984; Roland et al. 1986; Greenberg 1987), herbivores feeding near conspicuous damage caused by other species will be exposed to higher rates of attack (Heinrich and Collins 1983; Faeth 1990; but see Fowler and MacGarvin 1986; Bergelson and Lawton 1988). Many leaf-chewing herbivores go to great lengths to conceal evidence of their feeding (Heinrich 1979), while others, notably those that are unpalatable, often leave behind conspicuous damage (Heinrich 1979; Heinrich and Collins 1983; Heinrich, Chapter 7). In general, herbivores that feed inconspicuously should suffer lower discovery rates when they avoid plants or plant parts bearing conspicuous damage. Brower (1958) and Mauricio and Bowers (1990) suggested that some herbivores may widely disperse their feeding stages to reduce the likelihood of discovery by enemies.

The mutualisms between ants and honeydew-producing insects play a central role in the second, widespread type of enemy-mediated interaction. The common honeydew producers include various homopterans (Buckley 1987a,b), lycaenid caterpillars (Atsatt 1981a), and at least one cynipid wasp that induces a honeydew-

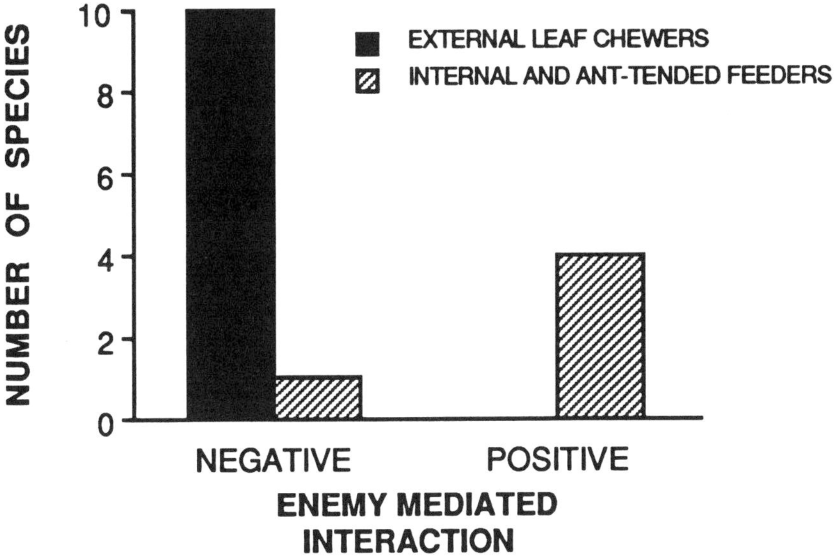

Figure 4.7. The tendency of enemy-mediated interactions involving honeydew producers and ants to affect negatively or positively other externally feeding leaf chewers and internally feeding or ant-tended herbivores.

producing gall (Washburn 1984). By attracting ants, the honeydew producers benefit by obtaining protection from natural enemies, but the ants incidentally remove other insects from the plant as well (Way 1983; Buckley 1987a,b). The impact of the honeydew-producing herbivore on other herbivores sharing its host plant depends both on the aggressiveness of the ant species tending the honeydew producer and the feeding behavior of the other herbivores. Honeydew producers attracting aggressive ants are more likely to become involved in enemy-mediated interactions than are those tended by relatively passive ants. Many ant/honeydew–producer associations are nonspecific, so that the aggressiveness of the attending ants may simply follow from which ant species happens to patrol the plant on which the honeydew producers have settled (Way 1963; Addicott 1979). As a consequence the intensity of enemy-mediated interactions involving honeydew producers may vary from plant to plant within a site.

Internally feeding insects, insects that are specialists, and insects that typically feed on nectary-bearing plants are all less vulnerable to attack by ants than are externally feeding generalists (Risch and Carroll 1982; Heads and Lawton 1985; Heads 1986; Schupp 1986; Buckley 1987b; Bernays and Carneline 1989). Such insects may actually benefit by associating with ant-tended herbivores, because the ants may protect them from natural enemies as effectively as they protect the honeydew producers (Fig. 4.7). For example, Fritz (1983) found that beetle larvae mining in locust leaves near colonies of an ant-tended membracid suffered lower levels of predation and parasitism than beetles mining far from a source of ants. Among some ant-tended lycaenid butterflies, females specifically oviposit on plants already patrolled by ants tending homopterans, thus ensuring their offspring of protection (Lamborn 1913; Atsatt 1981a; Smiley et al. 1988).

Provision of Shelters

Several studies have reported herbivorous insects that seek out and use shelters built by other species (Heie 1972; Akimoto 1981; Hajek and Dahlsten 1986). Carroll et al. (1979) observed late season tortricids and pyralids selectively ovipositing on or near leaf rolls that had been built by leaf-rolling caterpillars attacking oaks earlier in the growing season. Similar behaviors have been reported for the geometrid, *Rheumaptera hastata,* and the plutellid, *Homodaula anisocentra* (Rose and Lindquist 1982). The use by one insect of shelters built by another may prove to be extremely widespread; frequently leaf rolls in the field contain organisms other than the leaf roller (N. Cappuccino, personal communication, on sugar maple; H. Damman, personal observation, on goldenrod).

The role of shelters in mediating interactions between herbivorous insects straddles the boundary between enemy- and resource-mediated interactions. Insects feeding later in the growing season may avoid natural enemies by occupying shelters abandoned by insects feeding early in the season (Damman 1987; Atlegrin 1989), or give them access to a favorable microclimate (Willmer 1980; Berenbaum 1983) or a more easily eaten food (Lewis 1979; Dussourd and Eisner 1987). Halkka et al. (1977), investigating sharing of spittle masses by four spittlebug species, suggested that by sharing spittle masses the bugs lowered the cost of building and maintaining the spittle masses. In addition, newly hatched caterpillars may have difficulty in constructing shelters on their own, and the use of prefabricated shelters may be critical to their successful establishment on a plant (Damman 1987).

The Insects on Pawpaws: A Case Study

Pawpaws (*Asimina* spp., Annonaceae) reach their greatest diversity in the pine forests of Florida and adjacent Georgia and Alabama. In the Ocala National Forest of Florida, where I have studied the herbivore fauna of pawpaws, two freely hybridizing species, *A. speciosa* and *A. obovata,* grew as small, sclerophyllous shrubs on the poor, sandy soils supporting longleaf pine (*Pinus palustris*) savannas. At this site, pawpaws produced new growth synchronously during late March and early April, so that from late mid-May onward virtually all of the leaves were fully mature. The herbivore fauna associated with pawpaws was simple, being numerically dominated by two specialists: swallowtail butterfly caterpillars, *Eurytides marcellus,* which fed singly on young foliage, and caterpillars of the pyralid moth, *Omphalocera munroei,* which fed gregariously as leaf rollers on mature foliage. The remainder of the fauna consisted of a few generalist moths, beetles, homopterans, and grasshoppers all of which concentrated their damage on young leaves.

Direct Interactions

Direct interactions between the herbivores on pawpaws were rarely observed. Although damage to pawpaws was relatively severe, with 52% of the over 4000 leaves produced by 250 regularly censused plants being eaten or destroyed, direct interactions were minimized by temporal and spatial subdivision of the leaves by the herbivores. The majority of herbivores attacked young foliage, but did little damage, whereas *Omphalocera,* which alone consumed 38% of the annual leaf production, concentrated its damage on mature foliage, though it would accept any young foliage that it encountered (Fig. 4.8). In a Controlled experiment in which a typically sized group of 20 newly hatched *Omphalocera* larvae was placed on the same expanding shoot as a single newly hatched *Eurytides* larva, the *Omphalocera* caterpillars performed as well as when reared alone, whereas *Eurytides* ran out of food more rapidly when sharing a plant with *Omphalocera* (Table 4.7). Over the course of the experiment *Omphalocera* invariably won because its leaf-rolling behavior effectively denied *Eurytides* caterpillars access to the young leaves they needed. Under natural conditions, however, I never observed *Omphalocera* and *Eurytides* together on any of the census plants because the preference of ovipositing females for plants bearing mature and young leaves, respectively, effectively kept the two species apart.

Resource-Mediated Interactions

Resource-mediated interactions proved to be very important to the young-leaf feeders attacking pawpaws. Pawpaw foliage did not change in quality following simulated or natural herbivory, but plants frequently produced new growth in response to severe defoliation in early and midsummer (Damman 1989). Of the herbivores attacking pawpaws, only gregarious *Omphalocera* caterpillars inflicted damage sufficiently extensive to stimulate new growth (Damman 1989). Summer generations of *Eurytides* and the leaf-tying tortricids relied on the *Omphalocera*-induced young growth because little other appropriate foliage was available at this time. The various acridid grasshoppers also attacked regrowth following damage, but probably did not depend on this out-of-season young foliage because they could feed equally well on the succulent leaves of a diversity of other herbaceous plants. Cicadellids were unaffected by the damage-induced regrowth because they had just a single spring generation.

Enemy-Mediated Interactions

A number of predators attacked the various Lepidoptera feeding on pawpaws. Several species of vespid wasps and lycosid and salticid spiders represented

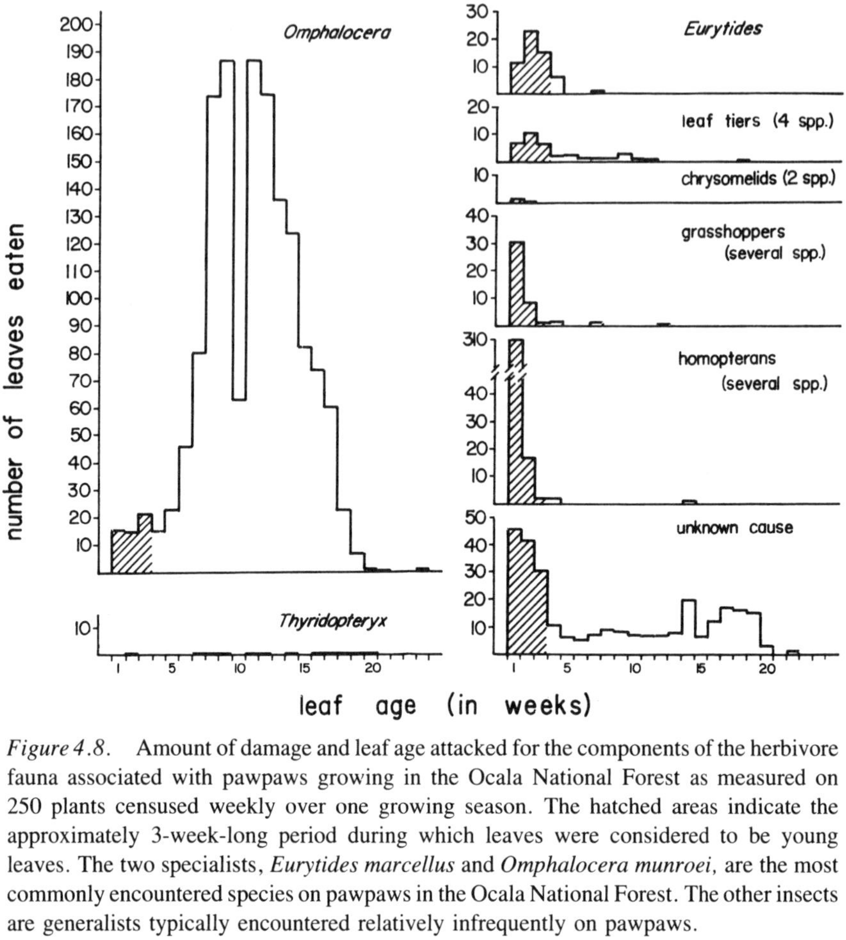

Figure 4.8. Amount of damage and leaf age attacked for the components of the herbivore fauna associated with pawpaws growing in the Ocala National Forest as measured on 250 plants censused weekly over one growing season. The hatched areas indicate the approximately 3-week-long period during which leaves were considered to be young leaves. The two specialists, *Eurytides marcellus* and *Omphalocera munroei,* are the most commonly encountered species on pawpaws in the Ocala National Forest. The other insects are generalists typically encountered relatively infrequently on pawpaws.

Table 4.7. The Potential for Direct Competition between 20 Newly Hatched Omphalocera *Larvae Placed onto Plants with Single Newly Hatched* Eurytides *Larvae on* Asimina *Growing in the Field*

	Days to deplete young growth			
	Alone	With competitor	$F_{1,8}$[a]	P
Eurytides	7.8	3.2	15.5	<0.005
Omphalocera	4.7	4.8	0.120	>0.75

[a]Analysis of variance for paired data.

common predators of *Eurytides, Omphalocera,* and the leaf-tying tortricids. Additional links connected *Eurytides* and *Omphalocera* larvae to herbivores feeding on other plant species. *Eurytides* larvae were parasitized by an ichneumonid wasp, *Trogus pennator,* which also attacked larvae of a second swallowtail species, *Papillo glaucus,* which fed on wild cherries (*Prunus* sp.) growing in the same habitat as the pawpaws. *Omphalocera* larvae were attacked by an unidentified sphecid wasp, that also commonly searched for leaf-rolls formed on nearby oaks.

Provision of Shelters

The leaf rolls constructed by *Omphalocera* larvae were central to their ability to avoid natural enemies (Damman 1987). At hatching, the groups of caterpillars have difficulty shaping the tough, mature foliage into leaf shelters capable of excluding predators. During this critical early stage of establishment on a host plant, the caterpillars showed a strong tendency to seek out naturally sheltered sites on the plant (Damman 1987). Frequently, the most readily available sheltered spots were leaf rolls abandoned by earlier generations of *Omphalocera* and leaves that had been deformed by cicadellid feeding.

Conclusions

In spite of the widespread emphasis on plant–herbivore and herbivore–natural enemy interactions, within-trophic-level interactions play an important role in structuring herbivore communities. Although the generally accepted conclusion that competition is less pervasive in herbivore communities than among organisms at other trophic levels holds true, our failure to recognize the impact of indirect interactions may limit our understanding the distribution of herbivorous insects. Indirect resource-, or enemy-mediated interactions not only comprise the majority of interactions between herbivorous insects, but also differ markedly from direct competition in that they frequently include positive, as well as negative, interactions. Although positive and indirect interactions have been reported in the past (e.g., Connell 1983, Strong et al. 1984), their potential importance in mediating interactions between herbivores has, except for recent reviews by Faeth (1987, 1988), been largely neglected.

Focusing so strongly on direct competition to the exclusion of other within-trophic-level interactions has colored the way we perceive the structure of herbivore communities. For example, Fritz et al. (1987) found that four gall-forming sawfly species were distributed over clones of willow, *Salix lasiolepis,* either independently of one another or in positive associations. They concluded that competition was not important in structuring this herbivore fauna, a contention experimentally supported in a subsequent study (Fritz and Price 1990). Fritz et al. (1987) explained the positive associations by invoking similar host–phenotype preferences by each of the four sawfly species. Maddox and Root (1990), in exploring the pattern in which the many insects that feed on goldenrod distributed themselves over genetically distinct clones, identified four groups of species that were consistently positively associated with one another. Although Maddox and Root briefly mention the possibility that direct competition might influence the pattern they observed, they concluded that the positive associations between species stem primarily from a preference for similar plant quality by the members of each of the four groups. The explanations offered by Fritz et al. (1987) and Maddox and Root (1990) were both based on between-trophic-level interactions and were completely consistent with the distributions that they observed. In both cases, however, the distributions could also be explained by positive resource- or enemy-mediated within-trophic-level interactions. The evolutionary interpretation of community structure based on between-trophic-level interactions implies that change in plant traits and insect responses to plant traits would influence the herbivore distribution pattern most strongly. In contrast, the evolutionary interpretation of community structure based on within-trophic-level interactions implies that change in plant traits would have a relatively minor affect on the herbivore distribution pattern.

Until we gain a better understanding of the mechanisms that underlie resource- and enemy-mediated interactions, these interactions will continue to seem dependent on the idiosyncrasies of the plants, herbivores, and natural enemies involved. As with the responses of insects to other poorly understood changes in habitat quality, the responses seem extremely variable (e.g., Larsson 1989). At the moment, good experimental field studies are sufficiently scarce that only the most pronounced patterns can be detected.

The experiments necessary to allow a confident assessment of the importance of direct and indirect, and positive and negative interactions in structuring herbivore communities will require a more inclusive approach than the studies of pairs of herbivores that now dominate the literature (Tilman 1987). To demonstrate that interactions between herbivores occur, field manipulations of entire plant-centered communities are required so as to capture the effects of plants and natural enemies on the herbivores. The experiments must run for a sufficiently long time to detect interactions between temporally separated herbivores and cover a sufficiently broad herbivore base to encompass the other herbivores sharing natural enemies with the focal species. Finally, the densities of the herbivores and their natural enemies, and the quality of the plants, must be directly relatable to those encountered in a natural setting. With this information in hand, it will become possible to address the relative importance of direct, resource-mediated, and enemy-mediated interactions in a particular system.

Acknowledgments

I thank Nancy Stamp, Stanley Faeth, and an anonymous reviewer, all of whom provided extremely valuable comments on earlier versions of this chapter. Their input was responsible for much of what is good in the current version. Naomi Cappuccino carefully went over a later draft to eliminate some of the many rough edges. The writing of this review was supported by a Natural Sciences and Engineering Research Council operating grant.

Literature Cited

Abrahamson, W. G., and McCrea, K. D. 1986. Nutrient and biomass allocation in *Solidago altissima:* Effects of two stem gallmakers, fertilization, and ramet isolation. Oecologia (Berl.) 68:174–180.

Addicott, J. F. 1978. Niche relationships among species of aphids feeding on fireweed. Can. J. Zool. 56:1837–1841.

Addicott, J. F. 1979. A multispecies aphid-ant association: Density dependence and species-specific effects. Can. J. Zool. 57:558–569.

Akimoto, S. 1981. Gall formation by *Eriosoma* fundatrices and gall parasitism in *Eriosoma yangi* (Homoptera, Pemphigidae). Kontyû, Tokyo 49:426–436.

Antolin, M. F., and Addicott, J. F. 1988. Habitat selection and colony survival of *Macrosiphon valeriani* Clarke (Homoptera: Aphididae). Ann. Entomol. Soc. Am. 81:245–251.

Askew, R. R. 1980. The diversity of insect communities in leaf-mines and plant galls. J. Anim. Ecol. 49:817–829.

Alsatt, P. R. 1981a. Ant-dependent-food plant selection by the mistletoe butterfly *Orgyris amaryllis* (Lycaenidae). Oecologia (Berl.) 48:60–63.

Alsatt, P. R. 1981b. Lycaenid butterflies and ants: Selection for enemy-free space. Am. Nat. 118:72–82.

Alsatt, P. R., and O'Dowd, D. J. 1976. Plant defense guilds. Science 193:24–29.

Atlegrin, O. 1989. Exclusion of birds from bilberry stands: Impact on insect larval density and damage to bilberry. Oecologia (Berl.) 79:136–139.

Auerbach, M., and Simberloff, D. 1984. Responses of leaf miners to atypical leaf production patterns. Ecol. Entomol. 9:361–376.

Baldwin, I. T. 1988a. Short-term damage-induced increases in tobacco alkaloids protect plants. Oecologia (Berl.) 75:367–370.

Baldwin, I. T. 1988b. The alkaloidal responses of wild tobacco to real and simulated herbivory. Oecologia (Berl.) 77:378–381.

Baldwin, I. T. 1990. Herbivory simulations in ecological research. Trends Ecol. Evol. 5:91–93.

Barbosa, P., and Wagner, M. R. 1989. Introduction to Forest and Tree Insects. Academic Press, New York.

Bender, E. A., Case, T. J., and Gilpin, M. E. 1984. Perturbation experiments in community ecology. Ecology 65:1–13.

Berenbaum, M. 1983. Coumarins and caterpillars: A case for coevolution. Evolution 37:163–179.

Bergelson, J., Fowler, S., and Hartley, S. 1986. The effects of foliage damage on casebearing moth larvae, *Coleophora serratella,* feeding on birch. Ecol. Entomol. 11:241–250.

Bergelson, J. M., and Lawton, J. H. 1988. Does foliage damage influence predation on the insect herbivores of birch? Ecology 69:434–445.

Bernays, E. A., and M. L. Carneline. 1989. Generalist caterpillar prey are more palatable than specialists for the generalist predator *Iridomyrmex humilis*. Oecologia (Berl.) 79:427–430.

Bess, H. A. 1936. The biology of *Leschenaultia exul* Townsend, a tachinid parasite of *Malacosoma americana* Fabricius and *Malacosoma disstria* Hubner. Ann. Entomol. Soc. Am. 29:593–613.

Blakely, N. R., and Dingle, H. 1978. Competition: Butterflies eliminate milkweed bugs from Caribbean island. Oecologia (Berl.) 37:133–136.

Brower, L. P. 1958. Bird predation and foodplant specificity in closely related pro-cryptic insects. Am. Nat. 92:183–187.

Bryant, J. P. 1981. Phytochemical deterrence of snowshoe hare browsing by adventitious shoots of four Alaskan trees. Science 213:889–890.

Buckley, R. 1987a. Ant-plant-Homoptera interactions. Adv. Ecol. Res. 16:53–85.

Buckley, R. C. 1987b. Interactions involving plants, Homoptera, and ants. Annu. Rev. Ecol. Syst. 18:111–135.

Bultman, T. L., and S. H. Faeth, 1985. Patterns of intra- and interspecific association in leaf-mining insects on three oak host species. Ecol. Entomol. 10:121–129.

Cappuccino, N. 1987. Comparative population dynamics of two goldenrod aphids: Spatial patterns and temporal constancy. Ecology 68:1634–1646.

Carne, P. B. 1965. Distribution of the eucalypt-defoliating sawfly *Perga affinis affinis* (Hymenoptera). Aust. J. Ecol. 13:593–612.

Carroll, C. R., and Hoffmann, C. A. 1980. Chemical feeding deterrent mobilized in response to insect herbivory and counteradaptation by *Epilachna tridecimnotata*. Science 209:414–416.

Carroll, M. R., Wooster, M. T., Kearby, W. H., and Allen, D. C. 1979. Biological observations on three oak leaftiers: *Psilocorsis querciella, P. reflexella,* and *P. cryptolechiella* in Massachusetts and Missouri. Ann. Entomol. Soc. Am. 72:441–447.

Cole, L. R. 1967. A study of the life-cycles and hosts of some Ichneumonidae attacking pupae of the green oak-leaf roller moth, *Tortrix viridana* (L.) (Lepidoptera: Tortricidae) in England. Trans. Roy. Entomol. Soc. (London) 119:267–281.

Coley, P. D., Bryant, J. P., and Chapin, F. S. 1985. Resource availability and plant antiherbivore defense. Science 230:895–899.

Compton, S. G., and Robertson, H. G. 1988. Complex interactions between mutualisms: Ants tending homopterans protect fig seeds and pollinators. Ecology 69:1302–1305.

Connell, J. H., 1983. On the prevalence and relative importance of interspecific competition: evidence from field experiments. Am. Nat. 122:661–696.

Crawley, M. J., and Nachapong, M. 1984. Facultative defences and specialist herbivores?

Cinnabar moth (*Tyria jacobaeae*) on the regrowth foliage of ragwort (*Senecio jacobaea*). Ecol. Entomol. 9:389–393.

Crawley, M. J., and Pattrasudhi, R. 1988. Interspecific competition between insect herbivores: Asymmetric competition between cinnabar moth and the ragwort seed-head fly. Ecol. Entomol. 13:243–249.

Cushman, J. H., and Addicott, J. F. 1989. Intra- and interspecific competition of mutualists: Ants as a limited and limiting resource for aphids. Oecologia (Berl.) 79:315–321.

Damman, H. 1987. Leaf quality and enemy avoidance by the larvae of a pyralid moth. Ecology 68:88–97.

Damman, H. 1989. Facilitative interactions between two lepidopteran herbivores of *Asimina*. Oecologia (Berl.) 78:214–219.

Danell, K., and Huss-Danell, K. 1985. Feeding by insects and hares on birches earlier affected by moose browsing. Oikos 41:75–81.

DeBach, P., Hendrickson, R. M., and Rose, M. 1978. Competitive displacement: Extinction of the yellow scale, *Aonidiella citrina* (Coq.) (Homoptera: Diaspididae), by its ecological homologue, the California red scale, *Aonidiella aurantii* (Mosk.) in southern California. Hilgardia 46:1–35.

Dempster, J. P. 1983. The natural control of populations of butterflies and moths. Biol. Rev. 58:461–481.

Denno, R. F., Raupp, M. J., and Tallamy, D. W. 1982. Organization of a guild of sap-feeding insects: equilibrium vs. nonequilibrium coexistence, pp. 151–181. In R. F. Denno and H. Dingle (eds.), Life History Patterns: Habitat and Geographic Variation. Springer-Verlag, New York.

Dethier, V. G. 1988. Induction and aversion-learning in polyphagous arctiid larvae (Lepidoptera) in an ecological setting. Can. Entomol. 120:125–131.

Dethier, V. G. 1989. Patterns of locomotion of polyphagous arctiid caterpillars in relation to foraging. Ecol. Entomol. 14:375–386.

Dixon, A. F. G., and Barlow, N. D. 1979. Population regulation in the lime aphid. Zool. J. Linn. Soc. 67:225–237.

Doutt, R. L., and Nakata, J. 1973. The *Rubus* leafhopper and its egg parasitoid: An endemic biotic system useful in grape-pest management. Environ. Entomol. 2:381–386.

Dussourd, D. E., and Eisner, T. 1987. Vein-cutting behavior: insect counterploy to the latex defense of plants. Science 237:898–901.

du Toit, J. T., Bryant, J. P., and Frisby, K. 1990. Regrowth and palatability of *Acacia* shoots following pruning by African savanna browsers. Ecology 71:149–154.

Edson, J. L. 1985. The influences of predation and resource subdivision on the coexistence of goldenrod aphids. Ecology 66:1736–1743.

Edwards, P. J., and Wratten, S. D. 1983. Wound induced defences in plants and their consequences for patterns of insect grazing. Oecologia (Berl.) 59:88–93.

Edwards, P. J., and Wratten, S. D. 1985. Induced plant defences against insect grazing: fact or artifact? Oikos 44:70–74.

Ehrlich, P. R., and Raven, P. H. 1964. Butterflies and plants: A study in coevolution. Evolution 18:586–608.

Ericsson, A., Hellqvist, C., Långström, B., Larsson, S., and Tenow, O. 1985. Effects of growth of simulated and induced shoot pruning by *Tomicus piniperda* as related to carbohydrate and nitrogen dynamics in Scots pine. J. Appl. Ecol. 22:105–124.

Evans, E. W. 1989. Interspecific interactions among phytophagous insects of tallgrass prairie: An experimental test. Ecology 70:435–444.

Faeth, S. H. 1985. Host leaf selection by leaf miners: interactions among three trophic levels. Ecology 66:870–875.

Faeth, S. H. 1986. Indirect interactions between temporally separated herbivores mediated by the host plant. Ecology 67:479–494.

Faeth, S. H. 1987. Community structure and folivorous insect outbreaks: The role of vertical and horizontal interactions, pp. 135–171. In P. Barbosa and J. C. Schultz (eds.), Insect Outbreaks. Academic Press, New York.

Faeth, S. H. 1988. Plant-mediated interactions between seasonal herbivores: Enough for evolution or coevolution?, pp. 391–414. In K. C. Spencer (ed.), Chemical Mediation of Coevolution. Academic Press, New York.

Faeth, S. H. 1990. Structural damage to oak leaves alters natural enemy attack on a leafminer. Entomol. Exp. Appl. 57:57–63.

Feeny, P., Blau, W. S., and Kareiva, P. M. 1985. Larval growth and survivorship of the black swallowtail butterfly in central New York. Ecol. Monogr. 55:167–187.

Fellows, D. P., and Heed, W. B. 1972. Factors affecting host plant selection in desert-adapted cactiphilic *Drosophila*. Ecology 53: 850–858.

Ferson, S., Downey, P., Klerks, P., Weissburg, M., Kroot, I., Stewart, S., Jacquez, G., Ssemakula, J., Malenky, J., and Anderson, K. 1986. Competing reviews, or why do Connell and Schoener disagree? Am. Nat. 127:571–576.

Finch, S. and Jones, T. H. 1989. An analysis of the deterrent effect of aphids on cabbage root fly (*Delia radicum*) egg laying. Ecol. Entomol. 14:387–391.

Fitt, G. P. 1984. Oviposition behaviour of two tephritid fruit flies, *Dacus tryoni* and *Dacus jarvisi,* as influenced by the presence of larvae in the host plant. Oecologia (Berl.) 62:37–46.

Fowler, S. V., and MacGarvin, M. 1986. The effects of leaf damage on the performance of insect herbivores on birch, *Betula pubescens*. J. Anim. Ecol. 55:565–573.

Fowler, S. V., and Lawton, J. H. 1985. Rapidly induced defenses and talking trees: The Devil's advocate position. Am. Nat. 126:181–195.

Fox, L. R., and P. A. Morrow. 1983. Estimates of damage by herbivorous insect on *Eucalyptus* trees. Aust. J. Ecol. 8:139–147.

Fritz, R. S. 1983. Ant protection of a host plant's defoliator: consequence of an ant-membracid mutualism. Ecology 64:789–797.

Fritz, R. S., and Price, P. W. 1990. A field test of interspecific competition on oviposition of gall-forming sawflies on willow. Ecology 71:99–106.

Fritz, R. S., Gaud, W. S., Sacchi, C. F., and Price, P. W. 1987. Patterns of intra- and interspecific association of gall-forming sawflies in relation to shoot size on their willow host plant. Oecologia (Berl.) 73:159–169.

Fritz, R. S., Sacchi, C. F., and Price, P. W. 1986. Competition versus host plant phenotype in species composition: Willow sawflies. Ecology 67:1608–1618.

Gange, A. C., and Brown, V. K. 1989. Effects of root herbivory by an insect on a foliar-feeding species mediated through changes in the host plant. Oecologia (Berl.) 56:38–42.

Gardner, T. R. 1938. Influence of feeding habits of *Tiphia vernalis* on the parasitization of the Japanese beetle. J. Econ. Entomol. 31:204–207.

Gibberd, R., Edwards, P. J., and Wratten, S. D. 1988. Wound-induced changes in the acceptability of tree foliage to Lepidoptera: within-leaf effects. Oikos 51:43–47.

Gibson, C., and Visser, M. 1982. Interspecific competition between two field populations of grass-feeding bugs. Ecol. Entomol. 7:61–67.

Gibson, C. W. D. 1976. The importance of foodplants for the distribution and abundance of some Stenodemini (Heteroptera: Miridae) of limestone grasslands. Oecologia (Berl.) 25:55–76.

Gibson, C. W. D. 1980. Niche use patterns among some Stenodemini (Heteroptera: Miridae) of limestone grassland, and an investigation of the possibility of interspecific competition between *Notostira elongata* Geoffroy and *Megaloceraea recticomis* Geoffroy. Oecologia (Berl.) 47:352–364.

Green, T. R., and Ryan, C. A. 1972. Wound-induced protease inhibitor in plant leaves: A possible defense mechanism against insects. Science 175:776–777.

Greenberg, R. 1987. Development of dead leaf foraging in a tropical migrant warbler. Ecology 68:130–141.

Greenberg, R., and J. Gradwohl. 1980. Leaf surface specializations of birds and *Arthropods* in a Panamanian forest. Oecologia (Berl.) 46:115–124.

Hairston, N. G., Smith, F. E., and Slobodkin, L. B. 1960. Community structure, population control, and competition. Am. Nat. 44:421–425.

Hajek, A. E., and D. L. Dahlsten. 1986. Coexistence of three species of leaf-feeding aphids (Homoptera) on *Betula pendula*. Oecologia (Berl.) 68:380–386.

Halkka, O., Raatikainen, M., Halkka, L., and Raatikainen, T. 1977. Coexistence of four species of spittle-producing Homoptera. Ann. Zool. Fennici 14:228–231.

Hanhimäki, S. 1989. Induced resistance in mountain birch: defence against leaf-chewing insect guild and herbivore competition. Oecologia (Berl.) 81:242–248.

Harrison, J. O. 1964. Factors affecting the abundance of Lepidoptera in banana plantations. Ecology 45:508–519.

Harrison, S., and Karban, R. 1986. Effects of an early-season folivorous moth on the success of a later-season species, mediated by a change in the quality of the shared host, *Lupinus arboreus* Sims. Oecologia (Berl.) 69:354–359.

Hartley, S. E. 1988. The inhibition of phenolic biosynthesis in damaged and undamaged birch foliage and its effect on insect herbivores. Oecologia (Berl.) 76:65–70.

Hartley, S. E., and Lawton, J. H. 1987. Effects of different types of damage on the chemistry of birch foliage, and the responses of birch feeding insects. Oecologia (Berl.) 74:432–437.

Hassell, M. P. 1968. The behavioural response of a tachinid fly (*Cyzenis albicans* Fall.) to its host, the winter moth (*Operophtera brumata* (L.)). J. Anim. Ecol. 37:627–639.

Hastings, A. 1987. Can competition be detected using species co-occurrence data? Ecology 68:117–123.

Haukioja, E. 1980. On the role of plant defences in the fluctuation of herbivore populations. Oikos 35:202–213.

Haukioja, E., and Neuvonen, S. 1987. Insect population dynamics and induction of plant resistance: the testing of hypotheses, pp. 411–432. In P. Barbosa and J. C. Schultz (eds.), Insect Outbreaks. Academic Press, New York.

Haukioja, E., and Niemelä, P. 1977. Retarded growth of a geometrid larva after mechanical damage to leaves of its host tree. Ann. Zool. Fennici 14:48–52.

Haukioja, E., and Niemelä, P. 1979. Birch leaves as a resource for herbivores: Seasonal occurrence of increased resistance in foliage after mechanical damage of adjacent leaves. Oecologia (Berl.) 39:151–159.

Hawkins, B. A. 1988. Foliar damage, parasitoids and indirect competition: a test using herbivores of birch. Ecol. Entomol. 13:301–308.

Hawkins, B. A., and R. D. Goeden. 1984. Organization of a parasitoid community associated with a complex of galls on *Atriplex* spp. in southern California. Ecol. Entomol. 9:271–292.

Hawkins, B. A., Askew, R. R., and Shaw, M. R. 1990. Influences of host feeding-niche and foodplant type on generalist and specialist parasitoids. Ecol. Entomol. 15:275–280.

Heads, P. A. 1986. Bracken, ants and extrafloral nectaries. IV. Do wood ants (*Formica lugubris*) protect plants against insect herbivores? J. Anim. Ecol. 55:795–809.

Heads, P. A., and Lawton, J. H. 1985. Bracken, ants and extrafloral nectaries. III. How insect herbivores avoid ant predation. Ecol. Entomol. 10:29–42.

Heichel, G. H., and Turner, N. C. 1976. Phenology and leaf growth of defoliated hardwood trees, pp. 31–40. In Anderson, J. F., and Kaya, H. K. (eds.), Perspectives in Forest Entomology. Academic Press, New York.

Heie, O. E. 1972. Bladlus på birk i Danmark (Hom., Aphidoidea). Entomol. Med. 40:81–105.

Heinrich, B. 1979. Foraging strategies of caterpillars. Oecologia (Berl.) 42:325–337.

Heinrich, B., and Collins, S. L. 1983. Caterpillar leaf damage and the game of hide-and-seek with birds. Ecology 64:592–602.

Hodson, A. C. 1981. The response of aspen (*Populus tremuloides*) to artificial defoliation. Great Lakes Entomol. 14:167–169.

Holt, R. D. 1984. Spatial heterogeneity, indirect interactions, and the coexistence of prey species. Am. Nat. 124:377–406.

Holt, R. D., and Kotler, B. P. 1987. Short-term apparent competition. Am. Nat. 130:412–430.

Horvitz, C. C., and Schemske, D. W. 1984. Effects of ants and an ant-tended herbivore on seed production of a neotropical herb. Ecology 65:1369–1378.

Hsaio, T. H. and Holdaway, F. G. 1966. Seasonal history and host synchronization of *Lydella grisescens* (Diptera: Tachinidae) in Minnesota. Ann. Entomol. Soc. Am. 59:125–133.

Hunter, M. D. 1987. Opposing effects of spring defoliation on late season oak caterpillars. Ecol. Entomol. 12:373–382.

Hunter, M. D. 1990. Differential susceptibility to variable plant phenology and its role in competition between two insect herbivores on oak. Ecol. Entomol. 15:401–408.

Hunter, M. D., and Willmer, P. G. 1989. The potential for interspecific competition between two abundant defoliators on oak: leaf damage and habitat quality. Ecol. Entomol. 14:267–277.

Jeffries, M. J., and Lawton, J. H. 1984. Enemy free space and the structure of ecological communities. Biol. J. Linn. Soc. 23:269–286.

Jervis, M. A. 1980. Ecological studies on the parasite complex associated with typhlocybine leafhoppers (Homoptera, Cicadellidae). Ecol. Entomol. 5:123–126.

Jones, T. H., Cole, R. A., and Finch, S. 1988. A cabbage root fly oviposition deterrent in the frass of garden pebble moth caterpillars. Entomol. Exp. App. 49:277–282.

Jutsum, A. R., Cherrett, J. M., and Fisher, M. 1981. Interactions between the fauna of citrus trees in Trinidad and the ants *Atta cephalotes* and *Azteca* sp. J. Appl. Ecol. 18:187–195.

Karban, R. 1986. Interspecific competition between folivorous insects on *Erigeron glaucus*. Ecology 67:1063–1072.

Karban, R. 1988. Resistance to beet armyworms (*Spodoptera exigua*) induced by exposure to spider mites (*Tetranychus turkestani*) in cotton. Am. Midl. Nat. 119:72–82.

Karban, R. 1989. Community organization of *Erigeron glaucus* folivores: effects of competition, predation, and host plant. Ecology 70:1028–1039.

Karban, R., and Myers, J. H. 1989. Induced plant responses to herbivory. Annu. Rev. Ecol. Syst. 20:331–348.

Karban, R., Adamchak, R., and Schnathorst, W. C. 1987. Induced resistance and interspecific competition between spider mites and a vascular wilt fungus. Science 2356:678–680.

Kareiva, P. 1982. Exclusion experiments and the competitive release of insects feeding on collards. Ecology 63:696–704.

Keiser, I., Kobayashi, R. M., Miyashita, D. H., Harris, E. J., Schneider, E. L., and Chambers, D. L. 1974. Suppression of Mediterranean fruit flies by oriental fruit flies in mixed infestations in guava. J. Econ. Entomol. 67:355–360.

Kidd, N. A. C., Lewis, G. B., and Howell, C. A. 1985. An association between two species of pine aphid, *Schizolachnus pineti* and *Eulachnus agilis*. Ecol. Entomol. 10:427–432.

Klijnstra, J. W. 1985. Interspecific egg load assessment of host plants by *Pieris rapae* butterflies. Entomol. Exp. Appl. 38:227–231.

Laine, K. J., and Niemelä, P. 1980. The influence of ants on the survival of mountain birches during an *Oporinia autumnata* (Lep., Geometridae) outbreak. Oecologia (Berl.) 47:39–42.

Lamborn, W. A. 1913. On the relationship between certain West African insects, especially ants, Lycaenidae, and Homoptera. Trans. Roy. Entomol. Soc. (London) 61:436–512.

Landsberg, J. 1990. Dieback of rural eucalypts: response of foliar dietary quality and herbivory to defoliation. Aust. J. Ecol. 14:88–96.

Larsson, S. 1989. Stressful times for plant stress—insect performance hypothesis. Oikos 56:277–283.

Lawton, J. H. 1982. Vacant niches and unsaturated communities: A comparison of bracken herbivores at sites on two continents. J. Anim. Ecol. 51:573–595.

Lawton, J. H. 1984. Non-competitive populations, non-convergent communities, and vacant niches: the herbivores of bracken, pp. 67–100. In D. R. Strong, D. Simberloff, L. G. Abele, and A. B. Thistle (eds.), Ecological Communities. Princeton University Press, Princeton, N.J.

Lawton, J. H. 1986. The effect of parasitoids on phytophagous insect communities, pp. 265–287. In J. Waage and D. Greathead (eds.), Insect Parasitoids. Academic Press, New York.

Lawton, J. H., and McNeill, S. 1979. Between the Devil and the deep blue sea: On the problem of being an herbivore, pp. 223–244. In R. M. Anderson, B. D. Turner, and L. R. Taylor (eds.), Population Dynamics. Blackwell Scientific, Oxford.

Lawton, J. H., and Strong, D. R. 1981. Community patterns and competition in folivorous insects. Am. Nat. 118:317–338.

Lessells, C. M. 1985. Parasitoid foraging: should parasitism be density dependent? J. Anim. Ecol. 54:27–41.

Lewis, A. C. 1979. Feeding preference for diseased and wilted sunflower in the grasshopper *Melanoplus differentialis*. Entomol. Exp. Appl. 26:202–207.

Lowman, M. D. 1982. Effects of different rates and methods of leaf area removal on rain forest seedlings of coachwood (*Ceratopetalum apetalum*). Aust. J. Bot. 30:477–483.

Maddox, G. D., and Root, R. B. 1990. Structure of the encounter between goldenrod (*Solidago altissima*) and its diverse insect fauna. Ecology 71:2115–2124.

Mattson, W. J. 1986. Competition of food between two principal cone insects of red pine, *Pinus resinosa*. Environ. Entomol. 15:88–92.

Mattson, W. J., Haack, R. A., Lawrence, R. K., and Herms, D. A. 1989. Do balsam twig aphids (Homoptera: Aphididae) lower tree susceptibility to spruce budworm? Can. Entomol. 121:93–103.

Mauricio, R., and Bowers, M. D. 1990. Do caterpillars disperse their damage?: Larval foraging behavior of two specialist herbivore, *Euphydryas phaeton* (Nymphalidae) and *Pieris rapae* (Pieridae). Ecol. Entomol. 15:153–161.

May, R. M. 1984. An overview: real and apparent patterns in community structure, pp.

3–16. In D. R. Strong, D. Simberloff, L. G. Abele, and A.B. Thistle (eds.), Ecological Communities. Princeton University Press, Princeton, NJ.

McClure, M. S. 1980. Competition between exotic species: Scale insects on hemlock. Ecology 61:1391–1401.

McClure, M. S. 1981. Effects of voltinism, interspecific competition and parasitism on the population dynamics of the hemlock scales, *Florinia extema* and *Tsugaspidiotus tsugae* (Homoptera: Diaspididae). Ecol. Entomol. 6:47–54.

McClure, M. S., and Price, P. W. 1975. Competition among sympatric *Erythroneura* leafhoppers (Homoptera: Cicadellidae) on American sycamore. Ecology 56:1388–1397.

McLain, D. K. 1981. Resource partitioning by three species of hemipteran herbivores on the basis of host plant density. Oecologia (Berl.) 48:414–417.

Messina, F. J. 1981. Plant protection as a consequence of an ant-membracid mutualism: Interactions on goldenrod (*Solidago* sp.). Ecology 62:1433–1440.

Messina, F. J., and J. A. A. Renwick, 1985. Ability of ovipositing seed beetles to discriminate between seeds with differing egg levels. Ecol. Entomol. 10:225–230.

Mopper, S., Whitham, T. G., and Price, P. W. 1990. Plant phenotype and interspecific competition between insects determine sawfly performance and density. Ecology 71:2135–2144.

Moran, N., and Whitham, T. G. 1990. Interspecific competition between root-feeding and leaf-galling aphids mediated by host-plant resistance. Ecology 71:1050–1058.

Myers, J. H., and Williams, K. S. 1987. Lack of short or long term inducible defenses in the red alder-western tent caterpillar system. Oikos 48:73–78.

Neuvonen, S., Haukioja, E., and Molarius, A. 1987. Delayed inducible resistance against a leaf-chewing insect in four deciduous tree species. Oecologia (Berl.) 74:363–369.

Neuvonen, S., Hanhimäki, S., Suomela, J., and Haukioja E. 1988. Early season damage to birch foliage affects the performance of a late season herbivore. Z. Ang. Entomol. 105:182–189.

Nickerson, J. C., Ralph Kay, C. A., and Buschman, L. L. 1977. The presence of *Spissistilus festinus* as a factor affecting egg predation by ants in soybeans. Florida Entomol. 60:193–199.

Nielsen, B. O. 1978. Above ground food resource and herbivory in a beech forest ecosystem. Oikos 31:273–279.

Niemelä, P. Tuomi, J., Mannila, R., and Ojula, P. 1984. The effect of previous damage on the quality of Scots pine foliage as food for diprionid sawflies. Z. Ang. Entomol. 98:33–43.

Odell, T. M. and P. A. Godwin. 1984. Host selection by *Blepharipa pratensis* (Meigen), a tachinid parasite of the gypsy moth, *Lymantria dispar* L. J. Chem. Ecol. 10:311–320.

Opler, P. A. 1974. Biology, ecology, and host specificity of Microlepidoptera associated with *Quercus agrifolia* (Fagaceae). Univ. Calif. Publ. Entomol. 75:1–83.

Perrin, R. M. 1975. The role of perennial stinging nettle, *Urtica dioica,* as a reservoir of beneficial natural enemies. J. Appl. Biol. 81:289–297.

Pilson, D., and Rausher, M. D. 1988. Clutch size adjustment by a swallowtail butterfly. Nature (London) 333:361–363.

Plakides, J. D. 1978. *Epiblema scudderiana* (Clemens) (Lepidoptera: Olethreutidae), a winter host reservoir for parasitic insects in southwestern Pennsylvania. J. N.Y. Entomol. Soc. 86:220–223.

Potter, D. A., and Redmond, C. T. 1989. Early spring defoliation, secondary leaf flush, and leafminer outbreaks on American holly. Oecologia (Berl.) 81:192–197.

Price, P. W., Bouton, C. E., Gross, P., McPheron, B. A., Thompson, J. N., and Weis, A. E. 1980. Interactions among three trophic levels: Influence of plants on interactions between insect herbivores and natural enemies. Annu. Rev. Ecol. Syst. 11:41–65.

Price, P. W., Westoby, M., Rice, B., Atsatt, P. A., Fritz, R. S., Thompson, J. N., and Mobley, K. 1986. Parasitic mediation in ecological interactions. Annu. Rev. Ecol. Syst. 17:487–505.

Pritchard, I. M., and James R. 1984. Leaf mines: their effect on leaf longevity. Oecologia (Berl.) 64:132–139.

Prokopy, R. J. 1981. Epideictic pheromones influencing spacing patterns of phytophagous insects, pp. 181–213. In D. A. Nordlund, R. Jones, and W. Lewis (eds.), Semiochemicals: Their Role in Pest Control. Wiley, New York.

Pullin, A. S. 1987. Changes in leaf quality following clipping, and regrowth of *Urtica dioica,* and consequences for a specialist insect herbivore, *Aglais urticae*. Oikos 49:39–45.

Rathcke, B. J. 1976. Competition and coexistence within a guild of herbivorous insects. Ecology 57:76–87.

Raupp, M. J., and Denno, R. F. 1984. The suitability of damaged willow leaves as food for the leaf beetle, *Plagiodera versicolora*. Ecol. Entomol. 9:443–448.

Raupp, M. J., Milan, F. R., Barbosa, P., and Leonhardt, B. A. 1986. Methylcyclopentanoid monoterpenes mediate interactions among insect herbivores. Science 232:1408–1410.

Risch, S. J., and Carroll, C. R. 1982. Effect of a keystone predaceous ant, *Solenopsis geminata,* on arthropods in a tropical agroecosystem. Ecology 63:1979–1983.

Rhoades, D. F. 1983. Responses of alder and willow to attack by tent caterpillars and webworms: evidence for pheromonal sensitivity of willows, pp. 55–68. In P. A. Hedin (ed.), Plant Resistance to Insects. American Chemical Society, Washington, D.C.

Rhoades, D. F. 1985. Offensive—defensive interactions between herbivores and plants: Their relevance in herbivore population dynamics and ecological theory. Am. Nat. 125:205–238.

Rockwood, L. L. 1974. Seasonal changes in the susceptibility of *Crescentia alata* to the flea beetle, *Oedionychus* sp. Ecology 55:142–148.

Roland, J., Hannon, S. J., and Smith, M. A. 1986. Foraging pattern of pine siskins and its influence on winter moth survival in an apple orchard. Oecologia (Berl.) 69:47–52.

Room, P. M. 1972. The fauna of the mistletoe *Tapinanthus bangvensis* (Engl. & K.

Krause) growing on cocoa in Ghana: Relationships between fauna and mistletoe. J. Anim. Ecol. 41:611–621.

Rose, A. H., and Lindquist, O. H. 1982. Insects of eastern hardwood trees. Can. For. Serv., For. Tech. Rep. 29:1–304.

Rothschild, G. H. L. 1971. The biology and ecology of rice-stem borers in Sarawak (Malaysian Borneo). J. Appl. Ecol. 8:287–322.

Rothschild, M., and Schoonhoven, L. M. 1977. Assessment of egg load by *Pieris brassicae* (Lepidoptera: Pieridae). Nature (London) 216:352–355.

Röttger, U. and Klingauf, F. 1976. Änderung im Stoffwechsel von Zuckerrübenblättern durch gefall mit *Pegomya betae* Curt. (Muscoidea: Anthomyidae). Z. Ang. Entomol. 98:33–43.

Ryan, C. A. 1983. Insect-induced chemical signals regulating natural plant protection responses, pp. 43–60. In R. F. Denno, and M. S. McClure (eds.), Variable Plants and Herbivores in Natural and Managed Systems. Academic Press, New York.

Schoener, T. W. 1983. Field experiments on interspecific competition. Am. Nat. 122:240–285.

Schultz, J. C. 1983. Habitat selection and foraging tactics of caterpillars in heterogeneous trees, pp. 61–90. In R. F. Denno and M. S. McClure (eds.), Variable Plants and Herbivores in Natural and Managed Systems. Academic Press, New York.

Schultz, J. C. 1988. Plant responses induced by herbivores. Trends Ecol. Evol. 3:45–49.

Schultz, J. C., and Baldwin, I. T. 1982. Oak leaf quality declines in response to defoliation by gypsy moth larvae. Science 217:149–151.

Schupp, E. W. 1986. *Azteca* protection of *Cecropia:* ant occupation benefits juvenile trees. Oecologia (Berl.) 70:379–385.

Shapiro, A. 1981. The pierid red egg syndrome. Am. Nat. 117:276–294.

Shearer, J. W. 1976. Effect of aggregations of aphids (*Periphyllus* spp.) on their size. Entomol. Exp. Appl. 20:179–182.

Sih, A., Crowley, P., McPeek, M., Petrunka, J., and Strohmeier, K. 1985. Predation, competition, and prey communities: a review of field experiments. Annu. Rev. Ecol. Syst. 16:269–311.

Singer, M. C., and Mandracchia, J. 1982. On the failure of two butterfly species to respond to the presence of conspecific eggs prior to oviposition. Ecol. Entomol. 7:327–330.

Skinner, G. J., and Whittaker, J. B. 1981. An experimental investigation of inter-relationships between wood-ant (*Formica rufa*) and some tree-canopy herbivores. J. Anim. Ecol. 50:313–326.

Smiley, J. T., Atsatt, P. R., and Pierce, N. E. 1988. Local distribution of the lycaenid butterfly, *Jalmenus evagorus*. In response to host ants and plants. Oecologia (Berl.) 76:416–422.

Sokal, R. R., and Rohlf, F. J. 1981. Biometry, 2nd ed. Freeman, New York.

Southwood, T. R. E., Brown, V. K., and Reader, P. M. 1986. Leaf palatability, life expectancy and herbivore damage. Oecologia (Berl.) 70:544–518.

Springett, B. P. 1978. On the ecological role of insects in Australian eucalypt forests. Aust. J. Ecol. 3:129–139.

Stamp, N. E. 1984. Effect of defoliation by checkerspot caterpillars (*Euphydryas phaeton*) and sawfly larvae (*Macrophya nigra* and *Tenthredo nigra*) on their host plants (*Chelone* spp.). Oecologia Berl.), 63:275–280.

Stamp, N. E. 1987. Availability of resources for predators of *Chelone* seeds and their parasitoids. Am. Midl. Nat. 117:265–279.

Starý, P. 1978. Seasonal relations between lucerne, red clover, wheat and barley agro-ecosystems through the aphids and parasitoids (Homoptera, Aphididae; Hymenoptera: Aphidiidae). Acta Entomol. Bohemoslov. 75:296–311.

Stiling, P. D. 1980. Competition and coexistence among *Eupteryx* leafhoppers (Hemiptera: Cicadellidae) occurring on stinging nettles. J. Anim. Ecol. 49:793–805.

Stiling, P. D. 1987. The frequency of density dependence on insect host-parasitoid systems. Ecology 68:844–856.

Stiling, P. D., and Strong, D. R. 1983. Weak competition among *Spartina* stem borers, by means of murder. Ecology 64:770–778.

Stiling, P. D., and Strong, D. R. 1984. Experimental density manipulation of stem boring insects: some evidence for interspecific competition. Ecology 65:1683–1685.

Stiling, P. D., Simberloff, D., and Anderson, L. C. 1987. Non-random distribution patterns of leaf miners on oak trees. Oecologia (Berl.) 74:102–105.

Strong, D. R. 1982a. Harmonious coexistence of hispine beetles on *Heliconia* in experimental and natural communities. Ecology 63:1039–1049.

Strong, D. R. 1982b. Potential interspecific competition and host specificity: hispine beetles on *Heliconia*. Ecol. Entomol. 7:217–220.

Strong, D. R. 1984. Exorcising the ghost of competition past: phytophagous insects, pp. 28–41. In D. R. Strong, D. Simberloff, L. G. Abele, and A. B. Thistle (eds.), Ecological Communities. Princeton University Press, Princeton, NJ.

Strong, D. R., Lawton, J. H., and Southwood, T. R. E. 1984. Insects on Plants. Harvard University Press, Cambridge, MA.

Tallamy, D. W. 1985. Squash beetle feeding behavior: An adaptation against induced cucurbit defenses. Ecology 66:1574–1579.

Tilman, D. 1987. The importance of the mechanisms of interspecific competition. Am. Nat. 129:769–774.

Tullock, G. 1971. The coal tit as a careful shopper. Am. Nat. 105:77–80.

Tuomi, J., Niemelä, Haukioja, E., Sirén, S., and Neuvonen, S. 1984. Nutrient stress: An explanation for plant anti-herbivore responses to defoliation. Oecologia (Berl.) 61:208–210.

Valentine, H. T., Wallner, W. E., and Wargo, P. W. 1983. Nutritional changes in host foliage during and after defoliation, and their relation to the weight of gypsy moth pupae. Oecologia (Berl.) 57:298–302.

Varley, G. C. 1947. The natural control of population balance in the knapweed gall-fly (*Urophora jacaena*). J. Anim. Ecol. 16:139–187.

Vinson, S. B. 1974. Biochemical coevolution between parasitoids and their hosts, pp. 14–48. In P. W. Price (ed.), Evolutionary Strategies of Parasitic Insects and Mites. Plenum, New York.

Waloff, N. 1968. Studies on the insect fauna on Scotch broom *Sarothamnus scoparius* (L.) Wimmer. Adv. Ecol. Res. 5:87–209.

Waring, G. L., and Price, P. W. 1989. Parasitoid pressure and the radiation of the gall forming group (Cecidomyiidae: *Asphondylia* spp.) on creosote bush (*Larrea tridentata*). Oecologia Berl.) 79:293–299.

Washburn, J. O. 1984. Mutualism between a cynipid gall wasp and ants. Ecology 65:654–656.

Watt, K. E. F. 1965. Community stability and the strategy of biological control. Can. Entomol. 97:887–895.

Way, M. J. 1963. Mutualism between ants and honeydew-producing Homoptera. Annu. Rev. Entomol. 8:307–344.

Webb, J. W., and Moran, V. C. 1978. The influence of the host plant on the population dynamics of *Acizzia russellae* (Homoptera: Psyllidae). Ecol. Entomol. 3:313–321.

West, C. 1985. Factors underlying the late seasonal appearance of the lepidopterous leaf-mining guild on oak. Ecol. Entomol. 10:111–120.

Whitham, T. G. 1987. Evolution of territoriality by herbivores in response to host plant defenses. Am. Zool. 27:359–369.

Whittaker, J. B. 1984. Response of sycamore (*Acer pseudoplatanus*) leaves to damage by a typhlocybine leaf hopper, *Ossiannilssonola callosa*. J. Ecol. 72:455–462.

Williams, K. S., and J. H. Myers. 1984. Previous herbivore attack of red alder may improve food quality for fall webworm larvae. Oecologia (Berl.) 63:166–170.

Willmer, P. G. 1980. The effects of fluctuating environment on the water relations of larval Lepidoptera. Ecol. Entomol. 5:271–292.

Wraten, S. D., Edwards, P. J., and Dunn, I. 1984. Wound-induced damages in the palatability of *Betula pubescens* and *B. pendula*. Oecologia (Berl.) 61:372–375.

Zwölfer, H. 1979. Strategies and counterstrategies in insect population systems competing for space and food in flower heads and plant galls. Fortschr. Zool. 57:558–569.

5

Invertebrate Predators and Caterpillar Foraging

Clytia B. Montllor and Elizabeth A. Bernays

Introduction

Several characteristics of lepidopteran feeding, such as host specificity, degree of gregariousness, degree of conspicuousness, and position or time of feeding are potentially influenced by selection pressure from natural enemies. Taken together for any given species, these and other related features define a foraging strategy. Presumably each of the diverse strategies seen in nature is a compromise between conflicting pressures, a balance of the costs and benefits incurred by feeding in particular ways on particular plants. The search for underlying unifying themes has driven much research, but it is becoming evident that the many forces involved in determining foraging behavior (Hassell and Southwood 1978) are differentially important in different systems (e.g., Stamp and Bowers 1990).

Invertebrate (arthropod) predators are recognized as important causes of mortality for lepidopteran larvae. But differential mortality of genotypes, and not just high mortality per se, is required for predators to be selective agents. Such differential mortality is difficult to document, therefore the degree to which invertebrates have shaped the evolution of host use and foraging patterns in caterpillars remains a largely unanswered question, although their potential to be important selective agents is often apparent (Heads and Lawton 1985). To address this question, we must know the degree to which arthropod predators choose prey items based on genetically determined behavioral, physiological, and morphological characteristics of the prey.

Heinrich (1979) hypothesized that some caterpillar foraging patterns, such as hiding during the day or clipping petioles of partially eaten leaves, might have evolved as a result of vertebrate predation pressure. The generality of this hypothesis remains largely untested, perhaps because it seems so intuitively sensible. However, it is likely that quantification of the behavior of many caterpillar species will reveal a multitude of foraging strategies, reflecting a variety of selection pressures, including predation by arthropods. As Heinrich (1979) and others (e.g., Mauricio and Bowers 1990) have suggested, a combination of strategies such as optimal foraging, predator avoidance, and thermoregulation is likely, and these will also depend on habitat and host plant.

Even where predation appears to have been a major selective force, contrasting pressures may occur. An interesting study of the pine looper, a geometrid that occurs in two color morphs, a common green form and a rare yellow form, showed that of the many potential selective forces that were studied, the only mortality factor that appeared to be different between morphs was predation (den Boer 1971). In experiments with captive Great Tits, birds took more later-instar yellow caterpillars, presumably because they were easier to see against the background of the host pine needles. Crab spiders (Thomisidae) took more yellow larvae, because they thrashed more on contact, but this preference did not hold for third instars, which did not thrash. In contrast, vespid wasps preferred green caterpillars, based on an unidentified chemical difference.

The variety of feeding strategies employed by caterpillars (Heinrich 1979; Schultz 1983), and the increasing attention being given to the influence of invertebrate predators on the evolution of their prey (e.g., Bernays 1988, 1989; Bernays and Graham 1988), has prompted this evaluation of the evidence for the importance of invertebrate predation on the shaping of caterpillar feeding strategies.

Impact of Invertebrate Predators on Caterpillars

Examples of the direct impact of invertebrate predators on populations of lepidopteran larvae are presented in Table 5.1. Only studies that included some measurement of rate of foraging or of caterpillar mortality due to predation are included. Although most of these studies do not address the question of whether arthropod predation might influence caterpillar foraging, they indicate that invertebrates are major causes of mortality in lepidopteran populations. If just one colony of the ant *Formica rufa* can collect 21,700 larvae in a single day (Strokov 1956; cited by Hölldobler and Wilson 1990), any slight changes in behavior of the prey that reduce the risk of being taken will surely be selected for.

Few studies have dealt with the potential indirect effect of predators on prey survivorship. Stamp and Bowers (1991) have estimated that 20% of the reduction in survivorship of buckmoth caterpillars (*Hemileuca lucina:* Saturniidae) was caused indirectly by the presence of *Polistes* wasps, which forced the caterpillars to feed on nutritionally inferior, shaded foliage (Stamp and Bowers 1988). Pentatomid predators, on the other hand, caused little indirect, but substantial direct, mortality on larvae of *Junonia coenia* (Nymphalidae) (Stamp and Bowers 1991). Given the evident importance of invertebrate predators as direct and, in addition, indirect mortality agents for caterpillars, we will now consider characteristics of these predators and prey that might affect the outcome of their interactions.

Table 5.1. Examples of Predation or Mortality Rates of Caterpillars due to Invertebrate Predators

Predator	Prey	Impact	Reference
Podisus maculiventris (Pentatomidae)	*Heliothis zea* *Trichoplusia ni* *Anticarsia gemmatalis* *Pseudoplusia includens* *Plathypena scabra*	Predation success: 12–88% of contacted larvae	Marston et al. (1978)
Podisus maculiventris Unid. hemipteran	*Halidisota caryae*	0.7–2.1 larvae/day	Lawrence (1990)
Podisus serieventris	*Choristoneura pinus* 4th–5th instar	~ 1 larva/day	Allen et al. (1970)
Orius insidiosus (Anthocoridae)	*Heliothis zea* *Spodoptera frugiperda*	Up to 20/24 hr Up to 16/24 hr (dependent on prey density and diet)	Isenhour et al. (1989)
Nabidae (3 spp.)	*Heliothis punctiger* (1st instar)	Up to > 1000 larvae/predator	Awan (1990)
Iridomyrmex (Formicidae)	*Pieris rapae*	Up to 70% reduction in larval populations (dependent on instar, prey density)	Jones (1987)
Formicidae (4 spp.) esp. *Formica, Camponotus*	*Lymantria dispar*	Contact in Field: every 20 min, probability of capture 4–20%	Weseloh (1989)

Table 5.1. *(continued)*

Predator	Prey	Impact	Reference
Formica rufa	*Operophthera brumata*	Mean density of larvae/shoot, with ants excluded: 1.2; ants allowed: 0.6	Skinner and Whittaker (1981)
Polistes spp. (Vespidae)	*Hemileuca lucina*	4–10 larvae/2 hr (dependent on larval group size)	Stamp and Bowers (1988)
Polistes	*Pieris rapae*	1–8 larvae/wasp/day	Gould and Jeanne (1984)
Polistes	*Manduca sexta*	Population reduced by 60%	Lawson et al. (1961)
Polistes	*Manduca sexta*	Population reduced by 68%	Rabb and Lawson (1957)
Vespula spp. (Vespidae)	*Heliothis virescens*	Mortality: 91%	Steward et al. (1988)
Vespula maculifrons	*Hyphantria cunea*	Up to 100% mortality by time of 5th instar	Morris (1972b)
Geocoris (Lygaeidae)	*Heliothis* spp (1st instar)	% reduction in 48 hr: 53%	Lopez et al. (1976)
Podisus (Pentatomidae)		25%	
Coleomegilla (Coccinellidae)		51%	
Anatis ocellata (Coccinellidae)	*Choristoneura pinus* 5th–6th instar	0.4 larva/day	Allen et al. (1970)
Arthropods	*Pieris rapae* (young instars)	50–60% mortality	Dempster (1967)
Spiders	*Spodoptera littoralis* neonates	Mortality up to 100/spider/day	Mansour et al. (1981)
Anthropods, especially spiders	*Orgyia pseudotsugata*	~ 47% of total losses	Mason and Torgersen (1983)

Characteristics of Invertebrate Predators

Many of the qualities of invertebrate predators relating to prey finding and acceptance have been generally discussed by Hagen (1987), Greany and Hagen (1981), and Hagen et al. (1976). Several characteristics of invertebrate predators distinguish them from their vertebrate counterparts and may be specifically related to their influence on the foraging behaviors of caterpillars. These include the relative ability of invertebrates compared to vertebrates to learn (e.g., aversion learning of aposematic patterns, or associative learning of cues such as leaf damage), the thermal constraints on searching for prey and other activities, and the relatively small size of invertebrate predators in relation to their prey. In addition, the behavioral/physiological susceptibilities of invertebrates to various chemical defenses of Lepidoptera might be expected to differ from those of vertebrates.

Further, invertebrate predators may be assumed to be more similar in behavior and physiology to their invertebrate (in this case, caterpillar) prey than are their vertebrate counterparts. However, arthropod predators of caterpillar larvae are a remarkably diverse group, and we should not overlook differences among invertebrate predators (e.g., Yeargan and Braman 1989), or between invertebrate predators and their prey (e.g., Evans 1982) in behavior and physiology.

Learning

Much recent work on the learning abilities of insects has focused on herbivores (Papaj and Prokopy 1989; Szentesi and Jermy 1990) and natural enemies, specifically parasitoids (e.g., Vet and Groenewald 1990). In contrast, very little work has been done on predatory insects with respect to their prey (Bernays 1991). Some early references to rejection of noxious prey by invertebrate predators are given in Berenbaum and Miliczky (1984). Learning in ants has been studied to a degree (Dejean 1988). Associative learning was suggested by Forsyth (1978, cited in Raveret Richter 1988) to account for observations of *Syncoeca* wasps biting open aborted buds under trees to extract insect larvae. At least two other species of vespid wasps (*Polybia* spp) learned and remembered where they had hunted successfully, and where they had left prey (Raveret Richter 1988, 1990), and other studies indicate an important role of learning in the successful foraging of the vespids *Polistes jadwigae* (Yamasaki et al. 1978; Takagi et al. 1980) and *Mischocyttarus flavitarsus* (Cornelius 1991).

The ability of predators to learn (e.g., aversion learning) is also particularly important in the context of adaptations on the part of prey, such as chemical defense and aposematism. Berenbaum and Miliczky (1984) showed that after encounters with cardenolide-containing prey (*Oncopeltus fasciatus,* Hemiptera), adult mantid predators learned to reject both distasteful and palatable prey, presumably based on visual cues alone. Juvenile mantids also learned to reject more quickly *O. fasciatus* at the second encounter, though contact was necessary before rejection took place (Paradise and Stamp 1991). Ants learned to avoid distasteful, aposematic prey (chrysomelid larvae), usually from a distance without antennal contact, after a single initial contact five days before (Dejean 1988). This memory lasted for at least 28 days. Bowers and Larin (1989) showed that ants learned to avoid a solution of sucrose and cycasin, a compound sequestered by the aposematic butterfly *Eumaeus atala* (Lycaenidae), after 10–20 min, although the solution was initially not deterrent; results were similar with ground butterfly bodies in suspension. However, when presented 24 hours later with similar solutions, the ants showed the same pattern of initial acceptance followed by rejection. In the context of Berenbaum and Miliczky's (1984) and Dejean's (1988) results, it is unclear whether ants might have responded differently, i.e., learned better, if presented with additional cues, perhaps visual or olfactory, at the same time.

The importance of aposematic characteristics per se to vertebrate predators has long been established; an assessment of the potential importance of aposematism to invertebrate predators has yet to be made. There is probably much more variability among invertebrates in their capacity to learn and remember than there is among vertebrates. For this reason it is important not to lump invertebrate predators together. For example, as visually hunting, diurnal predators capable of learning, wasps and birds may be more similar in predatory habits than wasps and other insects such as hemipterans.

Physiology

Ectothermy

Ectothermy is among the major factors that distinguish invertebrate predators, as well as parasitoids, from their vertebrate counterparts, and may make them more sensitive to environmental factors such as temperature or weather patterns. Physiological constraints on the predatory activities of invertebrates are likely to be similar in kind to those of their prey, complicating the interactions between predation and environmental factors. For example, fly larvae were more active (traveled further and burrowed more quickly in the ground) at higher temperatures. Ant predation on these larvae was higher as well at higher temperatures (Eskafi and Kolbe 1990), possibly due both to increased movement of prey as well as to direct effects on the predators. On the other hand, Evans (1982) described a case in which the activity of predatory stinkbugs at tents of *Malacosoma americanum* caterpillars was severely limited by a spring cold snap, while the caterpillars continued to feed and grow at a normal rate. Therefore, predatory and parasitic arthropods may be generally more vulnerable to such environmental setbacks than are vertebrate predators, as well as being in some cases more vulnerable than their prey, especially in the case of gregarious caterpillars that have a thermoregulatory advantage (Stamp, Chapter 15; Casey, Chapter 1).

Studies on the comparative effects of weather on different types of predators are few. Weseloh (1988) found that ant foraging was diminished under wet or humid conditions, as did Fowler and Roberts (1980). In terms of gypsy moth larvae (*Lymantria dispar*) removed, other predators, which included invertebrates, did not appear to be greatly affected by the environmental factors measured, including temperature and moisture, or by time of day (Weseloh 1988). However, few occasions of actual predation were observed, and many of the conclusions were inferential.

Phenology

Considerations of the phenology of some invertebrate predators have been discussed by Hagen et al. (1976). Phenological factors are especially important to host-specific predators and their prey (Tauber and Tauber 1987), and it is generally accepted that predatory arthropods tend to be generalists. Phenology may therefore be a more important factor in parasitoid–host interactions, since parasitoids tend to be more host-specific; host-range patterns of parasitoids of externally feeding Lepidoptera have been recently discussed by Sheehan (1991a).

Among predators, Evans and Eberhard (1970) described the solitary wasps as "host-specific." Though many solitary wasps specialize at the ordinal level, within the five tribes of Sphecidae reported by Bohart and Menke (1976) to take lepidopteran larvae, only two species could be found described as restricted to a particular family or genus. These were *Eremochares aureonata,* which preys on Notodontidae, and *Hoplammophila aemulans,* preying on the genus *Gonoclostera,* also in the Notodontidae. However, there are great gaps in the information available on the feeding habits of sphecids and other predators. Because of the great diversity of arthropod phenologies, it is not likely that the phenology of generalist predators would greatly affect caterpillar feeding strategies and, as far as we know, no ecological studies on the phenology of host-specific caterpillar-feeding predators have been published.

Susceptibility to Chemical Defense

The respective susceptibilities of vertebrate and invertebrate predators to chemical defenses of Lepidoptera have not been explicitly compared, although some differences might be expected. There is some information, however, on the relative activities of detoxicative enzymes of herbivorous compared to predatory arthropods (e.g., Mullin and Croft 1984). To date, the degree to which predatory arthropods can cope with plant secondary compounds ingested by their prey has been of less concern than their ability to deal with environmental toxins such as pesticides. Comparisons between phytophagous and predatory mites (Mullin et al. 1982), between *Trichoplusia ni* (Noctuidae) and *Chrysoperla carnea* (Chrysopidae) (Ishaaya and Casida 1980, 1981), and between various lepidopteran larvae and *Podisus maculiventris* (Pentatomidae) (Yu 1987) indicate that detoxification ability varies greatly. Duffey et al. (1986) have given some examples with parasitoids to support their assertion that a significant aspect of the interaction between herbivore and enemy is "the respective abilities of either the host or the parasitoid to tolerate allelochemical sequestrates or enzymatically detoxify them" (p. 49). We presume that this holds for predators as well. Though effects of herbivore-sequestered plant compounds on predator behavior and physiology have been documented (Dejean 1988; Bowers and Larin 1989; Berenbaum and Miliczky 1984; Paradise and Stamp 1991, 1992; Wink and Römer 1986; Malcolm 1989; Montllor et al. 1991), almost nothing is known generally about how predatory insects cope with high levels of potentially toxic compounds in their prey.

Predator Interactions

Predator pressures on caterpillar foraging might be complicated by competitive interactions between arthropod predators. The presence of congeners near a prey item attracted hunting individuals of two species of *Polybia* wasps (Raveret Richter 1990); subsequent behavior depended on the wasp species and prey size. On a large scale, the presence or abundance of a single predator species may greatly depress the number and diversity of both herbivores and other predators, as has been shown by selective removal of the fire ant *Solenopsis geminata* in an agroecosystem (Risch and Carroll 1982). Competitive interactions between arthropod predators may in turn be complicated by the fact that predators eat each other: for example, half of the observed prey of spiders in a peanut agroecosystem consisted of entomophagous arthropods (Agnew and Smith 1989). Thus predator pressures could be overestimated if their number alone is used to estimate impact.

Effects of Plant Characters on Predator Behavior

Plant characteristics that directly affect the foraging and/or success rate of natural enemies have been discussed by a number of authors (e.g., Price et al. 1980; Boethel and Eikenbary 1986; Nordlund et al. 1988; Hawkins et al. 1990; Sheehan 1991a), with an emphasis on parasitoids. Less attention has been paid to predators (Boethel and Eikenbary 1986; Hagen 1987; Barbosa 1988), mainly because they are largely generalists and have been thought to be less important in control of insect herbivores than the more specialized parasitoids.

The efficiency of arthropod predators has been shown to be affected by plant surface structure (Carter et al. 1984; Rasmy 1977; Obrycki 1986), plant diversity (e.g., Letourneau 1990), and plant chemistry (see Nordlund et al. 1988). The host plant is also known to affect aspects of caterpillar foraging such as movement (e.g., Mauricio and Bowers 1990; Lawrence 1990), rate of ingestion (Bernays

1986), and time spent feeding (Slansky and Scriber 1985). How these factors might be related to coincident effects on predatory arthropods is of interest.

Habitat

Habitat may influence numbers or types of predators present, as well as predator efficiency (Dempster 1969; Dempster et al. 1976; Jeanne 1979; Risch et al. 1982; Bugg et al. 1987; Riechert and Bishop 1990). A variety of characteristics may be responsible, e.g., plant architecture, and density (Letourneau 1990). Rates of ant predation have been shown to differ with latitude, type of vegetation, and on ground vs. on vegetation (Jeanne 1979). The distance of trees from colonies of the ant *Formica obscuripes* was a significant factor in determining the degree of predation by these ants on tent caterpillars (Tilman 1978).

Many arthropod predators also forage differentially on different plant parts. In forest habitats, predation of tethered gypsy moths due to a variety of invertebrate and vertebrate predators was greater lower in trees (highest mortality at 0.5 m). Vertebrate predators appeared to be more important on trunks than on leaves (Weseloh 1988). Ant predators of *Operophtera brumata* (Geometridae) were also apparently more active lower on trees (2 vs. 8 m) (Skinner and Whittaker 1981). Therefore the location of feeding and resting activities will subject caterpillars to potentially different degrees and kinds of predators (e.g., Weseloh 1989). It is interesting, in this context, that during the day only late instar gypsy moths rest in leaf litter, where ant activity is high; ant predation success on these larger instars is fairly low due to their defensive tactics (Weseloh 1989). Movement of caterpillars between host plants might also be strongly affected by the possibility of differential predation on the ground, though experimental evidence is lacking.

Plant Chemistry

Plant secondary chemistry and nutrients are potentially important mediators of predator behavior and of predator success. Plant chemistry may directly affect prey finding or mobility of predators on the plant; for example, predators may have to navigate around trichomes (Obrycki 1986). Some predators are attracted to plant odors. For example, Reid and Lampman (1989) found in the laboratory that extracts of corn silks contained attractants for the predatory anthocorid *Orius insidiosus*. Plant chemistry may also influence the predators indirectly: position and mobility of prey may be based on chemically mediated selection of feeding sites, and their palatability is also likely to be strongly influenced by plant chemistry. Because of the variety of these effects, they are discussed where relevant under separate sections below.

A particular aspect of plant chemistry which is of interest is the production of nutrient rewards. Examples include extrafloral (Bentley 1977) and floral (Dominguez et al. 1989) nectaries that have been shown to attract ants and wasps, respectively, often greatly reducing herbivory on plants offering them. However, in at least one case, it has been suggested that a pod-feeding caterpillar is fortuitously defended from its parasitoids by the presence of ants feeding at and defending nectaries (Koptur and Lawton 1988). In this case, when on plants with extrafloral nectaries, the pod-feeding habit is reinforced by protection from both potential predators (ants) and parasitoids.

Damage

Foliage damage left by feeding herbivores has been suggested to serve as a visual cue to vertebrate predators of caterpillars (Heinrich and Collins 1983; Heinrich, Chapter 7) and to provide chemical cues to parasitoids (e.g., Odell and Godwin 1984; Faeth 1985). Niemela and Tuomi (1987) have even suggested that the leaf shape of some plants might mimic feeding damage and thereby attract more natural enemies of herbivores.

Few experimental studies have been carried out with invertebrate predators, some of which (e.g., hunting wasps) might be expected to learn to recognize caterpillar feeding damage. Raveret Richter (1988) provided evidence that individuals of two species of naturally foraging *Polybia* wasps landed preferentially on leaf rolls made by caterpillars, or on damaged (>5% leaf area eaten) leaves, compared to undamaged leaves. Bergelson and Lawton (1988) investigated whether foliage damage might affect predation by increasing herbivore movement away from the damaged area. Movement of two caterpillar species, a coleophorid and a geometrid, on artificially damaged birch leaves did not increase, nor was predation by ants or birds affected by damage to foliage.

Plant and herbivore olfactory cues are known to be important in host finding by parasitoids (e.g., Nordlund et al. 1988; Whitman and Eller 1990). A recent report by Turlings et al. (1990) showed that a parasitic wasp used volatile terpenoids to find caterpillars of *Heliothis zea*. The terpenoids were released by corn plants, which had been fed upon by these caterpillars; artificially damaged plants were significantly less attractive to the parasitoids. Dicke and Sabelis (1988) and Dicke et al. (1990) had previously shown that predatory mites are attracted to volatiles presumed to be released by herbivore-damaged plants. Volatiles released from damaged plants are potentially important to arthropod predators of caterpillars as well. Cornelius (1991) has found that vespid wasps apparently use both visual and olfactory cues to find larvae of *Manduca sexta* on artificially damaged tobacco plants in the laboratory. Other invertebrate predators, such as anthocorids, respond to plant-derived volatiles (Reid and Lampman 1989) as well, and the phenomenon of using olfactory cues from damaged plants may simply be an extension of this attraction to plant odors.

Effects of Prey Characteristics on Predator Behavior

Some prey characteristics that influence invertebrate predator searching have been discussed by Hagen et al. (1976), Greany and Hagen (1981), and Hagen (1987).

Only a few predators of Lepidoptera have been studied in any detail. Invertebrate predators use a variety of searching modes to find their prey. The prey itself may not be necessary as a visually detectable presence if associated cues may be used. For example, caterpillar homogenates added to corn increased the numbers of a beetle predator and a hemipteran predator in the field for up to 6 days after application (Gross et al. 1985). In another study, vespid wasps hovered, walked, and landed preferentially on leaves that had been rubbed with caterpillar prey; a deodorized prey item was also attractive to foragers, but less so than the caterpillar odor (Raveret Richter 1988). Other vespid wasps, which removed more of a cryptic (green) color morph than of a more conspicuous (yellow) morph of a geometrid larva from host shoots, were also more attracted to shoots treated with green-larva hemolymph compared to yellow-larva hemolymph (den Boer 1971). Responses of parasitoids to lepidopteran frass have been well studied (reviewed in Nordlund et al. 1988). Similar studies with predators are lacking, though pentatomids were shown to follow artificial trails of hemolymph or frass of potential prey in the laboratory (McLain 1979). Physical and chemical attributes of caterpillars are known to be important in the success of invertebrate predators, and evidently affect invertebrate and vertebrate predators differently.

Physical Attributes

Relative Size

Size may be considered either a characteristic of predators or of their prey, since it is the relative size of these potential antagonists that is important. As a characteristic of the predator, small size is one of the major attributes distinguishing invertebrates from vertebrates. From the invertebrate predator's point of view, it may be dangerous to attack a large caterpillar. Morris (1963) described such encounters for a relatively small and "timid" predator, *Podisus maculiventris* (Pentatomidae): fifth-instar larvae of *Hyphantria cunea* apparently commonly grasped and even threw these predators, damaging their legs and antennae.

As a characteristic of prey, size may be an important factor determining the adaptiveness of other behaviors or morphologies. Interspecific differences in behavior of different instars may be related, for example, to a shift in predation pressure on different size classes of caterpillars. In terms of ecological and possibly evolutionary effects of invertebrate predators on caterpillar foraging, one might expect that there should be differential mortality of caterpillar instars due to small, arthropod predators compared to larger, vertebrate predators. Dempster (1967) showed this to be the case for *Pieris brassicae* in a crop ecosystem in which the small instars were taken predominantly by invertebrate predators, whereas the later instars were taken by birds. Weseloh (1988) gives circumstantial evidence for such differential predation on *Lymantria dispar* larvae, in which fewer large caterpillars were removed from screen cages that excluded larger (presumably, mammal and bird) predators, but small larvae were taken in similar numbers irrespective of cage mesh size. That invertebrate predators are restricted in terms of the size of lepidopteran larvae that they can attack successfully has been shown for ants preying on *Malacosoma americanum* (Tilman 1978) and gypsy moth, *Lymantria dispar* (Weseloh 1989), mirids preying on *Heliothis zea* (Cleveland 1987), carabids preying on *Euxoa ochrogaster* (Frank 1971), vespids preying on *Hemileuca lucina* (Stamp and Bowers 1988), coccinellids preying on *Battus philenor* (Stamp 1986), pentatomids attacking *Malacosoma californicum* (Iwao and Wellington 1970), and a range of predators feeding on *Heliothis* spp. (Lopez et al. 1976; Awan 1985, 1990). Many other examples of arthropod predation on small instars and/or bird predation on larger instars occur in the literature, including anecdotal remarks (e.g., Feichtinger and Reavey 1989), and observations coupled with circumstantial evidence (e.g., Bernays and Montllor 1989). Bird species may also differ in terms of the size of caterpillars that they will take. In a garden setting, although all stages of *Pieris rapae*, from eggs through the last larval instar, were taken by birds, different species accounted for mortality of the different instars (Baker 1970).

The degree to which larger caterpillars forage differently from earlier instars, given that they are protected from a certain amount of predation from invertebrates but may be more subject to predation from vertebrates, has not been much studied with differential predation in mind. Some ontological studies of shelter-making in lepidopteran larvae indicate that changes as larvae grow might be related to predation pressures. Ruehlmann et al. (1988) suggested that the different morphologies of feeding shelters of different instars of *Herpetogramma aeglealis* (Pyralidae) might in part be influenced by bird predation, though no experimental evidence was given. The leaf-tying *Achlya flavicornis* (Thyatiridae) moved longer distances between successive ties and fed further from current ties as they grew, and each new feeding area occurred just beyond the limit of a larva's previous feeding area (Feichtinger and Reavey 1989). It is also interesting to note that these larvae became more "selective" in later instars, i.e., ate a smaller proportion of leaves encountered as they foraged. This decrease in proportion of leaves eaten probably mainly reflects the fact that later instars fed further from their leaf ties, i.e., traveled further to feed. The relationship between feeding distance from ties and potential predation was not addressed.

The degree of dispersion of larvae may also change as larvae grow (examples in Vulinec 1990). Young instars of *Uresiphita reversalis* (Pyralidae) feed in groups within loosely tied leaves, and are subject to some predation by small predatory bugs and by wasps (Bernays and Montllor 1989); the larger, aposematic, and alkaloid-sequestering instars (Montllor et al. 1990) occur in dispersed groups, and are distasteful to a variety of arthropod predators, including hemipterans, hymenopterans, a neuropteran, and a salticid spider (Montllor et al. unpublished data). Last instar larvae usually occur singly, are almost certainly too large to be taken by invertebrates, and are probably also unpalatable to birds (Bernays and Montllor 1989). The decreased tendency to aggregate of later instar larvae of *Hemileuca lucina* may be related to their increased ability to defend themselves against arthropod enemies (Cornell et al. 1987). Larvae of *Hyphantria cunea* build colonial webs, but disperse in later instars to feed singly (Suzuki et al. 1980). Although invertebrate predators, including vespids, can prey on these larvae within their webs (Morris 1963, 1972b; Schaeffer 1977), aggregation in a web may confer protection to a degree not needed by older larvae. On the other hand, webs may be dangerous places for larger caterpillars, which are actually preferred by wasps and birds (Morris 1972b); such predators easily learn to return to a nest of caterpillars to feed. Other reasons for dispersing when larger may not involve the risk of predation, and include protection from potential food shortage (Suzuki et al. 1980).

Growth Rate

The rate at which larvae grow is affected by both genetic and environmental (e.g., diet and temperature) factors. It is generally believed that a more rapid rate of larval growth is beneficial in terms of avoidance of predation and parasitism (Price et al. 1980), though this has not been much studied (see Stamp and Bowers 1990). However, even in the absence of predation/parasitism, it has been suggested that for a group of pierid species those with higher growth rates had lower survival, due presumably to some unknown "cost" associated with high growth rates (Jones et al. 1987).

A slower growth rate that results from increased time spent in the molt, rather than growth time during the instar (Stamp 1990), may be especially risky for larvae because of their vulnerability during molting. A relatively slow growth rate need not lead to reduced final weight, but where a smaller body size at a given instar results, larvae would be available for longer and/or in additional instars, especially to small invertebrate predators (e.g., Isenhour et al. 1989).

Attaining a large size quickly may provide escape from some invertebrate predators and parasitoids, but Clancy and Price (1987) provided an example of a gall-forming sawfly larva for which rate of growth and likelihood of parasitization are positively correlated. Similarly, Damman (1987) found that slow growth and predator avoidance were positively correlated in another concealed feeder, a leaf-tying pyralid, because better shelters could be constructed on nutritionally inferior older leaves. However, there is no evidence to date that an externally feeding herbivore benefits from a slow growth rate.

Morphology

A variety of morphological features of caterpillars may be related to predation pressures, but for most there is no direct evidence linking the putative adaptation with selection by predators. For example, Bernays and Janzen (1988) suggested that differences in mandible morphology among some larval Lepidoptera might be related to maximization of ingestion rates (i.e., reduction of feeding time), which might in turn be selected for by visually hunting predators.

Other morphological attributes, such as cuticular spines and hairs, appear to be more obviously defensive, though not always consistently so. The relative costs and benefits to caterpillars of spines and hairs have been recently discussed by Sheehan (1991b). Although parasitism by generalist parasitoids appears to be especially high on hairy/spiny caterpillar species, such caterpillars often require special handling by other enemies such as birds and social wasps, and may be rejected outright by these predators. Advantages of hairs and spines not directly related to predation or parasitism include reduction of heat loss and increased absorption of solar radiation in some species (examples in Sheehan 1991b), which are important factors in maximizing growth rate. In addition, spiny/hairy morphology is correlated with gregariousness. Cuticular adaptations of this kind are therefore a particularly interesting feature to examine further with reference to the relative significance of invertebrate predators as selective agents. Other morphological features of caterpillars, such as mechanoreceptors, that are sensitive to the wingbeat frequency of predatory wasps (Tautz and Markl 1978), suggest that as selective agents such predators are indeed significant and perhaps underestimated.

Chemical Attributes

Chemical attributes of caterpillar prey are often related to their foraging strategies and have been shown to affect predator behavior. Examples of chemical defense of lepidopteran larvae are well known (Blum 1981; Brower 1984; Witz 1990; Bowers 1990). Witz (1990) has made a selective review of 20 years of literature on antipredator mechanisms of arthropods; of 35 studies on chemical defense of Lepidoptera, only seven specifically dealt with effects on invertebrate predators.

Many "chemically defended" Lepidoptera are aposematic, and store plant compounds that are known vertebrate toxins, such as cardenolides (e.g., Brower 1984), alkaloids (e.g., Boppré 1990; Rothschild et al. 1979; Kelly et al. 1987; Montllor et al. 1990), cyanogens (e.g., Franzl et al. 1988; Jones et al. 1962; Nahrstedt and Davis 1983), or azoxyglycosides (Bowers and Larin 1989). The significance of responses of invertebrate predators to sequestered "vertebrate" toxins or to warning coloration is not often explicitly discussed. Several studies have now shown that chemical defense of aposematic insects is effective against invertebrate predators, and that such predators can learn to subsequently avoid similar prey without contact (see section on Learning, above).

Lepidoptera larvae that are deterrent to invertebrate predators by virtue of ingested plant compounds include *Eumaeus atala* (Lycaenidae), which contain cycasin deterrent to ants (Bowers and Larin 1989), and *Uresiphita reversalis* (Pyralidae), which contain quinolizidine alkaloids deterrent to ants and wasps

(Montllor et al. 1991). That such plant-derived compounds may have adverse physiological effects on invertebrate predators has been shown by Berenbaum and Miliczky (1984) and Paradise and Stamp (1992), who documented a case of emesis and reduced growth, respectively, in mantid predators eating cardenolide-containing milkweed bugs. Web structure and predatory activities of spiders (Araneae) fed cardenolide-sequestering aphids were disrupted (Malcolm 1989), and predatory carabid beetles became narcotized after feeding on alkaloid-sequestering aphids (Wink and Römer 1986).

The implications of such ecological relationships to the evolution of feeding strategies is not always clear (e.g., see Guilford 1990). In a survey of 28 pairs of generalist and specialist caterpillars (total of 27 species), specialists were taken less frequently by foraging vespids than generalists (Bernays 1988). At least part of this phenomenon might be explained by the association of sequestration and specialization. Denno et al. (1990) suggested that specialization of a chrysomelid on a salicylate-rich willow species, from which they derived a defensive secretion, was a result of the reduced predation of larvae on these plants compared to those on a salicylate-poor species. Stamp (1992) gives an example of higher predation by wasps and stinkbugs on a non-cryptic specialist compared to a cryptic generalist on the same host plant. Both species sequestered iridoid glycosides from their plantain host, but the generalist species was killed less frequently due to its evasive movements and/or lack of apparency. However, experienced wasps apparently learned to reject the specialist, while the rate of predation on the cryptic generalist stayed constant. This result has interesting implications for the role of apparency and/or encounter rate in learning by predators.

Caterpillars may produce secretions (e.g., Honda 1983) or regurgitates (e.g., Peterson et al. 1987) that are deterrent to invertebrate predators. Such putatively defensive tactics, which are widespread among insects, are evidently important components of a given feeding strategy. The fact that invertebrate predators are among those repelled by sequestered, secreted, or regurgitated chemicals indicates that such predators may play a role in reinforcing feeding strategies, such as host-plant specialization, that incorporate such defenses. However, causality in such associations is far from determined. Potential predators may also be appeased by larval secretions, e.g., lycaenids and ants, and this often results in a mutualistic relationship. In this case, larval feeding strategies might then be shaped more by the mutualism than by association with particular host plants (Atsatt 1981).

Apart from the more obvious examples of the sequesterers discussed above, the effects of prey diet on their suitability as food for insect predators seem to be minimal where they have been investigated (e.g., Drummond et al. 1984; Isenhour et al. 1989). However Bernays (1988) noted that even cryptic species varied in acceptability to vespid wasps, and it might be fruitful to examine the role of ingested plant material. Work of Cornelius (1991) suggested that ants responded differently to the cryptic, oligophagous *Manduca sexta* (Sphingidae) and to the cryptic, polyphagous *Trichoplusia ni* (Noctuidae) depending on the caterpillars' diets.

Prey diet may affect predator success in search or capture. In laboratory experiments, *Spodoptera frugiperda* and *Heliothis zea* (Noctuidae) reared on resistant corn, or on a diet containing resistant silks, respectively, suffered increased mortality from an anthocorid predator compared to larvae on susceptible corn or diet without silk (Isenhour et al. 1989). A combination of increased movement and smaller size of larvae on the poorer diet appeared to be responsible.

Caterpillar Behavior: Foraging and Risks

Without conflicting pressures that require tradeoffs, caterpillars might be expected to prefer nutritionally superior foliage, on which their development and/or fecundity is optimized. Caterpillars have indeed been shown to select plants or plant parts that are easy to chew or digest, that are high in nitrogen and water, and low in allelochemicals (which even to specialists are often deterrent in high doses) (Slansky and Scriber 1985). In addition, larvae might be expected to eat maximally, especially during the warmth of the day, when rate of feeding and/or digestion will be higher (e.g., Stamp 1990), and to feed in the open, in sunshine, rather than in the shade (Stamp and Bowers 1988, 1990). In fact, caterpillars have many behaviors that appear to ensure their maximal performance in these regards. For example, positive phototaxis and preference for young leaves exhibited by many caterpillars leads them to feed on distal parts of plants, where they are more likely to feed on nutritionally superior foliage, in sunshine (e.g., Bernays and Montllor 1989). Such tactics should maximize growth rates.

However, there are many exceptions to these expectations, indicating that there are factors that constrain caterpillar foraging behavior to less than the nutritionally and physiologically ideal (e.g., Heinrich 1979; Damman 1987). One of these factors is "avoidance of risk," which is widely recognized as an important aspect of foraging in insects (Hassell and Southwood 1978). For example, Bentley and Benson (1988) describe behavior of both adult and larval *Heliconius* butterflies that they ascribe to selection pressure from ants. Eggs are laid on tendrils and larvae remain on the tendrils for at least the first instar, where they are better protected from ants, even though tendrils are not especially good food.

In the previous sections we have discussed attributes of invertebrate predators that make them able and likely to constrain caterpillar foraging. The relative importance of these predators compared to other factors such as nutrient quality of plants, plant secondary chemistry, temperature, and humidity in shaping feeding patterns has only recently been addressed experimentally (Damman 1987; Stamp and Bowers 1988, 1990). One complication is that it is often not clear which of many covarying plant characteristics might be influencing caterpillar behavior. For example, preferred foliage that is apparently nutritionally superior, e.g., young leaves of black cherry for *Malacosoma americanum* (Peterson 1987), may also prove to serve the caterpillars better in their defense against ant predators due to higher levels of cyanogens (Peterson et al. 1987). In such a case it would be interesting to assess the two factors for their relative effects on growth, predator avoidance, and survival.

In the following sections we will examine temporal and spatial aspects of feeding patterns in relation to benefits and risks to foraging caterpillars.

Movement

Movement is energetically costly, and potentially risky to prey. Therefore, the amount of movement engaged in during foraging is an important component of a feeding strategy. Movement away from feeding sites has been suggested to be advantageous for palatable prey whose predators use leaf damage as a cue in foraging (Heinrich 1979; Heinrich and Collins 1983; Heinrich, Chapter 7), in cases in which leaf damage decreases the suitability of nearby foliage (see Bergelson and Lawton 1988), or when leaf quality varies considerably among leaves (Schultz 1983). Movement within or between plants may also be affected directly by the presence of predators themselves (Riechert and Lockley 1984; Stamp and Bowers 1988, 1991).

The inherent riskiness of movement due to invertebrate predation has been shown repeatedly. Bergelson and Lawton (1988) found that search and capture times of *Apocheima pilosaria* (Geometridae) by ants were significantly shorter when prey were mobile on their birch hosts, but that prey movement had no effect on an avian predator's success. Other invertebrate predators have also been described as responsive to prey movement, albeit on a smaller scale and/or at closer range, e.g., the pentatomid *Podisus maculiventris* (Marston et al. 1978). Crab spiders (Thomisidae) took significantly more of a yellow color morph of *Bupalus piniarius* (Geometridae) because they moved more in response to the spider's initial contact than did green conspecifics (den Boer 1971).

On the other hand, movement of aposematic, toxic caterpillars may incur little risk; in fact, aposematic larvae often have distinct defensive movements that make them more conspicuous. Movement of aggregations may also tend to reduce risk. Mortality of group-feeding tussock moth larvae due to invertebrate predation, dominated by two hemipterans, was apparently reduced by about 10% for groups that moved compared to groups that were stationary (Lawrence 1990). Moving groups traveled an average of up to 15 cm/day, and though the presence of invertebrate predators did not affect the distance traveled, host plant species and group size did, with larger groups moving further. The reason for the variation in movements was not clear, therefore it is hard to assess the reasons for increased invertebrate predation on stationary groups.

Foraging Patterns

Caterpillars display a wide array of temporal and spatial feeding patterns (e.g., in the last decade, Damman 1987; Mauricio and Bowers 1990; Dethier 1988, 1989; Heinrich and Collins, 1983; Feichtinger and Reavy 1989; Stamp and Bowers 1988; Reynolds et al. 1986; Stamp 1984; Fitzgerald and Peterson 1988; Roessingh 1988; Tsubaki and Kitching 1986; examples in Schultz 1983; Lederhouse 1990). The influence of the presence or pressures of arthropod predators on foraging strategies has only occasionally been explicitly investigated.

One approach has been to examine the behavior of caterpillars in environments that differ in terms of riskiness. The degree to which caterpillars, or insects generally, can make behavioral decisions based on differing degrees of risk has not been extensively studied; both ants (see Lima and Dill 1990) and larval damselflies (Dionne et al. 1990) forage or hide differently in relation to the riskiness of the habitat. Stamp and Bowers (1988) found that the saturniid *Hemileuca lucina* responded to the presence of foraging vespids or to simulated attack by moving to the interior of the plant onto mature shaded foliage, where they grew more slowly and attained a lower weight (Stamp and Bowers 1990). They have also observed that larvae of the nymphalid *Junonia coenia* moved more between host plants in the presence of pentatomid predators (Stamp and Bowers 1991). Spider presence has been shown to cause abandonment of the plant altogether by caterpillars (Riechert and Lockley, 1984). Later instar *Papilio glaucus* (third instar and after) changed their foraging behavior depending on the host plant, appearing to forage in such a way as to minimize evidence of feeding damage (discussed in Lederhouse 1990); it is inferred that predators were able to use leaf damage as a cue in prey finding and that this shaped the behavior of the caterpillars.

A second approach in assessing predators as a factor shaping caterpillar foraging patterns has been to try to fit such patterns into adaptive syndromes. Heinrich (1979) and Heinrich and Collins (1983) made a case for the importance of visual predators, such as birds, in shaping foraging patterns of palatable cryptic vs. those of unpalatable aposematic caterpillars. They hypothesized that palatable caterpillars would be more likely to be cryptic, resting on the undersides of leaves, feeding at night, and moving away from their feeding damage, whereas distasteful caterpillars would be more conspicuous, feeding by day, without extensive movement either to reach, or to subsequently leave, feeding sites.

Although we do see such patterns among some species, many more need to be described. In a recent study in which larvae of two lepidopteran species, one aposematic and unpalatable and the other cryptic and palatable, were compared on their respective host plants, Mauricio and Bowers (1990) were not able to discern any pattern that was obviously related to predator avoidance. Both the aposematic, iridoid glycoside-sequestering larvae (*Euphydryas phaeton*), and the cryptic, palatable larvae (*Pieris rapae*) tended to forage similarly on their respective host plants, although *E. phaeton* did spend more time in the sun (i.e., more exposed) than *P. rapae,* as might be predicted given that it is an early spring feeder. The putatively chemically protected *E. phaeton* ate a small percentage of each leaf (usually less than 25%). In addition, larvae moved relatively long distances to feed, apparently contrary to the predictions of Heinrich (1979) for unpalatable prey of avian predators, though under more natural field conditions, *E. phaeton* is often sedentary (N. Stamp, personal communication). The influence of host plant factors on foraging of larvae was also pointed out by Mauricio and Bowers (1990): *P. rapae* larvae spent significantly more time moving and more time feeding on *Rhaphanus* than on *Brassica*.

In another study, five species of caterpillars were described as having very different feeding behaviors on soybean (Marston et al. 1978), from moving extensively and feeding little (*Heliothis zea*), to resting under leaves and feeding extensively (*Anticarsia gemmatalis*). Though *A. gemmatalis* suffered the lowest degree of mortality from a pentatomid predator, mortality was not related to encounter rates, which were similar for all five species, but rather to defensive behaviors of the caterpillars. Therefore, feeding pattern itself appeared to make little difference to the likelihood of being successfully preyed on by this particular invertebrate predator.

Of course, no one or two foraging strategies is likely to be able to deal with predation by "invertebrates," a very diverse group. One guild of arthropod predators in a corn agroecosystem comprised 30 taxa, in at least 10 families and two orders (Brust et al. 1986); another, in soybeans, consisted of at least 17 species (Godfrey et al. 1989), and a similar diversity of predators in rangelands was reported (in four orders of insects and spiders, Spangler and Macmahon 1990; and over 30 families in seven arthropod orders, Shaw et al. 1987). Less is known about invertebrate predator faunas in natural ecosystems, but they are also likely to be quite diverse (e.g., Mason and Torgersen 1983). As a group, birds may have more similar hunting modes, compared to invertebrate predators, though birds do differ in terms of size prey they take (Baker 1970), and may specialize on particular substrates, for example, upper vs. lower surfaces of leaves (Greenberg and Gradwohl 1980; Stamp and Wilkens, Chapter 9).

Temporal feeding strategies of caterpillars might be expected to be different for species or in habitats where the major predators are birds rather than invertebrates, since invertebrates are often active at night, while insectivorous birds are not. Though nighttime predation is not easily or often observed (e.g., Lawrence 1990), it can be considerable. Weseloh (1988) measured predation (due mostly to small mammals and ants) of gypsy moth larvae on an hourly basis, and found that it was about the same day or night. Dempster (1967) estimated that arthropod predation accounted for between about 50 and 60% of the mortality of young instars of *Pieris rapae,* and that it occurred mostly at night.

Retreats and Barriers

Other specialized behaviors of caterpillars include the building of shelters, such as webbing or tents, leaf rolls or ties (Fitzgerald, Chapter 11; Stamp and Wilkens, Chapter 9), the utilization of resting sites on plants or in litter, or feeding in internal tissues such as stems, in mines or in galls. That such physical barriers confer protection from predators is widely accepted (Lederhouse 1990) and although many instances of the breaching of caterpillar shelters by invertebrate predators exist (e.g., Morris 1972a, 1972b; Schaeffer 1977; Tilman 1978; Berisford and Tsao 1975), there is also evidence that survivorship is enhanced for sheltered larvae (Damman 1987). Sheltering often requires aggregation (discussed below). In complicated systems that might include nest building, group feeding, trail following, and recruitment (e.g., *Malacosoma americanum*), the relative importance of predation in forging such a life-style is very difficult to determine.

The importance of shelter as a defense has been shown by Damman (1987). In fact, feeding on nutritionally poor older leaves by the pyralid *Omphalocera munroei* appears to be tolerable because of their superiority for the construction of leaf ties, which provide protection from invertebrate predators, the major source of larval mortality. When given a choice between young and mature leaves, these larvae do not prefer the nutritionally superior younger leaves, as is common in many other insects. Hunter (1987) suggested that *Diurnea fagella* (Oecophoridae) is more common on damaged (regrowth) leaves of its oak host, in spite of poorer growth and survival on these leaves, because they can make shelters more quickly on this type of leaf. Predation was not measured in this study, and the relative importance of other advantages of leaf rolls could not be assessed. Leaf tying in *Achlya flavicornis* (Thyatiridae) (Feichtinger and Reavy 1989) and shelters constructed by a fern-feeding pyralid (Ruehlmann et al. 1988) were also suggested to be antipredator devices; both vertebrates and invertebrates were implicated, but, again, no actual data on predation were given.

Other reasons for sheltering have also been suggested, such as more suitable temperature for cutworms (*Agrotis segetum*), which find shelter in the ground (Esbjerg 1990), facilitation of thermoregulation on tents for *Malacosoma americanum* (Joos et al. 1988), or avoidance of phototoxicity by hypericin in the host plant of a dawn-feeding beetle larva, which hides during the day (Fields et al. 1990). Several leaf-rolling caterpillar species in Panama may benefit from the improved nutritional quality (lower tannin and higher water content), and reduced leaf toughness, possibly due to changes in light exposure, of rolled compared to unrolled leaves of their host plant (Sagers 1990). In these latter cases the relative importance of predators in reinforcing the sheltering habit has yet to be determined.

Gregariousness

Among the advantages and disadvantages of feeding in groups, avoidance and attraction of enemies have been suggested, respectively (see Lawrence 1990; Vulinec 1990). Theoretical and experimental considerations indicate that being clumped is advantageous if larvae are distasteful (e.g., Greenwood et al. 1989) and that being distasteful is an advantage if larvae are clumped (e.g., Sillen-Tullberg and Leimar 1988), and these two characteristics are indeed often associated throughout the Insecta. However, aggregations need not always be conspicuous, nor need their utility in predator avoidance be related to advertisement (Vulinec 1990). For example, groups may facilitate shelter building (e.g., Damman 1987), or increased growth rate (Stamp and Bowers 1990), both of which have been associated with predator avoidance.

The relative importance of aggregation as a defense against invertebrates compared to vertebrates is unclear, and probably dependent on the type of aggregation. Morris (1963) contrasted two types of predators of *Hyphantria cunea,* which feed as larvae in groups within silk nests. The first type included birds and wasps, which tended to remove larvae on the outside of the nest, and which might return repeatedly to the same nest. Being inside the nest would be of obvious advantage against this type of predator. The second type included hemipterans and spiders, which entered the nest and potentially completed their development inside. In this case, the density of caterpillars in the nest was always high enough that the functional response of a pentatomid predator (determined in the laboratory) was independent of larval density. Presumably, larger aggregations would be beneficial if they did not attract more predators of this type. In the case of another gregarious arctiid, *Halisidota caryae* (Lawrence 1990), larger aggregations did not attract more invertebrate predators than did smaller ones, and so the likelihood of being taken was less in a larger group. Accordingly, survivorship in groups of 125 was significantly greater than that in groups of 25 or less on three host tree species. Lawrence suggested that oviposition behavior may account for the fact that *H. caryae* in the field occur in quite variable group sizes (seven to 376 individuals), in spite of the apparent advantages of larger groups. In fact, the importance of oviposition in shaping caterpillar feeding patterns should not be overlooked (e.g., Moore et al. 1988). The vulnerability of ovipositing females to predators may indirectly influence subsequent patterns of herbivory which may otherwise appear to bear no relation to predation pressures.

Host Range

The wide variation in diet breadth of caterpillars has been a subject of increasing attention, because of the interest in what may be the selective factors influencing it. It is possible that predators play a role, and there are two studies that at least demonstrate that diet breadth is correlated with relative vulnerability to wasps and ants. Bernays (1988) showed that the vespid, *Mischocyttarus flavitarsus,* caused considerably more mortality among polyphagous species than among species with restricted diets, when these were tested in pairs with size and density controlled. Among those with narrow host range, species with warning coloration were better protected than those which were more cryptic, and in several of the species sequestered plant compounds were important. The results also indicate that visually hunting invertebrates may contribute to selection for aposematism. The ant, *Iridomyrmex humilis,* was similarly shown to prefer polyphagous caterpillars, and among specialist prey, to prefer those without warning coloration (Bernays and Cornelius 1989). In both of these studies the vulnerability of the relative polyphages was partly due to a greater level of acceptability to the predators after being found, and in the case of the more cryptic specialists, there is thus a tendency to have some chemical protection.

Two major questions arise from these studies of invertebrate predation on caterpillars. First, to what extent are differences in vulnerability of species with different diet breadths a cause or effect of the diet breadth? Second, when and why is a specialist better protected than a generalist when feeding on the same host plant? Although a pattern of increased vulnerability of generalist species to invertebrate predators has emerged from the work of Bernays (1988) and Bernays and Cornelius (1989), exceptions will certainly exist (e.g., Stamp 1992). Answers may be found by careful work on within-species differences in foraging strategies. If an individual polyphagous caterpillar retains a fidelity to a certain plant, is it better protected from predation than if it moves from one plant species to another? If so, this would provide a good reason for the phenomenon of induced preference, so common among Lepidoptera. Furthermore it would provide new reasons for the phenomenon of local food specialization of individuals or populations within some polyphagous species (Fox and Morrow 1981). It is tempting, if premature, to speculate that predators, including invertebrates, influence host use strategies in an evolutionary sense, as well as having direct effects on which plants are used or left at a particular time.

General Considerations and New Research

1. In many cases, invertebrate predators are not necessarily different in their effects on caterpillar behavior and evolution from vertebrate predators. As mobile, visual, day-feeding hunters with an ability to learn, wasps and birds may provide some similar constraints, although direct comparisons are needed. Similarly, small mammals and nocturnal invertebrate predators, such as ants and carabids, might be fruitfully compared.
2. Sucking predators, such as Hemiptera and neuropteran larvae, may provide very different challenges for caterpillar prey than mandibulate

predators, and may be able to circumvent the strategy of depositing defensive compounds in or on the cuticle (e.g., Montllor et al. 1991).

3. More information on whether certain predator types predominate in different biomes or habitats might tell us whether certain foraging strategies are related to the abundance of particular predators. For example, it is probable that birds are particularly important in woodlands, but little is known about the relative importance of different invertebrate predators in such habitats.
4. Little is known about the significance of invertebrates in the evolution of aposematism, or sequestration of defensive compounds by caterpillars. More information on the learning capability of wasps, as well as other visually hunting predators, coupled with a quantification of their relative importance as mortality factors would be worthwhile. Physiological studies of the effects of chemical "defenses" of caterpillars on invertebrate predators would also be enlightening.
5. To what extent are foraging strategies variable within a species and different according to particular predators present? Stamp and Bowers (1988) demonstrated dramatic effects of hunting wasps on foraging behavior of one species. It would be instructive to measure the intraspecific variation in caterpillar responses and whether other significant predators cause the same or different responses.
6. Above all, enlightenment will come from detailed and continuous observations of individuals in the field. A particular advantage of working with invertebrates is that such observations can more easily be made without disturbing the predators, but few such studies have been undertaken. Many casual observations on such details as caterpillar disturbance during feeding that result in temporary swinging down from the plant on silk threads need quantifying, lability of behaviors under different types of disturbance needs documentation, and whether the act of feeding is dangerous needs examination, since this is often assumed, and even proposed as a basis for mandibular adaptations that might increase ingestion rate (Bernays and Janzen 1988).

In conclusion, there are more questions than answers in relation to how invertebrate predators may constrain foraging by caterpillars, and a large and fascinating array of studies remain to be undertaken. It is clear, however, that invertebrates are very likely significant and underestimated factors shaping the foraging patterns of caterpillars.

Acknowledgments

Thanks to M. L. Cornelius, M. Raveret Richter, and N. E. Stamp for helpful suggestions. This work was supported in part by USDA-CSRS Grant 88-38300-3628 and NSF Grant BSR 8705014 to EAB.

Literature Cited

Agnew, C. W, and Smith, J. W., Jr. 1989. Ecology of spiders (Araneae) in a peanut agroecosystem. Environ. Entomol. 18:30–42.

Allen, D. C., Knight, F. B., and Foltz, J. L. 1970. Invertebrate predators of the jack-pine budworm, *Choristoneura pinus* in Michigan. Ann. Ent. Soc. Am. 63:59–64.

Atsatt, P. R. 1981. Lycaenid butterflies and ants: Selection for enemy-free space. Am. Nat. 118:638–654.

Awan, M. S. 1985. Anti-predator ploys of *Heliothis puntiger* (Lepidoptera:Noctuidae) caterpillars against the predator *Oechalia schellenbergii* (Hemiptera:Pentatomidae). Aust. J. Zool. 33:885–890.

Awan, M. S. 1990. Predation by three hemipterans: *Tropiconabis nigrolineates, Oechalla schellenbergii* and *Cermatulus nasalis,* on *Heliothis punctiger* larvae in two types of searching arenas. Entomophaga 35:203–210.

Baker, R. R. 1970. Bird predation as a selective pressure on immature stages of the cabbage butterflies, *Pieris rapae* and *P. brassicae*. J. Zool. 162:43–59.

Barbosa, P. 1988. Natural enemies and herbivore-plant interactions: influence of plant allelochemicals and host specificity, pp. 201–229. In P. Barbosa and D. K. Letourneau (eds.), Novel Aspects of Insect-Plant Interactions. Wiley, New York.

Bentley, B. L. 1977. Extrafloral nectaries and protection by pugnacious body guards. Annu. Rev. Ecol. Syst. 8:407–427.

Bentley, B. L., and Benson, W. W. 1988. The influence of ant foraging patterns on the behavior of herbivores, pp. 297–306. In J. C. Traeger (ed.), Advances in Myrmeoclogy. Brill Press, Leiden.

Berenbaum, M. R., and Miliczky, E. 1984. Mantids and milkweed bugs: Efficacy of aposematic coloration against invertebrate predators. Am. Midl. Nat. 111:64–68.

Bergelson, J. M., and Lawton, J. H. 1988. Does foliage damage influence predation on the insect herbivores of birch? Ecology 69:434–445.

Berisford, Y. C., and Tsao, C. H. 1975. Parasitism, predation, and disease in the bagworm, *Thyridopteryx ephemeraeformis* (Haworth) (Lepidoptera:Psychidae). Environ. Entomol. 4:549–554.

Bernays, E. A. 1986. Diet-induced head allometry among foliage-chewing insects and its importance for graminivores. Science 231:495–497.

Bernays, E. A. 1988. Host specificity in phytophagous insects: selection pressure from generalist predators. Entomol. Exp. Appl. 49:131–140.

Bernays, E. A. 1989. Host range in phytophagous insects: The potential role of generalist predators. Evol. Ecol. 3:299–311.

Bernays, E. A. 1992. Aversion learning and feeding. In D. R. Papaj and A. C. Lewis (eds.), Learning in Insects: Ecological and Evolutionary Perspectives. Chapman and Hall, New York (in press).

Bernays, E. A., and Cornelius, M. L. 1989. Generalist caterpillar prey are more palatable than specialists for the generalist predator *Iridomyrmex humilis*. Oecologia 79:427–430.

Bernays, E. A., and Graham, M. 1988. On the evolution of host specificity in phytophagous arthropods. Ecology 69:886–892.

Bernays, E. A., and Janzen, D. H. 1988. Saturniid and sphingid caterpillars: two ways to eat leaves. Ecology 69:1153–1160.

Bernays, E. A., and Montllor, C. B. 1989. Aposematism of *Uresiphita reversalis* larvae (Pyralidae). J. Lep. Soc. 43:261–273.

Blum, M. S. 1981. Chemical Defenses of Arthropods. Academic Press, New York.

Boethel, D. J., and Eikenbary, R. D. 1986. Interactions of Plant Resistance and Parasitoids and Predators of Insects. Ellis Horwood, Chichester.

Bohart, R. M., and Menke, A. S. 1976. Sphecid Wasps of the World. University of California Press, Berkeley.

Boppre, M. 1990. Lepidoptera and pyrrolizidine alkaloids. J. Chem. Ecol. 16:165–186.

Bowers, M. D. 1990. Recycling plant natural products for insect defense, pp. 353–386. In D. L. Evans and J. O. Schmidt (eds.), Insect Defenses, Adaptive Mechanisms and Strategies of Prey and Predators. SUNY Press, Albany.

Bowers, M. D., and Larin, Z. 1989. Acquired chemical defense in the lycaenid butterfly *Eumaeus atala*. J. Chem. Ecol. 15:1133–1146.

Brower, L. P. 1984. Chemical defense in butterflies, pp. 109–133. In R. I. Vane-Wright and P. R. Ackery (eds.), The Biology of Butterflies. Princeton University Press, Princeton.

Brust, G. E., Stinner, B. R., and McCartney, D. A. 1986. Predator activity and predation in corn agroecosystems. Environ. Entomol. 15:1017–1021.

Bugg, R. L., Ehler, L. E., and L. T. Wilson. 1987. Effect of common knotweed (*Polygonum aviculare*) on abundance and efficiency of insect predators of crop pests. Hilgardia 55:1–53.

Carter, M. C., Sutherland, D., and Dixon, A. F. G. 1984. Plant structure and the searching efficiency of coccinellid larvae. Oecologia 63:394–397.

Clancy, K. M., and Price, P. W. 1987. Rapid herbivore growth enhances enemy attack: sublethal plant defenses remain a paradox. Ecology 68:733–737.

Cleveland, T. 1987. Predation by tarnished plant bugs (Heteroptera: Miridae) of *Heliothis* (Lepidoptera:Noctuidae) eggs and larvae. Environ. Entomol. 16:37–40.

Cornelius, M. L. 1991. Tritrophic level interactions: effects of host plant traits on two predators of lepidopteran larvae, the Argentine ant *Iridomyrmex humilis* and the paper wasp *Mischocyttarus flavitarus*. Ph.D. Dissertation, Univ. of California, Berkeley.

Cornell, J. C., Stamp, N. E., and Bowers, M. D. 1987. Developmental change in aggregation, defense and escape behavior of buckmoth caterpillars, *Hemileuca lucina* (Saturniidae). Behav. Ecol. Sociobiol. 20:383–388.

Damman, H. 1987. Leaf quality and enemy avoidance by the larvae of a pyralid moth. Ecology 68:88–97.

Dejean, A. 1988. Memory effect on predatory behavior of *Odontomachus troglodytes* (Formicidae-Ponerinae). Behaviour 107:131–137.

Dempster, J. P. 1967. The control of *Pieris rapae* with DDT. I. The natural mortality of the young stages of *Pieris*. J. Appl. Ecol. 4:485–500.

Dempster, J. P. 1969. Some effects of weed control on the numbers of the small cabbage white (*Pieris rapae*) on Brussels sprouts. J. Appl. Ecol. 6:339–345.

Dempster, J. P., King, M. L., and Lakhani, K. H. 1976. The status of the swallowtail butterfly in Britain. Ecol. Entomol. 1:71–84.

den Boer, M. H. 1971. A colour polymorphism in caterpillars of *Bupalus piniarius* (L.) (Lepidoptera:Geometridae). Neth. J. Zool. 21:61–116.

Denno, R. F., Larsson, S., and Olmstead, K. L. 1990. Role of enemy-free space and plant quality in host-plant selection by willow beetles. Ecology 71:124–137.

Dethier, V. 1988. The feeding behavior of a polyphagous caterpillar (*Diacrisia virginica*) in its natural habitat. Can. J. Zool. 66:1280–1288.

Dethier, V. 1989. Patterns of locomotion of polyphagous arctiid caterpillars in relation to foraging. Ecol. Entomol. 14:375–386.

Dicke, M., and Sabelis, M. W. 1988. How plants obtain predatory mites as bodyguards. Neth. J. Zool. 38:148–165.

Dicke, M., van Beek, T. A., Posthumus, M. A., ben Dom, N., van Bokhoven, H., and de Groot, A. E. 1990. Isolation and identification of volatile kairomone that affects acarine predator-prey interactions: Involvement of host plant in its production. J. Chem. Ecol. 16:381–396.

Dionne, M., Butler, M., and Folt, C. 1990. Plant-specific expression of antipredator behavior by larval damselflies. Oecologia 83:371–377.

Dominguez, C. A., Dirzo, R., and Bullock, S. H. 1989. On the function of floral nectar in *Croton suberosus* (Euphorbiaceae). Oikos 56:107–114.

Drummond, F. A., James, R. L., Casagrande, R. A., and Faubert, H. 1984. Development and survival of *Podisus maculiventris* (Say) (Hemiptera: Pentatomidae), a predator of the Colorado potato beetle (Coleoptera:Chrysomelidae). Environ. Entomol. 13:1283–1286.

Duffey, S. S., Bloem, K. A., and Campbell, B. C. 1986. Consequences of sequestration of plant natural products in plant-insect-parasitoid interactions, pp. 31–60. In D. J. Boethel and R. D. Eikenbary (eds.), Interactions of Plant Resistance and Parasitoids and Predators of Insects. Ellis Horwood, Chichester.

Esbjerg, P. 1990. The significance of shelter for young cutworms (*Agrotis segetum*). Entomol. Exp. Appl. 54:97–100.

Eskafi, F. M., and Kolbe, M. M. 1990. Predation on larval and pupal *Ceratitis capitata* (Diptera: Tephritidae) by *Solenopsis geminata* (Hymenoptera: Formicidae) and other predators in Guatemala. Environ. Entomol. 19:148–153.

Evans, E. W. 1982. Influence of weather on predator/prey relations: stinkbugs and tent caterpillars. J. N.Y. Entomol. Soc. 90:241–246.

Evans, H. E., and Eberhard, M. J. W. 1970. The Wasps. University of Michigan Press, Ann Arbor.

Faeth, S. H. 1985. Host leaf selection by leafminers: Interactions among three trophic levels. Ecology 66:870–875.

Feichtinger, V. E., and Reavey, D. 1989. Changes in movement, tying and feeding patterns as caterpillars grow: The case of the yellow horned moth. Ecol. Entomol. 14:471–474.

Fields, P. G., Arnason, J. T., and Philogene, B. J. R. 1990. Behavioral and physical adaptations of three insects that feed on the phototoxic plant *Hypericum perforatum*. Can. J. Zool. 68:339–346.

Fitzgerald, T. D., and Peterson, S. C. 1988. Cooperative foraging and communication in caterpillars. BioScience 38:20–25.

Fowler, H. G., and Roberts, R. B. 1980. Foraging behavior of the carpenter ant, *Camponotus pennsylvanicus* (Hymenoptera:Formicidae) in New Jersey, USA. J. Kan. Entomol. Soc. 53:295–304.

Forsyth 1978. Cited by Raveret Richter, M. 1988. Prey hunting and interactions among social wasp (Hymenoptera: Vespidae) foragers and responses of caterpillars to hunting wasps. Ph.D. thesis, Cornell University, Ithaca, NY.

Fox, L. R., and Morrow, P. A. 1981. Specialization: Species property or local phenomenon? Science 211:887–893.

Frank, J. H. 1971. Carabidae (Coleoptera) as predators of the red-backed cutworm (Lepidoptera:Noctuidae) in central Alberta. Can. Entomol. 103:1039–1044.

Franzl, S., Naumann, C. M., and Nahrstedt, A. 1988. Cyanoglycoside storing cuticle of *Zygaena* larvae. Zoomorphology 108:183–190.

Godfrey, K. E., Whitcomb, W. H., and Stimac, J. L. 1989. Arthropod predators of velvetbean caterpillar, *Anticarsia gemmatalis* Hubner (Lepidoptera:Noctuidae), eggs and larvae. Environ. Entomol. 18:118–123.

Gould, W. P., and Jeanne, R. L. 1984. *Polistes* wasps (Hymenoptera:Vespidae) as control agents for lepidopteran cabbage pests. Environ. Entomol. 13:150–156.

Greany, P. D., and Hagen, K. S. 1981. Prey selection, pp. 121–135. In D. A. Nordlund, R. L. Jones, and W. J. Lewis (eds.), Semiochemicals, Their Role in Pest Control. Wiley, New York.

Greenberg, R., and Gradwohl, J. 1980. Leaf surface specializations of birds and arthropods in a Panamanian forest. Oecologia 46:115–124.

Greenwood, J. J. D., Cotton, P. A., and Wilson, D. M. 1989. Frequency-dependent selection on aposematic prey: Some experiments. Biol. J. Linn. Soc. 36:213–226.

Gross, H. R., Jr., Pair, S. D., and Jackson, R. D. 1985. Behavioral responses of primary entomophagous predators to larval homogenates of *Heliothis zea* and *Spodoptera frugiperda* (Lepidoptera:Noctuidae) in whorl-stage corn. Environ. Entomol. 14:360–364.

Guilford, T. 1990. The evolution of aposematism, pp. 23–61. In D. L. Evans and J. O. Schmidt (eds.), Insect Defenses, Adaptive Mechanisms and Strategies of Prey and Predators. SUNY Press, Albany.

Hegen, K. S. 1987. Nutritional ecology of terrestrial insect predators, pp. 533–577. In F.

Slansky, Jr. and J. G. Rodriguez (eds.), Nutritional Ecology of Insects, Mites, Spiders, and Related Invertebrates. Wiley, New York.

Hagen, K. S., Bombosch, S., and McMurtry, J. A. 1976. The biology and impact of predators, pp. 93–142. In C. B. Huffaker and P. S. Messenger (eds.), Theory and Practice of Biological Control. Academic Press, New York.

Hassell, M. P., and Southwood, T. R. E. 1978. Foraging strategies of insects. Annu. Rev. Ecol. Syst. 9:75–98.

Hawkins, B. A., Askew, R. R., and Shaw, M. R. 1990. Influences of host feeding, niche and foodplant type on generalist and specialist parasitoids. Ecol. Entomol. 15:275–280.

Heads, P. A., and Lawton, J. H. 1985. Bracken, ants and extrafloral nectaries. III. How insect herbivores avoid ant predation. Ecol. Entomol. 10:29–42.

Heinrich, B. 1979. Foraging strategies of caterpillars. Leaf damage and possible predator avoidance strategies. Oecologia 42:325–337.

Heinrich, B., and Collins, S. L. 1983. Caterpillar leaf damage, and the game of hide-and-seek with birds. Ecology 64:592–602.

Hölldobler, B. and Wilson, E. O. 1990. The Ants. Belknap Press of Harvard Univ. Press, Cambridge, MA.

Honda, K. 1983. Defensive potential of components of the larval osmeterial secretion of papilionid butterflies against ants. Physiol. Entomol. 8:173–179.

Hunter, M. 1987. Opposing effects of spring defoliation on late season oak caterpillars. Ecol. Entomol. 12:373–382.

Isenhour, D. J., Wiseman, B. R., and Layton, R. C. 1989. Enhanced predation by *Orius insidius* (Hemiptera:Anthocoridae) on larvae of *Heliothis zea* and *Spodoptera frugiperda* (Lepidoptera:Noctuidae) caused by prey feeding on resistant corn genotypes. Environ. Entomol. 18:418–422.

Isahaaya, I., and Casida, J. E. 1980. Properties and toxicological significance of esterases hydrolyzing permethrin and cypermethrin in *Trichoplusia ni* larval gut and integument. Pestic. Biochem. Physiol. 14:174–184.

Ishaaya, I., and Casida, J. E. 1981. Pyrethroid esterase(s) may contribute to natural pyrethroid tolerance of larvae of the common green lacewing. Environ. Entomol. 10:681–684.

Iwao, S., and Wellington, W. G. 1970. The influence of behavioral differences among tent-caterpillar larvae on predation by a pentatomid bug. Can. J. Zool. 48:896–898.

Jeanne, R. L. 1979. A latitudinal gradient in rates of ant predation. Ecology 60:1211–1224.

Jones, D. A., Parsons, J., and Rothschild, M. 1962. Released hydrocyanic acid from crushed tissues of all stages of the life cycle of species of Zygaeninae (Lepidoptera). Nature (London) 193:52–53.

Jones, R. E. 1987. Ants, parasitoids, and the cabbage butterfly *Pieris rapae*. J. Anim. Ecol. 56:739–749.

Jones, R. E., Nealis, V. G., Ives, P. M., and Scheermeyer, E. 1987. Seasonal and spatial variation in juvenile survival of the cabbage butterfly *Pieris rapae:* Evidence for patchy density dependence. J. Anim. Ecol. 56:723–738.

Joos, B., Casey, T. M., Fitzgerald, T. D., and Buttermer, W. A. 1988. Roles of the tent in behavioral thermoregulation of eastern tent caterpillars. Ecology 69:2004–2011.

Kelly, R. B., Seiber, J. N., Segall, D. D., and Brower, L. P. 1987. Pyrrolizidine alkaloids in overwintering monarch butterflies (*Danaus plexippus*) from Mexico. Experientia 43:943–946.

Koptur, S., and Lawton, J. H. 1988. Interactions among vetches bearing extrafloral nectaries, their biotic protective agents, and herbivores. Ecology 69:278–283.

Lawrence, W. S. 1990. The effects of group size and host species on development and survivorship of a gregarious caterpillar *Halisidota caryae* (Lepidoptera:Arctiidae). Ecol. Entomol. 15:53–62.

Lawson, F. R., Rabb, R. L., Guthrie, F. E., and Bowery, T. G. 1961. Studies of an integrated control system for hornworms on tobacco. J. Econ. Entomol. 54:93–97.

Lederhouse, R. C. 1990. Avoiding the hunt: primary defenses of lepidopteran caterpillars, pp. 175–189. In D. L. Evans and J. O. Schmidt (eds.), Insect Defenses, Adaptive Mechanisms and Strategies of Prey and Predators. SUNY Press, Albany.

Letourneau, D. K. 1990. Mechanisms of predator accumulation in a mixed crop system. Ecol. Entomol. 15:63–69.

Lima, S. L., and Dill, L. M. 1990. Behavioral decisions made under risk of predation: A review and prospectus. Can. J. Zool. 68:619–640.

Lopez, J. D., Ridgway, R. L., and Pinnell, R. E. 1976. Comparative efficacy of four insect predators of the bollworm and tobacco budworm. Environ. Entomol. 5:1160–1164.

Malcolm, S. M. 1989. Disruption of web structure and predatory behavior of a spider by plant-derived chemical defenses of an aposematic aphid. J. Chem. Ecol. 15:1699–1716.

Mansour, F., Rosen, D., and Shulov, A. 1981. Disturbing effect of a spider on larval aggregations of *Spodoptera littoralis*. Entomol. Exp. Appl. 29:234–237.

Marston, N. L., Schmidt, G. T., Biever, K. D., and Dickerson, W. A. 1978. Reaction of five species of soybean caterpillars to attack by the predator, *Podisus maculiventris*. Environ. Entomol. 7:53–56.

Mason, R. R., and Torgersen, T. R. 1983. Mortality of larvae in stocked cohorts of the Douglas-fir tussock moth, *Orgyia pseudotsugata* (Lepidoptera: Lymantriidae) Can. Entomol. 115:1119–1127.

Mauricio, R., and Bowers, M. D. 1990. Do caterpillars disperse their damage?: larval foraging behavior of two specialist herbivores, *Euphydryas phaeton* (Nymphalidae) and *Pieris rapae* (Pieridae). Ecol. Entomol. 15:153–161.

McLain, K. 1979. Terrestrial trail following by three species of predatory stinkbugs. Florida Entomol. 62:152–154.

Montllor, C. B., Bernays, E. A., and Barbehenn, R. V. 1990. Importance of quinolizidine alkaloids in the relationship between larvae of *Uresiphita reversalis* (Lepidoptera:Pyralidae) and a host plant, *Genista monspessulana*. J. Chem. Ecol. 16:1853–1865.

Montllor, C. B., Bernays, E. A., and Cornelius, M. L. 1991. Responses of two hymenopteran predators to surface chemistry of their prey: significance for an alkaloid–sequestering caterpillar. J. Chem. Ecol. 17:391–399.

Moore, L. V., Myers, J. H., and Eng, R. 1988. Western tent caterpillars prefer the sunny side of the tree, but why? Oikos 51:321–326.

Morris, R. F. 1963. The effect of predator age and prey defense on the functional response of *Podisus maculiventris* Say to the density of *Hyphantria cunea* Drury. Can. Entomol. 95:1009–1020.

Morris, R. F. 1972a. Predation by insects and spiders inhabiting colonial webs of *Hyphantria cunea*. Can. Entomol. 104:1197–1207.

Morris, R. F. 1972b. Predation by wasps, birds, and mammals on *Hyphantria cunea*. Can. Entomol. 104:1581–1591.

Mullin, C. A., and Croft, B. A. 1984. *Trans*-epoxide hydrolase: A key indicator enzyme for herbivory in arthropods. Experientia 40:176–178.

Mullin, C. A., Croft, B. A., Strickler, K., Matsumura, F., and Miller, J. R. 1982. Detoxification enzyme differences between a herbivorous and predatory mite. Science 217:1270–1272.

Nahrstedt, A., and Davis, R. H. 1983. Occurrence, variation and biosynthesis of the cyanogenic glycosides linamarin and lotaustralin in species of the Heliconini (Insecta: Lepidoptera). Comp. Biochem. Physiol. 75B:65–73.

Niemela, P., and Tuomi, J. 1987. Does the leaf morphology of some plants mimic caterpillar damage? Oikos 50:256–257.

Nordlund, D. A., Lewis, W. J., and Altieri, M. A. 1988. Influences of plant-produced allelochemicals on the host/prey selection behavior of entomophagous insects, pp. 65–90. In P. Barbosa and D. K. Letourneau (eds.), Novel Aspects of Insect-Plant Interactions. Wiley, New York.

Obrycki, J. J. 1986. The influence of foliar pubescence on entomophagous species, pp. 61–83. In D. J. Boethel and R. D. Eikenbary (eds.), Interactions of Plant Resistance and Parasitoids and Predators of Insects. Ellis Horwood, Chichester.

Odell, T. M., and Godwin, P. A. 1984. Host selection by *Blepharipa pratensis* (Meigen), a tachinid parasite of the gypsy moth, *Lymantria dispar* L. J. Chem. Ecol. 10:311–320.

Papaj, D. R., and Prokopy, R. J. 1989. Ecological and evolutionary aspects of learning in phytophagous insects. Annu. Rev. Entomol. 34:315–350.

Paradise, C. J., and Stamp, N. E. 1991. Prey recognition time of praying mantids and consequent survivorship of unpalatable prey. J. Insect Behav. 4:265–273.

Paradise, C. J., and Stamp, N. E. 1992. Episodes of unpalatable prey reduce consumption and growth of juvenile praying mantids. J. Insect Behav. (in press).

Peterson, S. C. 1987. Communication of leaf suitability by gregarious eastern tent caterpillars (*Malacosoma americanum*). Ecol. Entomol. 12:283–289.

Peterson, S. C., Johnson, N. D., and LeGuyader, J. L. 1987. Defensive regurgitation of allelochemicals derived from host cyanogenesis by eastern tent caterpillars. Ecology 68:1268–1272.

Price, P. W., Bouton, C. E., Gross, P., McPheron, B. A., Thompson, J. N., and Weiss, A. E. 1980. Interactions among three trophic levels: influence of plants on interactions between insect herbivores and natural enemies. Annu. Rev. Ecol. Syst. 11:41–65.

Rabb, R. L., and Lawson, F. R. 1957. Some factors influencing the predation of *Polistes* wasps on the tobacco hornworm. J. Econ. Entomol. 50:778–784.

Rasmy, A. H. 1977. Predatory efficiency and biology of the predatory mite, *Amblyseius gossipi* (Acarina:Phytoseidae) as affected by plant surfaces. Entomophaga 22:421–423.

Raveret Richter, M. 1988. Prey hunting and interactions among social wasp (Hymenoptera: Vespidae) foragers and responses of caterpillars to hunting wasps. Ph.D. thesis, Cornell University, New York.

Raveret Richter, M. 1990. Hunting social wasp interactions: influence of prey size, arrival order, and wasp species. Ecology 71:1018–1030.

Reid, C. D., and Lampman, R. L. 1989. Olfactory responses of *Orius insidiosus* (Hemiptera: Anthocoridae) to volatiles of corn silks. J. Chem. Ecol. 15:1109–1115.

Reynolds, S. E., Yeomans, M. R., and Timmins, W. A. 1986. The feeding behavior of caterpillars (*Manduca sexta*) on tobacco and artificial diet. Physiol. Entomol. 11:39–51.

Riechert, S. E., and Bishop, L. 1990. Prey control by an assemblage of generalist predators: spiders in garden test systems. Ecology 71:1441–1450.

Riechert, S. E., and Lockley T. 1984. Spiders as biological control agents. Annu. Rev. Entomol. 29:299–320.

Risch, S., and Carroll, C. 1982. Effect of a keystone predaceous ant, *Solenopsis geminata,* on arthropods in a tropical agroecosystem. Ecology 63:1979–1983.

Risch, S. J., Wrubel, R., and Andow, D. 1982. Foraging by a predaceous beetle, *Coleomegilla maculata* (Coleoptera: Coccinellidae), in a polyculture: Effects of plant density and diversity. Environ. Entomol. 11:949–950.

Roessingh, P. 1988. Trail marking and following by larvae of the small ermine moth *Yponomentua cagnagellus*. Ph.D. thesis, Agricultural University, Wageningen, The Netherlands.

Rothschild, M., Aplin, R. T., Cockrum, P. A., Edgar, J. A., Fairweather, P., and Lees, R. 1979. Pyrrolizidine alkaloids in arctiid moths (Lep.) with a discussion on host plant relationships and the role of these secondary plant substances in the Arctiidae. Biol. J. Linn. Soc. 12:305–326.

Ruehlmann, T. E., Matthews, R. W., and Matthews, J. R. 1988. Roles for structural and temporal shelter-changing by fern-feeding lepidopteran larvae. Oecologia 75:228–232.

Sagers, C. L. 1990. Herbivores manipulate host plant defenses. Bull. Ecol. Soc. Am. (Suppl.) 71:312.

Schaeffer, P. W. 1977. Attacking wasps, *Polistes* and *Therion,* penetrate silk nests of fall webworm. Environ. Entomol. 6:591.

Schultz, J. C. 1983. Habitat selection and foraging tactics of caterpillars in heterogeneous trees, pp. 61–90. In R. F. Denno and M. S. McClure (eds.), Variable Plants and Herbivores in Natural and Managed Systems. Academic Press, New York.

Shaw, P. B., Owens, J. C., Huddleston, E. W., and Richman, D. B. 1987. Role of arthropod predators in mortality of early instars of the range caterpillar, *Hemileuca oliviae* (Lepidoptera: Saturniidae), Environ. Entomol. 16:814–820.

Sheehan, W. 1991a. Host range patterns of hymenopteran parasitoids of exophytic lepidopteran folivores, pp. 209–248. In E. A. Bernays (ed.), Insect-Plant Interactions, Vol. 3. CRC Press, Boca Raton, FL.

Sheehan, W. 1991b. Caterpillar hairs and spines and the evolution of gregariousness in Lepidoptera. Unpublished manuscript.

Sillen-Tullberg, B., and Leimar, O. 1988. The evolution of gregariousness in distasteful insects as a defense against predators. Am. Nat. 132:723–734.

Skinner, G. J., and Whittaker, J. B. 1981. An experimental investigation of interrelationships between the wood-ant (*Formica rufa*) and some tree-canopy herbivores. J. Anim. Ecol. 50:313–326.

Slansky, F., and Scriber, J. M. 1985. Food consumption and utilization, pp. 87–163. In G. A. Kerkut and L. I. Gilbert (eds.), Comprehensive Insect Physiology Biochemistry and Pharmocology, Vol. 4. Pergamon Press, Oxford.

Spangler, S. M., and Macmahon, J. A. 1990. Arthropod faunas of monocultures and polycultures in reseeded rangelands. Environ. Entomol. 19:244–250.

Stamp, N. E. 1984. Foraging behavior of tawny emperor caterpillars (Nymphalidae:*Asterocampa clyton*). J. Lep. Soc. 38:186–191.

Stamp, N. E. 1986. Physical constraints of defense and response to invertebrate predators by pipevine caterpillars (*Battus philenor*:Papilionidae). J. Lep. Soc. 40:191–215.

Stamp, N. E. 1990. Growth versus molting time of caterpillars as a function of temperature, nutrient concentration and the phenolic rutin. Oecologia 82:107–113.

Stamp, N. E. 1992. Relative susceptibility to predation of two species of caterpillar on plantain. Oecologia (in press).

Stamp, N. E., and Bowers, M. D. 1988. Direct and indirect effects of predatory wasps (*Polistes* sp.: Vespidae) on gregarious caterpillars (*Hemileuca lucina:* Saturniidae). Oecologia 75:619–624.

Stamp, N. E., and Bowers, M. D. 1990. Variation in food quality and temperature constrain foraging of gregarious caterpillars. Ecology 71:1031–1039.

Stamp, N. E., and Bowers, M. D. 1991. Indirect effect on survivorship of caterpillars due to presence of invertebrate predators. Oecologia 88:325–330.

Steward, V. B., Smith, K. G., and Stephen, F. M. 1988. Predation by wasps on lepidopteran larvae in an Ozark forest canopy. Ecol. Entomol. 13:81–86.

Strokov, 1956. Cited by Holldobler, B., and Wilson, E. O. 1990. The Ants. Belknap Press of Harvard University Press, Cambridge, MA.

Suzuki, N., Kunimi, Y., Uematsu, S., and Kobayashi, K. 1980. Changes in spatial distribution pattern during the larval stage of the fall webworm, *Hyphantria cunea* Drury (Lepidoptera:Arctiidae). Res. Popul. Ecol. 22:273–283.

Szentesi, A., and Jermy, T. 1990. The role of experience in host plant choice by phytophagous insects, pp. 39–74. In E. A. Bernays (ed.), Insect-Plant Interactions, Vol. 2. CRC Press, Boca Raton.

Takagi, M., Hirose, Y., and Yamasaki, M. 1980. Prey-location learning in *Polistes jadwigae* Dalla Torre (Hymenoptera:Vespidae), field experiments on orientation. Kontyu 48:53–58.

Tauber, C. A., and Tauber, M. J. 1987. Food specificity in predaceous insects: a comparative ecophysiological and genetic study. Evol. Ecol. 1:175–186.

Tautz, J., and Markl, H. 1978. Caterpillars detect flying wasps by hairs sensitive to airborne vibration. Behav. Ecol. Sociobiol. 4:101–110.

Tilman, D. 1978. Cherries, ants and tent caterpillars: timing of nectar production in relation to susceptibility of caterpillars to ant predation. Ecology 59:686–692.

Tsubaki, Y., and Kitching, R. L. 1986. Central place foraging in larvae of the charaxine butterfly, *Polyura pyrrhus* (L.): A case study in a herbivore. J. Ethol. 4:59–68.

Turlings, T. C. J., Tumlinson, J. H., and Lewis, W. J. 1990. Exploitation of herbivore-induced plant odors by host-seeking parasitic wasps. Science 250:1251–1253.

Vet, L. E. M., and Groenewold, A. 1990. Sesiochemicals and learning in parasitoids. J. Chem. Ecol. 16:3119–3135.

Vulinec, K. 1990. Collective security: aggregation by insects as a defense, pp. 251–288. In D. L. Evans and J. O. Schmidt (eds.), Insect Defenses, Adaptive Mechanisms and Strategies of Prey and Predators. SUNY Press, Albany.

Weseloh, 1988. Effects of microhabitat, time of day, and weather on predation of gypsy moth larvae. Oecologia 77:250–254.

Weseloh, R. 1989. Simulation of predation by ants based on direct observations of attacks on gypsy moth larvae. Can. Entomol. 121:1069–1076.

Whitman, D. W., and Eller, F. J. 1990. Parasitic wasps orient to green leaf volatiles. Chemoecology 1:69–75.

Wink, M., and Romer, P. 1986. Acquired toxicity—the advantages of specializing on alkaloid-rich lupins to *Macrosiphon albifrons* (Aphidae). Naturwissenschaften 73:210–212.

Witz, B. 1990. Antipredator mechanisms in arthropods: A twenty year literature survey. Florida Entomol. 73:71–99.

Yamasaki, M., Hirose, Y., and Takagi, M. 1978. Repeated visits of *Polistes jadwigae* Dalla Torre (Hymenoptera:Vespidae) to its hunting site. Jpn. J. Appl. Entomol. Zool. 22:51–55.

Yeargan, K. V., and Braman, S. K. 1989. Comparative behavioral studies of indigenous hemipteran predators and hymenopteran parasites of the green cloverworm (Lepidoptera:Noctuidae). J. Kans. Entomol. Soc. 62:156–163.

Yu, S. J. 1987. Biochemical defense capacity in the spined soldier bug (*Podisus maculiventris*) and its lepidopterous prey. Pestic. Biochem. Physiol. 28:216–223.

6

Potential Effects of Parasitoids on the Evolution of Caterpillar Foraging Behavior

Ronald M. Weseloh

Introduction

Few relevant data exist on the subject of this chapter: potential effects of parasitoids on the evolution of caterpillar foraging behavior. Numerous studies have shown differences and similarities between parasitoid and host habitats and microhabitats, but few of these have emphasized parasitoid activities that might drive evolution of host behavior. Many selection pressures beside parasitoids impact on caterpillar foraging, as documented elsewhere in this volume. Identifying the specific contributions due to parasitoids may be difficult. Descriptive field studies can be helpful but cannot in general be definitive. Laboratory experiments can identify important features related to caterpillar foraging, and perhaps rule out alternative explanations such as host-plant defenses, but they are no substitute for field work. Careful field experiments can provide much insight about this subject, but few have been done.

The paucity of research does not mean that parasitoids have no effect. Parasitoids of Lepidoptera are numerous and ubiquitous. Hairston et al. (1960) and Lawton and Strong (1981) argue that because vegetation is seldom completely destroyed by herbivores, natural enemies are largely responsible for keeping populations of primary consumers in check [but see Murdoch (1966) for a contrary view]. It is thus tempting to conclude that parasitoids exert considerable selection pressure on their hosts. Although the lack of data make any discussion of this subject speculative, the idea deserves serious consideration. An understanding of how parasitoids influence caterpillar behavior could explain a great deal about foraging patterns seen in nature.

As one approach to this subject, I will examine unique characteristics of parasitoids to determine if any could cause changes in foraging behaviors of caterpillars. If they could, it might be possible to predict how caterpillar foraging would evolve. A comparison of parasitoid traits with known characteristics of caterpillar behavior could then provide insights about the roles of parasitoids in shaping foraging patterns. Later I will discuss selected parasitoid–host systems in greater detail.

Parasitoids are "parasite-like" because immatures exist as parasites in or on hosts. The host is killed or prevented from reproducing in almost all cases. Thus, in terms of population dynamics, parasitoids function as predators. The vast majority of parasitoids are either in the order Hymenoptera (Ichneumonoidea, Chalcidoidea, Proctotrupoidea, and related superfamilies) or in the dipteran family Tachinidae. The adult is almost always free-living and is usually the stage that locates hosts. Thus, effects on host foraging behavior will be caused by activities of adult parasitoids.

Parasitoids differ from predators in at least two ways: they are often much smaller than even invertebrate predators, and they must complete at least one generation for each generation of their host (unless a state of dormancy intervenes). These characteristics enable parasitoids to exploit microhabitats that would not support a predator and, owing to short generations, to evolve as fast or faster than their hosts. As a consequence they may become very specialized to microhabitats and/or host species (Lawton 1986).

In many cases, parasitoids seek out a specific habitat within which their hosts are likely to occur before searching for hosts themselves. Such host-habitat location is considered to be one of the primary categories in the host selection process of parasitoids (Vinson 1981). The habitats of almost all lepidopterous larvae are plants, a subset of which parasitoids may be attracted to. For instance, the ichneumonid *Cardiochiles nigriceps* parasitizes its host, *Heliothis virescens,* on cotton and tobacco but not on peanuts (Vinson 1981).

Parasitoids often respond specifically to chemicals produced by some of the plants on which their hosts feed, and they may restrict their activities to particular vegetation strata (Weseloh 1976d). They also use specific cues from hosts to locate them, including chemicals produced when plant leaves are damaged and odors from host feces (Vinson 1976). At least one parasitoid responds to silk produced by its preferred host, the gypsy moth, but not to silk of other Lepidoptera (Weseloh 1976c). Some predators may also have restricted habitat and host preferences, but it is clear that parasitoids are more specialized (Huffaker et al. 1971).

How will the extreme specialization of parasitoids modify the foraging patterns of caterpillars? Depending on selection pressures and evolutionary constraints, a parasitoid might follow a host into each new microhabitat that the latter invades or become specialized on a subset of microhabitats and perhaps widen its host range within that subset. The latter situation would permit some hosts to escape. For example, a very small parasitoid with a short ovipositor may have difficulty attacking caterpillars that bore in plant stems and so specialize on those that mine in leaves. Potential hosts that evolve into stem borers would not be exploited by that parasitoid. A comparison of host and parasitoid microhabitat preferences and constraints might indicate which of these scenarios is most likely. This comparison is done in the following section.

Comparison of Caterpillar and Parasitoid Habitats

A summary of caterpillar and parasitoid habitats is presented in Table 6.1. The information was taken from the literature, and neither host nor parasitoid habitats are necessarily well defined or consistent. In no case are habitats completely described, and many of the systems are artificial agroecosystems. Different studies have emphasized different aspects of host and parasitoid biologies. In some, parasitoid responses to plant odors and caterpillar feeding preferences are compared. In others, toxic effects of plant chemicals on parasitoids and hosts or foraging by parasitoids and hosts in various parts of plants are examined.

Nevertheless, a trend is evident: in 61% of the cases (14 of 23), hosts appear to have wider habitat preferences than do their parasitoids. Caterpillars tend to have a greater tolerance of host plant chemicals (which may become sequestered in the host and so kill immature parasitoids), or greater spatial distributions within plants, or greater ranges of plant associations than do their parasitoids. Thus, many parasitoids appear to be more specialized in habitat preferences, particularly with regard to association with host food plants, than are their caterpillar hosts. The relatively greater habitat generality of herbivores may enable them to evolve away from their parasitoids.

The patterns shown in Table 6.1 illustrate the concept of "enemy free space" as discussed by Lawton (1986). Arguing from a theoretical and general viewpoint, he emphasized that parasitoids are known to restrict their activities to certain parts of plants, microhabitats, or plant species, and that they often attack a number of taxonomically unrelated hosts that frequent the same habitat (see also Lawton 1978). This observation is not new. Cushman (1926) argued that many parasitic Hymenoptera, particularly those attacking concealed hosts, search a particular microhabitat, parasitizing any hosts that are there. The tendency is most pronounced in polyphagous parasitoïds, which may attack any hosts present in their microhabitat.

Most of the evidence, then, suggests that a plausible evolutionary route for lepidopterous larvae to escape parasitoids is to change habitats, such as by colonizing a new plant species. Whether this would actually occur and what the observed outcome would be depends in large part on the effectiveness of the parasitoids associated with the colonized plant.

At one extreme, parasitoid activity on some plants may cause caterpillars to avoid these plants and thus eventually to become more oligophagous. This is in contrast to a commonly accepted explanation for oligophagy, which involves only herbivore–plant interactions (see Strong et al. 1984 for discussion). Indeed,

Table 6.1. *Summary of Habitats of Caterpillar Hosts and Their Parasitoids*

Host	Host habitat	Parasitoid	Parasitoid habitat	Can host escape?	References
Differences in Chemical Tolerance					
Heliothis zea	Tolerates plants with tomatine	*Hyposoter exiguae*	Little tolerance for tomatine	Yes	Campbell and Duffey (1979)
Manduca sexta	Tolerates plants with nicotine	*Apanteles congregatus*	Little tolerance for tomatine	Yes	Gilmore (1938), Thurston and Fox (1972)
Microhabitat Differences					
Pieris rapae	*Brassica*—open and headed strains	*Apanteles glomeratus* *Phryxe vulgaris*	*Brassica*—open strains	Yes	Pimentel (1961)
Thyridopteryx ephemeraeformis	Females high in trees	*Itoplectis conquisitor*	Low in tree	Yes	Gross and Fritz (1982)
Bupalus piniarius	Top of needles	*Eucarcelia rutilla*	Top of needles	No	Herrebout (1960, 1969)
Choristoneura fumiferana	Upper tree crown	*Apanteles fumiferana*	Upper tree crown	No	Jaynes (1954), Miller (1959)
		Glypta fumiferanae	Whole tree	No	Jaynes (1954)
Choristoneura occidentalis	Upper tree crown	*Meteorus trachynotus*	Upper tree crown	No	Schmid (1981)
		Glypta fumiferanae	Whole tree	No	
Epinotia tedella	Shaded parts of trees	*Pimplopterus dubius*	Top of tree	Yes	Munster-Swendsen (1980)
		Apanteles tedellae	Low part of tree	No	
Adoxophyes orana	Outer branches	*Colpoclyeus florus*	Outer branches	No	Dijkstra (1986)

Table 6.1. (continued)

Host	Host habitat	Parasitoid	Parasitoid habitat	Can host escape?	References
Host Plant Differences					
Pieris napi	*Arabis* and other crucifers	*Apanteles glomeratus*	Crucifers—not *Arabis*	Yes	Ohsaki and Sato (1990)
Ostrinia nubilalis	Two corn strains	*Lydella griescens*	One corn strain	Yes	Franklin and Holdaway (1966)
O. nubilalis	Polyphagous	*Apanteles flavipes*	Corn	Yes	Inayatullah (1983)
O. nubilalis	Polyphagous	*Macrocentrus grandii*	Corn	Yes	Ding *et al.* (1989)
Rhyacionia buoliana	Pines	*Pimpla ruficollis*	Pines	No	Thorpe and Caudle (1938)
	Scot's, red pine	*Itoplectis conquisitor*	Scot's pine	Yes	Arthur (1962)
Heliothis zea and	Polyphagous	*Campoletis*	Sesame	Yes	Pair et al. (1982)
H. virescens	Polyphagous	*sonorensis*	Host plants restricted	Yes	Elzen et al. (1983)
Heliothis spp.	Polyphagous	*Microplitis croceipes*	Cotton	Yes	Mueller (1983)
H. armigera	Polyphagous	*Cotesia kazak*	Cotton	Yes	Jalali et al (1988)
Platyptilia carduidadyla	Artichoke	*Phaeogenes cynarae*	Artichoke	No	Bragg (1974)

Bernays and Graham (1988) argue that natural enemies may be responsible for the narrow host range preferences often seen in insect herbivores. They note that while many insects are deterred from feeding on various plants by secondary chemicals, the deterring chemicals are not really harmful and in many cases the plants are nutritionally suitable. However, the herbivores may be more vulnerable to predators and parasitoids on some plants, and so the herbivores have evolved to avoid those plants using the deterring chemicals as cues.

At the other extreme, a host could become more polyphagous by continually being forced to colonize new plant species. Perhaps at one level (e.g., among plant families) predators and parasitoids may act to select herbivores with narrow host ranges, and at a lower level (e.g., within plant families) parasitoids might force hosts to use a wider range of microhabitats or plant species. The outcome would be sensitive to local distributions of parasitoids on the plants and microhabitats involved, but detailed information on such distributions is too scarce for meaningful conclusions to be drawn at present. Thus, although a consideration of parasitoid habitats suggests ways in which herbivore foraging behavior might be influenced, testable predictions must await much greater knowledge about parasitoid ecology.

Ways of Avoiding Parasitoids

There are various ways that Lepidoptera could restrict their feeding to habitats offering the best protection from parasitoids:

1. *Feeding on Inconspicuous Hosts and Monophagy*. Inconspicuous and rare hosts may be difficult for parasitoids to find (Sato and Ohsaki 1987). A moth or butterfly colonizing such a plant may have such a selective advantage that it becomes monophagous, even though other plants are suitable for it (Bernays and Graham 1988).
2. *Exploitation of Ephemeral Habitats*. Butterflies or moths may be able to disperse and find plant hosts more quickly than parasitoids can find them. An example would be *Apanteles glomeratus,* whose host, *Pieris rapae,* is able to escape by continually colonizing new habitats (Ohsaki and Sato 1990).
3. *Use of Microhabitats that Parasitoids do not Exploit Readily*. Particularly in forests, the distributions of parasitoids and hosts often do not overlap completely. Numerous examples are given in Table 6.1.
4. *Escape in Time*. If a parasitoid is active only during the daytime, as are many parasitic Hymenoptera, a host may be able to avoid attack by feeding only at night. The gypsy moth, discussed below, is a good example. Other herbivore species may feed at irregular times either day

or night, but often this is attributed to pressure from predation by birds (Heinrich 1979; Heinrich and Collins 1983; Heinrich, Chapter 7).

5. *Confusion or Proliferation of Cues*. Caterpillars often have short feeding bouts and make many small, widely spaced feeding injuries. Such foraging patterns have usually been implicated as ways to confuse birds (Heinrich and Collins 1983), but they might also be effective against parasitoids that respond to damaged leaves (Hassell 1968, ODell and Godwin 1984).
6. *Association with Ants*. Aphids are often protected from natural enemies by their attending ants, and the same occurs with lycaenid caterpillars, as shown below (see also Baylis and Pierce, Chapter 12).
7. *Colonial Behavior*. Caterpillars feeding in groups may be able to better defend themselves from parasitoids. This behavior is also discussed below.

Impact of Parasitoids on Caterpillar Foraging

The Gypsy Moth and Its Parasitoids

Even when parasitoid habitat specialization occurs, it may still be difficult to show that parasitoids drive the evolution of foraging in caterpillars. Most Lepidoptera are attacked by many parasitoid species, each of which has difficult, possibly conflicting, effects on host behavior. Each parasitoid also is subject to its own selective pressures and evolutionary constraints, including competition with other parasitoids, that affect how it interacts with hosts. The gypsy moth–parasitoid system provides an example.

The gypsy moth, *Lymantria dispar,* is a very important forest defoliator in the northeastern United States. It is an imported insect, having arrived from Europe in 1868–1869 in the Boston area, and has been spreading ever since (McManus and McIntyre 1981). The pest is a typical outbreak species, with very low numbers in most years but producing periodic or quasiperiodic outbreaks lasting usually 2 to 3 years. It has one generation a year, and caterpillars hatch from overwintering eggs as tree leaves expand in spring. Larvae are present for about 2 months and feed on a wide variety of forest trees and shrubs, but especially species of oak. In Connecticut, pupae are formed usually by early July, and adults emerge a few weeks later. Adults mate and females lay eggs in late July.

Larval foraging behavior varies with age and population density. Young larvae are almost always present on leaves or small twigs of food plants, particularly in lower parts of trees and on shrubs (Ticehurst and Yendol 1989). At low population densities, about two-thirds of the fourth, fifth, and sixth instars generally rest during the day in bark fissures or under bark flaps on trees, and about one-third rest in the leaf litter (Campbell et al. 1975a). All ascend trees to feed at night

(Leonard 1970). However, at densities high enough for defoliation to occur, large larvae feed at any time and will migrate away from stripped food plants.

Many natural enemies were imported from Europe and Asia in a massive effort to control the gypsy moth early in this century. At present, 11 species of the imported parasitoids have become established in North America (Hoy 1976; Schaefer et al. 1989). For some the host–parasitoid relationships are known in some detail.

The Tachinid Compsilura concinnata

This tachinid fly was imported for control of the gypsy moth, but it is highly polyphagous and successfully attacks many lepidopterous larvae (Culver 1919). By placing gypsy moth larvae tethered by thread in various portions of trees, I was able to determine that *C. concinnata* attacks hosts primarily on the lower leaves (Weseloh 1982). At low population densities, large gypsy moth caterpillars are often on leaves only at night when *C. concinnata* is inactive. However, young caterpillars are most abundant on lower leaves of trees both day and night (Ticehurst and Yendol 1989), and so are vulnerable to this tachinid. Indeed, the parasitoid sometimes destroys many gypsy moth caterpillars when the latter's population densities are low (Elkinton and Gould 1989). But because the tachinid is dependent on other hosts to overwinter, and because it can produce only one generation each year in the gypsy moth (Weseloh 1982), it is not able to respond quickly to increasing gypsy moth numbers. As a consequence, *C. concinnata* rarely parasitizes more than a few percent of the caterpillars under outbreak conditions (personal observations).

Thus at low densities the large caterpillars escape in time, and at high densities the host population oversaturates the parasitoids. Many generalist predators are also effective at low gypsy moth populations (Smith and Lautenschlager 1981), but suffer the same disadvantage as *C. concinnata* at high densities. The gypsy moth and other outbreak species may have evolved adaptations that enable them to periodically escape such generalist natural enemies.

The Braconid Cotesia (=Apanteles) melanoscela

This parasitoid is a small braconid wasp that lays eggs in gypsy moth caterpillars. In the northeastern United States the gypsy moth is its only host. It overwinters as a mature larva inside a cocoon located on tree trunks and other objects, and emerges as an adult about the same time that gypsy moth eggs hatch. The female parasitoids attack very young caterpillars and their progeny develop without diapausing to form adults that attack somewhat larger caterpillars. The progeny of this generation overwinter.

Adult females are very good searchers. They are active throughout the forest canopy (Weseloh 1972), and find feeding hosts by responding specifically to chemical(s) in gypsy moth silk on leaves (Weseloh 1976c). When parasitoids were released into areas where gypsy moth population densities were low, parasitism rates increased substantially, showing that the parasitoid is able to locate hosts easily when the latter are scarce (Weseloh and Anderson 1975).

Because of its presence in the host habitat, good searching abilities and high reproductive potential, *C. melanoscela* should be able to exert substantial pressure on the evolution of caterpillar foraging, both at low and high population densities. However, parasitism levels rarely reach 20% for two reasons. Overwintering parasitoid larvae are attacked heavily by native hyperparasitoids (Weseloh 1978), and so adult *C. melanoscela* emerging in the spring are relatively scarce. The progeny they produce emerge when most caterpillars are fourth instars or larger, and adult parasitoids are not able to easily attack these because of their vigorous defensive behaviors (Weseloh 1976b). Also, fourth instars have already begun to abandon leaves during daylight hours when the parasitoid is active.

It is conceivable that *C. melanoscela* applies selection pressure for the rapid development of gypsy moth caterpillars. However, rapid development is also advantageous for other reasons, particularly because of declining foliage quality as leaves mature. Young caterpillars develop best on young foliage (Valentine 1981). Thus, rapid development may not be a primary response to selection pressure by this parasitoid.

Parasitism by *C. melanoscela* might also influence the diel periodicity of large caterpillars. However, if large gypsy moth larvae are driven from leaves by this parasitoid, then the tendency of young larvae to remain on leaves at all times must be accounted for. After all, the young larvae are the stages most susceptible to the parasitoid.

The restriction of small caterpillars to tree leaves could be caused by ants. *Formica* spp. are very common on the forest floor and are capable of destroying high proportions of young gypsy moth caterpillars in the litter. Larger larvae are not so susceptible, nor are larvae in trees, particularly those on leaves (Weseloh 1988, 1989a). The parasitization of young caterpillars by the small numbers of overwintering *C. melanoscela* that escape hyperparasitoids is probably low enough that caterpillars suffer less mortality by staying on leaves during the first two instars than if they took refuge on the ant-infested forest floor. The greatly increased populations of sccond generation *C. melanoscela* adults, along with parasitism by the tachinid, *C. concinnata,* may be threatening enough to drive caterpillars out of trees at a time when they are less susceptible to forest ants.

The Tachinid Parasitigena silvestris

This tachinid is also a specific parasitoid, and the gypsy moth is apparently its only host. It attacks large larvae, laying large, white eggs on them. *P. silvestris* has only one generation a year and overwinters within a puparium in the soil.

Again, by placing tethered gypsy moth larvae in various parts of forest trees,

I was able to show that *P. silvestris* attacks exposed hosts on tree trunks (i.e., those not hidden under bark flaps, etc.) rather than on leaves (Weseloh 1974). Its activity peaks in the afternoon and early evening (Weseloh 1976a).

Large gypsy moth larvae that remain in leaf litter during the day are generally unavailable to *P. silvestris*. For the caterpillars, such behavior is probably a compromise strategy. In the leaf litter they are vulnerable to a variety of generalist predators, mainly vertebrates such as forest mice and shrews, that feed on large larvae (Smith and Lautenschlager 1981). In the canopy they are vulnerable to *Compsilura* and *Cotesia,* as well as birds (Smith and Lautenshlager 1981). If caterpillars are able to find suitable concealed locations on trees, such as deep bark cracks or loose bark flaps, they tend to rest under these in preference to leaf litter. Here they are protected against generalist predators (Weseloh 1988), foliage-frequenting specific parasitoids, and *P. sylvestris*. Indeed, gypsy moths appear to survive best in forests where trees have many loose bark flaps and other resting locations above ground level (Campbell et al. 1975b). However, larvae must still climb trees around dusk in order to feed, and it is likely that *P. silvestris* attacks them heavily at that time because up to 90% of gypsy moth caterpillars may be parasitized in the years following the collapse of a heavy infestation (Sisojevic 1975; Weseloh personal observations). If larvae were continuously available to this parasitoid it is likely that many more would be destroyed. Thus, there is circumstantial evidence that selection pressures exerted by *P. silvestris* were influential in determining the foraging behavior of gypsy moths.

Discussion—Gypsy Moth Caterpillar Parasitoids

The gypsy moth example does not definitely demonstrate how parasitoids have shaped the foraging activities of this caterpillar. The available evidence is descriptive and suggestive, and confounding effects are present. The situation becomes even more complex when it is realized that the diel movements of caterpillars are also influenced by weather conditions. On heavily overcast or rainy days, many large gypsy moth larvae stay in the canopy (Weseloh 1987). The immediate cause is low light levels. Larvae in the laboratory exposed to daily cycles of high-intensity light migrated away from food when lights were on and toward food when lights were off, whereas those exposed to lower intensity light stayed near food continuously (Weseloh 1989b). The ultimate reason, however, could still be parasitoids if these are inactive under low light conditions (and especially during rain many are likely to be inactive). Then there would be no advantage for caterpillars to migrate to the litter where they may be attacked at any time by predators (Weseloh 1988).

The gypsy moth system does illustrate how natural enemies can interact to influence foraging behavior. The pattern of foraging of gypsy moth larvae cannot be appreciated by considering each natural enemy in isolation. *Cotesia* and *Compsilura* would discourage feeding during the day, but why do the larvae then rest in leaf litter rather than on tree trunks, particularly when the forest floor is dangerous? The answer may well be that *P. silvestris* is so effective at attacking caterpillars on trunks that they have little choice—either they go all the way to the litter or find hidden locations in trees when these are available.

For completeness, I must also mention another possibility. Campbell and Sloan (1976) suggested that the pattern of gypsy moth foraging may be a result of selection pressures in Europe and Asia where the insect originally evolved. The most prominent parasitoids of the gypsy moth caterpillar in North America were originally imported from Europe and Asia, and presumably the behavior of gypsy moth caterpillars is relevant in relation to these parasitoids. But Campbell and Sloan (1976) hypothesized that birds attacking caterpillars in the canopy are a much more important mortality agent in Europe than are small mammals foraging on the forest floor, whereas the opposite occurs in North America. Thus, as a consequence of being confronted with natural enemies it has not had time to adapt to, the foraging behavior of gypsy moth caterpillars in North America may be maladaptive. The gypsy moth has been studied much less in Europe than in the United States, however, so this interesting possibility cannot be verified.

The gypsy moth example shows how the behavior of one species of caterpillar is an integration of diverse and sometimes contradictory evolutionary pressures. The same must also be true for many other Lepidoptera, and this illustrates the necessity of tentative conclusions unless a study is truly exhaustive. Some systems are less complicated and amenable to analysis than others, however. The following case histories may be instructive.

Other Examples of the Impact of Parasitoids on Caterpillar Foraging

Plant Effects

Gross and Price (1988) worked with two gelechild leafminers, *Tildenia inconspicuella,* that feed on horsenettle (*Solanum carolinense*), and *T. georgei,* that feeds on groundcherry (*Physalis heterophylia*). The horsenettle miner is attacked by nine parasitoids and the other by only two. The groundcherry miner, like most other members of the genus, is mobile, and it ties and folds leaves. If disturbed, it often drops from the leaf by a thread. In contrast, the horsenettle miner is always concealed in a mine and so cannot escape when found by parasitoids. This is because the leaves of horsenettle are covered by stellate trichomes that prevent large instars of either species from reaching the leaf surface to feed. Only small larvae are able to fit in the spaces between the trichomes and so penetrate to the leaf surface where they can initiate a mine. As they grow, they become too large to begin another mine if they leave the first one, so of necessity they must stay in a single mine until development is complete.

The groundcherry miner is able to avoid parasitoids by a number of mechanisms. It uses microhabitats that parasitoids do not exploit readily (tied leaves), escapes in space (by using more than one leaf and abandoning a mine when disturbed), and leaves a profusion of cues (in the form of abandoned mines and tied leaves). Only two parasitoids attack it. In contrast, the horsenettle miner appears to have no way of avoiding parasitoids and is used by nine species. In this case the horsenettle miner occupies an "enemy-rich" space. Perhaps exploiting horsenettle is worth the risk nutritionally, because of reduced competition from other leafminers, or for other reasons. Or perhaps the horsenettle miner cannot readily colonize new host plant species. Most of the miners in this genus have the foraging habits of the leaf-tying groundcherry miner, and for them an "enemy-poor" space might be especially important. They may not exploit hosts such as horsenettle that would force them to change foraging behavior and so expose them to greater parasitism. The fact that only one species in this genus remains within leaves and thus has a vulnerable foraging behavior suggests that escaping parasitoids is a high priority in this group.

Pieris rapae *and Parasitoids*

Pimentel (1961) investigated the presence of *P. rapae* on *Brassica* varieties. More caterpillars were present on headed varieties (brussels sprouts and cabbage) than on open-leaf ones (kale, broccoli, and collards). Levels of parasitism caused by *Phryxe vulgaris* and *Apanteles glomeratus* were between 40 and 78% on headed varieties and 100% on open-leaf ones. Field observations showed that the parasitoids could find and parasitize hosts most easily on the open varieties because caterpillars could often escape in the crevices and folds of the headed varieties.

If larvae preferentially seek out hidden areas to escape parasitoids, then the natural enemies are constraining caterpillar foraging. In feeding tests, *P. rapae* caterpillars preferred the youngest leaves of cabbage, which constitute the head and wrapper, over the older, more open, frame leaves (Hoy and Shelton 1987). Thus, although there are many conceivable reasons for the development of this preference, including plant nutrition and response to thermal stress, it does appear that cabbage butterfly caterpillars actively seek out parts of plants where they are partially protected from parasitoids.

Another aspect of *P. rapae* foraging may also be important. Ohsaki and Sato (1990) noted that this butterfly primarily attacks cultivated crucifers, an ephemeral resource that is planted and harvested at least twice throughout the growing season. *A. glomeratus* attacks a much higher percentage of caterpillars in the first crop than the second. Evidently the removal of hosts at the first harvest forces the parasitoids to go elsewhere or die without reproducing. *P. rapae* then invades the growing second crop faster than the parasitoid does. By being able to search effectively for new host plants, *P. rapae* is thus able to escape the parasitoid.

Group Foraging

In at least some instances, group foraging by caterpillars or caterpillar-like larvae may have evolved as a defense against parasitoids. Tostowaryk (1971) worked with the diprionid sawflies, *Neodiprion swainei* and *N. pratti banksianae*. The larvae feed in tight groups. Parasitoids (*Spathimeigenia spinigera* [Tachinidae], *Lamachus* spp. [Ichneumonidae], and *Perilampus hyalinus* [Perilampidae]) attack a greater number of hosts at the periphery of the colony than in the middle (70 vs. 40%). Conceivably, group defense and consequently foraging evolved to facilitate protection from parasitoids and other invertebrate enemies.

Another example is provided by Baltimore checkerspot (*Euphydryas phaeton*) caterpillars, which construct communal webs on their food plant (Stamp 1982). First instars feed strictly within the webs. Older prediapause larvae are often found outside webs and use the shelter mainly at night, during cool weather, and when molting. Parasitoids, particularly *Apanteles euphydryidis*, attack first instar hosts by ovipositing through the web surface, but they attack larger hosts while they are outside. Second and third instars have long spines and vigorous defensive movements that seem to be effective against parasitoids. Thus, first instars apparently protect themselves against excessive parasitism by staying within webs when feeding. As first instars deplete food reserves inside the web, larger larvae are more likely to be exposed when they feed and increasingly use individual rather than group defenses to protect themselves.

These examples show that group foraging can be an effective means of protection against parasitoids. Because of collective action, groups can have more effective defensive movements than individuals, release greater amounts of defensive chemicals, or build more elaborate and effective shelters. However, it is an open question as to how extensively group foraging has evolved as a response to parasitoids. Group foraging is advantageous for other reasons as well, and may have developed without respect to or in spite of parasitoid activity. For example, Knapp and Casey (1986) note that eastern tent caterpillars, *Malacasoma americanum,* often bask on the outside of their webs in early spring to thermoregulate, even though this behavior may make them more vulnerable to tachinid parasitoids. In this case it is clear that parasitism was not the pressure that lead to the selection of group foraging. In other cases, however, parasitism may have modified the expression of communal living after the latter had developed, resulting in greater protection to the caterpillars.

Lycaenid Caterpillars and Ants

To my mind, lycaenid butterfly larvae give the clearest example of how caterpillars modify their foraging activities to escape parasitoids. Lycaenid butterfly larvae are generally associated with ants (Baylis and Pierce, Chapter 12).

Herbivorous lycaenid larvae often eat a variety of unrelated, sometimes nutritionally deficient, host plants, particularly those with nectar-producing flowers or extra-floral nectaries that attract ants (Atsatt 1981). For example, larvae of the tropical lycaenid, *Teratoneura isabellae,* remain in the foraging columns of ants on tree trunks, where they grow slowly by feeding on lichen, fungi, and algae (Farquharson 1921). Displaced larvae always return to the ants. The impression is that these caterpillars are exploiting a poor food source in return for a desirable association with ants. Furthermore, some adult lycaenids choose plants for oviposition based on numbers of ants present on them rather than on plant type (Pierce and Elger 1985). Other lycaenids produce sweet secretions and attractive chemicals that cause ants to attend them, and yet others have taken up carnivory and feed on ant-tended Homoptera or ant larvae in nests (Atsatt 1981).

The reason for ant associations in Lycaenidae is usually attributed to the protection ants provide against natural enemies. Atsatt (1981) presents numerous anecdotal references in which ants are described as repelling parasitoids from lycaenid caterpillars. Pierce and Mead (1981) report that parasitization of the lycaenid, *Glaucopsyche lygdamus,* was several times greater when ants where excluded from plants by sticky barriers than when they were not. Pierce and Easteal (1986) expanded these results by showing that *G. lygdamus* was also protected against pupal parasitism by ants, and that predation rates were high but similar for ant-tended and nontended larvae. Thus in this case ants were a protection against parasitoids only. Working with the lycaenid *Jalmenus evagoras,* Pierce et al. (1987) determined that young caterpillars actively sought out host trees on which ants occurred. Parasitization of pupae was zero if ants were present but over 90% if they were not. Interestingly, pupae lost more weight when they were ant-attended, probably because the ants induced the pupae to secrete attractive substances. In Lepidoptera, low-weight pupae produce smaller adults that usually produce fewer eggs than larger adults. The advantages to the butterfly of associating with ants must be large for such a decrease in reproductive potential to be tolerated. Clearly, parasitoids have had an enormous impact on the development of foraging behaviors of caterpillars in this family of butterflies.

Changes in Foraging after Parasitism

Caterpillar foraging may change after parasitism has occurred. For instance, a parasitized larva of the nymphalid butterfly, *Chlosyne harrisii,* is found on the top of herbs rather than lower where unparasitized larvae are common (Shapiro 1976). It is difficult to see how this would directly affect the evolution of caterpillar foraging, because a parasitized insect will almost never reproduce, although in some instances the parasitized host may survive (Clausen 1962; English-Loeb et al. 1990). Shapiro (1976) suggested that conspicuous parasitized larvae may be easily discovered by predators, thus killing the parasitoid and leading to lower rates of parasitism on other, presumably related, larvae. However, working with another nymphalid, *Euphydryas phaeton,* Stamp (1981) showed that parasitized caterpillars on the tops of plants were attacked less by hyperparasitoids than lower ones. In this case the parasitoid has apparently manipulated the parasitized host so as to decrease its own mortality. The scenario suggested by Shapiro (1976) remains a possibility for other systems; but, in light of Stamp's (1981) results, advantages to the host would have to be demonstrated in each case.

As mentioned above, caterpillars may sometimes survive parasitoid development. In one such case involving the tachinid parasitoid *Thelairia bryanti,* the host *Platyprepia virginalis* (Arctiidae) occasionally lived to adulthood and reproduced (English-Loeb et al. 1990). The developmental times of these parasitized larvae were lengthened considerably, thus prolonging foraging and probably having unspecified effects on the caterpillars' behavior. Such instances in which parasitized caterpillars survive to reproduce are probably very rare, however, and so it is unlikely that this phenomenon has much significance for changing foraging behavior of caterpillars.

Conclusions

Despite the speculative nature of this subject, I conclude that parasitoids can have profound effects on foraging behaviors of hosts. If one accepts the premise that parasitoids and other natural enemies are responsible for limiting populations of most herbivorous insects, then it follows that parasitoids may exert strong selective pressures on many aspects of host behavior, including foraging. It would then be necessary to demonstrate only that the foraging behavior of caterpillars protects them from parasitoids, as has been done in some of the studies detailed in this review. Many people would argue from general principles that parasitoids and predators do limit herbivore populations (Hairston et al. 1960; Lawton and Strong 1981), but others feel direct demonstration is necessary (Murdoch 1966). Greater habitat specialization by parasitoids than by their host is common (Table 6.1), providing opportunity for hosts to escape. However, the molding of host foraging behavior occurs in the midst of a multitude of other selection pressures, and so the effects of parasitoids may be subtle and difficult to appreciate fully.

The impact of parasitoids on foraging of caterpillars has been documented mainly where the foraging takes extreme forms, as with lycaenid larvae, or where foraging appears paradoxical, as with monophagous caterpillars that nevertheless can feed successfully on a variety of plants (Bernays and Graham 1988). I suspect that similar cases will be the ones best documented in the future.

To show that parasitoids are major influences on caterpillar foraging, one must first describe foraging patterns. This should generally be straightforward, as it would involve simple observation. Second, it would be necessary to demonstrate through field experiments whether this behavior eliminates or reduces parasitism levels significantly. Foraging behavior, or the consequences of foraging, could be selectively changed, and parasitism levels compared to controls. For lycaenids, this may involve only eliminating ants. For group foragers, it might involve isolating caterpillars or removing them from webs. Those larvae constructing leaf-shelters could be removed from the shelters. Host plants and host habitats could be changed. Times when hosts are exposed and hidden could be altered. Effective manipulations may require ingenuity and details will depend on the system being studied, but in principle this should be possible in many situations. Lastly, it would be necessary to assess the impact of other selection pressures on the foraging pattern in question. As I have attempted to demonstrate in the gypsy moth, parasitoids are only one factor that contribute to host foraging patterns. The pressures documented in other chapters of this book may also be important. It is unrealistic to expect that all evolutionary pressures in a natural system can be characterized, but at least obvious ones in addition to parasitism should be accounted for, as Pierce and Easteal (1986) have done for predators and Pierce et al. (1987) for effects on reproductive potential.

Where the impact of parasitoids on foraging of caterpillars can be documented, it is clear they do have diverse and profound effects. Additional research should indicate under what conditions parasitoids are particularly important in shaping host foraging patterns. These results will explain many aspects of caterpillar behavior if considered along with other selection pressures.

Acknowledgments

I would like to acknowledge the useful and extensive comments on this manuscript by two anonymous reviewers. My profound thanks as well to Nancy Stamp for her large amount of help and encouragement with this chapter.

Literature Cited

Arthur, A. P. 1962. Influence of host tree on abundance of *Itoplectis conquisitor* (Say) (Hymenoptera: Ichneumonidae), a polyphagous parasite of the European pine shoot moth, *Rhyacionia buoliana* (Schiff.) (Lepidoptera: Olethreutidae). Can. Entomol. 94:337–347.

Atsatt, P. R. 1981. Lycaenid butterflies and ants: Selection for enemy-free space. Am. Nat. 118:638–654.

Bernays, E., and Graham, M. 1988. On the evolution of host specificity in phytophagous arthropods. Ecology 69:886–892.

Bragg, D. E. 1974. Ecological and behavioral studies of *Phaeogenes cynarae:* Ecology; host specificity; searching and oviposition; and avoidance of superparasitism. Ann. Entomol. Soc. Am. 67:931–936.

Campbell, B. C., and Duffey, S. S. 1979. Tomatine and parasitic wasps: Potential incompatibility of plant antibiosis with biological control. Science 205:700–702.

Campbell, R. W., and Sloan, R. J. 1976. Influence of behavioral evolution on gypsy moth pupal survival in sparse populations. Environ. Entomol. 5:1211–1217.

Campbell, R. W., Hubbard, D. L., and Sloan, R. J. 1975a. Patterns of gypsy moth occurrence within a sparse and numerically stable population. Environ. Entomol. 4:535–542.

Campbell, R. W., Hubbard, D. L., and Sloan, R. J. 1975b. Location of gypsy moth pupae and subsequent pupal survival in sparse, stable populations. Environ. Entomol. 4:597–600.

Clausen, C. P. 1962. Entomophagous Insects. Hafner, New York, pp. 43–46.

Culver, J. J. 1919. A study of *Compsilura concinnata*, an imported tachinid parasite of the gipsy [sic] moth and the brown-tail moth. USDA Bull. 766, 1–27.

Cushman, R. A. 1926. Location of individual host versus systematic relation of host species as a determining factor in parasitic attack. Proc. Entomol. Soc. Wash. 28:5–6.

Dijkstra, L. J. 1986. Optimal selection and exploitation of hosts in the parasitic wasp *Colpoclypeus florus* (Hym., Eulophidae). Netherlands J. Zool. 36:171–301.

Ding, D., Swedenborg, P. D., and Jones, R. L. 1989. Plant odor preferences and learning in *Macrocentrus grandii* (Hymenoptera: Braconidae), a larval parasitoid of the European corn borer, *Ostrinia nubilalis* (Lepidoptera: Pyralidae). J. Kansas Entomol. Soc. 62:164–176.

Elkinton, J. S., and Gould, J. R. 1989. Are gypsy moth populations in North America regulated at low density?, pp. 233–249. In W. E. Wallner and K. A. McManus (eds.), Lymantriidae: A Comparison of Features of New and Old World Tussock Moths. USDA Forest Service, General Technical Report NE–123.

Elzen, G. W., Williams, H. J., and Vinson, S. B. 1983. Response by the parasitoid *Campoletis sonorensis* (Hymenoptera: Ichneumonidae) to chemicals (synomones) in plants: Implications for host habitat location. Environ. Entomol. 12:1872–1876.

English-Loeb, G. M., Karban, R., and Brody, A. K. 1990. Arctiid larvae survive attack by a tachinid parasitoid and produce viable offspring. Ecol. Entomol. 15:361–362.

Farquharson, C. O. 1921. Five year's observations (1914–1918) on the bionomics of southern Nigerian Insects, chiefly directed to the investigation of Lycaenidae, Diptera, and other insects to ants. Trans. Entomol. Soc. London, Parts III, IV:319–493.

Franklin, R. T., and Holdaway, F. G. 1966. A relationship of the plant to parasitism of European corn borer by the tachinid parasite *Lydella grisescens*. J. Econ. Entomol. 59:440–441.

Gilmore, J. U. 1938. Observations on the hornworms attacking tobacco in Tennessee and Kentucky. J. Econ. Entomol. 31:706–712.

Gross, P., and Price, P. W. 1988. Plant influences on parasitism of two leafminers: A test of enemy-free space. Ecology 69:1506–1516.

Gross, S. W., and Fritz, R. S. 1982. Differential stratification, movement and parasitism of sexes of the bagworm, *Thyridopteryx ephemeraeformis* on red cedar. Ecol. Entomol. 7:149–154.

Hairston, N. G., Smith, F. E., and Slobodkin, L. B. 1960. Community structure, population control, and competition. Am. Nat. 94:421–425.

Hassell, M. P. 1968. The behavioral response of a tachinid fly (*Cyzenis albicans* (Fall)) to its host, the winter moth (*Operophtera brumata* (L.)). J. Anim. Ecol. 37:627–639.

Heinrich, B. 1979. Foraging strategies of caterpillars. Oecologia 42:325–337.

Heinrich, B., and Collins, S. L. 1983. Caterpillar leaf damage, and the game of hide-and-seek with birds. Ecology 64:592–602.

Herrebout, W. M. 1960. Host selection in the parasitic fly, *Eucarcelia rutilla* VIII. Arch. Neerland. Zool. 13:626.

Herrebout, W. M. 1969. Some aspects of host selection in *Eucarcelia rutilla* VIII. (Diptera: Tachinidae). Netherlands J. Zool. 19:1–104.

Hoy, C. W., and Shelton, A. M. 1987. Feeding response of *Artogeia rapae* (Lepidoptera: Pieridae) and *Trichoplusia ni* (Lepidoptera: Noctuidae) to cabbage leaf age. Environ. Entomol. 16:680–682.

Hoy, M. A. 1976. Establishment of gypsy moth parasitoids in North America: An evaluation of possible reasons for establishment or non-establishment, pp. 215–232. In J. F. Anderson and H. K. Kaya (eds.), Perspectives in Forest Entomology. Academic Press, New York.

Huffaker, C. B., Messenger, P. S., and DeBach, P. 1971. The natural enemy component in natural control and the theory of biological control, pp. 16–67. In C. B. Huffaker (ed.), Biological Control. Plenum, New York.

Inayatullah, C. 1983. Host selection by *Apanteles flavipes* (Cameron) (Hymenoptera: Braconidae): Influence of host and host plant. J. Econ. Entomol. 76:1086–1087.

Jalali, S. K., Singh, S. P., Kumar, P., and Ballal, C. R. 1988. Influence of the food plants on the degree of parasitism of larvae of *Heliothis armigera* by *Cotesia kazak*. Entomophaga 33(1):65–71.

Jaynes, H. A. 1954. Parasitization of spruce budworm larvae at different crown heights by *Apanteles* and *Glypta*. J. Econ. Entomol. 47:355–356.

Knapp, R., and Casey, T. M. 1986. Thermal ecology, behavior, and growth of gypsy moth and eastern tent caterpillars. Ecology 65:598–608.

Lawton, J. H. 1978. Host-plant influences on insect diversity: The effects of time and space, pp. 105–125. In L. A. Mound and N. Waloff (eds.), Diversity of Insect Faunas. Sym. Roy. Entomol. Soc. London. 9.

Lawton, J. H. 1986. The effect of parasitoids on phytophagous insect communities, pp. 265–289. In J. Waage and D. Greathead (eds.), Insect Parasitoids. Academic Press, London.

Lawton, J. H., and Strong, D. R., Jr. 1981. Community patterns and competition in folivorous insects. Am. Nat. 118:317–338.

Leonard, D. E. 1970. Feeding rhythm of the gypsy moth. J. Econ. Entomol. 63:1454–1457.

McManus, M. L., and McIntyre, T. 1981. Introduction, pp. 1–7. In C. C. Doane and M. L. McManus (eds.), The Gypsy Moth: Research Toward Integrated Pest Management. USDA Forest Service Tech. Bull. 1584.

Miller, C. A. 1959. The interaction of the spruce budworm, *Choristoneura fumiferana* (Clem.), and the parasite *Apanteles fumiferanae* Vier. Can. Entomol. 91:457–476.

Mueller, T. F. 1983. The effect of plants on the host relations of a specialist parasitoid of *Heliothis* larvae. Entomol. Exp. Appl. 34:78–84.

Munster-Swendsen, M. 1980. The distribution in time and space of parasitism in *Epinotia tedella* (Cl.) (Lepidoptera, Tortricidae). Ecol. Entomol. 5:373–383.

Murdoch, W. W. 1966. "Community structure, population control, and competition"—a critique. Am. Nat. 100:219–226.

ODell, T. M., and Godwin, P. A. 1984. Host selection by *Blepharipa pratensis* (Meigen) a tachinid parasite of the gypsy moth, *Lymantria dispar* L. J. Chem. Ecol. 10:311–320.

Ohsaki, N., and Sato, Y. 1990. Avoidance mechanisms of three *Pieris* butterfly species against the parasitoid wasp *Apanteles glomeratus*. Ecol. Entomol. 15:169–176.

Pair, S. D., Laster, M. L., and Martin, D. F. 1982. Parasitoids of *Heliothis* spp. (Lepidoptera: Noctuidae) larvae in Mississippi associated with sesame interplantings in cotton, 1971–1974: Implications of host habitat interaction. Environ. Entomol. 11:509–512.

Pierce, N. E., and Easteal, S. 1986. The selective advantage of attendant ants for the larvae of a lycaenid butterfly, *Glaucopsyche lygdamus*. J. Anim. Ecol. 55:451–462.

Pierce, N. E., and Elgar, M. A. 1985. The influence of ants on host plant selection by *Jalmenus evagoras,* a myrmecophilous lycaenid butterfly. Behav. Ecol. Soc. 16:209–222.

Pierce, N. E., and Mead, P. S. 1981. Parasitoids as selective agents in the symbiosis between lycaenid butterfly larvae and ants. Science 211:1185–1187.

Pierce, N. E., Kitching, R. L., Buckley, R. C., Taylor, M. F. J., and Benbow, K. F. 1987. The costs and benefits of cooperation between the Australian lycaenid butterfly, *Jalmenus evagoras,* and its attendant ants. Behav. Ecol. Sociobiol. 21:237–248.

Pimentel, D. 1961. An evaluation of insect resistance in broccoli, brussels sprouts, cabbage, collards, and kale. J. Econ. Entomol. 54:156–158.

Sato, Y., and Ohsaki, N. 1987. Host-habitat location by *Apanteles glomeratus* and effect of food-plant exposure on host-parasitism. Ecol. Entomol. 12:291–297.

Schaefer, P. W., Fuester, R. W., Chianese, R. J., Rhoads, L. D., and Tichenor, R. B., Jr. 1989. Introduction and North American establishment of *Coccygomimus disparis* (Hymenoptera: Ichneumonidae), polyphagous pupal parasite of Lepidoptera, including gypsy moth. Environ. Entomol. 18:1117–1125.

Schmid, J. M. 1981. Distribution of western spruce budworm (Lepidoptera: Tortricidae) insect parasites in the crowns of host trees. Can. Entomol. 113:1101–1106.

Shapiro, A. M. 1976. Beau Geste? Am. Nat. 110:900–902.

Sisojevic, P. 1975. Population dynamics of tachinid parasites of the gypsy moth (*L. dispar* L.) during a gradation period. Plant Protection (Belgrade) 132:167–170.

Smith, H. R., and Lautenschlager, R. A. 1981. Gypsy moth predators, pp. 96–125. In

C. C. Doane and M. L. McManus (eds.), The Gypsy Moth: Research Toward Integrated Pest Management. USDA Forest Service Tech. Bull. 1584.

Stamp, N. E. 1981. Behavior of parasitized aposematic caterpillars: Advantageous to the parasitoid or the host? Am. Nat. 118:715–725.

Stamp, N. E. 1982. Behavioral interactions of parasitoids and the Baltimore checkerspot caterpillar (*Euphydryas phaeton*). Environ. Entomol. 11:100–104.

Strong, D. R., Lawton, J. H., Southwood, R. 1984. Insects on Plants: Community Patterns and Mechanisms. Harvard University Press, Cambridge, MA.

Thorpe, W. H., and Caudle, H. B. 1938. Studys of the olfactory responses of insect parasites to the food plant of their host. Parasitology 30:523–528.

Thurston, R., and Fox, P. M. 1972. Inhibition by nicotine of emergence of *Apanteles congregatus* from its host, the tobacco hornworm. Ann. Entomol. Soc. Am. 65:547–550.

Ticehurst, M., and Yendol, W. 1989. Distribution and abundance of early instar gypsy moth (Lepidoptera: Lymantriidae). Environ. Entomol. 18:459–464.

Tostowaryk, W. 1971. Relationships between parasitism and predation of diprionid sawflies. Ann. Entomol. Soc. Am. 64:1424–1427.

Valentine, H. T. 1981. A model of oak forest growth under gypsy moth influence, pp. 50–61. In C. C. Doane and M. L. McManus (eds.), The Gypsy Moth: Research Toward Integrated Pest Management. USDA Forest Service Tech. Bull. 1584.

Vinson, S. B. 1976. Host selection by insect parasitoids. Annu. Rev. Entomol. 21:109–133.

Vinson, S. B. 1981. Habitat location, pp. 51–77. In D. A. Nordlund, R. L. Jones, and W. J. Lewis (eds.), Semiochemicals: Their Role in Pest Control. Wiley, New York.

Weseloh, R. M. 1972. Spatial distribution of the gypsy moth [Lepidoptera: Lymantriidae] and some of its parasitoids within a forest environment. Entomophaga 17:339–351.

Weseloh, R. M. 1974. Host-related microhabitat preferences of the gypsy moth larval parasitoid, *Parasetigena agilis*. Environ. Entomol. 3:363–364.

Weseloh, R. M. 1976a. Diel periodicity and host selection, as measured by ovipositional behavior, of the gypsy moth parasite, *Parasetigena silvestris,* in Connecticut woodlands. Environ. Entomol. 5:514–516.

Weseloh, R. M. 1976b. Reduced effectiveness of the gypsy moth parasite, *Apanteles melanoscelus,* in Connecticut due to poor seasonal synchronization with its host. Environ. Entomol. 5:743–746.

Weseloh, R. M. 1976c. Behavioral responses of the parasite, *Apanteles melanoscelus,* to gypsy moth silk. Environ. Entomol. 5:1128–1132.

Weseloh, R. M. 1976d. Behavior of forest insect parasitoids, pp. 99–110. In J. F. Anderson and H. K. Kaya (eds.), Perspectives in Forest Entomology. Academic Press, New York.

Weseloh, R. M. 1978. Seasonal and spatial mortality patterns of *Apanteles melanoscelus* due to predators and gypsy moth hyperparasites. Environ. Entomol. 7:662–665.

Weseloh, R. M. 1982. Implications of tree microhabitat preferences of *Compsilura concin-*

nata (Diptera: Tachinidae) for its effectiveness as a gypsy moth parasitoid. Can. Entomol. 114:617–622.

Weseloh, R. M. 1987. Accuracy of gypsy moth (Lepidoptera: Lymantriidae) population estimates based on counts of larvae in artificial resting sites. Ann. Entomol. Soc. Am. 80:361–366.

Weseloh, R. M. 1988. Effects of microhabitat, time of day, and weather on predation of gypsy moth larvae. Oecologia 77:250–254.

Weseloh, R. M. 1989a. Simulation of predation by ants based on direct observations of attacks on gypsy moth larvae. Can. Entomol. 121:1069–1076.

Weseloh, R. M. 1989b. Behavioral responses of gypsy moth (Lepidoptera: Lymantriidae) larvae to abiotic environmental factors. Environ. Entomol. 18:361–367.

Weseloh, R. M., and J. F. Anderson. 1975. Inundative release of *Apanteles melanoscelus* against the gypsy moth. Environ. Entomol. 4:33–36.

7

How Avian Predators Constrain Caterpillar Foraging

Bernd Heinrich

Introduction

Caterpillars are potentially an ideal food for many insectivorous birds. They are slow moving. They do not have a large portion of their proteins tied up in indigestible hard parts that are necessary for exoskeletal support for rapid locomotion. They are also generally abundant enough to make specialization to hunt them economically worthwhile.

The majority of the world's approximately 100,000 species of Lepidoptera have specialized to feed from the most common and otherwise nearly uncontested food source available: foliage, primarily from trees. Given a larva's small body mass, tree foliage provides a caterpillar a nearly limitless resource under most situations, except those where populations have temporarily exceeded the ability of parasites and predators to control them. Birds have been identified as primary predators of numerous caterpillars (MacArthur 1959; Baker 1970; Homes et al. 1979), and many caterpillars have elaborate visually oriented displays (Jones 1932; Ford 1945; Greene 1989) and poison spines (Gilmer 1925; Turnbull 1979; Fourrie and Hull 1980). Consequently, it seems reasonable that these visually and mechanically oriented defenses of caterpillars are evolved from the selective pressure exerted by birds.

Although almost nothing is known of the evolutionary history of lepidopteran feeding behavior and probable bird predation on caterpillars, a rough chronology suggests a possible scenario. Arthropod feeding damage is apparent on Upper Carboniferous vegetation (Scott and Taylor 1983), but the Lepidoptera did not evolve until somewhat later, at approximately 100,000,000 years B.P., in the mid-Cretaceous (McKay 1970; Carpenter 1976), probably from caddisfly-like ancestors (Kukalova-Peck 1978). Although the ancestral bird *Archeopteryx* dates from the Jurassic, some 140,000,000 years B.P., the passerine birds who likely evolved to prey on caterpillars did not undergo extensive radiation until the Tertiary, some 50,000,000 years B.P. (Feduccia 1980). It is therefore likely that the radiation of lepidopteran species long preceded that of the birds, and indeed it may have helped to promote it. In any case, the selective pressure of, at least, some tens of million years of birds hunting caterpillars has undoubtedly been a potent selective pressure, and the morphology and feeding behavior of caterpillars now provide an indelible record of that evolutionary history.

Although bird predators have likely left the clearest and most obvious evolutionary imprint on caterpillars, caterpillars have had and still have other selective pressures acting on them that could result in compromises of given traits that we inspect in isolation.

As a first requirement for utilizing foliage, a caterpillar must be able to handle large leaves mechanically (Heinrich 1971), at the same time to use the leaves as a shield from solar radiation (Casey 1976, 1977; Greenberg and Gradwohl 1980) and/or to hide from birds (Thurston and Prachuabmoh 1971). Therefore, body exposure and leaf manipulation must be integrated to reduce exposure to predators and the elements.

Another major consideration on feeding behavior could be avoidance of the defenses that plants mount against herbivores (Rothschild 1973; Dussourd and Eisner 1987; Scriber and Ayres 1988). Plants may employ various mechanisms including mechanical barriers such as spines, chemical defenses, and leaf smoothness and toughness (Coley 1983) that limit access to feeding. Possibly leaf morphology and leaf spacing on the plant could also affect exposure to the herbivores to predators (Robinson and Holmes 1984) and in turn place selective pressure on the herbivores for mechanical skill in harvesting.

Third, caterpillars face a large array of invertebrate enemies, such as *Polistes* (Gould and Jeanne 1984; Stamp and Bowers 1988), *Vespula* (Morris 1972), *Polybia* (Raveret Richter 1990), sphecid (Evans 1987), ichneumonid (Damman 1986), and other wasps as well as flies (Turnbull 1979, Greene 1989). Relative to birds that employ remarkably detailed search images (Pietrewicz and Kamil 1979), *Vespula* and possibly other insect predators display very little visual discrimination (Heinrich 1984) and many species rely on scent in hunting (Vinson 1985; Wardle and Borden 1989), which birds do not. Scents of the caterpillars would tend to be very species specific and thus one might expect the invertebrate predators to be either generalists with considerably less detailed visual search images than those of birds, or that they can become specialists using scent as a cue. Since frass resulting from feeding would drop to the ground, it would usually not be a reliable cue in locating larvae.

Although the evolutionary pressure exerted by invertebrate predators is predictably host plant specific with regard to scent, it could be general with regard to immediate behavioral responses to mechanical stimulation (see Stamp and Bowers 1988). Once found by predators, caterpillar behavioral responses such as dropping and writhing, are thus likely shaped both by invertebrate and vertebrate predators.

In contrast, evading predators from the outset by appearance and by reducing scent could be shaped differently (Turnbull 1979) by avian predators and invertebrate enemies, respectively.

The numerous selective pressures acting simultaneously on any one trait in any one instar cannot be readily partitioned, although some deductions may still be possible. For example, the ichneumonid wasp *Trogus pennator* attacks swallowtail caterpillars exclusively (Heinrich 1962), yet the scent glands (osmeteria) that swallowtail caterpillars extrude when they are touched do not elicit an aversion reaction in the wasp (Damman 1986). It is therefore unlikely that this specialized organ has evolved (and is still retained) for defense against this and probably other large wasps, again pointing to birds as the major selective agent on these caterpillars.

Generalist predators with acute vision have the potential to exert considerable impact because numerous responses are available that could foil visual detection. Hiding cannot block scent, but it can block detection by vision. The range of movement, form, color, and activity time of larvae are all subject to selection by visually oriented predators. Given these considerations and the experimental observations that birds of many species converge to prey on the same species of caterpillars, accounting for perhaps most of their long-term mortality in northeastern deciduous woodland (Holmes et al. 1979, 1986), I here present the broad outlines of how bird predation may have shaped caterpillar morphology and feeding behavior.

Caterpillar Appearance

The visible appearance of caterpillars is one of the most obvious proofs of the all-pervasive role of *visually* oriented predators having shaped their evolution. There are few caterpillars in existence that are prized by birds whose appearance does not in some very obvious way bear the marks of having passed through a stringent visual filter of evolution (see Greene 1989). All of these adaptations are too numerous to mention, and the general point needs little documentation. Here I give only reminders of what every naturalist knows. For example, geometrid "stick" caterpillars resemble the specific twigs on which they perch (Fig. 7.1) in the finest detail of color, texture, and leaf scars and lentices. *Catocala* and numerous other genera of moth caterpillars resemble the bark of their host trees. Notodontid leaf-mimicking caterpillars may have "perfect" leaf blemishes and/or leaf edges (Fig. 7.2). Other caterpillars eaten by birds mimic catkins (Greene 1989) and other assorted debris, bird droppings, snakes, and poisonous amphibians (Fig. 7.3). Birds prey on salamanders, and mimics of poisonous species gain protection (Brodie and Brodie 1980) and thus presumably mimetic caterpillars of the same models are also protected from birds.

All of the visual appearances of caterpillars are backed by elaborate and very

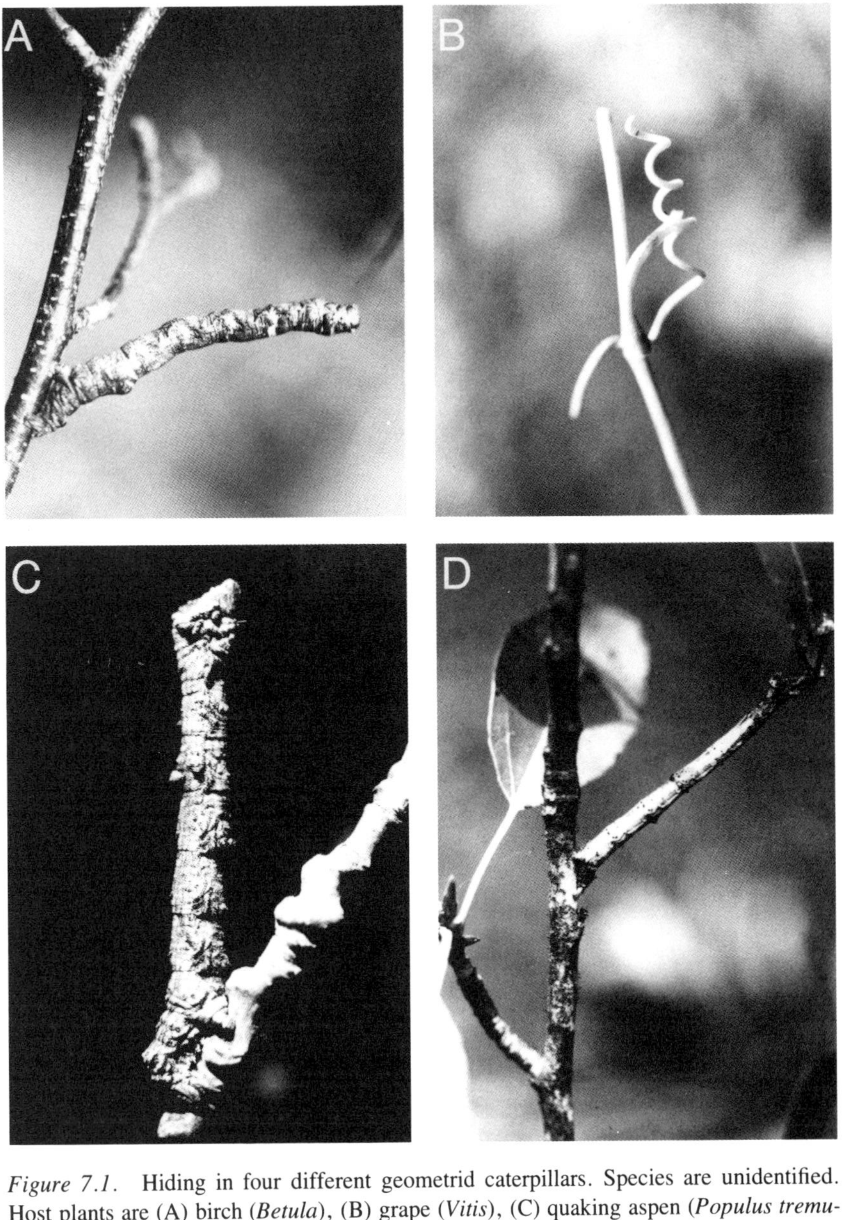

Figure 7.1. Hiding in four different geometrid caterpillars. Species are unidentified. Host plants are (A) birch (*Betula*), (B) grape (*Vitis*), (C) quaking aspen (*Populus tremuloides*), and (D) balsam poplar (*P. balsamifera*). Only the caterpillar on grape (B) moves and feeds in the daytime. It effectively mimics the leaf petiole.

Figure 7.2. Four different species of notodontid caterpillars mimicking leaf edges at which they feed. (A) *Heterocampa biundata* on birch, (B) *Nerice bidentata* on elm, (C) *Heterocampa guttivitta* on oak, and (D) *Schizura unicornis* on birch.

appropriate behaviors that are surely not meant for parasitic wasps and tachinid flies, which rely largely on scent to locate hosts or other insect enemies (Vinson 1985), and some of which must bump into prey to detect them (see Montllor and Bernays, Chapter 5).

In contrast to the above caterpillars that seem to have gone to extremes to make themselves seem what they really are not, there are others that make no special effort to hide themselves. Indeed, many are conspicuous from a great distance because of their prominently displayed bright and contrasting colors (Fig. 7.4). These caterpillars are usually well defended by noxious poisons, spines, or both (Jones 1932; Ford 1945), whereas members of the first group have neither poisons nor hairs nor spines (Rothschild 1973).

The above broad and obvious patterns of correlations in behavior, appearance, and defenses are helpful to keep in mind when considering the evolution of feeding behavior, because they help us identify the predators and selective agents involved. I shall argue that although insect predators and parasites use vision in

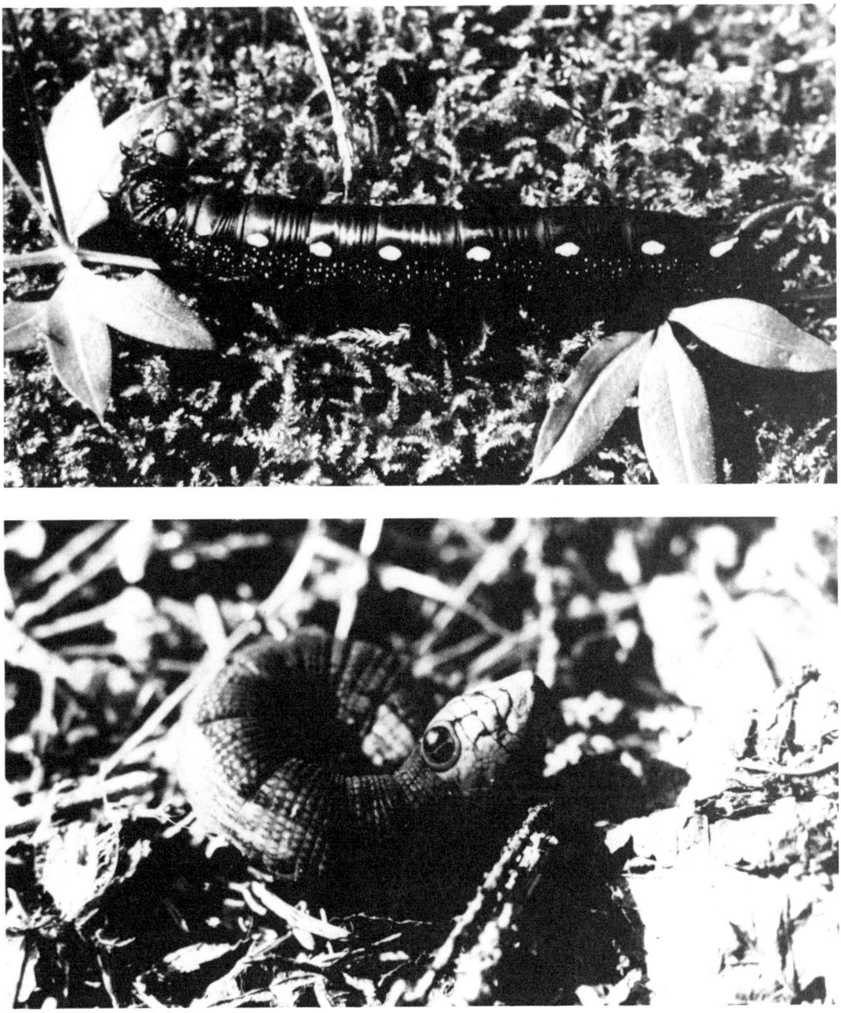

Figure 7.3. Galium sphinx caterpillar, *Celerio intermedia,* feeding on creeping *Galium* sp., a herb that grows in moist places. This black and yellow-spotted caterpillar may mimic the spotted salamander, *Ambystoma maculatum*. Lower picture shows posture that Abbots sphinx caterpillar, *Sphecodina abbotti,* takes in its wandering stage on the ground after leaving its food plant (grape) to pupate in the ground.

hunting prey (Morris 1972; Stamp and Bowers 1988), these agents are not likely to have been discriminating enough to shape the variety and precision of caterpillar appearance (Greene 1989) and the foraging behavior strongly associated with that appearance that I.will discuss in this paper. Predatory vespid wasps usually hunt by pouncing on leaf blemishes, nail heads, and almost anything else that visually

Figure 7.4. "Messy" feeding by conspicuous caterpillars that are unpalatable to birds. (A) Forest tent caterpillar, *Malacosoma disstria,* on basswood, (B) *Halysidota maculata* on red maple, (C) monarch butterfly caterpillar, *Danaus plexippus,* on milkweed, and (D) mourning cloak butterfly caterpillars, *Nymphalis antiopa,* on willow.

contrasts from the background (Heinrich 1984). It is highly unlikely that their behavior is discriminating enough to have caused the near perfect mimicry of natural objects that even humans have difficulty in distinguishing. Taken together, the above considerations suggest that birds, and not insects, have been the primary selective agents in those aspects of caterpillar design that relate to visual stimuli.

Throughout the remainder of this chapter I will attempt to ascertain the effect of avian foraging behavior on caterpillar foraging behavior through a comparative perspective of "palatable" vs. "unpalatable" species to birds. However, no strict categorization is realistic. I make the simplification only to detect broad patterns. In general, hairy, spiny, and brightly colored caterpillars are unpalatable, whereas smooth, cryptic species that hide are eaten by birds. As always, this assumes "all things being equal," which they seldom are. For example, Yellow-billed Cuckoos are well known to eat hairy caterpillars (Knapp and Casey 1986) not eaten by most other insectivorous birds. Chickadees, *Parus atricapillus,* accept without hesitation smooth-skinned heterocampid, geometrid, and small sphingid caterpillars, but bristly caterpillars (such as those of the tussock moth, *Heterocampus leucostigma;* most arctiids, including *Halysidota maculata;* the mourning cloak, *Nymphalis antiopa;* and the saturniid *Arisota rubicunda*) are not even pecked at

(Heinrich and Collins 1983). When hairy caterpillars are (rarely) taken by Red-eyed Vireos, *Vireo olivaceus*, they require long handling time to knock off the spines (Heinrich 1979) and are therefore only marginally palatable. Smooth caterpillars are swallowed immediately, and they are preferentially hunted and eaten. To some extent, therefore, palatable vs. unpalatable is a matter of degree, as are aposematic coloration, spininess, and hairiness.

Caterpillar Movement

One of the obvious cues whereby predators locate prey is by their movement. Movement should therefore vary conspicuously between palatable and unpalatable caterpillars. However, possible differences have not been quantitatively investigated.

Leaf toughness is a major constraint in caterpillar foraging, and many lowland tropical leaves are protected from caterpillars by being tough (Coley 1983). However, in addition, any one leaf is not uniformly tender or tough. Even a tender leaf generally has a tough rachis and tough supporting veins, and these structures are often not eaten by small as well as large unpalatable caterpillars. On the other hand, even moderately sized palatable caterpillars tend to consume them (Heinrich, personal observation). It seems that many unpalatable caterpillars are very "fussy" eaters, quickly moving on when encountering veins, midribs and other less suitable leaf tissue, whereas palatable caterpillars do not move from the spot, and keep right on eating. Indeed, in groups of spiny *Nymphalis* or hairy *Malacasoma* larvae there is often a continuous bustle as the caterpillars jostle for choice feeding spots (Heinrich, personal observation). Similarly, when these and other highly unpalatable caterpillars, such as those of some hairy arctiid moths, travel from one feeding spot to the next they appear to move in great haste. Perhaps they can "afford" to be choosy because the release from predator pressure allows them to travel.

In palatable caterpillars there are at least two patterns of movement: very fast or very slow. Notodontid larvae that disguise themselves as debris or parts of leaves stay at all times on the leaf (Fig. 7.2), and they are virtual sloths, always remaining still or moving only in very slow motion when moving on to another leaf. The second group does not move at all in the daylight hours, but under cover of darkness they move at a fast pace. The highly cryptic catocala moth caterpillars belong to this group. In the daytime they spend most or all of their time pressed motionless against a limb or a tree trunk, and at dusk and dawn they commute rapidly to and from their feeding site, sometimes racing many meters up and down the branches from their hiding place. Similarly, most of the large-bodied geometrid caterpillars that mimic sticks or branches (Fig. 7.1) do not feed or move about during daylight. Throughout the day they (at least the latter instars) remain rigid and extended like a stick by remaining firmly anchored by their claspers and by either grasping a leaf above them, or by suspending their anterior end by a small thread (Heinrich 1979). In contrast, small geometrid larvae feed both day and night. Apparently they are "cryptic" enough (except at very close range) without immobility. Alternately, long-term increased exposure time to parasites (by potentially not eating and growing) is not counterbalanced enough by immediate evasion of avian predators.

Feeding Time

In caterpillars growth rate is a relatively direct function of feeding rate, and the faster a caterpillar grows, the smaller the temporal window of vulnerability to both predators and parasites. No data are available that compare feeding (and growth) rates of palatable vs. unpalatable caterpillars to test whether predator pressure drives feeding rate. Nevertheless, the feeding and growth rates of some caterpillars are impressive. For example, a *Manduca sexta* (Sphingidae) caterpillar weighing 5.3 g can gain 2.5 g or about 50% body weight per day during its last larval instar (Heinrich, personal observation) when it appears to feed nearly continuously. The potential cost of feeding rapidly could be a decreased efficiency in food assimilation (Scriber and Slansky 1981), but again, data are lacking. A caterpillar has to feed to grow. But it presumably has the options to feed fast or slowly, continuously or in bursts, in the daytime and/or at night. Costs and tradeoffs for predator avoidance undoubtedly exist, but they remain to be elucidated (see Stamp and Wilkens, Chapter 9).

A caterpillar that is hiding motionless on a branch from a bird is not likely to be hidden from a parasitoid hunting by scent. Therefore, the primary threat that highly camouflaged night-feeding caterpillars are guarding against is *bird* predation. The caterpillar likely compromises growth rate (and *increases* the time it will be exposed to invertebrates) by not feeding for some 12–14 hours per day to reduce the bird threat.

Unpalatable caterpillars could potentially feed more freely and thus develop fast, although parasites could still exert pressure for reducing development time. Indeed, unpalatable caterpillars, having the option to feed in the daytime because they are not discriminated against by moving, could potentially have even faster developmental times than palatable caterpillars. Unfortunately, the extent to which caterpillars minimize exposure to parasites and/or predators by shortening development time via feeding regimens remains unknown.

Bird Recognition of Caterpillar Feeding Damage

I shall here examine the hypothesis that bird predation has shaped the way caterpillars eat leaves. This hypothesis is based on the preposition that caterpillar-hunting birds can recognize leaf damage and then use leaf damage as a cue in their search. Few data are available examining this hypothesis. Although the evidence is so far relatively consistent, much more work could be done in this area, which necessarily combines behavioral studies of birds with comparative studies of caterpillars.

The experimental evidence for the underlying assumption that birds recognize leaf damage from caterpillar feeding comes from two studies. In the first, Black-capped Chickadees, *Parus atricapillus,* were tested in a large outdoor aviary where hidden food (small green, extremely cryptic caterpillars) were offered on saplings with leaves that had been either naturally damaged or had holes punched into them. The chickadees quickly learned to preferentially search on the trees with damaged leaves (Fig. 7.5), visiting these trees in the feeding trials (when the caterpillars had been removed). Furthermore, after having been trained to search for hidden caterpillars on leaf-damaged trees of one species, they continued to search on that tree species even when offered several other tree species which also had leaf damage (Fig. 7.6). In summary, having been trained to search for caterpillars near leaf damage on one tree species, they did not necessarily generalize to search for caterpillars near leaf damage on other tree species. Such specificity would be highly useful because in the field many tree species have leaf damage but no palatable caterpillars at that time associated with it (Janzen 1988), while in others leaf damage is associated with palatable caterpillars (Fig. 7.7). To what extent and in what situations chickadees and other caterpillar-foraging birds may use leaf damage as a cue in the field is not known (but see Greenberg and Gradwohl 1980). The above study shows that at least chickadees have the capacity to make use of leaf damage left by caterpillars as a cue in their hunting behavior.

A second experimental study (Real et al. 1984) indicated that Blue Jays, *Cyanocitta cristata,* can differentiate quickly between projected photographic images of whole leaves vs. caterpillar-damaged leaves. Furthermore, the birds also readily differentiated between leaves that had been partially eaten by palatable caterpillars vs. those that had been partially eaten by unpalatable caterpillars (see later discussion).

Leaf Handling

Caterpillars face the physical problem of consuming sometimes large leaves on which they perch, without at the same time severing their perch from the plant. They could potentially avoid the problem of removing themselves from their food plant by only partially consuming a leaf before moving on to the next. But this response could then create two other problems: (1) waste of food and (2) leaving cues for avian predators to find them. The solutions to the leaf-handling problem undoubtedly vary depending on the size and shape of the leaves of the food plant, and the palatability of the caterpillar. The following examples show the range of leaf-handling options dictated by different contingencies.

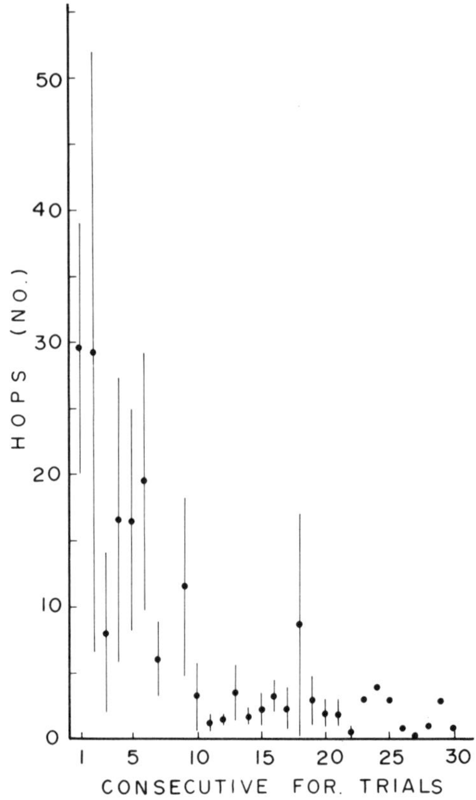

Figure 7.5. Tree choice of Black-capped chickadees, *Parus atricapillus*, over consecutive foraging trials when the birds were let into an aviary containing 10 small trees, 2 of which had damaged leaves. Numbers of hops here refer to those on the undamaged trees. Note that after about 10 foraging trials the chickadees no longer searched (hopped) on the trees with undamaged leaves. Instead they flew straight to one of the two trees with damaged leaves. The numbers of trials (birds) varied from 4 (values at left) to 1 (at right). Values are means ± SE. (From Heinrich and Collins 1983.)

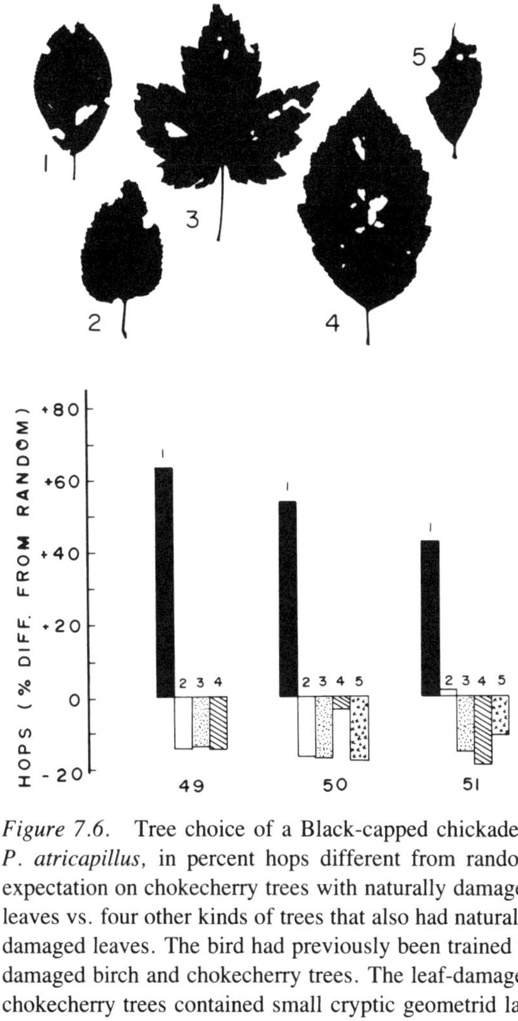

Figure 7.6. Tree choice of a Black-capped chickadee, *P. atricapillus,* in percent hops different from random expectation on chokecherry trees with naturally damaged leaves vs. four other kinds of trees that also had naturally damaged leaves. The bird had previously been trained to damaged birch and chokecherry trees. The leaf-damaged chokecherry trees contained small cryptic geometrid larvae in trials 49 and 50, but in trial 51 none of the trees contained food reward. 1, Chokecherry; 2, white birch; 3, red maple; 4, alder; 5, willow. In all three trials the percent hops on the experimental trees was significantly different ($P<0.001$) from random expectation. (From Heinrich and Collins 1983.)

Harvesting without waste may be the primary consideration while feeding on some leaves. Consider a pine sphinx larvae, *Lapara bombycoides,* harvesting lanceolate pine leaves that may be several times longer than its own body. If the caterpillar simply chewed into the nearest leaf tissue at the base of the pine needle where it is attached, then it would discard most (at least 95%) of each leaf it feeds from. However, the caterpillar has solved the problem of excessive wastage by

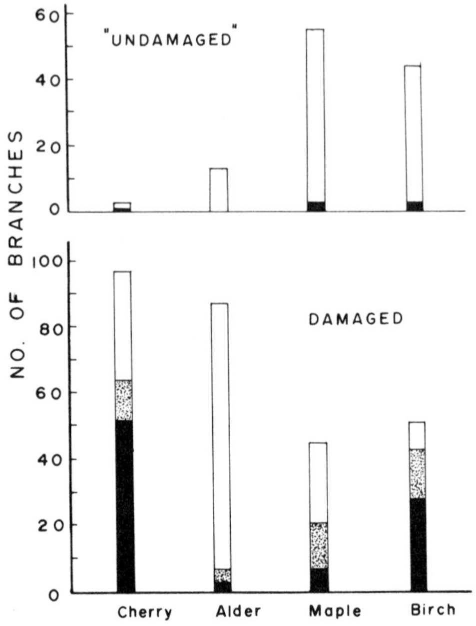

Figure 7.7. Number of branches ($1-m^2$ area) with damaged leaves (bottom) and "undamaged" leaves (top) out of a sample of 100 branches in each of four tree species on 26 and 27 July 1981. Open bars, no caterpillars visible. Stippled bars, number of branches with one to seven unpalatable caterpillars visible. Solid bars, number of branches with one to three palatable caterpillars visible. Note that (1) most cherries and alders had damaged leaves, and only half of the maples and birches showed leaf damage, and (2) there were 7 times more palatable caterpillars associated with damaged than with undamaged leaf patches of birch. (From Heinrich and Collins 1983.)

walking up each needle with its thoracic prolegs, while its clasper at the rear remains clamped near the needle's base. As a result, the needle bends into a bow. The caterpillar does not begin to feed until the tip of the needle reaches its mouth, and then it eats the bent needle down to the base.

Consider now fennel, *Foeniculum* sp., and various other plants of the carrot family that have filiform but highly branched leaves. In this case the anise swallowtail butterfly larvae, *Papilio zelicoan,* feeding on this plant at first do the same as the sphinx caterpillars on pine. They feed only from leaf tips. However, each time they reach a branch point in the leaf they momentarily stop feeding and walk with their thoracic legs up the main branch until they again encounter another leaf tip. In this way the whole highly branched leaf is consumed without snipping off or wasting any leaf tissue.

The later instars of the sphinx moth caterpillars, *Manduca sexta,* use a similar feeding behavior to that given above, although it is modified to allow them to handle large broad leaves of solanaceous plants (Heinrich 1971). Caterpillars remain attached near the petiole underneath leaves much larger than themselves, and they eat the entire leaf without wastage. Caterpillars pull leaf tissue toward themselves, walk forward with their thoracic legs while remaining attached posteriorally by their clasper, and then feed back toward the leaf base on the bowed leaf. (Early instar caterpillars are unable to bend leaves and they chew holes into the leaf instead, avoiding the leaf veins.)

The three examples above indicate the capacity of caterpillars to efficiently harvest leaf tissue without wastage. The efficient leaf harvesting is advantageous because it facilitates the high rate of food intake required to maintain a high growth rate and reduce developmental time. It also reduces the amount of visible feeding damage. In the case of *M. sexta* feeding on jimson weed (*Datura* sp.), one plant may not be enough for the many caterpillars often found on it in the field (Heinrich, personal observation). A feeding contest ensues, and those caterpillars slower in harvesting will be forced to move, to be exposed to predators and to face the uncertainty of finding another often very distant and rare food plant.

Another potential feeding constraint of leaf-eating caterpillars aside from efficient leaf handling concerns the plant's chemical defenses. For example, many insects (including monarch butterfly larvae, *Danaus plexippus*), feeding on plants with sticky latex surmount the latex exuding at the feeding wound by snipping the main supply veins (Dussourd and Eisner 1987; Dussourd, Chapter 3).

Plants could potentially also mobilize secondary compounds to the leaf area that is being damaged to induce the caterpillar to move to feed on a different portion of the plant. So far, however, the latter idea has not been proven. (Why should a tree care if its leaves are eaten from one limb vs. another except possibly to make the caterpillar move to make it more vulnerable to bird predation?) If localized mobilization of secondary compounds is responsible for caterpillar movement patterns, then unpalatable caterpillars should be the first to move, whereas palatable may be obliged to stay to avoid detection because of their locomotor movement. The strategy of staying put makes sense if movement increases the likelihood of predation, or if the caterpillars incorporate the plants' toxins for their own defense or otherwise tolerate the toxins.

Association with Leaf Damage

While feeding on broad leaves, caterpillars necessarily leave damage that could compromise their disguise no matter how good their mimicry and how little their

Figure 7.8. Messy feeding damage by caterpillars of the mourning cloak butterfly, *Nymphalis antiopa*. Left photo shows feeding by a single caterpillar placed on poplar. Right photo shows twigs of willow. Note the leaves are eaten from near the base of the petioles, and remnants left are primarily on the terminal portions of the leaves.

movement. If this leaf damage serves as a hunting cue to predators, then one might predict that palatable caterpillars have evolved means to reduce the apparent leaf damage or their association with it.

Unpalatable caterpillars can serve as a control because they do not show such presumptive behaviors. On a large tree with many leaves, a caterpillar could potentially feed selectively, leaving the tough and presumably less nutritious leaf veins. As a result of such selective feeding, leaf tatters would be left behind (Heinrich and Collins 1983). Indeed, unpalatable caterpillars characteristically leave leaf tatters that are often visible from a long distance (Heinrich 1979). Many unpalatable caterpillars leave not only the leaf veins, they also leave the tips of leaves and other hard-to-get portions. In general, they appear to be lackadaisical "messy" feeders (Figs. 7.4 and 7.8). Presumably some unpalatable caterpillars do die because they are conspicuous, so they probably receive some selective pressure to remain hidden. However, this selective pressure is likely countered by another or others, such as advantages of being exposed to sunlight, staying in groups, reducing movement, and longer feeding periods.

The feeding behavior of palatable caterpillars contrasts markedly with that of the unpalatable ones who presumably have been under less selective pressure by predators to avoid leaving evidence of their feeding damage (Heinrich 1979). Palatable caterpillars have one or a combination of several behaviors that act to dissociate them from their feeding damage. Few later instar palatable caterpillars make holes in leaves. One exception is the noctuid moth caterpillar, *Amphiphyra pyramindoides*. On the average these caterpillars eat only 5 cm^2 from a large (up to 300 cm^2) leaf of their food plant, basswood, *Tilia americana*. After having finished their short meal, they immediately leave and later feed on another leaf often a meter or more distant (Heinrich 1979). Perhaps this is a "shell-game" strategy where the predator would be confronted with many damaged leaves, but seldom with the perpetrator. Presumably this strategy of the caterpillar would be most appropriate when feeding on plants with very large leaves. The caterpillar could presumably complete nearly its entire larval development on a single leaf, but if it attempted to do so, the predator would with very high reliability find a caterpillar at any one damaged leaf. In that regard, caterpillars that leave a leaf soon after a feeding bout should be at an advantage.

More commonly, especially on trees with relatively small leaves, such as cherry, birch, or willow, the palatable caterpillars move very little but eat the entire or nearly the entire leaf. However, they do not accomplish this by chewing holes in the leaf. Instead, the leaves are pared down along the side so that partially eaten leaves merely look smaller than intact leaves (Figs. 7.2, 7.9,and 7.10). As already mentioned, at least some birds can readily learn to differentiate leaves chewed on by palatable vs. unpalatable caterpillars, but they have difficulty distinguishing whole leaves from those fed on by palatable caterpillars (Real et al. 1984).

The behavior of hiding feeding damage on leaves reaches perhaps its highest sophistication in a number of notodontid species where the larva's body itself mimics the leaf and "fills in" the place where leaf tissue has been removed. In one species, *Nerice bidentata,* the larva's back even has serrations mimicking the serrations of the leaf edge of its food plant, elm (Fig. 7.2B). In several species the larvae have discolorations that mimic leaf blemishes or their outline is broken up to create the effect of a partially dried leaf edge. The above combination of behavior and disguise allows these caterpillars to minimize travel and to feed both day and night.

Leaf Clipping

In the notodontid larvae discussed above, effective leaf mimicry depends on maintaining a high leaf/body ratio, so that the body can mimic a part of the leaf. The more that leaf tissue has been removed from the leaf, the more precarious is the caterpillars' camouflage. However, the caterpillars have an elegant solution to this problem. After having eaten part of the leaf, a caterpillar backs off down the leaf and chews through the petiole, thereby clipping off the leaf remnant. Some species even clip off the remaining leaf petioles at the base where they attach to the twig (Figs. 7.9 and 7.11).

Many palatable caterpillars snip off partially eaten leaves (Tables 7.1 and 7.2).

Figure 7.9. Examples of leaf handling. (A) A yellow-banded underwing caterpillar, *Catocala cerogana,* clipping off a basswood leaf remnant on which it has just finished feeding. (B) A waved sphinx caterpillar, *Ceratomia undulosa,* also clipping off a leaflet remnant just after finishing feeding. (C) Two *Viburnum* leaves trimmed down by the hummingbird clearwing caterpillar, *Hemaris thysbe.* (D) Feeding behavior of the bombyx sphinx, *Lapara bombycoides.*

Is this a behavior that fools predators, or is it a response to thwart the host plant's defenses (by removing the "receptors" in the damaged leaf tissue so that the plant does not "know" it is being eaten)? For leaf clipping to thwart the host plant, plants can send induced toxins only to damaged leaves and leaves adjacent to damaged leaf *blades,* and *not* to leaves near chewed petioles. However, there is no evidence that plants are indeed capable of sending toxins to specific and partially damaged leaves but not to leaves near defoliated petioles. Until such data are forthcoming, it seems more prudent to suppose that predation is more likely to shape this kind of feeding behavior (clipping partially eaten leaves) than plant physiology.

The leaf-trimming and leaf-clipping behaviors (Figs. 7.10, 7.12, and 7.13 BC) are not restricted to notodontids. They are also common in numerous other palatable moth caterpillars (e.g., Noctuidae, Sphingidae). In a number of these leaf-clipping species, the caterpillars feed only at night. In the daytime they hide away from the feeding site on a branch in the vicinity of where they fed. Such larval behaviors are explicable if birds learn to search near where they find leaf damage. The leaf clipping behavior is not consistent with the idea of reducing the apparent damage to plants so as to mobilize fewer toxins. If leaf clipping reduced plant defenses, unpalatable caterpillars (at least spiny and hairy ones—

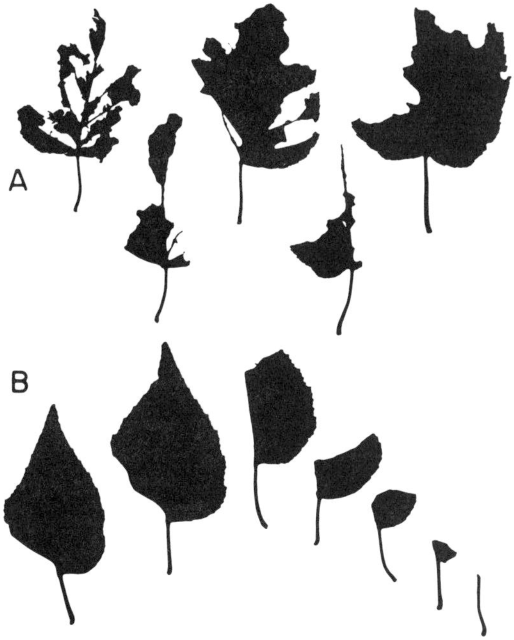

Figure 7.10. Representative leaf damage and feeding sequence on birch. (A) The spotted halysidota, *Halysidota maculata* (Arctiidae). (B) *Heterocampa* sp. (Notodontidae) showing a leaf trimmed and consumed in stages (left to right).

those not defended by toxins they sequester from the plant) should clip leaves but they do not (Table 7.1). The possibility remains that a caterpillar leaves scent on leaves where it has fed (Vinson 1985), so that by clipping off leaves it has fed from it is also reducing its scent trail (Faeth 1986). If so, then caterpillars that feed on herbs should have the same behavior. However, so far the leaf clipping behavior has been observed in 16 species, all of which forage on trees or vines (Table 7.2). I presume that ground-foraging birds are also important predators of larvae. But they are not likely to search for caterpillars by using leaf damage because (1) leaf damage there would not be readily visible, (2) caterpillars near the ground often hide in the leaf litter and soil so that leaf damage is not a reliable clue, and (3) relatively few if any ground-dwelling birds *specialize* on caterpillars, feeding mainly on other items.

Conclusions and Future Direction

I conclude that caterpillars' appearance, activity patterns, and feeding behaviors are all integrated by selective pressures exerted by the dominant constraint of

Figure 7.11. Examples of hiding. (A) The caterpillar of the waved sphinx, *Ceratomia undulosa,* on trimmed lateral ash leaflet, with terminal leaflet eaten and remnants clipped off. The same ash twig also shows the rachi and petioles of two leaves already processed. (B) The relict underwing caterpillar, *Catocala relicta,* hiding motionless (during daylight) on poplar twig away from feeding site. Note fake shadow mark across back of caterpillar. (C, D) An apparently ungrazed basswood twig, which on being turned over (right) reveals that three leaves are missing. They have been fed from and then clipped off by the yellow-banded underwing caterpillar, *Catocala cerogana.*

highly acute visual predators. However, although most of the data from a variety of perspectives all point to the same conclusion, the empirical evidence is still sparse. In particular, geographic and temporal trends need to be documented.

The effectiveness of leaf-clipping behavior is contingent on the background of existing leaf damage, and that leaf damage probably varies temporally. In the

Table 7.1. Feeding Behaviors of 30 Palatable and 13 Unpalatable Caterpillars on Trees[a]

	Trim leaves	Leave tatters	Eat whole leaf	Leaf partially eaten	Clip leaf remnants off
30 Palatable	30	1	25	6	16
13 Unpalatable	0	13	0	13	0

[a]Adapted from Heinrich and Collins (1983). The numbers in the columns show numbers of caterpillars (of the 30 and 13, respectively) showing the behavior. Note the behaviors are not necessarily exclusive.

Table 7.2. Feeding Behaviors of 30 Palatable and 13 Unpalatable Caterpillars as a Function of Food Plant[a]

	Trim leaves		Eat whole leaf		Clip off leaf remnants	
	Yes	No	Yes	No	Yes	No
Trees						
Fraxinus	4	1	3	2	4	1
Populus	7	2	6	2	6	4
Prunus	1	2	1	2	1	3
Betula	4	1	4	1	4	4
Acer	4	2	3	3	4	4
Herbs	3	4	3	4	0	7

[a]Adapted from Heinrich and Collins (1983). Herbs were *Asclepias syriaca, Aristolochia* sp., *Aster* sp., *Galium* sp., *Vitis* sp., *Datura* sp.

north-temperate regions, the trees flush out within about 2 weeks in May, and most of the insect damage could then be associated reliably with palatable prey. But after about 2 months, much leaf damage has accumulated, and leaf damage then becomes a much less reliable cue to indicate the presence of edible caterpillars.

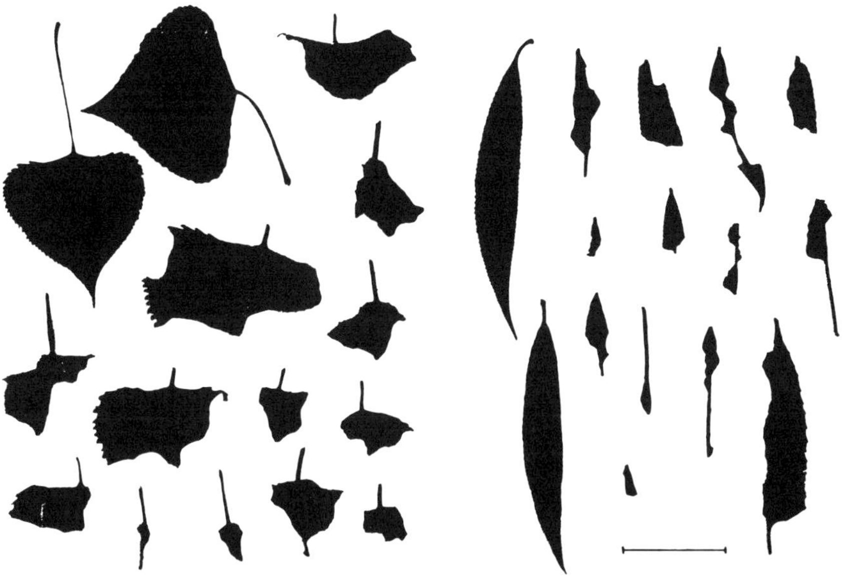

Figure 7.12. Clean feeding by caterpillars of the big poplar sphinx, *Pachysphinx modesta,* on cottonwood (left) showing leaf trimming and leaf clipping (note petiole stubs and long petioles on the two reference leaves). Pictures at right show feeding on willow, where caterpillars also clip off leaf remnants.

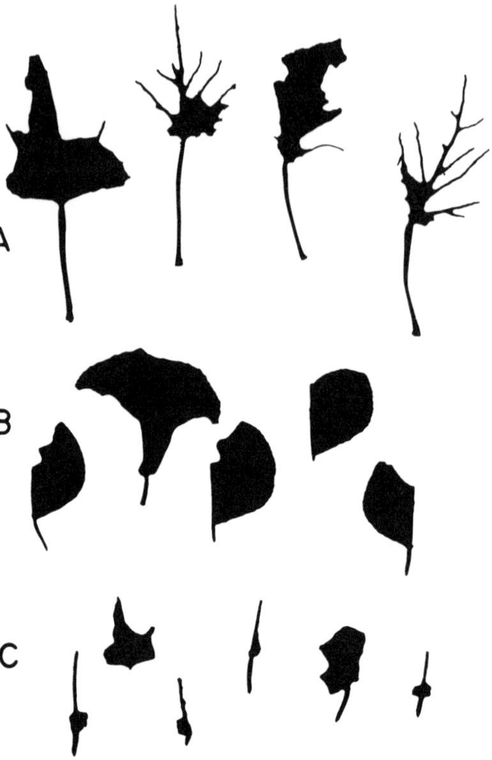

Figure 7.13. Leaf handling on poplar. (A) *Nymphalis antiopa* (Nymphalidae), showing "messy" feeding and no clipipng (long petioles). (B) *Zale* sp. (Noctuidae) showing trimming of the leaf followed by clipping. (C) The American dagger moth caterpillar, *Cronicta americana* (Noctuidae), showing leaf trimming followed by clipping off the shown leaf remnants.

Leaf-clipping behavior may vary geographically as well. For instance, the leaf-clipping phenomenon could be much less widespread in tropical forests. I made cursory observations of caterpillar feeding behavior at the beginning of the wet season (early June) in Costa Rica. After continuous searching for at least 1 day each at four sites (Santa Rosa National Park, Monte Verde, Manuel Antonio National Park, and Playa Hermosa) I found evidence of leaf clipping at all sites. However, by comparison with the great abundance of clipped leaves littering the forest floor in almost all the temperate woodlands of New England, the leaf clipping in Costa Rica seemed sporadic and rare. Of course, impressions are no substitute for data. But why might leaf clipping be less common in the tropics than in temperate forests?

As suggested by Janzen (1988), perhaps a higher level of defoliation (in Costa Rican dry forest) makes damaged leaves a poor indicator of the presence of a caterpillar. In the wet lowland tropics on the other hand, many plants hold their leaves for long periods, sometimes more than 2 years (Coley 1988), and such leaves are more resistant to predation (Coley 1983). The tough and chemically defended leaves of wet lowland tropics (Coley 1987) are likely to be used mainly by specialist insects that are chemically defended themselves, so that leaf clipping may not be necessary.

Finally, migrant birds move between the temperate and tropical habitats, and their foraging behavior changes accordingly (Greenberg 1987). But how such changes may impact on caterpillar communities and their ways of evading birds is unclear.

Acknowledgments

I give my sincere thanks to an anonymous reviewer and to N. E. Stamp for her patient and insightful editing. Writing of this manuscript was in part supported by NSF Grant BNS-8819705.

Literature Cited

Baker, R. R. 1970. Bird predation as a selective pressure on the immature stages of the cabbage butterflies, *Pieris rapae* and *P. brassicae*. J. Zool. Lond. 162:43–59.

Brodie, D. B., Jr., and Brodie, E. D., III. 1980. Differential avoidance of mimetic salamanders by free-ranging birds. Science 208:181–182.

Casey, T. M. 1976. Activity patterns, body temperature and thermal ecology of two desert caterpillars (Lepidoptera: Sphingidae). Ecology 56:485–497.

Casey, T. M. 1977. Physiological responses to temperature of caterpillars of desert populations of *Manduca sexta*. Comp. Biochem. Physiol. 57A:485–487.

Carpenter, F. M. 1976. Geological history and evolution of insects. Proc. XV Int. Congr. Entomol. Washington, D.C., pp. 63–70.

Coley, P. D. 1983. Herbivory and defensive characteristics of tree species in a lowland tropical forest. Ecol. Monogr. 53:209–233.

Coley, P. D. 1987. Interspecific variation in plant anti-herbivore properties: The role of habitat quality and rate of disturbance. New Phytol. 106:251–263.

Coley, P. D. 1988. Effects of plant growth rate and leaf lifetime on the amount and type of anti-herbivore defense. Oecologia 74:531–536.

Damman, H. 1986. The osmeterial glands of the swallowtail butterfly *Eurytides marcellus* as a defence against natural enemies. Ecol. Entomol. 11:261–265.

Dussourd, D. E., and Eisner, T. 1987. Vein-cutting behavior: Insect counterplay to the latex defense of plants. Science 237:898–901.

Evans, H. E. 1987. Observations of the prey and nests of *Podalonia occidentatis* Murray (Hymenoptera: Sphecidae). Pan-Pac. Entomol. 63:130–134.

Faeth, S. H. 1986. Indirect interactions between temporally separated herbivores mediated by the host plant. Ecology 67:479–494.

Feduccia, A. 1980. The Age of Birds. Harvard University Press, Cambridge, MA.

Ford, E. B. 1945. Butterflies, Collins, London.

Fourie, P. B., and Hull, P. R. 1980. Urticaria caused by the slug caterpillar *Latoia vivida* (Lepidoptera: Limacodidae). S. Afr. Tydskr. Dierk. 15:56.

Gilmer, P. M. 1925. A comparative study of the poison apparatus of certain lepidopterous larvae. Ann. Entomol. Soc. Am. 18:203–239.

Gould, W. P., and Jeanne, R. L. 1984. *Polistes* wasps (Hymenoptera: Vespidae) as control agents for lepidopterous cabbage pests. Environ. Entomol. 13:150–156.

Greenberg, R. 1987. Seasonal foraging specialization in the Worm-eating Warbler. Condor 89:158–168.

Greenberg, R., and Gradwohl, J. 1980. Leaf surface specializing birds and arthropods in a Panamanian forest. Oecologia 46:114–124.

Greene, E. 1989. A diet-induced developmental polymorphism in a caterpillar. Science 243:643–646.

Heinrich, G. H. 1962. Synopsis of nearctic Ichneumoninae Stenopneusticae with particular reference to the northeastern region (Hymenoptera). Part VII. Synopsis of the Trogini. Can. Entomol., Suppl. 29:805–886.

Heinrich, B. 1971. The effect of leaf geometry on the feeding behavior of the caterpillar of *Manduca sexta* (Sphingidae). Anim. Behav. 19:119–124.

Heinrich, B. 1979. Foraging strategies of caterpillars: leaf damage and possible predator avoidance strategies. Oecologia 42:325–337.

Heinrich, B. 1984. Strategies of thermoregulation and foraging in two vespid wasps, *Dolichovespula maculata* and *Vespula vulgaris*. J. Comp. Physiol. 154B:175–180.

Heinrich, B., and Collins, S. L. 1983. Caterpillar leaf damage, and the game of hide-and-seek with birds. Ecology 64:592–602.

Holmes, R. T., Schultz, J. C. and Nothnagle, P. 1979. Bird predation on forest insects: an exclosure experiment. Science 206:462–463.

Holmes, R. T., Sherry, T. W., and Sturges, F. W. 1986. Bird community dynamics in a temperate deciduous forest: Long-term trends at Hubbard Brook. Ecol. Monogr. 56:202–220.

Janzen, D. H. 1988. Ecological characterization of a Costa Rican dry forest caterpillar fauna. Biotropica 20:120–135.

Jones, F. M. 1932. Insect coloration and the relative acceptability to birds. Entomol. Soc. London Trans. 80:345–385.

Knapp, R., and Casey, T. M. 1986. Thermal ecology, behavior, and growth of gypsy moth and Eastern tent caterpillars. Ecology 67:598–608.

Kukalova-Peck, J. 1978. Origin and evolution of insect wings and their relation to metamorphosis, as documented by the fossil record. J. Morphol. 156:53–126.

MacArthur, R. H. 1959. On the breeding distribution patterns of North American migrant birds. Auk 76:318–325.

McKay, M. R. 1970. Lepidopteran in Cretaceous amber. Science 167:379–384.

Morris, R. F. 1972. Predation by wasps, birds, and mammals on *Hyphantria cunea*. Can. Entomol. 104:1581–1591.

Pietrewicz, A. T., and Kamil, A. C. 1979. Search image formation in the Blue Jay (*Cyanocitta cristata*). Science 204:1332–1333.

Raveret Richter, M. 1990. Hunting social wasp interactions: influence of prey size, arrival order, and wasp species. Ecology 71:1018–1030.

Real, R. G., Ianazzi, R., Kamil, A. C., and Heinrich, B. 1984. Discrimination and generalization of leaf damage by blue jays (*Cyanocitta cristata*). Anim. Learn. Behav. 12:202–208.

Robinson, S. K. and Holmes, R. T. 1984. Effects of plant species and foliage structure on the foraging behavior of forest birds. Auk 101:672–684.

Rothschild, M. 1973. Secondary substances and warning colouration in insects. In H. F. Van Emdem (ed.), Insect-Plant Relationships. Symp. Roy. Entomol. Soc. London 6:59–83.

Scott, A. C., and Taylor, T. N. 1983. Plant-animal interactions during the Upper Carboniferous. Bot. Rev. 49:259–307.

Scriber, J. M., and Slansky, F., Jr. 1981. The nutritional ecology of insects. Annu. Rev. Entomol. 26:183–211.

Scriber, J. M., and Ayres, M. P. 1988. Leaf chemistry as a defense against insects, pp. 117–123. In ISI Atlas of Science: Animal and Plant Sciences, Vol. 1.

Stamp, N. E., and Bowers, M. D. 1988. Direct and indirect effects of predatory wasps (*Polistes* sp.: Vespidae) on gregarious caterpillars (*Hemileuca lucina:* Saturniidae). Oecologia 75:619–624.

Thurston, R., and Prachuabmoh, O. 1971. Predation by birds on tobacco hornworm larvae infecting tobacco. J. Econ. Entomol. 64:1548–1549.

Turnbull, C. L. 1979. Adaptive changes in morphology and behavior of *Clossiana selene* larva (Lepidoptera: Nymphalidae). Entomol. News 90:125–130.

Vinson, S. B. 1985. The behavior of parasitoids, pp. 417–469. In G. A. Kerkut and L. I. Gilbert (eds.), Comprehensive Insect Physiology, Biochemistry, and Pharmacology, Vol. 9, Behaviour. Pergamon, Oxford.

Wardle, A. R., and Borden, J. H. 1989. Learning of an olfactory stimulus associated with a host microhabitat by *Exeristes roboyator*. Entomol. Exp. Appl. 52:271–279.

8

Why Body Size Matters to Caterpillars

Duncan Reavey

> The Caterpillar was the first to speak.
> "What size do you want to be?" it asked.
> "Oh, I'm not particular as to size," Alice hastily replied, "only one doesn't like changing so often you know."
> "I don't know," said the Caterpillar.
>
> *Alice's Adventures in Wonderland,* Lewis Carroll

Changes in Body Size

There is a spectacular change in body size as caterpillars grow. Larval weights for the few species for which data are available increase by 2200 to 3300 times between hatching and the completion of larval growth (Table 8.1). The variation among species is considerably greater; eggs of *Heliconius erato* (Heliconidae), for example, are 20 times the volume of eggs of *Colias eurytheme* (Pieridae), though adults of these two species are of a similar size (Labine 1968). But in all cases, with such a change in size, the world must seem a very different place to first and final instars.

Likewise, small and large species must face very different challenges. The difference in size between a newly hatched *Pterophorus pentadactyla* (Pterophoridae) caterpillar (0.022 mg, Reavey 1992) and a fully grown *Eacles imperialis* (Saturniidae) 25–30 g, Janzen 1984) spans six orders of magnitude. Both of these species feed externally on leaves; the smallest newly hatched leaf mining caterpillars are likely to be at least two orders of magnitude smaller still. Caterpillars of all sizes across this range are often present at the same time on the same food plant, but the ways in which they make use of it, and the ways in which they experience and conceive their environment, are likely to differ considerably.

In this chapter I will highlight the importance of body size as a major influence on the feeding ecology and behavior of caterpillars. First I will look at the effect of body size on some of the constraints on foraging patterns considered in other chapters of this book. This is the first time an attempt has been made to bring together information on the changes that occur as caterpillars grow. It has been a somewhat frustrating task because surprisingly few studies even *mention* body

Table 8.1. Changes in Size between Newly Hatched and Fully Grown Caterpillars

	Live weight			
	Newly hatched (mg)	Fully grown (mg)	Increase in size	Reference
Hyalophora cecropia (Saturniidae)	3.7	10200	2760×	Schroeder (1972)
Achlya flavicornis (Thyatiridae)	0.12	265	2200×	D. Reavey (unpublished data)
Eacles imperialis (Saturniidae)	0.1	25000–30000	2750–3300×	P. Z. Goldstein (unpublished data); Janzen (1984)

size, never mind consider its implications. Furthermore, although it is possible to begin to understand some of the pieces of the puzzle and sense their relative importance, there are few studies that give any idea of how they fit together in the real world.

Having gathered together a scattered and rather sparse literature, I will then consider three recent studies at the University of York, U.K., which demonstrate the impact of body size on the feeding habits of microlepidoptera, the feeding ecology and behavior of newly hatched caterpillars, and the foraging patterns of a leaf-tying caterpillar over its whole larval development. These highlight the value of studying patterns across large numbers of species and the need to study early as well as late instars because the ways they differ are not necessarily predictable. These same body size effects can also begin to give an insight into patterns in feeding ecology among species of different sizes which I will consider briefly.

I will restrict the discussion to the differences *between* instars, even though the increase in weight *within* an instar can be as much as ten (e.g., Williams 1980) or even 100-fold (e.g., Thomas and Wardlaw 1992). I do this because there are distinct patterns of physiological and behavioral change within each instar that deserve a chapter of their own, and because many developmental changes, in color or mouthpart structure, for example, occur only at ecdysis. Even so, some of the arguments, for instance about mobility or vulnerability to enemies, are likely to apply equally to changes within an instar. In this chapter I consider sawfly caterpillars as honorary Lepidoptera because of their ecological similarity, and I occasionally draw on information from other insects and even spiders where it seems appropriate and data from Lepidoptera are sparse or absent.

Effects of Body Size on Caterpillars—Metabolism, Physiology, Ecology, and Behavior

Physiological Performance and Metabolism

There are consistent changes in physiological performance as caterpillars grow (Schowalter et al. 1977; Larsson and Tenow 1979; Scriber and Slansky 1981;

Blake and Wagner 1984; Sehnal 1985; Slansky and Scriber 1985; Trichilo and Mack 1989). Relative metabolic rate is highest in the earliest instars and declines from then on, at least in part reflecting an accumulation of fat reserves that are metabolically less active. Possible exceptions are cases in which the first instar has to locate the food plant or overwinter and feeds very little, in which case the second instar is likely to have the highest metabolic rate because of the difference in activity (Sehnal 1985). The relative food consumption rate decreases, but the total amount consumed increases. Assimilation efficiency tends to decrease with body size: possible reasons are that reduced feeding selectivity by larger individuals means more indigestible fiber is consumed, and that their larger mouthparts mean they ingest food particles of a larger size, with a relatively small surface area exposed to digestive enzymes. Another is that larger caterpillars have a smaller gut surface area relative to body weight. Note, however, that although the gut surface area-to-volume ratio might decrease dramatically, it can be the *volume* of gut epithelial cells rather than the surface area of the lumen that matters in digestion, and that this might actually increase. For the gypsy moth *Lymantria dispar* (Lymantriidae), the gut surface area-to-volume ratio decreases to less than one-ninth of its original value between the first and fifth instars, but the relative volume of epithelial cells increases 13-fold, so in fact there is a net *increase* in the epithelial volume-to-gut-volume ratio (Ahmad 1986). Relative growth rate decreases with increasing body size. Although assimilation efficiency decreases, efficiency of conversion of digested food increases; together they determine the overall efficiency of conversion of ingested food, which is not related to size in any predictable way. Successive instars tend to be of longer duration because relative metabolic activity decreases while growth rate per instar remains broadly constant (Sehnal 1985). Toward the end of larval development, within the final instar, there is a change in preference for a diet high in protein to one high in carbohydrates as larvae switch from growth to the metabolic activity preceding pupation (Cohen et al. 1987; Zucoloto 1987).

It is clear that changes in performance as caterpillars grow determine their nutritional needs and reflect the quality and quantity of the food consumed. But how do these changes impinge on feeding and foraging behavior? Because smaller individuals show higher relative performance rates, it is possible that differences in food quality have most influence on the early instars (Taylor 1984; Slansky and Scriber 1985 and references therein; Krishna 1987). Indeed, the nitrogen level in the diet of first instars of *Samea multiplicalis* (Pyralidae) affects the duration of larval development and the number of instars, but the nitrogen level in the diet of subsequent instars has little influence (Taylor 1984). Thus the need to make the correct feeding choices could be greatest for early instars.

Furthermore, smaller larvae are possibly less able to compensate for poorer food quality. Late instars of the buckeye *Junonia coenia* (Nymphalidae) maintain their growth rate on low nitrogen foliage by increasing the amount of food they consume, but early instars cannot maintain their growth even with increased consumption (Fajer 1989). Although larvae can adjust their feeding rates to maximize nitrogen intake, hence growth rate (Miles et al. 1982), this is possible only within limits because utilization is less efficient when food is processed more quickly and there comes a point when increased feeding brings no benefits (Slansky and Feeny 1977). These limits are apparently narrower for early instars.

On a rather different note, the way larvae respond to photoperiod can depend on their instar. *Sasakia charonda* (Nymphalidae) hibernates in the fourth instar. For fourth instars, growth is slowed and diapause induced by short days, and diapause is delayed by long days. For third instars, however, growth is accelerated by short days and slowed by long days. But not only that; for one morph of the caterpillar, whether the caterpillar responds to photoperiod also depends on larval body weight (Kato 1989).

Feeding

A change in body size affects a caterpillar's feeding opportunities. Take, for example, size-related changes in feeding selectivity. Smaller mouthparts allow early instar caterpillars to make feeding choices that become impossible as they grow larger. Even within a leaf, tissues differ in nutritional value, toughness, and allelochemicals (Kimmerer and Potter 1987) and the smallest caterpillars can respond. First instars of *Plutella xylostella* (Plutellidae) feed as leaf miners on soft spongy mesophyll and ignore tougher palisade tissue; second instars, however, are external feeders and consume all the tissue except the upper epidermis (Salinas 1984). Young soybean loopers *Pseudoplusia includens* (Noctuidae) feed selectively on leaf tissues low in fiber and of high digestibility, but older larvae are less discriminating (Kogan and Cope 1974). First instar larvae of the common looper *Autographa precationis* (Noctuidae) scrape shallow cavities into the parenchyma of the leaf and eventually chew through to the opposite side; second and third instars chew across the entire thickness leaving schlerenchymatous tissue and veins, and later instars clip across all but the thickest veins (Khalsa et al. 1979). This reduction in feeding selectivity could be one reason why assimilation efficiency declines as caterpillars grow. *Neodiprion sertifer* larvae are selective only in the first few days after hatching and assimilation efficiency hardly changes subsequently; similarly, larvae of the pine beauty *Panolis flammea* consume whole needles after the second instar and from then on there is little change in assimilation efficiency (Oldiges 1959; Larsson and Tenow 1979). More fundamental changes in feeding habit, for example, from leaf mining to stem boring or external feeding, can occur during larval growth and are a striking demonstration of the influence of body size on ways of feeding. I examine these in more detail in Changes in Feeding Habit.

The size and strength of the mandibles appear to influence how effective caterpillars can be in circumventing sticky, toxic latex defenses within the leaves of some plants (Compton 1989). Some species feeding on the leaves of figs achieve this by cutting the leaf veins to sabotage the latex defenses before feeding. As *Aganais speciosa* (Hypsidae) caterpillars grow, there is a progression from no cutting behavior to chewing through small side veins and eventually to cutting through the midrib of the leaf. Only later instars of *Chrysodeixis acuta* (Noctuidae) were observed chewing through the midrib, and only larvae of the largest species studied, *Pseudoclanis postica* (Sphingidae), could cut the midrib in early instars. A further 10 species were examined on that same food plant but none showed such behavior; interestingly, all 10 are smaller than the three that cut the leaf veins (Compton 1989).

A change in selectivity can be paralleled by a change in mandible morphology as well as size. For instance, the mandibles of most notodontid larvae change from having a toothed cutting edge in the first instar to a smooth edge in later instars; first instars skeletonize the leaves leaving one epidermis intact whereas later instars cut through the full thickness (Godfrey et al. 1989). Gracillariids mine the leaf epidermis as sap drinkers in their earliest instars and their mouthparts are modified to cut the walls of sap-filled cells, but at the second or third ecdysis there is a radical restructuring of the mouthparts and from then on the larvae mine by chewing out parenchyma tissue (Heath and Emmet 1985).

Mandible morphology also determines the size and shape of food particles entering the gut. Saturniids have short, simple mandibles that cut the leaf, but only into large pieces. Only the cut edge of each fragment is digested, but this could be a good thing because it means that the bulk of the growth-inhibiting tannins do not enter the gut from within the leaf. As their mouthparts increase in size, the pieces of leaf are cut larger, reducing the efficiency of digestion. This is a possible reason why the mouthparts of saturniid larvae are relatively large when the caterpillars are small, and relatively small when the caterpillars are large (Bernays and Janzen 1988).

Some caterpillars switch food plant species as they grow (Reavey and Lawton 1991). Although this could be due to seasonal changes in plant quality, it could equally reflect changes in the ability of caterpillars to accept alternative food plants due to their increased body size. Caterpillars of the gypsy moth *Lymantria dispar* (Lymantriidae) often switch from plant to plant in the wild and may feed on two or more species during development. Larvae fed on coniferous species do not survive the first instar. But if larvae are first fed on a deciduous species and then switched to a conifer, adults have greater fecundity than if fed on the deciduous species alone. A possible reason is that the tough conifer needles prevent initiation of feeding by newly hatched larvae (Barbosa et al. 1986).

As caterpillars grow, there are certainly differences in the concentrations of some gut enzymes that might be important in the detoxification of secondary plant compounds, for instance, mixed function oxidases (Ahmad et al. 1986 and references therein), but how they relate to changing needs during development is uncertain (Gould 1984). For *Lymantria dispar,* the relative activity of mixed function oxidases increases more than 4-fold between the second and fifth instars.

Possible reasons are that it is linked to an increase in relative consumption rate (this species is unusual among caterpillars in this respect, see Physiological Performance and Metabolism), that there is a seasonal change in, or induction of, allelochemicals in the food plant, and that larger larvae are more likely to move to new plants with more or different allelochemicals (Ahmad 1986).

Caterpillars probably have the greatest potential difficulties with the physical act of feeding when they have just hatched, and this could be one factor affecting the size of newly hatched larvae. Eggs (and the head capsules of newly hatched larvae) of *Parnara guttata* (Hesperiidae) are small in the first and second generations when the larvae feed on tender leaves, but larger in the third generation when they feed on hard grasses (Nakasuji and Kimura 1984), and hesperiid species feeding on tougher grasses tend to lay larger eggs (Nakasuji 1987). I will look more closely at patterns in the feeding ecology and early larval behavior with respect to the size of newly hatched larvae in The Size of Newly Hatched Caterpillars.

Starvation

The greater energy reserves and lower relative (weight specific) metabolic rate associated with larger body size could mean that larger individuals are better able to endure starvation for longer periods of time. While this is supported for many different animal groups (e.g., Lindsey 1966; Millar and Hickling 1990), there is less evidence for invertebrates (Derr et al. 1981; Yuma 1984; Slansky and Scriber 1985; Palanichamy and Baskaran 1989). Among fifth instars of the small white *Pieris rapae* (Pieridae), larger individuals endure starvation for longer (Jones 1977). Similarly, larger larvae of the gypsy moth *Lymantria dispar* (Lymantriidae) resist starvation for longer than smaller individuals, though feeding history has an important influence (Stockhoff 1991). Among newly hatched individuals, differences in body weight have no effect on survival of several satyrid species (Karlsson and Wiklund 1984, 1985). However, larger larvae of the corn earworm *Heliothis armigera* (Noctuidae) are less eager than smaller larvae to begin feeding (Barber 1941) suggesting that they may be better able to resist starvation. The relationship across species between body size and survival time of newly hatched larvae is further considered in The Size of Newly Hatched Caterpillars.

Mobility

Mobility affects the distances over which caterpillars can forage, and thus their opportunities to make feeding choices, the ease of locating a new food plant (Rausher 1979; Cain et al. 1985), or dispersal to new feeding sites (Weiss et al. 1987), and the feasibility of moving a safe distance from conspicuous feeding damage between feeding bouts.

Larger individuals of the same species tend to move faster and further (e.g.,

Dethier 1959; Coshan 1974; Jones 1977; Rausher 1979; Holdren and Ehrlich 1982; Hansen et al. 1984b; Cain et al. 1985). For example, 2.5 mm newly hatched larvae of the yellow horned moth *Achlya flavicornis* (Thyatiridae) moved at 148 cm h^{-1} on the stem of a birch sapling, whereas 30 mm fifth instar larvae moved at 668 cm h^{-1} (Feichtinger and Reavey 1989). Although larger larvae move faster, their power and agility do not increase in proportion to weight (Enders 1976). Thus the relative energy cost per stride (but not necessarily per unit distance, which is what matters) will be less for smaller caterpillars.

The speed a caterpillar can move depends on the length of its stride; larger caterpillars can move faster because their stride is longer. But the absolute distance smaller and larger caterpillars have to move to cross a surface can differ considerably if the surface is anything other than completely smooth: small larvae closely follow the ups and downs of the surface that large caterpillars do not even notice. Weiss and Murphy (1988) highlight this in a simple fractal model. In their imaginary rough-textured grassland, 1-mm-long caterpillars travel 15.8 m to cover 1 linear meter, whereas 5-mm caterpillars travel 8.32 m and 30-mm caterpillars only 4.08 m because of the fractal nature of the surface. At an extreme, movement of small caterpillars might be prevented entirely by plant surface structures like leaf pubescence (e.g., Ramalho et al. 1984), which have little effect on larger individuals. Although larger caterpillars might be physically capable of moving faster and further, they might not necessarily do so if it increases vulnerability to enemies or the risk of separation from the food plant, or if it opens up no new feeding choices (see Changes in Movement).

"Silking," the suspension of caterpillars by silk threads, is probably only feasible for smaller individuals even though later instars can have stronger silk (Yeargan and Braman 1986). Larger caterpillars are simply too heavy for the silk to hold them. Indeed, among the newly hatched larvae of 12 species of British Lepidoptera, smaller species are more likely to silk (Reavey 1992). However, I know of no data that give an idea of the threshold size beyond which silking becomes impossible. Silking can be an escape route (e.g., Allen et al. 1970; Dempster 1971), prevent larvae falling accidentally from the plant, or be a means of dispersal associated with "ballooning" (e.g., Taylor and Reling 1986), but in each case it is likely to be effective only when the larvae are relatively small.

Body size determines the parts of the food plant a caterpillar can use for feeding, basking, or resting. Early instar large white *Pieris brassicae* (Pieridae) caterpillars tend to feed and rest on the lower surfaces of leaves, but when they grow larger they move onto the upper surface because they more easily fall from the lower surface when disturbed (Long 1955). Small crickets climb grass stems to reach a good position for basking, but heavier individuals cannot climb the stems and search instead for open ground or stones (Remmert 1985).

The Physical Environment

Environmental temperature determines caterpillar body temperature, which influences body processes and ultimately growth rate. In sunshine, body temperature rises above ambient temperature as solar radiation is absorbed. Smaller insects heat up and cool down more quickly because of their relatively large surface-to-volume ratio. However, larger insects reach a higher temperature excess at equilibrium because of their higher heat capacity, and this also means that they are less affected by short-term temperature changes (Casey 1981). In the arctic woolly bear *Gynaephora rossii* (Lymantriidae), for example, 40-mm caterpillars can have a temperature excess of >10°C compared to an excess of 3.2°C for those 20 mm long (Kevan et al. 1982). Wind decreases body temperature to a greater extent in smaller insects (Whitman 1987) and the rate of desiccation is likely to be greater for smaller insects (Stamp and Bowers cited by Cornell et al. 1987) because of their greater surface area-to-volume ratio.

The heterogeneity of its thermal environment depends on the size of the insect. To a smaller insect the environment is more patchy and pockets of closer to optimal temperature could be available that are too small to affect larger individuals (May 1985). Different sized individuals of several insect species are affected in different ways by the various microclimates to which they are exposed, even to the extent that the relationship can influence mating patterns (Larsson 1989). Furthermore, for several insect groups, larger individuals are less restricted to the lower leaf surface possibly because they are less influenced by constraints of microclimate (Greenberg and Gradwohl 1980), and a similar pattern might be expected for different sized individuals of the same species. Such body size–microclimate relationships are likely to have broad implications for caterpillars but have received all too little attention.

The optimum temperatures for insect growth and activity tend to differ from the environmental temperature and some caterpillars thermoregulate by their behavior (Casey, Chapter 1). Behavioral mechanisms can depend on body size. In some species only larger individuals bask openly on vegetation. Only fourth and later instar larvae of the marsh fritillary *Euphydryas aurinia* (Nymphalidae) move from their food plant low in the undergrowth to bask in the sun on bare ground or exposed vegetation (Porter 1982). Presumably less mobile early instars are less able to move about their environment to regulate their body temperature, and in any case the rewards in terms of temperature excess are not so high. Early instars of this species, however, are gregarious within a silken web. Webs reduce air movement while allowing the penetration of solar radiation, so may have a thermoregulatory role (Wellington 1974; Knapp and Casey 1986). Thus, smaller individuals may thermoregulate, but in a different way from when they are large.

Coloration could have an adaptive value in thermoregulation. Insects colored black reach a higher temperature excess more quickly than those of other colors, so it is not surprising that basking instars of *Euphydryas aurinia* are jet black whereas earlier instars are pale brown (Porter 1982). *Ctenucha virginica* (Arctiidae) caterpillars change color as they grow. In the autumn and spring, their hairs are mostly black, allowing the caterpillars to increase temperature excess, hence growth, in these cooler seasons; but, in the summer, hairs of the same individuals are mostly yellow, perhaps preventing overheating because yellow tends to reflect rather than absorb heat (Fields and McNeil 1988). However, large black individuals run a grave risk of overheating and this is a likely reason why the largest species of ladybird are not black (Stewart and Dixon 1989). For this reason we might predict that there are relatively few black species among the largest diurnal caterpillars and that some larger species change from black to a less absorbant color as they grow.

Susceptibility to cold might also be affected by body size. Larval mortality due to chilling is greater for earlier instars of the eastern tent caterpillar *Malacosoma americanum* (Lasiocampidae) (Mansingh 1974).

The effects of other climatic factors are also likely to be influenced by body size. The greatest mortality among first and second instar larvae of *Pieris rapae* in one study came from rainfall drowning or dislodging larvae; larger larvae are not usually affected by rain and, if dislodged, can usually find their way back to the food plant (Harcourt 1966). In the same way, it is usually only first and second instars of the black swallowtail *Papilio polyxenes* (Papilionidae) that suffer death from drowning (Blau 1980). Newly hatched sawfly larvae are small enough to become trapped in water or resin droplets which are unlikely to have any effect on larger larvae (Ghent 1960).

Enemies and Defense

The appearances and behaviors of caterpillars are likely in part to reflect pressures from enemies. Even if the pressures remain the same as caterpillars grow, the change in body size can require subtle or more radical changes just to maintain the same defense strategy. Take, for example, species feeding on pine needles (Herrebout et al. 1963). The young caterpillars closely resemble the needles, but as they grow they become conspicuously fatter. Their crypsis is maintained by the appearance of white stripes that break up the green background into green stripes never wider than a single needle. More white stripes appear as the caterpillars grow larger. Beyond a certain size it seems that even this is not enough. In the final instar of some of these species there is a wholesale change in color to a mottled brown, and larvae rest on the pine bark instead of the needles (Herrebout et al. 1963). Some species closely resemble bird droppings, but this mimicry is effective only over a certain size range. Gregarious caterpillars of *Trilocha* spp (Bombycidae) appear as a mass of droppings left by a bird roosting above; the solitary final instar larvae of *T. kolga* resembles the dropping of a larger bird or lizard while that of *T. obliquissima* is cryptic on the brown stems of the food plant (Edmunds 1974). The third and fourth instars of the convolvulus hawkmoth *Agrius convolvuli* (Sphingidae) rest on the stems of their food plant and are highly cryptic. The final instar caterpillar is brown and highly conspicuous on the food plant, but for most of the last instar it is hidden below ground during the day (Edmunds 1974). Whether this change occurs because it is no longer feasible to maintain the former strategy at a larger body size is unclear. The caterpillar of

Morpho peleides (Morphidae) changes its resting site as it grows because body size places a constraint on where it can hide. First and second instars rest on the undersides of leaves, second and third instars among larger leaves, and third to fifth instars on even larger leaves close to the ground or on branches. In fact, this change has an interesting consequence because it affects, albeit to a small extent, the time of day at which the caterpillars feed (Young 1972).

Different responses to the same enemy become feasible as body size changes. The parasitoid *Apanteles euphydryidis* attacks the Baltimore checkerspot *Euphydryas phaeton* (Nymphalidae) at all stages of caterpillar development. The parasitoid attacks different instars in different ways; the defenses of the caterpillars also change as they grow (Stamp 1982, 1984a). The web of first instars provides some protection and the gregarious caterpillars respond collectively to attack by head jerking or synchronous movement away from the enemy. Second to sixth instar larvae may be partly protected by spiny tubercules that, by the third instar, are of similar length to the ovipositor of the parasitoid. Similarly, early instars of *Malacosoma americana* are vulnerable to predation by ants but later instars are protected increasingly by a dense covering of hairs (Ayre and Hitchon 1968). Earlier instars of the green coverworm *Plathypena scabra* (Noctuidae) escape many predators by silking down from the food plant while later instars drop to the ground without silking (Yeargan and Braman 1986). But the parasitoid *Diolcogaster facetosa* can detect the silk threads and slides down them to reach second and third instar larvae. However, first instars are rarely attacked successfully because their weaker silk thread breaks under the parasitoid's weight. Defenses that depend on scale—like tubercules or spines sufficiently long to keep parasitoids at bay, or silking behavior—might well be feasible only at particular body sizes.

There is a further complication however. Caterpillars are likely to face different enemies as they grow. *Papilio xuthus* (Papilionidae), for example, suffers high mortality from predatory spiders, ants, and bugs in the first and second instars and from predatory wasps in later instars (Watanabe 1976, 1981). It is attacked by two different parasitoids, one in the second and third instars and one in the fourth and fifth instars; and predation by birds is high from the third instar onward.

In general, predators select prey that are small enough to be overcome easily but large enough to make the effort in time and energy worthwhile. In his detailed study of the feeding behavior of birds in pine woods, Tinbergen (1960) showed that small caterpillars are much scarcer in the diets of nestling tits than would be expected from their numbers in the field. The birds clearly prefer larger prey (see also Clark 1964; Mattson et al. 1968). Furthermore, as nestlings grow, parents select larger caterpillars for them. Interestingly the minimum acceptable size varies among prey species with, for example, tits rarely taking *Panolis flammea* less than 20 mm long, but frequently taking *Cacoecia piceana* (Tortricidae) of 12–15 mm (Tinbergen 1960). There is a complex relationship between caterpillar quality and size: some birds prefer the largest larvae even if they are of lower quality (Prop 1960). Others take small larvae even though larger are available (Baker 1970). Birds feeding on hairy caterpillars tend to feed on smaller rather than larger individuals (Witler and Kulman 1972), perhaps because the larger ones are too hairy for the birds to manage.

In the same way, invertebrate predators are usually more successful against prey smaller than themselves. Although certain predatory wasps attack all sizes of *Hemileuca lucina* (Saturniidae) caterpillars, the larger larvae are less likely to be killed (Stamp and Bowers 1988). Fifth instar eastern tent caterpillars *Malacosoma americanum* (Lasiocampidae), *Heliothis punctiger* (Noctuidae) and various sawflies are large and aggressive enough to defend themselves successfully against pentatomid bugs, but earlier instars are not (Iwao and Wellington 1970; Tostowaryk 1972; Awan 1985). In cannibalistic species, larger caterpillars tend to eat smaller ones (Semlitsch and West 1988). Only in the first 3 weeks of growth are eastern tent caterpillars small enough to be attacked by ants; cherry trees produce nectar to encourage these ants to forage over them, but only for the same 3 weeks in which the caterpillars are vulnerable (Tilman 1978).

It is less easy to generalize about the effect of body size on attack by parasitoids. If anything it seems that parasitoids tend to confine oviposition to one or two instars or at least to prefer particular instars (Hopper and King 1984; Hébert and Cloutier 1990 and references therein). Many species tend to lay in earlier rather than later instars.

Species that face different enemies as they grow might be expected to alter their behavior or appearance in response. The scotch argus *Erebia aethiops* (Satyridae) feeds day and night in its early instars, but in the final instar feeds only at night (Ford 1957). Contrast this with the small mountain ringlet *E. epiphron* that feeds only at night until the final instar when it also feeds by day (Carter and Hargreaves 1986). The two species have similar phenologies, broadly similar habitats, but different food plants. We can only speculate about the reasons for such changes in the timing of feeding. Pressures from enemies are likely to affect timing of larval activity, but so too are requirements for high humidity (at least for sawfly larvae, Benson 1950) or a suitable temperature for activity (e.g., Schultz 1983) and it may be difficult to reconcile conflicts between the need to avoid enemies and other selection pressures.

It is difficult to separate changes due to increasing body size from those due to changing pressures from enemies. Accompanying the striking change in appearance of several *Papilio* species from "bird dropping" penultimate instars to green, snake-like final instars are differences in the composition of pungent chemical secretions released when the caterpillars are disturbed (Burger et al. 1978; Honda 1980a, 1981). The composition of secretions does not change significantly in other papilionid genera that do not change in appearance (Honda 1980b). Penultimate and final instars of *Papilio anchisiades* (Papilionidae) do not differ in appearance to any significant degree but, still, the composition of their secretions differs considerably (Young et al. 1986). Whether these changes are a response to pressures from different enemies or are due simply to this particular mechanism becoming more functional as the larvae grow remains an intriguing question (Brower 1984; Young et al. 1986).

As its own body size and its potential enemies change, a caterpillar's direct behavioral response to attack might also differ. There are two active ways to respond: defense or escape. All instars of the pipe vine swallowtail *Battus philenor* (Papilionidae) attempt to defend themselves against coccinellids but only caterpillars larger than the predators are successful (Stamp 1986). Smaller caterpillars of *Heliothis punctiger* tend to drop from the food plant when attacked, but larger individuals respond to predators aggressively (Awan 1985). The buckmoth *Hemileuca lucina* (Saturniidae) is gregarious in its early instars and shows defensive behavior; when one larva is attacked, all respond by head rearing or thrashing (Cornell et al. 1987). Later instars are solitary and escape attack by dropping to the ground. When defense is unlikely to succeed, escape may be a more appropriate response (Stamp 1986) but the risks of leaving the food plant are considerable, especially for small caterpillars. On the other hand, for individuals large enough to defend themselves, the risks of escape are also reduced. Although these examples illustrate changes from escape to defense and from defense to escape, we have no indication which is more usual.

Group Size

Some caterpillar species are gregarious, and they feed, move, and rest in groups. Many are solitary. But others, for instance four out of seven gregarious species of British butterflies (Thomas and Lewington 1991), change from a gregarious to a solitary life-style as they grow. All sorts of possible explanations for different group sizes have been suggested (e.g., Stamp 1980; Young 1983). But it is interesting to look at these in the context of body size: changes in group size as caterpillars grow could reflect changes in selective pressures acting on the insects (Stamp 1977).

One possible advantage of gregarious behavior is facilitation of feeding, illustrated so well by Ghent's (1960) classic study of the jack pine sawfly *Neodiprion pratti banksinae* (Hymenoptera: Neodiprionidae). Newly hatched larvae have great difficulty in breaking through the tough, mature pine needle and beginning to feed; failing to feed is by far the heaviest cause of mortality for these larvae. But when one larva manages to initiate feeding, additional larvae join it and form a group feeding from the break the first larva has made. In the same way, groups of *Mechanitis isthmia* (Ithomiidae) are more effective than individual caterpillars at stripping the leaf surface of its dense layer of trichomes before feeding; these trichomes are no barrier to larger larvae, which consume them with the tissue (Young and Moffett 1979). In both cases aggregation facilitates feeding by small larvae but would seem unnecessary when the larvae are larger. Yet larvae of both species remain in groups until pupation.

Aggregation can improve thermoregulation. Caterpillars in close proximity are similar thermodynamically to a single larger organism with a relatively small surface area, so they can reach a higher temperature excess (Seymour 1974; Knapp and Casey 1986). Construction of silken webs or tents by gregarious species could have a thermoregulatory function (Knapp and Casey 1986). As caterpillars grow and individuals are more able to thermoregulate independently, dispersal from groups might be expected. This is indeed what happens in the case of *Euphydryas aurinia* (see The Physical Environment).

Aggregation could have advantages in defense (Bowers, Chapter 10; Fitzgerald, Chapter 11) and might change if defense requirements change. For instance, aggregation could enhance the effect of head thrashing behavior, chemical warning signals, or aposematic coloration for small larvae whereas larger individuals are able to achieve the same effect when alone (Ford 1957). The two studies that hint at the possibilities do not give firm evidence one way or the other. For *Hemileuca lucina,* larger group size reduces vulnerability to predation for second instars but not for third and fourth instars; yet this does not tell us much because the species is gregarious in all these stages (Stamp and Bowers 1988). A parasitoid attacks larvae of *Euphydryas aurinia* when they are gregarious and, later, when they are dispersed, so change in group size does not seem to be a response to this enemy (Porter 1982).

The change from gregarious to solitary behavior could simply reflect loss of individuals through death, the separation of individuals through escape behavior, or underlying difficulties of remaining in groups when caterpillars are larger. Feeding groups of *Neodiprion pratti banksinae* become smaller as fewer larger larvae are able to fit side by side around a pine needle (Ghent 1960). As food consumption increases, solitary caterpillars and those in smaller groups could benefit from spending less time moving to new food and more time feeding (Stamp 1977; Tsubaki and Shiotsu 1982).

Color

In some cases caterpillar coloration might have an adaptive value in thermoregulation (see The Physical Environment) or defense (see Enemies and Defense) but all too often its significance is unclear (Burtt 1981; Booth 1990). This is highlighted in species that change color as they grow. *Chlosyne lacinia* (Nymphalidae) caterpillars show no color patterns in their early gregarious instars but distinct patterns appear as the larvae become solitary (Drummond et al. 1970; Stamp 1977). Similarly, early instars of the silver Y *Plusia gamma* (Noctuidae) are leaf green but, in the fourth instar, a distinct pattern begins to appear and continues to darken (Long 1953). In this and other species, crowding of larvae leads to a darker color still (e.g., Hodjat 1970; Schneider 1973). Presumably, spectacular changes in color close to the end of larval development (Fuzeau-Braesch 1985 and references therein) are signs of an underlying developmental change rather than being adaptations for better camouflage, aposematic warning or whatever.

Foraging

It is the theme of this volume that foraging behavior is set within a framework of conflicting pressures, such as nutritional, defensive, and microclimatic. In preceding sections it has been possible to examine some of the ways in which caterpillar body size affects these constraints on foraging. But, ironically, it is much more difficult to see how body size affects foraging itself because it has scarcely ever been investigated. Indeed, all too few studies have considered caterpillar foraging behavior in detail even at a single stage of larval development. Descriptions of foraging behavior encompass spatial patterns of movement about the food plant and the habitat, temporal patterns in feeding, movement and other activities, and factors influencing feeding decisions. From the few studies that exist we can hardly begin to generalize on how these change as caterpillars grow.

There are distinct changes in the speed of movement about the habitat for some species. Caterpillars of Harris' checkerspot *Melitaea harrisii* (Nymphalidae) must move to new food plants when food is exhausted and the caterpillars reach them by random wandering. Although the closest plant might only be 66 cm away, newly hatched larvae can take 12 days to reach it and, not surprisingly, a high proportion do not make it. Larger larvae reach new plants much more quickly (Dethier 1959). Similarly, larger larvae of *Battus philenor* are more likely to locate a food plant, and to do so more quickly (Rausher 1979). There can also be qualitative changes in movement patterns as caterpillars grow, for instance, in their response to light. First and second instars of the variegated cutworm *Peridroma saucia* (Noctuidae) are strongly photopositive, third instars unresponsive to light, and fourth to sixth instars photonegative; in the field, early instars are to be found on their food plants by day, late instars are below the soil surface or in the litter, and third instars can be present in either location (Shields and Wyman 1984; see also Sullivan and Wellington 1953). For *Ectropis excursaria* (Geometridae), the strength but not the direction of the response to light becomes weaker as caterpillars develop (Mariath 1984). Whether the rules governing movement decisions change as larvae grow is unclear, but it seems likely that in many species they must, given the huge changes in selection pressures acting on caterpillars as larvae grow.

Daily timing of activity may also change. Mention has already been made of *Morpho peleides* and of *Erebia* spp. (See Enemies and Defense). In the same way, fourth and fifth instars of *Malacosoma americanum* have three distinct periods of activity at dawn, mid-afternoon, and dusk, but final instar caterpillars forage only at night (Fitzgerald et al. 1988). Early instar *Lymantria dispar* feed by day but from the fourth instar feed only by night (Leonard 1970). Two heliconiines feed at all times until their final instar when they tend to feed only at night (Alexander 1961).

Duration of different activities may change. The average length of feeding bouts for *Malacosoma americanum* caterpillars increases almost 4-fold between first and fifth instars (Fitzgerald 1980) and for the tobacco hornworm *Manduca sexta* (Sphingidae) the length of the feeding bout is correlated to body weight (Bowdan 1988). The length of feeding bout and the total time spent feeding each day by woolly bear caterpillars of *Diacrisia virginica* (Arctiidae) increases between third and fifth instars (Dethier 1988).

Range caterpillars *Hemileuca oliviae* (Saturniidae) move between distinct feeding and resting sites. As they become larger, they are more likely to rest on nonfood plant species, presumably reflecting a change in mobility. There are some differences in daily patterns of the timing of feeding, with fourth and fifth instars feeding in the morning and late afternoon but sixth instars also continuing to feed into the early afternoon (Mansen et al. 1984a,b). The total linear distance moved during each instar increases from virtually nothing in the first instar to 1–3 m each for second to fourth, 9 m for fifth, and 20 m for sixth. In early instars most movement is in mid-afternoon; from the fifth instar it is in mid-morning and dusk. Early instars are gregarious and late instars solitary. And food plant preferences change as caterpillars grow (Hansen et al. 1984a,b). Thus there is a clear picture of key aspects of larval behavior—when and where larvae move, though not the feeding choices they make—for this one species. But how the patterns fit into the framework of possible constraints discussed above, and whether the patterns are typical of other species, is at this stage wholly unknown.

A key aspect of foraging behavior is the feeding decisions caterpillars make. The question of how these decisions change as caterpillars grow, though seemingly one of the simplest to answer, remains unresolved. Feeding preferences may certainly change during development. Only new leaves are present when first instars of *Hemileuca lucina* are feeding, and in choice tests they show no preference for new or mature leaves; however, both new and mature leaves are present when third instars feed, and they show a clear preference for new leaves (Stamp and Bowers 1990). This suggests a change in feeding decision rules that parallels the change in food plant heterogeneity, though in this case it is not affected directly by body size. The area of the food plant accessible for foraging is likely to alter with a change in caterpillar mobility, but whether this opens up new feeding choices is another matter. Larvae could become more selective if they have access to a greater part of the plant, but if absolute consumption also increases they might have to become less choosy. Whether mobility really influences feeding choices is questionable. Take the tawny emperor *Asterocampa clyton* (Nymphalidae), for example. After hatching, caterpillars feed on adjacent leaves, eating 35–60% of the 3 or 4 closest leaves and apparently not selectively. They move a considerable distance (up to 1.2 m) prior to molting, molt, feed again on the 4–8 closest leaves, then move (this time up to 3.1 m) before molting to the third instar (Stamp 1984b). It is clear that larvae do not make the most of their mobility in hour by hour feeding choices, but whether the food resource is heterogeneous at a scale that matters to the caterpillars and whether they are selective when choosing a molting and feeding site is unclear. I discuss changes in the movement, tying, and feeding patterns of one caterpillar species on its natural food plant over its whole larval feeding period in Changes in Movement.

Whether it will be possible to construct "roadmaps" to caterpillar behavior (sensu Vander Meer 1987), in which aspects of behavior are found to be strongly correlated with body size, we will not know until it is attempted.

New Perspectives on the Relationship between Body Size and Feeding

Changes in Feeding Habit as Caterpillars Grow

Caterpillars of a great many species show striking changes in their feeding behavior as they grow. Early instars feed in one manner, for instance, by mining leaves, but later change to do something entirely different, such as feeding externally on the plant surface or concealed below leaves tied with silk. Among the British microlepidoptera, certainly the best known assemblage of smaller lepidopteran species in the world, 17.6% of the 1137 species for which there are data make a single marked change in their feeding habit some time during larval development. Such changes are interesting because they are likely to reflect changes in selective pressures operating on caterpillars as they grow.

A closer look at the changes these species make is intriguing because there are clear patterns to the directions of the changes (Gaston et al. 1991). Caterpillars that start out as leaf miners, concealed feeders, or external feeders are most likely to change their feeding habit, although none of those that begin as case bearers or gall formers do so. The most frequent habit changes are from leaf mining to case bearing, to spinning, tying, and rolling, and to external feeding, and from concealed feeding to case bearing and to spinning, tying, and rolling (Table 8.2). A likely explanation for the majority of these changes is based on body size: species that remain as leaf miners throughout larval development are significantly smaller as adults than those that start out as leaf miners but later change. Similarly, species remaining as concealed feeders throughout development are significantly smaller than species that later change, once case bearers are excluded from the analysis. This suggests that certain habits are best suited to caterpillars of certain sizes and that changes in habit might be required if growing caterpillars are to span certain size ranges.

Leaf mining and concealed feeding almost certainly constrain body size, and a size increase could force a change in behavior as caterpillars outgrow the plant structures they occupy. Many leaf miners, for instance, *Phyllonorycter* spp. (Gracillaridae), spin silk threads within the mine that open it up to a size large enough for the growing larva (Heath and Emmet 1985); this avoids the complexi-

Table 8.2. *Numbers of Species That Change Their Larval Feeding Habits from One Category to Another*[a]

	Late habit[b]						
	Leaf miners	Gall formers	Concealed feeders	Case bearers	Spinners, tiers, and rollers	External feeders	Underground
Early habit							
Leaf miners (9.27±3.76)	(242)	—	9	49	49	17	—
Gall formers (14.33±2.82)	—	(7)	—	—	—	—	—
Concealed feeders (14.74±4.70)	4	—	(234)	37	14	3	1
Case bearers (16.12±5.15)	—	—	—	(18)	—	—	—
Spinners, tiers, and rollers (17.44±4.80)	—	—	4	7	(392)	1	—
External feeders (19.52±6.09)	—	—	1	—	3	(32)	—
Underground (19.76±5.12)	—	—	1	—	—	—	(12)

[a]The order in which feeding habits are listed is the progression from small to large mean adult wing spans for species in each of the early habit categories; mean adult wing spans ± SD are given in parentheses. The main diagonal shows the number of species that do not change feeding habit. (After Gaston et al. 1991.)

[b]Leaf miners, species feeding in galleries within leaves. Gall formers, species forming galls in plant tissue, within which they feed. Concealed feeders, species concealed within the plant, but without specialized structures or external cues associated with their presence. Includes species that bore stems and roots and those that infest catkins, flower heads, seed pods, seeds, and buds, or are inquilines of galls. Case bearers, species constructing cases that are borne with them as they move about the plant between feeding sites. External feeders, species feeding on the outside of the plant, typically on the leaves. Does not include species that feed on the root surface. Web spinners, leaf tiers, and leaf rollers, species constructing simple structures externally on the plant by covering, tying, folding, or rolling plant structures with silk, and retaining the ability to move freely inside or outside these structures when feeding. Underground feeders, species that spend much of their time underground, but are not concealed within plant tissue.

ties of changing leaves or feeding habit. Others remain as internal feeders but move to larger plant structures, for instance from leaf to stem. Nine species in our study change from leaf mining to concealed feeding. However, 66 species move at some point during growth from the interior of the leaf to feed on the outside. Leaf mining has possible advantages of physical protection, better microclimate, and greater feeding selectivity and is perhaps preferred as a feeding habit so long as body size permits (Gaston et al. 1991).

Body size is also likely to affect the feasibility of other feeding habits. External feeders, for example, must deal with tough and nutritionally poor epidermal tissue before reaching the leaf interior and this might be easier for larger caterpillars with larger mouthparts (see Feeding). Spinning, tying, and rolling might only be effective when larvae are larger because physical manipulation of plant structures could become easier. Clearly there are many interesting possibilities. We suggest that, across species, there is a sequence of predominant feeding habits as body size increases (Fig. 8.1); the progression we see in the microlepidoptera is the first part of this sequence but unfortunately suitable data are not available to complete it for the macrolepidoptera. However, our observations indicate that many macrolepidoptera change from concealed feeding to spinning, tying, and rolling, and from spinning, tying, and rolling to external feeding. The largest of all caterpillars are external feeders and these very large individuals probably have little to gain from tying or rolling plant structures that are small relative to their own size (Gaston et al. 1991). Note that changes in feeding habit as caterpillars grow reflect differences in feeding habit among species of different sizes, an idea to which I will return.

The Size of Newly Hatched Caterpillars in Relation to Larval Feeding

It is probably newly hatched caterpillars that face the greatest feeding challenges, especially if they are less mobile (see Mobility), less tolerant of starvation (see Starvation), and less able to feed on physically tough foliage (see Feeding) than when they are larger. General relationships between the sizes of newly hatched larvae of different species and their feeding ecology and early larval behavior could reflect the importance of body size as a constraint on feeding. I examined the relationship between absolute egg size and aspects of feeding ecology for different lepidopteran families in different temperate regions using data in the literature, then extended the study to look at patterns in the behavior and survival of newly hatched larvae of 42 species of British Lepidoptera in relation to larval size and to food plant characteristics (Reavey 1992). I found strong links between the size of eggs (and therefore newly hatched larvae) and larval feeding ecology, and between feeding ecology and larval behavior, but there are surprisingly few links between larval size and behavior.

Within families, eggs of generalist feeders tend to be larger than those of specialists. A similar relationship between adult size and feeding specificity is well

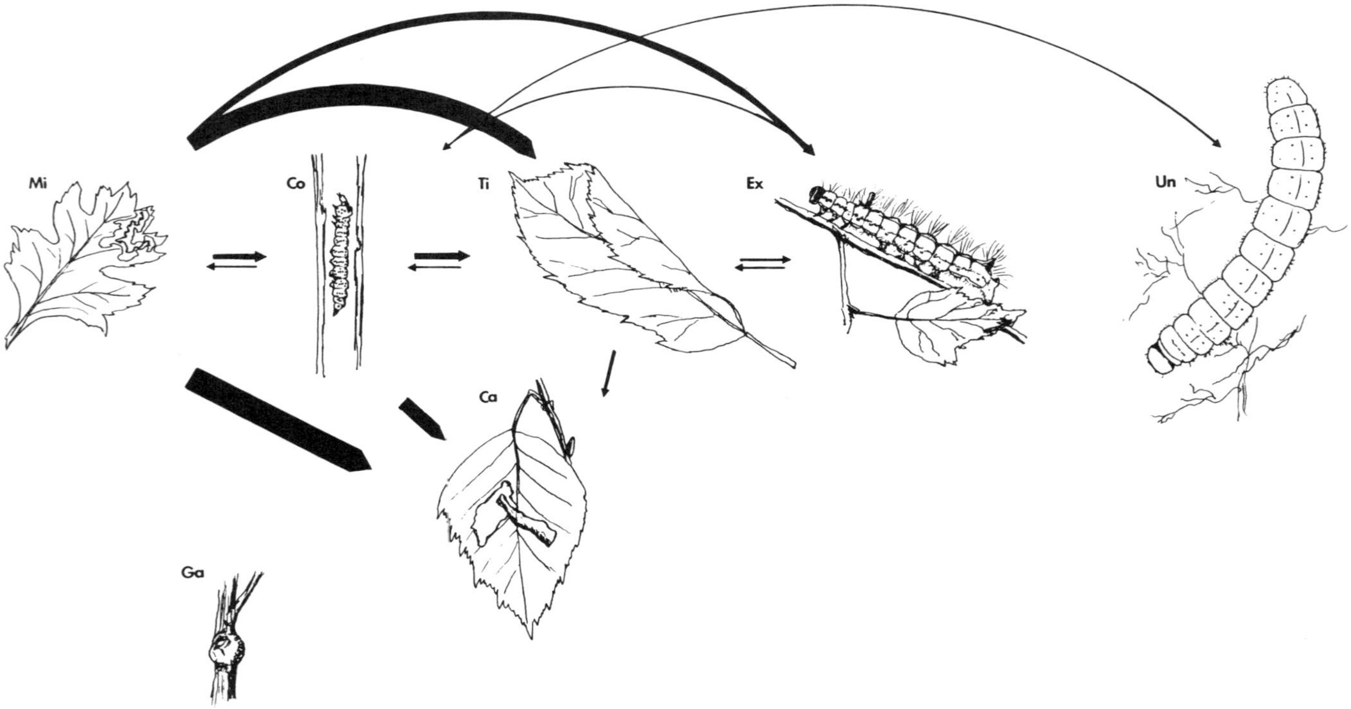

Figure 8.1. The sequence of predominant feeding habits for species of different sizes. Body size increases from left to right. Arrows show switches in feeding habit as caterpillars grow; the thickness of each arrow indicates the relative number of species switching. Feeding habits are Mi, leaf miners; Ga, gall formers; Co, concealed feeders; Ca, case bearers; Ti, spinners, tiers, and rollers; Ex, external feeders; Un, underground. (Illustration: Simon Reavey.)

known—large bodied species are more polyphagous—though this relationship is not well understood (Gaston and Reavey 1989 and references therein). Generalist feeding could be a response to unpredictable food plant availability, so one possibility is that larger larvae have an advantage if they more readily survive starvation or more effectively disperse to find food (see Starvation; Mobility). However, my work with newly hatched larvae shows no significant differences between larval survival, speed of movement, or silking behavior and feeding specificity.

Within families, eggs of woody plant feeders are larger than those of grass feeders, which are larger than eggs of herb feeders. Again, there is a similar relationship between adult size and food plant growth form (Gaston and Reavey 1989 and references therein). The pattern could indicate an adaptation to the feeding requirements of the young larvae. Leaves of woody plants tend to be tougher than those of herbs (e.g., Morrow 1983) and larger larvae could have larger mouthparts more suited to this tough food. Furthermore, leaves of woody plants are nutritionally poorer than leaves of herbs (e.g., Slansky and Scriber 1985), and larger individuals are more able to make use of these foods because their relative metabolic demands decrease with body size. It is possible, then, that having large newly hatched larvae allows a species to avoid potential difficulties of a small body size or at least to outgrow them more quickly.

Eggs laid on the plant part that is eaten tend to be smaller than those laid away from it. But, although we might expect larvae of species laying away from their food to be more active or active for longer than larvae that have food immediately available on hatching, I find no evidence that this is the case.

Among 38 species with a size range of two orders of magnitude I found no correlation between larval size and survival time. There is considerable variation among species, with a mean survival time for newly hatched larvae kept without food at 15°C of 1.0 to 20.0 days. The species surviving longest, the common footman *Eilema lurideola* (Arctiidae), was the only lichen feeder we studied. Lichens are especially poor nutritionally (Lawrey 1987), and lichen feeders probably have a low metabolic rate to counter this. Nor was there a correlation between larval size and speed of movement. The smallest species, *Pterophorus pentadactyla* (Pterophoridae) (0.022 mg), moved at 34 cm h^{-1} and the fastest, the obscure wainscot *Mythimna obsoleta* (Noctuidae) (0.17 mg) at 268 cm h^{-1}. It is likely that differences among species are adaptations to feeding needs related to accessibility or predictability of the food source, though this is something that is hard to quantify. However, it is not possible to say to what extent body size per se, through its influence on mobility or on the ability of larvae to tolerate starvation, is a constraint on larval behavior.

Changes in Movement, Tying, and Feeding Patterns as Caterpillars Grow

Although many laboratory and some field studies have examined caterpillar foraging behavior over short periods, virtually none traces the movements and feeding behavior of individual caterpillars on their natural food plants over their whole larval period. Perhaps this is not surprising given the logistical difficulties of following recognizable individuals for considerable distances around a complex habitat for several weeks. However, we have been able to follow the movements, tying, and feeding activities of leaf tying caterpillars of the yellow horned moth *Achlya flavicornis* (Thyatiridae) throughout their feeding stages (Feichtinger and Reavey 1989).

Achlya flavicornis is a specialist feeder on bushes and trees of birch *Betula* spp. Eggs are laid before the first flush of leaves, and newly hatched larvae hide within the sheaths of newly opened buds when they are not feeding. After a few days, each constructs its first leaf tie by folding one edge of a leaf across the upper surface towards or beyond the midrib and strongly attaching it with silk. Subsequent ties are made in the same way. Caterpillars leave their ties to feed but always return between feeding bouts.

Newly hatched caterpillars were placed on potted birch trees and each day we recorded the location of each caterpillar and the presence of new feeding damage and newly constructed ties. Larvae were not disturbed in any way and it was possible to recognize individuals without marking them because their movements on the tree were so restricted. They remained on the same trees throughout their development.

During their five instars, caterpillars constructed 4–7 ties. There was no link between the timings of tying and ecdysis, and there was great variability in the duration a tie was occupied. One possible reason for larvae moving is that they outgrow their ties. Larvae construct larger ties as they grow, but each leaf differs subtly in size, conformation, and toughness so there is likely to be variation in the size of their new ties and the time they remain suitable. Other reasons for moving could be deterioration in the food quality of adjacent leaves, perhaps induced by their own feeding (D. Reavey unpublished data), or escape from their conspicuous feeding damage that might be a cue to predators.

As the caterpillars grow, they move greater distances between successive ties (Fig. 8.2) but, considering their ability to move long distances very quickly, their movements are surprisingly limited. Newly hatched larvae can move 148 cm h^{-1} but the distance from the first to second tie is on average 10 cm; fifth instars can move at 668 cm h^{-1} but their ties are on average 51 cm apart. Indeed, when larvae construct new ties, the distance they move is only just sufficient to ensure there is no overlap between the feeding areas around the old and new ties. Larger caterpillars tend to feed further away from their current ties (Fig. 8.2) and so they encounter more leaves as they move about the tree. They also become more selective: first instars feed on 75% of leaves they encounter, fifth instars on 47%. This increased selectivity parallels an increase in heterogeneity within the tree; at leaf flush only young leaves are present, but later there are leaves of a wide range of ages, levels of herbivore damage, degree of shade from the sun, and so on. Even so, most larvae, most of the time, do not move far from their ties.

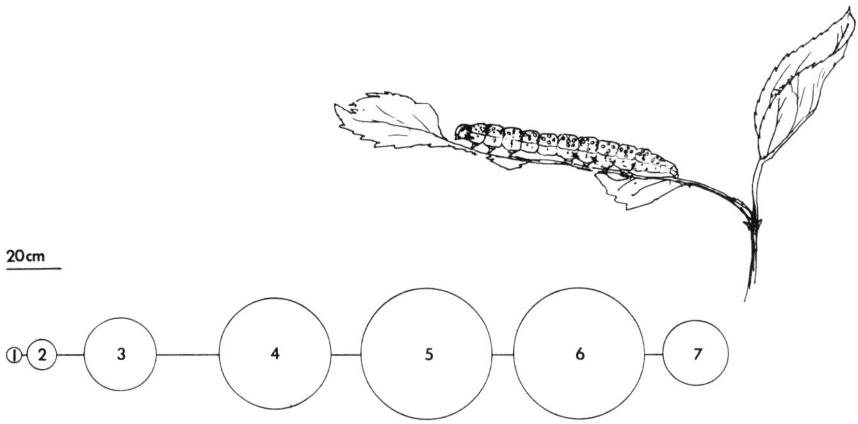

Figure 8.2. The spatial relationship between successive ties and foraging areas of *Achlya flavicornis*. Larvae construct four to seven ties during their five instar development. In the center of each circle is the tie. The size of each circle represents the distance from the tie at which larvae forage. The distance of separation between successive ties is drawn to the same scale (From Feichtinger and Reavey 1989; illustration: Simon Reavey.)

Perhaps there is a high cost to losing the tie, in terms of wasted time, increased risk of predation or additional silk synthesis. Alternatively, it is possible that foraging at greater distances is simply not worthwhile because food quality is relatively homogeneous and moving greater distances opens up no new feeding choices.

Thus there are changes in tying and feeding distances and in feeding selectivity as caterpillars of *A. flavicornis* grow. It would be interesting to see if such patterns are characteristic of other leaf-tying species. It would also be intriguing to obtain comparable data for birch feeders that are not constrained by the need to construct a tie and to return to it after each feeding bout.

Body Size as a Constraint on Caterpillar Feeding Ecology

A consideration of body size as a constraint on feeding would not be complete without a brief mention of factors affecting the sizes caterpillars start and finish, and the amount they grow in between. Much is known of the proximate factors that determine egg size (e.g., adult size, number of eggs laid previously) and pupal size (e.g., food availability and quality, crowding, time available for feeding) but there is no need to cover these here. More interesting in the context of this chapter are the constraints that a particular starting or finishing size might place on feeding. The sizes of newly hatched larvae, fully grown larvae, and adults are not likely to be independent of each other. Thus evolutionary pressures for a particular body size at one of these stages will influence body size at all others too. To what degree, then, does each stage constrain the others? The patterns of change as caterpillars grow already discussed give some initial clues.

The advantages of large or small species size are difficult to assess for Lepidoptera. For instance, large adult size could reduce vulnerability to environmental perturbation, for example, in weather or food availability, and increase the ability to disperse (e.g., Derr et al. 1981) but small species could fit more generations into the year (Schoener and Janzen 1968; Gaston and Reavey 1989). Each of these may impose selective pressures for a particular body size. For caterpillars, large size could buffer against physiological stress, perhaps a reason why generalists tend to be larger than specialists (Wasserman and Mitter 1978) and could permit use of poorer quality food resources, perhaps a reason why tree feeders tend to be larger than herb feeders (Mattson 1977), whereas small species could make use of food sources that are not large or persistent enough to permit growth to a large size (Opler 1978; Scriber 1978). Thus the impact of caterpillar feeding considerations on adult size is likely to be considerable. The question, then, is whether these pressures are strongest when the larvae are small or when they grow larger. At this stage the evidence suggests the former, as body size seems to place real constraints on the feeding choices of newly hatched larvae (see Feeding; The Size of Newly Hatched Caterpillars). If the use of a particular food plant requires newly hatched larvae to be a particular minimum size, it could require adults to be large enough to produce an adequate number of large eggs. Equally, if a particular feeding niche requires larvae to be very small at hatching, adult size may be influenced accordingly.

It is therefore clearly important to look to the early instars in assessing factors affecting patterns in body size and feeding ecology. We need some idea too of just how closely linked are the starting and finishing sizes and how much flexibility there is in the number of instars and growth per instar. This will give an indication of the degree to which pressures on early instars are liable to affect final body size. Equally, there may well be situations in which adult body size is of overwhelming concern, for instance, for migration. Then, larval feeding ecology and size may be strongly constrained. Here are virtually unexplored problems ripe for investigation.

The Search for Patterns

In this chapter I have suggested ways in which foraging and constraints on foraging might be affected by body size. What is missing in most cases is an indication of the significance of each of the possibilities in the real world. For example, it would be useful to know the proportion of species showing thermoregulatory or gregarious behavior that change their life-styles as they grow. It would be valuable to have some idea of the numbers of species that change food plant species as they grow and of the nature of the changes they make.

And so on. The emphasis has been—and remains—on individual case studies illustrating particular points in a spectacular way rather than on seeking robust patterns and broad trends shown by large numbers of species. Furthermore, although we can gain some insight of the importance of body size by examining changes of individuals as they grow, it would be even more revealing to look closely at patterns among species of different sizes. The general effects of body size on nutritional considerations and feeding habit, for example, are similar for individuals as they grow and for species of different sizes.

Acknowledgments

My particular thanks go to Kevin Gaston, John Lawton, Graciela Valladares, and Veronika Feichtinger, ecologists of widely differing body sizes, for their ideas and suggestions throughout this work. Chris Rees and Jeremy Thomas gave helpful comments on the chapter. I was supported by a Kennedy Scholarship at Harvard University while writing this chapter.

Literature cited

Ahmad, S. 1986. Enzymatic adaptations of herbivorous insects and mites to phytochemicals. J. Chem. Ecol. 12:533–560.

Ahmad, S., Brattsten, L. B., Mullin, C. A., and Yu, S. J. 1986. Enzymes involved in the metabolism of plant allelochemicals, pp. 73–151. In L. B. Brattsten, and S. Ahmad (eds.), Molecular Aspects of Insect-Plant Associations. Plenum, New York.

Alexander, A. J. 1961. A study of the biology and behavior of the caterpillars, pupae and emerging butterflies of the subfamily Heliconiinae in Trinidad, West Indies. Part I. Some aspects of larval behavior. Zoologica 46:1–24.

Allen, D. C., Knight, F. B., and Foltz, J. L. 1970. Invertebrate predators of the jack-pine budworm, *Choristoneura pinus,* in Michigan. Ann. Entomol. Soc. Am. 63:59–64.

Awan, M. S. 1985. Anti-predator ploys of *Heliothis punctiger* (Lepidoptera: Noctuidae) caterpillars against the predator *Oechalia schellenbergii* (Hemiptera: Pentatomidae). Aust. J. Zool. 33:885–890.

Ayre, G. L., and Hitchon, D. E. 1968. The predation of tent caterpillars, *Malacosoma americana* (Lepidoptera: Lasiocampidae) by ants (Hymenoptera: Formicidae). Can. Entomol. 100:823–826.

Baker, R. R. 1970. Bird predation as a selective pressure on the immature stages of the cabbage butterflies, *Pieris rapae* and *P. brassicae*. J. Zool. London 162:43–59.

Barber, G. W. 1941. Observations on the egg and newly hatched larva of the corn ear worm on corn silk. J. Econ. Entomol. 34:451–456.

Barbosa, P., Martinat, P., and Waldvogel, M. 1986. Development, fecundity and survival of the herbivore *Lymantria dispar* and the number of plant species in its diet. Ecol. Entomol. 11:1–6.

Benson, R. B. 1950. An introduction to the natural history of British sawflies (Hymenoptera Symphyta). Trans. Soc. Br. Entomol. 10:45–142.

Bernays, E. A., and Janzen, D. H. 1988. Saturniid and sphingid caterpillars: Two ways to eat leaves. Ecology 69:1153–1160.

Blake, E. A., and Wagner, M. R. 1984. Effect of sex and instar on food consumption, nutritional indices, and foliage wasting by the western spruce budworm, *Choristoneura occidentalis*. Environ. Entomol. 13:1634–1638.

Blau, W. S. 1980. The effect of environmental disturbance on a tropical butterfly population. Ecology 61:1005–1012.

Booth, C. L. 1990. Evolutionary significance of ontogenetic colour change in animals. Biol. J. Linn. Soc. 40:125–163.

Bowdan, E. 1988. Microstructure of feeding by tobacco hornworm caterpillars, *Manduca sexta*. Entomol. Exp. Appl. 47:127–136.

Brower, L. P. 1984. Chemical defence in butterflies, pp. 109–134. In R. I. Vane-Wright and P. R. Ackery (eds.), The Biology of Butterflies. Symp. R. Entomol. Soc. Lond. 11. Academic Press, Orlando.

Burger, B. V., Röth, M., LeRoux, M., Spies, H. S. C., Truter, V., and Geertsema, H. 1978. The chemical nature of the defensive larval secretion of the citrus swallowtail, *Papilio demodocus*. J. Insect Physiol. 24:803–805.

Burtt, E. H., Jr. 1981. The adaptiveness of animal colours. BioScience 31:723–729.

Cain, M. L., Eccleston, J., and Kareiva, P. M. 1985. The influence of food plant dispersion on caterpillar searching success. Ecol. Entomol. 10:1–7.

Carter, D. J., and Hargreaves, B. 1986. A Field Guide to Caterpillars of Butterflies and Moths in Britain and Europe. Collins, London.

Casey, T. M. 1981. Behavioral mechanisms of thermoregulation, pp. 79–114. In B. Heinrich (ed.), Insect Thermoregulation. Wiley, New York.

Clark, L. R. 1964. Predation by birds in relation to the population density of *Cardiaspina albitextura* (Psyllidae). Aust. J. Zool. 12:349–361.

Cohen, R. W., Waldbauer, G. P., Friedman, S., and Schiff, N. M. 1987. Nutrient self-selection by *Heliothis zea* larvae: A time-lapse film study. Entomol. Exp. Appl. 44:65–73.

Compton, S. G. 1989. Sabotage of latex defences by caterpillars feeding on fig trees. S. Afr. J. Sci. 85:605–606.

Cornell, J. C., Stamp, N. E., and Bowers, M. D. 1987. Developmental changes in aggregation, defense and escape behavior of buckmoth caterpillars, *Hemileuca lucina* (Saturniidae). Behav. Ecol. Sociobiol. 20:383–388.

Coshan, P. F. 1974. The biology of *Coleophora serratella* (L.) (Lepidoptera: Coleophoridae). Trans. R. Entomol. Soc. London 126:169–188.

Dempster, J. P. 1971. The population ecology of the cinnabar moth *Tyria jacobaeae* L. (Lepidoptera, Arctiidae). Oecologia 7:26–67.

Derr, J. A., Alden, B., and Dingle, H. 1981. Insect life histories in relation to migration, body size, and host plant array: A comparative study of *Dysdercus*. J. Anim. Ecol. 50:181–193.

Dethier, V. G. 1959. Food-plant distribution and density and larval dispersal as factors affecting insect populations. Can. Entomol. 91:581–596.

Dethier, V. G. 1988. The feeding behavior of a polyphagous caterpillar (*Diacrisia virginica*) in its natural habitat. Can. J. Zool. 66:1280–1288.

Drummond, B. A., III, Bush, G. L., and Emmel, T. C. 1970. The biology and laboratory culture of *Chlosyne lacinia* Geyer (Nymphalidae). J. Lepid. Soc. 24:135–142.

Edmunds, M. 1974. Defence in Animals. Longman, London.

Enders, F. 1976. Size, food-finding, and Dyar's constant. Environ. Entomol. 5:1–10.

Fajer, E. D. 1989. The effects of enriched Co_2 atmospheres on plant-insect herbivore interactions: growth responses of larvae of the specialist butterfly, *Junonia coenia* (Lepidoptera: Nymphalidae). Oecologia 81:514–520.

Feichtinger, V. E., and Reavey, D. 1989. Changes in movement, tying and feeding patterns as caterpillars grow: the case of the yellow horned moth. Ecol. Entomol. 14:471–474.

Fields, P. G., and McNeil, J. N. 1988. The importance of seasonal variation in hair coloration for thermoregulation of *Ctenucha virginica* larvae (Lepidoptera: Arctiidae). Physiol. Entomol. 13:165–175.

Fitzgerald, T. D. 1980. An analysis of daily foraging patterns of laboratory colonies of the eastern tent caterpillar, *Malacosoma americanum* (Lepidoptera: Lasiocampidae), recorded photoelectronically. Can. Entomol. 112:731–738.

Fitzgerald, T. D., Casey, T., and Joos, B. 1988. Daily foraging schedule of field colonies of the eastern tent caterpillar *Malacosoma americanum*. Oecologia 76:574–578.

Ford, E. B. 1957. Butterflies. Collins, London.

Fuzeau-Braesch, S. 1985. Colour changes, pp. 549–589. In G. A. Kerkut and L. I. Gilbert (eds.), Comprehensive Insect Physiology, Biochemistry and Pharmacology, Vol. 9. Pergamon, Oxford.

Gaston, K. J., and Reavey, D. 1989. Patterns in the life histories and feeding strategies of British macrolepidoptera. Biol. J. Linn. Soc. 37:367–381.

Gaston, K. J., Reavey, D., and Valladares, G. R. 1991. Changes in feeding habit as caterpillars grow. Ecol. Entomol. 16:339–344.

Ghent, A. 1960. A study of the group-feeding behaviour of larvae of the jack pine sawfly, *Neodiprion pratti banksinae* Roh. Behaviour 16:110–148.

Godfrey, G. L., Miller, J. S., and Carter, D. J. 1989. Two mouthpart modifications in larval Notodontidae (Lepidoptera): Their taxonomic distributions and putative functions. J. N.Y. Entomol. Soc. 97:455–470.

Gould, F. 1984. Mixed function oxidases and herbivore polyphagy: The devil's advocate position. Ecol. Entomol. 9:29–34.

Greenberg, R., and Gradwohl, J. 1980. Leaf surface specializations of birds and arthropods in a Panamanian forest. Oecologia 46:115–124.

Hansen, J. D., Ludwig, J. A., Owens, J. C., and Huddleston, E. W. 1984a. Motility, feeding, and molting in larvae of the range caterpillar, *Hemileuca oliviae* (Lepidoptera: Saturniidae). Environ. Entomol. 13:45–51.

Hansen, J. D., Ludwig, J. A., Owens, J. C., and Huddleston, E. W. 1984b. Larval movement of the range caterpillar, *Hemileuca oliviae* (Lepidoptera: Saturniidae). Environ. Entomol. 13:415–420.

Harcourt, D. G. 1966. Major factors in survival of the immature stages of *Pieris rapae* (L.) Can. Entomol. 98:653–662.

Heath, J., and Emmet, A. M. (eds.) 1985. The Moths and Butterflies of Great Britain and Ireland. Vol. 2. Cossidae-Heliodinidae. Harley, Great Horkesley, Essex.

Hebert, C., and Cloutier, C. 1990. Host instar as a determinant of preference and suitability for two parasitoids attacking late instars of the spruce budworm (Lepidoptera: Tortricidae). Ann. Entomol. Soc. Am. 83:734–741.

Herrebout, W. M., Kuyten, P. J., and De Ruiter, L. 1963. Observations on colour patterns and behaviour of caterpillars feeding on Scots pine with a discussion of their possible functional significance. Arch. Néerl. Zool. 15:315–357.

Hodjat, S. H. 1970. Effect of crowding on colour, size and larval activity of *Spodoptera littoralis* (Lepidoptera: Noctuidae). Entomol. Exp. Appl. 13:97–106.

Holdren, C. E., and Ehrlich, P. R. 1982. Ecological determinants of food plant choice in the checkerspot butterfly *Euphydryas editha* in Colorado. Oecologia 52:417–423.

Honda, K. 1980a. Volatile constituents of larval osmeterial secretions in *Papilio protenor demetrius*. J. Insect Physiol. 26:39–45.

Honda, K. 1980b. Osmeterial secretions of papilionid larvae in the genera *Luehdorfia*, *Graphium* and *Atrophaneura* (Lepidoptera). Insect Biochem. 10:583–588.

Honda, K. 1981. Larval osmeterial secretions of the swallowtails (*Papilio*). J. Chem. Ecol. 7:1089–1113.

Hopper, K. R., and King, E. G. 1984. Preference of *Microplitis croceipes* (Hymenoptera: Braconidae) for instars and species of *Heliothis* (Lepidoptera: Noctuidae). Environ. Entomol. 13:1145–1150.

Iwao, S., and Wellington, W. G. 1970. The influence of behavioral differences among tent-caterpillar larvae on predation by a pentatomid bug. Can. J. Zool. 48:896–898.

Janzen, D. H. 1984. How to be a tropical big moth: Santa Rosa saturniids and sphingids. Oxford Stud. Evol. Biol. 1:85–140.

Jones, R. E. 1977. Search behaviour: a study of three caterpillar species. Behaviour 60:237–259.

Karlsson, B., and Wiklund, C. 1984. Egg weight variation and lack of correlation between egg weight and offspring fitness in the wall brown butterfly *Lasiommata megera*. Oikos 43:376–385.

Karlsson, B., and Wiklund, C. 1985. Egg weight variation in relation to egg mortality and starvation endurance of newly hatched larvae in some satyrid butterflies. Ecol. Entomol. 10:205–211.

Kato, Y. 1989. Role of photoperiod in larval growth of *Sasakia charonda* (Lepidoptera: Nymphalidae). Jpn. J. Entomol. 57:221–230.

Kevan, P. G., Jensen, T. S., and Shorthouse, J. D. 1982. Body temperatures and behavioral thermoregulation of high arctic woolly-bear caterpillars and pupae (*Gynaephora rossii,* Lymantriidae: Lepidoptera) and the importance of sunshine. Arctic Alpine Res. 14:125–136.

Khalsa, M. S., Kogan, M., and Luckmann, W. H. 1979. *Autographa precationis* in relation to soybean: Life history, and food intake and utilization under controlled conditions. Environ. Entomol. 8:117–122.

Kimmerer, T. W., and Potter, D. A. 1987. Nutritional quality of specific leaf tissues and selective feeding by a specialist leafminer. Oecologia 71:548–551.

Knapp, R., and Casey, T. M. 1986. Thermal ecology, behavior, and growth of gypsy moth and eastern tent caterpillars. Ecology 67:598–608.

Kogan, M., and Cope, D. 1974. Feeding and nutrition of insects associated with soybeans. 3. Food intake, utilization, and growth in the soybean looper, *Pseudoplusia includens.* Ann. Entomol. Soc. Am. 67:66–72.

Krishna, S. S. 1987. Nutritional modulation of reproduction in two phytophagous insect pests. Proc. Indian Acad. Sci. (Anim. Sci.) 96:153–169.

Labine, P. A. 1968. The population biology of the butterfly, *Euphydryas editha.* VIII. Oviposition and its relation to patterns of oviposition in other butterflies. Evolution 22:799–805.

Larsson, F. K. 1989. Insect mating patterns explained by microclimatic variables. J. Therm. Biol. 14:155–157.

Larsson, S., and Tenow, O. 1979. Utilization of dry matter and bioelements in larvae of *Neodiprion sertifer* Geoffr. (Hym., Diprionidae) feeding on Scots pine (*Pinus sylvestris* L.). Oecologia 43:157–172.

Lawrey, J. D. 1987. Nutritional ecology of lichen/moss arthropods, pp. 209–233. In F. Slansky, Jr. and J. G. Rodriguez (eds.), Nutritional Ecology of Insects, Mites, Spiders, and Related Invertebrates. Wiley, New York.

Leonard, D. E. 1970. Feeding rhythm in larvae of the gypsy moth. J. Econ. Entomol. 63:1454–1457.

Lindsey, C. C. 1966. Body sizes of poikilotherm vertebrates at different latitudes. Evolution 20:456–465.

Long, D. B. 1953. Effects of population density on larvae of Lepidoptera. Trans. R. Entomol. Soc. London 104:543–585.

Long, D. B. 1955. Observations on sub-social behaviour in two species of lepidopterous larvae, *Pieris brassicae* L. and *Plusia gamma* L. Trans R. Entomol. Soc. London 106:421–436.

Mansingh, A. 1974. Studies on insect dormancy. II. Relationship of cold-hardiness to diapause and quiescence in the eastern tent caterpillar, *Malacosoma americanum* (Fab.), (Lasiocampidae: Lepidoptera). Can. J. Zool. 52:629–637.

Mariath, H. A. 1984. Factors affecting the dispersive behaviour of larvae of an Australian geometrid moth. Entomol. Exp. Appl. 35:159–167.

Mattson, W. J. 1977. Size and abundance of forest Lepidoptera in relation to host plant resources. Coll. Int. C.N.R.S. 265:429–441.

Mattson, W. J., Knight, F. B., Allen, D. C., and Foltz, J. L. 1968. Vertebrate predation on the jack-pine budworm in Michigan. J. Econ. Entomol. 61:229–234.

May, M. L. 1985. Thermoregulation, pp. 507–522. In G. A. Kerkut and L. I. Gilbert (eds.), Comprehensive Insect Physiology, Biochemistry and Pharmacology, Vol. 4. Pergamon, Oxford.

Miles, P. W., Aspinall, D., and Correll, A. T. 1982. The performance of two chewing insects on water-stressed food plants in relation to changes in their chemical composition. Aust. J. Zool. 30:347–355.

Millar, J. S., and Hickling, G. J. 1990. Fasting endurance and the evolution of mammalian body size. Func. Ecol. 4:5–12.

Morrow, P. A. 1983. The role of sclerophyllous leaves in determining insect grazing damage. Ecol. Studies 43:509–524.

Nakasuji, F. 1987. Egg size of skippers (Lepidoptera: Hesperiidae) in relation to their host specificity and to leaf toughness of host plants. Ecol. Res. 2:175–183.

Nakasuji, F., and Kimura, M. 1984. Seasonal polymorphism of egg size in a migrant skipper, *Parnara guttata guttata* (Lepidoptera, Hesperiidae). Kontyû, Tokyo 52:253–259.

Oldiges, H. 1959. Der Einflua der Temperatur auf Stoffwechsel und Eiproduktion von Lepidopteren. Z. Angew Entomol. 44:115–166.

Opler, P. A. 1978. Interaction of plant life history components as related to arboreal herbivory, pp. 23–31. In G. G. Montgomery (ed.), Ecology of Arboreal Folivores. Smithsonian Institute, Washington, D.C.

Palanichamy, S., and Baskaran, P. 1989. Effects of starvation and dehydration in the tropical spider (*Cyrtophora cicricola*). Environ. Ecol. 7:407–411.

Porter, K. 1982. Basking behaviour in larvae of the butterfly *Euphydryas aurinia*. Oikos 38:308–312.

Prop, N. 1960. Protection against birds and parasites in some species of tenthredinid larvae. Arch. Néerl. Zool. 13:380–447.

Ramalho, F. S., Parrott, W. L., Jenkins, J. N., and McCarty, J. C., Jr. 1984. Effects of cotton leaf trichomes on the mobility of newly hatched tobacco budworms (Lepidoptera: Noctuidae). J. Econ. Entomol. 77:619–621.

Rausher, M. D. 1979. Egg recognition: Its advantage to a butterfly. Anim. Behav. 27:1034–1040.

Reavey, D. 1992. Egg size in Lepidoptera and its relation to larval feeding. J. Zool. London 227:277–297.

Reavey, D., and Lawton, J. H. 1991. Larval contribution to fitness in leaf-eating insects, pp. 293–329. In W. J. Bailey and J. Ridsdill-Smith (eds.), Reproductive Behaviour of Insects: Individuals and Populations. Chapman and Hall, London.

Remmert, H. 1985. Crickets in sunshine. Oecologia 68:29–33.

Salinas, P. J. 1984. Studies on the behavior of the larvae of *Plutella xylostella* (Linnaeus) (Lepidoptera: Plutellidae), a world pest of cruciferous crops: Normal and "spacing" behavior. Turrialba 34:77–84.

Schneider, G. 1973. Über den Einflua verschiedener Umweltfaktoren auf den Färbungs-polyphänismus der Raupen des tropisch-amerikanischen Schwärmers *Erinnyis ello* L. (Lepidopt., Sphingid.). Oecologia 11:351–370.

Schoener, T. W., and Janzen, D. H. 1968. Notes on environmental determinants of tropical versus temperate insect size patterns. Am. Nat. 102:207–224.

Schowalter, T. D., Whitford, W. G., and Turner, R. B. 1977. Bioenergetics of the range caterpillar, *Hemileuca oliviae* (Ckll.). Oecologia 28:153–161.

Schroeder, L. A. 1972. Energy budget of cecropia moths, *Platysamia cecropia* (Lepidoptera: Saturniidae), fed lilac leaves. Ann. Entomol. Soc. Am. 65:367–372.

Schultz, J. C. 1983. Habitat selection and foraging tactics of caterpillars in heterogeneous trees, pp. 61–90. In R. F. Denno and M. S. McClure (eds.), Variable Plants and Herbivores in Natural and Managed Systems. Academic Press, New York.

Scriber, J. M. 1978. The effects of larval feeding specialization and plant growth form on the consumption and utilization of plant biomass and nitrogen: an ecological consideration. Entomol. Exp. Appl. 24:694–710.

Scriber, J. M., and Slansky, F., Jr. 1981. The nutritional ecology of immature insects. Ann. Rev. Entomol. 26:183–211.

Sehnal, F. 1985. Growth and life cycles, pp. 1–86. In G. A. Kerkut and L. I. Gilbert (eds.), Comprehensive Insect Physiology, Biochemistry and Pharmacology, Vol. 2. Pergamon, Oxford.

Semlitsch, R. D., and West, C. A. 1988. Size-dependent cannibalism in noctuid caterpillars. Oecologia 77:286–288.

Seymour, R. S. 1974. Convective and evaporative cooling in sawfly larvae. J. Insect Physiol. 20:2447–2457.

Shields, E. J., and Wyman, J. A. 1984. Responses of variegated cutworm (Lepidoptera: Noctuidae) to various light levels. Ann. Entomol. Soc. Am. 77:152–154.

Slansky, F., Jr., and Feeny, P. 1977. Stabilization of the rate of nitrogen accumulation by larvae of the cabbage butterfly on wild and cultivated food plants. Ecol. Monogr. 47:209–228.

Slansky, F., Jr., and Scriber, J. M. 1985. Food consumption and utilization, pp. 87–163. In G. A. Kerkut and L. I. Gilbert (eds.), Comprehensive Insect Physiology, Biochemistry and Pharmacology, Vol. 4. Pergamon, Oxford.

Stamp, N. 1977. Aggregation behavior of *Chlosyne lacinia* larvae (Nymphalidae). J. Lepid. Soc. 31:35–40.

Stamp, N. E. 1980. Egg deposition patterns in butterflies: why do some species cluster their eggs rather than deposit them singly? Am. Nat. 115:367–380.

Stamp, N. E. 1982. Behavioral interactions of parasitoids and Baltimore checkerspot caterpillars (*Euphydryas phaeton*). Environ. Entomol. 11:100–104.

Stamp, N. E. 1984a. Interactions of parasitoids and checkerspot caterpillars *Euphydryas* spp. (Nymphalidae). J. Res. Lepid. 23:2–18.

Stamp, N. E. 1984b. Foraging behavior of tawny emperor caterpillars (Nymphalidae: *Asterocampa clyton*). J. Lepid. Soc. 38:186–191.

Stamp, N. E. 1986. Physical constraints of defense and response to invertebrate predators by pipevine caterpillars (*Battus philenor:* Papilionidae). J. Lepid. Soc. 40:191–205.

Stamp, N. E., and Bowers, M. D. 1988. Direct and indirect effects of predatory wasps (*Polistes* sp.: Vespidae) on gregarious caterpillars (*Hemileuca lucina:* Saturniidae). Oecologia 75:619–624.

Stamp, N. E., and Bowers, M. D. 1990. Variation in food quality and temperature constrain foraging of gregarious caterpillars. Ecology 71:1031–1039.

Stewart, L. A., and Dixon, A. F. G. 1989. Why big species of ladybird beetles are not melanic. Func. Ecol 3:165–177.

Stockhoff, B. A. 1991. Starvation resistance of gypsy moth, *Lymantria dispar* (L.) (Lepidoptera: Lymantriidae): tradeoffs among growth, body size, and survival. Oecologia 88:422–429.

Sullivan, C. R., and Wellington, W. G. 1953. The light reactions of larvae of the tent caterpillars, *Malacosoma disstria* Hbn., *M. americanum* (Fab.), and *M. pluviale* (Dyar) (Lepidoptera: Lasiocampidae). Can. Entomol. 85:297–310.

Taylor, M. F. J. 1984. The dependence of development and fecundity of *Samea multiplicalis* on early larval nitrogen intake. J. Insect Physiol. 30:779–785.

Taylor, R. A. J., and Reling, D. 1986. Density/height profile and long-range dispersal of first-instar gypsy moth (Lepidoptera: Lymantriidae). Environ. Entomol. 15:431–435.

Thomas, J., and Lewington, R. 1991. The Butterflies of Britain and Ireland. Dorling Rindersley, London.

Thomas, J. A., and Wardlaw, J. C. 1992. The capacity of a *Myrmica* ant nest to support a predacious species of *Maculinea* butterfly. Oecologia (in press).

Tilman, D. 1978. Cherries, ants and tent caterpillars: Timing of nectar production in relation to susceptibility of caterpillars to ant predation. Ecology 59:686–692.

Tinbergen, L. 1960. The natural control of insects in pinewoods. I. Factors influencing the intensity of predation by songbirds. Arch. Néer. Zool. 13:265–343.

Tostowaryk, W. 1972. The effect of prey defense on the functional response of *Podisus modestus* (Hemiptera: Pentatomidae) to densities of the sawflies *Neodiprion swainei* and *N. pratti banksianae* (Hymenoptera: Neodiprionidae). Can. Entomol. 104:61–69.

Trichilo, P. J., and Mack, T. P. 1989. Soybean leaf consumption by the soybean looper (Lepidoptera: Noctuidae) as a function of temperature, instar and larval weight. J. Econ. Entomol. 82:633–638.

Tsubaki, Y., and Shiotsu, Y. 1982. Group feeding as a strategy for exploiting food resources in the burnet moth *Pryeria sinica*. Oecologia 55:12–20.

Vander Meer, R. K. 1987. Per cent emergent weight: a roadmap to adult rhinoceros beetle, *Oryctes rhinoceros,* behaviour. J. Insect Physiol. 33:437–441.

Wasserman, S. S., and Mitter, C. 1978. The relationship of body size to breadth of diet in some Lepidoptera. Ecol. Entomol. 3:155–160.

Watanabe, M. 1976. A preliminary study on population dynamics of the swallowtail butterfly, *Papilio xuthus* L. in a deforested area. Res. Pop. Ecol. 17:200–210.

Watanabe, M. 1981. Population dynamics of the swallowtail butterfly, *Papilio xuthus* L., in a deforested area. Res. Pop. Ecol. 23:74–93.

Weiss, S. B., and Murphy, D. D. 1988. Fractal geometry and caterpillar dispersal: Or how many inches can inchworms inch? Func. Ecol. 2:116–118.

Weiss, S. B., White, R. R., Murphy, D. D., and Ehrlich, P. R. 1987. Growth and dispersal of larvae of the checkerspot butterfly *Euphydryas editha*. Oikos 50:161–166.

Wellington, W. G. 1974. Tents and tactics of caterpillars. Nat. Hist. 83(1):64–72.

Whitman, D. W. 1987. Thermoregulation and daily activity patterns in a black desert grasshopper, *Taeniopoda eques*. Anim. Behav. 35:1814–1826.

Williams, C. M. 1980. Growth in insects, pp. 369–383. In M. Locke and D. S. Smith (eds.), Insect Biology in the Future. Academic Press, New York.

Witler, J. A., and Kulman, H. M. 1972. A review of the parasites and predators of tent caterpillars (*Malacosoma* spp.) in North America. Agric. Exp. Sta. Univ. Minnesota, Tech. Bull. 289.

Yeargan, K. V., and Braman, S. K. 1986. Life history of the parasite *Diolcogaster facetosa* (Weed) (Hymenoptera: Braconidae) and its behavioral adaptation to the defensive response of a lepidopteran host. Ann. Entomol. Soc. Am. 79:1029–1033.

Young, A. M. 1972. Adaptive strategies of feeding and predator avoidance in the larvae of the butterfly *Morpho peleides limpida* (Lepidoptera: Morphoidae). J. N.Y. Entomol. Soc. 80:66–82.

Young, A. M. 1983. On the evolution of egg placement and gregariousness of caterpillars in the Lepidoptera. Acta Biotheor. 32:43–60.

Young, A. M., and Moffett, M. W. 1979. Studies on the population biology of the tropical butterfly *Mechanitis isthmia* in Costa Rica. Am. Midl. Nat. 101:309–319.

Young, A. M., Blum, M. S., Fales, H. M., and Bian, Z. 1986. Natural history and ecological chemistry of the neotropical butterfly *Papilio anchisiades* (Papilionidae). J. Lepid. Soc. 40:36–53.

Yuma, M. 1984. Egg size and viability of the firefly, *Luciola cruciata* (Coleoptera, Lampyridae). Kontyû, Tokyo 52:615–629.

Zucoloto, F. S. 1987. Feeding habits of *Ceratitis capitata* (Diptera: Tephritidae): Can larvae recognise a nutritionally effective diet? J. Insect Physiol. 33:349–353.

PART II

Ecological and Evolutionary Consequences: Caterpillar Life-Styles

The foraging patterns among caterpillar species are diverse. Some caterpillars spend considerable time moving among feeding sites, never staying long enough to finish a leaf, whereas others are relatively sedentary (e.g., leaf miners). Some feed both during daylight and at night, whereas others restrict their feeding mainly to nighttime. Some caterpillars are messy eaters (i.e., feeding sites are obvious with leaves left in tatters) and thus those larvae are conspicuous; in contrast, others neatly trim leaves. Some caterpillars are mutualists. A few species are carnivorous.

Caterpillars vary in many other ways as well. Color pattern ranges from blending into the background to mimicking inedible objects to bright, seemingly conspicuous signals. Larval size varies tremendously within and among species. Some caterpillars are thermal regulators, moving among microsites in response to changes in temperature, which thus allows them to maintain near optimal body temperatures for growth and survival, but others are thermal conformers.

Such diversity in traits affecting foraging exists despite the seemingly simple set of constraints on larvae: caterpillars, within their limitations as ectotherms, must avoid their natural enemies while obtaining food for growth. Even though the diversity in foraging behavior and other traits may seem somewhat bewildering, there are some striking patterns, which we refer to here as caterpillar life-styles. A caterpillar life-style arises from an array of traits that presumably reflects a set of constraints, i.e., the phylogenetic, abiotic, and biotic factors that shape the foraging pattern of the caterpillar. Some degree of crypsis is a critical element of every caterpillar's life (Stamp and Wilkens, Chapter 9). However, the degree of crypsis among species varies from relatively unapparent on any scale to conspicuous within a predator's perceptual field, in which case the caterpillar may be warningly colored. Caterpillars that are warningly colored may also be unpalatable, which presumably lessens the force of some constraints on foraging but may increase that of others (Bowers, Chapter 10). Another interesting contrast is that of solitary versus gregarious life-styles. Solitary caterpillars tend to exhibit a large degree of crypsis. In contrast, gregarious larvae tend to be conspicuous and able to exploit food resources quickly. Being gregarious may reduce some constraints on foraging but impose others (Fitzgerald, Chapter 11). Many caterpillars use host plants with floral and extrafloral nectaries or end up sharing a host plant with honey-dew producing homopterans, which may attract ants, major caterpillar predators. Some caterpillars effectively exploit that kind of relationship through the mutualism of being "nectaries" for the ants, which then protect them (Baylis and Pierce, Chapter 12). Caterpillars can also be categorized as specialist versus generalist feeders, or as exposed versus enclosed feeders. These classifications will also be dealt with to some extent in the following chapters.

The objective of this section is to examine caterpillar life-styles in terms of the cryptic, aposematic, gregarious, and mutualistic patterns. These life-styles represent some evolutionary "solutions" to similar overall problems. Thus, the contrasts among these patterns are instructive for our developing a better understanding of the ecology and evolution of caterpillars and can serve as a framework for quantitatively investigating these different life-styles.

9

On the Cryptic Side of Life: Being Unapparent to Enemies and the Consequences for Foraging and Growth of Caterpillars

Nancy E. Stamp and Richard T. Wilkens

Introduction

Cryptic coloration and behavior in insect herbivores presumably reflect selection pressure from predators (Cott 1940; Robinson 1969; Edmunds 1974; Heinrich 1979; Lederhouse 1990). Various studies provide a glimpse of the nature of that process (e.g., Pietrewicz and Kamil 1981; Greenberg 1987a; Ito and Higashi 1991a). For instance, some evidence exists for the evolution of caterpillar coloration driven by predators. Differing susceptibilities of yellow and green morphs of larvae of a geometrid species to various predators (den Boer 1971) indicate ways in which predators may drive the evolution of prey color and pattern. Other cases are also suggestive in this regard (de Ruiter 1952; Clarke et al. 1963; Herrebout et al. 1963; Curio 1970), but in general studies on the effectiveness of cryptically colored caterpillars in evading predators have been incomplete (e.g., major assumptions untested, lack of adequate controls, statistical problems). There are fewer examples of the presence of predators contributing to altered and cryptic behavior of herbivorous prey. Several studies of aquatic carnivorous insects show them taking refuge in different microhabitats in the presence of their predators (Sih 1987, and references therein), but there are few clear examples of insect herbivores shifting microhabitats in direct response to predators with demonstrated consequences for survivorship (but see Stamp and Bowers 1988, 1990, 1991 for one case, which is discussed later). Nonetheless, there is considerable indirect evidence that cryptic coloration and behavior reflect the need to avoid predators and thus play an important part in the foraging patterns of insect herbivores.

We will use the term crypsis to mean the appearance, behavior, and other characteristics that make an organism less apparent to its enemies (Cott 1940; Robinson 1969; Edmunds 1974; Endler 1986). Being cryptic is for the most part a primary defensive mechanism (the means by which the prey avoids attack due to the predator ignoring it; Robinson 1969) that occurs regardless of the presence or absence of predators (Edmunds 1974). Within the category of crypsis, prey may exhibit a general resemblance to their background, referred to as eucrypsis, or they may masquerade as some inedible item, such as a leaf or bird dropping. In the former case, the model is the background and thus the appearance of the prey relative to its background is especially important, whereas in masquerading, the model is an inedible object and thus matching the background is not critical (Endler 1981). If a predator does encounter or attack the prey, the detected prey may return to a cryptic (and thus undetected) state by its behavior (e.g., by dropping off the leaf onto a background that it matches). So some secondary defensive mechanisms (Edmunds 1974) (i.e., those providing escape) contribute to crypsis. In this chapter, we focus on the primary line of defense, crypsis (both eucrypsis and masquerade), and one aspect of the secondary line of defense, escape that makes the prey unapparent.

Rather than classifying insects as cryptic versus warningly colored, we take the position that probably all larvae are cryptic at some scale relevant to their predators, which we will explore more fully below. Thus, our view is that warningly colored larvae are likely cryptic to their predators at least some and perhaps much of the time, with most of the exceptions due to the predators approaching close enough to detect such prey. The first line of defense for any animal ought to minimize the risk of predation by avoiding encounter or confrontation with its enemies (Endler 1986; Lederhouse 1990).

Our objectives here are (1) to discuss how perception and detection by different kinds of enemies (birds, invertebrate predators, and parasitoids) may shape cryptic coloration and behavior in caterpillars, (2) given such perception and detection by enemies, to examine the ways in which caterpillars can be cryptic, (3) to explore the consequences of caterpillar crypsis for foraging and growth, and (4) to offer some guidelines for future studies of crypsis.

Perception and Detection by Caterpillar Enemies

Crypsis to Aposematism: A Continuum

Crypsis is a function of scale. Both the distance between the predator and prey and the distance between the prey and its background affect the discriminatory abilities of the predator (Endler 1978). For instance, for each predator species, the anatomy of its eyes and the distance to prey will determine whether, in its perception, color patches of the prey blend together or can be distinguished separately (Endler 1978). Thus, the prey's color pattern should match the background as the predator sees it, blended or in fine detail, reflecting the average distance over which the prey is sighted. The degree to which prey match their background relative to a predator's perception can be measured, at least for some systems (e.g., Endler 1984). Prey apparency then is a function of the prey's cryptic traits, nearness of an enemy, and the enemy's perceptive abilities relative to the prey's background, including the chemical and sound backgrounds.

Regarding caterpillar color per se, it is important to recognize that (1) it may have some other function for an animal than as a signal to another organism, or even have no function, and (2) the perception of color depends on the conditions for perception, including light intensity, light quality, distance, and perceptual abilities of the animal. As an example of the first case, colors that are associated with warning coloration, such as yellow, red, white, and black and combinations of these, may be beneficial for thermoregulation and may or may not serve as a signal about palatability to the perceiver. For instance, *Ctenucha virginica* larvae that are covered with black and yellow hairs in the spring and fall can obtain higher body temperatures than caterpillars of the summer form with predominantly yellow hair, which reflects more light (Fields and McNeil 1988). Therefore, the color pattern of yellow and black does not necessarily reflect selection by predators for warning coloration of prey. As for the second point, Endler (1978) describes a situation that illustrates the effect of distance between the prey and its background on the predator's visual acuity and thus perception of color; orange spots on a fish may match the blending in the predator's eyes of the red and yellow patches in the coral reef background. Thus, there is the possible fallacy of assuming that bright coloration, such as orange spots on prey, reflects warning coloration.

Coloration in caterpillars may be similarly deceptive to observers. We human observers must be very careful about classifying, for instance, coloration and pattern of an organism as warningly colored (deliberately conspicuous), thus indicating either noxious or unpalatable properties, or mimicking organisms with such characteristics. To be classified as aposematic (unpalatable *and* warningly colored), it is necessary to demonstrate that (1) the organism is unpalatable or noxious *and* (2) its coloration is associated by the predator with the unpalatable or noxious qualities of the prey as indicated by altering the predator's behavior toward the prey. An organism with warning coloration but that is not noxious or unpalatable is bluffing. If it is a good mimic, its apparency to predators should be similar to that of its model.

An important point here is that, for their average background, probably most insect herbivores are cryptic the majority of the time relative to the set of enemies commonly preying on them. Even insect herbivores that are considered warningly colored or aposematic may be apparent only on the near scale and be quite cryptic at a distance (Edmunds 1974; Papageorgis 1975; Rothschild 1975; Jarvi et al. 1981; Lederhouse 1990). For example, while feeding on their asclepiad host plants that contain and thus allow them to sequester cardiac glycosides (Reichstein et al. 1968), monarch caterpillars (*Danaus plexippus*), with their yellow, white, and black stripes, are conspicuous when viewed up close but blend into their background, especially as they usually feed on the undersurface of leaves, which are lighter in color than the top surface (Stamp, personal observation). Thus, the average distance between the predator and prey and average background of the prey influence the prey's apparency (or lack of it).

Factors Affecting the Probability of Encounter between Prey and Predator

Although we cannot readily determine the probability of encounter of [or average distance between, sensu Endler (1978)] caterpillars and their enemies, we can estimate the distances beyond which prey are probably undetected, by examining reactive or attack distances. Reactive distance is that point from which the predator clearly orients to the prey; attack distance is that point from which the predator attempts contact. Reactive distance in some cases may be the same as attack distance. Detection distance is that point from which the predator detects the prey, but in contrast to reactive and attack distances, it is not measurable in the environment because we cannot be sure when the predator actually detects the prey. Of course, reactive and attack distances do not necessarily equal detection distance, but they probably provide a fairly good approximation. For example, under natural conditions reactive distances of avian insectivores are in general less than 1 m (Eckhardt 1979; Greenberg and Gradwohl 1980; Robinson and Holmes 1982). At such distances experienced predators may correctly detect cryptic prey 90% of the time, whereas at distances of 4.8 m, they may find prey only 50% of the time (Pietrewicz and Kamil 1977). Overall, caterpillars should benefit when the average distance between them and their predators exceeds the reactive or attack distance of the predator. Anything that increases the average distance between prey and predator should increase the likelihood that the prey is undetected.

In this section we consider the probability of encounter between caterpillars and their major enemies, birds, invertebrate predators, and parasitoids, as a function of the enemies' perception and detection of cryptic prey, to establish a comprehensive picture of these enemies as a selective force for cryptic traits.

Avian Predators and Caterpillars

The probability of encounter between caterpillars and avian predators may be affected by various factors, the most important of which are (1) the foraging mode of the predator, (2) leaf morphology and plant architecture, (3) searching and attack rates of the predator, (4) degrees of stereotypic behavior and learning of the predator, (5) prey density and/or biomass, and (6) behavior of the prey.

Foraging Mode of Predator

The probability of encounter between insectivorous birds and caterpillars depends on the birds' foraging modes. Gleaners actively search by hopping along branches and usually attack stationary insects at short range (Eckhardt 1979;

Robinson and Holmes 1982). For example, gleaning warblers in the Rocky Mountains of Colorado exhibited attack distances less than 0.6 m (Eckhardt 1979). Attack distance of gleaners depends on whether the insects are on the upper or lower surfaces of leaves. For instance, the average attack distance for a gleaning warbler in Panamanian lowland forest was 10 cm on upper leaf surfaces in front of them in contrast to 29 cm for lower leaf surfaces above them (Greenberg and Gradwohl 1980). Hoverers survey an area while perching, pick prey from a substrate while in flight, and thus usually attack insects from greater distances (Greenberg and Gradwohl 1980; Robinson and Holmes 1982). Seven of 11 species of foliage-dwelling insectivorous birds in a temperate forest predominantly used the hovering tactic; hover flight distances (equivalent to reactive distances) ranged from 0.3 to 1.6 m (Robinson and Holmes 1982). There are other foraging modes that are relevant to capturing caterpillars, such as terrestrial leapers (from ground to glean low foliage) (Greenberg and Gradwohl 1980) and ground foragers (e.g., thrushes) that may encounter larvae that have left host plants (Holmes et al. 1986). Some gleaning warblers specialize in dislodging or frightening prey by shaking leaves as they hop along branches (Morton 1980). Foliage gleaning birds that have rictal bristles frequently chase down prey that drop from leaves and are probably must successful in obtaining such prey when there is no foliage directly below the prey to obscure the bird's view (Morton 1980). Another tactic is hanging, in which a bird flies to a leaf or twig and while hanging from it uncurls a rolled leaf (Robinson and Holmes 1982), which itself serves as a cue to the bird (Greenberg 1987a,b). But reactive distances for these modes are unknown.

Leaf Morphology and Plant Architecture

The number and arrangement of leaves along branches determines in part how many leaves can be searched, foraging speed, and how easily prey can be captured (Morton 1980; Robinson and Holmes 1982). For example, in the canopy layer, vireos (*Vireo philadelphicus*) foraged more efficiently, and had higher prey attack rates and shorter attack distances on white ash and yellow birch than on sugar maple and beech (Robinson and Holmes 1984). Ash and birch have multilayered foliage (more open canopy with leaves distributed fairly evenly within them) in contrast to sugar maple and beech, which have monolayered foliage (Horn 1971; Robinson and Holmes 1984). Birds seem to adjust for the number and arrangement of leaves by changing the rate of flight between perches and the rate of hops while searching rather than by altering the search radius or attack distance, which may reflect inherent limitations on the birds' perceptual abilities (Robinson and Holmes 1984).

Leaf size, shape, and petiole length had little effect on foraging behavior of hoverers (Robinson and Holmes 1984). However, birds that gleaned prey from leaves showed stronger preferences for tree species (Holmes and Robinson 1981) and presumably are affected by such tree species differences (Jackson 1979).

Some gleaning species consistently changed their foraging pattern from favoring upper leaf surfaces in beech (where 15% of the beech lepidopteran larvae were) to favoring lower leaf surfaces in sugar maple (where 62% of the maple larvae were), which probably reflects the influence of these tree species on the foraging behavior of the birds (Holmes and Schultz 1988).

These patterns suggest that cryptic strategies of prey should vary among tree species. For example, tiger swallowtail caterpillars (*Papilio glaucus*) exhibit different patterns of feeding on various host trees, such as greater frequency of petiole-clipping for large-leafed host plants (tuliptree and sassafras) than for small-leaved host plants (black cherry and white ash), which in both cases would seem to reduce cues for predators (Lederhouse 1990).

Searching and Attack Rates

Speed of a searching predator influences the likelihood that the predator will perceive prey within its searching radius. For example, the vireo *Vireo solitarius* searches slowly on inner branches, foliage, and trunks, exhibiting a low prey attack rate of 1.4 per minute, but captures a high proportion of relatively large prey (Robinson and Holmes 1982). Of the species that heavily utilized caterpillars in a temperate forest, the number of leaves that would have to have been searched (i.e., included in the predator's searching radius) to yield their attack rates ranged between 465 and 1707 per minute, with the solitary vireo having the lowest value and the others having rates at least double that. Such high leaf searching rates for the majority of the bird species coupled with the often long distances from which leaves are searched means that birds must miss or pass over many small and cryptic prey (Robinson and Holmes 1982).

Studies indicate that as prey become more cryptic the search speed of predators may decline (Smith 1974; Goss-Custard 1977). In theory, when a predator searches for two prey types that differ in crypticity, the optimal search rate decreases with declining relative density of the more conspicuous prey, and this in turn may increase the probability of the predator detecting the more cryptic prey (Gendron and Staddon 1983). With increasing relative density of the more conspicuous prey, the optimal search rate increases and that may reduce the probability of the predator finding the more cryptic prey, despite its absolute abundance.

The probability of encounter between predator and prey may also be affected by changes in the predator's attack rate and territorial coverage. In the breeding season, attack rates are much higher during nonsinging periods than singing (territorial defensive) periods (Eckhardt 1979). When singing, birds occur at different heights in the foliage than they do when they are feeding, and foliage density may be less or greater around the singing sites (Hunter 1980; Welty and Baptista 1988). Also, singing sites change through the season as different tree species begin to leaf out (Hunter 1980). So a change in the attack rate due to territorial behavior coupled with the different coverage of the territory during the defensive periods probably contributes to more variation in risk of predator encounter among the caterpillars within the territory. Singing is greatest at dawn, somewhat less at dusk, and least during the afternoon (Welty and Baptista 1988). Some caterpillar species migrate at dawn to diurnal refuges in bark and leaf litter and at dusk to nocturnal feeding sites in the foliage (Schultz 1983; Fitzgerald et al. 1988; Weseloh 1989). Because of the overlap in timing of avian predators' need for territorial defense and the daily larval migrations, the probability of encounter between prey and predator may be less than expected for prey in transit.

The state of the predator's hunger may also affect the search and attack rates. Hungrier animals tend to be more persistent (Charnov 1976) and less selective (Snyderman 1983). Thus, the encounter rate between prey and hungrier predators may be greater than that between prey and relatively well-fed predators.

In sum, searching speed, attack rate, and hunger of a bird vary and consequently may influence the probability of caterpillar prey being encountered.

Degree of Learning and Stereotypic Behavior

The probability of encounter with prey may be increased by the predator's learning. Avian predators can learn through experience where and on what they can feed most profitably (e.g., Alcock 1973; Smith and Sweatman 1974; Erichsen et al. 1980). But foraging on cryptic prey necessitates spending time discriminating them from the background (Erichsen et al. 1980; Pietrewicz and Kamil 1981). Specialists, such as those that show an innate foraging preference for dead curled leaves hanging on forest understory plants, must actively examine and manipulate such hiding places of their arthropod prey, which is time consuming, and thus, the birds benefit from juvenile exploratory learning with these kinds of prey locations (Greenberg 1987a). Birds can learn to use other cues, such as leaf damage, to locate prey (Heinrich and Collins 1983). For instance, black-capped chickadees (*Parus atricapillus*) can learn to discriminate among leaf shapes (and thus tree species), which allows them to use leaf shape as a cue to locating prey (Heinrich and Collins 1983). But empirical tests show that a predator will stop looking for an otherwise profitable prey that has become sufficiently cryptic (Erichsen et al. 1980). Therefore, avian predation should favor caterpillar traits that decrease predictability of the caterpillar in time, space, and appearance, and consequently traits that maximize the amount of learning required by the predator (Morse 1980).

Substrate-restricted searching modes are commonly used by birds in neotropical forests (Munn and Terborgh 1979; Morton 1980; Gradwohl and Greenberg 1984). Whether a predator specializes within a microhabitat or forages more generally may also influence the probability of encounter with a particular caterpillar species. Insectivorous birds that forage in a broad array of microhabitats are likely to investigate novel situations, whereas those that specialize on certain microhabitats tend to be wary of novel situations (Greenberg 1983).

Prey Density

Absolute and relative densities of prey may affect the probability of a predator encountering a particular prey species. Over time the densities of arthropods among and within tree species vary, and birds respond accordingly (Holmes et al. 1986; Greenberg 1987b; Holmes and Schultz 1988). For instance, in one study yellow birch had higher arthropod densities than the other major tree species, and correspondingly all of the 10 insectivorous birds in that temperate forest favored yellow birch at that time (Holmes and Robinson 1981).

Densities of prey also vary due to periodic outbreaks, and that may influence patterns of selective pressure by predators on cryptic prey. For example, over 16 years in a temperate forest, there were three outbreaks (1968–1971, *Heterocampa guttivitta*; 1982–1983, *Itame pustularia*; and 1982, unidentified geometrid) involving different sets of tree species (Holmes et al. 1986). The insectivorous birds responded by concentrating their foraging on the trees harboring the outbreak insects. Patterns of insect outbreaks followed by increases in bird populations suggest that forest birds may experience prolonged periods of food limitation punctuated by brief periods of superabundant food (Janzen 1980, 1988; Holmes et al. 1986; Holmes 1990).

These patterns may mean that selection for crypsis is greater at two times for a species that occasionally exhibits an outbreak phase. For a species in an outbreak phase, selection for crypsis should be high just past the peak of the outbreak, due to a numerical response by predators lagging behind the increase in the outbreak species and the birds' familiarity with the outbreak species (Fig. 9.1). For nonoutbreak species and outbreak species currently in a nonoutbreak phase, selection for crypsis should be high after the outbreak of another species, when predator numbers are still high but the outbreak species is becoming less common (Fig. 9.1). An experiment in a temperate forest showed that 18–63% of the understory caterpillars were captured by birds during a nonoutbreak period (Holmes et al. 1979), which contrasts markedly with less than 1% of forest caterpillars killed by birds during outbreaks (Morris et al. 1958; Buckner and Turnock 1965).

Even if populations of territorial bird species are relatively low, migrants may provide fairly regular and intense selective pressure on caterpillars. Migrants can take advantage of the annual rise in prey and of local outbreaks in the spring. One spring study in temperate forest showed that 75–98% of the foliage invertebrates were lepidopteran larvae and that small larvae (< 15 mm) and in particular leaf rollers were preyed on disproportionately by such migrants, which ate 1.2–1.7 times their own weight in caterpillars per day (Graber and Graber 1983).

Birds may alter their foraging mode relative to prey availability. For instance, the warbler *Helmitheros vermivorus* switched from spending 78% of its searching time inspecting dead curled leaves in the understory during its overwintering in lowland neotropical forests to spending 75% of its time gleaning prey off live

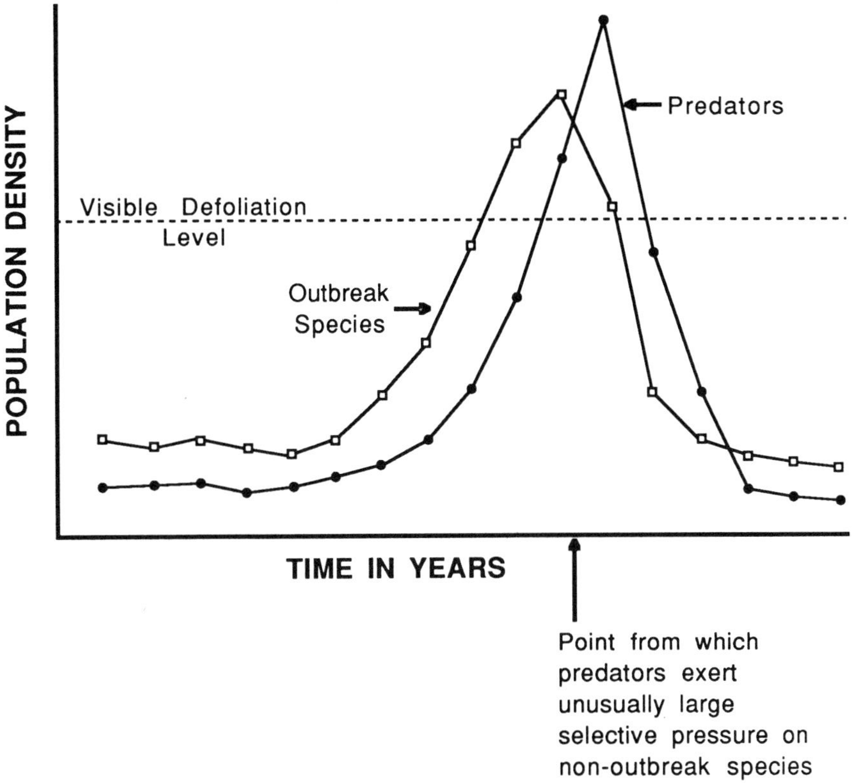

Figure 9.1. Patterns of population density for an outbreak lepidopteran species and its predators over several years. (Modified from Martinat 1987.)

leaves during its breeding period in temperate forests (Greenberg 1987b). Arthropod biomass in the dead, curled leaves versus on live leaves was 153:1 during the warbler's overwintering period in tropical forest, whereas during its breeding season in temperate forest the largest arthropods (34% of which were caterpillars) were on leaf surfaces during the day. Examining leaf curls takes extra time; for example, another specialist on leaf curls spent on average 2–7 seconds per curl (Gradwohl and Greenberg 1984). Thus, a switch from the less specialized mode of gleaning in the breeding period to the more specialized mode of hanging and investigating curled leaves on the overwintering site reflects the changing profitability of such foraging modes in the respective habitats.

Shifting relative abundance of prey may also influence detection and learning by predators. In a laboratory experiment, Pietrewicz and Kamil (1981) found that experienced blue jays (*Cyanocitta cristata*) improved in their detection of a cryptic prey type (*Catocala* moth) against a natural background with 4–5 successive encounters. In contrast to that pattern, jays showed no improvement in detecting prey when two different cryptic prey types (two *Catocala* species) were offered in random order.

Behavior of Prey

Prey behavior may affect the probability of encounter between prey and predator. Cues such as leaves rolled by caterpillars may be detected at some distance, at least up to a few meters if the line of vision is clear, but use of cues such as leaf mining require close inspection (Heinrich and Collins 1983).

The probability of encounter may depend on how caterpillars behave on being disturbed. Green sawfly (*Croesus varus*) larvae move to new feeding sites when the branch they are on is disturbed. (Boevé 1991). Under field conditions, gregarious larvae (*Hemileuca lucina*) react defensively, for example, by falling off the host plant when humans get within a meter and especially if direct sunlight is blocked (Stamp, personal observation). Other caterpillars respond to disturbance up to a half meter away (Tautz and Markl 1978). In these cases, a caterpillar may move before a predator has detected it or, once an encounter between the predator and prey occurs, the caterpillar may escape successfully, for example, by dropping off the leaf, and thus become unapparent once again, at least in part by increasing its distance from the predator.

Effective defense by ectotherms (such as caterpillars) may be subject to thermal conditions, in which case prey may benefit by exhibiting different behaviors under different thermal conditions. For instance, cooler lizards cannot move as rapidly as warmer lizards, and in one case did not allow predators to approach as closely before fleeing (Rand 1964) and in another stood their ground and were more aggressive (Hertz et al. 1982). To what extent temperature affects predator-caterpillar interactions is unknown and may depend on the sensitivity of the caterpillars to disturbances that could signal the presence of enemies. As discussed above, some caterpillars react to potential threats at distances that are comparable to the lizard-predator situation.

Another factor that may affect the probability of encounter between predator and prey is the relationship between degree of crypsis and behavior of the prey. Cryptic lizards allow predators to approach more closely before fleeing than noncryptic lizards (Heatwole 1968). To what extent caterpillars recognize the degree to which they match their substrate and respond accordingly in the presence of predators is largely unknown. Catkin-like caterpillars moved to catkins when placed on leaves or twigs and stayed put when placed on catkins; likewise twig-like caterpillars choose twigs for resting sites (Greene 1989). In a choice test, sphingid *Amphion floridensis* larvae that were green more frequently rested under green leaves than pink larvae, whereas pink larvae more often rested on woody stems (Fink 1989). But for some caterpillars, matching the substrate may be primarily achieved through fixed behaviors. The reflectance quality of leaves of the host plant can determine, by way of diet, the color that caterpillars develop (Grayson and Edmunds 1989) and consequently staying put on leaves would ensure a high degree of background matching. For bark-colored larvae, negative phototaxy may promote them to seek refuge on bark, in bark crevices, or on the ground during the daylight hours (Weseloh 1989).

In sum, birds probably have been a major selective pressure for cryptic coloration and behavior of caterpillars. Birds can learn to recognize cryptic prey, but developing an effective search image may be compromised by foraging conditions (e.g., lighting, plant architecture). Furthermore, given the constraints on birds' foraging (e.g., meeting their energy requirements, territorial defense, avoiding their own predators), they must forage quickly and with some attention focused elsewhere (on neighbors and predators). Consequently, cryptic caterpillars are more likely to escape predation than noncryptic prey.

Invertebrate Predators and Caterpillars

The probability of encounter between caterpillars and invertebrate predators is affected by the same factors that influence that of caterpillars and avian predators. A variety of invertebrate predators attack caterpillars: carabid and coccinellid beetles; predatory wasps; ants; anthrocorid, pentatomid, reduviid, and nabid bugs; and spiders are the most important (DeBach 1974; Coppel and Mertins 1977; Montllor and Bernays, Chapter 5). The morphologies and behaviors among the invertebrate predators vary to a much greater extent than they do among insectivorous birds. Consequently, reactive distance, which serves as our estimate of the distance beyond which the prey are undetected, varies considerably as well. Many invertebrate predators have a visual field of about 1–2 cm and frequently have to make physical contact to orient to and attack prey (coccinellids, Banks 1957; Dixon 1959; ants, Swynnerton 1915; Myers and Campbell 1976; coccinellids and ants, Allen et al. 1970). However, other predators, such as salticid spiders, which have a field of vision up to about 80 cm (Forster 1982) and predatory wasps with a field of vision up to 25–75 cm (Spradbery 1973), may have reactive distances that approach and even rival those of insectivorous birds.

Sufficient data are available to make comparisons among predatory stinkbugs (Pentatomidae), ants (Formicidae), and one group of predatory wasps (Vespidae). All three groups are major predators of caterpillars that have been used as biological control agents (Adlung 1966; Lopez et al. 1976; Risch and Carroll 1982; Saks and Carroll 1980; Gould and Jeanne 1984, and references within these). Most of these predators are generalists. For instance, caterpillars of at least 15 lepidopteran families are prey of a wood ant (*Formica lugubris*, McNeil et al. 1978). Over 50 caterpillar species are listed as prey of *Podisus* stinkbugs (Mukerji and LeRoux 1965; Tostowaryk 1971; Lopez et al. 1976). Likewise, over 50 caterpillar species are listed as prey of *Polistes* wasps in North Carolina, with 87–100% of wasp prey being caterpillars, depending on the year and species of *Polistes* (Rabb 1960).

Predatory stinkbugs use visual, chemical, and tactile cues to locate prey while walking on a plant (McLain 1979; Awan et al. 1989). *Podisus maculiventris* search over more area and longer at low prey density than at high density but do not respond to damaged leaves or intensively re-search areas in which they were previously successful (Wiedenmann and O'Neil 1991). Plant architecture and density influence colonization by predators (Letourneau 1990) and plant characteristics affect the searching efficiency of various insect predators. For example, trichomes impede the progress of predators (Scopes 1969) and smooth surfaces increase the likelihood of predators falling off the plant (Carter et al. 1984). It is likely that plant characteristics affect stinkbugs, too.

Stinkbugs react to moving prey at distances up to 10 cm, but their reactive distance to immobile prey is considerably less, with detection often seeming to occur at contact (Tostowaryk 1972). Often caterpillars do not respond to the approach of a predator (Marston et al. 1978). After finding and orienting to prey, stinkbugs may spend as much as an hour stealthily approaching a caterpillar (Morris 1963; Mukerji and LeRoux 1965; Tostowaryk 1971). Different species of caterpillars vary in their response to attack by stinkbugs; some drop off the plant, whereas others defend themselves in place (Marston et al. 1978; Awan 1985). Stinkbugs and caterpillars are often of similar size, and the prey can injure the stinkbugs (Morris 1963; Tostowaryk 1972). A relatively low kill rate by stinkbugs is at least in part a function of the defensive behaviors of larvae. Under laboratory conditions, stinkbugs (*P. maculiventris*) killed 0.9 restrained fall webworms (*Hyphantria cunea*) per day but only 0.5 defensive larvae per day (Morris 1963). At relatively low prey densities under field conditions, stinkbugs would have had to search on average 3264 leaves per day to meet a predation rate of 0.4 larva per day (O'Neil 1989).

However, stinkbugs often locate gregarious caterpillars and sawfly larvae and then remain near the aggregation (on average 20 cm away, Tostowaryk 1971) for hours to days, periodically attacking and feeding (Tostowaryk 1971; Morris 1972a; Evans 1983). Gregarious behavior by caterpillars may increase their risk of being attacked by stinkbugs. Under these conditions, the stinkbugs may spend less than 1% of their time searching for prey and often exhibit a sit-and-wait predator strategy (Tostowaryk 1971). Furthermore, stinkbugs tend to aggregate at nests of gregarious caterpillars (Evans 1983). This aggregation by predators probably accounts for Morris' (1972a) finding that a disproportionate percentage of fall webworm nests had no stinkbugs. Thus, there may be considerable variation in probability of encounter between prey and predator among webworm sites.

Another factor that may influence the probability of encounter between prey and predator is the age of the predator. Second, third, and fourth instar stinkbugs search more actively and attack more aggressively than fifth instar and adult stinkbugs (*Podisus modestus*) (Tostowaryk 1971). Furthermore, once a prey is encountered, young nymphs continue to attack even after several failures whereas adults usually retreat if they are unsuccessful with the first attack. Also, as adults get older, the rate of predation decreases (Morris 1963).

Vespid wasps exhibit two foraging tactics, hovering around plants, presumably making a visual inspection, and walking on plants and the ground, apparently responding largely to chemical cues (Takagi et al. 1980; Raveret Richter and Jeanne 1985; Stamp and Bowers 1988; Stamp 1992). Vespid wasps rely on visual cues to locate prey more than stinkbugs (Morris 1963, 1972b). The foraging intensity of *Polistes* wasps increases with increasing densities of prey (Nakasuji et al. 1976), especially when prey are large (Morris 1972b). When they locate high densities of prey, wasps make repeated visits to those sites (Yamasaki et al. 1978; Steward et al. 1988). Once prey are located, they are dispatched quickly; the wasp pounces on the prey, bites repeatedly, and carries off the balled-up prey or a piece of it to provision the larvae at the nest (Raveret Richter and Jeanne 1985; Stamp and Bowers 1988). The behavior and size of gregarious prey determines how effective they are at escaping once wasps have found them (Stamp and Bowers 1988). Caterpillars that have been recently attacked unsuccessfully (harassed) by wasps may spend much of their time at the base of host plants and away from the crowns where the wasps most frequently inspect the plants. Thus, the larvae may effectively decrease the risk of predation but at the expense of having to remain in the shade and feed on mature leaves, which reduces their growth rate and survivorship (Stamp and Bowers 1988, 1990, 1991).

Many ant species prey on caterpillars to some extent (Finnegan, 1974). *Formica* ants of temperate forests are especially important predators of defoliating caterpillars (Adlung 1966; McNeil et al. 1978; Laine and Niemala 1980). The rate at which ants leave the nest and search for food is a function of temperature and hunger of the colony (Taylor 1977; Rissing 1982; Stradling 1987). Over the course of a day, there are peaks of foraging activity (Stradling 1987). In temperate forests, some foraging occurs at night when it is warm enough (McNeil et al. 1978; Skinner 1980). Although sight may be used for navigation, ants probably cannot see prey more than 2 cm away, and thus they rely mainly on chemical cues (Stradling 1987). Because detection by visual and chemical cues is likely to be limited to short distances, scouting seems to be for the most part random (Stradling 1987; Traniello 1989). In many species, scouts may recruit nestmates to food items (Traniello 1989); recruitment rate is higher when the food patch is closer to the nest, the diameter of the patch is greater, the resource is more abundant, and prey size is larger (Taylor 1977). As a consequence, prey located lower on plants, especially on trees, are found first (Skinner and Whittaker 1981), and thus prey higher up may avoid detection by ants (Campbell and Torgersen 1983). On average, ant densities decline with distance from the nest (Laine and Niemela 1980), and correspondingly prey survivorship is greater with increasing distance from the ant colony (e.g., tent caterpillars *Malacosoma americanum* at 25 m or more survived, Tilman 1978; geometrid *Oporinia autumnata* at 20 m or more, Laine and Niemala 1980).

If honey-dew producing homopterans are present on plants, they may be tended by ants and thus increase the activity of the ants on those plants (Stradling 1987; Traniello 1989). *Formica* ants may tend aphids up to 200 m from the nest (Brian 1955), and thus be a threat to caterpillars near ant-tended aphids. In a temperate forest in the spring, ant activity was greatest on sycamore, but through the summer, it declined on sycamore and ash and increased on oak and pine, reflecting the changing distribution of ant-tended aphids (Skinner 1980). With increasing homopteran–ant interaction on plants, the risk of predation for caterpillars on those plants should increase. Studies have shown a reduction in numbers of caterpillars on plants with ant-tended aphids compared to plants without ants (Skinner and Whittaker 1981; Heads and Lawton 1985; Ito and Higashi 1991b). *Formica* ants have main routes to homopteran colonies and to other productive foraging areas, and workers that survive the winter retain some fidelity to these routes (Rosengren 1971). Such collective learning and memory may facilitate quick response by ants to resource availability and therefore promote a relatively high risk of predation for caterpillars yearly in localized areas.

Overall, some invertebrate predators, such as wasps, may act as selective agents for cryptic coloration in caterpillars. Predatory wasps and other invertebrate predators probably promote cryptic behavior of caterpillars at both the primary and secondary defensive levels. The spatial and temporal distribution of ant-tended homopterans may influence not only the distribution of caterpillars but, by way of the ants, contribute to differential selection for cryptic traits in caterpillars.

Parasitoids and Caterpillars

Chemical cues play a major role at every level of host location by parasitoids, beginning with odors of the plant species used by hosts (Vinson 1976, 1985). The caterpillar hosts may use a wider range of host plants than the parasitoid responds to (Weseloh, Chapter 6, and references therein), which would mean lower risk of parasitism when a caterpillar uses a plant species the parasitoid is unlikely to search than that for plant species the parasitoid usually searches. Once the habitat of the host plants of the hosts has been located, some parasitoids search for hosts by flying among host plants about 2–5 cm from the substrate and walking on host plants while examining surfaces with the antennae (Vinson 1977). Parasitoids often orient to particular plant parts (Vinson 1976, 1985). They may be attracted visually and chemically to plant damage (Hassell 1968; Sugimoto et al. 1988; Faeth 1990; Turlings et al. 1990). Injured plant tissue, such as that damaged by caterpillars, may yield different odors or amounts of volatile phytochemicals (Turlings and Tumlinson 1991). However, plant characteristics may interfere with host location (Vinson 1976; Obrycki 1986). For example, high trichome density reduced the walking speed of parasitoids (Hulspas-Jordaan and van Lenteren 1978), and sticky trichomes trapped parasitoids (Rabb and Bradley 1968).

In contrast to the longer range volatile chemicals of plants and perhaps damaged tissue, odors of host insects may be detected or at least responded to only at short distances (e.g., 2–20 cm away; Hendry et al. 1973; Schmidt 1974). Parasitoids may use silk, frass, and mandibular gland products of hosts to locate hosts (Vinson 1968; Weseloh 1977; Waage 1978; Dmoch et al. 1985), but a continuous trail may not be necessary (Vinson 1977). Response to frass, silk, and feeding damage of hosts increases with previous experience with these cues, and as a result experienced parasitoids are more active than less experienced ones (Dmoch et al. 1985). Parasitoids may respond to host odors by concentrating their search efforts in those areas (Vinson 1977). Thus, areas with low host densities may receive disproportionately less coverage, and the risk of parasitism may be much less than in areas with higher host densities.

Just as birds, predatory wasps, and stinkbugs concentrate their foraging at high density caterpillar sites and consequently increase the probability of encounter between prey and themselves, so do parasitoids. Parasitoids of the gregarious web-maker *Hyphantria cunea* remained on or near webs up to several days, periodically searching the webs for larvae to attack (Morris 1963). Large webs of the Baltimore checkerspot (*Euphydryas phaeton*) had a disproportionately greater number of parasitoids attending them, which resulted in a higher level of parasitism than on smaller webs (Stamp 1982). But aggregative response to high host densities does not always result in a greater percentage of parasitism (Morrison and Strong 1980; Waage 1983).

Once it is located, the host may exhibit escape behaviors that increase the distance between it and the parasitoid and thus increase host unapparency. For instance, green cloverworms (*Plathypena scabra*) frequently reacted to disturbance by parasitoids by dropping from the leaves, with younger larvae dropping on silk threads and older larvae often dropping without the threads (Yeargan and Braman 1989a). The parasitoid, *Cotesia marginiventris*, which attacks first instars, captured larvae before they dropped half the time, and dove after the dropping larvae, capturing 73% of those. A second parasitoid, *Diolcogaster facetosa*, which attacks second and third instars, rarely caught larvae before they dropped; these wasps slowly walked down the silk thread after the larvae. A third parasitoid, *Rogas nolophanae*, which primarily attacks third instars, captured 57% of the larvae before they dropped, and dove after those that dropped, although not always immediately. The hyperparasitoid *Mesochorus discitergus*, which utilizes the three parasitoid species and thus attacks first through fourth instar cloverworms, exhibited different tactics depending on the instar (Yeargan and Braman 1989a,b). They usually captured second instar larvae by either hanging from the edge of the leaf and reeling in the caterpillar by the silk thread or walking part way down the silk and then pulling the silk toward them. Third and fourth instar larvae were captured before dropping 40–44% of the time, and those that dropped were pulled up by their silk to the wasps. Some studies report that occasionally parasitoids wait for larvae that have dropped on silk threads to climb back to the plant (Waage 1983; Yeargan and Braman 1989a). Although dropping on silk threads may have several advantages, such as increasing the distance between the caterpillar and its predator, removing the larva from the predator's radius of detection, and ease with which the caterpillar can return to its host plant, parasitoids may often circumvent this caterpillar defense, due to their relatively small size (and thus weight) and ability to track chemical cues (such as those associated with silk).

For the most part, parasitoids probably select for cryptic behavior and fewer olfactory cues by their hosts. Cryptic behavior is especially important for primary defense because parasitoids may be able to circumvent a caterpillar's secondary defensive mechanisms, such as dropping off a leaf.

Sum Effects of Caterpillar Enemies

In examining crypticity in caterpillars, it is important to keep in mind that an array of enemies has shaped the caterpillar traits. Avian predators rely primarily on vision and perhaps to a lesser extent on hearing to detect prey. Invertebrate predators rely mainly on vision, smell, and touch. Parasitoids rely primarily on smell but also use vision, touch, and perhaps hearing to locate hosts. Therefore, selective pressure on caterpillars by birds should be greatest for prey traits that reduce visual cues. Selective pressure by parasitoids should be greatest for host traits that reduce and interfere with chemical cues. Reflecting the cues that it uses, an invertebrate predator is likely to promote the reduction of visual cues and/or chemical cues. Furthermore, small, young caterpillars are preyed on by certain enemies, and the larger, older caterpillars are attacked by another array of enemies. Presumably such changes in predator pressure account for color and behavioral changes among caterpillar instars (e.g., Nentwig 1985; Cornell et al. 1987).

Overall, caterpillars are likely to benefit most by a suite of traits that facilitate not being seen, smelled, and heard up to a meter or so and by being sufficiently unpredictable in time and space to have an average distance between themselves and their enemies greater than the enemy's reactive distance and consequently have a low probability of encounter with enemies.

Factors Affecting the Similarity between Prey and Their Background

Another critical element affecting detection of prey by predators is the similarity between prey and their background (or the prey's average background, sensu Endler 1978). Plants provide the primary backdrop for caterpillars. The average background of the prey may be affected by host plant range of the prey, degree of synchrony between the phenologies of the prey and the plant parts it uses, and whether larvae defoliate their host plant. Factors not directly associated with the host plant may also be important, including thermal conditions and pattern of behavior of the prey.

Many populations of caterpillars feed on several host plant species. For instance, the woolly bear caterpillar *Diacrisia virginica* feeds on a variety of herbaceous plants (Dethier 1988), with perhaps as many as 100 species serving as host plants (Tietz 1972). Even populations of aposematic species may oviposit on and eat more than one host plant species (e.g., depending on the *Euphydryas editha* population, 2 of 7 potential host plant species are used, 3 of 6 and 4 of 5; Thomas et al. 1990). Populations of larvae that use more than one host plant species have a broader array of backgrounds to blend into than those from populations using one host plant species. Selection for crypsis ought to reduce host plant range (Fox and Morrow 1981), which may explain in part why larvae tested in the laboratory often can habituate to or are not deterred by allelochemicals that they might be expected to encounter among host plants (Bernays and Graham 1988). That is, predators more than plant chemistry may have shaped host plant range.

Insect herbivores that are either in close synchrony with the phenology of their host plant (e.g., that regularly hatch right after budburst, Du Merle 1988; Potter and Redmond 1989) or can tolerate asynchrony (e.g., by constructing refuges and withstanding starvation until food is available, Hunter 1990) will be less apparent, because they do not have to move around looking for suitable food, and therefore need to match a smaller array of backgrounds than those without these traits. Species that seldom or never defoliate their host plants and so may not have to leave the host plant until pupation will not need to blend into as wide a set of backgrounds as those that frequently have to leave host plants in search of a new supply of food. Many gregarious species commonly defoliate their host plants (Dethier 1959a,b; Rausher et al. 1981; Tsubaki and Shiotsu 1982; van der Meijden 1979; Stamp 1984a; Pullin 1987) and thus presumably experience a relatively wider array of backgrounds than solitary species.

Thermal conditions may influence movement of caterpillars and thus affect their set of backgrounds. Caterpillars that are thermal conformers (i.e., lacking behavioral thermoregulation, such as tobacco hornworms, *Manduca sexta*, Casey 1976) may have fewer backgrounds into which they need to blend than behavioral thermoregulators that move frequently to adjust their body temperatures (e.g., white-lined sphinx moth larvae *Hyles lineata*, Casey 1976; tent caterpillars, *Malacosoma americanum*, Knapp and Casey 1986). Thermal conformers may also frequent microhabitats with little or no direct sunlight (Casey 1976, Knapp and Casey 1986). In that case, they would be subject to relatively uniform low-intensity lighting with low color diversity and that should favor cryptic patterns with less contrast and that are less colorful (Endler 1978).

In contrast, behavioral thermoregulators bask and seek shade, as needed, to approach their optimal body temperature (Casey 1976, Knapp and Casey 1986). In that case, they may often be in, or move among, microhabitats that vary more in light intensity than those of thermal conformers. Use of such microhabitats would favor color patterns in prey that have many strongly contrasting, brightly colored patches (Endler 1978). The high contrast in the background is likely to make capture of such prey difficult because the predator's eyes will not be able to accommodate quickly enough to the changes in contrast between shady patches and sunflecks (Young 1971; Papageorgis 1975; Lederhouse 1990). In forests, most sunflecks are small relative to leaf size and less than a minute in length, with 1–10 second sunflecks often clustered together (Chazdon 1986). Such short duration of sunflecks is the general pattern for tropical and temperate forests (Hutchinson and Matt 1977; Pearcy 1983; Chazdon 1986), whereas sunfleck duration is somewhat longer in subalpine forests and varies with tree species (Young and Smith 1979). But even though sunflecks are small in size and duration, they may cause sufficient contrast in lighting to interfere with detection of prey, especially if the prey body has contrasting colors. However, when it is cloudy, brightly colored prey are likely to be more conspicuous to the predator (Endler 1978).

Alterations in prey behavior may also affect the range of backgrounds that they experience. Some caterpillars have fixed behavioral patterns by which they move at dawn and dusk among microhabitats, such as from nocturnal feeding sites on leaves to diurnal resting sites on tree branches and trunks or on the ground (Herrebout et al. 1963; Leonard 1970; Schultz 1983; Fitzgerald et al. 1988; Weseloh 1989). The visual and chemical backgrounds among these microhabitats may vary markedly, although the visual differences may be minimized if most of the long-distance movement occurs when light intensities are low and if one of the microhabitats is used almost exclusively at night, at least for some of the instars. Caterpillars tend to leave their host plant and seek pupation sites elsewhere. Sometimes caterpillars that appear cryptically colored on their host plant are conspicuous to predators on the ground (e.g., *Pieris rapae*), and often caterpillars that appeared conspicuous on their host plant seem less so on the ground (*Pieris brassicae*, Baker 1970). Induced behavior in prey by the presence of predators may also increase the range of backgrounds the prey experience. For instance, green caterpillars that drop off a leaf in response to a predator or parasitoid may land on the ground where they may be quite conspicuous in coloration (e.g., Geometridae, Evans 1986) and by movement (Allen et al. 1970).

In addition, the predator's experience with the prey relative to the prey's background is important. Blue jays (*Cyanocitta cristata*) required additional learning to detect a previously recognized prey type (*Catocala* moth) offered against a new background and had even more difficulty with a novel prey type (another *Catocala* species; Pietrewicz and Kamil 1981). Pietrewicz and Kamil (1981) concluded that a predator may have to become very familiar with prey's appearance before it can detect the prey when it is cryptic against a natural background; they refer to this phenomenon as development of a search image. But it may also reflect the necessity of having to learn to recognize different features of the prey relative to different backgrounds. For instance, against some backgrounds, recognizing prey shape may be sufficient for detection of prey and thus prey color pattern may be irrelevant and therefore ignored, but against other backgrounds recognition of prey color pattern may be necessary.

Effects of Sets of Enemies

Although we cannot ascertain to what extent each of a caterpillar's enemies have contributed to its appearance and behavior, it is possible to identify how sets of enemies or circumstances might select for patterns of predator avoidance. Here we consider the following sets of enemies: (1) parasitoids, (2) ants versus other invertebrate enemies, (3) avian and invertebrate predators, and (4) sum of vertebrate and invertebrate enemies.

1. Generalist and specialist parasitoids may act as selective agents for somewhat different host traits. Parasitoids that prevent further development of their hosts may be generalists, and they account for the majority of parasitoids on British insect herbivores that live within plant tissue; in contrast, parasitoids that permit further development of their hosts tend to be specialists and make up most of the parasitoids on the externally feeding herbivores (Hawkins et al. 1990). This pattern suggests that generalist parasitoids may have played an important role in shaping endophytic host habits, whereas specialist parasitoids may have contributed to molding the exophytic host patterns.
2. Ants may negatively affect other invertebrate predators and parasitoids and, consequently, influence caterpillar traits. As ants patrol plants where they are tending honey-dew-producing homopterans or plant nectaries, they may interfere with and even exclude invertebrate predators (Laine and Niemala 1980; Skinner and Whittaker 1981; Risch and Carroll 1982) and parasitoids (Washburn 1984, Koptur 1985). Therefore on ant plants, any insect herbivores that can avoid or are avoided by ants may often forage undeterred by other invertebrate enemies.

   Thus, in microhabitats where ants and caterpillars cooccur, ants can act as selective agents for caterpillar traits that protect against ants but that may or may not facilitate evading other caterpillar enemies in the absence of the ants. For instance, shelter-using herbivores (such as leaf tiers, leaf miners, gall makers, borers) and chemically defended herbivores (e.g., various cryptically colored sawfly larvae) often occur at higher densities on plants where ants are tending aphids or extrafloral nectaries than on plants without these extra ant resources (Fowler and MacGarvin 1985; Heads and Lawton 1985; Heads 1986).

3. The level to which birds prey on invertebrate predators as well as caterpillars may influence the extent to which caterpillars need to be cryptic to particular predators. Predators, such as birds and predatory wasps, that feed on invertebrate predators as well as the caterpillar prey of those predators may cause a shift in the community of both invertebrate predators and caterpillars toward more cryptic forms at relatively low densities. Although this kind of effect of top predators has been shown in freshwater littoral communities (Blois-Heulin et al. 1990), such an effect of birds and wasps on the terrestrial invertebrate community has not been investigated. Such a shift toward more cryptic forms might result in a different array of cryptic traits favored in a particular caterpillar species than in a community in which birds feed only on caterpillars. In the latter case, invertebrate predators densities would be higher and therefore exert more selective pressure for cryptic traits that help caterpillars evade invertebrate predators.
4. The members of the set of enemies of a caterpillar species are likely to change as caterpillars grow, which may promote changes in larval appearance and behavior (Reavey, Chapter 8). But as caterpillars get larger, it may be more difficult to blend into the background (e.g., feeding damage and associated cues that enemies use become more abundant). The sum effect of a set of enemies then may select not so much for alterations that reflect particular kinds of enemies, but rather select for increasingly all-purpose cryptic behavior by caterpillars, to the point of their changing from exposed feeding in early stages to enclosed feeding in latter stages. For example, young *Pieris rapae* larvae eat exposed on outer cabbage leaves, but larger (older) larvae move to the center of the plant presumably in response to the combination of parasitoids and avian and invertebrate predators (Harcourt 1966). This change in behavior occurs even though it places caterpillars at greater risk to pathogens. As the larvae become concentrated in the center of plants, the likelihood of death by pathogens increases (e.g., from 0% while on outer leaves to 34% within the plant, Harcourt 1966). This finding indicates that a set of enemies (predators and parasitoids) may exert considerable selective pressure for a cryptic trait (in this case enclosed feeding).

Ways to Be Cryptic

Reducing Cues

Caterpillars can be cryptic in any number of ways, which are not mutually exclusive. Here we outline ways in which caterpillars may reduce cues to predators and then provide a few examples to illustrate how a set of traits may reduce most cues on the scale at which most enemies operate. Several reviews offer more extensive examples than we do here of crypsis in general and how cryptic traits may reduce cues utilized by predators (Cott 1940; Robinson 1969; Edmunds 1974; Heinrich, Chapter 7).

Visual Cues

Larval Color

Color vision has been reported for many insects, including invertebrate predators and parasitoids (Vinson 1976; Chapman 1982), with wavelengths ranging from 300 to 650 nm and peaks of activity at near-ultraviolet and blue-green (Matthews and Matthews 1978). Color vision in birds, which perceive wavelengths similar to humans (380–760 nm, Schmidt-Nielsen 1983; Ali and Klyne 1985), is especially well developed (e.g., with some detection of color hue and intensity, Welty and Baptista 1988). Thus, caterpillars may benefit in general by matching their background colors (Baker 1970, Mariath 1982).

The basic colors of externally feeding caterpillars can be broadly classified as those that make it difficult to see them when they are on leaves (e.g., green) and those that make it difficult to see when they are adjacent to or off leaves (e.g., yellow, pink, red, brown, gray, black). Holmes and Schultz (1988) found that 78–91% of the caterpillars in a northern hardwood forest were on leaves, with 62–75% on the undersurfaces, and most of these were some shade of green as opposed to brown or gray. Being green is often a function of the presence of yellow carotenoids from food and blue bile in the hemolymph (Fuzeau-Braesch 1985). Internal feeders in general lack pigmentation; gut contents of leaf material may make them greenish. Other coloration is due to synthesis of pigments, sequestration of plant substances, products of metabolism, and structural properties of cuticular substances (Fuzeau-Braesch 1985). Most larvae found on twigs and petioles tend to match their substrate in color, pattern, and shape (e.g., Geometridae, Holmes and Schultz 1988). Birds have difficulty discriminating such caterpillars from twigs (Mariath 1982). Relative to larvae found on leaves, such caterpillars may be underutilized by birds (Holmes and Schultz 1988). Tent caterpillars (*Malacosoma* spp.) have spots and stripes that reflect ultraviolet light, which may camouflage them from arthropod enemies (Byers 1975), and caterpillars hidden in UV-reflecting leaves and flowers may be unapparent as well (Silberglied 1979).

Many caterpillar species are polymorphic or variable in coloration. This variation may be under genetic control, or determined by the host plant or plant parts eaten (Clarke et al. 1963; den Boer 1971; Grayson and Edmunds 1989; Greene 1989) and by temperature (Imura 1982). In monomorphic caterpillars, it is not surprising that color and pattern match the typical resting background. In some polymorphic caterpillars, color morphs have been shown to differ in their resting site preferences (Fink 1989; Greene 1989). For example, in an experiment with the sphingid *Amphion floridensis*, green larvae rested under leaves on 69% of the censuses, whereas pink larvae were found there only 16% of the time (Fink 1989).

As caterpillars develop, their color and pattern may change, presumably reflecting how the set of enemies perceiving and attacking them changes (Cott 1940). For example, a tropical caterpillar (*Oxytenis naemia*: Oxytenidae) resembles droppings of small birds in the first through third instar; the fourth instar resembles oily feces of some other kind of animal; the fifth instar is green, reddish-brown, or pink and looks like a fallen, rolled leaf, which when disturbed shows two eye-spots and acts "snake-like" (Nentwig 1985). Snake-like appearance and behavior may startle a predator just long enough for the caterpillar to then make a quick escape, by dropping (e.g., Vaughan 1983).

Leaf Alteration

Caterpillars may alter leaves in the process of eating and making shelters, and these alterations may serve as cues to enemies. Some caterpillars exhibit behaviors that presumably reduce such cues, for example, by leaving smooth feeding edges, clipping petioles of partially eaten leaves, and moving frequently to new feeding sites (Heinrich 1979; Heinrich and Collins 1983; Heinrich, Chapter 7). These behaviors may vary with larval age and host plant species (i.e., size of leaf used; Lederhouse 1990).

Movement

During feeding and resting bouts, remaining still when enemies are present may often be advantageous because many predators respond to movement by prey (Rilling et al. 1959; Herzog and Burghardt 1974). A common response to disturbance by caterpillars is to freeze unless further disturbed (Minnich 1935; Tautz and Markl 1978). When perception of the predator by prey is limited, as it is with caterpillars due to their perceptual abilities and screening vegetation, the prey should flee (Ydenberg and Dill 1986). Dropping off the host plant, a common defensive response by caterpillars (e.g., Allen et al. 1970; Awan 1985; Stamp 1984b; Heads and Lawton 1985; Cornell et al. 1987; Yeargan and Braman 1989a), may reduce feeding rate and place the prey at greater risk due to desiccation, other predators (Dethier 1959a; Allen et al. 1970) or more of the same predator species (e.g., ants, Fowler and MacGarvin 1985). However because "flightier" prey will lose feeding time relative to bolder prey, there is an advantage to staying put (Ydenberg and Dill 1986), provided that the prey is undetected by the predator or can defend itself.

Although most of a caterpillar's time is spent feeding and resting, some time is spent moving among feeding and resting sites. Movement may be between leaves on different branches (Young 1972; Schultz 1983; Stamp 1984b), between leaves and bark (Herrebout et al. 1963; Leonard 1970), or off and on the host plant (Leonard 1970; McCorkle and Hammond 1988). The degree to which larvae move among such sites may depend on their age (and thus size) (Reavey, Chapter 8) and on when they eat. For instance, some *Catocala* spp. feed day and night in the first instar and rest on leaves between feeding bouts, but after that they only eat at night and between feedings rest on twigs (Johnson 1984).

How well caterpillars match the background when they are moving depends on their speed, color, and pattern, and the predator's flicker fusion frequency (the rate of flicker at which light appears continuous). For example, if a caterpillar can move rapidly enough so that the travel of color patches across the predator's visual field is faster than the flicker fusion frequency, then the color patches will blend together as they do in projected movie film (Endler 1978). In rapidly moving snakes, annulated coloring or vertical banding, even when bright and contrasting in color, may blend to match the background (Jackson et al. 1976; Pough 1976). When such prey are not moving quite fast enough for flicker fusion to occur in the predator, the prey may be perceived to be moving faster than they actual are (Diener et al. 1976), which may confuse the predator, especially if direction is erratic (Endler 1978).

For the most part, caterpillars move slowly. Birds have flicker fusion frequencies ranging from 10 to 100 cycles per second (Ali and Klyne 1985). Some insects, including wasps, have much higher flicker fusion rates (up to 250 cycles per second) than endothermic vertebrates and thus they can perceive the form of a prey item even when it is flying quickly (Mazokhin-Porshnyakov 1969). With decreasing light intensity, flicker fusion frequency declines (Ali and Klyne 1985). Therefore, caterpillars moving fast enough to exceed the flicker fusion frequency of a predator may only occur under certain conditions: (1) for invertebrate enemies with limited vision, (2) interactions between especially fast caterpillars (e.g., many arctiids, Carter and Hargreaves 1986; Stamp 1992) and some of their enemies, (3) under dim light, and (4) after encounter with enemies, when caterpillars respond by dropping off the host plant.

Olfactory Cues

Odors that direct a predator or parasitoid to a caterpillar are disadvantageous to the prey. But release of some chemicals may be unavoidable (e.g., those associated with mandibular secretions and frass, induced response by host plants) or may serve as pheromones (i.e., affecting conspecifics, such as feeding-site-recruiting pheromones, Fitzgerald, Chapter 11) and allomones (e.g., volatile secretions from osmeteria that deter some but not all enemies; Honda 1983; Chow and Tsai 1989; Leslie and Berenbaum 1990). Because birds have a poorly developed sense of smell (Welty and Baptista 1988), odor primarily serves as a cue for insect predators and parasitoids. Caterpillars may benefit by tossing frass away (Rawlins 1984; Usher 1984; Friedlander 1986; Lederhouse 1990) because they eliminate both visual and olfactory cues. However, for internal and shelter feeders, getting rid of frass (Mopper et al. 1984) may simply reduce microbial contamination of food. Also, without an enemy having the ability to detect minute quantities of specific chemicals, volatile cues are likely with increasing distance to be less reliable than visual cues. Therefore, limiting olfactory cues may be advantageous mainly against some invertebrate enemies, in particular parasitoids.

Audible Cues

As insect larvae eat and move around, they make sounds that may be audible to predators and parasitoids (and also to humans, Lawrence 1981; Sugimoto et al. 1988). Predators and parasitoids may use such sounds as cues to locate prey or hosts, especially ones that are concealed within plant tissue (e.g., Ryan and Rudinsky 1962; Lawrence 1981; Sugimoto et al. 1988). Because such sounds may be unavoidable, caterpillars may benefit by limiting activities that are relatively noisy to periods when parasitoids and predators, especially birds, are not foraging (e.g., at night), but that may increase their risk to nocturnal predators. With increasing distance, sound will be a less reliable cue to predators.

Tactile Cues

It is difficult to determine whether a predator's or parasitoid's response on contact with the prey/host reflects chemical or tactile sensitivities. Some predators seem unable to detect motionless prey on physical encounter (e.g., crab spiders with pine loopers *Bupalus piniarius*, den Boer 1971). Some coccinellid predators walk so quickly that they run over solitary prey apparently without detecting them (Frazer et al. 1981). Correspondingly, prey behavior may consist of playing dead when disturbed (e.g., as some mycophagous caterpillars do, Rawlins 1984).

Some Examples of Cryptic Life-Styles

Some organisms are usually unapparent even when predators are close, and when apparent they are not aposematic and do not mimic unpalatable organisms. It is these, with a largely cryptic life-style, that we will focus on in this section. Such caterpillars may or may not be palatable (e.g., Turner 1984; Heads and Lawton 1985), but usually they are. These caterpillars may be external feeders, living openly on host plants or within shelters they have constructed from silk and leaves. Other larvae may spend most or even all of their time within host plant tissue. Caterpillars that spend most of their time in refuges include gall makers, leaf miners, leaf rollers, and leaf tiers. Here we present some examples of cryptic life-styles.

Leaf Miners

Lepidopteran leaf miners are relatively small larvae that often spend their entire developmental period in one or sometimes a few leaves (Needham et al. 1928). The advantages of leaf mining may include avoiding unpalatable tissue within leaves (Hagen and Chabot 1986; Kimmerer and Potter 1987), protection from desiccation, and concealment from predators (Needham et al. 1928).

The major factors that affect predation rate of leaf miners are density of leaf miners, type of predator, and nearness to ant-tended homopterans. For example, less than 10% of the 6078 identifiable caterpillars gathered by ants in Canadian forests were leaf miners (McNeil et al. 1978). When leaf miner densities are high, considerable mortality may occur due to birds (30–62%, Owen 1975; 14–19%, Itamies and Ojanen 1977; 39%, Heads and Lawton 1983). These levels of predation are similar to those by ants on externally feeding larvae and leaf rollers (Ito and Higashi 1991a), and probably by some vespid wasps (Krombein 1967) and birds, too, since some specialize on leaf miners (Werner and Sherry 1987). But leaf characteristics (e.g., waxiness, prickles) may interfere with birds attacking leaf miners (Owen 1975). Furthermore, it may only be energetically feasible for birds to forage for leaf miners when 10% or more of the leaves are mined (Itamies and Ojanen 1977). Densities of leaf miners in eastern North America are usually less than 10%, which may mean that under those conditions invertebrate predators contribute more to leaf miner mortality than birds (Faeth 1980). However, parasitoids may be the most important enemy of leaf miners (Needham et al. 1928). In various studies the majority of leaf miners were killed by parasitoids (Askew 1968; Askew and Shaw 1979; Kato 1984). One leaf miner pattern, branching or crossing of mines within a leaf, may result in leading a parasitoid in the wrong direction long enough for it to give up searching that leaf (Kato 1984). Some studies have found little or no predation on leaf miners when they are on plants where ants are responding to extrafloral nectaries (Heads 1986; Heads and Lawton 1985) or honey-dew-producing homopterans (Fowler and MacGarvin 1985). Parasitism can also be less when ants are present (Sato and Higashi 1987). In the presence of aphid-tending ants, leaf miners fared better than leaf tiers and exposed feeders (Fowler and MacGarvin 1985).

Overall, leaf mining as a cryptic life-style seems to depend on erratic densities and on enclosed feeding and small final size, which yield few cues to predators. Use of a host plant with ant-tended homopterans may be advantageous at least when leaf miner densities are low.

Leaf Rollers and Leaf Tiers

Leaf rollers (using one leaf at a time) and leaf tiers (using several leaves at a time) tie leaves around themselves with silk and then feed from within that construction. Leaf rollers and leaf tiers usually construct more than one shelter over their developmental period. For example, a fern-feeding caterpillar (*Herpetogramma aeglealis*) builds on average 5 shelters of 3 distinct types (Ruehlmann et al. 1988). Shelter building and switching usually occur at night. The shelters of leaf rollers and leaf tiers often have at least two entries/exists, and larvae will drop out of one when the other end is probed (Stamp, personal observation). Usually the densities of leaf rollers and leaf tiers are low, but outbreaks do occur (Mumma and Zettle 1977).

The advantages of living within these shelters may include obtaining body temperatures above ambient during the day (Henson 1958a), and avoiding desiccation (Henson 1958b), direct sunlight (Berenbaum 1978; Sandberg and Berenbaum 1989), and enemies (Ruehlmann et al. 1988). Some vespid wasps prey heavily on leaf tiers and leaf rollers (Krombein 1967). Some birds specialize on curled leaves, which require more time and skill per leaf than gleaning and hovering (Gradwohl and Greenberg 1984; Greenberg 1987a,b). Parasitism accounts for much of the mortality of these caterpillars (19–32%, Mayer and Beirne 1974; 18–63%, Mumma and Zettle 1977). Of the identifiable caterpillars collected by ants, one study found that 71% were leaf rollers and leaf tiers (McNeil et al. 1978), but in another study, leaf tiers in the presence of ants fared better than exposed feeders (Fowler and MacGarvin 1985). Leafrolling allowed one pyralid caterpillar species (*Acrospila gastralis*) to escape ants patroling plants that reward them (Vasconcelos 1991).

Success as a leaf roller or leaf tier may depend primarily on (1) the shelter blending into the background with relatively little tell-tale leaf damage and frass, (2) tight construction of the shelter, (3) dispersing the evidence, by moving periodically and thus constructing several shelters over the developmental period, and (4) confusing or startling predators by quickly exiting shelters. Once again, sharing a host plant with an ant-tended homopteran or feeding on a plant protected by ants may also play a role provided the caterpillar shelter is invulnerable to ants.

Exposed Feeders

A large proportion of lepidopteran larvae feed externally. It is often their damage to plants that is the most conspicuous among the three categories discussed in this section. Such feeding damage may provide cues to natural enemies (Heinrich 1979; Heinrich and Collins 1983) and correspondingly exposed feeders may benefit by eliminating cues at feeding sites and being relatively mobile. Their color patterns vary from generally eucryptic (e.g., green foliage-feeding caterpillars) to masquerade (e.g., twig-mimicking caterpillars). Some are eucryptic at a distance and aposematic at close range (e.g., bright colored, striped monarch caterpillars). Of the exposed foliage-feeding caterpillar-like guild on striped maple (*Acer pennsylvanicum*), 48% of the species were eucryptic in general, 48% were masqueraders, and 4% conspicuous on the near scale (Marquis and Passoa 1989; R. Marquis, unpublished data). In terms of numbers of larvae, 88% were eucryptic in general, 10% were masqueraders and 2% were conspicuous. At night the larvae were almost always on leaves (top or bottom). But during the daytime, green larvae were on the underside or edges of leaves, twig mimics, regardless of color, were off leaves (on petioles and twigs), and the conspicuously colored larvae showed no preference.

When exposed feeders are not feeding and, in particular, when they are molting or in diapause, they may seek refuge by tying leaves together (Stamp 1984b), by hiding in bark or soil crevices, or by moving into the leaf litter (Esbjerg 1988; Weseloh 1989). Even cryptic hidden feeders may be found by predators. For example, only the diapausing codling moth larvae that were best concealed under bark flaps escaped predation by birds, which intensified their searching on finding prey and revisited the site throughout the winter (Solomon and Glen 1979).

Compared to leaf rollers, leaf tiers, and leaf miners, exposed feeders move around more and over the larval development are likely to travel greater distances (e.g., enclosed feeders, Ruehlmann et al. 1988, vs. exposed feeders, Dethier 1959a,b, 1988). Such mobility may allow exposed feeders to find high quality food, but whether they experience better foraging conditions by their mobility than enclosed feeders and without increased risk to predation is not clear. Exposed feeders may be quite susceptible to environmental extremes, such as cold-snaps and drought. Unfavorable environmental conditions may prolong molting (Ayres and MacLean 1987; Stamp 1990), which is probably a particularly vulnerable stage for caterpillars (Rabb and Lawson 1957; Morris 1972a), especially those that are not enclosed. However, exposed feeders may thermoregulate behaviorally and consequently spend more time near their optimal temperature for growth (Casey 1976; Knapp and Casey 1986). Exposed feeders may be especially vulnerable to natural enemies. Mortality of exposed feeders can be relatively high compared to enclosed feeders (Fowler and MacGarvin 1985). But exposed feeders may also respond to local densities of enemies, by short-term hiding and local movement (e.g., buckmoth *Hemileuca lucina* caterpillars, Stamp and Bowers 1988; sawfly *Croesus varus* larvae, Boeve 1991).

Consequences of a Cryptic Life-Style

Here we address the questions: How does crypsis constrain foraging patterns of caterpillars? Is there a cost? If so, how might the cost, for example, of eucrypsis compare to the cost of the foraging pattern of an aposematic species? Presumably a cryptic life-style may incur a cost to species that exhibit it. Cryptic caterpillars that forage only at times and in ways that reduce apparency may grow slowly compared to larvae that may forage wherever and whenever they choose because they are aposematic (Herrebout et al. 1963; Schultz 1983). Such slow growth for cryptic caterpillars may reflect the food they use, thermal conditions, and the need to avoid enemies.

In general, growing relatively quickly is probably advantageous for caterpillars (Scriber and Slansky 1981; Slansky and Scriber 1985). The only examples for which the cost of growing slowly appears worth bearing involve enclosed feeders (Clancy and Price 1987; Damman 1987). For instance, pyralid caterpillars *Omphalocera munroei* feed on mature leaves despite better growth on new leaves; the tied mature leaves are more effective in providing protection against enemies (Damman 1987).

Growth rate may be increased by thermoregulatory behavior, such that more time is spent at the caterpillar species' optimal temperature than is possible for a thermal conformer (Casey 1976; Knapp and Casey 1986; Casey, Chapter 1). The optimal temperature facilitates digestive and growth processes (Sherman and Watt 1973; Casey 1976; Knapp and Casey 1986). Thermoregulatory behavior and some degree of conspicuousness on the near scale often occur together, by the caterpillar choosing direct sunlight at least part of the time (Seymour 1974; Casey 1976; Knapp and Casey 1986; Mattson and Scriber 1987). In contrast, probably most eucryptic caterpillars and masqueraders are thermal conformers; the need to avoid being conspicuous to enemies is likely to override the advantages of thermoregulatory behavior.

Thus, the critical measure of the effectiveness of crypsis is not solely the cost, but the cost:benefit ratio. Theoretical work has demonstrated that for a juvenile animal to maximize its fitness, it must choose habitats that minimize the ratio between mortality (μ) and growth rate (g) for each size class (Werner and Gilliam 1984). Figure 9.2 shows six possible microhabitat choices available to a caterpillar and the predicted μ/g for each. If we assume that a microhabitat in which the caterpillar can be exposed on its host plant provides a thermal and nutritional optimum because the caterpillar can bask while feeding on the plant parts of its preference (e.g., young leaves), the ideal choice for the caterpillar will be microhabitat 1, which provides a high growth rate and low mortality (due to lack of enemies). By using microhabitat 1 as a reference point, the cost of predator avoidance by crypsis and aposematism can be assessed.

Aposematic (and, by that definition, unpalatable) caterpillars often do not seek shelters, at least in later instars, and they may feed on new foliage often at branch tips and bask. Therefore they tend to be quite exposed. These unpalatable caterpillars may be able to utilize microhabitat 2, even though enemies are present (Fig. 9.2). Comparison of the μ/g ratios for aposematic caterpillars in microhabitats 1 and 2 would indicate the relative costs of aposematism.

However, aposematic caterpillars may alter their behavior in the presence of predators that are undeterred by their defenses and thus potentially reduce their apparency to predators. A study of buckmoth caterpillars (*Hemileuca lucina*, which are black with urticating spines and thus aposematic) showed that they sought refuges in the interior of the host plant when predatory wasps were present (Stamp and Bowers 1988; corresponding to microhabitats 4 and 6 in Fig. 9.2). The differences in food quality and temperature between the original feeding sites

	Predator Absent	Predator Present
Exposed Sites	# 1 $\frac{\mu}{g} = \frac{++}{+++}$	# 2 $\frac{\mu}{g} = \frac{++++}{+++}$
Indirect Sunlight but on Leaves	# 3 $\frac{\mu}{g} = \frac{++}{++}$	# 4 $\frac{\mu}{g} = \frac{+++}{++}$
Off Leaves	# 5 $\frac{\mu}{g} = \frac{++}{+}$	# 6 $\frac{\mu}{g} = \frac{++}{+}$

Figure 9.2. Potential microhabitats that caterpillars might experience. μ = mortality rate and g = growth rate. A greater number of "+"s indicates a larger relative value. Microhabitat 1 offers access to high quality food and direct sunlight and therefore the possibility of ideal conditions for growth. Presumably aposematic caterpillars can exploit microhabitat 2 better than eucryptic larvae and thus green caterpillars are more common in microhabitats 3 and 4, and bark-colored caterpillars and twig mimics more common in microhabitats 5 and 6.

(analogous to microhabitat 1 in Fig. 9.2) and their refuges (microhabitats 4 and 6), in turn, reduced larval growth rate and survivorship (Stamp and Bowers 1988, 1990, 1991). Presumably, the predation on larvae feeding outside these refuges (corresponding to microhabitat 2 in Fig. 9.2) would have been high (e.g., perhaps as much as 100% mortality), though this was not examined empirically.

Compared to aposematic caterpillars, eucryptic green caterpillars and bird-dropping masqueraders are less exposed. They inhabit microhabitats 3 and 4, with and without enemies (Fig. 9.2). Presumably the growth rates of any particular eucryptic species are similar between these microhabitats. But the presence of predators may have an indirect effect on larval growth rate if larvae detect enemies and consequently move more often. For example, green sawfly larvae *Croesus varus* migrate from feeding sites when their leaves are disturbed (Boevé 1991);

with frequent interruption of their eating, developmental time may be prolonged. Comparison of the μ/g ratios for aposematic species and eucryptic species could indicate under what conditions the ratios are similar. For instance, if eucryptic caterpillars spend half their time in each of their microhabitats (3 and 4), they would have an average μ/g value of 1.25 (using the values in Fig. 9.2), which would be equivalent to aposematic caterpillars that spend equal time among microhabitats 1, 2, 4, and 6 (Fig. 9.2).

In contrast to green caterpillars and bird-dropping mimics, eucryptic nongreen caterpillars and twig masqueraders would be in microhabitats 3 and 4 at night to feed but spend more than half of their time (daytime) in microhabitats 5 and 6, off leaves. If they spend equal time among these microhabitats, the μ/g or cost:benefit ratio is more than twice that of the other caterpillar life-styles, but it may be that they are subject to much less predation than depicted in Figure 9.2. In a 24-hour test, green *Amphion floridensis* caterpillars gained more weight than pink larvae, which was due to the green larvae eating more than pink caterpillars (which rest away from the leaves in the daytime), and in another outdoor experiment, temperatures of these morphs did not differ (Fink 1989). Since wild females produce both green and pink offspring (Fink 1989), there must be some situations that favor pink larvae, whereby the mortality of pink larvae (away from feeding sites during the daytime, i.e., microhabitat 6) is much less than that of green caterpillars (microhabitat 4), presumably due to differential effects of predators.

Microhabitat use may be complicated by the effect of temperature on pathogens and parasites of caterpillars. High temperature enables some insects to resist viral infection (Tanada 1967), and silkworms (*Bombyx mori*) can overcome a flacherie virus if they are warm enough (37°C, Inoue and Tanada 1977). Although caterpillars have not been tested in a thermal gradient, other invertebrates chose a higher ambient temperature when they were infected and consequently exhibited higher survivorship and fecundity than infected animals kept at the lower temperatures chosen by uninfected animals (Bronstein and Conner 1984; Louis et al. 1986; Boorstein and Ewald 1987; McClain et al. 1988). So infected eucryptic and masquerading caterpillars may benefit by seeking exposed sites (microhabitats 1 and 2 in Fig. 9.2) for thermal therapy. However, the reduction in mortality due to pathogens may be offset by increased mortality due to other enemies (microhabitat 2).

Presumably eucrypsis, masquerade, and aposematism reduce the μ/g ratio in the presence of enemies compared to that of noncryptic palatable species, but is one of these three defensive mechanisms more effective than the others at minimizing the ratio? One way to investigate the costs and benefits of various degrees of crypsis would be to compare a guild of foliage-feeding caterpillars concurrently sharing a host plant in terms of their life-style characteristics (such as exposed versus enclosed feeders) and categories of crypsis (eucrypsis, masquerade, and conspicuous on the near scale) with their performance (e.g., relative consumption rate, relative growth rate, developmental time, and final size) and survivorship in the presence and absence of enemies. Similar analysis of a taxonomic group in which there were contrasting life-styles would be useful.

Croesus varus and *C. septentrionalis* sawfly larvae, which are cryptic and aposematic, respectively, may serve as an example. They both have eversible ventroabdominal glands with secretions that deter ants and they are moderately distasteful to birds (great tits), with the aposematic species somewhat more noxious than the cryptic species (Boevé and Pasteels 1985; Boevé 1991). These sawfly species often occupy the same host plant and use it in similar ways (Boevé 1991; J.-L. Boevé, personal communication). But there are some striking differences in behavior. When a branch or leaf is disturbed, the cryptic species leave their feeding site and eventually settle at different sites on the plant without reaggregating, whereas the aposematic species show no such response (Boevé 1991). Although the cryptic species takes fewer days to complete larval development than the aposematic species (14 vs. 18 days; J.-L. Boevé, unpublished data), the aposematic species achieves a larger size (Boevé 1991). Our estimates of relative growth rate of the cryptic species are larger than those of the aposematic species (0.136 vs. 0.128 mm/day using head–capsule width as an indicator of growth and 1.64 vs. 1.39 mm/day using body length; from Fig. 3 in Boevé 1991). Some data suggest that the cryptic species is more susceptible to predators and parasitoids than the aposematic species (Boevé 1991). Thus, the μ/g ratios of these two species may be similar.

Much to our dismay, we have found no published data for any set of species for which we could provide a more quantitative analysis here. Comparison of such species that are related, concurrently using the same host plant and exhibiting similar chemical defense would contribute significantly to increasing our understanding of the factors affecting the cost:benefit ratio of eucrypsis versus aposematism. Any such comparison should simulate environmental conditions (e.g., for a summer guild, maintenance of mean day and night summer temperatures, summer photoperiod, and host plant material of the appropriate age).

Despite the lack of explicit examination, by way of empirical and comparative studies, of the costs and benefits of various degrees of crypsis in caterpillars, we can make some predictions. The null hypothesis is that the cost:benefit ratio for one life-style (e.g., eucryptic in general) equals that of another (e.g., aposematic), and therefore that such contrasts simply reflect different ways to achieve the same end. For instance, the comparison of the eucryptic and aposematic *Croesus* species discussed above supports that hypothesis. However, there is some evidence to suggest that cost:benefit ratios may vary significantly. For some species, the consequences of crypsis are slow growth compared to more apparent species (Herrebout et al. 1963; Schultz 1983) and, if that is coupled with greater mortality, the μ/g ratio of those cryptic species may be greater than that of aposematic species using the same host plant.

In sum, crypsis can have a cost and there are potential conditions in which it may be difficult for cryptic caterpillars to minimize the cost:benefit ratio. Under what conditions one defensive mode is more effective than another in reducing the μ/g ratio is unknown. A methodical examination of related caterpillars or a guild of caterpillars utilizing the same host plant species could demonstrate the relative costs and benefits of eucrypsis vs. aposematism and of eucrypsis vs. masquerade.

Conclusions and Future Directions

Here we suggest some different approaches for investigating the ecology and, by inference, the evolution of crypsis in caterpillars.

Examining Predator and Prey One-on-One

Cryptic characters for a caterpillar consist mainly of behavior, shape, and coloration. Endler (1986) outlines two ways to test the hypothesis that a given trait functions as a defense against a predator: (1) tests of differential survival of individuals with and without the trait (e.g., painting prey a different color, removing spines, comparing individuals that differ in expression of a trait, altering the background), and (2) predictions about the details of the defense (e.g., a change from prey immobility to dropping off the host plant on the basis of the predator's speed, size, and danger, or relative to temperature, degree of crypsis, previous experience, and stage of prey). The former method should be conducted along with the latter because only the former method demonstrates that the mechanism actually works (Endler 1986). Although a number of studies have used the first method (e.g., den Boer 1971; Mariath 1982), none has taken the second approach to investigate antipredator behavior in caterpillars, which would indicate the conditions that favor a particular trait.

Some behaviors that appear to lead to predator avoidance, such as moving from feeding sites in tree foliage at dawn to spend the daylight hours in crevices in bark, may be fixed genetically and thus Endler's method 1 is appropriate. In this case, for example, tethering caterpillars in the various microhabitats, as Weseloh (1988) did with gypsy moth caterpillars, would indicate their susceptibility to predators among the microhabitats.

However, caterpillars often exhibit some degree of flexibility in their behavior. When populations are small, later instar gypsy moth larvae rest during the day in bark crevices and leaf litter but stay in the canopy when populations are large (Weseloh, Chapter 6). Early instar cutworms (*Argotis segetum*) rest in dry soil between feeding bouts but stay on the plant when soil is wet or when they cannot burrow into the substrate (Esbjerg 1988, 1990). Buckmoth caterpillars (*Hemileuca lucina*) abandon basking and feeding on new leaves to move into the interior of the host plants when their aggregations are attacked by predatory wasps (Stamp and Bowers 1988). When caterpillars exhibit some flexibility in behavior and distribution among microhabitats relative to the presence and absence of predators, Endler's method 2, in addition to 1, would be especially informative.

Main's (1987) procedure to evaluate the mechanisms of predator avoidance is basically a combination of these two methods. It entails (1) examining the microhabitat distribution and behavioral repertoire of the prey in the absence of the predator, (2) determining whether the prey modified their behavior or distribution in the presence of the predator, (3) assessing whether behavior and distribution in response to the predator changed with prey size, (4) determining whether behavioral and spatial alterations in response to the predator reduced predation, and (5) assessing whether the prey color and pattern limited predation. Although Main's (1987) system involved pinfish preying on shrimp inhabiting seagrass, it may provide a particularly instructive example to those who wish to examine crypsis in insect herbivores.

Tackling a Caterpillar's Array of Antipredator Defenses

Crypsis, as the primary line of defense, and escape that makes the prey unapparent are just part of the multicomponent defensive repertoire of prey. In some cases, it may be possible to identify the likely selective agent for a particular trait. More often the multiple antipredator traits of prey are likely to reflect the combination of historical factors, evolutionary time lags, differential genetic lability, and ontogenetic carryover (Pearson 1989). The possibilities of multiplicative or synergistic effects and factors that might be counterselective (e.g., characters associated with thermoregulation) further complicate attempts to correlate selective agents with antipredator traits (Pearson 1989).

Pearson (1989) outlines a program for investigating the role of multiple antipredator traits, which includes (1) determining *all* the potential predators, (2) identifying the rate of susceptibility of *all* stages of the prey, (3) validating the target of each antipredator trait, (4) determining the reduction of predation due to each trait, and (5) ascertaining alternative functions of the antipredator traits. We have been unable to find any study of cryptic traits in caterpillars that has addressed each of Pearson's recommendations. Most studies, but only to a limited extent, have dealt with points 2–4 by, for example, assessing susceptibility of prey on natural versus unnatural backgrounds to the most likely type of predator for a particular trait (e.g., an insectivorous bird's response to twig-mimic caterpillars, Mariath 1982). However, if we are to understand the ecology and evolution of antipredator characters and, in particular, of crypsis, we need studies that methodically examine predator-prey interactions as indicated above.

Conclusions

All caterpillars are cryptic to some degree. The average distance between a caterpillar and its enemies is critical to the caterpillar's unapparency; that average distance varies considerably due to characteristics of the enemy and the caterpillar, as well as to environmental conditions. For most caterpillars (i.e., the nonmasqueraders), average background is also quite important and is mainly a function of caterpillar behavior and environmental conditions. The set of traits that facilitates being unapparent to enemies presumably reflects a compromise between avoiding a diverse array of parasitoids and invertebrate and vertebrate predators and obtaining sufficient and appropriate food under fluctuating environmental conditions. How different life-styles compare in terms of the cost:benefit ratio (e.g., mortality rate:growth rate ratio) is unknown. Rigorous examination of the costs and benefits of various kinds of crypsis in caterpillars, as we have outlined, would significantly advance, from a field full of ideas and anecdotal information and so lacking in data, our understanding of crypsis, the major antipredator mechanism of insect herbivores.

Acknowledgments

We thank D. Bowers, H. Damman, L. Fink, C. Montllor, and R. Weseloh for their insightful comments on this manuscript and J.-L. Boevé and R. Marquis for providing unpublished data. NES was supported by NSF Grant BSR-8906259.

Literature Cited

Adlung, K.G. 1966. A critical evaluation of the European research on the use of red woodants (*Formica rufa* group) for the protection of forests against harmful insects. Zeit. Ange. Entomol. 57:167–189.

Alcock, J. 1973. Cues used in searching for food by red-winged blackbirds (*Agelaius phoeniceus*). Behaviour 46:174–188.

Ali, M.A., and Klyne, M.A. 1985. Vision in Vertebrates. Plenum Press, New York.

Allen, D.C., Knight, F.B., and Foltz, J.L. 1970. Invertebrate predators of the jackpine budworm, *Choristoneura pinus*, in Michigan. Ann. Entomol. Soc. Am. 63:59–64.

Askew, R.R. 1968. A survey of leaf-miners and their parasites on *Laburnum*. Trans. R. Entomol. Soc. London 120:1–37.

Askew, R.R., and Shaw, M.R. 1979. Mortality factors affecting the leaf-mining stages of *Phyllonorycter* (Lepidoptera: Gracillariidae) on oak and birch. 1. Analysis of the mortality factors. Zool. J. Linn. Soc. 67:31–49.

Awan, M.S. 1985. Anti-predator ploys of *Heliothis punctiger* (Lepidoptera: Noctuidae) caterpillars against the predator *Oechalia schellenbergii* (Hemiptera: Pentatomidae). Aust. J. Zool. 33:885–890.

Awan, M.S., Wilson, L.T., and Hoffman, M.P. 1989. Prey location by *Oechalia schellembergii*. Entomol. Exp. Appl. 51:225–231.

Ayres, M.P., and Maclean, S.F., Jr. 1987. Molt as a component of insect development:

Galerucella sagittariae (Chrysomelidae) and *Epirrita autumnata* (Geometridae). Oikos 48:273–279.

Baker, R.R. 1970. Bird predation as a selective pressure on the immature stages of the cabbage butterflies, *Pieris rapae* and *P. brassicae*. J. Zool. London 162:43–59.

Banks, C.J. 1957. The behaviour of individual coccinellid larvae on plants. Br. J. Anim. Behav. 5:12–24.

Berenbaum, M. 1978. Toxicity of a furanocoumarin to armyworms: A case of biosynthetic escape from insect herbivores. Science 201:532–534.

Bernays, E. and Graham, M. 1988. On the evolution of host specificity in phytophagous arthropods. Ecology 69:886–892.

Blois-Heulin, C., Crowley, P.H., Arrington, M., and Johnson, D.M. 1990. Direct and indirect effects of predators on the dominant invertebrates of two freshwater littoral communities. Oecologia 84:295–306.

Boer, M.H. den 1971. A colour-polymorphism in caterpillars of *Bupalus piniarius* (L.) (Lepidoptera: Geometridae). Neth. J. Zool. 21:61–116.

Boevé, J.-L. 1991. Gregariousness, field distribution and defence in the sawfly larvae *Croesus varus* and *C. septentrionalis* (Hymenoptera, Tenthredinidae). Oecologia 85:440–446.

Boevé, J.-L., and Pasteels, J.M. 1985. Modes of defense in nematine sawfly larvae: efficiency against ants and birds. J. Chem. Ecol. 11:1019–1036.

Brian, M.V. 1955. Food collection by a Scottish ant community. J. Anim. Ecol. 24:336–351.

Bronstein, S.M., and Conner W.E. 1984. Endotoxin-induced behavioural fever in the Madagascar cockroach, *Gromphadorhina portentosa*. J. Insect Physiol. 30:327–330.

Boorstein, S.M., and Ewald, P.W. 1987. Costs and benefits of behavioral fever in *Melanoplus sanguinipes* infected by *Nosema acridophagus*. Physiol. Zool. 60:586–595.

Buckner, C.H., and Turnock, W.J. 1965. Avian predation on the larch sawfly, *Pristiphora erichsonii* (Htg.), (Hymenoptera: Tenthredinidae). Ecology 46:223–236.

Byers, J.R. 1975. Tyndall blue and surface white of tent caterpillars, *Malacosoma* spp. J. Insect Physiol. 21:401–415.

Campbell, R.W., and Torgersen, T.R. 1983. Effect of branch height on predation of western spruce budworm (Lepidoptera: Tortricidae) pupae by birds and ants. Environ. Entomol. 12:697–699.

Carter, D.J., and Hargreaves, B. 1986. A Field Guide to Caterpillars of Butterflies and Moths in Britain and Europe. Collins, London.

Carter, M.C., Sutherland, D., and Dixon, A.F.G. 1984. Plant structure and the searching efficiency of coccinellid larvae Oecologia 63:394–397.

Casey, T.M. 1976. Activity patterns, body temperature and thermal ecology in two desert caterpillars (Lepidoptera: Sphingidae). Ecology 57:485–497.

Chapman, R.F. 1982. The Insects: Structure and Function, 3rd. ed. Harvard University Press, Cambridge.

Charnov, E. 1976. Optimal foraging: An attack strategy of a mantid. Am. Nat. 110:141–151.

Chazdon, R.L. 1986. Light variation and carbon gain in rain forest understorey palms. J. Ecology 74:995–1012.

Chow, Y.S., and Tsai, R.S. 1989. Protective chemicals in caterpillar survival. Experientia 45:390–392.

Clancy, K.M., and Price, P.W. 1987. Rapid herbivore growth enhances enemy attack: Sublethal plant defenses remain a paradox. Ecology 68:733–737.

Clarke, C.A., Dickson, C.G.C., and Sheppard, P.M. 1963. Larval color pattern in *Papilio demodocus*. Evolution 17:130–137.

Coppel, H.C., and Mertins, J.W. 1977. Biological Insect Pest Suppression. Springer-Verlag, Berlin.

Cornell, J.C., Stamp, N.E., and Bowers, M.D. 1987. Developmental change in aggregation, defense and escape behavior of buckmoth caterpillars, *Hemileuca lucina* (Saturniidae). Behav. Ecol. Sociobiol. 20:383–388.

Cott, H.B. 1940. Adaptive Coloration in Animals. Methuen, London.

Curio, E. 1970. Validity of the selective coefficient of a behaviour trait in hawkmoth larvae. Nature (London) 228:382.

Damman, H. 1987. Leaf quality and enemy avoidance by the larvae of a pyralid moth. Ecology 68:88–97.

DeBach, P. 1974. Biological control by Natural Enemies. Cambridge University Press, London.

Dethier, V.G. 1959a. Egg-laying habits of Lepidoptera in relation to available food. Can. Entomol. 91:554–561.

Dethier, V.G. 1959b. Food-plant distribution and density and larval dispersal as factors affecting insect populations. Can. Entomol. 91:581–596.

Dethier, V.G. 1988. The feeding behavior of a polyphagous caterpillar (*Diacrisia virginica*) in its natural habitat. Can. J. Zool. 66:1280–1288.

Diener, H.C., Wist, E.R., Dichgans, J., and Brant, T. 1976. The spatial frequency effect on perceived velocity. Vis. Res. 16:169–176.

Dixon, A.F.G. 1959. An experimental study of the searching behaviour of the predatory coccinellid beetle *Adalia decempunctata* (L.). J. Anim. Ecol. 28:259–281.

Dmoch, J., Lewis, W.J., Martin, P.B., and Nordlund, D.A. 1985. Role of host-produced stimuli and learning in host selection behavior of *Cotesia* (=*Apanteles*) *marginiventris* (Cresson). J. Chem. Ecol. 11:453–463.

Du Merle, P. 1988. Phenological resistance of oaks to the green oak leafroller, *Tortrix viridana* (Lepidoptera: Tortricidae), pp. 215–226. In W.J. Mattson, J. Levieux, and C. Bernard-Vagan (eds.), Mechanisms of Woody Plant Defense against Insects. Search for Pattern. Springer-Verlag, Berlin.

Eckhardt, R.C. 1979. The adaptive syndromes of two guilds of insectivorous birds in the Colorado Rocky mountains. Ecol. Monogr. 49:129–149.

Edmunds, M. 1974. Defence in Animals: A Survey of Anti-predator Defences. Longman, Essex.

Endler, J.A. 1978. A predator's view of animal color patterns. Evol. Biol. 11:319–364.

Endler, J.A. 1981. An overview of the relationships between mimicry and crypsis. Biol. J. Linn. Soc. 16:25–31.

Endler, J.A. 1984. Progressive background matching in moths, and a quantitative measure of crypsis. Biol. J. Linn. Soc. 22:187–231.

Endler, J.A. 1986. Defense against predators, pp. 109–134. In M.E. Feder and G.V. Lauder (eds.), Predator-Prey Relationships: Perspectives and Approaches from the Study of Lower Vertebrates. University of Chicago Press, Chicago.

Erichsen, J.T., Krebs, J.R., and Houston, A.I. 1980. Optimal foraging and cryptic prey. J. Anim. Ecol. 49:271–276.

Esbjerg, P. 1988. Behaviour of 1st- and 2nd-instar cutworms (*Agrotis segetum*) (Lep., Noctuidae): The influence of soil moisture. J. Appl. Entomol. 105:295–302.

Esbjerg, P. 1990. The significance of shelter for young cutworms (*Agrotis segetum*). Entomol. Exp. Appl. 54:97–100.

Evans, D.L. 1986. Anti-predatory autecology in the geometrid larvae of *Larentia clavaria pallidata*. Entomol. Exp. Appl. 40:209–214.

Evans, E.W. 1983. Niche relations of predatory stinkbugs (*Podisus* spp., Pentatomidae) attacking tent caterpillars (*Malacosoma americanum*, Lasiocampidae). Am. Midl. Nat. 109:316–323.

Faeth, S.H. 1980. Invertebrate predation of leaf-miners at low densities. Ecol. Entomol. 5:111–114.

Faeth, S.H. 1990. Structural damage to oak leaves alters natural enemy attack on a leafminer. Entomol. Exp. Appl. 57:57–63.

Fields, P.G., and McNeil, J.N. 1988. The importance of seasonal variation in hair coloration for thermoregulation of *Ctenucha virginica* larvae (Lepidoptera: Arctiidae). Physiol. Entomol. 13:165–175.

Fink, L.S. 1989. Color polymorphism in sphingid caterpillars (Lepidoptera: Sphingidae). Ph.D. dissertation. University of Florida, Gainesville.

Finnegan, R.J. 1974. Ants as predators of forest pests. Entomophaga (Mem. Hors. Ser.) 7:53–59.

Fitzgerald, T.D., Casey T. and Joos, B. 1988. Daily foraging schedule of field colonies of the eastern tent caterpillar *Malacosoma americanum*. Oecologia 76:574–578.

Forster, L.M. 1982. Vision and prey-catching strategies in jumping spiders. Am. Sci. 70:165–175.

Fowler, S.V., and MacGarvin, M. 1985. The impact of hairy wood ants, *Formica lugubris*, on the guild structure of herbivorous insects on birch, *Betula pubescens*. J. Anim. Ecol. 54:847–855.

Fox, L.R., and Morrow, P.A. 1981. Specialization: Species property or local phenomenon? Science 211:887–893.

Frazer, B.D., Gilbert, N., Ives, P.M., and Raworth, D.A. 1981. Predation of aphids by coccinellid larvae. Can. Entomol. 113:1043–1046.

Friedlander, T. 1986. Taxonomy, phylogeny, and biogeography of *Asterocampa* Rober 1916 (Lepidoptera, Nymphalidae, Apaturinae). J. Res. Lepid. 25:215–338.

Fuzeau-Braesch, S. 1985. Colour changes, pp. 549–58. In G.A. Kerkut and L.I. Gilbert (eds.), Comprehensive Insect Physiology, Biochemistry and Pharmacology, Vol. 9, Behaviour. Pergamon, Oxford.

Gendron, R.P., and Staddon, J.E.R. 1983. Searching for cryptic prey: The effect of search rate. Am. Nat. 121:172–186.

Goss-Custard, J.D. 1977. Optimal foraging and the size selection of worms by redshank, *Tringa totanus*, in the field. Anim. Behav. 25:10–29.

Gould, W.P., and Jeanne, R.L. 1984. *Polistes* wasps (Hymenoptera: Vespidae) as control agents for lepidopteran cabbage pests. Environ. Entomol. 13:150–156.

Graber, J.W., and Graber, R.R. 1983. Feeding rates of warblers in spring. Condor 85:139–150.

Gradwohl, J.A., and Greenberg, R. 1984. Search behavior of the checker-throated antwren foraging in aerial leaf litter. Beh. Ecol. Sociobiol. 15:281–285.

Grayson, J., and Edmunds, M. 1989. The causes of colour and colour change in caterpillars of the popular and eyed hawkmoths (*Laothoe populi* and *Smerinthus ocellata*). Biol. J. Linn. Soc. 37:263–279.

Greenberg, R. 1983. The role of neophobia in determining the degree of foraging specialization in some migrant warblers. Am. Nat. 122:444–453.

Greenberg, R. 1987a. Development of dead leaf foraging in a tropical migrant warbler. Ecology 68:130–141.

Greenberg, R. 1987b. Seasonal foraging specialization in the worm-eating warbler. Condor 89:158–168.

Greenberg, R., and Gradwohl, J. 1980. Leaf surface specializations of birds and arthropods in a Panamanian forest. Oecologia 46:115–124.

Greene, E. 1989. A diet-induced developmental polymorphism in a caterpillar. Science 243:643–646.

Hagen, R.H. and Chabot, J.F. 1986. Leaf anatomy of maples (*Acer*) and host use by Lepidoptera larvae. Oikos 47:335–345.

Harcourt, D.G. 1966. Major factors in survival of the immature stages of *Pieris rapae* L. Can. Entomol. 98:653–662.

Hassell, M.P. 1968. The behavioral response of a tachinid fly [*Cyzenis albicans* (Fall.)] to its host, the winter moth [*Operophtera brumata* (L.)]. J. Anim. Ecol. 37:627–639.

Hawkins B.A., Askew, R.R., and Shaw M.R. 1990. Influences of host feeding-niche and foodplant type on generalist and specialist parasitoids. Ecol. Entomol. 15:275–280.

Heads, P.A. 1986. Bracken, ants and extrafloral nectaries. IV. Do wood ants (*Formica lugubris*) protect the plant against insect herbivores? J. Anim. Ecol. 55:795–809.

Heads, P.A., and Lawton, J.H. 1983. Studies on the natural enemy complex of the holly leaf-miner: The effects of scale on the detection of aggregative responses and the implications for biological control. Oikos 40:267–276.

Heads, P.A., and Lawton, J.H. 1985. Bracken, ants and extrafloral nectaries. III. How insect herbivores avoid ant predation. Ecol. Entomol. 10:29–42.

Heatwole, H. 1968. Relationship of escape behavior and camouflage in anoline lizards. Copeia *1968*:109–113.

Henson, W.R. 1958a. The effects of radiation on the habitat temperatures of some poplar-inhabiting insects. Can. J. Zool. 36:463–478.

Henson, W.R. 1958b. Some ecological implications of the leaf-rolling habit in *Compsolechia niveopulvella* Chamb. Can. J. Zool. 36:809–818.

Heinrich, B. 1979. Foraging strategies of caterpillars: leaf damage and possible predator avoidance strategies. Oecologia 42:325–337.

Heinrich, B., and Collins, S. 1983. Caterpillar leaf damage and the game of hide-and-seek with birds. Ecology 64:592–602.

Hendry, L.B., Greany, P.D., and Gill, R.J. 1973. Kaironome mediated host-finding behavior in the parasitic wasp *Orgilus lepidus*. Entomol. Exp. Appl. 16:471–477.

Herrebout, W.M., Kuyten, P.J., and de Ruiter, L. 1963. Observations on colour patterns and behaviour of caterpillars feeding on Scots pine. Arch. Neerl. Zool. 15:315–357.

Hertz, P.E., Huey, R.B., and Nevo, E. 1982. Fight versus flight: Body temperature influences defensive responses of lizards. Anim. Behav. 30:676–679.

Herzog, H.A., Jr., and Burghardt, G.M. 1974. Prey movement and predatory behavior of juvenile western yellow-bellied racers, *Coluber constrictor mormon*. Herpetologica 30:285–289.

Holmes, R.T. 1990. Ecological and evolutionary impacts of bird predation on forest insects: an overview. Studies Avian Biol. 13:6–13.

Holmes, R.T., and Robinson, S.K. 1981. Tree species preferences of foraging insectivorous birds in a northern hardwoods forest. Oecologia 48:31–35.

Holmes, R.T., and Schultz, J.C. 1988. Food availability for forest birds: effects of prey distribution and abundance on bird foraging. Can. J. Zool. 66:720–728.

Holmes, R.T., Schultz, J.C., and Nothnagle, P. 1979. Bird predation on forest insects: an exclosure experiment. Science 206:462–463.

Holmes, R.T., Sherry, T.W., and Sturges, F.W. 1986. Bird community dynamics in a temperate deciduous forest: long-term trends at Hubbard Brook. Ecol. Monogr. 56:201–220.

Honda, K. 1983. Defensive potential of components of the larval osmeterial secretion of papilionid butterflies against ants. Physiol. Entomol. 8:173–179.

Horn, H.S. 1971. Adaptive Geometry of Trees. Princeton University Press, Princeton.

Hulspas-Jordaan, P.M., and van Lenteren, J.C. 1978. The relationship between host-plant leaf structure and parasitization efficiency of the parasitic wasp *Encarsia formosa* Gahan (Hymenoptera: Aphelinidae). Med. Fac. Landbouww. Rijksuniv. Gent. 43:431–440.

Hunter, M.D. 1990. Differential susceptibility to variable plant phenology and its role in competition between two insect herbivores on oak. Ecol. Entomol. 15:401–408.

Hunter, M.L. 1980. Microhabitat selection for singing and other behaviour in great tits, *Parus major*: Some visual and acoustical considerations. Anim. Behav. 28:468–475.

Hutchinson, B.A., and Matt, D.R. 1977. The distribution of solar radiation within a deciduous forest. Ecol. Monogr. 47:185–207.

Imura. O. 1982. Studies on the colour variation in larvae of *Ephestia kuhniella* Zeller (Lepidoptera: Phycitidae) II. Effect of environmental factors on larval pigmentation. Appl. Entomol. Zool. 17:52–59.

Inoue, H., and Tanada, Y. 1977. Thermal therapy of the flacherie virus disease in the silkworm, *Bombyx mori*. J. Invert. Pathol. 29:63–68.

Itamies, J., and Ojanen, M. 1977. Autumn predation of *Parus major* and *P. montanus* upon two species of *Lithocolletis* (Lepidoptera, Lithocolletidae). Ann. Zool. Fennici 14:235–241.

Ito, F., and Higashi, S. 1991a. Variance of ant effects on the different life forms of moth caterpillars. J. Anim. Ecol. 60:327–334.

Ito, F., and Higashi, S. 1991b. An indirect mutualism between oaks and wood ants via aphids. J. Anim. Ecol. 60:463–470.

Janzen, D.H. 1980. Heterogeneity of potential food abundance for tropical small land birds, pp. 545–552. In A. Keast and E.S. Morton (eds.), Migrant Birds in the Neotropics: Ecology, Behavior, Distribution, and Conservation. Smithsonian Inst. Press, Washington, D.C.

Janzen, D.H. 1988. Ecological characterization of a Costa Rican dry forest caterpillar fauna. Biotropica 20:120–135.

Jackson, J.A. 1979. Tree surfaces as foraging substrates for insectivorous birds, pp. 69–93. In J.G. Dickson, R.N. Conner, R.R. Fleet, J.C. Kroll, and J.A. Jackson (eds.), The Role of Insectivorous Birds in Forest Ecosystems. Academic Press, New York.

Jackson, J.F., Ingram, W. III, and Campbell, H.W. 1976. The dorsal pigmentation of snakes as an anti-predator strategy: A multivariate approach. Am. Nat. 110:1029–1053.

Jarvi, T., Sillen-Tullberg, B., and Wiklund, C. 1981. The cost of being aposematic. An experimental study of predation on larvae of *Papilio machaon* by the great tit *Parus major*. Oikos 36:267–272.

Johnson, J.W. 1984. The immature stages of six California *Catocala* (Lepidoptera: Noctuidae). J. Res. Lepid. 23:303–327.

Kato, M. 1984. Mining pattern of the honeysuckle leaf-miner *Phytomyza lonicerae*. Res. Popul. Ecol. 26:84–96.

Kimmerer, T.W., and Potter, D.A. 1987. Nutritional quality of specific leaf tissues and selective feeding by a specialist leafminer. Oecologia 71:548–551.

Knapp, R., and Casey, T.M. 1986. Thermal ecology, behavior, and growth of gypsy moth and eastern tent caterpillars. Ecology 67:598–608.

Koptur, S. 1985. Alternative defenses against herbivores in *Inga* (Fabaceae: Mimosoideae) over an elevational gradient. Ecology 66:1639–1650.

Krombein, K.V. 1967. Trap-Nesting Wasps and Bees: Life Histories, Nests, and Associates. Smithsonian Press, Washington, D.C.

Laine, K.J., and Niemela, P. 1980. The influence of ants on the survival of mountain birches during an *Oporinia autumnata* (Lep., Geometridae) outbreak. Oecologia 47:39–42.

Lawrence, P.O. 1981. Host vibration—a cue to host location by the parasite, *Biosteres longicaudatus*. Oecologia 48:249–251.

Lederhouse, R.C. 1990. Avoiding the hunt: primary defenses of lepidopteran caterpillars, pp. 175–189. In D.L. Evans and J.O. Schmidt (eds.), Insect Defense: Adaptive Mechanisms and Strategies of Prey and Predators. State University of New York Press, Albany.

Leonard, D.E. 1970. Feeding rhythm of the gypsy moth. J. Econ. Entomol. 63:1454–1457.

Leslie, A.J., and Berenbaum, M.R. 1990. Role of the osmeterial gland in swallowtail larvae (Papilionidae) in defense against an avian predator. J. Lepid. Soc. 44:245–251.

Letourneau, D.K. 1990. Mechanisms of predator accumulation in a mixed crop system. Ecol. Entomol. 15:63–69.

Lopez, J.D.Jr., Ridgway, R.L., and Pinnell, R.E. 1976. Comparative efficacy of four insect predators of the bollworm and tobacco budworm. Environ. Entomol. 5:1160–1164.

Louis, C., Jourdan, M., and Cabanac, M. 1986. Behavioral fever and therapy in a rickettsia-infected Orthoptera. Am. J. Physiol. 250:R991–R995.

Main, K.L. 1987. Predator avoidance in seagrass meadows: Prey behavior, microhabitat selection, and cryptic coloration. Ecology 68:170–180.

Mariath, H.A. 1982. Experiments on the selection against different colour morphs of a twig caterpillar by insectivorous birds. Zeit. Tierpsychol. 60:135–145.

Marquis, R.J., and Passoa, S. 1989. Seasonal diversity and abundance of the herbivore fauna of striped maple *Acer pennsylvanicum* L. (Aceraceae) in western Virginia. Am. Midl. Nat. 122:313–320.

Marston, N.L., Schmidt, G.T., Biever, K.D., and Dickerson, W.A. 1978. Reaction of live species of soybean caterpillars to attack by the predator, *Podisus maculiventris*. Environ. Entomol. 7:53–56.

Martinat, P.J. 1987. The role of climatic variation and weather in forest insect outbreaks, pp. 241–268. In P. Barbosa and J.C. Schultz (eds.), Insect Outbreaks. Academic Press, New York.

Matthews, R.W., and Matthews, J.R. 1978. Insect Behavior. Wiley, New York.

Mattson, W.J., and Scriber, J.M. 1987. Nutritional ecology of insect folivores of woody plants: nitrogen, water, fiber, and mineral considerations, pp. 105–146. In F. Slansky, Jr. and J.G. Rodriguez (eds.), Nutritional Ecology of Insects, Mites, Spiders and Related Invertebrates. Wiley, New York.

Mayer, D.F. and Beirne, B.P. 1974. Aspects of the ecology of apple leaf rollers (Lepidoptera: Tortricidae) in the Okanagan Valley, British Columbia. Can. Entomol. 106:349–352.

Mazokhin-Porshnyakov, G.A. 1969. Insect Vision. Plenum, New York.

McClain, E., Magnuson, P., and Warner, S.J. 1988. Behavioural fever in a Namib Desert tenebrionid beetle, *Onymacris plana*. J. Insect Physiol. 34:279–284.

McCorkle, D.V., and Hammond, P.C. 1988. Biology of *Speyeria zerene hippolyta* (Nymphalidae) in a marine-modified environment. J. Lepid. Soc. 42:184–195.

McLain, K. 1979. Terrestrial trail-following by three species of predatory stink bugs. Florida Entomol. 62:152–153.

McNeil, J.N., Delisle, J., and Finnegan, R.J. 1978. Seasonal predatory activity of the introduced red wood ant, *Formica lugubris* (Hymenoptera: Formicidae) at Valcartier, Quebec, in 1976. Can. Entomol. 110:85–90.

Meijden, E. van der. 1979. Herbivore exploitation of a fugitive plant species: Local survival and extinction of the cinnabar moth and ragwort in a heterogeneous environment. Oecologia 42:307–404.

Minnich, D.E. 1935. The responses of caterpillars to sound. J. Exp. Zool. 72:439–453.

Mopper, S., Faeth, S.H., Boecklen, W.J., and Simberloff, D.S. 1984. Host-specific variation in leaf miner population dynamics: Effects on density, natural enemies and behaviour of *Stilbosis quadricustatella* (Lepidoptera: Cosmopterigidae). Ecol. Entomol. 9:169–177.

Morris, R.F. 1963. The effect of predator age and prey defense on the functional response of *Podisus maculiventris* Say to the density of *Hyphantria cunea* Drury. Can. Entomol. 95:1009–1020.

Morris, R.F. 1972a. Predation by insects and spiders inhabiting colonial webs of *Hyphantria cunea*. Can. Entomol. 104:1197–1207.

Morris, R.F. 1972a. Predation by wasps, birds, and mammals on *Hyphantria cunea*. Can. Entomol. 104:1581–1591.

Morris, R.F., Cheshire, W.F., Miller, C.A., and Mott, D.G. 1958. The numerical response of avian and mammalian predators during a gradation of spruce budworm. Ecology 39:487–494.

Morrison, G., and Strong, D.R. 1980. Spatial variations in host density and the intensity of parasitism: Some empirical examples. Environ. Entomol. 9:149–152.

Morse, D.H. 1980. Behavioral Mechanisms in Ecology. Harvard University Press, Cambridge, MA.

Morton, E.S. 1980. Adaptations to seasonal changes by migrant land birds in the Panama Canal Zone, pp. 437–453. In A. Keast and E.S. Morton (eds.), Migrant Birds in the Neotropics: Ecology, Behavior, Distribution and Conservation. Smithsonian Inst. Press, Washington, D.C.

Mukerji, M.K., and LeRoux, E.J. 1965. Laboratory rearing of a Quebec strain of the pentatomid predator, *Podisus maculiventris* (Say) (Hemiptera: Pentatomidae). Phytoprotection 46:40–60.

Mumma, R.O., and Zettle, A.S. 1977. Larval and pupal parasites of the oak leafroller, *Archips semiferanus*. Environ. Entomol. 6:601–605.

Munn, C.A. and Terborgh, J.W. 1979. Multi-species territoriality in neotropical foraging flocks. Condor 81:338–347.

Myers, J.H., and Campbell, B.J. 1976. Predation by carpenter ants: A deterrent to the spread of cinnabar moth. J. Entomol. Soc. Br. Col. 73:7–9.

Nakasuji, F., Yamanaka, H., and Kiritani, K. 1976. Predation of larvae of the tobacco cutworm *Spodoptera litura* (Lepidoptera: Noctuidae) by *Polistes* wasps. Kontyu 44:205–213.

Needham, J.G., Frost, S.W., and Tothill, B.H. 1928. Leaf-mining insects. Williams & Wilkens, Baltimore.

Nentwig, W. 1985. A tropical caterpillar that mimics faeces, leaves and a snake (Lepidoptera: Oxytenidae: *Oxytenis naemia*. J. Res. Lepid. 24:136–141.

Obrycki, JJ. 1986. The influence of foliar pubescence on entomophagous species, pp. 61–83. In D.J. Boethel and R.D. Eikenbary (eds.), Interactions of Plant Resistance and Parasitoids and Predators of Insects. Halsted, New York.

O'Neil, R.J. 1989. Comparison of laboratory and field measurements of the functional response of *Podisus maculiventris* (Heteroptera: Pentatomidae). J. Kan. Entomol. Soc. 62:148–155.

Owen, D.F. 1975. The efficiency of blue tits *Parus caeruleus* preying on larvae of *Phytomyza ilicis*. Ibis 117:515–516.

Papageorgis, C. 1975. Mimicry in neotropical butterflies. Am. Sci. 63:522–532.

Pearcy, R.W. 1983. The light environment and growth of C_3 and C_4 tree species in the understory of a Hawaiian forest. Oecologia 58:19–25.

Pearson, D.L. 1989. What is the adaptive significance of multicomponent defensive repertoires? Oikos 54:251–253.

Pietrewicz, A.T., and Kamil, A.C. 1977. Visual detection of cryptic prey by blue jays (*Cyanocitta cristata*). Science 195:580–582.

Pietrewicz, A.T., and Kamil, A.C. 1981. Search images and the detection of cryptic prey: an operant approach, pp. 311–331. In A.C. Kamil and T.D. Sargent (eds.), Foraging Behavior: Ecological, Ethological, and Physiological Approaches. Garland STPM Press, New York.

Potter, D.A., and Redmond, C.T. 1989. Early spring defoliation, secondary leaf flush, and leafminer outbreaks on American holly. Oecologia 81:192–197.

Pough, F.H. 1976. Multiple cryptic effects of crossbanded and ringed patterns in snakes. Copeia *1976*:834–836.

Pullin, A.S. 1987. Changes in leaf quality following clipping and regrowth of *Urtica dioica*, and consequences for a specialist insect herbivore, *Aglais urticae*. Oikos 49:39–45.

Rabb, R.L. 1960. Biological studies of *Polistes* in North Carolina (Hymenoptera: Vespidae). Ann. Entomol. Soc. Am. 53:111–121.

Rabb, R.L., and Bradley, J.R. 1968. The influence of host plants on parasitism of eggs of the tobacco hornworm. J. Econ. Entomol. 61:1249–1252.

Rabb, R.L., and Lawson, F.R. 1957. Some factors influencing the predation of *Polistes* wasps on the tobacco hornworm. J. Econ. Entomol. 50:778–784.

Rand, A.S. 1964. Inverse relationship between temperature and shyness in the lizard *Anolis lineatopus*. Ecology 45:863–864.

Rausher, M.D., MacKay, D.A., and Singer, M.C. 1981. Pre- and post-alighting host discrimination by *Euphydryas edithia* butterflies: The behavioral mechanisms causing clumped distributions of egg clusters. Anim. Beh. 29:1220–1228.

Raveret Richter, M.A., and Jeanne, R.L. 1985. Predatory behavior of *Polybia sericea* (Olivier), a tropical social wasp (Hymenoptera: Vespidae). Behav. Ecol. Sociobiol. 16:165–170.

Rawlins, J.E. 1984. Mycophagy in Lepidoptera, pp. 382–423. In Q. Wheeler and M. Blackwell (eds.), Fungus-Insect Relationships. Columbia University Press, New York.

Reichstein, T., von Euw, J., Parsons, J.A., and Rothschild M. 1968. Heart poisons in the monarch butterfly. Science 161:861–866.

Rilling, S., Mittelstaedt, H., and Roeder, K.D. 1959. Prey recognition in the praying mantis. Behaviour 14:164–184.

Risch, S.J., and Carroll, C.R. 1982. Effect of a keystone predacious ant, *Solenopsis geminata*, on arthropods in a tropical agroecosystem. Ecology 63:1979–1983.

Rissing, S.W. 1982. Foraging velocity of seed-harvestor ants *Veromessor pergandei* (Hymenoptera: Formicidae). Environ. Entomol. 11:905–907.

Robinson, M.H. 1969. Defenses against visually hunting predators. Evol. Biol. 3:225–259.

Robinson, S.K., and Holmes, R.T. 1982. Foraging behavior of forest birds: the relationships among search tactics, diet, and habitat structure. Ecology 63:1918–1931.

Robinson, S.K., and Holmes, R.T. Effects of plant species and foliage structure on the foraging behavior of forest birds. Auk 101:672–684.

Rosengren, R. 1971. Route fidelity, visual memory and recruitment behaviour in foraging wood ants of the genus *Formica* (Hymenoptera: Formicidae). Acta Zool. Fennica 133:1–106.

Rothschild, M. 1975. Remarks on carotenoids in the evolution of signals, pp. 20–50. In L.E. Gilbert and P.H. Raven (eds.), Coevolution of Animals and Plants. University of Texas Press, Austin.

Ruehlmann, T.E., Matthews, R.W., and Matthews, J.R. 1988. Roles for structural and temporal shelter-changing by fern-feeding lepidopteran larvae. Oecologia 75:228–232.

Ruiter, L. de 1952. Some experiments on the camouflage of stick caterpillars. Behaviour 4:222–232.

Ryan, R.B., and Rudinsky, J.A. 1962. Biology and habits of the Douglas-fir beetle parasite, *Coeloides brunneri* Viereck (Hymenoptera: Braconidae), in western Oregon. Can. Entomol. 94:748–763.

Saks, M.E., and Carroll, C.R. 1980. Ant foraging activity in tropical agro-ecosystems. Agro-Ecosystems 6:177–188.

Sandberg, S.L., and Berenbaum, M.R. 1989. Leaf-tying by tortricid larvae as an adaptation for feeding on phototoxic *Hypericum perforatum*. J. Chem. Ecol. 15:875–885.

Sato, H., and Higashi, S. 1987. Bionomics of *Phyllonorycter* (Lepidoptera, Gracillariidae) on *Quercus*. II. Effects of ants. Ecol. Res. 2:53–60.

Schmidt, G.T. 1974. Host-acceptance behaviour of *Campoletis sonorensis* toward *Heliothis zea*. Ann. Entomol. Soc. Am. 67:835–844.

Schmidt-Nielsen, K. 1983. Animal Physiology: Adaptation and Environment, 3rd ed. Cambridge University Press, Cambridge.

Schultz, J.C. 1983. Habitat selection and foraging tactics of caterpillars in heterogeneous trees, pp. 61–90. In R.F. Denno and M.S. McClure (eds.), Variable Plants and Herbivores in Natural and Managed Systems. Academic Press, New York.

Scopes, N.E.A. 1969. The potential of *Chrysopa carnea* as a biological control agent of *Myzus persicae* on glasshouse chrysanthemums. Ann. Appl. Biol. 64:433–439.

Scriber, J.M., and Slansky, F., Jr. 1981. The nutritional ecology of immature insects. Annu. Rev. Entomol. 26:183–211.

Seymour, R.S. 1974. Convective and evaporative cooling in the sawfly larvae. J. Insect Physiol. 20:2447–2457.

Sherman, P.W., and Watt, W.B. 1973. The thermal ecology of some *Colias* butterflies. J. Comp. Physiol. 83:25–40.

Sih, A. 1987. Predators and prey lifestyles: an evolutionary and ecological overview, pp. 203–224. In W.C. Kerfoot and A. Sih (eds.), Predation: Direct and Indirect Impacts on Aquatic Communities. University Press New England, Hanover, NH.

Silberglied, R.E. 1979. Communication in the ultraviolet. Annu. Rev. Entomol. 10:373–398.

Skinner, G.J. 1980. The feeding habits of the wood-ant, *Formica rufa* (Hymenoptera: Formicidae), in limestone woodland in north-west England. J. Anim. Ecol. 49:417–433.

Skinner, G.J., and Whittaker, J.B. 1981. An experimental investigation of inter-relationships between the wood-ant (*Formica ryfa*) and some tree-canopy herbivores. J. Anim. Ecol. 50:313–326.

Slansky, F., Jr., and Scriber, J.M. 1985. Food consumption and utilization, pp. 87–163. In G.A. Kerkut and L.I. Gilbert (eds)., Comprehensive Insect Physiology, Biochemistry and Pharmacology, Vol. 4. Pergamon, Oxford.

Smith, J.N.M. 1974. The food searching behaviour of two European thrushes. II. The adaptiveness of the search patterns. Behaviour 49:1–61.

Smith, J.N.M., and Sweatman, H.P.A. 1974. Food-searching behavior of titmice in patchy environments. Ecology 55:1216–1232.

Snyderman, M. 1983. Optimal prey selection: the effects of food deprivation. Behav. Anal. Lett. 3:131–147.

Solomon, M.E., and Glen, D.M. 1979. Prey density and rates of predation by tits (*Parus* spp.) on larvae of codling moth (*Cydia pomonella*) under bark. J. Appl. Ecol. 16:49–59.

Spradbery, J.P. 1973. Wasps: An Account of the Biology and Natural History of Solitary and Social Wasps. University of Washington Press, Seattle.

Stamp, N.E. 1982. Searching behaviour of parasitoids for web-making caterpillars: A test of optimal searching theory. J. Anim. Ecol. 52:387–395.

Stamp, N.E. 1984a. Effect of defoliation by checkerspot caterpillars (*Euphydryas phaeton*) and sawfly larvae (*Macrophya nigra* and *Tenthredo grandis*) on their host plants (*Chelone* spp.). Oecologia 63:275–280.

Stamp, N.E. 1984b. Foraging behavior of tawny emperor caterpillars (Nymphalidae: *Asterocampa clyton*). J. Lepid. Soc. 38:186–191.

Stamp, N.E. 1990. Growth versus molting time of caterpillars as a function of temperature, nutrient concentration and the phenolic rutin. Oecologia 82:107–113.

Stamp, N.E. 1992. Relative susceptibility to predation of two species of caterpillar on plantain. Oecologia: in press.

Stamp, N.E., and Bowers, M.D. 1988. Direct and indirect effects of predatory wasps (*Polistes* sp.: Vespidae) on gregarious caterpillars (*Hemileuca lucina*: Saturniidae). Oecologia 75:619–624.

Stamp, N.E., and Bowers, M.D. 1990. Variation in food quality and temperature constrain foraging of gregarious caterpillars. Ecology 71:1031–1039.

Stamp, N.E., and Bowers, M.D. 1991. Indirect effect on survivorship of caterpillars due to presence of invertebrate predators. Oecologia 88:325–330.

Steward, V.B., Smith, K.G., and Stephen, F.M. 1988. Predation by wasps on lepidopteran larvae in an Ozark forest canopy. Ecol. Entomol. 13:81–86.

Stradling, D.J. 1987. Nutritional ecology of ants, pp. 927–969. In F. Slansky, Jr. and J. Rodriguez (eds.), Nutritional Ecology of Insects, Mites, Spiders, and Related Invertebrates. Wiley, New York.

Sugimoto, T., Shimono, Y., Hata, Y., Nakai, A., and Yahara, M. 1988. Foraging for patchily-distributed leaf-miners by the parasitoid, *Dapsilarthra rufiventris* (Hymenoptera:Braconidae). III. Visual and acoustic cues to a close range patch-location. Appl. Entomol. Zool. 23:113–121.

Swynnerton, C.F.M. 1915. Experiments on some carnivorous insects, especially the driver ant *Dorylus* and with butterflies' eggs as prey. Trans. Entomol. Soc. London 1915:317–350.

Takagi, M., Hirose, Y., and Yamasaki, M. 1980. Prey-location learning in *Polistes jadwigae* Dalla Torre (Hymenoptera, Vespidae), field experiments on orientation. Kontyu 48:53–58.

Tanada, Y. 1967. Effect of high temperatures on the resistance of insects to infectious diseases. J. Sericult. Sci. Jpn. 36:333–339.

Tautz, J. and Markl, H. 1978. Caterpillars detect flying wasps by hairs sensitive to airborne vibration. Behav. Ecol. Sociobiol. 4:101–110.

Taylor, F. 1977. Foraging behavior of ants: Experiments with two species of myrmecine ants. Behav. Ecol. Sociobiol. 2:147–167.

Thomas, C.D., Vasco, D., Singer, M.C., Ng, D., White, R.R., and Hinkley, D. 1990. Diet divergence in two sympatric congeneric butterflies: community or species level phenomenon? Evol. Ecol. 4:62–74.

Tietz, H.M. 1972. An Index to the Described Life Histories, Early Stages and Hosts of the Macrolepidoptera of the Continental United States and Canada, Vol. II. Allyn Museum, Sarasota, FL.

Tilman, D. 1978. Cherries, ants and tent caterpillars: Timing of nectar production in relation to susceptibility of caterpillars to ant predation. Ecology 64:1411–1422.

Tostowaryk, W. 1971. Life history and behavior of *Podisus modestus* (Hemiptera: Pentatomidae) in boreal forest in Quebec. Can. Entomol. 103:662–673.

Tostowaryk, W. 1972. The effect of prey defense on the functional response of *Podisus modestus* (Hemiptera: Pentatomidae) to densities of the sawflies (*Neodiprion swainei* and *N. pratti banksianae* (Hymenoptera: Neodiprionidae). Can. Entomol. 104:61–69.

Traniello, J.F.A. 1989. Foraging strategies of ants. Annu. Rev. Entomol. 34:191–210.

Tsubaki, Y., and Shiotsu, Y. 1982. Group feeding as a strategy for exploiting food resources in the burnet moth *Pryeria sinica*. Oecologia 55:12–20.

Turlings, T.C., and Tumlinson, J.H., 1991. Do parasitoids use herbivore-induced plant chemical defenses to locate hosts? Florida Entomol. 74:42–50.

Turlings, T.C., Tumlinson, J.H., and Lewis, W.J. 1990. Exploitation of herbivore-induced plant odors by host-seeking parasitic wasps. Science 250:1251–1253.

Turner, J.R.G. 1984. Mimicry: The palatability spectrum and its consequences, pp. 141–161. In R.I. Vane-Wright and P.R. Ackery (eds.), The Biology of Butterflies. Academic Press, London.

Usher, B.F. 1984. Housecleaning behavior of an herbivorous caterpillar: Selective and behavioral implications of frass-throwing by *Pieris rapae* larvae. Ph.D. dissertation, Cornell University, Ithaca, New York.

Vasconcelos, H.L. 1991. Mutualism between *Maieta guianensis* Aubl., a myrmecophytic melastome, and one of its ant inhabitants: ant protection against insect herbivores. Oecologia 87:295–298.

Vaughan, F.A. 1983. Startle responses of blue jays to visual stimuli presented during feeding. Anim. Behav. 31:385–396.

Vinson, S.B. 1968. Source of a substance in *Heliothis virescens* that elicits a searching response in its habitual parasite, *Cardiochiles nigriceps*. Ann. Entomol. Soc. Am. 61:8–10.

Vinson, S.B. 1976. Host selection by insect parasitoids. Annu. Rev. Entomol. 21:109–133.

Vinson, S.B. 1977. Behavioral chemicals in the augmentation of natural enemies, pp. 237–279. In R.L Ridgway and S.B. Vinson (eds.), Biological Control by Augmentation of Natural Enemies. Plenum, New York.

Vinson, S.B. 1985. The behavior of parasitoids, pp. 417–469. In G.A. Kerkut and L.I. Gilbert (eds.), Comprehensive Insect Physiology, Biochemistry and Pharmacology, Vol. 9, Behaviour. Pergamon, Oxford.

Waage, J.K. 1978. Arrestment responses of the parasitoid, *Nemeritis canescens*, to a contact chemical produced by its host, *Plodia interpunctella*. Physiol. Entomol. 3:135–146.

Waage, J.K. 1983. Aggregation in field parasitoid populations: foraging time allocation by a population of *Diadegma* (Hymenoptera: Ichneumonidae). Ecol. Entomol. 8:447–453.

Washburn, J.O. 1984. Mutualism between a cynipid gall wasp and ants. Ecology 65:654–656.

Welty, J.C., and Baptista, L. 1988. The Life of Birds, 4th ed. Saunders, Philadelphia.

Werner, E.E., and Gilliam, J.F. 1984. The ontogenetic niche and species interactions in size-structured populations. Annu. Rev. Ecol. Syst. 15:393–425.

Werner, T.K., and Sherry, T.W. 1987. Behavioral feeding specialization in *Pinaroloxias inornata*, the "Darwin's Finch" of Cocos Island, Costa Rica. Proc. Natl. Acad. Sci. U.S.A. 84:5506–5510.

Weseloh, R.M. 1977. Behavioral responses of the parasite, *Apanteles melanoscelus*, to gypsy moth silk. Environ. Entomol. 5:1128–1132.

Weseloh, R.M. 1988. Effects of microhabitat, time of day, and weather on predation of gypsy moth larvae. Oecologia 77:250–254.

Weseloh, R.M. 1989. Behavioral responses of gypsy moth (Lepidoptera: Lymantriidae) larvae to abiotic environmental factors. Environ. Entomol. 18:361–367.

Wiedenmann, R.N., and O'Neil, R.J. 1991. Searching behavior and time budgets of the predator *Podisus maculiventris*. Entomol. Exp. Appl. 60:83–93.

Yamasaki, M., Hirose, Y. and Takagi, M. 1978. Repeated visits of *Polistes jadwigae* Dalla Torre (Hymenoptera: Vespidae) to its hunting site. J. Appl. Entomol. Zool. 22:51–55 (in Japanese with English summary).

Yeargan, K.V., and Braman, S.K. 1989a. Comparative behavioral studies of indigenous hemipteran predators and hymenopteran parasites of the green cloverworm (Lepidoptera: Noctuidae). J. Kansas Entomol. Soc. 62:156–163.

Yeargan, K.V., and Braman, S.K. 1989b. Life history of the hyperparasitoid *Mesochorus discitergus* (Hymenoptera: Ichneumonidae) and tactics used to overcome the defensive behavior of the green cloverworm (Lepidoptera: Noctuidae). Ann. Entomol. Soc. Am. 82:393–398.

Ydenberg, R.C., and Dill, L.M. 1986. The economics of fleeing from predators. Adv. Study Behav. 16:229–249.

Young, A.M. 1971. Wing coloration and reflectance in *Morpho* butterflies as related to reproductive behavior and escape from avian predators. Oecologia 7:209–222.

Young, A.M. 1972. Adaptive strategies of feeding and predator-avoidance in the larvae of the neotropical butterfly, *Morpho peleides limpida* (Lepidoptera: Morphidae). J. N.Y. Entomol. Soc. 80:66–82.

Young, D.R., and Smith, W.K. 1979. Influence of sunflecks on the temperature and water relations of two subalpine understory congeners. Oecologia 43:195–205.

10

Aposematic Caterpillars: Life-Styles of the Warningly Colored and Unpalatable

M. Deane Bowers

Introduction

Caterpillars have an impressive array of defenses to avoid being eaten. For example, they may blend in extremely well with their background (Cott 1940; Edmunds 1974, 1990), they may mimic potential predators of their own predators (Nentwig 1985; Pough 1988), they may construct shelters (Fitzgerald 1980; Fitzgerald and Willer 1983; Damman 1987), and they may be unpalatable due to urticating hairs (Kawamoto and Kumada 1984), defensive glands such as osmeteria (Honda 1983; Damman 1986), or chemicals sequestered from their host plants (Duffey 1980; Blum 1983; Bowers 1990). Insects that are unpalatable are particularly interesting in that they not only use their bad taste or unpleasant odor as a defense, but they usually also advertise this defense to would-be predators by attributes such as conspicuous coloration, gregariousness, and sedentary behavior.

Possession of unpalatable qualities coupled with advertisement of those qualities has many consequences for the life history features, population biology, physiology, and foraging behavior of aposematic caterpillars. For example, aposematic caterpillars need not hide from most potential predators and so, on a daily basis, may be able to spend more time feeding than cryptic caterpillars that must behave in ways that reduce their susceptibility to predators. Sequestration of defense compounds from larval host plants may require particular physiological adaptations by larvae to ingest, accumulate, and store those compounds (Brattsten 1986; Bowers 1992). In addition, unpalatable caterpillar species may be mimicked by palatable caterpillar species, and the presence of varying proportions of palatable look-alikes could affect the efficacy of aposematism, and the population dynamics of both model and mimic.

Conspicuously colored, unpalatable insects have been referred to as warningly colored or aposematic. The terms warning coloration and aposematism are often used synonymously, but their meanings are quite different. The term *aposematic* was first defined by Poulton (1890) as "an appearance which warns off enemies because it denotes something unpleasant or dangerous, or which directs the attention of an enemy to some specially defended or merely non-vital part, or which warns off other individuals of the same species" (Poulton 1890, table following p. 338). Current appropriate usage of the term aposematic employs the first part of Poulton's definition ("an appearance which warns off enemies because it denotes something unpleasant or dangerous"), and thus links an unpleasant or toxic quality of a prey (*unpalatability*) with an advertisement of this feature (*warning coloration*). It is this definition that is used in this chapter. The term *warning coloration* refers to coloration that enhances the conspicuousness of an individual. A conspicuously colored animal is therefore not necessarily aposematic. For example, a brightly colored mimic may be said to exhibit warning coloration, but because it is palatable, it is not aposematic. *Unpalatability* is used in this chapter to refer to a noxious taste or odor that facilitates rejection by predators (Brower 1984).

Prior to Poulton (1890) defining aposematism, Charles Darwin had considered aposematic caterpillars to be an enigma in his investigation of bright coloration in animals, which he explained as important in courtship, and thus reproductive success (Darwin 1878). Because caterpillars are the immature stages of Lepidoptera, and thus do not reproduce, brightly colored larvae did not fit with Darwin's theory. Puzzled, Darwin asked Alfred Russell Wallace if he could suggest any explanation, and Wallace hypothesized that these caterpillars were protected by a nauseous taste or smell and that they advertised this by their bright colors (Wallace 1867, pp. 146–148 in Marchant 1975). Darwin responded very positively to Wallace's suggestion: "You are the man to apply to in a difficulty. I never heard anything more ingenious than your suggestion, and I hope you may be able to prove it true" (Darwin 1867, p. 148 in Marchant 1975). This exchange sparked several investigators to perform feeding trials with many different species of caterpillars (Poulton 1890 and references therein), and these trials showed that, indeed, brightly colored caterpillars were usually rejected by potential predators.

We have progressed a long way since Darwin and Poulton began writing about warning coloration and unpalatability in caterpillars (and other insects). For example, experiments by many researchers have shown that some species of Lepidoptera are unpalatable, and that this unpalatability protects them against potential predators. In addition, advances in analytical chemical techniques have permitted isolation, identification, and quantification of chemical compounds involved in unpalatability of many insect species. The evolution of unpalatability, warning coloration, and mimicry has been addressed from both theoretical and empirical perspectives. The goal of this chapter is to examine the scope of our progress in these areas from the perspective of caterpillar (rather than adult)

Table 10.1. Families of Lepidoptera That Have Species with Stinging or Irritating Hairs or Spines and the Life Stage in Which the Defense Occurs[a]

Family	Life stage
Anthelidae	Larvae
Arctiidae	Larvae
Eupterotidae	Larvae (common) and adults (rare)
Lasiocampidae	Larvae
Limacodidae	Larvae
Lymantriidae	Larvae (common) and adults (rare)
Megalopygidae	Larvae
Noctuidae	Larvae
Nolidae	Larvae
Nymphalidae	Larvae
Saturniidae	Larvae (common) and adults (rare)
Thaumetopoedae	Larvae
Zygaenidae	Larvae

[a]From Kawamoto and Kumuda (1984).

biology, to suggest some questions that remain in the study of unpalatable caterpillars, and to indicate where future research in these areas may take us.

Mechanisms of Unpalatability and Their Consequences

Unpalatability in caterpillars may be due to any of several mechanisms: (1) stinging or irritating hairs or spines, (2) osmeteria and other eversible glands, (3) regurgitation, (4) presence of toxic leaf material in the gut, and (5) sequestration of phytochemicals from the host plant. Use of a particular mode of defense by aposematic caterpillars may have consequences for their foraging behavior, as well as for their life history patterns in general. In addition, there may be costs (as well as the benefit of predator deterrence) to the employment of certain defenses instead of others.

Urticating Hairs and Spines

Urticating (irritating), stinging, or venomous hairs are found on the larvae of many species of Lepidoptera. Twelve families of moths and one family of butterflies have been reported to have representatives with such hairs or spines (Maschwitz and Kloft 1971; Pesce and Delgado 1971; Kawamoto and Kumada, 1984) (Table 10.1). Poisoning by these caterpillars is called "erucism" (from the Latin "*eruca*" for caterpillar), or "lepidopterism." Reactions to these hairs by humans are generally an allergic respiratory response to hairs present in the air, and pain, coupled with dermatitis from contact with poisonous hairs or setae

(Kawamoto and Kumada 1984). Allergic respiratory responses are usually due to the physical irritation of mucosal membranes in nose and lungs. Although the reactions of potential predators, such as toads, lizards, and birds, have been relatively unstudied compared to human response, such predators may reject urticating caterpillars (Bellows et al. 1982), apparently in response to pain and irritation caused by the hairs and spines of the larvae.

Pain, irritation, or dermatitis are caused by chemical compounds produced by specialized poison cells associated with the hairs or spines (Kawamoto and Kumada 1984). As far as is known, these venoms are produced *de novo* by the caterpillars and are not derived from compounds found in the hostplant (Kawamoto and Kumada 1984). The specific components of the venoms of different species vary substantially (Rothschild et al. 1970; Kawamoto and Kumada 1984). Although the complete chemical composition of these venoms has not been determined, some components have been identified. Many lepidopteran venoms contain histamine, or histamine-like substances (Kawamoto and Kumada 1984). Other components include various proteins, but the specific identity of these proteins is unknown (Kawamoto and Kumada 1984).

These venomous spines or hairs may effectively deter vertebrate predators, such as birds, lizards, or rodents (Root 1966; Bellows et al. 1982). For example, larvae of the range caterpillar (*Hemileuca oliviae*, Saturniidae), which have urticating spines, were avoided by lizards and various rodent species found in range caterpillar habitats, whereas the pupae, which lack these spines, were freely eaten (Bellows et al. 1982).

The caterpillar's hairs or spines may also have other functions. They may act as a physical barrier to prevent parasitism or predation by small invertebrate predators (Langston 1957). For example, the spines on larvae of many checkerspot species (*Euphydryas*, Nymphalidae), although not venomous, may be long enough to prevent oviposition by a common parasitoid, *Apanteles euphydriidis* (Braconidae) (Stamp 1982). Hairs and spines may also aid in thermoregulation, by enhancing heat gain in basking species (Porter 1982; Knapp and Casey 1986), or by reducing heat loss (Casey and Hegel 1981). They may also serve as a means of detecting predators and parasitoids (Tautz and Markl 1978).

Although hairs and spines may reduce predation and parasitism by some natural enemies (Heinrich and Collins 1983; Stamp 1982), they may be ineffective against others (Furuta 1983; Stamp and Bowers 1988). For example, the social wasps, *Polistes fuscatus* and *P. dominulus* (Vespidae), which collect caterpillars to feed to the young developing at the nest (Evans and West-Eberhard 1970), can bypass this defense. When they attack caterpillars of *Hemileuca lucina* (Saturniidae), which have urticating spines, the wasps bite off the spines before cutting the caterpillar up into pieces of a size that they can carry (M. Raveret-Richter, personal communication). Birds are responsible for substantial mortality in other hairy caterpillar species, including *Hyphantria cunea* (Arctiidae) (Morris 1972) and *Orgyia pseudotsugata* (Lymantriidae) (Mason and Torgersen 1983), so hairs and spines may not always be effective defenses.

Hairs and spines compose a substantial proportion of the body mass, and are shed at each molt. The exuvium of a hairy caterpillar species was measured as 9% of the gutless dry biomass, whereas that of a smooth caterpillar was only 1% of the gutless dry biomass (Sheehan 1992). Hairs and spines thus represent a substantial energetic investment by the caterpillar.

Osmeteria and Other Defensive Glands

Osmeteria are eversible epidermal glands found in larvae of certain species of swallowtail butterflies (Papilionidae). The gland, located on the dorsal surface between the head and first thoracic segment, is everted when the caterpillar is disturbed, and the volatile products may repel some predators (Eisner and Meinwald 1965; Eisner et al. 1970; Honda 1980a,b, 1981, 1983; Wiklund and Sillen-Tullberg 1985; Damman 1986). Osmeterial secretions include a diversity of terpenoids (Honda 1980a,b, 1981). These compounds have a strong, spicey aroma that may repel vertebrates (e.g., Wiklund and Sillen-Tullberg 1985), and the gland may also be wiped on smaller invertebrate predators and parasitoids (Eisner and Meinwald 1965; Eisner 1970; Damman 1986).

Osmeteria may be particularly effective against invertebrate predators, such as ants and praying mantids (Honda 1983; Damman 1986; Chow and Tsai 1989). Larvae of the zebra swallowtail, *Eurytides marcellus*, were able to fend off ants and small spiders by using their osmeteria (Damman 1986). If the larvae were prevented from everting their osmeteria by the application of Liquid Paper© to block the gland opening, survivorship declined relative to larvae able to use their osmeteria (Damman 1986). Honda (1983) found that certain of the components of the larval osmeterial secretions of papilionid caterpillars were toxic to two species of ants, while others were merely repellent. The praying mantis, *Hirodula patellifera*, avoided larvae of *Papilio memnon heronus* (Papilionidae) in feeding trials, apparently due to the osmeterial secretions (Chow and Tsai 1989). However, osmeteria may not be effective against all types of predators. For example, osmeteria of *Papilio protenor* were ineffective against *Polistes* wasps or the sparrow, *Passer montanus saturatus* (Honda 1983), and larvae of *Papilio machaon* were not protected by their osmeteria in feeding trials with birds (*Parus major*) (Jarvi et al. 1981).

Other caterpillar species have eversible glands that produce defensive chemicals. Larvae of most notodontid moth species have an eversible cervical gland, or adenosma (Weller 1987), located ventrally between the head and prothoracic legs (Godfrey and Appleby 1987; Weller 1987; Miller 1991). This gland produces a mixture of ketones and aqueous formic acid, that is sprayed when the caterpillar is disturbed (Poulton 1887; Herrick and Detwiler 1919; Roth and Eisner 1962;

Eisner et al. 1972). Chemical analyses showed that last instar larvae of *Schizura concinna* (Notodontidae), that had not been previously disturbed, could eject 0.05 g of secretion, containing about 40% formic acid (Poulton 1887). The composition of the secretion appears to be unaffected by the plant that the caterpillar is reared on (Poulton 1887). The spray from a 2.5-cm caterpillar was ejected over an area of about 12.5 by 17.5 cm (Herrick and Detwiler 1919). Larvae appear able to direct the spray in the direction of a disturbance (Poulton 1887; Herrick and Detwiler 1919; Eisner et al. 1972).

Although relatively few observations have been made on the ability of the glandular secretion of notodontid larvae to deter predators and parasitoids, those that have been made suggest that the caterpillars can effectively repel attackers. *Heterocampa manteo* (Notodontidae) larvae sprayed a mixture of formic acid and ketones that repelled lycosid spiders (Eisner et al. 1972). The acid spray of *Schizura leptinoides* (Notodontidae) was deterrent to blue-jays, lizards, and toads (Eisner et al. 1972). In *Litodonta hydromeli* (Notodontidae), the gland's secretion repelled one ant species, *Atta texana*, but not another, *Camponotus* sp. (Weller 1987).

Caterpillars of some species of Noctuidae also have a cervical gland (Marti and Rogers 1988). The gland in noctuid larvae is generally a smooth tube (Marti and Rogers 1988), compared to the bifid structure of the gland in members of the Notodontidae (Weller 1987; Miller 1991). Although by analogy with the cervical gland of the notodontids, the function of the gland in noctuid larvae would appear to be defensive, there have been no experiments to determine this. This chemical composition of the gland products is unknown (Marti and Rogers 1988).

Repeated use of the osmeterium or other eversible gland in defense depletes the supply of compounds. As a consequence, attack by natural enemies may exhaust the defensive secretions to such a degree that they are ineffective. For example, larvae of *S. concinna* produced decreasing amounts of formic acid in response to a series of simulated attacks, and on the seventh attack were unable to produce any secretion at all (Herrick and Detwiler 1919). Trials with freshly field-collected larvae of *H. manteo* demonstrated that 3 to 10 glandular ejections could be elicited from individual larvae (Eisner et al. 1972).

Regurgitation

Regurgitation is a potent defense of many lepidopterous larvae (Eisner et al. 1974; Blum 1981; Stamp 1982; Cornell et al. 1987; Peterson et al. 1987), although the phenomenon has been better documented in sawfly larvae (Prop 1960; Eisner et al. 1974; Morrow et al. 1976). Caterpillars and sawfly larvae may respond to the attack of invertebrate predators, such as ants or wasps (Stamp 1982; Peterson et al. 1987), or to simulated attack (Cornell et al. 1987), by regurgitating. The regurgitant of the caterpillars may contain defensive chemicals derived from the larval host plant. Eastern tent caterpillars, *Malacosoma americanum* (Lasiocampidae), regurgitate a fluid containing hydrogen cyanide and benzaldehyde, which effectively repels ants (Peterson et al. 1987). These compounds result from the action of plant enzymes on cyanogenic glycosides present in leaves of the host plant, *Prunus serotina* (Rosaceae) (Peterson et al. 1987). Larvae of the Monarch butterfly (*Danaus plexippus*, Nymphalidae) regurgitate copious amounts of fluid when disturbed, and this fluid is rich in cardenolides, which are derived from the larval host plant (Nishio 1983). The regurgitant of catalpa sphinx caterpillars, *Ceratomia catalpae* (Saturniidae), contains iridoid glycosides derived from the hostplant (Bowers and Collinge, unpublished data).

One consequence of regurgitation is the loss of the defensive material. However, the regurgitant may also contain nutritional components from the plant material, as well as fluid, that are lost when the material is discharged. Although in some cases the regurgitant may be reimbibed if it is not used (Cornell et al. 1987), it is usually expelled. Thus there may be a cost (loss of defense compounds, fluid, and nutritional components) to regurgitating. For example, sawfly larvae (*Neodiprion sertifer*) that had their regurgitated droplets removed grew significantly less than those that were allowed to reabsorb their droplets (Bjorkman and Larsson 1991). These droplets contain primarily resin acids derived from the host plant, Scots pine (*Pinus sylvestris*). Larvae of the catalpa sphinx, *Ceratomia catalpae* (Sphingidae), that feed on species of *Catalpa* (Bignoniaceae), which contain iridoid glycosides, regurgitate when disturbed (Bowers, personal observation). The regurgitant contains at least two iridoid glycosides from the host plant, catalpol and catalposide (Bowers, Collinge, and Lawrence, unpublished data). In an experiment to determine whether regurgitation affected larval growth, over a period of 9 days, newly molted third instar caterpillars that were pinched with forceps twice a day to simulate attack from a predator, and induce regurgitation, gained about half the weight of control caterpillars that were observed but not pinched [control larvae, $N = 13$, mean biomass gained $= 962.13$ mg, $SE = 140.5$, regurgitating larvae, $N = 9$, mean biomass gained $= 557.97$ mg, $SE = 120.83$, t test, $t(20) = 2.42$, $P < 0.05$] (Bowers and Collinge, unpublished data). Thus, both sawfly larvae and catalpa sphinx caterpillars that repeatedly defended themselves by regurgitating paid the price of reduced growth for using this mode of defense.

Presence of Toxic Host Plant Compounds in the Gut

The presence of chemical compounds in the host plant material in the caterpillar's gut may provide some protection against predation (Brower 1984; Janzen 1984). If the compounds occur solely in the gut, then it is likely that the caterpillar must be killed for the predator to experience any deterrent effect of the compounds, unless these larvae commonly regurgitate (see Regurgitation). If the compounds are moved out of the gut to the hemolymph or other site of storage, such as the fat body or cuticle (i.e., they are sequestered, see Sequestration), then they may be concentrated in the caterpillar at levels much higher than those at which they occur in the host plant (Roeske et al. 1976; Brower 1984; Bowers and Collinge 1992). Thus the presence of host plant compounds in the gut that are potentially toxic or distasteful to caterpillar predators may provide some protection, but will probably not provide the same degree of protection that they would if they were sequestered in hemolymph, cuticle, or defensive gland.

Although some aposematic caterpillars may be chemically defended only by the chemical compounds in the plant material in the gut, there has been no experimental work to demonstrate this. One anecdotal example illustrates the possibility. Mopane worms, larvae of the saturniid moth, *Gonimbrasia belina*, are brightly colored white with red, black and yellow markings, and are gregarious (Brandon 1987). These larvae feed on the legume, *Colophospermum mopane* (Fabaceae), and become quite large, 8 to 10 cm in length and 1.5 cm in diameter. They are used as food in parts of South Africa, and may be eaten fresh or dried (Quin 1959; Dreyer and Wehmeyer 1982; Brandon 1987). The preparation of these caterpillars for consumption suggests that their bright coloration indicates that they are unpalatable due to the presence of toxins in the food material in the gut. Traditionally, the gut contents of the fresh caterpillars are squeezed out by hand (Quin 1959; Dreyer and Wehmeyer 1982), and the caterpillars are then roasted and eaten, or dried and stored (Quin 1959). Without chemical analyses of the caterpillars and their host plant, as well as feeding trials with potential consumers, it is impossible to say whether the the larvae are in fact protected by toxic compounds in the food in the gut.

For aposematic caterpillars, their warning coloration may serve the function of advertisement of their noxious qualities, but if the defense compounds are confined to the gut, the caterpillar must be killed for the predator to experience the effects of the defense compound. Therefore aposematic larvae that are unpalatable due to gut contents only, would be predicted to be gregarious, since kin selection must be invoked to explain selection for a trait for which the individual possessing it must be killed.

Cryptic as well as aposematic caterpillars could be protected by the presence of toxic host plant chemicals in the gut. Janzen (1984) suggested that some species of cryptic sphingids at Santa Rosa National Park in Costa Rica may be avoided by vertebrate predators because of this mechanism. However, as with the aposematic caterpillars, such a larva would probably be killed, since the gut material must be tasted for the deterrence or toxicity of the host plant compounds to be effective.

Since little (if any) chemical or experimental work has been done examining the potential of toxic host plant compounds in the gut as a defense of aposematic or cryptic caterpillars, this would be a productive area to pursue. Chemical analyses of gut contents, feeding tests with potential predators, and comparisons of foraging behavior, defense behavior, and susceptibility to predation of aposematic and cryptic species that may use toxic gut contents as a defense are all unexplored areas of enquiry.

Sequestration of Host Plant Chemicals

Most research on acquired chemical defenses of insects, and especially Lepidoptera, has concentrated on the adult stage. It is primarily this stage that has been studied from the perspective of effectiveness of chemical defense (Brower 1984; Rothschild 1985), the evolution of warning coloration (Sillen-Tullberg 1988), and mimicry (Bates 1862; Gilbert 1983). However, in most cases it is the larval stage that ingests, processes, and stores the defensive chemicals acquired from the host plant. The ingestion of pyrrolizidine alkaloids by adult danaines and ithomiines is an exception to this generalization (Brown 1984; Ackery and Vane-Wright 1986).

The term *sequestration* has been used in a variety of contexts (see Duffey 1980). Here, I use it to mean the acquisition and accumulation of defensive chemicals in the insect's body, as compared to the *de novo* synthesis of defensive compounds (Duffey 1980; Bowers 1992). These chemicals can be acquired from the larval host plant (Duffey 1980, Brower 1984; Brattsten 1986; Bowers 1992) or from other dietary sources (e.g., butterfly acquisition of pyrrolizidine alkaloids from nectar (Brown 1984; Ackery and Vane-Wright 1986)). Compounds that are ingested, but pass out of the insect's body without leaving the gut are not considered to be sequestered. Compounds that pass out of the gut and are found in the hemolymph, cuticle, fat body, or other organ are sequestered.

Processing of Host Plant Chemicals

Although many species of aposematic caterpillars are unpalatable due to the chemical compounds that they sequester from their host plants, many insect species feed on plants containing potential defensive chemicals that they do not sequester (Rothschild 1972; Isman et al. 1977; Blum 1983; Nishio 1983; Bowers and Puttick 1986; Brattsten 1986; Rowell-Rahier and Pasteels 1986). In such species, the chemicals do not accumulate during the larval stage, nor are they found in the pupae or adults. In these caterpillars, the compounds may be eliminated relatively intact (Self et al. 1964; Duffey 1980; Blum 1983; Brattsten 1986), or they may be metabolized or broken down in some way (Duffey 1980; Blum 1983; Brattsten 1986; Gardner and Stermitz 1988).

Compounds that are sequestered are largely immune to the attack of detoxification enzymes in the larval gut (Terriere 1984; Ahmad et al. 1986; Brattsten 1986). They pass through the gut wall and into the hemolymph, which may require transport by carrier molecules (Duffey 1980; Wink and Schneider 1988). The hemolymph itself may be the site of storage, and the compounds may be effective against potential predators due to reflex bleeding (Happ and Eisner 1961; Eisner 1970; Blum 1981; Boros et al. 1991), or when the animal is wounded. Defense compounds may also be transported via the hemolymph to sites of storage in the cuticle, fat body, or in specialized organs such as the dorsolateral spaces found in milkweed bugs (Duffey and Scudder 1974; Duffey et al. 1978). In some larvae, defense compounds may be stored in a foregut diverticulum, and they are regurgitated when the insect is disturbed (Eisner et al. 1974; Morrow et al. 1976; Nishio 1983; Larsson et al. 1986; see Regurgitation).

The tissues of unpalatable insects that sequester chemical compounds from their host plants may be immune to, or protected from, potential toxic effects of those compounds. However, nonadapted insect species may be adversely affected through a variety of mechanisms (Berenbaum 1986). For example, increasing doses of iridoid glycosides in the diet are toxic to some species of generalist caterpillars (Puttick and Bowers 1988), but caterpillars that specialize on iridoid glycosides, such as larvae of the buckeye, *Junonia coenia* (Nymphalidae) feed with impunity on diets with doses of iridoids at levels at or above those found in their host plants (Bowers and Puttick 1989). At a physiological level, the tissues of monarchs (*Danaus plexippus*, Nymphalidae) and milkweed bugs (*Oncopeltus fasciatus*, Lygaeidae), that specialize on plants containing cardenolides, were immune to the effects of those compounds on sodium and potassium transport, while those of nonspecialists were not (Vaughan and Jungreis 1977; Moore and Scudder 1986).

Nonetheless, for caterpillars that sequester chemical compounds from their hostplants to use in their own defense, there may be a cost of acquiring and sequestering these compounds. This cost is certainly manifest in the relegation of resources that are used in transport or storage of defense compounds that might otherwise be used for different functions. But, is there also a cost at the level of an insect's fitness, such that individuals that are better protected by having larger amounts of defense compounds pay a price for having those compounds? In the absence of predation, insects higher in defense compounds may have a lower fitness than individuals that have less, because of physiological costs associated with processing and sequestration of relatively high amounts of defense compounds; but in the presence of predation, the palatable individuals may be killed and eaten, thus their fitness would be zero. Attempts to determine a cost of chemical defense have met with mixed success (see review in Bowers 1992). In some species there is an indication of cost, whereas in others there is not. Because it is the larval stage that acquires and processes the chemical compounds, perhaps study of caterpillars rather than adults would be most revealing in this regard.

Patterns of Chemical Defense in Caterpillars versus Other Life Stages

In holometabolous insect species, such as the Lepidoptera, there may be very different ecological and evolutionary forces acting on the different life stages, egg, larva, pupa, and adult. These forces may be important in the evolution of aposematism in these four life stages. In some lepidopteran species the larvae are warningly colored, but the pupae and adults are not, whereas in other species both larvae and adults are warningly colored. In only very few species, such as members of the butterfly genus *Eumaeus* (Lycaenidae) (DeVries 1976; Bowers and Larin 1989) are all three of these life stages warningly colored. In an experiment to investigate why most species of Lepidoptera have cryptic pupae even though the larvae and adults are aposematic, Wiklund and Sillen-Tullberg (1985) found that pupae are generally killed when they are attacked, and caterpillars and adults usually survive.

Selection pressures on caterpillars may be quite different than those on other life stages (Brower 1984; Gilbert 1984). Differences in strategies for predator avoidance and escape among these life stages, coupled with differences in physiology and behavior among larva, pupa, and adult may act in concert to determine what kind(s) of defenses are employed. For example, eggs and larvae may experience predation primarily by invertebrates, whereas pupae and adults are preyed on primarily by vertebrates (Brower 1984; Gilbert 1984). In addition to avoiding predation and dealing with the environment, caterpillars must cope primarily with finding food, while adults must find mates and oviposition sites as well. Although caterpillars may move around on (Jones 1977; Dethier 1988, 1989; Mauricio and Bowers 1990) or leave the host plant (Dethier 1988), they are often relatively sedentary, especially compared to the winged adult. Pupae, however, are completely immobile, and have little possibility of behavioral defenses, although some pupae may wiggle or squeak when disturbed (Hinton 1948; Downey and Allyn 1973). In addition, the habitats of these three life stages (larva, pupa, adult) may be quite different. Larvae feed on the host plant, but pupation may occur some distance from the host plant or in the soil, and adults may fly considerable distances in search of food, mates, or oviposition sites.

A comparison of patterns of chemical defense in species in which only larvae are aposematic with those in which larvae and adults are aposematic (indicating that the defensive compounds are retained through the pupal stage and into the adult stage) may provide some information about the factors affecting the evolution of aposematism in these different stages, and most importantly here, for the larvae. Larvae of the buckeye butterfly, *Junonia coenia* (Nymphalidae), and the Catalpa sphinx, *Ceratomia catalpae* (Sphingidae), sequester iridoid glycosides from their host plants, but they do not retain them through to the adult stage (Bowers and Puttick 1986, Table 10.2). Larvae of *C. catalpae* are warningly colored black, white, and yellow, and are gregarious, while the adults are cryptically colored gray and brown and are solitary and palatable (Bowers and Farley 1990). Larvae contain 15% dry weight iridoid glycosides, while pupae are very low in iridoid glycosides and adults are devoid of these compounds (Table 10.2). The buckeye, *J. coenia*, shows a similar pattern. Larvae may contain as much

Table 10.2. Total Iridoid Glycoside Content (Mean ± SE) of Different Life Stages of Three Lepidopterans That Sequester Iridoid Glycosides

Species	Host plant	Life stage	*N*	Iridoid content (% dry weight)
Junonia coenia[a]	*Plantago lanceolata*	Newly molted I_4	5	2.88 (± 0.73)
	(Plantaginaceae)	Newly molted I_5	16	6.83 (± 0.71)
		Pupa	14	0.19 (± 0.04)
		Adult	10	0
Ceratomia catalpae[b]	*Catalpa speciosa*	Newly molted I_3	12	14.53 (± 0.50)
	(Bignoniaceae)	Newly molted I_4	12	1.32 (± 0.38)
		Pupa	10	0.26 (± 0.03)
		Adult	10	0
Euphydryas phaeton[c]	*Chelone glabra*	Newly molted I_7	5	15.50 (± 0.48)
	(Scrophulariaceae)	Pupa	6	7.60 (± 0.42)
		Adult	15	1.15 (± 0.30)

[a]Data from Bowers and Collinge (1992).

[b]Data from Bowers, Collinge, and Lawrence (unpublished).

[c]Data from Bowers and Collinge (unpublished).

as 12% dry weight iridoid glycosides, pupae contain less than 0.2% iridoid glycosides and adults are devoid of these compounds (Bowers and Collinge 1992, Table 10.2). Thus, the larvae are protected from potential predators by the sequestered compounds, but the adults are not.

Checkerspots of the genus *Euphydryas* (Nymphalidae) also sequester iridoid glycosides during larval feeding (Bowers and Puttick 1986; Bowers 1991). However, in these species the compounds are retained through to the adult stage (Bowers and Puttick 1986; Belofsky et al. 1989; Bowers 1992, Table 10.2), rendering larvae, pupae, and adults unpalatable (Bowers, 1980, 1981; Bowers and Farley 1990). However, the host plant species may affect both the amounts and kinds of iridoid glycosides that are sequestered (Bowers 1980, 1981; Bowers and Puttick 1986; Belofsky et al. 1989; Bowers and Farley 1990; Bowers 1991), which in turn may affect the relative palatability or unpalatability of an individual caterpillar (Bowers 1991, 1992).

Although only members of the genus *Euphydryas* retain iridoid glycosides through to the adult stage, in all three taxa (*J. coenia, C. catalpae*, and *Euphydryas* species), there is a decrease in the concentration of iridoid glycosides from larval to pupal to adult stages (Table 10.2). For example, in *E. phaeton*, the concentration of iridoid glycosides drops by more than 50% between larva and pupa, and again by about 75% between pupa and adult (Table 10.2).

The drop in concentration of defense compounds from larval to pupal to adult stages may reflect differences in the selective factors operating on these larvae versus pupae and adults. The larvae are relatively immobile compared to the adults, and, in the three taxa described above, could be very susceptible to predators that have some tolerance to their defenses. Thus very high concentrations of iridiod glycosides in the larvae may be necessary to deter predators or parasitoids. The pupal stage is cryptic in all these species. The combination of crypsis as an antipredator mechanism, with the potential for autotoxic effects during metamorphosis, may act in concert to select for a decrease in chemical content of the pupal stage. The pupal stage may present an evolutionarily difficult barrier to the storage of defensive chemicals through to the adult stage. During the pupal stage, there is massive reorganization of the internal organs, accompanied by high levels of cell growth and division. This high level of metabolic activity may render insects in the pupal stage particularly sensitive to autotoxicity of compounds that were used as larval defenses, and previously existing barriers to these toxic effects may be changing or lost during this period (Bowers 1992).

Adult butterflies and moths can fly, therefore they have an additional means of defense against potential predators and chemical defenses may be less important to adults relative to the more sedentary larvae. These differences in strategies for predator avoidance and escape between the three life stages, larva, pupa, and adult, may contribute to differences in the amount of chemical defense observed in these lepidopterans that sequester iridoid glycosides, and may be important for other unpalatable species as well.

Multiple Defenses

In some species, more than a single defense may be employed. For example, many aposematic caterpillars are not only unpalatable, but also regurgitate on small predators and parasitoids (Stamp 1982; Nishio 1983). Some cryptic caterpillars may be somewhat unpalatable due to the presence of toxic host plant compounds in the gut (Brower 1984; Janzen 1984). Larvae of the buckmoth, *Hemileuca lucina* (Saturniidae), are covered with stinging hairs, but may also regurgitate when disturbed (Cornell et al. 1987). There may be several advantages to caterpillars to having multiple antipredator defenses (Pearson 1989): (1) some characters may function together to maximize protection, such as the frequent cooccurrence of aposematism and gregariousness, (2) certain antipredator characters may be targeted toward different foraging phases of a predator (Endler 1986) such as cryptic swallowtail larvae having osmeteria, and (3) different defenses may be targeted against different types of predators (Stamp 1982, 1986; Cornell et al. 1987; Pearson 1989).

In addition to their primary line of defense, be it sequestered chemicals, stinging hairs, or osmeteria, many aposematic caterpillars demonstrate a rather impressive repertoire of defensive behaviors that may enhance their chemical arsenal (Tostowaryk 1971, 1972; Stamp 1982, 1986; Cornell et al. 1987; Peterson et al. 1987). Caterpillars may bite, thrash and/or fall off the plant (Stamp 1982, 1986; Cornell et al. 1987; Peterson et al. 1987). The size of the caterpillar relative to its predators may determine what type of defense is employed (Stamp 1986; Cornell et al. 1987; Reavey, Chapter 8). For example, small larvae of aposematic species may bite and regurgitate, whereas larger larvae may thrash and drop off the host plant (Cornell et al. 1987).

Advertisement of Unpalatability

Whether unpalatability is due to stinging hairs, defensive glands, or defensive chemicals from the host plant, aposematic caterpillars often exhibit several distinctive characteristics that serve to advertise their noxious qualities to potential predators. These characteristics include morphological modifications such as warning coloration, and behaviors such as aggregation.

Advertising Strategies

One method of advertising unpalatable qualities is by conspicuous, or warning, coloration. Warning colors include red, orange, yellow, black, and white. The predominance of these colors in making up the patterns of unpalatable caterpillars (and unpalatable insects in general) may be due to several reasons (Turner 1984). There may be an innate tendency for vertebrate predators to avoid warning colors (Coppinger 1970; Smith 1975; Schuler and Hesse 1985), or they may learn more quickly to avoid prey that have such coloration (e.g., Gibson 1980). Warning colors may also enhance learning by promoting easy recognition of unpalatable prey (Turner 1984). Furthermore, these conspicuous colors may reduce mistakes by experienced predators (Guilford 1986).

Not all brightly colored caterpillars are unpalatable, however; some brightly colored caterpillars may in fact be palatable Batesian mimics (see Mimicry). Nor are all unpalatable caterpillars warningly colored. For example, the larvae of moths in the genus *Zygaena* (Arctiidae) are cryptic, yet quite toxic to potential predators, and, interestingly, the adults are warningly colored (Turner 1984).

Many aposematic caterpillars are also gregarious (Fisher 1930; Cott 1940; Stamp 1980; Boeve and Pasteels 1985; Sillen-Tullberg 1988; Vulinec 1990; Fitzgerald, Chapter 11) (Fig. 10.1). Aggregation itself may be a defense in some insects that are palatable. High-density patches of palatable aphid prey resulted in a lower capture rate by syrphid larvae predators than a lower density patch (Kidd 1982). Larger group sizes may also facilitate early detection of predators (Treherne and Foster 1980, 1981, 1982; Vulinec 1990) and may increase the efficacy of defensive behavior (Vulinec 1990). There may also be a dilution effect, in which an individual is less likely to be attacked when in a group than when alone, once it has been detected by the predator (Hamilton 1971;

Figure 10.1 Aggregated aposematic caterpillars. Clockwise from upper left: (1) *Hemileuca lucina* (Saturniidae), molting from instar III to instar IV. These black larvae have urticating spines and are quite conspicuous as they feed and bask at branch tips of the host plant, *Spiraea latifolia* (Rosaceae). (2) *Euphydryas chalcedona* (Nymphalidae), instar V. These dark brown larvae have orange spots and their spines are non-urticating. They are quite conspicuous against the green leaves of their host plant, *Scrophularia californica* (Scrophulariaceae). The larvae are unpalatable due to sequestration of iridoid glycosides. (3) *Euphydryas phaeton* (Nymphalidae), instars V and VI. These aggregating larvae are orange with black stripes and nonurticating spines and are unpalatable due to sequestration of iridoid glycosides. Their coloration renders them very conspicuous on their host plant, *Chelone glabra* (Scrophulariaceae). (4) *Chlosyne nycteis* (Nymphalidae), instar III. These brown larvae aggregate, but their palatability has never been tested. They feed on species of *Helianthus* (Asteraceae).

Treisman 1975; Turner and Pitcher 1986; Sillen-Tullberg and Leimar 1988; Vulinec 1990).

For aposematic caterpillars (and other aposematic insects), aggregation may increase the effectiveness of the warning signal (Cott 1940; Edmunds 1974). Aggregation may also increase the efficacy of a defense by increasing its magnitude (Aldrich and Blum 1978), such as more stinging hairs or a larger discharge of a chemical defense. In addition, the presence of other unpalatable individuals in the immediate vicinity of the attacked individual may immediately reinforce avoidance learning in predators (Sillen-Tullberg and Leimar 1988; Guilford 1990). Although many species of aposematic caterpillars are gregarious (Table 11.1 in Vulinec 1990), some are solitary (Ackery and Vane-Wright 1986; Vulinec 1990). For example, although larvae of danaine butterflies are warningly colored and most are unpalatable due to sequestration of cardenolides from the host plants, they are usually solitary (Ackery and Vane-Wright 1986).

Advertising Costs

Advertising may have its costs, however. Warning coloration may not always be an advantage. Warningly colored caterpillars presumably are very conspicuous to potential predators, and thus more easily discovered than cryptic prey (Guilford 1988, 1990). Therefore once the defense is breached, or if the degree of chemical defense is relatively low, warningly colored caterpillars may be extremely vulnerable. For example, conspicuousness may increase contact and capture rate of warningly colored caterpillars by predators tolerant of their defenses (Guilford 1990).

For species with aggregated larvae, discovery of an aggregation by a predator that can cope with the defense may dramatically reduce parental fitness because a large proportion of the offspring are killed. In addition, predators and parasitoids may intensify a search in an area after finding a prey item there (Yamasuki et al. 1978; Heinrich and Collins 1983; Vulinec 1990), leading to increased chance of discovery of additional aggregations in a particular area.

Other Characteristics Often Associated with Aposematism

Aposematic insects are reported to have very tough bodies, allowing them to survive handling and tasting by predators (Bates 1862; Carpenter 1942; Jones 1932, 1934; Boyden 1976; Jarvi et al. 1981; Wiklund and Jarvi 1982). Most caterpillars, however, may not fit this description. Although some caterpillar species may have tough flexible cuticles that can withstand the beak of an attacking bird (Wiklund and Sillen Tullberg 1985) most caterpillars have very soft bodies that may be easily pierced by a sharp beak or jaws. Those species that employ stinging spines as a defense may deter predators before they are damaged. However, species that sequester host plant chemicals may be killed or, if the cuticle is ruptured, may die of desiccation, even if they are eventually not eaten by the predator (personal observation). In some caterpillars, it appears that the outer cuticle contains deterrent substances so that the caterpillar need not be damaged to be avoided (Bernays and Cornelius 1989; Montllor et al. 1991).

Based on life history observations of adult saturniid moths, Blest (1963) suggested that aposematic insect species would be long-lived relative to cryptic, palatable species. Long-lived, aposematic individuals would help predators learn to avoid other similar individuals, and therefore increase the chances of survival for those individuals. Long-lived cryptic individuals would be detrimental to others in the population because they would be exposed to predators for a longer period of time, enabling predators to learn to find other similarly colored cryptic individuals. There have been no experiments to test Blest's (1963) hypothesis, and no such analyses for the larval stages.

Foraging Behavior of Aposematic Caterpillars

The foraging behavior of caterpillars reflects the importance of three major factors: acquiring nutrients, avoiding predation, and coping with abiotic conditions (Table 10.3). Characteristics of the host plant, such as intraplant variation in secondary compounds (Schultz 1983; Edwards and Wratten 1985), leaf morphology (Mauricio and Bowers 1990), and plant architecture (Lawton 1986), may influence caterpillar foraging behavior. These factors do not act independently, but in concert, to determine the ecology and the evolution of foraging behavior in caterpillars.

In most lepidopteran species, it is the caterpillar that acquires the resources that are then translated into adult fitness, and there is a strong positive correlation between pupal weight of a caterpillar and its adult fecundity (Scriber and Slansky 1981; Reavey and Lawton 1991). In some taxa, adult Lepidoptera do not feed at all (notably certain members of the family Saturniidae), and thus all nutritional intake occurs during the larval stages. Therefore adult fitness may depend heavily on larval foraging behavior and an individual caterpillar's decisions about what, where, and when to eat.

Aposematic caterpillars that are unpalatable due to compounds they sequester from their host plant have a fourth component to their foraging behavior: acquisition of those defense compounds (Table 10.3). Although cryptic or mimetic caterpillars must cope with chemical compounds in their host plants, there is no enhanced survival on plants that provide more of those compounds (and indeed, there may be reduced survival). Therefore, although the effects of some constraints on foraging behavior are the same for aposematic caterpillars and cryptic or mimetic caterpillars, the effects of other constraints differ (Table 10.3). In the following sections, I will compare and contrast aposematic caterpillars with those that do not exploit this mode of defense.

Coping with the Abiotic Environment

Larval physiology and response to abiotic conditions may in part determine the foraging behavior of aposematic caterpillars (as well as other caterpillars) (Table 10.3) (see also Casey, Chapter 1, Stamp, Chapter 15, and Kukal, Chapter 16). For example, by increasing body temperature, basking may enhance growth rates by increasing digestion rates and efficiency (Sherman and Watt 1973; Casey 1976; Stamp and Bowers 1990a). Dark colors that aid in absorption of radiant energy may also make caterpillars quite conspicuous (Stamp and Bowers 1990b; Sheehan 1992), and may function as warning colors in some species. Aggregation

Table 10.3. The Major Factors with Which Aposematic Caterpillars Must Cope, and How They Are Likely to Fare Compared to Cryptic, Palatable Caterpillars and Warningly Colored, Palatable (Mimetic) Caterpillars

Problem or constraint	Solution	Potential cost	How are the consequences different for aposematic larvae, compared with cryptic or mimetic larvae?
Coping with abiotic conditions			
Unfavorable in the short term			
Too hot	Seek shade	Less nutritious food	Probably not different
Too cold	Bask	More conspicuous to natural enemies	Aposematic larvae protected, cryptic larvae more vulnerable, and mimetic larvae susceptible to naive predators
Unfavorable in the long term	Diapause or aestivation	Freezing or desiccation, susceptibility to natural enemies	Abiotic factors have similar effects on all types of larvae, but aposematic larvae protected from natural enemies
Obtaining nutrients			
Low quality food	Selectivity in food choice (e.g., choose new leaves)	Increased movement and susceptibility to natural enemies	Aposematic larvae protected, others more susceptible to natural enemies
Induced defenses	Move to avoid	Susceptibility to natural enemies	Aposematic larvae protected, others more susceptible to natural enemies
Competition	Leave plant or plant part	Starvation, susceptibility to natural enemies	Aposematic larvae that are gregarious may have more intraspecific competition, but they are protected from natural enemies. Starvation—a potential problem for all types of larvae
Avoiding predation	Temporal separation	Feeding at less optimal times (e.g., at night)	Aposematic larvae protected from natural enemies because of their unpalatability, mimetic larvae protected by their resemblance to the unpalatable model, so most important for cryptic larvae

Table 10.3. (continued)

Problem or constraint	Solution	Potential cost	How are the consequences different for aposematic larvae, compared with cryptic or mimetic larvae?
Avoiding predation *(continued)*			
	Spatial separation	Shelter building, which takes time and resources	Aposematic larvae protected from natural enemies because of their unpalatability, mimetic larvae protected by their resemblance to the unpalatable models, so most important for cryptic larvae
	Defensive behavior	Time, separation from host plant, possible injury when drop from host plant	Defensive behavior enhanced by unpalatability
	Unpalatability	Processing and storing defense compounds	Aposematic larvae must process and store; mimetic larvae are protected by their resemblance to the unpalatable model, but do not have to store the compounds, cryptic larvae must process but do not store
	Warning coloration	Susceptibility to naive and chemically tolerant natural enemies	Aposematic larvae may be rejected and survive attack, but mimetic larvae would be eaten
Obtaining defense chemicals	*De novo* synthesis, or feed on plants or plant parts high in defense chemicals	More conspicuous or vulnerable to natural enemies	Aposematic larvae protected from natural enemies unless the enemies are tolerant of the defense

may aid in elevating body temperature in basking caterpillars (Stamp and Bowers 1990b) and therefore may be an important component of feeding behavior in gregarious, unpalatable caterpillars. Thus predation and the abiotic environment may both select for dark coloration in unpalatable, gregarious caterpillars that bask.

Excessive heat may pose a particular problem for darkly colored aposematic caterpillars, and may affect their foraging behavior and defenses differently than more lightly colored larvae. For example, the black, gregarious larvae of *Hemileuca lucina* (Saturniidae) respond to high temperatures by retreating to the shaded interior of their host plant (Stamp and Bowers 1990b). In the shade, aposematic caterpillars in general, and these black larvae in particular, may be less conspicuous to predators and parasitoids, and thus less susceptible to natural enemies that can overcome their defenses. However, they may also be more easily mistaken for acceptable prey by predators that find them unpalatable, if they are not as conspicuous as when in their typical locations. In addition, leaves in the shade are more likely to be older leaves that are probably lower in nutrients, water, and, perhaps most importantly, the chemicals potentially used for the insects' defense (McKey 1979; Scriber and Slansky 1981; Stamp and Bowers 1988, 1990a).

Acquiring Nutrients

The nutritional requirements of most caterpillars are quite similar (Mattson 1980; Scriber and Slansky 1981), although different caterpillar species may vary in their ability to mechanically (Bernays and Janzen 1988) or physiologically (Slansky and Scriber 1985) process food and thus to extract nutrients from leaves. For most caterpillar species studied, water content, nitrogen content, and allelochemical content of the food determine larval and, ultimately, adult fitness (Slansky, Chapter 2, and references therein). Aposematic caterpillar species probably do not differ from cryptic species in their ability to find host plants and utilize the nutrients in their hostplants [but see the distinction between "stealthy" (usually cryptic) and "opportunistic" (often aposematic) caterpillars, Rhoades 1985]. However, acquisition of nutrients may be affected by differences between aposematic versus cryptic or mimetic caterpillars in their susceptibility to predators (e.g., Heinrich 1979; Heinrich and Collins 1983; Montllor and Bernays, Chapter 5; Heinrich, Chapter 7; Stamp and Wilkens, Chapter 9). For example, cryptic caterpillar species are usually solitary and may restrict their foraging to nighttime (Heinrich 1979), which may limit their foraging range and prevent them from moving to higher quality food, whereas aposematic caterpillars need not limit foraging to nighttime (Table 10.3).

Induction of plant chemical defenses may change the quality of the plant material as food for caterpillars (Edwards and Wratten 1983; Schultz 1983; Baldwin 1988; Rossiter et al. 1988; Haukioja, Chapter 13). The major effect is probably to decrease the food quality by increasing fiber, phenolics, and other secondary chemicals, while at the same time decreasing water content (Baldwin 1988; Rossiter et al. 1988; Haukioja, Chapter 13). Such a change may cause caterpillars to move more (Edwards and Wratten 1983; Schultz 1983), thus making them more susceptible to predators and parasitoids (but see Bergelson and Lawton 1988). Because many aposematic caterpillar species are also gregarious, they may have a different effect on induction of plant defenses than solitary caterpillars. Rhoades (1985) suggests that gregarious caterpillars may be able to overwhelm the host plant defenses by employing mass attack. However, there may also be a higher likelihood of aggregated caterpillars inducing chemical changes in the host plant because of greater feeding in a limited area on the host plant. These alternatives have not been experimentally investigated. If the defensive chemicals used by the caterpillars are induced by feeding, induction of higher amounts of these chemicals may in fact be an advantage. These larvae may not have to move as much to obtain higher amounts of defense chemicals, which they would usually obtain by feeding on parts of plants, such as new leaves, which generally provide those higher amounts. Instead, the larvae could concentrate feeding activity on certain areas of the host plant that are responding by inducing chemical changes.

Many aposematic caterpillars are gregarious, and this may be a potential liability to the acquisition of nutrients because larger numbers of individuals may more quickly exhaust the available food, resulting in intraspecific competition (Dethier 1959). This is probably more evident in aposematic larvae that acquire their defense compounds from herbaceous plants, rather than aposematic larvae feeding on tree species. Defoliation of the host plant may result in larvae leaving the host plant when they are relatively small, and lead to a greater mortality due to starvation because of an inability to find a new host plant and increased susceptibility to natural enemies (Dethier 1959; Rausher 1981; Holdren and Ehrlich 1982). Because most aposematic caterpillars are monophagous or oligophagous (Stamp 1980), it is probably more difficult for them to find an acceptable host plant than it would be for individuals of a more polyphagous species, making starvation more likely.

Avoiding Predation

Patterns of larval feeding may also be related to strategies of predator avoidance (Heinrich 1979, Chapter 7; Montllor and Bernays, Chapter 5) (Table 10.3). Based on his observations of caterpillar foraging patterns of both aposematic and cryptic caterpillar species, Heinrich (1979) suggested that their foraging strategies were primarily a function of predator avoidance. He found that cryptic, palatable caterpillars did the following: (1) restricted themselves to the underside of leaves at all times, (2) foraged at night, (3) moved away from damaged leaves, and (4) snipped off partially eaten leaves. In contrast, aposematic caterpillars

Table 10.4. Comparison of Foraging Behavior of the Cryptic, Palatable Larvae of Pieris rapae *and the Aposematic Larvae of* Euphydryas phaeton[a]

Behavior	*E. phaeton*	*P. rapae*	Significantly different ($P < 0.05$)
Distance travelled/day (range per individual) (cm)	25–50	23–83	No
Observations in direct sunlight (%)	92	75	Yes
Observations on leaf undersides when in the shade (%)	4	5	No
Observations feeding (%)	14	11	No
Observations moving (%)	18	15	No
Observations stationary (%)	82	85	No
Observations feeding was followed by a change in position on the host plant	54	57 or 74[b]	No

[a]Data are from Mauricio and Bowers (1990).

[b]On broccoli, *Brassica oleracea* and radish, *Raphanus sativus*, respectively.

did not restrict foraging to leaf undersides or to the nighttime, did not move away from damage, and did not snip damaged leaves (Heinrich 1979). Others have suggested that foraging patterns of caterpillars are not related to predator avoidance, but rather are a response to variation in the chemical content of the host plant (e.g., Edwards and Wratten 1983). One prediction of this latter hypothesis is that cryptic, palatable caterpillars and aposematic caterpillars forage in the same way.

A comparison of the foraging behavior of a cryptic palatable caterpillar, the cabbage white, *Pieris rapae* (Pieridae), and an aposematic caterpillar, the Baltimore checkerspot, *Euphydryas phaeton* (Nymphalidae), showed that the foraging behavior of these two species on their host plants and in the absence of predators, was quite similar (Mauricio and Bowers 1990) (Table 10.4). Larvae of the two species spent similar proportions of their time feeding, moving or stationary. They also changed position after feeding a similar percentage of time. *E. phaeton* larvae did spend a greater proportion of their time in direct sunlight then *P. rapae* larvae, but both species spent most of their time in sunlight. Thus in these specialist caterpillars, fixed morphological strategies of predator avoidance (crypsis versus aposematism) either resulted in convergent foraging behavior in the absence of predators, or were less important in determining foraging patterns than other factors.

Although warningly colored, unpalatable caterpillars may be less frequently attacked by most predators and parasitoids, they may provide an abundant food supply for natural enemies that are undeterred by their defenses (e.g., Stamp and Bowers 1988). For example, in a series of field experiments, predatory wasps, *Polistes dominulus* and *P. fuscatus* (Vespidae), killed and transported back to their nests 77 to 99% of *Hemileuca lucina* (Saturniidae) larvae (Stamp and Bowers 1988). These larvae are gregarious, warningly colored black, and defended by stinging spines. However, the spines did not deter the wasps, which bit the spines off before transporting the caterpillars (M. Raveret-Richter, personal communication).

Nonetheless, many predators are, in fact, deterred by caterpillar defenses. For example, larvae of the Monarch butterfly, *D. plexippus* (Nymphalidae) (Brower 1984), and the Baltimore checkerspot, *E. phaeton* (Nymphalidae) (Bowers 1980), were avoided in feeding tests with birds. Ants, *Iridomyrmex humilis*, avoided aposematic prey in choice tests using 36 species of caterpillars (Bernays and Cornelius 1989). *Polistes dominulus* wasps attacked, tasted, then rejected, larvae of the catalpa sphinx (*Ceratomia catalpae*, Sphingidae) (personal observation), which are warningly colored and sequester iridoid glycosides (Table 10.2). The osmeteria of swallowtail caterpillars are effective at repelling small invertebrate predators (Eisner et al. 1970; Honda 1981; Damman 1986). Possession of urticating spines or hairs may be the basis for rejection of certain caterpillars by foraging birds (Heinrich and Collins 1983).

Natural enemies may not only kill caterpillars, but may have indirect effects on larval behavior that then reduce survival (Stamp and Bowers 1988, 1991). For example, predator attack may cause caterpillars to alter their foraging behavior, which may then affect their ability to bask, aggregate, or feed on food of the highest quality (Stamp and Bowers 1988). For gregarious, warningly colored caterpillars, which are relatively easy for predators and parasitoids to find, indirect effects of predators and parasitoids may significantly reduce survival due to alteration of foraging behavior and the consequences thereof (Stamp and Bowers 1991).

Acquiring Defense Compounds

For aposematic caterpillar species that sequester defense compounds from their host plant, there is a fourth component to foraging patterns: acquiring those defense compounds (Table 10.3). The chemical content of the host plant often determines the degree to which a caterpillar and (in some species) the adult can be protected from potential predators. Variation in host plant chemicals available for use in caterpillar defense may occur among host plant species, populations, individuals, as well as among organs within an individual plant (Dolinger et al. 1973; McKey 1979; Roeske et al. 1976; Brower et al. 1982, 1984; Seiber et al. 1986; Harris et al. 1986; Bowers 1988). Similarly, caterpillars (and, in some species, adults) within a single population may vary substantially in degree of unpalatability depending on the plant species, individual, or organ on which the larvae fed (Brower et al. 1967; Bowers 1980, 1981; Brower 1984; Gardner and Stermitz 1988; Belovsky et al. 1989; L'Empereur and Stermitz 1990a,b; Bowers et al. 1992). Thus the decision (or lack thereof) by a caterpillar about which individual plant or plant part to feed on may be critical in determining its degree of unpalatability.

Although the initial feeding site of a newly hatched caterpillar is determined by the ovipositing female, larger caterpillars may move substantial distances (Dethier 1959, 1988; Rausher 1979; Schultz 1983; Hansen et al. 1984; Stamp 1984; Bergelson and Lawton 1988; Mauricio and Bowers 1990), and so may choose plants or plant parts relatively high in the chemicals the caterpillars use for defense. However, plant parts that are highest in defensive chemicals may also be highest in nutrients (McKey 1979). Experiments distinguishing between the importance of nutrients versus chemicals in diet choice of aposematic caterpillars are lacking, but are needed to determine whether caterpillars can and do choose plants or plant parts highest in defensive chemicals, independently of the nutrient content of those plants or plant parts.

Although detailed observations of the foraging behavior of aposematic caterpillars are uncommon, studies of the foraging behavior of two unpalatable caterpillar species showed that they chose plant parts highest in the chemicals that they sequester for their own defense. Larvae of *Euphydryas phaeton* (Nymphalidae) are gregarious, warningly colored orange and black, and sequester iridoid glycosides from their host plants, which make them unpalatable to potential predators (Bowers 1980, 1981; Bowers and Puttick 1986; Belovsky et al. 1989). Observations of feeding behavior of these larvae on a host plant, *Plantago lanceolata* (Plantaginaceae), showed that they ate the new leaves first (Bowers, Schmitt, and Collinge, unpublished data). Similarly, observations of larvae of the buckeye, *Junonia coenia* (Nymphalidae), which also sequester iridoid glycosides, showed that they preferred to feed on new leaves (Stamp and Bowers 1992). New leaves of *P. lanceolata* are higher in iridoid glycosides and nitrogen, a critical nutrient for caterpillars (Scriber and Slansky 1981; Tabashnik and Slansky 1987), than mature leaves (Bowers and Stamp 1992). Therefore it is impossible to determine whether choice of new leaves was due to high nitrogen content or high iridoid glycoside content. As a consequence of feeding on these new leaves, larvae may grow faster due to the high nitrogen content (Scriber and Slansky 1981; Slansky and Rodrieguez 1987), and also have a higher iridoid glycoside content (Bowers and Cooney, unpublished data), which may better protect them from natural enemies.

To distinguish between the relative importance of nutrients compared to iridoid glycosides (or other plant phytochemicals) in determining diet choice, experiments in which nutrient and chemical content of the diet can be decoupled and manipulated independently are necessary. Such diets could then be offered to larvae in paired choice experiments to determine whether nutrients or phytochemicals determined diet choice. For example, Lincoln et al. (1982) showed that the negative effects of a flavonoid resin on caterpillar growth of *Euphydryas chalcedona* (Nymphalidae) could be ameliorated by increasing the amounts of protein in an artificial diet.

Mimicry of Aposematic Caterpillars

Although mimicry of the adult stages of Lepidoptera has been extensively studied (Bates 1862; Gilbert 1983; Brower 1988; Turner 1988), it has been much less frequently considered in the larvae. The three types of mimicry in caterpillars that I consider here are (1) automimicry (or Browerian mimicry, Pasteur 1982), which occurs within a species and reflects differences in palatability among individuals within a single species or population, (2) Batesian mimicry, the resemblance of a palatable mimic to an unpalatable model of a different species, and (3) Mullerian mimicry, the resemblance of two or more unpalatable species to each other.

Automimicry in larvae of unpalatable species, as in the adults (Bowers 1988), is probably widespread, if not ubiquitous. For species that derive their defense compounds from their host plant(s), the variation in the chemical content of different hostplant species, individuals, and plant parts is reflected in the variation among individual adult butterflies in the amount and kind of defensive chemicals they contain (Brower et al. 1968; Roeske et al. 1976; Gardner and Stermitz 1988; Belofsky et al. 1989; L'Empereur and Stermitz 1990a,b; Bowers 1991). Since it is the larval stage that acquires these compounds, similar degrees of variation should and, in the few species studied, do exist in the larvae (Roeske et al. 1976; Bowers 1991, Table 10.2).

Batesian mimicry in caterpillars is considered rare (Turner 1984) or even nonexistent (Sillen-Tullberg 1988). However, there have been few investigations of this phenomenon. In most species in which the adults exhibit Batesian mimicry, such as the monarch, *Danaus plexippus*, and the Viceroy, *Limenitis archippus* (Brower 1958a; but see Ritland and Brower 1991), or the pipevine swallowtail, *Battus philenor*, and its various mimics (Brower 1958b), the larvae do not resemble each other at all.

There are, however, potential examples of Batesian mimicry in caterpillars. One example is the larvae of the relatively palatable butterfly, *Chlosyne harrisii* (Nymphalidae) that appear to mimic the larvae of the checkerspot. *Euphydryas phaeton* (Nymphalidae) (Bowers 1983), which are unpalatable and emetic (Bowers 1980) due to sequestration of iridoid glycosides (Bowers 1988). The larvae of both these species are orange with black spines and they occur in the same habitats (Bowers 1983). In the host plant literature, this resemblance has resulted in the mistaken attribution of the host plant of *C. harrisii, Aster umbellatus* (Asteraceae), as a host plant of *E. phaeton*, probably due to misidentification of the larvae. The adults of *C. harrisii* are palatable (Bowers 1983), whereas *E. phaeton* adults are unpalatable and emetic (Bowers 1980, 1983). The color pattern of the undersides of the adults of both species are similar, and blue-jays learn to avoid adults of *C. harrisii* after experience with the unpalatable *E. phaeton* (Bowers 1983). Although there is Batesian mimicry of adult *E. phaeton* by adult

C. harrissii, the palatability of the larvae of *C. harrisii* has never been tested. It is entirely possible that the larvae of *C. harrissii* may be unpalatable in their own right. If this were the case, then larvae of *E. phaeton* and *C. harrisii* would be Mullerian mimics, whereas the adults are Batesian mimics.

Another putative example of Batesian mimicry in caterpillars occurs between the unpalatable larva of the swallowtail butterfly, *Papilio memnon heronus* and the palatable larva of the puss moth, *Cerura erminea menciana* (Chow and Tsai 1989). The swallowtail larvae have a red, eversible osmeterium, while the puss moth larvae have a red, eversible tube on their posterior end. To human observers, this resemblance suggested that the larvae of these two species were mimics. Experiments with mantids showed that the swallowtail larvae were unpalatable and the puss moth larvae were palatable, thus suggesting that they were Batesian rather than Mullerian mimics. However, these experiments also showed that this apparent mimicry did not protect the puss moth larvae from being eaten by the mantids (Chow and Tsai 1989). This may not, therefore, be an example of Batesian mimicry in caterpillars.

Larvae with urticating spines may also serve as the unpalatable models in Batesian mimicry complexes. Palatable caterpillars may be spiny, but the spines are not urticating. For example, Janzen (1984) suggested that among the larval Saturniidae at Santa Rosa, Costa Rica, the saturniines and some ceratocampines were palatable mimics of the unpalatable hemileucine larvae. All the hemileucines at Santa Rosa (and most hemileucine species elsewhere) had extremely urticating spines, whereas the saturniines and ceratocampines look similar but the spines were harmless (Janzen 1984).

Sawfly larvae might provide examples of Batesian mimicry (Pasteels 1976), but have not been well studied in this regard. One possible example is larvae of *Nematus melanopsis* and *N. saliceti*, which are warningly colored and look similar to human observers (Pasteels 1976). The putative model, *N. melanopsis*, secretes substantial amounts of two defense compounds, benzaldehyde and 2-heptenal, from a row of ventral glands. The suggested mimic, *N. saliceti*, is sympatric with the model, and less common, and has glands that are small and odorless, apparently having lost their defensive function. Experiments to test the efficacy of this mimicry are needed.

Mullerian mimicry has also been documented among lepidopterous caterpillars. Larvae of two geometrids, *Meris alticola* and *Neoterpes graefiaria*, feed on *Penstemon* species (Scrophulariaceae) (Stermitz et al. 1988). Larvae of both species are black and white with yellow-orange spots (Poole 1970), sequester iridoid glycosides (Stermitz et al. 1988), and are probably unpalatable, since other caterpillar species that sequester iridoid glycosides are unpalatable (Bowers 1980). These larvae are also sympatric (Stermitz et al. 1988). Their similar appearances, sequestration of iridoid glycosides, and sympatry suggest that these two caterpillar species are Mullerian mimics (Stermitz et al. 1988). Feeding experiments with potential predators are needed to establish the unpalatability of

Table 10.5. Factors Affecting the Evolution of Coloration in Different Life Stages of Lepidoptera[a]

	Importance for		
Factor	Larva	Pupa	Adult
Visually hunting predators			
Vertebrate	++	+++	+++
Invertebrate	+++	+	+
Parasitoids	+++	+	○
Abiotic factors	++	++	++
Mate recognition	○	○	+++
Intraspecific recognition (other than mating)	○	○	++

[a]+++, very important; ++, important; +, somewhat important; ○, not important.

these larvae, and the ability of predators to avoid larvae of one species after experience with the other.

There are, indeed, relatively few documented examples of mimicry in caterpillars. Although part of this paucity may be due to a concentration of observations and experiments on adult (rather than larval) protective coloration, it appears to be a real phenomenon. Why might there be fewer instances of Batesian and Mullerian mimicry in caterpillars than in adults? A comparison of the importance of the factors contributing to selection for coloration in insects suggests that some factors may have more or less impact on larvae than on adults (Table 10.5). Coloration in adult Lepidoptera may be critical in mate recognition and intraspecific interactions among males (Silberglied 1984), but coloration for conspecific recognition is not at all important for larvae, which do not mate. Parasitoids are important selective agents for caterpillar traits, but larval coloration may not serve as an important cue for parasitoid selection; parasitoids find their prey primarily by chemical means (Vinson 1983). Visually hunting vertebrate predators are probably important selective agents for lepidopteran characters, including coloration, but relative to the evolution of coloration in the Lepidoptera, they may be most influential for the adult stage (Holmes et al. 1979; Dempster 1983; Brower 1984; Gilbert 1984; Bowers et al. 1985).

Mobility of the various life stages of lepidopteran species, as well as of the primary predators on those life stages, may also contribute to the relative paucity of examples of mimicry in caterpillars. For example, because caterpillars are less mobile than adults, and, generally, invertebrate predators less mobile than vertebrate predators, a closer physical association of model and mimic caterpillars (e.g., sharing an individual host plant) might be necessary for the evolution of mimetic resemblance in caterpillars, instead of the kind of general habitat association of model and mimic adults. However, sharing an individual host plant may have serious costs, such as a greater risk of inducing plant defenses or running out of food.

Conclusions and Future Directions

As the existence of this volume attests, the study of caterpillar biology has brought us a greater understanding of many areas within the much broader fields of genetics, physiology, behavior, ecology, and evolution. Despite an apparent human predeliction for studying adult Lepidoptera, and especially adult butterflies, because of their aesthetic as well as scientific appeal, an understanding of lepidopteran and, more generally, insect biology and its contribution to these other fields is impossible without a focus on the larval stage as well. For aposematic Lepidoptera, I have tried to illustrate the importance of study of the larval stage in understanding the evolution of warning coloration, acquisition of defense compounds, predator avoidance, and mimetic resemblance or lack thereof.

From the perspective of the study of aposematic caterpillars, I view the following as particularly productive areas of future enquiry.

1. *The use or disuse of host plant-derived compounds in the gut as a defense*. Although this has been suggested as a potential defense of both cryptic and aposematic caterpillars (Brower 1984; Janzen 1984), it has been little studied. For example, we know little about how the environment of the insect gut affects the compounds potentially used for defense, and we have no idea about how high the concentrations of these compounds can be in the insect gut. There have been no feeding experiments with predators to determine whether feeding on a toxic plant and presence of those compounds only in the gut are sufficient to defend a caterpillar from potential predators.
2. *Unpalatability and mimicry in caterpillars*. In many lepidopteran species, only the adults have been examined for unpalatability or mimicry. A thorough understanding of the ecology and evolution of unpalatability requires the examination of the larval stage as well. For example, is the larva of the Frangipani sphinx, *Pseudosphinx tetrio*, unpalatable and therefore aposematic, or is it a palatable mimic of the coral snake? The larvae are brightly colored black, yellow and red, whereas the adults are mottled gray and appear to be cryptic. These larvae may be unpalatable due to sequestration of defense compounds from their host plants in the Apocynaceae, in which case they are aposematic (Dinther 1956; Janzen 1983; Santiago-Blay 1985) or else they may be mimics of coral snakes (Janzen 1980), and thus not aposematic.

   Because some species may sequester defense chemicals in the larval stage but not retain them through to the adult stage, there may be many caterpillar species that are unpalatable to a wide range of predators, even though the adult stage is palatable to natural enemies. A more thorough examination of the palatability or unpalatability of caterpillars that feed on host plants containing compounds they could sequester would be extremely revealing in this regard.

3. *Use of multiple defenses*. Aposematic caterpillars may enhance the effectiveness of their unpalatability by regurgitation, defensive behaviors, or reflex bleeding. Some caterpillars may be aposematic under some circumstances and relatively cryptic under others (Edmunds 1974; Papageorgis 1975 for adult butterflies; Stamp and Wilkens, Chapter 9). Different defenses may be directed against different natural enemies, or may be effective only under certain conditions, such as attack by predatory wasps versus parasitoids (Cornell et al. 1987), or attack at night versus during the day.
4. *The evolution of aposematism in larval versus adult Lepidoptera*. There are two features of aposematism, presence of distasteful or toxic qualities and advertisement of those qualities. The evolution of the ability of caterpillars to sequester defense compounds from their host plants and in some species to retain them through the pupal to the adult stage has not been well studied. Indeed, we are just beginning to understand caterpillar palatability and chemical defense, whereas these properties are relatively well known for adults. Clearly, caterpillars deserve more attention. Although we certainly have a better understanding of aposematic caterpillars than did Darwin when he wrote to Wallace asking him to explain their bright coloration, there is still much we can learn.

Acknowledgments

I thank Sharon Collinge and Nancy Stamp for comments on the manuscript. Research in my laboratory on aposematic caterpillars has been supported by the National Science Foundation and the Whitehall Foundation.

Literature Cited

Ackery, P.R., and Vane-Wright, R.I. 1986. Milkweed Butterflies. Cornell University Press, Ithaca, NY.

Ahmad, S., Brattsten, L.B., Mullin, C.A., and Yu, S.J. 1986. Enzymes involved in the metabolism of plant allelochemicals, pp. 211–255. *In* L.B. Brattsten and S. Ahmad (eds.), Molecular Aspects of Insect-Plant Associations. Plenum, New York.

Aldrich, J.R., and Blum, M.S. 1978. Aposematic aggregation of a bug and formation of aggregations. Biotropica 10:58–61.

Baldwin, I. T. 1988. Short-term damage-induced alkaloids protect plants. Oecologia 75:367–370.

Bates, H. W. 1862. Contributions to an insect fauna of the Amazon valley. Trans. Linn. Soc. London 23:495–566.

Bellows, T.S., Jr., Owens, J.C., and Huddleston, E.W. 1982. Predation of range caterpillar, *Hemileuca oliviae* (Lepidoptera: Saturniidae) at various stages of development by different species of rodents in New Mexico during 1980. Environ. Entomol. 11:1211–1215.

Belofsky, G., Bowers, M.D., Janzen, S., and Stermitz, F.R. 1989. Iridoid glycosides of *Aureolaria flava* and their sequestration by *Euphydryas phaeton* butterflies. Phytochemistry 28:1601–1604.

Berenbaum, M. 1986. Post-ingestive effects of phytochemicals on insects: on Paracelsus and plant products, pp. 123–154. In T.A. Miller and J. Miller (eds.), Insect-Plant Interactions. Springer-Verlag, New York.

Bergelson, J.M., and Lawton, J.H. 1988. Does foliage damage influence predation on the insect herbivores of birch? Ecology 69:434–445.

Bernays, E.A., and Cornelius, M.L. 1989. Generalist caterpillar prey are more palatable than specialists for the generalist predator *Iridomyrmex humilis*. Oecologia 79:427–430.

Bernays, E.A., and Janzen, D.H. 1988. Saturniid and sphingid caterpillars: Two ways to eat leaves. Ecology 69:1153–1160.

Björkman, C., and Larsson, S. 1991. Pine sawfly defence and variation in host plant resin acids: a trade-off with growth. Ecol. Entomol. 16:283–289.

Blest, A.D. 1963. Longevity, palatability, and natural selection in five species of New World saturniid moths. Nature (London) 197:1183–1186.

Blum, M. S. 1981. Chemical Defenses of Arthropods. Academic Press, New York.

Blum, M. S. 1983. Detoxication, deactivation, and utilization of plant compounds by insects, pp. 265–275. In Hedin, P.A. (ed.), Plant Resistance to Insects. ACS Symposium Series 208. American Chemical Society, Washington, D.C.

Boevé, J.-L., and Pasteels, J.M. 1985. Modes of defense in nematine sawfly larvae: Efficacy against ants and birds. J. Chem. Ecol. 11:1019–1036.

Boros, C.A., Stermitz, F.R., and McFarland, N. 1991. Processing of the iridoid glycoside antirrhinoside from *Maurandya antirrhiniflora* (Scrophulariaceae) by *Meris paradoxa* (Geometridae) and *Lepolys* species (Noctuidae). J. Chem. Ecol. 17:1123–1133.

Bowers, M.D. 1980. Unpalatability as a defense strategy of *Euphydryas phaeton* (Lepidoptera: Nymphalidae). Evolution 34:586–600.

Bowers, M.D. 1981. Unpalatability as a defense strategy of western checkerspot butterflies (*Euphydryas*, Nymphalidae). Evolution 35:367–375.

Bowers, M.D. 1983. Mimicry in North American checkerspot butterflies: *Euphydryas phaeton* and *Chlosyne harrisii* (Nymphalidae). Ecol. Entomol. 8:1–8.

Bowers, M.D. 1988. Plant allelochemistry and mimicry, pp. 273–311. In P. Barbosa and D. Letourneau (eds.), Novel Aspects of Insect-Plant Interactions. Wiley, New York.

Bowers, M.D. 1990. Recycling plant natural products for insect defense, pp. 353–386. In D.L. Evans and J.O. Schmidt (eds.), Insect Defenses. State University of New York Press, Albany, NY.

Bowers, M.D. 1991. The iridoid glycosides, pp. 297–325. In G. Rosenthal (ed.), Herbivores: Their interactions with Secondary Plant Metabolites. Vol. 1, the Chemical Players. Academic Press, New York.

Bowers, M.D. 1992. Unpalatability and the cost of chemical defense in insects, pp. 216–244. In B. Roitberg and M.B. Isman (eds.), Chemical Ecology of Insects: An Evolutionary Approach. Chapman and Hall, New York.

Bowers, M.D., and Collinge, S.K. 1992. Fate of iridoid glycosides in different life stages of the buckeye, *Junonia coenia* (Lepidoptera: Nymphalidae). J. Chem. Ecol. 18:817–831.

Bowers, M.D., and Farley, S. 1990. The behaviour of gray jays (*Perisoreus canadensis*) toward palatable and unpalatable Lepidoptera. Anim. Behav. 39:699–705.

Bowers, M.D., and Larin, Z. 1989. Acquired chemical defense in the lycaenid butterfly, *Eumaeus atala* (Lycaenidae). J. Chem. Ecol. 15:1133–1146.

Bowers, M.D., and Puttick, G.M. 1986. The fate of ingested iridoid glycosides in lepidopteran herbivores. J. Chem. Ecol. 12:169–178.

Bowers, M.D., and Puttick, G.M. 1989. Iridoid glycosides and insect feeding preferences: Gypsy moths (*Lymantria dispar*, Lymantriidae) and buckeyes (*Junonia coenia*, Nymphalidae). Ecol. Entomol. 14:247–256.

Bowers, M.D., and Stamp, N.E. 1992. Chemical variation within and between individuals of *Plantago lanceolata* (Plantaginaceae). J. Chem. Ecol. 18:985–995.

Bowers, M.D., Brown, I.L., and Wheye, D. 1985. Bird predation as a selective agent in a butterfly population. Evolution 39:93–103.

Bowers, M.D., Stamp, N.E., and Collinge, S.K. 1992. Early stage of host range expansion by a specialist herbivore, *Euphydryas phaeton* (Nymphalidae). Ecology 73:526–536.

Boyden, T.C. 1976. Butterfly palatability and mimicry: experiments with *Ameiva* lizards. Evolution 30:73–81.

Brandon, H. 1987. The snack that crawls. Internatl. Wildlife 17(2):16–21.

Brattsten, L.B. 1986. Fate of ingested plant allelochemicals in herbivorous insects, pp. 211–255. In L.B. Brattsten and S. Ahmad (eds.), Molecular Aspects of Insect Plant Associations. Plenum, New York

Brower, J.V.Z. 1958a. Experimental studies of mimicry in some North American butterflies. I. The monarch *Danaus plexippus* and viceroy *Limenitis archippus*. Evolution 12:32–47.

Brower, J.V.Z. 1958b. Experimental studies of mimicry in some North American butterflies. II. *Battus philenor* and *Papilio troilus, P. polyxenes* and *P. glaucus*. Evolution 12:273–285.

Brower, L.P. 1984. Chemical defense in butterflies, pp. 109–132. In R.I. Vane-Wright and P.R. Ackery (eds.), The Biology of Butterflies. Academic Press, London.

Brower, L.P. 1988. Avian predation on the monarch butterfly and its implications for mimicry theory. Am. Nat. 131:S4–S6.

Brower, L.P., Brower, J.V.Z., and Corvino, J.M. 1967. Plant poisons in a terrestrial food chain. Proc. Natl. Acad. Sci. U.S.A. 57:893–898.

Brower, L.P., Ryerson, W.N., Coppinger, L.L., and Glazier, S.C. 1968. Ecological chemistry and the palatability spectrum. Science 161:1349–1351.

Brower, L.P., Seiber, J.N., Nelson, C.J., Lynch, S.P., and Tuskes, P.M. 1982. Plant determined variation in the cardenolide content, thin-layer chromatography profiles, and emetic potency of Monarch butterflies, *Danaus plexippus*, reared on the milkweed, *Asclepias eriocarpa* in California. J. Chem. Ecol. 8:579–633.

Brower, L.P., Seiber, J.N., Nelson, C.J., Lynch, S.P., and Holland, M.M. 1984. Plant determined variation in the cardenolide content, thin-layer chromatography profiles, and emetic potency of monarch butterflies, *Danaus plexippus*, reared on the milkweed *Asclepias speciosa* in California J. Chem. Ecol. 10:601–639.

Brown, K.S. 1984. Adult-obtained pyrrolizidine alkaloids defend ithomiine butterflies against a spider predator. Nature (London) 309:707–709.

Carpenter, G.D.H. 1942. Observations and experiments in Africa by the late C.F.M. Swynnerton on wild birds eating butterflies and the preference shown. Proc. Linn. Soc. London 154:10–46.

Casey, T.M. 1976. Activity patterns, body temperature and thermal ecology of two desert caterpillars. Ecology 57:485–497.

Casey, T.M., and Hegel, J.R. 1981. Caterpillar setae: Insulation for an ectotherm. Science 214:1131–1133.

Chow, Y.S., and Tsai, R.S. 1989. Protective chemicals in caterpillar survival. Experientia 45:390–192.

Coppinger, R.P. 1970. The effect of experience and novelty on avian feeding behavior with reference to the evolution of warning coloration in butterflies. II. Reactions of naive birds to novel insects. Am. Nat. 104:323–335.

Cornell, J.C., Stamp, N.E., and Bowers, M.D. 1987. Developmental change in aggregation, defense and escape behavior of buckmoth caterpillars, *Hemileuca lucina* (Saturniidae). Behav. Ecol. Sociobiol. 20:383–388.

Cott, H.B. 1940. Adaptive Coloration in Animals. Methuen, London.

Damman, H. 1986. The osmeterial glands of the swallowtail butterfly *Eurytides marcellus* as a defense against natural enemies. Ecol. Entomol. 11:261–265.

Damman, H. 1987. Leaf quality and enemy avoidance by the larvae of a pyralid moth. Ecology 68:88–97.

Darwin, C.D. 1878. The Descent of Man, Part ii. Sexual Selection. London.

Dempster, J.P. 1983. The natural control of butterflies and moths. Biol. Rev. 58:461–481.

Dethier, V.G. 1959. Food-plant distribution and density as factors affecting insect populations. Can. Entomol. 91:581–596.

Dethier, V.G. 1988. The feeding behavior of a polyphagous caterpillar (*Diacrisia virginica*) in its natural habitat. Can. J. Zool. 66:1280–1288.

Dethier, V.G. 1989. Patterns of locomotion of polyphagous arctiid caterpillars in relation to foraging. Ecol. Entomol. 14:375–386.

DeVries, P. J. 1976. Notes on the behavior of *Eumaeus minyas*, an aposematic lycaenid butterfly. Brenesia 10:269–270.

Dinther, J.B.M. 1956. Three noxious hornworms in Surinam. Entomol. Ber. 16:12–15.

Dolinger, P.M., Ehrlich, P.R., Fitch, W.L., and Breedlove, D.E. 1973. Alkaloid and predation patterns in Colorado lupine populations. Oecologia 13:191–204.

Downey, J.C., and Allyn, A.C. 1973. Butterfly ultrastructure. I. Sound production and associated abdominal structures in pupae of Lycaenidae and Riodinadae. Bull. Allyn Mus. 14:1–48.

Dreyer, J.J., and Wehmeyer, A.S. 1982. On the nutritive value of mopanie worms. S. Afri. J. Sci. 78:33–35.

Duffey, S.S. 1980. Sequestration of plant natural products by insects. Annu. Rev. Entomol. 25:447–477.

Duffey, S.S., and Scudder, G.G.E. 1974. Cardiac glycosides in *Oncopeltus fasciatus* (Dallas). I. The uptake and distribution of natural cardenolides in the body. Can. J. Zool. 52:283–290.

Duffey, S.S., Blum, M.S., Isman, M., and Scudder, G.G.E. 1978. Cardiac glycosides: A physical system for their sequestration by the milkweed bug. J. Insect Physiol. 24:639–645.

Edmunds, M. 1974. Defence in Animals: A Survey of Antipredator Defences. Longman, Essex, England.

Edmunds, M. 1990. The evolution of cryptic coloration, pp. 3–22. In D.L. Evans and J.O. Schmidt (eds.), Insect Defenses. State University of New York Press, Albany, NY.

Edwards, P.J., and Wratten, S.D. 1983. Wound induced defenses in plants and their consequences for patterns of insect grazing. Oecologia 59:88–93.

Edwards, P.J., and Wratten, S.D. 1985. Induced plant defences against insect grazing: Fact or artefact? Oikos 44:70–74.

Eisner, T. 1970. Chemical defense against predation in arthropods, pp. 157–217. In E. Sondheimer and J.B. Simeone (eds.), Chemical Ecology. Academic Press, New York.

Eisner, T., and Meinwald, J. 1965. Defensive secretion of a caterpillar (*Papilio*). Science 150:1733–1735.

Eisner, T., Pliske, T.E., Ikeda, M., Owen, D.F., Vazquez, L., Perez, H., Franclemont, J.G., and Meinwald, J. 1970. Defense mechanisms of arthropods XXVII. Osmeterial secretion of papilionid caterpillars (*Baronia, Papilio, Eurytides*). Ann. Entomol. Soc. Am. 63:914–915.

Eisner, T., Kluge, A.F., Carrel, J.C., and Meinwald, J. 1972. Defense mechanisms of arthropods. XXXIV. Formic acid and acyclic ketones in the spray of a caterpillar. Ann. Entomol. Soc. Am. 65:765–766.

Eisner, T., Johnessee, J.S., Carrel, J., Hendry, L.B., and J. Meinwald. 1974. Defensive use by an insect of a plant resin. Science 184:996–999.

Endler, J.A. 1986. Defense against predators, pp. 109–134. In M.E. Feder and G.V.

Lauder (eds.), Predator-Prey Relationships: Perspectives and Approaches from the Study of Lower Vertebrates. University of Chicago Press, Chicago, IL.

Evans, H.E., and West-Eberhard, M.J. 1970. The Wasps. University of Michigan Press, Ann Arbor.

Fisher, R.A. 1930. The Genetical Theory of Natural Selection. Clarendon Press, Oxford.

Fitzgerald, T.D. 1980. An analysis of daily foraging patterns of laboratory colonies of the eastern tent caterpillar, *Malacosoma americanum* (Lepidoptera: Lasiocampidae) recorded photoelectronically. Can. Entomol. 112:731–738.

Fitzgerald, T.D., and Willer, D.E. 1983. Tent-building behavior of the eastern tent caterpillar *Malacosoma americanum* (Lepidoptera: Lasiocampidae). J. Kan. Entomol. Soc. 56:20–31.

Furuta, K. 1983. Behavioral response of the Japanese paper wasp (*Polistes jadwigae* Dalla Torre: Hymenoptera: Vespidae) to the gypsy moth (*Lymantria dispar* L.: Lepidoptera: Lymantriidae). App. Entomol. Zool. 18:464–474.

Gardner, D.R., and Stermitz, F.R. 1988. Hostplant utilization and iridoid glycoside sequestration by *Euphydryas anicia* (Lepidoptera: Nymphalidae). J. Chem. Ecol. 14:2147–2168.

Gibson, D.O. 1980. The role of escape in mimicry and polymorphism: I. The response of captive birds to artificial prey. Biol. J. Linn. Soc. 14:201–214.

Gilbert, L.E. 1983. Coevolution and mimicry, pp. 263–281. In D.J. Futuyma and M.S. Slatkin (eds.), Coevolution. Sinauer, Sunderland, MA.

Gilbert, L.E. 1984. The biology of butterfly communities, pp. 41–53. In R.I. Vane-Wright and P.R. Ackery (eds.), The Biology of Butterflies. Academic Press, New York.

Godfrey, G.L., and Appleby, J.E. 1987. Notodontidae (Noctuoidea), pp. 524–533. In F. W. Stehr (ed.), Immature Insects. Kendall/Hunt, Dubuque, IA.

Guilford, T. 1986. How do "warning colours" work? Conspicuousness may reduce recognition errors in experienced predators. Anim. Behav. 34:286–288.

Guilford, T. 1988. The evolution of conspicuous coloration. Am. Nat. 131:S7–S21.

Guilford, T. 1990. The evolution of aposematism, pp. 23–61. In J.O. Schmidt and D.L. Evans (eds.), Insect Defenses. State University of New York Press, Albany, NY.

Hamilton, W.D. 1971. Geometry for the selfish herd. J. Theor. Biol. 31:295–311.

Hansen, J.D., Ludwig, J.A., Owens, J.C., and Huddleston, E.W. 1984. Larval movement of the range caterpillar, *Hemileuca oliviae*. (Lepidoptera: Saturniidae). Environ. Entomol. 13:415–420.

Happ, G.M., and Eisner, T. 1961. Hemorrhage in a coccinellid beetle and its repellent effect on ants. Science 132:329–331.

Harris, G.H., Stermitz, F.R., and Jing, W. 1986. Iridoids and alkaloids from *Castilleja* (Scrophulariaceae) host plants for *Platyptilia pica* (Lepidoptera: Pterophoridae): Rhexifoline content of *P. pica*. Biochem. Syst. Ecol. 14:499–504.

Heinrich, B. 1979. Foraging strategies of caterpillars: Leaf damage and possible predator avoidance strategies. Oecologia 42:325–337.

Heinrich, B., and Collins, S.L. 1983. Caterpillar leaf damage and the game of hide and seek with birds. Ecology 64:592–602.

Herrick, G.W., and Detwiler, J.D. 1919. Notes on the repugnatorial glands of certain notodontid caterpillars. Ann. Entomol. Soc. Am. 12:44–48.

Hinton, H.E. 1948. Sound production in lepidopterous pupae. Entomologist 81:254–269.

Holdren, C.E., and Ehrlich, P.R. 1982. Ecological determinants of food plant choice in the checkerspot butterfly, *Euphydryas editha* in Colorado. Oecologia 52:417–423.

Holmes. R.T., Schultz, J.C., and Nothnagle, P. 1979. Bird predation on forest insects: An exclosure experiment. Science 206:462–463.

Honda, K. 1980a. Volatile constituents of larval osmeterial secretions in *Papilio protenor demetrius*. J. Insect Physiol. 26:39–45.

Honda, K. 1980b. Osmeterial secretions of papilionid larvae in the genera *Luehdorfia, Graphium* and *Atrophaneura* (Lepidoptera). Insect Biochem. 10:583–588.

Honda, K. 1981. Larval osmeterial secretions of the swallowtails (*Papilio*). J. Chem. Ecol. 7:1089–1113.

Honda, K. 1983. Defensive potential of components of the larval osmeterial secretion of papilionid butterflies against ants. Physiol. Entomol. 8:173–179.

Isman, M.B., Duffey, S.S., and Scudder, G.G.E. 1977. Cardenolide content of some leaf-and stem-feeding insects on temperate North American milkweeds (*Asclepias* spp.). Can. J. Zool. 55:1024–1028.

Janzen, D.H. 1980. Two potential coral snake mimics in a tropical deciduous forest. Biotropica 12:77–78.

Janzen, D.H. 1983. *Pseudosphinx tetrio*, 754–765. In D.G. Janzen (ed.), Costa Rican Natural History. University of Chicago Press, Chicago IL.

Janzen, D.H. 1984. Two ways to be a tropical big moth: Santa Rosa saturniids and sphingids. Oxford Surveys Evol. Biol. 1:85–140.

Jarvi, T., Sillen-Tullberg, B., and Wiklund, C. 1981. The cost of being aposematic. An experimental study of predation on larvae of *Papilio machaon* by the great tit, *Parus major*. Oikos 36:267–272.

Jones, F.M. 1932. Insect coloration and the relative acceptability of insects to birds. Trans. R. Entomol. Soc. London 80:345–385.

Jones, F.M. 1934. Further experiments on colouration and relative acceptability of insects to birds. Trans. R. Entomol. Soc. London 82:443–453.

Jones, R.E. 1977. Search behaviour: A study of three caterpillar species. Behaviour 60:237–259.

Kawamoto, F., and Kumada, N. 1984. Biology and venoms of Lepidoptera, pp. 291–330. In A.T. Tu (ed.), Handbook of Natural Toxins, Vol. 2. Insect Poisons, Allergens, and Other Invertebrate Venoms. Dekker, New York.

Kidd, N.A.C. 1982. Predator avoidance as a result of aggregation in the grey pine aphid. *Schizolachnus pineti*. J. Anim. Ecol. 51:397–412.

Knapp, R., and Casey, T.M. 1986. Thermal ecology, behavior, and growth of gypsy moth and eastern tent caterpillars. Ecology 67:598–608.

L'Empereur, K.M., and Stermitz, F.R. 1990a. Iridoid glycoside content of *Euphydryas anicia* (Lepidoptera: Nymphalidae) and its major hostplant, *Besseya plantaginea* (Scrophulariaceae), at a high plains Colorado site. J. Chem. Ecol. 16:187–197.

L'Empereur, K.M., and Stermitz, F.R. 1990b. Iridoid glycoside metabolism and sequestration by *Poladryas minuta* (Lepidoptera: Nymphalidae) feeding on *Penstemon virgatus* (Scrophulariaceae). J. Chem. Ecol. 16:1495–1506.

Langston, R.L. 1957. A synopsis of hymenopterous parasites of *Malacosoma* in California (Lepidoptera: Lasiocampidae). Univ. Calif. Publ. Entomol. 14:1–50.

Larsson, S., Bjorkman, C., and Gref, R. 1986. Responses of *Neodiprion sertifer* (Hym., Diprionidae) larvae to variation in needle resin acid concentration in Scots pine. Oecologia 70:77–84.

Lawton, J.H. 1986. Surface availability and insect community structure: The effects of architecture and fractal dimension of plants, pp. 317–331. In B. Juniper and T.R.E. Southwood (eds.), Insects and the Plant Surface. Arnold, London.

Lincoln, D.E., Newton, T.S., Ehrlich, P.R., and Williams, K.S. 1982. Coevolution of the checkerspot butterfly *Euphydryas chalcedona* and its larval food plant *Diplacus aurantiacus*: Response to protein and leaf resin. Oecologia 52:216–223.

Marchant, J. 1975. Alfred Russel Wallace: Letters and Reminiscences. Arno Press, New York.

Marti, O.G., and Rogers, C.E. 1988. Anatomy of the ventral eversible gland of fall armyworm, *Spodoptera frugiperda* (Lepidoptera: Noctuidae), larvae. Ann. Entomol. Soc. Am. 81:308–317.

Maschwitz, U.W.J., and Kloft, W. 1971. Morphology and function of the venom apparatus of insects: Bees, wasps, ants, and caterpillars, pp. 1–60. In W. Bucherl and E.E. Buckley (eds.), Venomous Animals and their Venoms. Academic Press, New York.

Mason, R., and Torgersen, T. 1983. Mortalilty of larvae in stocked cohorts of the Douglas-fir tussock moth, *Orgyia pseudotsugata* (Lepidoptera: Lynamtriidae). Can. Entomol. 104:1119–1127.

Mattson, W.J. 1980. Herbivory in relation to plant nitrogen content. Annu. Rev. Ecol. Syst. 11:119–161.

Mauricio, R., and Bowers, M.D. 1990. Do caterpillars disperse their damage?: Larval foraging behaviour of two specialist herbivores, *Euphydryas phaeton* (Nymphalidae) and *Pieris rapae* (Pieridae). Ecol. Entomol. 15:153–161.

McKey, D. 1979. The distribution of secondary compounds within plants, pp. 55–133. In G. Rosenthal and D.H. Janzen (eds.), Herbivores: Their Interactions with Plant Secondary Metabolites. Academic Press, New York.

Miller, J.S. 1991. Cladistics and classification of the Notodontidae (Lepidoptera: Noctuoidea) based on larval and adult morphology. Bull. Am. Mus. Nat. Hist. 204:1–230.

Montllor, C.B., Bernays, E.A., and Cornelius, M.L. 1991. Responses of two hymenopteran predators to surface chemistry of their prey: Significance for an alkaloid-sequestering caterpillar. J. Chem. Ecol. 17:391–400.

Moore, L.V., and Scudder, G.G.E. 1986. Ouabain-resistant Na, K-ATPases and cardenolide tolerance in the large milkweed bug, *Oncopeltus fasciatus*. J. Insect Physiol. 32:27–33.

Morris, R. 1972. Predation by wasps, birds and mammals on *Hyphantria cunea*. Can. Entomol. 104:1581–1589.

Morrow, P.A., Bellas, T.E., and Eisner, T. 1976. Eucalyptus oils in the defensive oral discharge of Australian sawfly larvae (Hymenoptera: Pergidae). Oecologia 24:193–206.

Nentwig, W. 1985. A tropical caterpillar that mimics faeces, leaves and a snake (Lepidoptera: Oxytenidae: *Oxytenis naemia*). J. Res. Lepid. 24:136–141.

Nishio, S. 1983. The fates and adaptive significance of cardenolides sequestered by larvae of *Danaus plexippus* (L.) and *Cycnia inopinatus* (Hy. Edwards). Ph. D. thesis. University Microfilms, University of Georgia, Athens, GA.

Papageorgis, C. 1975. Mimicry in neotropical butterflies. Am. Sci. 63:522–532.

Pasteels, J.M. 1976. Evolutionary aspects in chemical ecology and chemical communication. Internatl. Cong. Entomol. Proc. 15:281–293.

Pasteur, G. 1982. A classificatory review of mimicry systems. Annu. Rev. Ecol. Syst. 13:169–199.

Pearson, D.L. 1989. What is the adaptive significance of multicomponent defensive repertoires? Oikos 54:251–253.

Pesce, H., and Delgado, A. 1971. Poisoning from adult moths and caterpillars, pp. 119–156. In W. Bucherl and E.E. Buckley (eds.), Venomous Animals and their Venoms, Vol. 3. Academic Press, New York.

Peterson, S., Johnson, N.D., and LeGuyader, J.L. 1987. Defensive regurgitation of allelochemicals derived from host cyanogenesis by eastern tent caterpillars. Ecology 68:1268–1272.

Poole, R.W. 1970. Convergent evolution in the larvae of two *Penstemon*-feeding geometrids (Lepidoptera: Geometridae). J. Kan. Entomol. Soc. 43:292–297.

Porter, K. 1982. Basking behaviour in larvae of the butterfly *Euphydryas aurinia*. Oikos 38:308–312.

Pough, F.H. 1988. Mimicry of vertebrates: Are the rules different? Am. Nat. 131:S67–S102.

Poulton, E.B. 1887. The secretion of pure aqueous formic acid by lepidopterous larvae for the purpose of defence. Br. Assoc. Adv. Sci. Rept. 57:765–766.

Poulton, E.B. 1890. The Colours of Animals. Appleton, New York.

Prop, N. 1960. Protection against birds and parasites in some species of tenthredinid larvae. Neth. J. Zool. 13:380–447.

Puttick, G.M., and Bowers, M.D. 1988. Effect of qualitative and quantitative variation in allelochemicals on a generalist insect: Iridoid glycosides and the southern armyworm. J. Chem. Ecol. 154:335–351.

Quin, P.J. 1959. Foods and Feeding Habits of the Pedi. Witwaterstrand University Press, Johannesburg, South Africa.

Rausher, M.D. 1979. Egg recognition: Its advantage to a butterfly. Anim. Behav. 27:1034–1040.

Rausher, M.D. 1981. Host plant selection by *Battus philenor* butterflies: The roles of predation, nutrition, and plant chemistry. Ecol. Monog. 51:1–20.

Reavey, D., and Lawton, J.H. 1991. Larval fitness in leaf-eating insects, pp. 293–329. In W. Bailey and T.J. Ridsdill Smith (eds.), Reproductive Behaviour in Insects: Individuals and Populations. Chapman and Hall, New York.

Rhoades, D.F. 1985. Offensive-defensive interactions between herbivores and plants: Their relevance in herbivore population dynamics and ecological theory. Am. Nat. 125:205–238.

Ritland, D.B., and Brower, L.P. 1991. The viceroy butterfly is not a batesian mimic. Nature (London) 350:497–498.

Roeske, C.N., Seiber, J.S., Brower, L.P., and Moffitt, C.M. 1976. Milkweed cardenolides and their comparative processing by monarch butterflies (*Danaus plexippus*). Rec. Adv. Phytochem. 10:93–167.

Root, R.B. 1966. The avian response to a population outbreak of the tent caterpillar, *Malacosoma constrictum* (Stretch) (Lepidoptera: Lasiocampidae). Pan-Pac. Entomol. 42:48–52.

Rossiter, M.C., Schultz, J.C., and Baldwin, I.T. 1988. Relationships among defoliation, red oak phenolics, and gypsy moth growth and reproduction. Ecology 69:267–277.

Roth, L.M., and Eisner, T. 1962. Chemical defenses of arthropods. Annu. Rev. Entomol. 7:107–136.

Rothschild, M. 1972. Secondary plant substances and warning colouration in insects, pp. 59–83. In H.F. Van Emden (ed.), Insect/Plant Relationships. Wiley, New York.

Rothschild, M. 1985. British aposematic Lepidoptera, pp. 9–62. In J. Heath and A.M. Emmet (eds.), The Moths and Butterflies of Great Britain and Ireland. Harley Books, Essex, England.

Rothschild, M., Reichstein, T., von Euw, J., Aplin, R.T., and Harman, R.R.M. 1970. Toxic Lepidoptera. Toxicon 8:293 299.

Rowell-Rahier, M., and Pasteels, J.M. 1986. Economics of chemical defense in Chrysomelinae. J. Chem. Ecol. 12:1189–1203.

Santiago-Blay, J.A. 1985. Notes on *Pseudosphinx tetrio* (L.) (Sphingidae) in Puerto Rico. J. Lepid. Soc. 39:208–214.

Schuler, W., and Hesse, E. 1985. On the function of warning coloration: A black and yellow pattern inhibits prey-attack by naive domestic chicks. Behav. Ecol. Sociobiol. 16:249–255.

Schultz, J.C. 1983. Habitat selection and foraging tactics of caterpillars in heterogeneous trees, pp. 61–90. In R.F. Denno and M.S. McClure (eds.), Variable Plants and Herbivores in Natural and Managed Systems. Academic Press, New York.

Scriber, J.M., and Slansky, F., Jr. 1981. The nutritional ecology of immature insects. Annu. Rev. Entomol. 26:183–211.

Seiber, J.N., Brower, L.P., Lee, S.M., McChesney, M.M., Cheung, H.T.A., Nelson, C.J., and Watson, T.R. 1986. Cardenolide connection between overwintering monarch butterflies from Mexico and their larval food plant, *Asclepias syriaca*. J. Chem. Ecol. 12:1157–1170.

Self, L.S., Guthrie, F.E., and Hodgson, E. 1964. Adaptations of tobacco hornworms to the ingestion of nicotine. J. Insect Physiol. 10:907–914.

Sheehan, W. 1992. Caterpillar defenses, conspicuous consumption and the evolution of gregariousness in Lepidoptera. Unpublished manuscript.

Sherman, P.W., and Watt, W.B. 1973. The thermal ecology of some *Colias* butterflies. J. Comp. Physiol. 83:25–40.

Silberglied, R.E. 1984. Visual communication and sexual selection in butterflies, pp. 207–223. In R.I. Vane-Wright and P.R. Ackery (eds.), The Biology of Butterflies. Academic Press, New York.

Sillen-Tullberg, B. 1988. Evolution of gregariousness in aposematic butterfly larvae: A phylogenetic analysis. Evolution 42:293–305.

Sillen-Tullberg, B., and Leimar, O. 1988. The evolution of gregariousness in distasteful insects as a defense against predators. Am. Nat. 132:723–734.

Slansky, F., Jr., and Rodrieguez, J.G. (eds.). 1987. Nutritional Ecology of Insects, Mites, Spiders, and Related Invertebrates. Wiley, New York.

Slansky, F., Jr., and Scriber, J.M. 1985. Food consumption and utilization pp. 87–163. In G.A. Kerkut and L. Gilbert (eds.), Comprehensive Insect Physiology, Biochemistry and Pharmacology, Vol. 4. Pergamon, Oxford.

Smith, S.M. 1975. Innate recognition of coral snake pattern by a possible avian predator. Science 187:759–760.

Stamp, N.E. 1980. Egg deposition patterns in butterflies: Why do some species cluster their eggs rather than deposit them singly? Am. Nat. 115:367–380.

Stamp, N.E. 1982. Behavioral interactions of parasitoids and Baltimore checkerspot caterpillars (*Euphydryas phaeton*). Environ. Entomol. 11:100–104.

Stamp, N.E. 1984. Foraging behavior of tawny emperor caterpillars *Asterocampa clyton* (Nymphalidae). J. Lepid. Soc. 40:191–205.

Stamp, N.E. 1986. Physical constraints of defense and response to invertebrate predators by pipevine caterpillars (*Battus philenor*: Papilionidae). J. Lepid. Soc. 40:191–205.

Stamp, N.E., and Bowers, M.D. 1988. Direct and indirect effects of predatory wasps (*Polistes sp*.: Vespidae) on gregarious caterpillars (*Hemileuca lucina*: Saturniidae). Oecologia 75:619–624.

Stamp, N.E., and Bowers, M.D. 1990a. Variation in food quality and temperature constrain foraging of gregarious caterpillars. Ecology 71:1031–1039.

Stamp, N.E., and Bowers, M.D. 1990b. Body temperature, behavior and growth of early-spring caterpillars (*Hemileuca lucina*, Saturniidae). J. Lepid. Soc. 44:143–153.

Stamp, N.E., and Bowers, M.D. 1991. Indirect effect on survivorship of caterpillars due to presence of invertebrate predators. Oecologia 88:325–330.

Stamp, N.E., and Bowers, M.D. 1992. Behaviour of specialist and generalist caterpillars on plantain (*Plantago lanceolata*). Ecol. Entomol. 17:(in press).

Stermitz, F.R., Gardner, D.R., and McFarland, N. 1988. Iridoid glycoside sequestration by two aposematic *Penstemon*-feeding geometrid larvae. J. Chem. Ecol. 14:435–441.

Tabashnik, B., and Slansky, F., Jr. 1987. Nutritional ecology of forb foliage-chewing insects, pp. 71–103. In F. Slansky, Jr. and J.G. Rodrieguez (eds.), Nutritional Ecology of Insects, Mites, Spiders, and Related Invertebrates. Wiley, New York.

Tautz, J., and Markl, H. 1978. Caterpillars detect flying wasps by hairs sensitive to airborne vibration. Behav. Ecol. Sociobiol. 4:101–110.

Terrierre, L.C. 1984. Induction of detoxication enzymes in insects. Annu. Rev. Entomol. 29:71–88.

Tostowaryk, W. 1971. Life history and behavior of *Podisus modestus* (Hemiptera: Pentatomidae) in a boreal forest in Quebec. Can. Entomol. 103:662–673.

Tostowaryk, W. 1972. The effect of prey defense on the functional response of *Podisus modestus* (Hemiptera: Pentatomidae) to densities of the sawflies *Neodiprion swainei* and *N. pratti banksianae* (Hymenoptera: Neodiprionidae). Can. Entomol. 104:61–69.

Treherne, J.E., and Foster, W.A. 1980. The effects of group size on predator avoidance in a marine insect. Anim. Behav. 1119–1122.

Treherne, J.E., and Foster, W.A. 1981. Group transmission of predator avoidance in a marine insect: The Trafalgar effect. Anim. Behav. 29:911–917.

Treherne, J.E., and Foster, W.A. 1982. Group size and anti-predator strategies in a marine insect. Anim. Behav. 30:536–542.

Treisman, M. 1975. Predation and the evolution of gregariousness. I. Models for concealment and evasion. Anim. Behav. 23:779–800.

Turner, G.F., and Pitcher, T.J. 1986. Attack abatement: a model for group protection by combined avoidance and dilution. Am. Nat. 128:228–240.

Turner, J.R.G. 1984. Mimicry: The palatability spectrum and its consequences, pp. 141–161. In R.I. Vane-Wright and P.R. Ackery (eds.), The biology of Butterflies. Academic Press, London.

Turner, J.R.G. 1988. The evolution of mimicry: A solution to the problem of punctuated equilibrium. Am. Nat. 131:S42–S66.

Vaughan, G.L., and Jungreis, A.M. 1977. Insensitivity of lepidopteran tissues to ouabain: Physiological mechanisms for protection from cardiac glycosides. J. Insect Physiol. 23:585–589.

Vinson, S.B. 1983. Parasitoid-host relationship, pp. 205–233. In W.J. Bell and R.T. Carde (eds.), Chemical Ecology of Insects. Sinauer, Sunderland, MA.

Vulinec, K. 1990. Collective security: Aggregation by insects as a defense, pp. 251–288. In D.L. Evans and J.O. Schmidt (eds.), Insect Defenses. State University of New York Press, Albany NY.

Weller, S.J. 1987. *Litodonta hydromeli* Harvey (Notodontidae): Description of life stages. J. Lepid. Soc, 41:187–194.

Wiklund, C., and Jarvi, T. 1982. Survival of distasteful insects after being attacked by naive birds: A reappraisal of the theory of aposematic coloration evolving through individual selection. Evolution 36:998–1002.

Wiklund, C., and Sillen-Tullberg, B. 1985. Why distasteful butterflies have aposematic larvae and adults, but cryptic pupae: Evidence from predation experiments on the monarch and the European swallowtail. Evolution 39:1155–1158.

Wink, M., and Schneider, D. 1988. Carrier-mediated uptake of pyrrolizidine alkaloids in larvae of the aposematic and alkaloid-exploiting moth *Creatonotos*. Naturwissenschaften 75:524–525.

Yamasuki, M., Hirose, Y., and Takagi, M. 1978. Repeated visits of *Polistes jadwigae* Dalla Torre (Hymenoptera: Vespidae) to its hunting site. J. App. Entomol. Zool. 22:51–55 (in Japanese with English summary).

11

Sociality in Caterpillars

T. D. Fitzgerald

Introduction

Assemblages of caterpillars, unlike ants, termites, and other eusocial insects, may or may not constitute adaptive networks of cooperating individuals. Nonco-operative assemblages occur when populations of normally solitary species reach epizootic proportions forcing individuals to forage in proximity. When such conditions are encountered frequently enough during the evolutionary life of a species, normally solitary individuals are likely to be genetically prepared to switch to behavioral patterns that enable them to cope with intense intraspecific competition. Gypsy moth larvae, for example, rest on the bole of the tree at low density but when crowded stay near their food (Leonard 1970). When compared to low-density populations, crowded army worms (*Leucania separta*) are more conspicuously colored, feed during the day rather than at night, are more active and more responsive to disturbance, feed on a broader range of food plants, and show a greater tolerance for starvation (Iwao 1968 and references therein).

Distinct from these and other such "facultatively gregarious" species, some species of caterpillars are genetically programmed to aggregate for a portion or for all of their larval life, and show well-defined patterns of integrated behavior. Approximately 5% of the larvae of North American butterflies form aggregates containing 10 or more caterpillars (Stamp 1980). The larvae of moths are less well known, but analysis of life history data collected during surveys of Canadian forest species showed that 7.7% of the larvae of the 392 species studied aggregate in groups containing eight or more caterpillars for at least part of the larval stage (Herbert 1983). Similar analysis of data for moths found on the British Isles indicates that 4.1% of 783 species lay their eggs in batches (Herbert 1983), but since moth larvae may disperse immediately after hatching, this value probably overestimates the number of species that remain gregarious as larvae.

In spite of their relatively small numbers compared to solitary species, social caterpillars are particularly successful, often achieving numerical dominance in the communities they inhabit. Approximately 70% of the forest species that periodically achieve outbreak status are batch layers; of these, over half are gregarious as larvae (Nothnagle and Schultz 1987). The reason why so many irruptive species are gregarious is unclear but when conditions are just right, social species may have particularly high survival rates due to enhanced predator defense, thermoregulation, and foraging efficiency (Nothnagle and Schultz 1987).

It is the purpose of this chapter to review what is known of the foraging behavior of these "obligatorially gregarious" species, and to consider the extent to which their behavior enhances fitness. Emphasis is placed on behavioral patterns associated with chemical communication and cooperative foraging, social facilitation in feeding, cooperative shelter building, thermoregulation, and antipredator defense. Frequent reference is made to studies of the tent caterpillars (*Malacosoma*), which are the best known of the social species (Fitzgerald and Peterson 1988).

While biologists consider a broad diversity of vertebrate and invertebrate organisms to be social, the term has had much more restrictive usage among entomologists. The social insects are generally considered to be those that have overlapping generations, cooperative brood care, and division of labor: ants, termites, some bees, and wasps. Although gregarious caterpillars may exhibit adaptive group behaviors even more integrated than those of many groups of vertebrates considered to be social (e.g., fish schools, Wittenberg 1981), gregarious caterpillars are considered by entomologists to be only "presocial" (Eickwort 1981), a term that tends to obscure the true nature of these associations (Fitzgerald and Peterson 1988). In this chapter, species of obligatorially gregarious caterpillars are referred to as social organisms in the larger sense of the term.

Aggregation in Caterpillars

Colony Size

Aggregates of social caterpillars vary in size from only a few individuals to assemblages containing hundreds or even thousands of larvae. Variation in colony size among species may be largely attributable to differing life histories and niche-associated constraints, but significant within-species variation in colony size also exists. Many studies have shown that the number of eggs brought to maturity by a given female is highly dependent on larval nutrition. Thus, egg masses of the semelparous eastern tent caterpillar may contain as few as 76 eggs or as many as 434 (Stacey et al. 1975). Colony size in some iteroparous butterflies is attributable in part to their ability to vary the size of their egg clutches in apparent response to host density (Courtney 1984). The butterfly *Ascia monuste,* for example, lays single eggs on some species of its host plants and batches containing up to 50 eggs on others (Nielsen and Nielsen 1950).

While the maximum size of a sibling cohort is ultimately limited by female fecundity, large multicolony assemblages have been reported for a number of species including the caterpillars *M. americanum* (Fitzgerald and Willer 1983), *Euphydryas phaeton* (Stamp 1981a), *Yponomeuta cagnagellus, Archips cerasivoranus* (Fitzgerald, personal observation), *Brassolis isthmia* (Dunn 1917), and the sawflies *Neodiprion swainei* and *N. pratti banksianae* (Tostowaryk 1972). In some cases multicolony aggregates occur because females choose oviposition sites adjacent to other egg masses (Fitzgerald and Willer 1983; Stamp 1981a; Tsubaki 1981), suggesting that under some circumstances larval fitness may be greater in colonies of a larger size than that generated by a single egg mass. Multicolony aggregates may also arise when mobile colonies encounter each other while foraging and subsequently aggregate at a common site (Berger 1906; Fitzgerald and Willer 1983). Although some aspects of the behavioral repertoire of social caterpillars may eventually be shown to be the result of kin selection, there is at present no evidence that caterpillars discriminate between siblings and nonsiblings.

Survivorship of insects in colonies is influenced by colony size, the modal number of individuals in a colony being a compromise value determined by the various selective forces that shape the genetic structure of the population (Wilson 1975). Factors bearing on survivorship of caterpillars that may be influenced by colony size include the ability of the caterpillars to locate sufficient food and to feed efficiently, the need to defend against predators and parasitoids, the necessity of constructing shelters or to otherwise physically manipulate the host, and the need to maintain a body temperature conducive to growth and development. The number of larvae required to facilitate these functions is likely to vary at different points in the life history of the colony and optimal colony size may vary accordingly (Stamp 1981a).

In the pine webworm, the early instars feed solitarily while the last three instars aggregate (Hertel and Benjamin 1979). This, however, appears to be the exception. Maturing colonies typically break up into smaller aggregates or lose the aggregative tendency completely, suggesting that for many species the major benefits of sociality occur during the early stages of a colony's growth. Increased demand for food as caterpillars grow appears to be one of the most common factors leading to increased dispersion of the caterpillars in maturing colonies (Porter 1982; Fitzgerald et al. 1988). The last instar of the forest tent caterpillar *M. disstria,* for example, consumes approximately 82% of all the aspen foliage consumed by all instars combined (Hodson 1941). Although colonies of the eastern tent caterpillar often disband during the penultimate stadium, when food is plentiful the caterpillars continue to aggregate in their communal tent until fully grown (Fitzgerald et al. 1988).

Aggregation Sites and Shelter Building

Caterpillars that forage together may aggregate on or in communally built silk or leaf shelters, aggregate partially hidden by bark or leaf surfaces, or form conspicuous aggregates in the open. Communal shelter builders include tent makers such as *M. americanum* and *M. californicum;* the fall webworm, *Hyphantria cunea;* the ermine moth, *Yponomeuta cagnagellus* and leaf tiers and folders such as the ugly nest caterpillar, *A. cerasivoranus* and the cherry scallop-shell caterpillar, *Hydria prunivorata.* The dull-colored early instars of *M. disstria* aggregate inconspicuously under leaves or against matching bark surfaces, but the later instars are more conspicuously colored and aggregate in the open on the mainstream of the tree. Aposematically colored species, such as the spiny elm caterpillar *Nymphalis antiopa* (Sharplin 1964) and larvae of the cinnabar moth *Callimorpha jacobaeae* (Myers and Campbell 1976), aggregate gregariously in the sunlit areas of their host plants. Most leaf mining caterpillars feed solitarily, but in *Cameraria cincinnatiella,* a miner of white oak leaves, as many as a dozen individuals may share a single mine (Johnson and Lyon 1988).

By acting collectively, caterpillars are able to construct shelters that can both lessen the impact of predation and parasitism and create a microclimate conducive to growth and development. Silk, alone or in combination with leaves, forms the basis for most of the structures caterpillars create. Leaf-shelter-building caterpillars harness forces generated by stretching silk strands before attaching them to the leaf surfaces (Fitzgerald et al. 1991). Successive strands exert miniscule forces as they contract, and eventually draw the leaves into compact structures. Studies of spinning aggregates of *A. cerasivoranus* show that the collective effect of the strands of many caterpillars working together is required to pull both leaves and branches into a shelter (Fitzgerald et al. 1991). Damman (1987) found that groups containing at least 20 larvae of the pawpaw caterpillar, *Omphalocera munroei,* were needed to construct leaf shelters from the older leaves of the host.

Among the most elaborate shelters constructed by the group effort of caterpillars are the silk tents formed by tent caterpillars and other lasiocampids. The shelters facilitate thermoregulation and predator avoidance (see below) and provide secure attachment sites allowing the aggregate to endure prolonged periods of cold and even subfreezing weather without falling from the tree. Unlike the shelters of most species, the tents of these caterpillars typically stand apart from the colony's food supply and serve as resting sites between bouts of feeding. The air entrapped in these shelters, like the silk and leaf shelters constructed by *A. cerasivoranus, H. cunea,* and *E. phaeton,* has above ambient moisture content. This higher humidity is thought to facilitate the development of the caterpillars, and may be particularly important when caterpillars molt (Morris and Fulton 1970, Stamp 1982).

Foraging Patterns

Although considerable variation exists in the foraging patterns of caterpillars, gregarious species can be broadly classified as patch-restricted foragers, nomadic

Table 11.1. Foraging Patterns of Some Species of Social Caterpillars[a]

Species	Common name	Foraging pattern	Reference
Tetralopha robustella	Pine webworm	C?	Hertel and Benjamin (1979)
Malacosoma spp.	Tent caterpillars	C	Many
Brassolis isthmia	Coconut caterpillar	C	Dunn (1917); Young (1985)
Eriogaster amygdali	Almond tent caterpillar	C	Talhouk (1975)
Eriogaster lanestris	Small eggar moth	C	Balfour-Browne (1933)
Hylesia lineata	—	C	Janzen (1984)
Ichthyura inclusa	Poplar tentmaker	C?	Johnson and Lyon (1988)
Malacosoma disstria	Forest tent caterpillar	N	Fitzgerald and Costa (1986)
Asterocampa clyton	Tawny emperor caterpillar	N	Stamp (1984b)
Datana ministra	Yellownecked caterpillar	N?	Johnson and Lyon (1988)
Nymphalis antiopa	Spiny elm caterpillar	N	Sharplin (1964)
Hemileuca lucina	Buckmoth caterpillar	N	Cornell et al. (1988)
Mechanitis isthmia	—	N	Young and Moffett (1979)
Morpho peleides limpida	—	N	Young (1972)
Pryeria sinica	Burnet moth	N	Tsubaki (1981)
Papilio anchisiades	—	N?	Young et al. (1986)
Datana integerrima	Walnut caterpillar	N?	Hixon (1941)
Lophocampa argentata	Silverspotted tiger moth	P	Duncan (1982)
Hyphantria cunea	Fall webworm	P/C	Berger (1906)
Archips cerasivoranus	Ugly nest caterpillar	P	Personal observation
Yponomeuta cagnagella	Ermine moth caterpillar	P	Hoebeke (1987)
Homadaula anisocentra	Mimosa webworm	P	Webster and St. George (1947)
Euphydryas phaeton	Baltimore checkerspot caterpillar	P?	Stamp (1981a)
Hydria prunivorata	Cherry scallop-shell caterpillar	P	Schultz and Allen (1975)
Omphalocera munroei	Pawpaw caterpillar	P	Damman (1987)
Chlosyne harrisii	—	P	Dethier (1959a)
Euchaetes egle	Dogbane caterpillar	P	Dethier (1959b)
Thaumetopoea pityocampa	Pine processionary	P/C	Balfour-Browne (1933)
Pieris brassicae	Large white butterfly	P	Long (1955)
Artona funeralis	—	P/N?	Mizuta (1968)
Homaledra saballella	Palmskeletionizer	P	Johnson and Lyon (1988)

[a]P, patch-restricted forager; N, nomadic forager; C, central-place forager. P/C indicates that as colony matures larvae switch from patch-restricted to central-place foraging.

foragers, central-place foragers, or as some combination of these (Fitzgerald and Peterson 1988, Table 11.1).

Patch-Restricted Foraging

Patch-restricted foragers obtain all of the food required during the social phase of their larval development from the leaves found in a single contiguous patch or from several such closely spaced patches. The foraging arena is typically well defined by a protective silk envelope or by leaves bound together. On large trees, patches usually consist of the leaves found on a part of a branch, an entire branch, or several closely situated branches. But on small trees and herbaceous plants the entire host may eventually be enveloped.

The ugly nest caterpillar *A. cerasivoranus* is a typical patch-restricted forager. The caterpillars are gregarious from eclosion and move from their egg mass, located at the base of the host sapling, to the uppermost branches. Here they envelop a few leaves with silk and feed from within the enclosure. As the colony matures the larvae pull more leaves into the shelter until eventually the entire top of a small sapling may be bound tightly together. Patch restricted foragers often complete the gregarious phase of their development at the initial feeding site, but colonies may move to a new, noncontiguous, section of the host tree if the original site is exhausted. Some species, such as the fall webworm, *H. cunea,* may switch to central-place foraging (see below) when the original patch is depleted, and move between a shelter constructed at the original site and distant feeding sites (Berger 1906).

Nomadic Foraging

Colonies of caterpillars that forage nomadically establish only temporary resting sites and make frequent moves from one patch to another. In one study, the nomadic forest tent caterpillar fed for approximately 2 days at a single site before moving to a new patch which, depending on the instar, was located 45 to 145 cm from the previous site (Fitzgerald and Costa 1986). Larvae of the tawny emperor butterfly, *Asterocampa clyton,* were observed to move up to 310 cm within a tree to establish new feeding sites (Stamp 1984a). For both species, the caterpillars passed by many potential feeding sites during their migrations.

The depletion of a patch may prompt nomadic caterpillars to move but successive patches may be only partially utilized. Forest tent caterpillars feeding on maple, for example, left from 50 to 62% of the foliage uneaten (Fitzgerald and Costa 1986). *A. clyton* larvae consumed from 34 to 79% of the foliage of hackberry before abandoning a site (Stamp 1984a). Nomadic foragers are typically unrestrained by webs or other shelters, though, like the larvae of *A. clyton* (Stamp 1983, 1984a), they may construct temporary molting shelters and confine their movement to silk trails when traveling to new sites.

Central-Place Foraging

Central-place foragers form a permanent or semipermanent shelter from which they launch intermittent forays to distant sites in search of food. Between bouts of feeding the caterpillars rest at the shelter. Tent caterpillars (*Malacosoma*) and other lasiocampids such as *Eriogaster lanestris* (Balfour-Browne 1933), *E. amygdali* (Talhouk 1975), *Gloveria howardi,* and *Eutachyptera psidii* (Franclemont 1973) make forays from large, conspicuous tents. Other species such as

Hylesia lineata (Janzen 1984) move from silk mats spun on the bark of the host tree, or like the coconut caterpillar *B. isthmia* (Dunn 1917; Young 1985), from shelters constructed by webbing together the foliage of the tree. Conspicuous silk pathways typically lead from these shelters to foraging patches.

Evolution of Foraging Patterns

A number of factors might influence the evolution of foraging patterns. Distasteful and aposematically colored caterpillars may forage freely over the plant surface, whereas more palatable species may need to construct leaf or silk shelters to defend against parasitoids and predators. Species that invest heavily in such shelters are likely to be predisposed to limit their foraging to the confines of the shelter or to brief episodic forays, often under the cover of darkness, to more distant sites. Conversely, caterpillars that are susceptible to disease agents might experience strong selection pressure to periodically abandon feeding and resting sites, leaving behind weakened and infectious siblings (Stehr and Cook 1968). Selective pressure from particularly effective predators or parasitoids that are attracted to heavily used resting sites or to damage caused by feeding (Heinrich 1979; Stamp 1984b) may also favor colony mobility. Basking species may need to move frequently to optimize their exposure to the sun (Porter 1982).

Characteristics of the food supply are likely to play a major role in the evolution of foraging patterns. Colonies of finicky feeders that forage in trees with high meristematic activity may need to move frequently from patch to patch to obtain sufficient food of a particular quality to satisfy colony demand. In contrast, caterpillars that feed in trees bearing mature leaves of relatively consistent quality may have little incentive to favor one patch over another. The pronounced difference in the foraging patterns of *M. americanum,* which feeds in the spring and favors young partially expanded leaves, and *H. cunea,* which feeds in late summer and fall on near-senescent leaves, may be largely attributable to marked seasonal variation in the extent of leaf heterogeneity of their common host black cherry, *Prunus serotina* (Fitzgerald and Peterson 1988). Variation in leaf quality in some tree species may also result from the production of feeding deterrents in response to herbivory (Schultz 1983), forcing caterpillars to abandon repeatedly partially consumed patches.

Adaptive Significance of Sociality

Nearly four decades of study have shown that for a wide variety of social caterpillars, isolated individuals or individuals kept in small groups suffer disproportionate mortality compared to those in larger groups (see, for example, Long 1953; Hosoya 1956; Dethier 1959a; Ghent 1960; Mizuta 1968; Lyons 1962; Sugimoto 1962; Henson 1965; Iwao 1968; Watanabe and Umeye 1968; Shiga 1976; Stamp 1980, 1981a; Tsubaki 1981). For the most part, these and other investigators have found the grouped caterpillars of social species to have an advantage over solitary siblings in foraging, thermoregulating, and defense against predators and parasitoids.

Facilitation of Foraging

Social Facilitation

A major apparent benefit to aggregation in many social caterpillars involves social facilitation. This facilitation occurs when a pattern of behavior is initiated or the pace or frequency of the pattern increases as a consequence of the presence or actions of another individual (Wilson 1975). In almost all cases of feeding facilitation reported among caterpillars, such facilitation is most pronounced during the first stadium.

Ghent (1960) showed that under laboratory conditions, hungry first instar larvae of the sawfly *N. pratti* were attracted to the feeding site of siblings that had succeeded in penetrating the tough needles of the Jack pine. Attracted larvae enlarged these locations to establish their own feeding sites alongside the "first biter." Because of this facilitation in establishing a feeding site, first instar larvae survived better in groups than they did when reared solitarily. Once a larva had established its own feeding site, the presence of siblings had no significant effect on the rate of larval growth. Disproportionate mortality among isolated as compared to grouped early instar caterpillars attributed to this "establishment mortality" has also been reported for a number of other species, notably *N. swainei* (Lyons 1962), *N. sertifer* (Henson 1965), *Euproctis pseudoconspersa* (Hosoya 1956), *Artona funeralis* (Sugimoto 1962, Mizuta 1968), *H. cunea* (Watanabe and Umeya 1968), and *M. neustrium* (Shiga 1976).

Some studies indicate that starvation is not the sole reason for this differential mortality. Henson (1965) suggested that isolated larvae of *N. sertifer* are vulnerable because they are less able than aggregated caterpillars to withstand even slight departures from optimal temperature and humidity. Presumably, enhanced feeding that occurs in groups allows the newly emerged caterpillars to gain enough water to withstand desiccating conditions. Stamp and Bowers (1990a) suggested that mortality of isolated *H. lucinia* caterpillars was attributable to strong selection pressure for aggregation, leading isolated larvae to search restlessly for the group. Although isolated caterpillars, which were initially placed on the food, established feeding sites and fed, they wandered extensively and often became isolated from the food supply. In his studies of Neotropical butterfly larvae, Young (1983) argued that a major incentive for gregariousness may be the need to overcome structural plant defenses. Groups of larvae of *Mechanitis isthmia,* for example, were observed to have more success in removing the fine trichomes protecting the leaf of solanaceous host plants than solitary larvae (Young and Moffet 1979). The advantage of group feeding, however, accrued primarily to the early instars, since the later instars were not effected by the trichomes. Stamp (1981a) found that although there was no evidence of social facilitation of feeding in *E. phaeton,* the presence of siblings facilitates movement to feeding sites by the first instar caterpillars, probably because the caterpillars follow and reinforce the silk trails deposited by siblings.

It has also been suggested that the increased agitation that occurs in dense aggregates may stir caterpillars to activity causing them to feed more often and to grow more rapidly than those kept alone. Long (1953) found that when larvae of *Plusia gamma,* which form loose aggregates in nature, are reared in isolation, they grow slower than caterpillars kept in groups, a difference he attributed to an increase in overall activity among crowded larvae. Preliminary studies show, however, that in *M. americanum* aggregation may limit individual activity. Colonies of this species exhibit a program of feeding en masse three to four times each day (Fitzgerald 1980; Fitzgerald et al. 1988; Casey et al. 1988). In contrast, when caterpillars are reared individually on host branches in the laboratory they feed up to 12 times a day (Fitzgerald unpublished data). Although the effect of this feeding regimen on growth and development has not yet been assessed, these preliminary findings indicate that rather than stimulating feeding activity, colony life may constrain individual activity. Casey et al. (1988) suggest that strong selection pressure for synchronous foraging may prevent caterpillars from following a foraging schedule dictated by individual hunger level. The proximate mechanism for this constraint is unknown but may involve tactile or chemical cues associated with resting siblings.

Polyethism

The foraging behavior of some social caterpillars has been purported to be influenced by individual differences in larval behavior. Wellington (1957) was the first to report that colonies of tent caterpillars contained both independent (type I) and dependent (type II) larvae. He argued that the success of a colony was largely attributable to the occurrence of the independent caterpillars, which comprised from 0 to 38% of a colony. In contrast, type II caterpillars were relatively sluggish and largely incapable of independent movement, but followed trails established by type I larvae, or travelled in close formation with other type II caterpillars. These differences in larval activity were thought to be permanent and attributable to variation in the amount of yolk deposited in the individual eggs of an egg mass (Wellington 1965). Adequately provisioned eggs produced type I individuals; those with less yolk produced type II larvae. It was held that the mixing of larval types led to a weak division of labor. Type I larvae kept the colony in an overall state of activity and established trails to new food finds; the sluggish type II caterpillars were said to be largely responsible for building the tent and maintaining colony cohesiveness. Wellington (1957) argued that activity differences among larvae also had an important bearing on the population dynamics of the caterpillar. Adults resulting from type II larvae were small and too weak to disperse. These moths produced more local colonies with increasingly larger proportions of unfit caterpillars. This, combined with successful emigration of fitter adults resulting from type I larvae, led to population declines at the original sites of infestation.

Despite the considerable attention that continues to be drawn to these early studies (see for example Barbosa and Baltensweiler 1987), work by other investigators failed to provide convincing evidence in support of adaptive polyethism in tent caterpillars. Shiga (1979) reported that individual differences in the Japanese tent caterpillar *M. neustrium testacea* were less obvious than those reported by Wellington for *M. c. pluviale,* and was not an important factor in the population dynamics of this species. Laux (1962) and Franz and Laux (1964) reported that the larvae of *M. neustrium* were not consistently active or inactive, but shifted from one category to another during the course of separate tests. Greenblatt (1974) and Greenblatt and Witter (1976) came to similar conclusions in their studies of *M. americanum* and *M. disstria*. Myers (1978) also reported that, contrary to previous reports, no apparent relationship existed between the order in which eggs were laid and the activity level of the larvae of *M. c. pluviale*. A reanalysis of Wellington's (1965) data failed to support the contention that maternal nutrition influences the proportions of larvae that exhibit different activity levels within egg masses (Papaj and Rausher 1983). After reviewing studies on tent caterpillars spanning two decades, Papaj and Rausher (1983) concluded that the whole question of adaptive polyethism in tent caterpillars needs to be reexamined using more appropriate behavioral assays and statistical techniques.

In a more recent study, Edgerly and Fitzgerald (1982) showed that larval activity levels are normally distributed in *M. americanum* and that exploration is a group function. There are no apparent leaders nor any indication of adaptive polyethism in this species. In one of the few investigations of polyethism in other social caterpillars, Cornell et al. (1988) undertook a study to determine if larval activity varied within groups of *Hemileuca lucinia*. Their study differed from others in that larvae were given distinctive marks to facilitate identification, but kept together in groups. They found that in this species larvae in the vanguard during any one foraging bout tended to stay there. In about half of their tests, activity ranks of caterpillars were consistent from one test to another so that some individuals were apparent leaders for a series of foraging bouts. However, ranking of larvae remained consistent from one instar to the next in less than 25% of the tests, indicating that in this species leadership is a transient phenomenon.

Chemical Communication and Cooperative Foraging

The caterpillars of some moths and butterflies use pheromones to mark trails and foraging arenas. The markers appear to form the proximate basis of colony cohesion and facilitate movement between resting and feeding sites (Fitzgerald and Peterson 1988). In a few species, they have been shown to underlie more elaborate systems of recruitment communication. Although preliminary surveys suggest that the use of chemical markers by caterpillars is not uncommon (Fitzgerald and Peterson 1988, and references therein), only a handful of species have been studied, and only a single chemical marker has been identified.

Trail-based foraging behavior has been investigated in *Thaumetopoea* (Fabre 1916), *Pieris* (Wojtusiak 1929, 1930; Long 1955), *Yponomeuta* (Kalkowski 1958, 1966; Roessingh et al. 1988; Roessingh 1989, 1990), *Hyphantria* (Masaki and Umeva 1977), *Archips* (Fitzgerald and Edgerly 1979a; Fitzgerald unpublished data), *Eriogaster* (Weyh and Maschwitz 1978), *Chlosyne* (Bush 1969), *Hemileuca* (Capinera 1980), and *Malacosoma* (Fitzgerald 1976; Fitzgerald and Peterson 1983). In addition, solitary species may move back and forth between resting and feeding sites (Heinrich 1979), and in the few cases that have been investigated solitary caterpillars have been shown to follow trails (McManus and Smith 1972; Weyh and Maschwitz 1982). It has been suggested that the marking behavior of solitary species may have provided an initial stimulus for aggregation among siblings of some species, facilitating the evolution of sociality (Fitzgerald and Peterson 1988).

Although trail following in lepidopterous larvae may to some extent depend on tactile components of silk trails (Fitzgerald and Edgerly 1982; Roessingh 1989), in all cases investigated the proximate basis for trail following has been shown to be largely chemical in nature. Weyh and Maschwitz (1978, 1982) found that unidentified chemical markers are secreted along with the silk trails of *E. lanestris* and *Iphiclides podalirius*. A crude extract of the anterior portion of the larvae of *Hemileuca olivae,* which contains the silk glands, elicits trail following in this species (Capinera 1980). Chemical trail factors occur in the silk of the gregarious leaf tier *A. cerasivoranus* (Fitzgerald, unpublished data) and *Y. cagnagellus* (Roessingh 1989, 1990). Other species such as *Euphydryas phaeton* (Stamp 1982), *Chlosyne lacinia* (Stamp 1977), *Asterocampa clyton* (Stamp 1984a), and *Pieris brassicae* (Long 1955) follow silk trails but it has not yet been established that a chemical trail marker is involved. Attempts to isolate the chemical components of silk responsible for trail following have been largely unsuccessful (Capinera 1980; Roessingh 1989, 1990).

Pheromones may also facilitate foraging in colonies of sawflies. Henson (1965) observed the movement of unfed first instar larvae of *N. sertifer* to the feeding site of a successful sibling. He reported preliminary studies that showed that while mashed foliage or mashed larvae were not attractive to unfed larvae, the actively feeding insect was. This result suggests the possibility that an aggregating pheromone may be involved, though none has yet been demonstrated. In another study, Prop (1960) observed that the larvae of *N. sertifer* and *Diprion pini* migrated over distances up to 2 m following defoliation of their initial feeding site. The investigator's observations indicate that the larvae deposit scent trails as they proceed over new branch surfaces and that the trail serves as an orientational cue for the siblings. Scent marking enables the larvae to reassemble en masse at a new feeding site. The nature of the trail marking stimulus was not determined and there have been no subsequent studies of trail marking either in these species or in other species of sawflies.

Recruitment Communication in Tent Caterpillars

The most sophisticated systems of trail-based foraging so far described for social caterpillars occur among the tent caterpillars (*Malacosoma*). The basis for trail marking in these insects differs from that of all other species so far investigated. Caterpillars lay down a chemical trail marker, secreted from the sternum at the posterior tip of the abdomenen, as they move over previously unmarked branches in search of food (Fitzgerald and Edgerly 1982). An active component of the trail of the eastern tent caterpillar was recently identified as 5β-cholestane-3,24-dione and shown to elicit trail following at threshold concentrations of 10^{-11} to 10^{-12} g/mm of trail (Crump et al. 1987). Laboratory and field studies demonstrated that the forest and eastern tent caterpillars follow each other's trails (Fitzgerald and Edgerly 1979a) and the forest tent caterpillar responds to the synthetic pheromone of the eastern tent caterpillar (Crump et al. 1987). The European tent caterpillar *M. neustrium* also responds positively to this compound (Peterson 1988), suggesting that the structure of the pheromone may be conserved at the generic level. Bilateral ablation of the maxillary palps leads to loss in the ability of the caterpillars to follow trails, showing that the receptors for the trail pheromone are located at this site (Roessingh et al. 1988).

As they return to the tent, caterpillars that have fed to repletion overmark exploratory trails with a recruitment marker. Overmarked trails are more effective in eliciting trail following than exploratory trails and serve to lead hungry tentmates directly to food finds. The trails are extended onto the surface of the tent and the structure serves as a center of communication. Studies of *M. neustrium* show that this species also produces distinct exploratory and recruitment trails (Peterson 1988), but it is not presently known for any species how exploratory and recruitment trails differ.

Recruitment to food in the eastern tent caterpillar (*M. americanum*) is an elective process contingent on individual assessment of the quality of feeding sites (Fitzgerald and Peterson 1983). Caterpillars that feed at crowded sites typically do not recruit tentmates. In addition, caterpillars allowed to feed on either young or aged leaves of host tree species consume comparable quantities of both, but recruit significantly more tentmates to young leaves. Peterson (1987) showed that caterpillars fed young leaves of the preferred host black cherry grew faster and had higher pupal weights than those fed aged leaves. Survival was also significantly greater among larvae fed young leaves. Thus, recruitment communication in this insect appears to facilitate growth and development by enabling colonies to locate efficiently food of optimal quality. Although recruitment occurs during each foraging bout, it may play an increasingly important role as young leaves become patchily distributed and host trees near defoliation. Caterpillars that leave defoliated host trees in search of food may even succeed in recruiting a whole colony over the ground to a newly discovered host tree (Fitzgerald and Edgerly 1979b).

The communication system of the eastern tent caterpillar is sufficiently flexible to allow the caterpillars to recruit to suboptimal food sources when preferred food is not available (Fitzgerald and Peterson 1983; Peterson 1986b). When reared on their natural host, black cherry, caterpillars recruited little or not at all after feeding for the first time on either sweet cherry or scarlet oak. But when caterpillars were reared on either sweet cherry or scarlet oak, they laid recruitment trails to these nonhost species. These results indicate that after defoliating their host tree, colonies require a period of deprivation and adjustment before accepting (and recruiting to) a suboptimal food plant. No matter what the previous rearing history, however, caterpillars always fed on and recruited preferentially to black cherry when given a choice between it and a nonhost species.

Trail-based communication and foraging behavior of the forest tent caterpillar (*M. disstria*), a sympatric congener of the eastern tent caterpillar, has also been investigated (Fitzgerald and Costa 1986). The behavior of this species differs from that of the eastern tent caterpillar in that it is nomadic and ranges widely in search of food. The caterpillars rest gregariously between bouts of feeding on silk mats, which are periodically abandoned (Stehr and Cook 1968). Caterpillars mark trails that are homologs of the exploratory and recruitment trails of the eastern tent caterpillar, but in the forest tent caterpillar recruitment trails serve to facilitate aggregation rather than to recruit siblings to food finds. Fitzgerald and Costa (1986) showed that this species lacks the finely tuned system of recruitment communication employed by the eastern tent caterpillar.

Social caterpillars exhibit a distinctly lower grade of colony organization than the ants and clearly lack their behavioral flexibility. Although the pine webworm is reported to bring pieces of needles back to its nest (Hertel and Benjamin 1979), there is no evidence that any caterpillar either stores food or distributes it to siblings. Unlike ants, tent caterpillars follow recruitment trails only when hungry. Individuals analogous to scout ants, which reinforce recruitment trails without carrying food (Jaffe 1980), do not occur in colonies of tent caterpillars.

Nonetheless, tent caterpillars share important details of their foraging ecology with ants. They have evolved convergent foraging strategies involving trail-based systems of chemical communication that facilitate group foraging through information sharing. Both forage from trunk trails or from fixed bases for patchily distributed resources and employ chemical systems involving the active secretion of minute quantities of pheromones that serve to recruit siblings to food finds. Unlike tent caterpillars, many species of ants depend on visual cues to maintain contact with their nests and do not lay down exploratory trails. Army ants are an exception. These blind or semiblind nomadic predators rely on chemically based exploratory trails to relocate their bivouacs and to hold foraging columns together (Chadab and Rettenmeyer 1975; Topoff et al. 1980). The most evolutionarily advanced trail systems employed by ants are those in which the trail pheromone alone is adequate to both stimulate and orient trail following; direct physical interaction between individuals is not required (Höllodobler 1978). The recruiting insect, however, may enhance recruitment by alerting and mobilizing nestmates (Wilson 1962; Chadab and Rettenmeyer 1975; Hölldobler 1978; Topoff et al. 1980). The trail system of the eastern tent caterpillar resembles these systems of mass recruitment in that chemical cues alone are adequate to elicit recruitment. Likewise, there is no evidence that direct physical contact between caterpillars plays any role in recruitment.

Thermoregulation

A well-documented advantage to aggregation in caterpillars is the facilitation of thermoregulation by group basking. At the elevated T_bs (body temperatures) they achieve by basking, caterpillars may be more effective in escaping predators and parasitoids. Evans (1982), and Porter (1983) show, for example, that when conditions are sunny but cool, basking caterpillars may have higher levels of activity and develop more rapidly than their nonbasking predators and parasitoids. In addition, group basking may allow siblings to develop at similar rates, which in some sawflies has been shown to facilitate cooperative burrowing (Seymour 1974). Caterpillars that bask clearly grow faster (Casey et al. 1988) allowing them to exploit rapidly growing host trees while foliage is at peak nutritional quality, or lowest in deterrents (Stamp and Bowers 1988, 1990c). Faster development also results in caterpillars spending less time exposed to predators, parasitoids, and disease agents, allowing them to avoid the build up of these mortality factors as the season progresses (Evans 1982; Porter 1982, 1983; Stamp and Bowers 1988, 1990a).

Some early studies [Fintelmann 1839 (cited in Prop 1960), Mosebach-Pukowski 1938] suggested that clustered caterpillars might generate enough metabolic heat to warm the cluster above ambient temperature, but subsequent studies have failed to reveal an endothermic component in caterpillar thermoregulation (Knapp and Casey 1986). Nonetheless, caterpillars that aggregate to bask appear to be particularly effective thermoregulators. Stamp and Bowers (1990b) showed that larvae of the moth *H. lucinia* exposed to full sunlight were able to elevate their T_bs up to 5°C above T_a (air temperature). Under partly cloudy conditions, aggregated caterpillars were able to sustain their T_bs for longer periods than isolated larvae. Faster growth at higher body temperatures allowed these spring feeding caterpillars to complete most of their larval development while host leaves were still of high quality (Stamp and Bowers 1990c). Porter (1982) showed that larval aggregates of the dark colored *Euphydryas aurinia* were able to elevate their temperature far above ambient by basking in the sun. The gain is due, in part, to insulation against lateral heat loss by the closely packed bodies of siblings.

Studies of the gregarious sawfly *Perga dorsalis* (Seymour 1974) showed a gain in temperature attributable to apparent reduction in convection when larvae aggregated to bask. The larvae of this species are also able to cool below T_a by smearing their bodies with a fluid excreted from the anus when air temperatures exceed about 37°C, a procedure that is markedly enhanced by aggregation. Seymour (1974) observed that isolated larvae had difficulty directing the fluid onto their own body but when aggregated, fluid otherwise wasted, was smeared on adjacent siblings.

Thermoregulation has been studied extensively in the tent caterpillars. These insects are active at a time of the year when the average daily T_a is often below the minimum temperature required for their growth and development (Hodson 1941; Knapp and Casey 1986). Nonetheless, the caterpillars are highly active foragers and achieve T_bs necessary to process the food stored in their guts by basking in the sun (Casey et al. 1988). Both the tent and gregariousness appear to play an important role in thermoregulation.

Average body temperature excesses ($T_b - T_a$) reported for eastern tent caterpillars show that they can achieve a broad range of temperatures by basking at different locations on the tent. In one study, models of solitary caterpillars placed on branches in the sun achieved mean temperature excesses of 0.4–11.2°C whereas models placed on the outside of the tent achieved temperature excesses of 3.5–27.5°C (Joos et al. 1988). The gain in T_b over caterpillars that basked off the tent occurred because the caterpillars rested within the boundary layer of the tent, a layer of unstirred air that effectively reduces convective heat loss. In addition, the tent itself gains heat, and provides a warmer substrate than the branches of the tree. Indeed, models placed on the underside of the tent, completely shaded from direct sunlight, achieved temperature excesses of 2.5–15.7°C solely by conducting heat away from the tent.

Air currents can result in substantial loss of heat to basking caterpillars, particularly when a strong wing is present. But the long setae of caterpillars serve as effective barriers to currents, and can greatly reduce convective heat loss (Casey and Hegel 1981). Studies with models of tent caterpillars show that loss of heat due to such convection can be further reduced by aggregation (Joos et al. 1988). Models simulating aggregated caterpillars resting on the southeast face of the tent achieved temperature excesses of 6.5–43.9°C. It was estimated that approximately one-third of the temperature excess of these caterpillars is attributable to aggregation, one-third to direct exposure to sunlight, and from 14 to 47%, depending on orientation to incident sunlight, to the high ambient temperature and boundary layer of the tent (Joos et al. 1988).

The mapping of the thermal habitat of the eastern tent caterpillar with models shows that not only can caterpillars attain physiological optimal T_b, but that they must take measures to avoid temperatures that exceed their upper physiological threshold. Caterpillars move from site to site to achieve or maintain T_b, and they actively avoid temperature extremes. Body temperatures of living caterpillars, measured with a thermocouple needle, or by infrared thermography, are much narrower than those of models (Joos et al. 1988). In one instance, the range of T_bs of caterpillar models was 25–55°C, whereas living caterpillars under similar conditions had T_bs of 40–48°C. When T_bs exceeded about 40°C, the caterpillars ceased to aggregate in the sunlight and rested in the shade, either inside of the tent or on the underside of the structure. Under extreme conditions, the caterpillars hang by their abdominal prolegs from the shaded side of the tent, keeping their bodies clear of the structure, to maximize convective heat loss.

The tent of the eastern tent caterpillar functions like a greenhouse, trapping solar radiation while greatly reducing convective heat loss by blocking the wind. Temperature excesses of tents ($T_{tent} - T_a$) measured with thermocouples buried at various locations within the structure vary from as little as 4°C to as much as 23°C. Sites deep within the tent are warmer than peripheral sites, and the side of the tent facing the sun is warmer than the shaded side. The temperature gradient within 39 tents exposed to full sunlight was 7.3 ± 3.1°C, providing a thermally heterogeneous environment within which caterpillars can thermoregulate (Joos et al. 1988).

The earlier instars of the eastern tent caterpillar typically bask just beneath the outer layer of the tent (Knapp and Casey 1986). During cool spring mornings, the small caterpillars warm rapidly, gaining the heat necessary to initiate activity and to process food collected during the morning foray. Although the layered structure of the tent might be expected to provide some insulation, helping caterpillars to retain T_bs during sunless intervals, there is little indication of such an effect. Tents, and the caterpillars in them, cool to the ambient temperature shortly after the sun sets. Carlberg (1980) reported similar rapid cooling of the tent of *E. lanestris* in the evening.

Tent caterpillars clearly orient their tents to take advantage of solar radiation (Balfour-Browne 1933; Carlberg 1980). Compass bearings taken perpendicular to the most developed face of 69 tents of the eastern tent caterpillar had an average facing direction of 122 ± 27.4° (Fitzgerald and Willer 1983). The southeast orientation of these tents probably reflects the direct influence of the position of the sun during the morning and afternoon on the spinning behavior of the photopositive caterpillars. Laboratory studies showed that caterpillars concentrate their spinning activity on the most intensely illuminated side of their tent, so that the most developed face of the structure lies perpendicular to the light source (Fitzgerald and Willer 1983).

Group Defense against Predators and Parasitoids

Crowding in Space and Time

The probability that any one individual in an aggregation will be killed by an attacking predator decreases as the size of the colony increases. Individuals may benefit from such passive defense if predators are not attracted to grouped prey in disproportionate numbers and the rate of attack by an individual predator is independent of colony size (see Vulinec 1990 for a review of passive defense models). Individuals may also gain protection from predators by attempting to surround themselves with others (Hamilton 1971). Tostowaryk (1971), for example, found that sawfly larvae that lie on the periphery of the aggregate are approximately twice as likely to be attacked as those toward the center. In addition, crowding may serve to confuse or distract predators, particularly if caterpillars are very active.

Caterpillars that make brief, intermittent forays from protective structures to feed may crowd predators and parasitoids in time as well as space and further reduce overall colony losses. Tent caterpillars follow such a pattern of intermittent foraging. Under field conditions colonies have three peaks of activity off the tent each day. Movement to and from food during these forays is typically en masse, particularly during the early stadia. Two of these forays also occur largely under the cover of darkness, a pattern that may have evolved to limit contact with day-active predators such as birds (Fitzgerald et al. 1988).

Influence of Colony Size

Lawrence (1990) investigated the effect of group size on growth rate and survivorship of the larvae of *Halisidota caryae*. Caterpillars were established under field conditions in groups of 10, 25, or 125. The caterpillars in the larger group not only grew more rapidly, probably due to a facilitation of feeding, but also had a higher survival rate. Differences in survival among groups were attributed largely to differential mortality from hemipteran predators. Predators were equally abundant for all group sizes so that the number of predators per caterpillar decreased with group size with a ninefold difference in the number of predators per caterpillar occurring between the smallest and largest groups. Larger groups also moved over greater distances and groups that moved the most had fewer predators per group than more stationary groups. A more rapid development of caterpillars in large groups may also limit predation by reducing the duration of the vulnerable larval stage.

Morris (1972a, 1976) showed that for populations of *Hyphantria cunea,* in which egg mass size varies considerably, larger egg masses produced a higher proportion of surviving fifth instar larvae than did smaller clusters. An average of only 27% of larvae from egg masses containing 400 eggs survived while 53% of the larvae from egg masses containing 700 eggs survived. In this same study, the average size of egg masses was approximately 500, thus colonies of greater than average size had the highest survivorship. The reason for these differences is at least partially attributable to parasitism. Morris (1976) found that the size of a colony had no bearing on whether it was eventually located by a parasitoid, but for all of eight species of parasitoids studied the proportion of larvae parasitized was significantly less in larger colonies. Caterpillars in larger colonies, however, are not always better off than those in smaller colonies. Stamp (1981a) found that artificially constructed colonies of *E. phaeton* caterpillars of a size typical of those arising from a single egg mass had a significantly lower rate of parasitism than larger colonies. Thus, for this species, the habit of some females to place their egg mass adjacent to another results in large multicolony aggregates that suffer increased parasitism.

Tostowaryk (1972) studied the functional response of the pentatomid predator *Podisus modestus* to varying densities of two species of gregarious sawflies. Although predators typically specialize to an increasing degree as the abundance of a particular prey item increases, his studies show that the rate of predation declined as the size of the sawfly aggregate increased. This result he attributed to a disturbance effect that alerted larvae near to an attacked sibling. The ensuing group defense reaction frequently resulted in the predator being bathed in a sticky liquid regurgitated by the larvae. Smeared predators sometimes had their appendages stuck to their bodies and had more difficulty moving, slowing down the rate of subsequent attacks.

Shelters

Silk and leaf shelters appear only partially effective in protecting caterpillars from birds and invertebrate predators. Some birds only take caterpillars that lie on the outside of the web but others tear the web apart and can completely destroy colonies (Mosebach-Pukowski 1938). Knapp and Casey (1986) found that the yellow-billed cuckoo destroyed all of the tent caterpillar colonies in three of their four study sites. Morris (1972b) found that the web of the fall webworm did offer some protection from birds, and some species of wasps took only caterpillars that ventured outside of the web. But colonies were also attacked by *Vespula* wasps that entered the web, and the combined effects of wasp and bird predation often resulted in complete loss of the colonies. In one study in Japan, webs covered with mesh to exclude both birds and wasps suffered 10 to 20% mortality, whereas unprotected webs suffered losses greater than 95% (Ito 1977). Similar studies by Shiga (1979) and Filip and Dirzo (1985) showed that when tents of *M. neustrium* and *M. incurvum aztecum* were caged to exclude parasitoids and predators, there was a 4- to 15-fold increase in larval survival. Thus, for these species, the web appears to provide only limited protection against predators and parasitoids.

When they rest deep within the tent, caterpillars are secure from braconid wasps and tachinid flies (Fitzgerald personal observation) but some insects, such as the predaceous stinkbug *Podisus placidus,* may enter the tent (Evans 1983). The small tents of the early instars of the eastern tent caterpillar appear to afford little protection from ants. Ayre and Hitchon (1968) found that the ant *Formica obscuripes* was a particularly effective predator of eastern tent caterpillars. Once discovered, ants continued to visit tents until, in many cases, the entire colony was destroyed. Green and Sullivan (1950) also reported that colonies of the forest tent caterpillar, which rest on silk mats in the open, were readily decimated once located by foraging ants. The ants systematically removed individuals from the assemblage with no apparent behavioral response by the remainder of the colony. In laboratory experiments, aggregation did not protect second instar larvae of the range caterpillar *Hemileuca oliviae* from predation by ants. Ants, however, were observed to have more difficulty pulling caterpillars from silk resting mats formed by the group than from more flimsy silk laid down by individuals (Capinera 1980).

Stamp (1982) found that the first instar larvae of *E. phaeton* spent more time enclosed by their web than the later instars. She attributed this to the relative vulnerability of the smaller caterpillars to predators, but the ready availability of food within the structure and elevated humidity may have also been contributing factors. Later instars molted within the deeper layers of the tent where they were least vulnerable to predators. When parasitic wasps attacked colonies, caterpillars were most vulnerable when they were just under the surface of the silk where they were partially constrained and susceptible to being probed with the parasitoid's ovipositor. Outside of the tent, various defensive maneuvers, such as thrashing and regurgitation, often involving more than one caterpillar, effectively repelled the wasps (Stamp 1984b).

The gregarious larvae of *Omphalocera munroei* develop more rapidly on young leaves of the pawpaw but are better protected from wasp predators when they feed from within leaf shelters constructed from older and tougher leaves (Damman 1987). Shelters constructed from older leaves retain their shape as they are fed on whereas younger leaves wilt and are less effective in protecting the caterpillars from predators. When allowed to forage freely, the caterpillars trade off development time for protection and feed preferentially on the older leaves.

Antipredator Group Displays

Although isolated caterpillars commonly respond defensively to attack by predators and parasitoids, simultaneous displays involving a number of caterpillars are likely to amplify the deterrence value of the behavior. No alarm pheromones have yet been described from social caterpillars, but visual and tactile cues associated with defensive display may alert nearby siblings and cause them to join in the display. Defensive group responses by caterpillars have been reported from a number of species of caterpillars.

Prop (1960) studied five species of sawflies and observed that the two gregarious species, *N. sertifer* and *D. pini,* produced more effective predator-deterring displays than the four solitary species. Group display behavior of the gregarious sawflies included head jerking and stretching responses not seen in the solitary species. Prop (1960) found that the displays were effective in frightening naive avian predators and in deterring oviposition by an ichneumonid parasitoid.

Counter ploys by parasitoids to the defensive behavior of grouped larvae have been little studied but Prop (1960) found that a tachinid fly parasitoid of the sawflies was not affected by the displays since it employed a sit-and-wait tactic followed by sudden attack and rapid oviposition that failed to incite the larvae until it was too late.

Tent-inhabiting caterpillars resting on the outside of their tent respond with a defensive body-flicking display when threatened (Ancona 1930; Morris 1963; Myers and Smith 1978). Such displays, which may set up vibrations in the tent, appear to alert nearby siblings, and the behavior radiates rapidly through the assemblage. Stamp (1984b) demonstrated that a general web disturbance accomplished by jabbing the silk web of *E. phaeton* with a blunt probe caused larvae distant from the disturbance to move into the tent and engage in head-jerking displays. *Euphydryas phaeton* larvae at the scene of a wasp-caterpillar encounter also engaged in group body thrashing and regurgitation. In some cases thrashing was vigorous enough that it resulted in wasps being knocked away (Stamp 1982; Stamp 1984b), though the main value of such display is probably in causing confusion and deterring oviposition.

During encounters with their enemies, gregarious caterpillars may repel intruders by regurgitating droplets of liquid containing distasteful or toxic compounds. The larvae of *A. cerasivoranus* respond aggressively to intruders by exuding droplets of the juices of choke cherry from their mouths as they strike at them (Fitzgerald personal observation). Stamp (1984b) noted that similar behavior in *E. phaeton* caused wasps to spend extended periods grooming after encounters. Peterson (1986a) and Peterson at al. (1987) studied defensive regurgitation by the larvae of *M. americanum* in the laboratory. Regurgitated liquid collected from caterpillars that had just fed contained both benzaldehyde and HCN at concentrations identical to those of the cyanogenic host *Prunus serotina*. Predaceous ants were found to be repelled at concentrations far below those found in the regurgitate droplets. Indeed, when food items were treated with the toxin, feeding ants became cateleptic and in some cases were irreversibly poisoned. Selective group foraging on the young leaves of the host (see above) allows the caterpillars to obtain large amounts of the toxin, since these leaves have the highest cyanogenic potential. The effectiveness of this tactic has not yet been assessed under field conditions, and it is not known if siblings near to a disturbance are also stimulated to regurgitate.

A curious instance of group response to predation was reported by Green and Sullivan (1950). They observed that pentatomid predators of the eastern tent caterpillar are sometimes fatally webbed to the surface of the tent by small groups of caterpillars that become agitated when they encounter the predator. Except for this single instance, the defensive use of silk during encounters with small parasitoids and predators is unknown but may be more common than presently recognized.

It has been hypothesized that the behavior of some individuals (that also have a probability of low relative fitness) within groups may make them especially susceptible to enemies and thus increase the likelihood that siblings escape (Smith Trail 1980; Shapiro 1976). Wellington (1960), for example, suggested that sluggish caterpillars in colonies of the western tent caterpillar (*M. californicum*) were particularly attractive to parasitoids and served as sinks, favoring more active siblings. Larvae of the eastern tent caterpillar parasitized by tachinids or braconids often settle on the outsides of the tent and individuals have been observed to have as many as 23 tachinid eggs attached to them (Fitzgerald unpublished data).

Stamp (1981b) tested the hypothesis that *E. phaeton* caterpillars parasitized by *Apanteles* wasps moved to exposed resting sites to attract hyperparasitoids. Reduction in *Apanteles* numbers by hyperparasitoids would favor late instar siblings of *E. phaeton* which are hunted by the newly emerged *Apanteles* wasps. Hyperparasitoids of *Apanteles,* however, attacked fewer exposed larvae than those placed experimentally in less exposed locations. Stamp (1981b) concluded that parasitized larvae were in exposed locations not to advertise but because they were fastened there by the cocoons of the *Apanteles* parasitoids, possibly to facilitate eventual mating of newly emerged wasps, although such locations might also facilitate faster development by the parasitoids (Porter 1983).

Aposematism

Gregarious caterpillars often forage conspicuously and are commonly warningly colored. While solitary caterpillars can also be warningly colored and distasteful, the relationship is particularly well developed among gregarious species. Stamp (1980) showed that egg clustering in butterflies is usually associated with warning coloration and distastefulness. Less is known of the situation in moths, but Nothnagle and Schultz (1987) found that two-thirds of the species of gregarious moth caterpillars they investigated had conspicuous, defensive features. Gregariousness in distasteful and warningly colored species might favor the survival of caterpillars in a number of ways (see review by Guilford 1990). Gregariousness may act like warning coloration itself, enhancing the conspicuousness of the aggregate. Predators may use proximity as a cue, avoiding caterpillars near to one that makes them sick. When there is a delayed toxic effect, aggregation increases the probability that subsequent prey eaten are of the same type as the one causing the sickness.

Some distasteful caterpillars may further enhance their conspicuousness by constructing highly visible shelters. Thus, although the early instars of tent caterpillars are dark-bodied and not particularly visible against the bark of the tree, the tents they construct can often be seen at distances of tens or even hundreds of meters (Fitzgerald personal observation). The colonies, however, may not be completely protected. Thus, while the larger larvae of tent caterpillars are usually not eaten by most birds because of their hairiness, smaller caterpillars may be eaten by as many as 60 different species of birds (Witter and Kulman 1972).

Conspicuously colored caterpillars are typically avoided by birds (see Wiklund and Järvi 1982). The value of aposematism and distastefulness is much less clear in the case of the invertebrate predators of caterpillars. Young (1983), for example, noted that distasteful and warning colored Neotropical caterpillars are little protected from such predators. The 150 species of parasitoids and predators that attack North American species of tent caterpillars are apparently oblivious to the properties that deter birds, though a number of these avoid the larva and specialize on the egg and pupa (Witter and Kulman 1972).

Fisher (1958) considered the relationship between gregariousness, distastefulness, and warning coloration in caterpillars. While recognizing that solitary caterpillars may be distasteful and warningly colored, and that differential survival of individuals sampled by predators could lead to the evolution of distastefulness, he favored a kin selection model to explain the origins of distastefulness. He suggested that even though a conspicuously colored and distasteful caterpillar may be fatally wounded by a bird, the sickened bird would thenceforth avoid the victim's nearby siblings and thus favor the evolution of the trait. He argued that distastefulness preceded the evolution of warning coloration and that gregariousness was a prerequisite for the evolution of distastefulness.

More recent studies have given additional support to the possibility that individual caterpillars can survive attack by predators (Boyden 1976; Järvi et al. 1981; Wiklund and Järvi 1982) and that individual selection can give rise to aposematic coloration (Sillen-Tullenberg and Bryant 1983). The current consensus appears to be that both individual selection and kin selection play a role in the evolution of aposematism, but the relative importance of each remains to be determined (Guilford 1990). A recent phylogenetic analysis indicates that in some gregarious butterflies, distastefulness may have evolved before gregariousness (Sillen-Tullenberg 1988). In cases where the evolution of aposematism and gregariousness could be separated, aposematism appears to have evolved before gregariousness, leading Sillen-Tullenberg (1988) to conclude that kin selection was of minor importance in the evolution of distastefulness and aposematism. The study involved only a small number of species and suffers, to some extent, from an incomplete knowledge of the relationship between apparent aposematism and distastefulness.

Evolution of Caterpillar Societies

The genetic systems underlying sociality in caterpillars affect both egg and larval dispersal patterns. Although larvae of the batch-laying pine webworm are reported to feed solitarily in the early instars, then aggregate later, there appears to be no instance where larvae that are products of dispersed, solitary eggs regularly find each other and form adaptive aggregates. Thus, all social caterpillars appear to be dirived from egg clusters. It is not true, however, that all egg clusters give rise to adaptive aggregates of larvae. In moths, siblings arising from a common egg mass may either disperse on eclosion or aggregate. Species that disperse are numerous and include such common forms as the gypsy moth *Lymantria dispar*, the fall cankerworm *Alsophila pomentaria* and the tussock moth *Orygia leucostigma*. The relationship between batch laying and larval gregariousness appears less flexible in butterflies. Nearly all of the North American species that lay eggs in batches have larvae that are gregarious for at least part of the larval stage (Stamp 1980).

Most investigators that have dealt with the evolution of aggregation in caterpillars have been interested chiefly in butterflies and have concentrated on elucidating the origins of the batch laying habit (see Stamp 1980; Chew and Robbins 1983; Young 1983; Courtney 1984). Stamp (1980) and Young (1983) suggest that cluster oviposition is particularly likely to arise when ovipositional plants are patchily distributed. Since encounters with larval food plants are rare, females are likely to deposit more than one egg on the hosts they discover. Selection pressure from predators that capture flying females, inclement weather that limits egg-laying opportunities, scarcity of males requiring females to search extensively for mates, and scarcity of adult food plants may also favor cluster oviposition (Stamp 1980). Clustering may render eggs less vulnerable to desiccation and to predation (particularly if distasteful and warningly colored) than solitary eggs (Stamp 1980).

It has been suggested that the apparent benefits accruing to aggregated larvae (the major emphasis of this review) are consequences rather than causes of the evolution of egg clustering. Courtney (1984) argued that adults of batch laying butterflies have higher realized fecundity than butterflies that lay eggs singly, and suggested that selection pressure for realized fecundity, in itself, is the chief factor underlying the evolution of cluster oviposition. Accordingly, cluster oviposition arises first and sets the stage for the evolution of larval adaptations that could lead initially to the maintenance of contact and eventually to sociality. To the extent that larval success depends on cluster size, however, the evolution of fitness-enhancing patterns of cooperative behavior among siblings could be expected to contribute to the evolution of adult ovipositional patterns. Moreover, Fitzgerald and Costa (1986) and Fitzgerald and Peterson (1988) suggested an explicit evolutionary pathway whereby advantages accruing to siblings that rest in proximity after encountering each other as they wandered over the plant might favor females that deposit eggs close enough to facilitate such encounters.

Acknowledgments

I thank Karen Clark, Peter Ducey, and Nancy Stamp for their helpful comments on the manuscript.

Literature Cited

Ancona, L. 1930. Biologia de la *Clistocampa azteca* Neum. Ann. Instit. Biol. Univ. Mexico 1:215–225.

Ayre, G. L., and Hitchon, D. E. 1968. The predation of tent caterpillars *Malacosoma americana* (Lepidoptera: Lasiocampidae) by ants. (Hymenoptera: Formicidae). Can. Entomol. 100:823–826.

Balfour-Browne, F. 1925. The evolution of social life among caterpillars, pp. 334–339. In Proceedings of the Third International Congress on Entomology, Zurich, Switzerland.

Balfour-Browne, F. 1933. The life-history of the "smaller eggar moth," *Eriogaster lanestris* L. Proc. Zool. Soc. London 161–180.

Barbosa, P., and Baltensweiler, W. 1987. Phenotypic plasticity and herbivore outbreaks, pp. 469–503. In J. C. Schultz and P. Barbosa (eds.), Insect Outbreaks. Academic Press, New York.

Berger, E. W. 1906. Observations upon the migration, feeding, and nesting habits of the fall webworm (*Hyphantria cunea* Dru.). Bull. 60, Bur. of Entomol. USDA: 41–51.

Boyden, T. C. 1976. Butterfly palatability and mimicry: Experiments with *Ameiva* lizards. Evolution 30:73–81.

Bush, G. L. 1969. Trail laying by larvae of *Chlosyne lacinia*. Ann. Entomol. Soc. Am. 62:674–675.

Capinera, J. L. 1980. A trail pheromone from the silk produced by larvae of the range caterpillar *Hemileuca olivae* (Lepidoptera: Saturniidae) and observations on aggregation behavior. J. Chem. Ecol. 3:655–644.

Carlberg, U. 1980. Larval biology of *Eriogster lanestris* (Lepidoptera, Lasiocampidae) in S. W. Finland. Notulae Entomol. 60:65–72.

Casey, T. M., and Hegel, J. R. 1981. Caterpillar setae: Insulation for an ectotherm. Science. 214:1131–1133.

Casey, T. M., Joos, B., Fitzgerald, T. D., Yurlona, M., and Young, P. 1988. Synchronized group foraging, thermoregulation, and growth of eastern tent caterpillars in relation to microclimate. Physiol. Zool. 61(4):372–377.

Chadab, R., and Rettenmeyer, C. W. 1975. Mass recruitment by army ants. Science 188:1124–1125.

Chew, F. S., and Robbins, R. K. 1983. Egg laying in butterflies. Symp. R. Entomol. Soc. London 11:65–79.

Cornell, J. C., Stamp, N. E., and Bowers, M. D. 1988. Variation and developmental change in activity of gregarious caterpillars *Hemileuca lucina* (Saturniidae). Psyche 95:45–58.

Courtney, S. P. 1984. The evolution of egg clustering by butterflies and other insects. Am. Nat. 123:276–281.

Crump, D., Silverstein, R. M., Williams, H. J., and Fitzgerald, T. D. 1987. Identification of the trail pheromone of the eastern tent caterpillar *Malacosoma americanum* (Lepidoptera: Lasiocampidae). J. Chem. Ecol. 13:397–402.

Damman, H. 1987. Leaf quality and enemy avoidance by the larvae of a pyralid moth. Ecology 68:88–97.

Dethier V. 1959a. Food-plant distribution and density and larval dispersal as factors affecting insect populations. Can. Entomol. 91:581–596.

Dethier, V. 1959b. Egg-laying habits of Lepidoptera in relation to available food. Can. Entomol. 91:554–561.

Duncan, R. W. 1982. Silver spotted tiger moth. Pacific For. Res. Centre, Can. Forestry Serv. Pest Leaflet 5.

Dunn, L. H. 1917. The coconut tree caterpillar (*Brassolis isthmia*) of Panama. J. Econ. Entomol. 10:473–488.

Edgerly, J. S., and Fitzgerald, T. D. 1982. An investigation of behavioral variability within colonies of the eastern tent caterpillar *Malacosoma americanum* (Lepidoptera: Lasiocampidae). J. Kan. Entomol. Soc. 55:145–155.

Eickwort, G. C. 1981. Presocial insects, pp. 199–279. In H. R. Hermann (ed.), Social Insects, Vol. II. Academic Press, New York.

Evans, E. W. 1982. Influence of weather on predator/prey relations: stinkbugs and tent caterpillars. J. N.Y. Entomol. Soc. 90:241–246.

Evans, E. W. 1983. Niche relations of predatory stinkbugs (*Podisus* spp., Pentatomidae) attacking tent caterpillar (*Malacosoma americanum,* Lasiocampidae). Am. Midl. Nat. 109:316–323.

Fabre, J. H. 1916. The Life of the Caterpillar. Dodd Mead, New York.

Filip, V., and Dirzo, R. 1985. Tabla de vida gusano de bolsa *Malacosoma incurvum* var. *aztecum* Neumogen (Lepidoptera: Lasiocampidae) en Xochimilco, D. F. Mexico. Folia Entomol. Mex. 66:31–45.

Fisher, R. A. 1958. The Genetical Theory of Natural Selection, 2nd ed. Dover, New York.

Fitzgerald, T. D. 1976. Trail marking by larvae of the eastern tent caterpillar. Science 194:961–993.

Fitzgerald, T. D. 1980. An analysis of daily foraging patterns of laboratory colonies of the eastern tent caterpillar, *Malacosoma americanum* (Lepidoptera: Lasiocampidae), recorded photoelectronically. Can. Entomol. 112:731–738.

Fitzgerald, T. D., and Costa, J. T. 1986. Trail-based communication and foraging behavior of young colonies of the forest tent caterpillar *Malacosoma disstria* Hubn. (Lepidoptera: Lasiocampidae). Ann. Entomol. Soc. Am. 79:999–1007.

Fitzgerald, T. D., and Edgerly, J. S. 1979a. Specificity of trail markers of forest and eastern tent caterpillars. J. Chem. Ecol. 5:564–574.

Fitzgerald, T. D., and Edgerly, J. S. 1979b. Exploration and recruitment in field colonies of the eastern tent caterpillar. J. Georgia Entomol. Soc. 14:312–314.

Fitzgerald, T. D., and Edgerly, J. S. 1982. Site of secretion of the trail marker of the eastern tent caterpillar. J. Chem. Ecol. 8:31–39.

Fitzgerald, T. D., and Peterson, S. C. 1983. Elective recruitment communication by the eastern tent caterpillar (*Malacosoma americanum*). Anim. Behav. 31:417–442.

Fitzgerald, T. D., and Peterson, S. C. 1988. Cooperative foraging and communication in social caterpillars. BioScience 38(1):20–25.

Fitzgerald, T. D., and Willer, D. E. 1983. Tent building behavior of the eastern tent caterpillar *Malacosoma americanum* (Lepidoptera: Lasiocampidae). J. Kan. Entomol. Soc. 56:20–31.

Fitzgerald, T. D., Casey, T. M., and Joos, B. 1988. Daily foraging schedule of field colonies of the eastern tent caterpillar *Malacosoma americanum*. Oecologia 76:574–578.

Fitzgerald, T. D., Clark, K. L., Vanderpool, R., and Phillips, C. 1991. Leaf-shelter-building caterpillars harness forces generated by axial retraction on stretched and wetted silk. J. Insect Behav. 4:21–32.

Franclemont, J. 1973. Mimallonoidea and Bombycoidea: Apatelodidae, Bombycidae, Lasiocampidae. In The Moths of America North of Mexico. Fasc 20. Part 1. Curwen Press, London.

Franz, J. M., and Laux, W. 1964. Individual differences in *Malacosoma neustria* (L.). Proc. XII Int. Cong. Entomol. 393–394.

Ghent, A. W. 1960. A study of the group-feeding behaviour of larvae of the jack pine sawfly *Neodiprion pratti banksianae* Roh. Behaviour 16:110–148.

Green, G. W., and Sullivan, C. R. 1950. Ants attacking larvae of the forest tent caterpillar, *Malacosoma disstria* Hbn. (Lepidoptera: Lasiocampidae). Can. Entomol. 82:194–195.

Greenblatt, J. A. 1974. Behavioral studies on tent caterpillars, M.S. thesis. University of Michigan, Ann Arbor.

Greenblatt, J. A., and Witter, J. A. 1976. Behavioral studies on *Malacosoma disstria* (Lepidoptera: Lasiocampidae). Can. Entomol. 108: 1225–1228.

Guliford, T. 1990. The evolution of aposematism, pp. 23–62. In D. L. Evans and J. O. Schmidt (eds.), Insect Defenses. State University of New York Press, Albany.

Hamilton, 1971. Geometry for the selfish herd. J. Theor. Biol. 31:295–311.

Heinrich, B. 1979. Foraging strategies of caterpillars: Leaf damage and possible predator avoidance strategies. Oecologia (Berlin) 42:325–337.

Henson, W. R. 1965. Individual rearing of the larvae of *Neodiprion sertifer* (Geoffroy) (Hymenoptera: Diprionidae). Can. Entomol. 97:773–779.

Herbert, P. D. N. 1983. Egg dispersal patterns and adult feeding behaviour in the Lepidoptera. Can. Entomol. 115:1477–1481.

Hertel, T., and Benjamin, D. M. 1979. Biology of the pine webworm in Florida slash pine plantations. Ann. Entomol. Soc. Am. 72:816–819.

Hixon, E. 1941. The walnut *Datana*. Oklahoma A&M Coll. Exp. Stn. Bull. B-246. 29 pp.

Hodson, A. C. 1941. An ecological study of the forest tent caterpillar *Malacosoma disstria* Hbn. in Northern Minnesota. Univ. Minn. Agric. Expt. Stn. Tech. Bull. 158.

Hoebeke, R. E. 1987. *Yponomeuta cagnagella* (Lepidoptera; Yponomeutidae): A palearctic ermine moth in the United States, with notes in its recognition, seasonal history, and habits. Ann. Entomol. Soc. Am. 80:462–467.

Hölldobler, B. 1978. Ethological aspects of chemical communication in ants, pp. 75–115. In J. S. Rosenblatt, R. A. Hinde, C. Beer, and M. Bushnel (eds.), Advances in the Study of Animal Behavior, Vol. 8, Academic Press, New York.

Hosoya, J. 1956. Notes on the biology of the tea-tussock-moth *Euproctis pseudoconspersa* Strand. Jpn. J. Sanit. Zool. 7:77–82.

Ito, Y. 1977. Birth and death, pp. 101–127. In T. Hidaka (ed.), Adaptation and Speciation in the Fall Webworm. Kodansha, Tokyo.

Iwao, S. 1968. Some effects of grouping in lepidopterous insects. Centre National de la Recherche Scientifique Colloq. Int. No. 173:185–210.

Jaffe, K. 1980. Theoretical analysis of the communication system for chemical mass recruitment in ants. J. Theor. Biol. 84:589–609.

Janzen, D. H. 1984. Natural history of *Hylesia lineata* (Saturniidae: Hemileucinae) in Santa Rosa National Park, Costa Rica. J. Kan. Entomol. Soc. 57:490–514.

Järvi, T. B., Sillen-Tullberg, B., and Wiklund, C. 1981. The cost of being aposematic. An experimental study of predation on larvae of *Papilio machaon* by the great tit *Parus major*. Oikos 36:267–272.

Johnson, W. T., and Lyon, H. H. 1988. Insects That Feed on Trees and Shrubs, 2nd ed. Comstock, Ithaca, NY.

Joos, B., Casey, T. M., Fitzgerald, T. D., and Buttemer, W. A. 1988. Roles of the tent in behavioral thermoregulation of eastern tent caterpillars. Ecology 69:2004–2011.

Kalkowski, W. 1958. Investigations on territorial orientation during ontogenic development in *Hyponmeuta evonymellus* L., Lepidoptera, Yponmeutidae, part II. Folia Biol. (Cracow) 6:245–263.

Kalkowski, W. 1966. Feeding orientation of caterpillars during ontogenic development in *Hyponomeuta evonymellus* L., Lepidoptera, Hyponomeutidae. Folia Biol. (Cracow) 14:23–46.

Knapp, R., and Casey, T. M. 1986. Thermal ecology, behavior, and growth of gypsy moth and eastern tent caterpillars. Ecology 67:598–608.

Laux, W. 1962. Individuelle Unterschiede in Verhalted und Leistung des Ringelspinners, *Malacosoma neustria* (L.). Ztsche, Angew. Zool. 49:465–525.

Lawrence, W. S. 1990. The effects of group size and host species on development and survivorship of a gregarious caterpillar *Halisidota caryae* (Lepidoptera: Arctiidae). Ecol. Entomol. 15:53–62.

Leonard, D. E. 1970. Feeding rhythms in larvae of the gypsy moth. J. Econ. Entomol. 63:1454–1457.

Long, D. B. 1953. Effects of population density on larvae of Lepidoptera. Trans. R. Entomol. Soc. London 104:533–585.

Long, D. B. 1955. Observations on subsocial behaviour in two species of lepidopterous larvae, *Pieris brassicae* L. and *Plusia gamma* L. Trans. R. Entomol. Soc. London 106:421–437.

Lyons, L. A. 1962. The effect of aggregation on egg and larval survival in *Neodiprion swainei* Midd. (Hymenoptera: Diprionidae). Can. Entomol. 94:49–58.

Masaki, S., and Umeya, K. 1977. Larval life, pp. 13–29. In T. Hidaka (ed.), Adaptation and Speciation in the Fall Webworm. Kodansha, Tokyo.

McManus, M. L., and Smith, H. R. 1972. Importance of silk trails in the diel behavior of late instars of the gypsy moth. Environ. Entomol. 1:793–795.

Mizuta, K. 1968. The effect of larval aggregation upon survival, development, adult longevity and fecundity of a zygaenid moth *Artona funeralis* Butler. Bull. Hiroshima Agric. Coll. 3:97–107.

Morris, R. F. 1963. The effect of predator age and prey defense on the functional response of *Podisus maculiventris* Say to the density of *Hyphantria cunea* Drury. Can. Entomol. 95:1009–1020.

Morris, R. F. 1972a. Fecundity and colony size in natural populations of *Hyphantria cunea*. Can. Entomol. 104:399–409.

Morris, R. F. 1972b. Predation by wasps, birds, and mammals on *Hyphantria cunea*. Can. Entomol. 104:1581–1591.

Morris, R. F. 1976. Relation of parasite attack to the colonial habit of *Hyphantria cunea*. Can. Entomol. 108:833–836.

Morris, R. F., and Fulton, W. C. 1970. Models for the development and survival of *Hyphantria cunea* in relation to temperature and humidity. Mem. Ent. Soc. Can. 70.

Mosebach-Pukowski, E. 1938. Uber die raupengesellschaften von *Vanessa io* und *Vanessa urticae*. Z. Morphol. Okol. Tiere 33:358–380.

Myers, J. H. 1978. A search for behavioural variation in first laid eggs of the western tent caterpillar and an attempt to prevent a population decline. Can. J. Zool. 56:2359–2363.

Myers, J. H., and Campbell, B. J. 1976. Distribution and dispersal in populations capable of resource depletion: A field study on cinnabar moth. Oecologia (Berlin) 24:7–20.

Myers, J. H., and Smith, J. N. M. 1978. Head flicking by tent caterpillars: A defensive response to parasite sounds. Can. J. Zool. 56:1628–1631.

Nielsen, E., and Nielsen, A. 1950. Contributions towards the knowledge of the migration of butterflies. Am. Mus. Novit., No. 1471.

Nothnagle, P. J., and Schultz, J. C. 1987. What is a forest pest?, pp. 59–80. In J. C. Schultz and P. Barbosa (eds.), Insect Outbreaks. Academic Press, New York.

Papaj, D. R., and Rausher, M. D. 1983. Individual variation in host location by phytophagous insects, pp. 77–124. In S. Ahmad (ed.), Herbivorous Insects: Host-Seeking Behavior and Mechanisms. Academic Press, New York.

Peterson, S. C. 1986a. Breakdown products of cyanogenesis: Repellancy and toxicity to predatory ants. Naturwissenschaften 73:627–628.

Peterson, S. C. 1986b. Host specificity of trail marking to foliage by eastern tent caterpillars, *Malacosoma americanum*. Entomol. Exp. Appl. 42:91–96.

Peterson, S. C. 1987. Communication of leaf suitability by gregarious eastern tent caterpillars (*Malacosoma americanum*). Ecol. Entomol. 12:283–289.

Peterson, S. C. 1988. Chemical trail marking and following by caterpillars of *Malacosoma neustria*. J. Chem. Ecol. 14:815–823.

Peterson, S. C., Johnson, N. D., and LeGuyader, J. L. 1987. Defensive regurgitation of allelochemicals derived from host cyanogenesis by eastern tent caterpillars. Ecology 68:1268–1272.

Porter, K. 1982. Basking behaviour in larvae of the butterfly *Euphydryas aurina*. Oikos 38:308–312.

Porter, K. 1983. Multivoltinism in *Apanteles bignellii* and the influence of weather on synchronization with its host *Euphydryas aurinia*. Entomol. Exp. Appl. 34:155–162.

Prop, N. 1960. Protection against birds and parasites in some species of tenthredinid larvae. Arch. Neerl. Zool. 13:380–447.

Roessingh, P. 1989. The trail following behaviour of *Yponomeuta cagnagellus*. Entomol. Exp. Appl. 51:49–57.

Roessingh, P. 1990. Chemical trail marker from silk of *Yponomeuta cagnagellus*. J. Chem. Ecol. 16:2203–2216.

Roessingh, P., Peterson, S. C., and Fitzgerald, T. D. 1988. The sensory basis of trail following in some lepidopterous larvae: Contact chemoreception. Physiol. Entomol. 13:219–224.

Schultz, D. E., and Allen, D. C. 1975. Biology and descriptions of the cherry scallop moth, *Hydria prunivorata* (Lepidoptera: Geometridae). Can. Entomol. 107:99–106.

Schultz, J. C. 1983. Tree tactics. Nat. Hist. 92:12–25.

Seymour, R. 1974. Convective and evaporative cooling in sawfly larvae. J. Insect Physiol. 20:2447–2457.

Shapiro, A. M. 1976. Beau geste? Am. Nat. 110:900–902.

Sharplin, J. 1964. The mourning cloak butterfly. University of Alberta, Department of Entomology Leaflet 475.

Shiga, M. 1976. A quantitative study on food consumption and growth of the tent caterpillar *Malacosoma neustria testacea* Motschulsky (Lepidoptera: Lasiocampidae). Bull. Fruit Tree Res. Stn. A 3:67–86.

Shiga, M. 1979. Population dynamics of *Malacosoma neustria testacea* (Lepidoptera, Lasiocampidae). Bull. Fruit Tree Res. Stn. A 6:59–168.

Sillen-Tullenberg, B. 1988. Evolution of gregariousness in aposematic butterfly larvae: a phylogenetic analysis. Evolution 42:293–305.

Sillen-Tullenberg, B., and Bryant, E. H. 1983. The evolution of aposematic coloration in distasteful prey: An individual selection model. Evolution 37:993–1000.

Smith-Trail, D. 1980. Behavioral interactions between parasites and hosts: Host suicide and the evolution of complex life cycles. Am. Nat. 116:71–91.

Stacey, L., Roe, R., and Williams, K. 1975. Mortality of eggs and pharate larvae of the eastern tent caterpillar *Malacosoma americana* (F.) (Lepidoptera: Lasiocampidae). J. Kan. Entomol. Soc. 48:521–523.

Stamp, N. E. 1977. Aggregation behavior of *Chlosyne lacinia* (Nymphalidae). J. Lepid Soc. 31:35–40.

Stamp, N. E. 1980. Egg deposition patterns in butterflies: why do some species cluster their eggs rather than deposit them singly? Am. Nat. 115:367–380.

Stamp, N. E. 1981a. Effect of group size on parasitism in a natural population of the Baltimore checkerspot, *Euphydryas phaeton*. Oecologia (Berlin) 49:201–206.

Stamp, N. E. 1981b. Behavior of parasitized aposematic caterpillars: Advantage to the parasitoid or host? Am. Nat. 118:715–725.

Stamp, N. E. 1982. Behavioral interactions of parasitoids and Baltimore checkerspot caterpillars *Euphydryas phaeton*. Environ. Entomol. 11:100–104.

Stamp, N. E. 1983. Overwintering aggregations of hackberry caterpillars (*Asterocampa clyton:* Nymphalidae). J. Lepid. Soc. 37:145.

Stamp, N. E. 1984a. Foraging behavior of tawny emperor caterpillars (Nymphalidae: *Asterocampa clyton*). J. Lepid. Soc. 38:186–191.

Stamp, N. E. 1984b. Interactions of parasitoids and checkerspot caterpillars *Euphydryas* spp. (Nymphalidae). J. Res. Lepid. 23:2–18.

Stamp, N. E., and Bowers, M. D. 1988. Direct and indirect effects of predatory wasps (*Polistes* sp.: Vespidae) on gregarious caterpillars (*Hemileuca lucina:* Saturniidae). Oecologia 75:619–624.

Stamp, N. E., and Bowers, M. D. 1990a. Variation in food quality and temperature constrain foraging of gregarious caterpillars. Ecology 71:1031–1039.

Stamp, N. E., and Bowers, M. D. 1990b. Body temperature, behavior and growth of early-spring caterpillars (*Hemileuca lucina:* Saturniidae). J. Lepid. Soc. 44:143–155.

Stamp, N. E., and Bowers, M. D. 1990c. Phenology of nutritional differences between new and mature leaves and its effect on caterpillar growth. Ecol. Entomol. 15:447–454.

Stehr, F. W., and Cook, E. F. 1968. A revision of the genus *Malacosoma* Hubner in North America (Lepidoptera: Lasiocampidae): Systematics, biology, immatures, and parasites. Smithsonian Institution, U.S. Nat. Mus. Bull. 276.

Sugimoto, T. 1962. Influences of individuals in aggregation or isolation on the development and survival of larvae of *Artona funeralis*. Jpn. J. Appl. Ent. Zool. 6:196–199.

Talhouk, A. M. 1975. Contribution to the knowledge of almond pests in East Mediterranean countries. I. Notes on *Eriogaster amygdali* Wilts. (Lepid., Lasiocampidae) with a description of a new subspecies by E. P. Wiltshire. Z. Angew. Entomol. 78:306–312.

Topoff, H., Mirenda, J., Droual, R., and Herrick, S. 1980. Behavioural ecology of mass recruitment in the army ant *Neivamyrmex nigrescens*. Anim. Behav. 28:779–789.

Tostowaryk, W. 1971. Relationship between parasitism and predation of diprionid sawflies. Ann. Entomol. Soc. Am. 64(6):1424–1427.

Tostowaryk, W. 1972. The effect of prey defense and the functional response of *Podisus modestus* (Hemiptera: Pentatomidae) to densities of the sawflies *Neodiprion swainei* and *N. pratti banksianae* (Hymenoptera: Neodiprionidae). Can. Entomol. 104:61–69.

Tsubaki, Y. 1981. Some beneficial effects of aggregation in young larvae of *Pryeria sinica* Moore (Lepidoptera: Zygaenidae). Res. Popul. Ecol. 23:156–167.

Vulinec, K. 1990. Collective security: aggregation by insects in defense, pp. 251–288.

In D. L. Evans and J. O. Schmidt (eds.), Insect Defenses. State University of New York Press, Albany.

Watanabe, N., and Umeya, K. 1968. Biology of *Hyphantria cunea* Drury (Lepidoptera; Arctiidae) in Japan. IV. Effects of group size on survival and growth of larvae. Jpn. Plant Prot. Serv. Res. Bull. 6:1–6.

Webster, H. V., and St. George, R. A. 1947. Life history and control of the webworm, *Homadaula albizziae*. J. Econ. Entomol. 40:546–553.

Wellington, W. G. 1957. Individual differences as a factor in population dynamics: The development of a problem. Can. J. Zool. 35:293–323.

Wellington, W. G. 1960. Qualitative changes in natural populations during periods of abundance. Can. J. Zool. 38:289–314.

Wellington, W. G. 1965. Some maternal influences on progeny quality in the western tent caterpillar, *Malacosoma pluviale* (Dyar). Can. Entomol. 97:1–14.

Weyh, R., and Maschwitz, U. 1978. Trail substance in larvae of *Eriogaster lanestris* L. Naturwissenshaften 65:64.

Weyh, R., and Maschwitz, U. 1982. Individual trail marking by larvae of the scarce swallowtail *Iphiclides podalirius* L. (Lepidoptera: Papilionidae). Oecologia (Berlin) 52:415–416.

Wiklund, C., and Järvi, T. 1982. Survival of distasteful insects after being attacked by naive birds: A reappraisal of the theory of aposematic coloration evolving through individual selection. Evolution 36:998–1002.

Wilson, E. O. 1962. Chemical communication among workers of the fire ant *Solenopsis saevissima* (Fr. Smith): 1. The organization of mass foraging; 2. An informational analysis of the odor trail; 3. The experimental induction of social response. Anim. Behav. 10:134–164.

Wilson, E. O. 1975. Sociobiology: The New Synthesis. Belknap Press, Cambridge, MA.

Wittenberg, J. F. 1981. Animal Social Bchavior. Wadsworth, Belmont, CA.

Witter, J. A., and Kulman, H. M. 1972. A review of the parasites and predators of tent caterpillars (*Malacosoma* spp.) in North America. Agr. Expt. Sta. Univ. Minn. Tech. Bull. 289.

Wojtusiak, R. J. 1929. Über die Raumorientierung bei *Pieris* Raupen. Bull. Int. Acad. Polon. Sc. B. II:59–66.

Wojtusiak, R. J. 1930. Weitere Untersuchungen über die Raumorientierung bei Kohlweisslingraupen. Bull. Int. Acad. Polon. Sc. B. II:629–655.

Young, A. M. 1972. Adaptive strategies of feeding and predator avoidance in the larvae of the neotropical butterfly, *Morpho peleides* (Lepidoptera: Morphidae). J. N.Y. Entomol. Soc. 80:66–82.

Young, A. M. 1983. On the evolution of egg placement and gregariousness of caterpillars in the Lepidoptera. Acta Biother. 32:43–60.

Young, A. M. 1985. Natural history notes on *Brassolis isthmia* Bates (Lepidoptera: Nymphalidae) in northeastern Costa Rica. J. Res. Lepid. 24:385–392.

Young, A. M., and M. W. Moffett. 1979. Studies on the population biology of the tropical butterfly *Mechanitis isthmia* in Costa Rica. Am. Midl. Nat. 101:309–319.

Young, A. M., Blum, M. S., Fales, H. H., and Bian, Z. 1986. Natural history and ecological chemistry of the neotropical butterfly *Papilio anchisiades* (Papilionidae). J. Lepid. Soc. 40:36–53.

12

The Effects of Ant Mutualism on the Foraging and Diet of Lycaenid Caterpillars

Matthew Baylis and Naomi E. Pierce

Introduction

Larvae of species in at least 10 families of the Lepidoptera associate with ants (Hinton 1951; Maschwitz et al. 1986), and of these the great majority are members of the Lycaenidae. About half of the Lycaenidae whose life histories have been described are myrmecophilous (Pierce 1987 cites records for 433 species from 6 biogeographic regions) and, with more information, this proportion is likely to be even greater (e.g., Fiedler 1991 provides thorough data for 118 species from Europe and North Africa alone). Ant association is also commonly found among species of the closely related taxon, the riodinids, although the frequency may not be as great as in the Lycaenidae (DeVries 1990b). Since together, these groups comprise about 30% of the some 17,280 species of butterflies estimated to occur worldwide (Shields 1989), larval association with ants is clearly a significant component of butterfly ecology.

Lycaenid–ant associations can be parasitic, commensal, or mutualistic (Hinton 1951; Atsatt 1981b; Cottrell 1984; Pierce 1987). Mutualisms in which the fitness of each partner is increased by the action of the other appear to be by far the most common type of interaction: the larvae of many species have specialized glands that visibly secrete droplets of food that are consumed by ants. Not only do attendant ants not attack the larvae themselves, but they often appear to protect them against other, potential enemies. Although some degree of chemical deception of ants by lycaenids is possible, it seems likely that ants in the majority of these relationships are harvesting substantial food rewards and are not being fooled by the caterpillars. To date, quantitative evidence assessing the costs and benefits for each partner in these associations is available for only a few lycaenid–ant and riodinid–ant systems (Ross 1966; Pierce and Mead 1981; Horvitz and Schemske 1984; Pierce and Easteal 1986; Pierce et al. 1987; DeVries 1988,

1990a, b; Fiedler and Maschwitz 1988, 1989a, b; Nash 1990; Savignano 1990; Baylis and Pierce 1991b). Our discussion here concentrates mainly on these mutualistic associations.

This chapter will examine the nature of the nutritional hurdles imposed by lycaenid–ant mutualisms and the responses of lycaenids to those challenges. We will begin with a brief overview of some of the effects that ants may have had on the dietary choices of lycaenid butterflies. We will then focus our discussion on *Jalmenus evagoras,* the species that has been the subject of our research in this area, and evaluate some of the physiological constraints imposed by the need to secrete rewards for ants as well as evidence for physiological and behavioral responses to these constraints.

Ants involved in lycaenid–ant interactions typically obtain their food from a variety of sources, and no relationships have been described in which attendant ants rely entirely on the secretions of lycaenid larvae for sustenance. However, lycaenids vary considerably in the strength of their associations with attendant ants. In some species, larvae appear to appease ants but have no other direct interactions with them; in others, perhaps the majority, larvae not only appease ants, but are intermittently tended by any of a large number of different species; and in still others, larvae are continuously tended by a single species of ant. This variation makes it difficult to generalize about the effects of ants on lycaenid foraging. Moreover, despite a valuable higher classification of the family which helps to distinguish many of the main taxa (Eliot 1973), the absence of a cladistic phylogenetic analysis restricts our ability to perform informative comparative studies across species.

There are, nevertheless, a number of common themes in lycaenid–ant interactions. Myrmecophilous lycaenid larvae have remarkably thick cuticles that are contoured in ways that protect the vital organs against occasional bites by attendant ants (Malicky 1970). The pupae and sometimes the larvae of many species stridulate, and these vibrations may serve as important communication signals for their attendant ants (Downey 1966; Downey and Allyn 1978; DeVries 1990a).

The association is primarily mediated through several specialized exocrine glands that secrete substances that appease ants, and in many species, reward them. The composition of these secretions has been partially analyzed for only a few species, and none has been fully characterized for all possible chemical components (Henning 1983b). This is particularly true for the pair of lateral tentacular organs found on the eighth abdominal segment of many species which secrete volatile chemicals that appear to mimic ant alarm signals (Claassens and Dickson 1977; Fiedler and Maschwitz 1987). Yamaguchi (1988) mentions that dendrolasin may be involved in interactions between the Japanese species, *Shirozua jonasi* and its attendant ants, and points out that dendrolasin can also function as a warning pheromone for ants such as *Lasius fulginosis*.

Secretions from the "dorsal organ" found on the seventh abdominal segment of many species were examined from larvae of *Polyommatus (Lysandra) hispana,* and found to contain largely carbohydrates, including fructose, glucose, sucrose, and trehalose in total concentrations ranging from about 13 to 19% by weight, and only trace amounts of the amino acid methionine (Maschwitz et al. 1975). Secretions from the dorsal organ of the Australian lycaenid, *Jalmenus evagoras,* contained sucrose, fructose, and glucose in concentrations of about 10%, as well as the amino acid serine, in concentrations that varied diurnally from about 20 to 40 mM (Pierce 1989). Amino acids were also detected in secretions from the many single-celled glands scattered in the epidermis of the larvae, called "pore cupolae" organs (Malicky 1969), and similar structures in the pupae (Kitching 1983), although these glands are likely to secrete additional compounds (Henning 1983b). The dorsal organ secretions of two congenors of *J. evagoras* produced similar amounts of amino acids, whereas those of a third produced aqueous solutions of a small, unidentified polypeptide (Pierce 1989). Among the Riodinidae, glands analogous to the dorsal organ, the "tentacle nectary organs" of *Thisbe irenea* were found to contain 10% sugar solutions and relatively high concentrations of a number of amino acids (DeVries 1988; DeVries and Baker 1989).

Lycaenid secretions differ in at least one significant way from the honeydew of Homoptera, with which they are frequently compared. Homopteran honeydew is excreted as a byproduct of feeding on plant phloem and xylem (Way 1963). Homopterans are known to modify aspects of their honeydew in response to attendant ants, including the rate of production, and the manner in which the honeydew is presented (Mittler 1958; Banks and Nixon 1958; Auclair 1963; Dixon 1985; Letourneau and Choe 1987). However, since it is essentially an excrement, honeydew is likely to be inexpensive for the homopterans to produce, and they produce it whether ants are there to receive it or not.

In contrast, lycaenids must synthesize their secretions actively. Amino acid secretions are metabolically costly to produce since they require diverting valuable protein resources from growth and development to defense: not surprisingly, lycaenids typically present these secretions only on solicitation by attendant ants. Indeed, larvae of *J. evagoras* raised without ants pupate at a larger size than those raised with ants (Pierce et al. 1987). It is difficult to make meaningful comparisons between the composition of these lepidopteran secretions and that of homopteran honeydew since so few species have been analyzed. However, lycaenid and riodinid secretions appear to contain lower concentrations of carbohydrates than most honeydews, and in some cases, higher concentrations of amino acids (Auclair 1963; Dixon 1985).

Lycaenid larvae benefit from ant associations in at least two ways. First, by secreting chemicals that appease ants, they are protected against the ants themselves that might otherwise be threatening predators (Malicky 1970). Second, experiments with several species have shown that attendant ants protect lycaenid larvae from predators and parasitoids (e.g., Pierce and Easteal 1986; Pierce et al. 1987; DeVries 1991; Fiedler and Maschwitz 1988, 1989a, b; Savignano 1990).

For example, the presence of attending ants was estimated to make a four to 12-fold difference in survival to pupation of larvae of the North American lycaenid, *Glaucopsyche lygdamus* (Pierce and Easteal 1986). The degree of protection afforded by attendant ants depends on characteristics such as the species of attendant ants, the density of predators and parasitoids in the habitat, and the phenologies of the different interacting parties (e.g., Pierce and Mead 1981; Bristow 1984).

Ant-associated lycaenid larvae thus live with unusual dietary constraints: not only must they consume sufficient food for their own development, but they must additionally supply food to their ant guard. These nutritional constraints depend in part on the nature of the rewards provided to the ants, and these in turn depend on the dietary preferences and alternative food sources available to the ants. For example, a lycaenid larva competing with homopterans and other nectar-secreting sources for the attention of ant mutualists may be under considerable pressure to produce rewards that are more attractive to attendant ants than the carbohydrates found in most honeydews. This may explain why the secretions of *J. evagoras* contain concentrated free amino acids as well as simple sugars. These amino acids can be shown to act as powerful phagostimulants for the attendant ant species, and may ensure that the larvae are continuously tended by ants, even when other honeydew-secreting insects are present (Pierce 1989).

Although amino acid secretions may be expensive to produce, populations of *J. evagoras* whose ants are experimentally excluded are preyed on so heavily that they cannot survive, meaning that the attraction and provision of an ant guard is, for this species, mandatory (Pierce et al. 1987). So, while Lawton's and McNeill (1979) herbivorous insect was caught between the devil of plant defenses and the deep blue sea of malnutrition, a myrmecophilous lycaenid caterpillar must contend not only with the same devil, but with a sea that is even deeper and bluer because of the additional nutritional demands imposed by its ant guard.

The Effect of Ant Protection on Lycaenid Diets

Protection by ants may have direct effects on the spatial and temporal foraging patterns of lycaenid larvae. Myrmecophilous lycaenid larvae occupy what has been called "enemy-free space" (Lawton 1978; Atsatt 1981b; Jeffries and Lawton 1984), areas in which the threat of predation and parasitism has been reduced because of the activities of attendant ants. Thus larvae may be able to feed in places, or at times of day, which would not be possible without ant protection. Feeding on the terminal foliage where leaves are more nutritious, and feeding during the day despite increased visibility to predators can be of considerable advantage to lycaenid larvae: the consumption of more nutritious food, and consumption throughout the day, permit shorter development times (Slansky and Scriber 1985). A good example of this is provided by *J. evagoras*. Larvae of *J.*

evagoras feed on acacia trees and are tended by ants in the genus *Iridomyrmex*. The larvae aggregate, and feed both during the day and night, often forming clearly visible congregations.

However, such apparency is rather exceptional among lycaenids. On the opposite extreme, protection by ants may also encourage behavioral crypsis among lycaenids such as the Australian species *Ogyris genoveva:* larvae of this butterfly shelter during the day in "byres" or earthen "corrals" constructed at the base of trees by their attendant ants, species of *Camponotus,* and emerge only at night to feed on the foliage of mistletoe hanging in the host trees (Common and Waterhouse 1981). Similarly, the ant-tended larvae of the Adonis Blue, *Polyommatus (Lysandra) bellargus,* have crepuscular activity patterns (Thomas 1983).

The foraging patterns adopted by different species of ant-associated lycaenids must depend in part on the activity patterns and the quality of protection offered by their ant associates. Lycaenids whose attendant ants are assiduous tenders, diurnally active, and possess relatively large colony sizes and/or well-developed systems of mass recruitment may be able to feed more openly. In contrast, those whose attendant ants are relatively weak tenders, nocturnal or crepuscular foragers, and possess small colony sizes and/or poor means of mass recruitment might resort to a more cryptic mode of foraging.

Foraging lycaenids may be influenced not only by the temporal activities of their attendant ants, but also by their spatial distributions. Lycaenids that associate with ants are more likely to lay their eggs in clusters than their untended counterparts, and the larvae and pupae are also more likely to aggregate (Kitching 1981). Feeding in aggregations may bring benefits in terms of attracting a larger, more efficient ant guard and facilitating foraging and handling time, especially for young larvae (Pierce et al. 1987). However, it can also incur costs: larvae of *J. evagoras* frequently consume all the available foliage on their host plants. Larvae must either find a new host plant or starve, although in some cases, feeding on extrafloral nectar or honeydew from homopterans may allow larvae on defoliated host plants to persist for prolonged periods.

Since selection favors myrmecophilous lycaenids that feed in places where the presence of associated ants reduces the density of potential enemies, attendant ants may also affect lycaenid host plant choice (Atsatt 1981b). Thus lycaenids may be more likely to feed on plants with extrafloral nectaries or a homopteran fauna, both of which serve to attract ants to the plant. Larvae of the Lipteninae, a subfamily of the Lycaenidae, are associated with ant columns on tree trunks, and feed on lichens, fungi, and algae found under the bark of trees. Characteristics of the host plants themselves, such as their growth form and/or relative degree of "apparency," may further influence the ant environment and thereby the lycaenid fauna (Malicky 1969; Atsatt 1981b).

Proximity to ants that farm a variety of nectar-producing sources has presumably led to the behavior seen among many lycaenid and riodinid larvae of imbibing nectar themselves, including extrafloral nectar and homopteran honeydew (Cottrell 1984). This nectar feeding is a curious sight, and has led to the belief that these caterpillars are parasitizing ant–plant and ant–homopteran interactions, possibly as a means of reducing the cost of producing ant attractants themselves (Horvitz and Schemske 1984; Maschwitz et al. 1984; Horvitz et al. 1987; DeVries and Baker 1989). For example, larvae of *Lachnocnema bibulus* feed on the honeydew of jassids, membracids, and psyllids, and larvae of *Shirozua jonasi* drink the honeydew of coccids and aphids (Hinton 1951). The larvae of *J. evagoras* and some of its congenors frequently drink extrafloral nectar from the host plant as well as homopteran honeydew, although this behavior is more often observed when ants are in attendance and disappears almost entirely when ants have been experimentally excluded (Pierce and Elgar 1985; unpublished observations). Larvae of the Japanese species, *Niphanda fusca,* feed primarily on regurgitations from their host ants (Fukuda et al. 1984). Moreover, the carnivorous consumption of homopterans themselves is widespread in the Lycaenidae: Cottrell (1984) cites examples in eleven genera from four subfamilies.

A number of lycaenid species live in, or directly around, ant nests or shelters. Some of these species maintain herbivorous diets: larvae of *Hypochrysops apollo,* which live in ant nests made in the ant plant *Myrmecodia beccarii,* feed on internal plant tissues (Common and Waterhouse 1981), and *Anthene emolus* lives in nests woven from the leaves of the tree *Saraca thaipingensis* by the weaver ant *Oecophylla smaragdina,* and feeds on the leaves of the tree (Fiedler and Maschwitz 1989b). This trend has been extended beyond mutualism to myrmecophagy by the parasitic species *Liphyra brassolis,* whose larvae prey on the ant brood within the nests of the weaver ants (Dodd 1902). These larvae possess an impressive array of adaptations for preying on ants, including an exceptionally thick and sclerotized cuticle, a fringe of hairs along the ventral side that appear to aid the larvae in clamping down on the substrate and resisting attempts by ants to expose the vulnerable ventral surface, as well as larvae antennae-like structures that are used in finding ant brood. The majority of other lycaenid species that parasitize ants do so by chemically mimicking aspects of their recognition signals, thereby fooling the ants into accepting them into the brood chamber of the nest where they then set about devouring the brood (Cottrell 1984; Thomas et al. 1989; Fiedler 1990b).

The necessity for myrmecophilous lycaenids to be found in, or near to ant nests or foraging places may act to limit the diversity of possible diets available to such lycaenids. However, myrmecophily may have also promoted diversification of lycaenid diets, not only to nectar feeding and to the carnivorous extremes exhibited by homopterophagous and myrmecophagous species, but also to a wider range of host plants among the strictly herbivorous taxa. Ant-dependent oviposition appears to be relatively common among the strongly myrmecophilous species. Caged females of *Ogyris amaryllis* laid larger egg batches on branches where they encountered ants than on those where ants had been excluded (Atsatt 1981a). When presented with trees bearing attendant ants and larvae, and trees with just larvae, females from one population of *J. evagoras* laid 86% of their eggs on the trees with ants and larvae (Pierce and Elgar 1985). Indeed, females could be induced to lay eggs on wooden dowling in the field, provided that pupae of *J. evagoras* and attendant ants were present on these artificial substrates (Atsatt, Pierce, and Smiley unpublished data). Pierce and Elgar (1985) cite 46 species of lycaenids, from 29 genera in 5 subfamilies, in which ant-dependent oviposition has been suspected or described.

Oviposition "mistakes" induced when appropriate attendant ants occur on plants other than the customary host species might therefore be especially common among ant tended lycaenids. For example, the larvae of *Hypochrysops ignitus* feed on at least 12 different plant families, but they are tended only by the ant *Iridomyrmex nitidus* (Common and Waterhouse 1981). Depending on the relative costs and benefits and the different species involved, lycaenids that have a mobile ant defense may also be able to shift hosts more readily than species that rely on sequestering chemical defenses from their food plants. Pierce and Elgar (1985) compared the diet breadth of 282 species of ant-associated and non-ant-associated lycaenids and found that, in general, lycaenids that are tended by ants feed on a greater number of host plant families and genera than their non-ant-attended counterparts. Myrmecophily may not be the only reason for increased diet breadth (Fiedler 1990a); however, the existence of ant-dependent oviposition provides a plausible mechanism promoting host expansion in myrmecophilous taxa.

It is tempting to speculate from these observations that ant association may have a paradoxical effect on the host plant range of lycaenids: ant dependence might simultaneously restrict the number of adequate feeding niches and facilitate host range expansion. Ant-induced host plant sampling may also help to explain the wide taxonomic diversity of food sources utilized by the Lycaenidae compared with other families of butterflies (Ehrlich and Raven 1965). The diets of herbivorous lycaenids include fungi, blue-green algae, lichens, ferns, cycads, conifers, bamboos, mistletoes, oaks, and legumes (Atsatt 1981b; Henning 1983a; Pierce 1987).

Because they decide where to lay eggs, female butterflies by and large determine the ant environment of the larvae. However, the juveniles of *J. evagoras* aggregate, and ovipositing females are attracted to their own juveniles as well as to ants (Pierce and Elgar 1985). Thus responses by the larvae to the presence or absence of ants can influence both the immediate ant environment, and the ant environment of the next larval generation. For example, larvae on plants where ant densities are high pupate in higher locations than larvae on plants where ant densities are low (Carper 1989). Presumably both the presence of larvae (indicating successful survival), and the relative degree of visibility of their pupation sites provide valuable feedback to females about the ant environment of potential host plants.

Presumably this is one of the mechanisms by which species specificity has arisen in certain mutualistic lycaenid–ant interactions. If females oviposit in response to the presence of conspecifics, then any ant that is a good enough tender that the larvae survive will enjoy an enhanced level of oviposition on trees where it is tending. If a particular ant species is an unusually proficient tender, then over time, the ant species itself, even in the absence of conspecific juveniles, might be sufficient to elicit enhanced levels of oviposition on appropriate host plant species. As long as female lycaenids lay at least some of their eggs on host plants, either not occupied by ants and/or visited by other species of ants, then this mechanism would allow them to sample the available ant fauna on a continuous basis. Should the ant fauna change for any reason (i.e., the "customary" ant associate become scarce, and/or a new species arise that is a better tender than the original associate), selection would cause the butterflies to change their affiliation accordingly.

Constraints Imposed by Secretion

Nitrogen and water are particularly important currencies in the nutrition of herbivores (McNeill and Southwood 1978; Strong et al. 1984). The growth rates and feeding efficiencies of herbivorous insects are strongly correlated with the relative ratio of nitrogen and water in their diet (Scriber 1977; Scriber and Slansky 1981; Slansky and Scriber 1985). Phytophagous ant-associated lycaenid larvae face these dietary constraints as well as having to secrete to ants aqueous solutions of carbohydrates and proteins. The amount secreted can be considerable. Sixty-two larvae of *J. evagoras* on a single tree were found to provide for ants approximately 400 mg dry biomass over a 24-hour period (Pierce et al. 1987). A larva of the European lycaenid, *Polyommatus (Lysandra) coridon* was estimated to produce 22–44 μl of solution in its lifetime, containing approximately 3.5–7.0 mg of dry biomass (Fiedler and Maschwitz 1988). The total volume produced by a single final instar of *Anthene emolus* is at least 80 μl, which, assuming a carbohydrate concentration of 15% (Maschwitz et al. 1975), contains approximately 12.7 mg of carbohydrate (Fiedler and Maschwitz 1989b).

Variation in the quality of lycaenid secretions is doubtlessly generated by a complex set of interacting variables, including the strength of predation and/or parasitism in a habitat, the degree of dependence of the lycaenid larvae on attendant ants for defense, the nature of the host plant, the dietary requirements of the ants, and the presence of alternative food resources for the ants. Selection should favor individual lycaenids that receive the maximum benefit of ant defense in exchange for the minimum cost of ant attraction. Thus we might expect to find considerable flexibility and variability in the nature of the secretions of different species and even of separate populations of the same species found in different habitats.

The costs of secretion can have far-reaching consequences. To investigate some of these consequences, we examined whether larvae of *J. evagoras* could

Table 12.1. Summary of the Effect of Ants on the Nutrition and Growth of Fifth Instar Larvae of J. evagoras[a]

	With ants[b]	Without ants[b]
Size at start of final instar	$<$	$>$
Development time	=	=
Relative consumption rate	=	=
AD	=	=
Relative growth rate	$<$	$>$
Size at start of pupal instar	$<<$	$>>$

[a]AD, approximate digestibility, a measure of the proportion of ingested food that is digested. From Baylis (1989) and Baylis and Pierce (1991a).

[b]$<$, significantly smaller; $>$, significantly greater; =, no significant difference.

compensate physiologically for the dietary cost of maintaining an ant guard. In other insects, compensation for changes in diet quality can occur by consuming more food, consuming food of a higher nutritional quality, altering digestive efficiencies (reviewed by Simpson and Simpson 1989; also see Fiedler 1990a), and/or extending development time (Mattson 1980).

We reared *J. evagoras* larvae from eggs under uniform conditions and fed them on cuttings from potted *Acacia* plants (Baylis 1989; Baylis and Pierce 1992). Half of the larvae were reared with attendant ants from a laboratory colony of *Iridomyrmex (anceps* species group); the other half were untended. The development time, relative consumption rate and approximate digestibility (AD; a measure of the proportion of ingested food that is digested) of final instar larvae were measured using the gravimetric method of Waldbauer (1968). The presence or absence of ants did not have a significant effect on any of the three compensatory methods examined (Table 12.1). As a consequence, final instar larvae reared with ants had a significantly lower growth rate than larvae reared alone (Baylis 1989; Baylis and Pierce 1992). This result agrees with the previous finding that larvae of *J. evagoras* reared in the laboratory on trees from which ants have been excluded are larger than those reared on trees with ants (Pierce et al. 1987).

It thus appears that larvae of *J. evagoras* do not compensate for the nutrient loss to ants; they simply bear the loss by growing less. Reduced growth represents a real cost to *J. evagoras:* both the lifetime mating success of males and the fecundity of females are positively correlated with relative adult size (Elgar and Pierce 1988).

The inability of larvae of *J. evagoras* to compensate for the nutrient loss to ants suggests that the nutrients, particularly nitrogen and water, may be especially limiting in the survival and growth of the larvae. There is good evidence that this is the case. In a further experiment, young, potted seedlings of *Acacia decurrens* were either given water containing a nitrogenous fertilizer, or water alone. The foliage of plants treated with fertilizer had a higher nitrogen content than the unfertilized plants, although this covaried with a number of other nutrients found

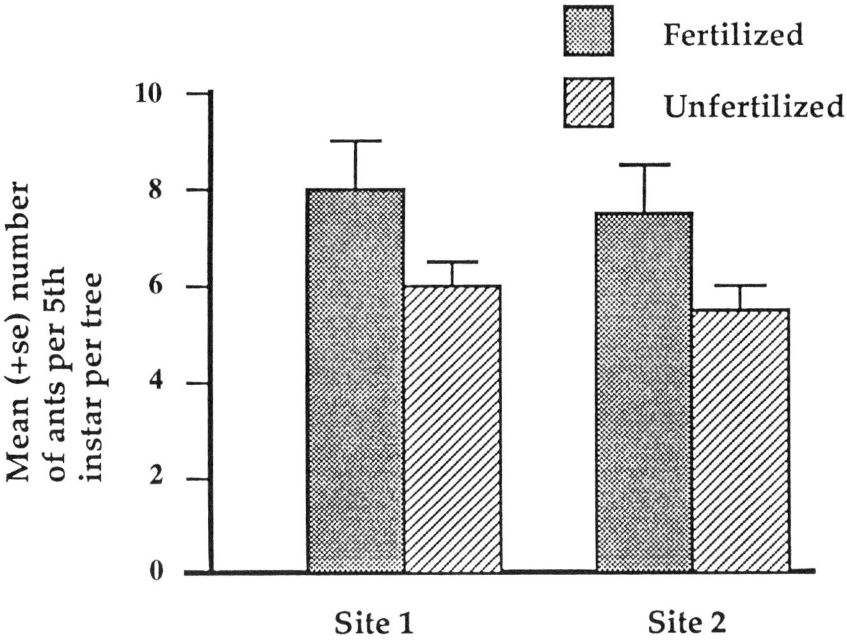

Figure 12.1. The Effect of Host Plant Quality on Ant Attendance of Larvae of *J. evagoras*. Mean number of ants (± SE) tending fifth instar larvae feeding on potted host plants that had been treated with nitrogenous fertilizer (solid columns) or not treated with fertilizer (hatched columns). Measurements were taken for ants tending larvae at two field sites in Mount Nebo, Queensland, Australia.

in the plants (Baylis and Pierce 1991a). Under field conditions, final instar larvae of *J. evagoras* feeding on the plants treated with fertilizer attracted a larger ant guard than those feeding on unfertilized plants (Fig. 12.1). In the absence of caterpillars, ants showed no preference for either plant type.

These results indicate that the ant attractants secreted by larvae of *J. evagoras* vary as a function of diet quality. Not only did caterpillars on plants treated with fertilizer attract a larger ant guard, but they also survived better in the field over a 10-day period than did larvae on unfertilized plants (Fig. 12.2). Control larvae reared in a screened bush house from which tending ants and predators were excluded survived equally on both types of plant. We therefore attribute the increase in survival rate to the attraction of a larger ant guard (Baylis and Pierce 1991a).

Thus host plant quality indirectly influences the survivorship of larvae of *J. evagoras* through altering the level of ant defense. As with other phytophagous insects, host plant quality also directly affects larval growth (Baylis 1989). Pupae of *J. evagoras* were collected from 82 trees of *Acacia melanoxylon* at Ebor in

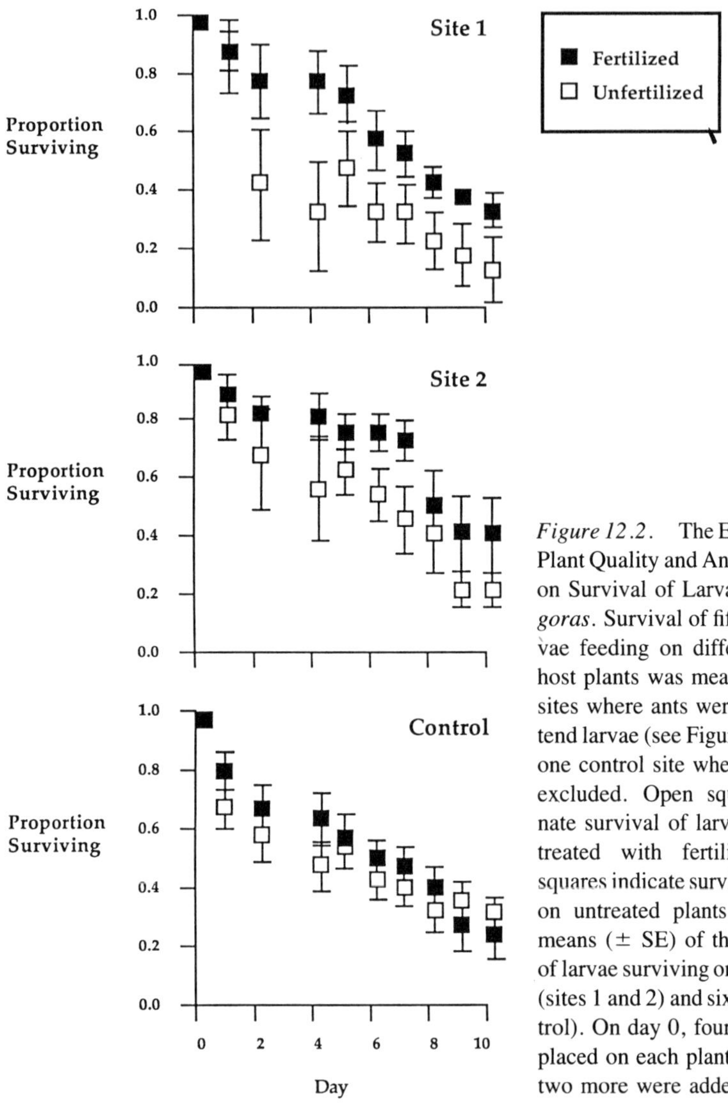

Figure 12.2. The Effect of Host Plant Quality and Ant Attendance on Survival of Larvae of *J. evagoras*. Survival of fifth instar larvae feeding on different quality host plants was measured at two sites where ants were allowed to tend larvae (see Figure 12.1), and one control site where ants were excluded. Open squares designate survival of larvae on plants treated with fertilizer; closed squares indicate survival of larvae on untreated plants. Points are means (± SE) of the proportion of larvae surviving on three plants (sites 1 and 2) and six plants (control). On day 0, four larvae were placed on each plant, and day 4, two more were added.

New South Wales, Australia. Although all the trees were of the same species, they were of two types: 58% of the trees had both young and mature leaves (YM), while the remaining 42% of the trees had only mature leaves (M). We found that both the young and mature leaves of the YM trees had higher nitrogen and water contents than the mature leaves of the M trees (Table 12.2). This suggests that the larvae that fed on the YM trees had a diet of higher nutritional quality than those that fed on M trees. Not surprisingly, the adults reared from the pupae

Table 12.2. The relationship between Percent Nitrogen and Percent Water Content of Young and Mature Leaves, and the Weights of Adults of J. evagoras. *Data are shown for YM trees (those with both young and mature leaves) and M trees (those with just mature leaves)*[a]

	YM trees			M trees			Unpaired
	Mean	SD	*n*	Mean	SD	*n*	*t* value
Nitrogen (dry weight) (%)							
Y	3.02a	0.56	48				
M	2.70a	0.32	48	2.50	0.35	34	2.81**
Water (%)							
Y	67.7b	3.2	48				
M	59.5b	4.5	48	56.4	4.0	34	3.13**
Male mass (mg)	86.1	22.4	29	57.6	21.6	25	4.85***
Female mass (mg)	163.8	48.4	37	84.2	9.2	24	6.46***

[a]Adults weights are means per tree. All data were log transformed prior to analysis by t test. Samples for a and b were taken from the same tree, and differences between means were analyzed by paired t test; a, df = 47, t = 3.72***; b, df 47, t = 9.72***. From Baylis (1989).

**$P < 0.01$.

***$P < 0.001$

collected from YM trees were larger than those reared from pupae collected from M trees.

Females of *J. evagoras* accordingly use plant quality as a cue in oviposition. Mated females placed in a cage containing plants that had been treated with fertilizer and plants that had not preferred to lay egg batches on the former (Baylis and Pierce 1991a).

Presumably because of the extreme dietary demands of ant association, lycaenids are likely to be able to respond not only to intraspecific variability in host plant quality, but to variability between individual plant parts, as well as to differences between species. Lycaenids are well known for feeding on nitrogen-rich, or water- and nitrogen-rich parts of plants such as flowers, terminal foliage, buds, shoots, and seed pods, although the degree to which this behavior correlates with myrmecophily is not known (Mattson 1980; Robbins and Aiello 1982; Thomas 1985; Milton 1990). However, larvae of the North American species, *Glaucopsyche lygdamus,* are more attractive to ants when feeding on seed pods of their legume host plants than when feeding on other parts of the plant (Pierce and Easteal 1986). Moreover, larvae of the European species, *Polyommatus icarus,* were more attractive to ants when reared on several species of herbaceous Fabaceae than on foliage of the tree *Robinia pseudacacia* (Fiedler 1990a). Fiedler points out that the relative stability of the interaction must therefore be mediated in part through the host plant.

A comparison of the food plants of 297 lycaenid species revealed that ant-tended lycaenids are more likely to feed on legumes, nitrogen-fixing nonleguminous plants and mistletoes than are their untended counterparts (Pierce 1985). An accurate evaluation of the extent to which particular ecological factors may have given rise to this pattern is impossible in the absence of a phylogeny for the group. Nevertheless, it is tempting to conclude that ants have been of central importance in shaping the host plant preferences of myrmecophilous lycaenids: legumes and nitrogen-fixing nonleguminous plants can convert atmospheric nitrogen to ammonia, which can subsequently be assimilated by the plant. These plants may thus be considered nitrogen-rich, or at least less variable in their nitrogen content over time because of their dependency on the relatively invariable atmospheric nitrogen level than on the variable soil nitrogen level. In addition, legumes frequently have extrafloral nectaries that attract ants and can act as a source of moisture for larvae (Pierce 1985; Baylis 1989). Similarly, mistletoes, which are parasitic and accordingly able to absorb selectively high quality nutrients from their host trees, may be more stable in their nitrogen and water levels than their hosts (Pierce 1985; Baylis 1989).

Conclusions

Mutualistic associations with ants have imposed several constraints on the dietary choice of the Lycaenidae. First, to obtain protection from ants, lycaenids must live in, or near to, ant-foraging trails or nests. Second, the secretion of often considerable amounts of aqueous proteins to ants, which can exact a considerable cost to larval growth, makes these nutrients of special importance in the diets of lycaenids. Severe nutritional requirements are likely to have led to the predilection, among herbivorous lycaenids, for nitrogen- and water-rich food plants such as legumes, and for the nitrogen- and water-rich parts of those food plants. These dietary preferences, coupled with the proximity of ants and ant-tended homopterans, may have facilitated the shift among some species to nectar feeding and to carnivory (Hinton 1951; Cottrell 1984).

Although myrmecophily imposes certain constraints on lycaenid foraging patterns, it may also act to diversify the range of possible diets. Ant-dependent oviposition has been frequently observed in myrmecophilous lycaenids, and this may encourage a high rate of host plant shifting. This, in turn, might be expected to promote polyphagy among myrmecophilous lycaenids, and may have been an important mechanism generating the startling diversity of diets consumed by extant species of Lycaenidae today.

Acknowledgments

We thank Andrew Berry, Catherine Bristow, Roger Kitching, David Nash, Steve Simpson, Nancy Stamp, and Diane Wagner for their contribution to the ideas presented in this chapter.

Literature Cited

Atsatt, P. R. 1981a. Ant-dependent food plant selection by the mistletoe butterfly *Ogyris amaryllis* (Lycaenidae). Oecologia (Berlin) 48:60–63.

Atsatt, P. R. 1981b. Lycaenid butterflies and ants: Selection for enemy-free space. Am. Nat. 118:638–654.

Auclair, J. L. 1963. Aphid feeding and nutrition. Annu. Rev. Entomol. 8:439–490.

Banks, C. J., and Nixon, H. L. 1958. Effects of the ant, *Lasius niger* L., on the feeding and excretion of the bean aphid, *Aphis fabae* Scop. J. Exp. Biol. 35:703–711.

Baylis, M. 1989. The role of nutrition in an ant-lycaenid-host plant interaction. D. Phil. thesis, University of Oxford.

Baylis, M., and Pierce, N. E. 1991a. The effect of host-plant quality on the survival of larvae and oviposition by adults of an ant-tended lycaenid butterfly, *Jalmenus evagoras*. Ecol. Entomol. 16:1–9.

Baylis, M., and Pierce, N. E. 1992. Lack of compensation by fifth instar larvae of the myrmecophilous lycaenid butterfly, *Jalmenus evagoras*, for the loss of nutrients to ants. Physiol. Entomol. 17:107–114.

Bristow, C. M. 1984. Differential benefits from ant attendance to two species of Homoptera on New York ironweed. J. Anim. Ecol. 53:715–726.

Carper, E. R. 1989. The effects of varying levels of ant attendance on the aggregation behavior and survivorship of larvae of *Jalmenus evagoras* (Lepidoptera: Lycaenidae). Senior thesis, Princeton University.

Claassens, A. J. M., and Dickson, C. G. C. 1977. A study of the myrmecophilous behaviour of the immature stages of *Aloides thyra* (L.) (Lepidoptera: Lycaenidae) with general reference to the function of the retractile tubercles and with additional notes on the general biology of the species. Entomol. Rec. J. Variat. 89:225–231.

Common, I. F. B., and Waterhouse, D. F. 1981. Butterflies of Australia, 2nd ed. Angus and Robertson, Sydney.

Cottrell, C. B. 1984. Aphytophagy in butterflies: Its relationship to myrmecophily. Zool. J. Linn. Soc. 80:1–57.

DeVries, P. J. 1987. Ecological aspects of ant association and hostplant use in a riodinid butterfly. Ph.D. thesis, University of Texas, Austin.

DeVries, P. J. 1988. The ant associated larval organs of *Thisbe irenea* (Riodinidaae) and their effects on attending ants. Zool. J. Linn. Soc. 94:379–393.

DeVries, P. J. 1990a. Enhancement of symbioses between butterfly caterpillars and ants by vibrational communication. Science 248:1104–1106.

DeVries, P. J. 1990b. Evolutionary and ecological patterns in myrmecophilous riodinid butterflies, pp. 143–147. In C. M. Huxley and D. R. Cutler (eds.), Ant-Plant Interactions. Oxford University Press, Oxford.

DeVries, P. J. 1991. Mutualism between *Thisbe irenea* larvae and ants, and the role of ant ecology in the evolution of myrmecophilous butterflies. Biol. J. Linn. Soc. 43:179–195.

DeVries, P. J., and Baker, I. 1989. Butterfly exploitation of a plant-ant mutualism: adding insult to herbivory. J. N.Y. Entomol. Soc. 97:332–340.

DeVries, P. J., Harvey, D. J., and Kitching, I. J. 1986. The ant associated epidermal organs on the larva of the lycaenid butterfly *Curetis regula* Evans. J. Nat. Hist. 20:621–633.

Dixon, A. F. G. 1985. Aphid Ecology. Blackie, New York.

Dodd, A. P. 1902. Contribution to the life history of *Liphyra brassolis* Westw. Entomology 35:153–156; 184–188.

Downey, J. C. 1962. Myrmecophily in *Plebejus (Icaricia) icarioides* (Lepidoptera: Lycaenidae). Entomol. News 73:57–66.

Downey, J. C. 1966. Sound production in pupae of Lycaenidae. J. Lep. Soc. 20:129–155.

Downey, J. C., and Ally, A. C., Jr. 1978. Sounds produced in pupae of Lycaenidae. Bull. Allyn Mus. 48:1–14.

Ehrlich, P. R., and Raven, P. 1965. Butterflies and plants: A study in coevolution. Evolution 18:596–604.

Elgar, M. A., and Pierce, N. E. 1988. Mating success and fecundity in an ant-tended lycaenid butterfly, pp. 59–75. In T. H. Clutton-Brock (ed.), Reproductive Success: Studies of Individual Variation in Contrasting Breeding Systems. University of Chicago Press, Chicago.

Eliot, J. N. 1973. The higher classification of the Lycaenidae (Lepidoptera): A tentative arrangement. Bull. Br. Mus. (Nat. Hist.) 28:371–505.

Fiedler, K. 1990a. Effects of larval diet on myrmecophilous qualities of *Polyommatus icarus* caterpillars (Lepidoptera: Lycaenidae). Oecologia (Berlin) 83:284–287.

Fiedler, K. 1990b. New information on the biology of *Maculinea nausithous* and *M. teleius* (Lepidoptera: Lycaenidae). Nota Lepidopt. 12:246–256.

Fiedler, K. 1991. European and North West African Lycaenidae and their associations with ants. J. Res. Lepidopt. 239–258.

Fiedler, K., and Maschwitz, U. 1987. Functional analysis of the myrmecophilous relationships between ants (Hymenoptera: Formicidae) and lycaenids (Lepidoptera: Lycaenidae) III. New aspects of the function of the retractile tentacular organs of lycaenid larvae. Zool. Beitr. 31:409–416.

Fiedler, K., and Maschwitz, U. 1988. Functional analysis of the myrmecophilous relationships between ants (Hymenoptera: Formicidae) and lycaenids (Lepidoptera: Lycaenidae) II. Lycaenid larvae as trophobiotic partners of ants—a quantitative approach. Oecologia (Berlin) 75:204–206.

Fiedler, K., and Maschwitz, U. 1989a. Functional analysis of the myrmecophilous relationships between ants (Hymenoptera: Formicidae) and lycaenids (Lepidoptera: Lycaenidae). I. Release of food recruitment in ants by lycaenid larvae and pupae. Ethology 80:71–80.

Fiedler, K., and Maschwitz, U. 1989b. The symbiosis between the weaver ant, *Oecophylla smaragdina,* and *Anthene emolus,* an obligate myrmecophilous lycaenid butterfly. J. Nat. Hist. 23:833–846.

Fukuda, H., Hama, E., Kuzuya, T., Takahashi, A., Takahashi, M., Tanaka, B., Tanaka, H., Wakabayashi, M., and Watanabe, Y. 1984. The Life Histories of Butterflies in Japan, Vol. 3. Hoikusha, Osaka.

Henning, S. F. 1983a. Biological groups within the Lycaenidae (Lepidoptera). J. Entomol. Soc. South. Afr. 46:65–85.

Henning, S. F. 1983b. Chemical communication between lycaenid larvae (Lepidoptera: Lycaenidae) and ants (Hymenoptera: Formicidae). J. Entomol. Soc. South. Afr. 46:341–366.

Hinton, H. E. 1951. Myrmecophilous Lycaenidae and other Lepidoptera—a summary. Proc. Transact. South London Entomol. Nat. Hist. Soc. 1949–50 111–175.

Horvitz, C. C., and Schemske, D. W. 1984. Effects of ant-mutualists and an ant-sequestering herbivore on seed production of a tropical herb, *Calathea ovandensis* (Marantceae). Ecology 65:1369–1378.

Horvitz, C. C., Turnbull, C., and Harvey, D. J. 1987. Biology of immature *Eurybia elvina* (Lepidoptera: Riodinidae), a myrmecophilous metalmark butterfly. Ann. Entomol. Soc. Am. 80:513–519.

Jeffries, M. J., and Lawton, J. H. 1984. Enemy free space and the structure of ecological communities. Biol. J. Linn. Soc. 23:269–286.

Kitching, R. L. 1981. Egg clustering and the southern hemisphere lycaenids: Comments on a paper by NE Stamp. Am. Nat. 118:423–425.

Kitching, R. L. 1983. Myrmecophilous organs of the larvae of the lycaenid butterfly *Jalmenus evagoras* (Donovan). J. Nat. Hist. 17:471–481.

Lawton, J. H. 1978. Host plant influences on insect diversity: the effects of space and time, pp. 105–125. In L. A. Mound and N. Waloff (eds.), Diversity of Insect Faunas. 9th Symposium of the Royal Entomological Society (London). Blackwell, London.

Lawton, J. H., and McNeill, S. 1979. Between the devil and the deep blue sea: on the problems of being a herbivore, pp. 223–244. In R. M. Anderson, B. D. Turner, and L. R. Taylor (eds.), Population Dynamics. Blackwell, Oxford.

Letourneau, D. K., and Choe, J. C. 1987. Homopteran attendance by wasps and ants: The stochastic nature of interactions. Psyche 94:81–91.

McNeill, S., and Southwood, T. R. E. 1978. The role of nitrogen in the development of insect/plant relationships, pp. 77–98. In J. B. Harborne (ed.), Biochemical Aspects of Plant and Animal Coevolution. Annual Proceedings of the Phytochemical Society of Europe, number 15. Academic Press, London.

Malicky, H. 1969. Versuch einer analyse der okologischen beziehungen zwischen Lycaeniden (Lepidoptera) und Formiciden (Hymenoptera). Tijd. Entomol. 112:213–298.

Malicky, H. 1970. New aspects of the association between lycaenid larvae (Lycaenidae) and ants (Formicidae, Hymenoptera). J. Lep. Soc. 24:190–202.

Maschwitz, U., Wust, M., and Schurian, K. 1975. Blaulingsraupen als Zuckerlieferanten fur Ameisen. Oecologia (Berlin) 18:17–21.

Maschwitz, U., Schroth, M., Hanel, H., and Pong, T. Y. 1984. Lycaenids parasitising symbiotic ant-plant partnerships. Oecologia (Berlin) 64:78–80.

Maschwitz, U., Dumpert, K., and Tuck, K. R. 1986. Ants feeding on anal exudate from tortricid larvae: a new type of trophobiosis. J. Nat. Hist. 20:1041–1050.

Mattson, W. J. 1980. Herbivory in relation to plant nitrogen content. Annu. Rev. Ecol. Syst. 11:119–161.

Milton, S. J. 1990. A lycaenid butterfly (*Anthene amarah* Guerin) selects unseasonal young *Acacia* shoots for oviposition. South Afr. J. Zool. 25:83–85.

Mittler, T. E. 1958. Studies on the feeding and nutrition of *Tuberolachnus salignus* (Gmelin) (Homoptera, Aphididae). II. The nitrogen and sugar composition of ingested phloem sap and excreted honeydew. J. Exp. Biol. 35:74–84.

Nash, D. R. 1990. Cost-benefit analysis of a mutualism between lycaenid butterflies and ants. D. Phil. thesis, University of Oxford.

Pierce, N. E. 1985. Lycaenid butterflies and ants: Selection for nitrogen fixing and other protein rich food plants. Am. Nat. 125:888–895.

Pierce, N. E. 1987. The evolution and biogeography of associations between lycaenid butterflies and ants. Oxford Surv. Evol. Biol. 4:89–116.

Pierce, N. E. 1989. Butterfly-ant mutualisms, pp. 299–324. In P. J. Grubb and J. Whittaker (eds.), Towards a More Exact Ecology. Blackwell, Oxford.

Pierce, N. E., and Easteal, S. 1986. The selective advantage of attendant ants for the larvae of a lycaenid butterfly, *Glaucopsyche lygdamus*. J. Anim. Ecol. 55:451–462.

Pierce, N. E., and Elgar, M. A. 1985. The influence of ants on host plant selection by *Jalmenus evagoras,* a myrmecophilous lycaenid butterfly. Behav. Ecol. Sociobiol. 16:209–222.

Pierce, N. E., and Mead, P. S. 1981. Parasitoids as selective agents in the symbiosis between lycaenid butterfly caterpillars and ants. Science 211:1185–1187.

Pierce, N. E., and Young, W. R. 1986. Lycaenid butterflies and ants: Two-species stable equilibria in mutualistic, commensal and parasitic interactions. Am. Nat. 1128:216–227.

Pierce, N. E., Kitching, R. L., Buckley, R. C., Taylor, M. F. J., and Benbow, K. 1987. The costs and benefits of cooperation for the Australian lycaenid butterfly, *Jalmenus evagoras* and its attendant ants. Behav. Ecol. Sociobiol. 21:237–248.

Robbins, R. K., and Aiello, A. 1982. Foodplant and oviposition records for Panamanian Lycaenidae and Riodinidae. J. Lep. Soc. 36:65–75.

Ross, G. N. 1966. Life history studies on Mexican butterflies. IV. The ecology and ethology of *Anatole rossi,* a myrmecophilous metalmark (Lepidoptera, Riodinidae). Ann. Entomol. Soc. Am. 59:985–1004.

Savignano, D. 1990. Associations between ants and larvae of the Karner blue. Ph.D. thesis, University of Texas, Austin.

Scriber, J. M. 1977. Limiting effects of low leaf-water content on the nitrogen utilization, energy budget, and larval growth of *Hyalophora cecropia*. Oecologia (Berlin) 28:269–287.

Scriber, J. M., and Slansky, F. Jr. 1981. The nutritional ecology of immature insects. Annu. Rev. Entomol. 26:183–211.

Shields, O. 1989. World numbers of butterflies. J. Lep. Soc. 43:178–183.

Simpson, S. J., and Simpson, C. L. 1989. The mechanisms of nutritional compensation by phytophagous insects pp. 111–160. In E. A. Bernays (ed.), Insect-Plant Interactions, Vol. 2. CRC Press, Boca Raton, FL.

Slansky, F., Jr., and Scriber, J. M. 1985. Food consumption and utilization, pp. 82–163. In G. A. Kerkut and L. I. Gilbert (eds.), Comprehensive Insect Physiology, Biochemistry and Pharmacology, Vol. 4. Pergamon, Oxford.

Strong, D. R., Lawton, J. H., and Southwood, T. R. E. 1984. Insects on Plants: Community Patterns and Mechanisms. Blackwell, Oxford.

Thomas, C. D. 1985. Specializations and polyphagy of *Plebejus argus* (Lepidoptera: Lycaenidae) in North Wales. Ecol. Entomol. 10:325–340.

Thomas, J. A. 1983. The ecology and conservation of *Lysandra bellargus* (Lepidoptera: Lycaenidae) in Britain. J. Appl. Ecol. 20:59–83.

Thomas, J. A., Elmes, G. W., Wardlaw, J. C., and Woyciechowski, M. 1989. Host specificity among *Maculinea* butterflies in *Myrmica* ant nests. Oecologia (Berlin) 79:452–457.

Vane Wright, R. I. 1978. Ecological and behavioural origins of diversity in butterflies pp. 56–59. In L. A. Mound and N. Waloff (eds.), Diversity of Insect Faunas, 9th Symposium of the Royal Entomological Society (London). Blackwell, London.

Waldbauer, G. P. 1968. The consumption and utilization of food by insects. Adv. Insect Physiol. 5:229–289.

Way, M. J. 1963. Mutualism between ants and honeydew producing Homoptera. Annu. Rev. Entomol. 8:307–344.

Yamaguchi, S. (1988) The Life Histories of Five Myrmecophilous Lycaenid Butterflies of Japan. Kodansha, Tokyo.

PART III

Environmental Variation in Time and Space

The objective of this section is to explore the effects of environmental variation in time and space on foraging patterns of caterpillars. Caterpillars face a tradeoff between maximizing adult fecundity through attaining maximal pupal size and having to develop quickly before food quality deteriorates too much and predators locate them. Damage to plants by herbivores may induce plant defenses, which may affect subsequent generations of herbivores. So these aspects of environmental variation through time can influence the foraging patterns and population dynamics of insect herbivores (Haukioja, Chapter 13). Most caterpillars contend with some degree of seasonality in their food resources and in predator pressures. In dry tropical forest, changes in abiotic and biotic conditions over the rainy period result in temporal (through diapause) or spatial (through migration) absence of Lepidoptera during the dry period (Janzen, Chapter 14). In arctic regions, basking behavior by caterpillars (thus they can approach their optimal temperature for growth) is critical because it facilitates acquisition of nutrients before the inevitable seasonal deterioration of food quality and sufficient larval growth before parasitoid activity becomes overwhelming (Kukal, Chapter 16). In addition to the effects of seasonality on food quality and population sizes of enemies, temperature directly affects the impact of predators and food quality on feeding, growth and survivorship of herbivores (Stamp, Chapter 15).

Study of foraging patterns of caterpillars has contributed to our general understanding of natural processes (e.g., how various factors constrain foraging by animals) and in particular to development of plant–herbivore theory, which focuses on an explanation of the degree of host plant specificity of insect herbivores. This knowledge of insect herbivores should be useful in insect pest management. However, agroecosystems differ greatly, both abiotically and biotically, from natural ecosystems. These differences are of such magnitude that they are likely to influence the foraging patterns of caterpillars and their insect enemies in significant ways (Barbosa, Chapter 17).

13

Effects of Food and Predation on Population Dynamics

Erkki Haukioja

Introduction

Factors that operate at the individual level of insect herbivores, such as food, predation, competition, and abiotic factors, generate a pattern of population dynamics. Members of extant populations originate from populations that have avoided extinction. Logically there have to be mechanisms that regulate population densities to positive values, but below numbers destroying possibilities for continuing existence. The regulating factors have to be density dependent at least temporally at some spatial scale (but see Murdoch et al. 1985; Hanski 1990). During the last decade there has been considerable debate concerning the existence of density dependency in field populations (Dempster 1983; Stiling 1987, 1988), and about methodological and biological difficulties in demonstrating density dependence (Hassell 1985; Dempster and Pollard 1986; Murdoch and Reeve 1987; Mountford 1988; Turchin 1990; Hanski 1990). When density dependent factors are assumed to operate, or are found, the logical next question concerns the biological characteristics of the regulative factors (Stiling 1987), whether they function directly or in combination with other factors (Roland 1988), and with or without lags (Turchin 1990; Hanski 1990).

To explain patterns in density, all the factors affecting herbivore density, by definition, should be taken into account (Haukioja et al. 1983; Strong et al. 1984). However, in this chapter I deal only with the effects of food and predation, and the interactions between them, as ecological and evolutionary constraints for herbivorous larvae, with repercussions for population dynamics. My examples mainly refer to forest caterpillars.

Studying each of the factors requires specialized methods, and a great deal of tedious work. Not surprisingly, seldom are all of the factors studied in detail for a particular population. Interactions among the factors may be important, too.

The effect of predation, for instance, is modified by the herbivores' choice of food and shelter. High levels of predation may determine to what extent high quality patches of food are available to caterpillars. Difficulties in studying the effects of single factors for population dynamics are even more exaggerated when studying the interactions of these factors.

Predation has long been regarded as the primary factor influencing insect population dynamics (Varley and Gradwell 1970; Strong et al. 1984; Hassell 1985). Its true importance is difficult to measure although occurrence of predation is relatively easy to observe: you can often see that a larva was parasitized or predated. The same is true with food quantity: lack of food can be observed. But the role of food quality in population dynamics of insect herbivores is often more elusive. For instance, poor food quality cannot usually be readily observed and may simply lead to disappearance of larvae, or to small adults with low fecundity (Neuvonen and Haukioja 1984; Haukioja et al. 1985; Lawton 1986; Zalucki and Brower 1992).

Effects of Food on Population Dynamics

Food availability influences population dynamics because food quality and quantity affect both survivorship and growth of larvae and, via the size of the pupa, the future fecundity of adults. Food plants also are the substrate of larval activity and can modify the consequences of predation.

Variation in Food Quality

Plant quality varies for genetic, developmental, and environmental reasons (e.g., Mattson 1980; McKey 1979). High among-plant variation in food quality occurs, with variability quantified by chemical measures (contents of nutricnts and allclochemicals), and by physical measures (like leaf toughness) (Coley 1983).

The degree of genetic variability of plants, both inter- and intraspecifically, has important consequences for herbivores, which have been discussed elsewhere (Rosenthal and Janzen 1979; Denno and McClure 1983; Strong et al. 1984). Variation in plant quality due to developmental reasons is often large; young foliage usually is rich in nutrients and sugars and often low in digestibility reducing compounds; the opposite characteristics are true for mature foliage (Feeny 1970; Haukioja et al. 1978; Rausher 1981).

Abiotic environmentally caused changes in foliage quality have large potential effects on insects. In particular, low nitrogen content of foliage in stressed trees is thought to be critical for survival of young larvae (White 1984). However, although stressed trees often are suitable diets for insects (Brodbeck and Strong 1987; Mattson and Haack 1987; Bazzaz et al. 1987; Chapin et al. 1987), variability in the effects on insect herbivores is common (Watt 1988; Coleman and Jones 1988; Jones and Coleman 1988; Larsson 1989; English-Loeb 1989; Price 1991).

Among environmentally caused biotic variation in foliage quality, insect-induced variation is especially interesting because it may be density dependent, and therefore offers a simple potential causal link to regulation of herbivore populations (Berryman 1987; Haukioja and Neuvonen 1987; Edelstein-Keshet and Rausher 1989). Insect feeding may induce changes in foliage quality, such that food quality may decline (because it then is lower in nutrients and higher in certain allelochemicals, e.g., Benz 1974; Schultz and Baldwin 1982; Tuomi et al. 1984; Clausen et al. 1989) or it may provide better nutrition for herbivores (Danell and Huss-Danell 1985; Wagner and Evans 1985; Haukioja et al. 1990). In the former case, the change in quality is called induced defense or induced resistance, the latter has been called resource regulation (Craig et al. 1986) or induced amelioration (Haukioja 1990). Furthermore, induced effects may operate with or without time lags (Haukioja 1982). Therefore they have the potential capacity either to stabilize or destabilize herbivore population density, and may even contribute to population outbreaks or regular cycles (Haukioja and Neuvonen 1987).

Effects of Food on Growth, Future Fecundity, and Mortality of Caterpillars

Physical and chemical plant traits do not affect population dynamics of insects directly, but through several intermediate steps operating both at the level of individuals and of populations. At the individual herbivore level, plant traits influence the searching and feeding behavior of caterpillars. Usually caterpillars search for more suitable diets, which has its dangers because the search may fail, or may make them susceptible to predation (Wratten et al. 1990; Edwards et al. 1991). Caterpillars may totally refuse to eat certain types of food, simply because they do not get suitable stimuli from the food (Schoonhoven 1982). Furthermore, larvae may circumvent potential constraints posed by plants, by detoxifying allelochemicals or becoming conditioned to them (Brattsten 1979; Dethier 1988). Competition and behavioral consequences of larval crowding may modify the effects of plant availability for individuals (Gruys 1970; Haukioja et al. 1988b).

Possibilities for insects to acclimatize or adapt to variable food availability obviously depend on predictability of the variation. Some forms of variation are highly predictable; for instance, in seasonal environments there is plenty of young foliage available in spring, but less later in the season. Chemically the variation from tender spring leaves to mature or senescing fall leaves is so large that it may halve the growth rate of larvae (Rausher 1981). The reason for the reduction in growth rate is reduced consumption of mature foliage, not reduced conversion of plant tissue to larval biomass (Rausher 1981; Lawson et al. 1984). Among diprionid sawfly larvae on pine, a spring-feeding species cannot survive on new needles due to their high content of allelochemicals (Ikeda et al 1977), but late-season species use the new needles without difficulties (Niemelä et al. 1991). In mountain birch, *Betula pubescens* ssp. *tortuosa*, insect growth rates in early summer are at least twice those late in the season and, even within the larval period of the early season species like autumnal moth, *Epirrita autumnata*, growth rates decrease steeply (Haukioja et al. 1978; Ayres and MacLean 1987; Senn et al. unpublished data). The difference in larval growth rates on early and mid-season birch leaves exceeds the differences in larval growth rates among plant species used by *Epirrita* (Neuvonen et al. 1987). Furthermore, heterophyllous trees (like birches and poplars), which even in late summer have both mature leaves in short shoots and still growing young foliage in long shoots, have a higher late-season species richness of lepidopteran fauna than other tree species, such as bird cherry, completing production of new leaves in spring (Niemelä and Haukioja 1982).

Other forms of variation in foliage availability are more stochastic. For instance, intratree chemical and physical variability among even-aged leaves is little studied but can be large (e.g., Schultz et al. 1982; Marquis 1988) and affect herbivores (Niemelä et al. 1984; Haukioja and Neuvonen 1985a; Edwards et al. 1990).

Food availability, including quality, largely determines both the final size of the caterpillar, and pupal size, and time needed to achieve that. Fecundity is positively correlated with pupal size. Pupal size is especially critical for species that do not feed as adults; all the resources for reproduction have to be obtained during the larval period.

Growth requires time. Other things being equal, achieving a large pupal weight takes longer than a smaller pupal weight. Consequently, pupating at a lower weight may be advantageous for three reasons: (1) it reduces the time that larvae are vulnerable to larval predation, (2) it increases the probability that the larvae successfully pupate before foliage is too old, and (3) at high larval densities, it increases the likelihood that foliage will be available for the whole larval period. Therefore, the length of the larval period has to be a compromise between the benefit of size-related fecundity and the costs of increased mortality risks associated with longer larval periods. But to shorten one life stage to proceed to the next to avoid predation does not help unless the risk of predation per time unit is lower in the next stage. The general assumption that larval period is more risky to predation than pupal time seems realistic (see Berryman 1988), although the pupal period may be quite vulnerable, too (Tenow 1963; Hanski and Parviainen 1985; Hanski 1987; Roland 1988).

Next I examine the relationship between larval weight, which is strongly correlated with fecundity, and length of the larval period in the autumnal moth using mountain birch. Caterpillars of the moth periodically defoliate birch forests at the northern tree line (Tenow 1972), in a strongly seasonal environment. The short growing season, lasting 3 months or less, causes rapid seasonal changes in foliage quality (Haukioja et al. 1978). Thus, the autumnal moth–birch system is particularly useful to examine how conflicting selective pressures on caterpillars affect population dynamics.

Because adult autumnal moths do not need food for laying eggs, their potential reproductive capacity depends on resources obtained by the larvae. (Here I ignore possible complications caused by other size-dependent factors, such as dispersal, avoidance of predation, but see Reavey, Chapter 8.) The critical pupal weight, under which the individual is unable to reproduce successfully, is 40–45 mg (Haukioja and Neuvonen 1985b). Weights above that limit have different importance for the sexes. For a female, every extra milligram in weight adds to the clutch size by, on average, 2.7 eggs. For a male, and even if multiple matings occur, simply exceeding the minimum limit is sufficient for successful reproduction (Haukioja and Neuvonen 1985b).

Larval growth in *Epirrita autumnata* is very rapid: on high quality diets, they may double their weight in a day (Haukioja and Niemelä 1974). For larvae reared under identical conditions on birches with different histories and treatments (previous defoliation, etc.), food quality may cause 27-fold differences in pupal weights (Haukioja and Neuvonen 1987). However, as mentioned above, the seasonal decline in foliage quality is very rapid, and thresholds in larval acceptance of aging leaves are obvious: at the end of the leaf expansion period, leaves may mature so much within 1 or 2 days that larvae refuse to eat them (Ayres and MacLean 1987; Suomela personal communication).

If the daily survival probability is lower in the larval stage than in the pupal stage, growing a day more is beneficial only if the growth and, correspondingly, fecundity increment is high enough to outweigh the probability of mortality with the additional days. If the risk of mortality per unit of time is equal in the larval and pupal stages or lower in the larval stage, extra growth—when possible—is beneficial for autumnal moth females but not necessarily for males. However, growth cannot be high for long periods because of the rapid seasonal deterioration in foliage quality; as leaf quality declines, larvae are deterred from feeding (Ayres and MacLean 1987) and start to loose weight.

The tradeoff between continued growth versus pupation suggests positive correlations between the size (and fecundity) and duration of the larval period. However, in reality such correlations between *E. autumnata* larval period and pupal weight may be either positive or negative (Ayres et al. 1987). The positive correlation occurs when broods are compared and thus is obviously genetically determined, with larger size associated with high fecundity and achieved by growing longer. But when tree-specific mean larval periods and pupal weights are compared, the correlation is negative. The reason is intertree variation in foliage quality: in some trees it is possible to achieve high weights rapidly, while in poor quality trees larval weights remain low even after growing a prolonged time.

Effects of Food on Insect Population Dynamics

There is no doubt that food availability may modify population dynamics, both the average population density and fluctuations around the mean. Foliage quality may directly affect larval mortality, because of toxicity, low content of nutrients, and feeding deterrence. On the other hand, food availability is likely to be a regulative factor only if it is density dependent. If it functions in a density dependent way, it is crucial whether the effect is direct or if there are time lags (Haukioja 1982; Rhoades 1985; Haukioja and Neuvonen 1987).

In data from field populations, the possible regulative role of food availability may be difficult to separate from other factors that simultaneously affect insect population dynamics (see, e.g., Dempster 1983; Hassell 1985; Dempster and Pollard 1986; Stiling 1988; Hassell et al. 1989 as examples of different opinions). What has been possible to do is to demonstrate that food availability introduces such strong delayed effects that it is potentially an important factor in the fluctuation of some insect populations.

Interactions between *E. autumnata* and mountain birch serve as an example of the magnitude of food-related lags in the population dynamics of the moth. Numbers of *E. autumnata* larvae fluctuate cyclically at 9 to 10 year intervals, and at the highest peaks, larvae completely destroy all the foliage (Tenow 1972; Haukioja et al. 1988a). Therefore, food quantity is important: shortage of foliage sets the limit to larval densities. Foliage quality in mountain birch is also quite variable (Haukioja et al. 1978). Constitutive resistance among birch stands is measurable but has not been well documented (Haukioja et al. 1990). Delayed forms of insect-induced resistance in birch foliage are more interesting for population dynamics because that kind of resistance modifies moth performance with time lags (Haukioja and Neuvonen 1985a; Haukioja et al. 1985, 1988a; Neuvonen and Haukioja 1991; Ruohomäki et al. 1992). Furthermore, because delayed induced resistance in birch foliage is triggered by larval feeding, it is density dependent, and may have time lags of 2 to 4 years (Haukioja et al. 1988a, 1990), long enough to be relevant for the population dynamics of the autumnal moth. Most data about the effects of induced resistance in birch quality on the moths are based on enclosure and laboratory experiments, so that predation was excluded.

Compared with larvae feeding on foliage of untreated control trees, the growth of larvae on foliage of trees defoliated in the previous season was retarded: mean pupal weights were 20–30% lower, and survival was up to 50% lower for those larvae reared on previously treated trees (Haukioja et al. 1985). These effects reduced egg production by 70–78% below that in control trees. Importantly, the magnitude of those effects on the moths and thus the enhanced tree resistance was carried over to the next growth seasons, and therefore to the next insect generations.

Not all forms of induced resistance may influence herbivore population dynamics. Rapidly induced forms of resistance may be so common that insects always have to cope with them (Fowler and Lawton 1985). Furthermore, the effects within a tree are hard to study because damage by larvae in a few hours induces resistance in adjacent control foliage also (Neuvonen and Haukioja 1991). Birch foliage can develop induced resistance so rapidly, within hours (Wratten et al.

1984) or days (Haukioja and Niemelä 1979; Neuvonen and Haukioja 1991), that its effects are encountered by the same insect generation. In this case, they do not cause time lags in the population dynamics but reduce the potential reproductive capacity of the current generation. Damaging the foliage of mountain birch a day or two before offering the leaves to larvae resulted in slower growth on damaged leaves and nearby undamaged leaves that on control foliage without previous damage of even nearby leaves (Haukioja and Niemelä 1979; Neuvonen and Haukioja 1991). Because the larval period was longer on foliage adjacent to damaged leaves than on control foliage, the pupal weights were only slightly but consistently (from 0 to 14%) lower on the former (Haukioja and Niemelä 1979; Haukioja and Hanhimäki 1985). Effects on survivorship varied among trials but usually were slightly lower on foliage from damaged trees than from control trees (Haukioja and Niemelä 1979). The combined effects of rapidly induced resistance on survival and fecundity decreased the reproductive potential of *E. autumnata* by as much as 22% (Haukioja and Neuvonen 1987).

Insect-induced changes in foliage quality of the host plant seem to be similar in another well studied species, the larch budmoth, *Zeiraphera diniana*. Like *E. autumnata*, the larch budmoth also has regular cycles of 9 to 10 year periodicity (Baltensweiler and Fischlin 1988). Benz (1974) showed that the length of larch needles, as well as their nitrogen content, declined and the fiber content increased in years following defoliations. In laboratory experiments, growth and survival of budmoth larvae on such foliage were poor. Fischlin and Baltensweiler (1979) modeled the cyclic fluctuation in budmoth density by including into the model only the insect-induced changes in the quantity and quality of larch foliage. Their simulations realistically mimicked the numerical behavior of the actual moth populations.

The above examples may not be representative for insects in general, but they are representative of forest insects, which often are chosen for study objects because of their tendency to show population outbreaks (Mason 1987). Strong delayed density dependency has been demonstrated in many forest insects (Turchin 1990), contrary to more random samples of moths (Hanski 1990). But it is not clear yet to what extent the outbreak tendency in forest lepidopterans is affected by their host plants, or if other features of forests contribute to outbreaks.

Effects of Predation on Population Dynamics

Predation has traditionally been regarded as an extremely important factor in influencing herbivore population dynamics (Varley and Gradwell 1970; Strong et al. 1984; Hassell 1985). The importance of predation has been demonstrated by both analytical methods and by experiments. The most widely used analytical method has been key-factor analysis (Varley and Gradwell 1960; Podoler and Rogers 1975), which partitions total mortality into cause–effect subcomponents to find the best correlation of the subcomponents with total mortality. The most extensive experimental results come from projects of biological control (Huffaker 1971; Huffaker et al. 1984).

In biological control, predators are added to the system, and results show that predation indeed may affect the density of herbivores. However, only some biological control projects succeed, in spite of choosing the most probable organisms. These results indicate that among predatory species attacking herbivorous larvae, only some of them have traits that make them effective in regulating herbivore populations. Luck (1990) listed the following traits characterizing successful biological control agents: synchrony with the host population, a high intrinsic rate of increase, high searching efficiency, intraspecific competition among the agents, and significant dispersal ability. But the characteristics that make for successful agents are not completely clear. In particular the degree to which dietary specialism by predators and stable prey equilibrums contribute to regulation of herbivore populations has been questioned (Murdoch et al. 1985).

Experimental treatments to demonstrate the role of predation usually use a tactic opposite to biological control: they try to exclude predators or parasites. Such experiments have demonstrated that herbivore populations grow better in the absence of predators (Faeth and Simberloff 1981; Skinner and Whittaker 1981; Mills 1990; Vasconcelos 1991). Similarly, exclusion of vertebrate predators, mainly birds, results in greater herbivore abundance (Holmes et al. 1979; for further examples, see Price 1987).

Gould et al. (1990) reported another type of predation experiment. What makes it especially interesting is that they modified herbivore densities. They artificially created eight gypsy moth (*Lymantria dispar*) populations with about 20-fold difference in density. They found a strong positive density-dependent reduction caused by parasitoids. However, different species of parasitoids showed different density dependence: from positive to negative.

Modeling studies demonstrate the potential regulative power of predation for insect populations. Analyses of cases of biological control indicate that assumptions of those models are not always met, and there obviously are multiple possibilities for effective control by predation (Murdoch et al. 1985). Biological control shows that long-term, even permanent, effects on prey populations are possible. Unfortunately, experiments do not usually cover long enough periods to give reliable data for natural populations (Price 1987).

Interactions between Food and Predation

Food availability for caterpillars may modify the caterpillars' value to predators, and similarly, the risk of predation may modify foraging habits of caterpillars and therefore what kind of food is really available to them. In addition to effects of plant traits on predation of herbivores, consequences via the whole herbivore guild may be important (Price 1986). For instance, the richer and more dense the plant-specific fauna of herbivores, the better strategy it is for a predator to search first for a specific host plant instead of a specific herbivore.

Predation Risks as Modified by Food

Perhaps the simplest ways in which food availability can modify predation risks occurs when plant availability is so low that development of herbivores is retarded (see Use of Food as Modified by Predation) or the herbivore has to move over considerable distances. Increased movement makes larvae more vulnerable to predation (Wratten et al. 1988). Retarded larval development by way of limited food availability also provides a logical explanation for the high diversity in secondary chemistry among and within plants, and for highly dispersed damage by herbivores within and among plants (Schultz 1983; Wratten et al. 1988; Edwards et al. 1991).

Location of larvae by a predator may be facilitated by passive consequences of feeding on the host plant, like the almost inevitable presence of feeding damage on leaves. Feeding damage is used by birds and other visually oriented predators to locate their prey (Heinrich and Collins 1983; Faeth 1990; Heinrich, Chapter 7). A larva may avoid being located by feeding damage in two ways. It can move frequently, which may make it more susceptible to predators. Bergelson and Lawton (1988) found increased vulnerability of moving larvae of the geometrid *Apocheima pilosaria* to ants, but not to titmice. But other studies did not reveal increased predation due to larval movement (Bergelson et al. 1986; Fowler and MacGarvin 1986; Mauricio and Bowers 1990). Second, the larva may camouflage feeding damage by cutting the damaged leaf from its petiole (Heinrich and Collins 1983). The latter is possible only for large larvae and may also be an adaptation against induced plant defenses (Edwards and Wanjura 1989; but see Weinstein 1990).

Volatile compounds released by plants when they are eaten may attract predators and parasitoids. Such compounds may be passive products of crushed tissue, just like visual feeding signs, but plants may also produce kairomones, or true infochemicals (sensu Dicke and Sabelis 1988), that attract predators. Such kairomones may be favored by natural selection because potentially they can facilitate finding of herbivores by their enemies, and thereby entice predators to be the plants' bodyguards (Dicke et al. 1990). Such compounds may even be released by uninfested plants in the neighborhood of infested ones (Takabayashi et al. 1991). Similarly, some plants provide food for predators, like ants and wasps, which may protect the plant against herbivores (Koptur 1979; Laine and Niemelä 1980; Risch and Rickson 1981; Mackay 1991; Cushman 1991). Concentration of, as well as content of nutrients in, extrafloral nectaries may be inducible through insect herbivory (Karban 1983; Smith et al. 1990).

Food availability modifies growth of a caterpillar. This modification may affect the caterpillar's suitability for the next trophic level (Price et al. 1980), and may even reverse the effect of plant allelochemicals (Duffey et al. 1986; Clancy and Price 1987; Taper and Case 1987). Food availability to herbivores is especially relevant in the case of parasitism. Small hosts are less favorable for parasitoids because they produce less fit offspring (or more males) (Vinson 1984). On the other hand, caterpillars from high quality food plants may be able to resist parasitoids more effectively (Price et al. 1980; Loader and Damman 1991).

Some of the most dramatic examples of plant effects on the third trophic level are noxious taste of the herbivore, or avoidance of larvae because they contain plant-derived toxic metabolites. For example, the monarch butterfly uses sequestered cardiac glycosides from the host plant *Asclepias* as a defense against its parasitoids (Smith 1978; Malcolm and Brower 1989; Zalucki et al. 1990). In the same way, leaf beetles utilized willow-derived compounds (Pasteels et al. 1988; Rowell-Rahier and Pasteels 1990), and sawfly larvae used food-derived resins to defend themselves against ants (Björkman and Larsson 1991). Success in predator avoidance on high-resin diets may cancel the direct negative influences that such diets cause for the larvae (Larsson et al. 1986).

Use of Food as Modified by Predation

High risk of predation makes some sources of otherwise usable food unavailable. As discussed above, this may happen if only some plants, or parts of them, are chemically suitable, and therefore herbivores become more predictable for predators (Schultz 1983; Wratten et al. 1988; Edwards et al. 1991).

On an evolutionary time scale, enemy-free space has been assumed to modify niche preferences of herbivores by restricting the number of herbivore species that can live in a specific habitat, or on a plant species (Lawton and Strong 1981; Bernays and Graham 1988). However, results may vary depending on the specific features of the tested systems. Denno et al. (1990) found a leaf beetle species that derived protective allelochemicals from a host plant for which the ovipositing females showed a preference but that was not optimal for larval growth; but another species of beetle did not use plant-derived defenses, and in this species oviposition preference and larval growth were well correlated. Gross and Price (1988) tested the enemy-free hypothesis by using two leaf miner species of the genus *Tildenia*. They found apparent selection toward enemy-rich space. Morphological traits of the host plant species affected the number of leaf miner species using the plants, in a way that increased access to and mobility on the host plant species that contributed to greater vulnerability to parasitoids.

Simultaneous Effects of Food and Predation on Population Dynamics of Herbivores

An analysis of field data of the relative contributions of food and predators as factors influencing larval performance would contribute to our understanding the dynamics of insect populations. However, for several reasons, relating both to limitations of analyses, and to the nature of the data, such analysis is difficult (e.g., Royama 1981; Price 1987; Stiling 1987; Bernays 1988; Bernays and Cornelius 1989; Hassell et al. 1989; Hanski 1990). Below I briefly discuss the types of existing data, and methods used in analysis of that data.

Empirical Research

The best data would result from studies measuring the fates of individuals, and causes of their failures and successes (Kyi et al. 1991). Such analyses can best be accomplished with sedentary larvae like leaf miners or gall makers (Taper et al. 1986; Hawkins 1988; Price and Pschorn-Walcher 1988; Cornell 1990; Auerbach 1991).

In leaf miners, premature or natural abscission of leaves is often an important mortality factor, sometimes much more important than predation (e.g., Askew and Shaw 1979; Simberloff and Stiling 1987; Auerback and Simberloff 1989; Faeth 1990a). Also intraspecific competition is important as indicated by an increased mortality rate in multiply mined leaves (e.g., Bultman and Faeth 1985; Faeth 1990a).

Clancy and Price (1989) partitioned larval mortality causes in a galling sawfly, *Pontonia* sp., to plant resistance, intraspecific competition, and predation. Natural enemies were responsible for deaths twice as often as plant-related causes and six times as often as competition.

Karban (1989) studied the relative importance of the host plant, *Erigeron glaucus* (by differences among clones), predation, and intra- and interspecific competition on plume moth larvae (*Platypilia williamsii*) over 3 years. Caging (which excluded bird predation) was the single significant factor in each year, and that factor did not interact with the other factors examined. Interestingly, competition was significant for another herbivore species, a spittlebug, and host clone was significant for a third herbivore, a thrips species. Therefore, the three insects were affected by different factors, and none of the important factors interacted significantly with other biotic factors (Karban 1989).

In the pine beauty moth, *Panolis flammea*, food plant quality modifies performance of larvae but still site-specific differences in predation rate could easily counteract the effects of the food plants (Watt and Leather 1988; Watt 1990).

Summarizing, both food and predation are important factors affecting population dynamics of Lepidoptera, but the foremost feature in real populations is the tremendous variation in the relative magnitude of these factors from system to system.

Analytical and Simulation Methods

The most widely available data consist of life tables that can be developed from any large-scale population study. Key-factor analysis offers a method to find the relative contributions of plants versus predators (Varley and Gradwell 1960). However, results of such analyses are only as good as the data analyzed. A major problem in key-factor analysis is classifying the true causes of death. Predation sometimes can be observed, but possible plant effects usually cannot be classified as anything other than disappearance of larvae, or reduced fecundity, and these effects may be confounded by thermal conditions.

Another approach is to simulate the simultaneous effects of food availability and predation on density and numerical trends in the herbivore species. In this way, with some initial assumptions, it is possible to show that both predation and food availability can strongly modify population patterns (Hassell 1985; Edelstein-Keshet and Rausher 1989). However, the value of this approach is compromised because different levels of predation and food availability may produce the same numerical outcome, and therefore make it difficult to draw conclusions about the relative effects of food availability and predators on natural populations of herbivores. For instance, van den Bos and Rabbinge (1976) studied the cyclic fluctuation of *Zeiraphera diniana* and found that all three trophic levels—the host plant, the moth, and predators—had to be included into the model for a good fit to actual patterns of the moth populations. With basically the same data, Fischlin and Baltensweiler (1979) demonstrated that models that produce realistic numerical trends for *Zeiraphera diniana* can be constructed by taking into account only herbivore-induced alterations in foliage quality in the course of the herbivore cycle. Making the total picture for *Zeiraphera diniana* even more confusing, Anderson and May (1980), using the same data, proposed that the budmoth cycles were driven by pathogens alone. But Baltensweiler and Fischlin (1988) then reported that pathogens were not associated with collapse of the moth populations for the latest cycle peak, unlike the previous declines of the peak densities on which Anderson and May (1980) based their model.

Demonstration of density dependence in field populations is problematic, but it is much more difficult to show which factors are causally responsible for population dynamics. Effects of even moderate variations in relative strengths of predation and food are hard to show and they may still be swamped by stochastic (abiotic) factors influencing population dynamics. Furthermore, different combinations of factors may regulate an insect population at different times (see Sinclair 1989). Only better field data will remedy the problem: to demonstrate the effects of all the major factors modifying mortality and fecundity. That can most easily be done in economically important forest defoliators (Berryman 1988), but that approach biases the conclusions with examples of outbreak species.

Conclusions

Appropriate food availability is essential for the growth of insect populations simply because food modifies reproductive potential and survivorship of insects. Predation, instead, mainly affects survivorship of insects.

Simultaneous effects of food and predation are simple at some levels. For instance, for populations of herbivores to increase, food must allow a high enough rate of increase that it exceeds the toll of all mortality factors, including predation. For food or predation to regulate population density, the factor has to function, at least temporally, in a density-dependent fashion. Both food availability, quantity and quality, and predation can function in a density-dependent way (Haukioja et al. 1983). Furthermore, both food availability and predation create time lags that may further modify population dynamics (Haukioja and Neuvonen 1987; Turchin 1990).

However, in actuality, effects of food and predation on herbivores are difficult to separate. Their interactions (Price et al. 1980; Whitman 1988; Barbosa 1988) are even more difficult to ascertain than the main effects. So far data about the relative efficacy of each of the factors in driving population dynamics have been accumulated mainly for two types of caterpillars: those that are economically important and those that are easy to study. The former have a higher than average tendency to outbreak (see, e.g., Berryman 1988), and the latter are often relatively large or exceptionally sessile.

Modeling of the effects of predation and food for insect populations should serve two purposes. First, models should help locate inconsistencies in the logic of the relative importance or weighing of factors and their interactions and, by generating predictions, test the models validity. Second, the models should generate sufficiently reliable predictions that decisions for controling pest populations could be based on them. Current models of insect population dynamics clearly serve the first goal but as yet do not serve the second goal well (Murdoch et al. 1985).

General models of population dynamics describe consequences of presumably important causal factors. They show that a certain combination of major factors can produce a specific outcome. Which factors are important are usually extracted from life table data. Usually, the analyses do not prove that it is just a particular set of factors that operates in the field. There are inherent difficulties in such analyses, and often no density-dependent factors are found (Stiling 1987, 1988). These results may reflect a number of technical problems (Turchin 1990, Hanski 1990), but they may also include the possibility that the data base do not describe relevant factors (Roland 1988), or simply that the time period of analysis is too short (Hassell et al. 1989). Furthermore, the models may identify in some situations a single factor of importance, when actually several factors control the population, but none of them is important all the time (Hilborn and Stearns 1982). These problems do not negate the value of general models in analyzing dynamic properties of the systems and ascertaining the sensitivity of these properties to different factors. But the models may still be of limited value in generation of truly predictive conclusions (Murdoch et al. 1985).

The complexity of the dynamics of natural populations may really be higher than generally anticipated. For instance, Duffey et al. (1986) and Broadway et al. (1986) emphasized that the toxicity of an allelochemical varies depending on the availability and kind of nutritive factors in the herbivore's diet. Accordingly, Duffey et al. (1986) stressed that we have to know the specific interactive details of the system: the dynamics of toxin, nutrient, herbivore, predator, and parasite levels. If most systems require such detailed and complex data to model them, our search for a general explanatory theory of population dynamics may be fruitless (Haukioja et al. 1983; Myers 1988). We cannot exclude the possibility that we must determine the specific ecological characters of the host plant, the prey, and the predator in each case. However, it seems likely that there are common patterns among systems that would allow at least some degree of modeling and therefore yield predictive power. Revealing the significant patterns is the real goal of research on insect population dynamics.

Acknowledgments

I wish to thank Peter Turchin, an anonymous reviewer, and especially Nancy Stamp for constructive criticism on the manuscript. Financial support was granted by the Academy of Finland.

Literature Cited

Anderson, R. M., and May, R. M. 1980. Infectious diseases and population cycles of forest insects. Science 210:658–661.

Askew, R.R., and Shaw, M.R. 1979. Mortality factors affecting the leaf-mining stages of *Phyllonorycter* (Lepidoptera: Gracillariidae) on oak and birch. I. Analysis of the mortality factors. Zool. J. Linn. Soc. 67:31–49.

Auerbach, M. 1991. Relative impact of interactions within and between trophic levels during an insect outbreak. Ecology 72:1599–1608.

Auerbach, M., and Simberloff, D. 1989. Oviposition site preference and larval mortality in a leaf-mining moth. Ecol. Entomol. 14:131–141.

Ayres, M. P., and MacLean, S. F. J. 1987. Development of birch leaves and the growth energetics of *Epirrita autumnata* (Geometridae). Ecology 68:558–568.

Ayres, M. P., Suomela, J., and MacLean, S. F. J. 1987. Growth performance of *Epirrita autumnata* (Lepidoptera: Geometridae) on mountain birch: trees, broods, and tree*brood interactions. Oecologia 74:450–457.

Barbosa, P. 1988. Natural enemies and herbivore-plant interactions: Influence of plant allelochemicals and host specificity, pp. 201–229. *In* P. Barbosa and D. K. Letourneau (eds.), Novel Aspects of Insect-Plant Interactions, Wiley, New York.

Baltensweiler, W., and Fischlin, A. 1988. The larch budmoth in the Alps, pp 331–351. In A.A. Berryman (ed.), Dynamics of Forest Insect Populations. Plenum, New York.

Bazzaz, F. A., Chiariello, N. R., Coley, P. D., and Pitelka, L. F. 1987. Allocating resources to reproduction and defense. BioScience 37:58–67.

Benz, G. 1974. Negative Rückkoppelung durch Raum- und Nahrungskonkurrenz sowie zyklische Veränderung der Nahrungsgrundlage als Regelprinzip in der Populationsdynamik des Grauen Lärchenwicklers, *Zeiraphera diniana* (Guenée) (Lep., Tortricidae). Z. Angew. Entomol. 76:196–228.

Bergelson, J.M., and Lawton, J.H. 1988. Does foliage damage influence predation on the insect herbivores of birch? Ecology 69:434–445.

Bergelson, J., Fowler, S., and Hartley, S. 1986. The effects of foliage damage on casebearing moth larvae, *Coleophora serratella*, feeding on birch. Ecol. Entomol. 11:241–250.

Bernays, E. A. 1988. Host specificity in phytophagous insects: selection pressure from generalist predators. Entomol. Exp. Appl. 49:131–140.

Bernays, E. A., and Cornelius, M. L. 1989. Generalist caterpillar prey are more palatable than specialist for the generalist predator *Iridomyrmex humilis*. Oecologia 79:427–430.

Bernays, E. A., and Graham, M. 1988. On the evolution of host specificity in phytophagous arthropods. Ecology 69:886–892.

Berryman, A.A. 1987. The theory and classification of outbreaks, pp. 3–30. In P. Barbosa and J. Schultz (eds.), Insect Outbreaks. Academic Press, San Diego.

Berryman, A. A. (ed.) 1988. Dynamics of Forest Insect Populations. Patterns, Causes, Implications. Plenum, New York.

Björkman, C., and Larsson, S. 1991. Pine sawfly defence and variation in host plant resin acids: A trade-off with growth. Ecol. Entomol. 16:283–289.

Bos, J., van den, and Rabbinge, R. 1976. Simulation of the Fluctuations of the Grey Larch Bud Moth. Pudoc, Wageningen.

Brattsten, L. B. 1979. Biochemical defense mechanisms in herbivores against plant allelochemicals, pp. 199–270. In G. A. Rosenthal and D. H. Janzen (eds.), Herbivores: Their Interactions with Secondary Plant Metabolites. Academic Press, New York.

Broadway, R.M., Duffey, S.S., Pearce, G., and Ryan, C.A. 1986. Plant protein inhibitors: A defense against herbivorous insects? Entomol. Exp. Appl. 41:33–38.

Brodbeck, B., and Strong, D. 1987. Amino acid nutrition of herbivorous insects and stress to host plants, pp. 347–364. In P. Barbosa and J. C. Schultz (eds.), Insect Outbreaks. Academic Press, San Diego.

Bultman, T.L., and Faeth, S.H. 1985. Patterns of intra- and interspecific association in leaf-mining insects on three oak host species. Ecol. Entomol. 10:121–129.

Chapin, F. S. I. I. I., Bloom, A. J., Field, C. B., and Waring, R. H. 1987. Plant responses to multiple environmental factors. BioScience 37:49–57.

Clancy, K. M., and Price, P. W. 1987. Rapid herbivore growth enhances enemy attack: Sublethal plant defenses remain a paradox. Ecology 68:733–737.

Clancy, K.M., and Price, P.W. 1989. Effects of plant resistance, competition, and enemies on a leaf-galling sawfly (Hymenoptera: Tenthredinidae). Environ. Entomol. 18:284–290.

Clausen, T. P., Reichardt, P. B., Bryant, J. P., Werner, R. A., Post, K., and Frisby, K. 1989. Chemical model for short-term induction in quaking aspen (*Populus tremuloides*) foliage against herbivores. J. Chem. Ecol. 15:2335–2346.

Coleman, J. S., and Jones, C. G. 1988. Plant stress and insect performance: Cottonwood, ozone and a leaf beetle. Oecologia 76:57–61.

Coley, P. D. 1983. Herbivory and defensive characteristics of tree species in a lowland tropical forest. Ecol. Monogr. 53:209–233.

Cornell, H. V. 1990. Survivorship, life history, and concealment: A comparison of leaf miners and gall formers. Am. Nat. 136:581–597.

Craig, T.P., Price, P.W., and Itami, J.K. 1986. Resource regulation by a stem-galling sawfly on the arroyo willow. Ecology 67:419–425.

Cushman, H. 1991. Host-plant mediation of insect mutualisms. Variable outcomes in herbivore-ant interactions. Oikos 61:138–142.

Danell, K., and Huss-Danell, K. 1985. Feeding by insects and hares on birches earlier affected by moose browsing. Oikos 44:75–81.

Dempster, J.P. 1983. The natural control of populations of butterflies and moths. Biol. Rev. 58:461–481.

Dempster, J.P., and Pollard, E. 1986. Spatial heterogeneity, stochasticity and the detection of density dependence in animal populations. Oikos 46:413–416.

Denno, R. F., and McClure, M. S. 1983. Variable Plants and Herbivores in Natural and Managed Systems. Academic Press, New York.

Denno, R.F., Larsson, S., and Olmstead, K.L. 1990. Role of enemy-free space and plant quality in host-plant selection by willow beetles. Ecology 71:124–137.

Dethier, V. G. 1988. Induction and aversion-learning in phytophagous arctiid larvae (Lepidoptera) in an ecological setting. Can. Entomol. 120:125–131.

Dicke, M., and Sabelis, M.W. 1988. Infochemical terminology: Should it be based on cost-benefit analysis rather than origin of compounds? Funct. Ecol. 2:131–138.

Dicke, M., Sabelis, M.W., Takabayashi, J., Bruin, J., and Posthumus, M. A. 1990. Plant strategies of manipulating predator-prey interactions through allelochemicals: Prospects for application in pest control. J. Chem. Ecol. 16:3091–3118.

Duffey, S.S., Bloem, K.A., and Campbell, B.C. 1986. Consequences of sequestration of plant natural products in plant-insect-parasitoid interactions, pp. 31–60. In D.J. Boethel, and R.D. Eikenbary, (eds.), Interactions of Plant Resistance and Parasitoids and Predators of Insects. Ellis Horwood, Somerset.

Edelstein-Keshet, L., and Rausher, M. D. 1989. The effects of inducible plant defenses on herbivore populations. I. Mobile herbivores in continuous time. Am. Nat. 133:787–810.

Edwards, P. J., Wratten, S. D., and Gibberd, R. 1991. The impact of inducible phytochemicals on food selection by insect herbivores and its consequences for the distribution of grazing damage, pp. 205–221. In D. W. Tallamy and M. J. Raupp (eds.), Phytochemical Induction by Herbivores. Wiley, New York.

Edwards, P.B., and Wanjura, W.J. 1989. Eucalypt-feeding insects bite off more than they can chew: Sabotage of induced defences? Oikos 54:246–248.

Edwards, P. B., Wanjura, W. J., and Brown, W. V. 1990. Mosaic resistance in plants. Nature (London) 347:434.

English-Loeb, G. M. 1989. Nonlinear responses of spider mites to drought-stressed host plants. Ecol. Entomol. 14:45–55.

Faeth, S.H. 1990a. Aggregation of a leafminer, *Cameraria* sp. nov. (Davis): Consequences and causes. J. Anim. Ecol. 59:569–586.

Faeth, S. H. 1990b. Structural damage to oak leaves alters natural enemy attack on a leafminer. Entomol. Exp. Appl. 57:57–63.

Faeth, S.H., and Simberloff, D. 1981. Population regulation of a leaf-mining insect, *Cameraria* sp. nov. at increased field densities. Ecology 62:620–624.

Feeny, P. 1970. Seasonal changes in oak leaf tannins and nutrients as a cause of spring feeding by winter moth caterpillars. Ecology 51:565–581.

Fischlin, A., and Baltensweiler, W. 1979. System analysis of the larch bud moth system. Part 1: the larch–larch bud moth relationship. Mitt. Schweiz. Entomol. Ges. 52:273–289.

Fowler, S. V., and Lawton, J. H. 1985. Rapidly induced defenses and talking trees: The devil's advocate position. Am. Nat. 126:181–195.

Fowler, S. V., and MacGarvin, M. 1986. The effects of leaf damage on the performance of insect herbivores on birch, *Betula pubescens*. J. Anim. Ecol. 55:565–573.

Gould, J.R., Elkinton, J.S., and Wallner, W.E. 1990. Density-dependent suppression of experimentally created gypsy moth, *Lymantria dispar* (Lepidoptera: Lymantriidae), populations by natural enemies. J. Anim. Ecol. 59:213–233.

Gross, P., and Price, P.W. 1988. Plant influences on parasitism of two leafminers: A test of enemy-free space. Ecology 69:1506–1516.

Gruys, P. 1970. Growth of *Bupalus piniarius* (Lepidoptera, Geometridae) in relation to larval population density. Centre Agricult. Publ. Doc. Wageningen.

Hanski, I. 1987. Pine sawfly population dynamics: Patterns, processes, problems. Oikos 50:327–335.

Hanski, I. 1990. Density dependence, regulation and variability in animal populations. Phil. Trans. R. Soc. London B 330:141–150.

Hanski, I., and Parviainen, P. 1985. Cocoon predation by small mammals, and pine sawfly population dynamics. Oikos 45:125–136.

Hassell, M. P. 1985. Insect natural enemies as regulating factors. J. Anim. Ecol. 54:323–334.

Hassell, M. P., Latto, J., and May, R. M. 1989. Seeing the wood for the trees: detecting density dependence from existing life-table studies. J. Anim. Ecol. 58:883–892.

Haukioja, E. 1982. Inducible defences of white birch to a geometrid defoliator, *Epirrita autumnata*, pp. 199–203. In J. H. Visser and A. K. Minks (eds.) Proceedings of the 5th International Symposium on Insect-Plant Relationships. PUDOC, Wageningen.

Haukioja, E. 1990. Positive and negative feedbacks in insect-plant interactions, pp. 113–122. In A.D. Watt, S.R. Leather, N.A.C. Kidd, and M. Hunter (eds.), Population Dynamics of Forest Insects. Intercept, Andover.

Haukioja, E., and Hanhimäki, S. 1985. Rapid wound-induced resistance in white birch (*Betula pubescens*) foliage to the geometrid *Epirrita autumnata*, a comparison of trees and moths within and outside the outbreak range of the moth. Oecologia 65:223–228.

Haukioja, E., and Neuvonen, S. 1985a. Induced long-term resistance in birch foliage against defoliators: defensive or incidental? Ecology 66: 1303–1308.

Haukioja, E., and Neuvonen, S. 1985b. The relationship between size and reproductive potential in male and female *Epirrita autumnata* (Lep., Geometridae). Ecol. Entomol. 10:267–270.

Haukioja, E., and Neuvonen, S. 1987. Insect population dynamics and induction of plant resistance: the testing of hypotheses, pp. 411–432. In P. Barbosa, and J. Schultz (eds.), Insect Outbreaks. Academic Press, San Diego.

Haukioja, E., and Niemelä, P. 1974. Growth and energy requirements of the larvae of *Dineura viridiorsata* (Retz.) (Hym., Tenthredinidae) and *Oporinia autumnata* (Bkh.) (Lep., Geometridae) feeding on birch. Ann. Zool. Fennici 11:207–211.

Haukioja, E., and Niemelä, P. 1979. Birch leaves as a resource for herbivores: Seasonal occurrence of increased resistance in foliage after mechanical damage to adjacent leaves. Oecologia 39:151–150.

Haukioja, E., Niemelä, P., Iso-Iivari, L., Ojala, H., and Aro, E. -M. 1978. Birch leaves as a resource for herbivores. I. Variation in the suitability of leaves. Rep. Kevo Subarct. Res. Stat. 14:5–12.

Haukioja, E., Kapiainen, K., Niemelä, P., and Tuomi, J. 1983. Plant availability hypothesis and other explanations of herbivore cycles: Complementary or exclusive alternatives. Oikos 40:419–432.

Haukioja, E., Suomela, J., and Neuvonen, S. 1985. Long-term inducible resistance in birch foliage: Triggering cues and efficacy on a defoliator. Oecologia 65:363–369.

Haukioja, E., Neuvonen, S., Hanhimäki, S., and Niemelä, P. 1988a. The autumnal moth in Fennoscandia, pp. 163–178. In A. A. Berryman (ed.), Dynamics of Forest Insect Populations. Patterns, Causes, Implications. Plenum, New York.

Haukioja, E., Pakarinen, E., Niemelä, P., and Iso-Iivari, L. 1988b. Crowding-triggered phenotypic responses alleviate consequences of crowding in *Epirrita autumnata*. Oecologia 75:549–558.

Haukioja, E., Ruohomäki, K., Senn, J., Suomela, J., and Walls, M. 1990. Consequences of herbivory in the mountain birch (*Betula pubescens* ssp *tortuosa*): Importance of the functional organization of the tree. Oecologia 82:238–247.

Hawkins, B. A. 1988. Do galls protect endophytic herbivores from parasitoids? A comparison of galling and non-galling Diptera. Ecol. Entomol. 13:473–477.

Heinrich, B., and Collins, S. L. 1983. Caterpillar leaf damage, and the game of hide-and-seek with birds. Ecology 64:592–602.

Hilborn, R., and Stearns, S. C. 1982. On inference in ecology and evolutionary biology: The problem of multiple causes. Acta Biotheor. 31:145–164.

Holmes, R.T., Schultz, J.C., and Nothnagle, P. 1979. Bird predation on forest insects: An exclosure experiment. Science 206:462–463.

Huffaker, C.B. (ed.) 1971. Biological Control. Plenum, New York.

Huffaker, C.B., Dahlsten, D. L., Janzen, D. H., and Kennedy, G.G. 1984. Insect influences in the regulation of plant populations and communities, pp. 659–691. In C.B. Huffaker (ed.) Ecological Entomology. Wiley, New York.

Ikeda, T., Matsumura, F., and Benjamin, D. M. 1977. Chemical basis for feeding adaptation of pine sawflies *Neodiprion rugifrons* and *Neodiprion swainei*. Science 197:497–499.

Jones, C. G., and Coleman, J. S. 1988. Plant stress and insect behavior: Cottonwood, ozone and the feeding and oviposition preference of a beetle. Oecologia 76:51–56.

Karban, R. 1983. Induced responses of cherry trees to periodical cicada oviposition. Oecologia 59:226–231.

Karban, R. 1989. Community organization of *Erigeron glaucus* folivores: Effects of competition, predation, and host plant. Ecology 70:1028–1039.

Koptur, S. 1979. Facultative mutualism between weedy vetches bearing extrafloral nectaries and weedy ants in California. Am. J. Bot. 66:1016–1020.

Kyi, A., Zalucki, M. P. and Titmarsh, I. J. 1991. An experimental study of early stage survival of *Helicoverpa armigera* (Lepidoptera: Noctuidae) on cotton. Bull. Entomol. Res. 81:263–271.

Laine, K., and Niemelä, P. 1980. The influence of ants on the survival of mountain birches during an *Oporinia autumnata* (Lepidoptera) outbreak. Oecologia 47:39–42.

Larsson, S. 1989. Stressful times for the plant stress-insect performance hypothesis. Oikos 56:277–283.

Larsson, S., Björkman, C., and Gref, R. 1986. Responses of *Neodiprion sertifer* (Hym., Diprionidae) larvae to variation in needle resin acid concentration in Scots pine. Oecologia 70:77–84.

Lawson, D. L., Merritt, R. W., Martin, M. M., Martin, J. S., and Kukor, J. J. 1984. The nutritional ecology of *Alsophila pometaria* and *Anisota senatoria* feeding on early- and late-season oak leaves. Entomol. Exp. Appl. 35:105–114.

Lawton, J. H. 1986. Food-shortage in the midst of apparent plenty: the case for birch-feeding insects, pp. 219–228. In H. H. W. Velthuis (ed.), Proceedings of the 3rd European Congress of Entomology, Amsterdam.

Lawton, J.H., and Strong, D.R. Jr. 1981. Community patterns and competition in folivorous insects. Am. Nat. 118:317–338.

Loader, C. and Damman, H. 1991. Nitrogen content of food plants and vulnerability of *Pieris rapae* to natural enemies. Ecology 72:1586–1590.

Luck, R. F. 1990. Evaluation of natural enemies for biological control: A behavioural approach. Trends Ecol. Evol. 5:196–199.

MacKay, D. A. 1991. The effects of ants on herbivory and herbivore numbers on foliage of the mallee eucalypt, *Eucalyptus incrassata* Labill. Aust. J. Ecol. 16:471–483.

Malcolm, S. B. and Brower, L. P. 1989. Evolutionary and ecological implications of cardeonolide sequestration in the monarch butterfly. Experientia 45:284–295.

Marquis, R. J. 1988. Intra-crown variation in leaf herbivory and seed production in striped maple, *Acer pensylvanicum* L. (Aceraceae). Oecologia 77:51–55.

Mason, R. R. 1987. Nonoutbreak species of forest lepidoptera, pp. 31–57. In P. Barbosa and J. C. Schultz (eds.), Insect Outbreaks. Academic Press, San Diego.

Mattson, W. J. 1980. Herbivory in relation to plant nitrogen content. Annu. Rev. Ecol. Syst. 11:119–161.

Mattson, W. J., and Haack, R. A. 1987. The role of drought stress in provoking outbreaks of phytophagous insects, pp. 365–407. In P. Barbosa and J. C. Schultz (eds.), Insect Outbreaks. Academic Press, San Diego.

Mauricio, R. and Bowers, M. D. 1990. Do caterpillars disperse their damage?: Larval foraging behaviour of two specialist herbivores, *Euphydryas phaeton* (Nymphalidae) and *Pieris rapae* (Pieridae). Ecol. Entomol. 15:153–161.

McKey, D. 1979. The distribution of secondary compounds within plants, pp. 55–133. In G. A. Rosenthal and D. H. Janzen (eds.), Herbivores. Their Interaction with Secondary Plant Metabolites. Academic Press, New York.

Mills, N.J. 1990. Are parasitoids of significance in endemic populations of forest defoliators? Some experimental observations from Gypsy Moth, *Lymantria dispar* (Lepidoptera: Lymantriidae), pp. 265–274. In A.D. Watt, S.R. Leather, N.A.C. Kidd, and M. Hunter (eds.), Population Dynamics of Forest Insects. Intercept, Andover.

Mountford, M. D. 1988. Population regulation, density dependence, and heterogeneity. J. Anim. Ecol. 57:845–858.

Murdoch, W. W. and Reeve, J. D. 1987. Aggregation of parasitoids and the detection of density dependence in field populations. Oikos 50:137–141.

Murdoch, W. W., Chesson, J. and Chesson, P. L. 1985. Biological control in theory and practice. Am. Nat. 125:344–366.

Myers, J. H. 1988. Can a general hypothesis explain population cycles of forest lepidoptera? Adv. Ecol. Res. 18:179–242.

Neuvonen, S., and Haukioja, E. 1984. Low nutritive quality as defence against herbivores: Induced responses in birch. Oecologia 63:71–74.

Neuvonen, S., and Haukioja, E. 1991. The effects of inducible resistance in host foliage on birch-feeding herbivores, pp. 277–291. In D. W. Tallamy and M. J. Raupp (eds.), Phytochemical Induction by Herbivores. Wiley, New York.

Neuvonen, S., Haukioja, E., and Molarius, A. 1987. Delayed induced resistance against a leaf-chewing insect in four deciduous tree species. Oecologia 74:363–369.

Niemelä, P., and Haukioja, E. 1982. Seasonal patterns in species richness of herbivores: Macrolepidopteran larvae of Finnish deciduous trees. Ecol. Entomol. 7:169–175.

Niemelä, P., Tuomi, J., and Sirén, S. 1984. Selective herbivory on mosaic leaves of variegated *Acer pseudoplatanus*. Experientia 40:1433–1434.

Niemelä, P., Tuomi, J., and Lojander, T. 1991. Defoliation of the Scots pine and performance of diprionid sawflies. J. Anim. Ecol. 60:683–692.

Pasteels, J. M., Rowell-Rahier, M. and Raupp, M. J. 1988. Plant-derived defense in chrysomelid beetles, pp. 235–272. In P. Barbosa and D. K. Letourneau (eds.), Novel Aspects of Insect-Plant Interactions. John Wiley & Sons, New York.

Podoler, H., and Rogers, D. 1975. A new method for the identification of key factors from life-table data. J. Anim. Ecol. 44:85–115.

Price, P.W. 1986. Ecological aspects of host plant resistance and biological control: Interactions among three trophic levels, pp. 11–30. In D.J. Boethel, and R.D. Eikenbary (eds.), Interactions of Plant Resistance and Parasitoids and Predators of Insects. Ellis Horwood, Frome, Somerset.

Price, P.W. 1987. The role of natural enemies in insect populations, pp. 287–312. In P. Barbosa and J. Schultz (eds.), Insect Outbreaks. Academic Press, San Diego.

Price, P. W. 1991. The plant vigor hypothesis and herbivore attack. Oikos 62:244–251.

Price, P. W., Bouton, C. E., Gross, P., McPheron, B. A., Thompson, J. N., and Weiss, A. E. 1980. Interactions among three trophic levels: Influence of plants on interactions between insect herbivores and natural enemies. Annu. Rev. Ecol. Syst. 11:41–65.

Price, P. W. and Pschorn-Walcher, H. 1988. Are galling insects better protected against parasitoids than exposed feeders?: A test using tenthredinid sawflies. Ecol. Entomol. 13:195–205.

Rausher, M. D. 1981. Host plant selection by *Battus philenor* butterflies: The roles of predation, nutrition, and plant chemistry. Ecol. Monogr. 51:1–20.

Rhoades, D. F. 1985. Offensive-defensive interactions between herbivores and plants: Their relevance to herbivore population dynamics and community theory. Am. Nat. 125:205–238.

Risch, S. J. and Rickson, F. R. 1981. Mutualism in which ants must be present before plants produce food bodies. Nature 291:149–150.

Roland, J. 1988. Decline in winter moth populations in North America: Direct versus indirect effects of introduced parasites. J. Anim. Ecol. 57:523–531.

Rosenthal, G.A., and Janzen, D.H. (eds.) 1979. Herbivores. Their Interaction with Secondary Plant Metabolites. Academic Press, New York.

Rowell-Rahier, M. and Pasteels, J. M. 1990. Chemical specialization on toxic plants provides increased protection from natural enemies. Symp. Biol. Hung. 39:343–347.

Royama, T. 1981. Evaluation of mortality factors in insect life table analysis. Ecol. Monogr. 51:495–505.

Ruohomäki, K., Hanhimäki, S., Haukioja, E., Iso-Iivari, L., Neuvonen, S., Niemelä, P., and Suomela, J. 1992. Variability in the efficacy of delayed inducible resistance in mountain birch. Entomol. Exp. Appl. 62:107–115.

Schoohoven, L.M. 1982. Biological aspects of antifeedants. Entomol. Exp. Appl. 31:57–69.

Schultz, J.C. 1983. Impact of variable plant defensive chemistry on susceptibility of insects to natural enemies, pp. 37–54. In E. Hedin (ed.), Mechanisms of Plant Resistance to Insects. American Chemical Society, Washington, D.C.

Schultz, J.C., and Baldwin, I.T. 1982. Oak leaf quality declines in response to defoliation by gypsy moth larvae. Science 217:149–151.

Schultz, J.C., Nothnagle, P.J., and Baldwin, I.T. 1982. Seasonal and individual variation in leaf quality of two northern hardwoods tree species. Am. J. Bot. 69 (5):753–759.

Simberloff, D., and Stiling, P. 1987. Larval dispersion and survivorship in a leaf-mining moth. Ecology 68:1647–1657.

Sinclair, A.R.A. 1989. Population regulation in animals, pp. 197–241. In J.M. Cherret (ed.), Ecological Concepts. Blackwell, Oxford.

Skinner, G. J. and Whittaker, J. B. 1981. An experimental investigation of inter-relationships between the wood ant *Formica rufa* and some tree canopy herbivores. J. Anim. Ecol. 50:313–326.

Smith, D.A.S. 1978. Cardiac glycosides in *Danaus chrysippus* (L.) provide some protection against an insect parasitoid. Experientia 34:844–846.

Smith, L.L., Lanza, J., and Smith, G.C. 1990. Amino acid concentrations in extrafloral nectar of *Impatiens sultani* increase after simulated herbivory. Ecology 71:107–115.

Stiling, P. 1987. The frequency of density dependence in insect host-parasitoid systems. Ecology 68:844–856.

Stiling, P. 1988. Density dependent processes and key factors in insect populations. J. Anim. Ecol. 57:581–594.

Strong, D.R., Lawton, J.H., and Southwood, R. 1984. Insects on plants. Community Patterns and Mechanisms. Blackwell, Southampton.

Takabayashi, J., Dicke, M., and Posthumus, M. A. 1991. Induction of indirect defence against spider-mites in uninfested lima bean leaves. Phytochemistry 30:1459–1462.

Taper, M. L. and Case, T. J. 1987. Interactions between oak tannins and parasite community structure: Unexpected benefits to cynipid gall-wasps. Oecologia 71:254–261.

Taper, M. L., Zimmerman, E. M. and Case, T. J. 1986. Sources of mortality for a cynipid gall-wasp (*Dryocosmus dubiosus* [Hymenoptera: Cynipidae]) : the importance of the tannin/fungus interaction. Oecologia 68:437–445.

Tenow, O. 1963. Leaf-eating insects on the mountain birch at Abisko (Swedish Lapland) with notes on bionomics and parasites. Zool. Bidr. Uppsala 35:545–570.

Tenow, O. 1972. The outbreaks of *Oporinia autumnata* Bkh. and *Operophthera* spp. (Lep., Geometridae) in the Scandinavian mountain chain and northern Finland 1862–1968. Zool. Bidr. Uppsala, Suppl. 2:1–107.

Tuomi, J., Niemelä, P., Haukioja, E., Sirén, S., and Neuvonen, S. 1984. Nutrient stress: An explanation for plant anti-herbivore responses to defoliation. Oecologia 61:208–210.

Turchin, P. 1990. Rarity of density dependence or population regulation with lags? Nature 344:660–663.

Varley, G. C., and Gradwell, G. R. 1960. Key factors in population studies. J. Anim. Ecol. 29:399–401.

Varley, G. C., and Gradwell, G. R. 1970. Recent advances in insect population dynamics. Annu. Rev. Entomol. 15:1–24.

Vasconcelos, H. L. 1991. Mutualism between *Maieta quianensis* Aubl, a myrmecophytic melastome, and one of its ant inhabitants—Ant protection against insect herbivores. Oecologia 87:295–298.

Vinson, S.B. 1984. Parasitoid-host relationship, pp. 205–233. In W.J. Bell and R.T. Cardé (eds.), Chemical Ecology of Insects. Chapman and Hall, Cambridge.

Wagner, M. R., and Evans, P. D. 1985. Defoliation increases nutritional quality and allelochemics of pine seedlings. Oecologia 67:235–237.

Watt, A. D. 1988. Effects of stress-induced changes in plant quality and host-plant species on the population dynamics of the pine beauty moth in Scotland: Partial life tables of natural and manipulated populations. J. Appl. Ecol. 25:209–221.

Watt, A. D. 1990. The consequences of natural, stress-induced and damage-induced differences in tree foliage on the population dynamics of the pine beauty moth, pp. 157–168. In A. D. Watt, S. R. Leather, M. D. Hunter, and N. A. C. Kidd (eds.), Population Dynamics of Forest Insects. Intercept, Andover.

Watt, A. D., and Leather, S. R. 1988. The pine beauty in Scottish lodgepole pine plantations, pp.243–266. In A. A. Berryman (ed.), Dynamics of Forest Insect Populations. Patterns, Causes, Implications. Plenum, New York.

Weinstein, P. 1990. Leaf petiole chewing and the sabotage of induced defences. Oikos 58:231–233.

White, T.C.R. 1984. The abundance of invertebrate herbivores in relation to the availability of nitrogen in stressed food plants. Oecologia 63:121–144.

Whitman, D. W. 1988. Allelochemical interactions among plants, herbivores, and their predators, pp. 11–64. In P. Barbosa and D. K. Letourneau (eds.), Novel Aspects of Insect-Plant Interactions. John Wiley & Sons, New York.

Wratten, S. D., Edwards, P. J., and Dunn, I. 1984. Wound-induced changes in the palatability of *Betula pubescens* and *Betula pendula*. Oecologia 61:372–375.

Wratten, S. D., Edwards, P. J., and Winder, L. 1988. Insect herbivory in relation to dynamic changes in host plant quality. Biol. J. Linn. Soc. 35:339–350.

Wratten, S. D., Edwards, P. J., and Barker, A. M. 1990. Consequences of rapid feeding-induced changes in trees for the plant and the insect: Individuals and populations, pp. 137–145. In A. D. Watt, S. R. Leather, M. D. Hunter, and N. A. C. Kidd (eds.), Population Dynamics of Forest Insects. Intercept, Andover.

Zalucki, M. P. and Brower, L. P. 1992. Survival of first instar larvae of *Danaus plexippus* in relation to cardiac glycoside and latex content of *Asclepias humistrata*. Chemoecology (in press)

Zalucki, M. P., Brower, L. P. and Malcolm, S. B. 1990. Oviposition by *Danaus plexippus* in relation to cardenolide content of three *Asclepias* species in the southeastern U.S.A. Ecol. Entomol. 15:231–240.

14

Caterpillar Seasonality in a Costa Rican Dry Forest

Daniel H. Janzen

Introduction

Caterpillars—taken here to mean the larvae of Lepidoptera—are not uniformly or randomly present in species or abundance in tropical lowland habitats. Instead, it is evident to any field naturalist that tropical caterpillar species and biomass are frequently correlated with seasonal phenomena, either directly or through their relationships with other organisms displaying seasonality. Indeed, other tropical insects are highly seasonal (e.g., Wolda 1988; Tauber et al. 1986; Nummelin 1989; Janzen 1973, 1983a,b, 1987a,c; Tanaka et al. 1987; Paarman and Stork 1987; Winston 1980) and there is no reason to expect caterpillars to be any different.

Caterpillar seasonality is particularly evident in tropical dry forest. This vegetation type once covered at least half of the terrestrial tropics (e.g., Murphy and Lugo 1986). However, it has been largely altered or replaced by agriculture, forestry, and animal husbandry (e.g., Uhl and Buschbacher 1985; Janzen 1988a,b). Ironically, the tropical dry forest caterpillar fauna is not of recent interest because portions of it are in danger of extinction (which it is), but rather because as tropical dryland agroforestry ecosystems begin to reacquire diverse cropping systems and even regenerate wildland vegetation, the caterpillar fauna becomes conspicuous as pests, hosts for parasitoids, food for valued vertebrates, and even as a source of genetic and chemical biodiversity for commercial exploitation.

A literature review of all the ways that tropical dry forest caterpillars have been associated with seasonal data or processes (e.g., Odendaal 1990; Jones 1987; Chippendale and Mahmalji 1987) would be a rewarding exercise, but it is not the intent of this chapter. Instead, my goal is to discuss descriptively a few aspects of the seasonality of a particular tropical dry forest caterpillar fauna, that of the eastern end of Santa Rosa National Park in the Guanacaste Conservation Area in northwestern Costa Rica (Janzen 1988b). I pass up a literature review because the act of studying this caterpillar fauna, and taking the steps necessary to ensure its long-term survival (Allen 1988; Janzen 1988b,c, 1989a, 1991, 1992a; Janzen and Hallwachs 1992b; Tangley 1990), has proven to be incompatible with the time investment necessary for a literature review. However, I do know that there is no published study or set of studies of a tropical caterpillar fauna with which the Santa Rosa caterpillar fauna can be contrasted. I justify a personal focus on the biology of *this* site by noting that it is large, complex, and probably representative of what once covered much of the neotropical dry forest, that its caterpillar fauna is taxonomically better known than is that of any other tropical wildland site, and that I am quite familiar with it.

I attempt to portray some of the seasonality of the Santa Rosa caterpillar fauna through generalizations as it appears to me at this time, sprinkled with illustrative examples. There are lifetimes of work ahead to put means and variances on these and other generalizations, and test the hypotheses advanced. I obviously do not have the time to examine more than a few examples in detail. However, there is an intensive effort to develop the Guanacaste Conservation Area (as with the other seven Areas de Conservación) as a huge biological station conserved into perpetuity (Janzen 1991; Janzen and Hallwachs 1992b), and to institutionalize the conservation of Costa Rica's biodiversity through nondestructive use under the auspices of the Instituto Nacional de Biodiversidad or INBio (Tangley 1990; Janzen 1991). This act should create a climate for later and less rushed study of the ACG caterpillar fauna into perpetuity, along with a host of other studies.

The salient seasonal feature of the Santa Rosa dry forest caterpillar fauna is that it fluctuates enormously in biomass and proportional species composition within the year. While this general pattern is repeated annually, its intensity is highly variable, owing to both biological and climate variation.

Materials and Methods

References

Much of what I report here about the caterpillars of Santa Rosa's dry forest has not been published before. Rather than repeatedly state "Janzen, unpublished field notes," I adopt the convention of viewing the entire chapter as previously unpublished field notes and commentary, unless otherwise stated.

Study Site

The study site is the general region of the administration area in the southeastern end of Santa Rosa National Park (Parque Nacional Santa Rosa) in the Guanacaste

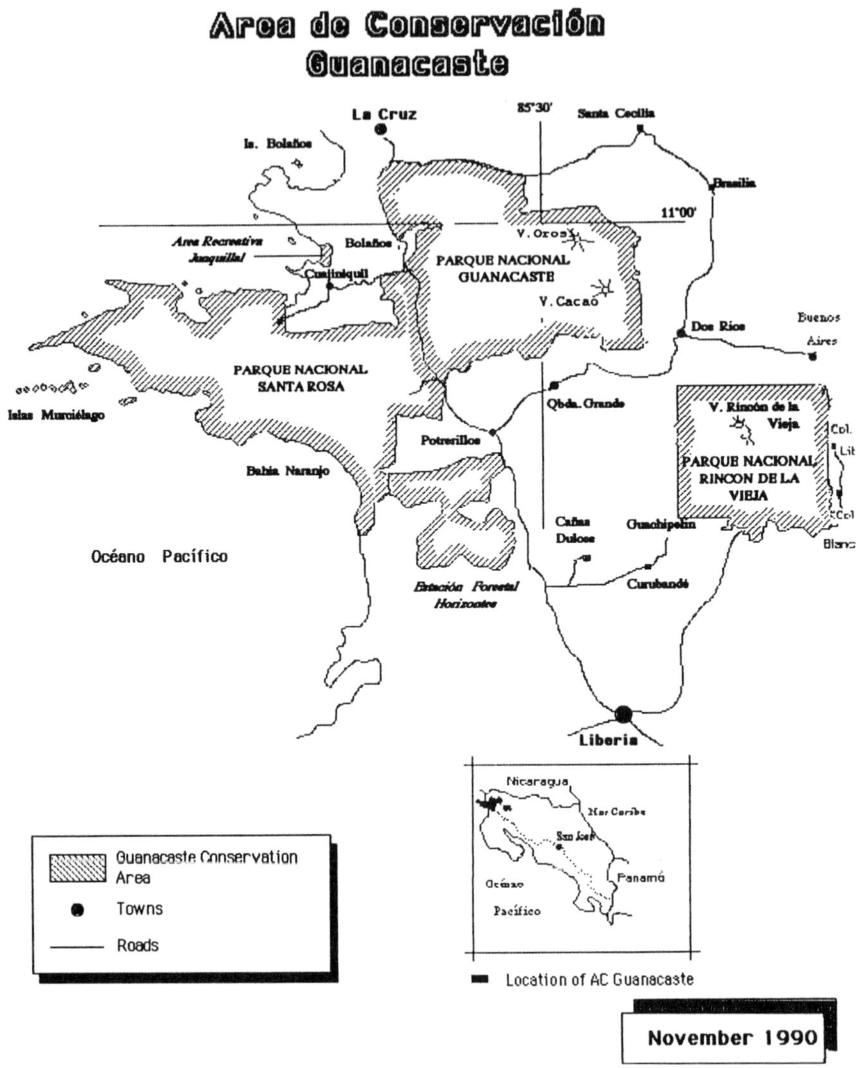

Figure 14.1 The Area de Conservación Guanacaste (ACG), northwestern Guanacaste Province, Costa Rica.

Conservation Area (ACG, or Area de Conservación Guanacaste). The study site will henceforth be referred to as "Santa Rosa."

The 104,000-ha ACG contains Santa Rosa National Park, Guanacaste National Park, Rincon de la Vieja National Park, the Junquillal Recreation Area, and Horizontes Forest Experimental Station (Fig. 14-1). The study site is approximately 5000 ha of tropical dry forest remnants at 200–300 m elevation, imbedded in a patch of approximately 60,000 ha of dry forest remnants extending from the Pacific Ocean to the western foothills of the Cordillera Guanacaste (Volcan Orosí, Volcan Cacao, Volcan Rincon de la Vieja). In these foothills, the dry forest blurs into the wetter rainforests that extend then eastward to the Caribbean.

Virtually all the ACG dry forest has been cut, cleared, selectively logged, pastured, and/or farmed one or more times by European-style agrobusiness during the past four centuries. Prior to this it was occupied by indigenous peoples for at least 10,000 years (see Janzen and Martin 1982). It thus comprises an extremely complex mosaic of secondary succession ranging from 1 to 400-plus years in age. Dotted through this mosaic are a few small patches of forest with a structure similar to that of the original forest. Despite this extensive disturbance, there is no hint of the recent extinction of any dry forest species of animal or plant from the ACG other than the scarlet macaw (*Ara macaw*) and the extinctions associated with the Pleistocene megafaunal extinctions (see Janzen and Martin 1982).

However, the relative densities of the animals and plants in the ACG dry forest—and thus the qualitative and quantitative traits of many of their interations—are (at present) clearly not those that would occur in this site were it never to have been touched by indigenous and European societies after the Pleistocene extinctions. This ecological situation applies to all neotropical dry forest north of the Panama canal. Of the interactions reported here, all are to some degree modifications of what they would be in a pristine world. However, the more the interaction depends on direct genetic traits, the less the modification. For example, whether a species of caterpillar snips or chews leaves (Bernays and Janzen 1988) is quite unaffected by whether the caterpillar is feeding early or late in the rainy season, but the relative abundance of caterpillars using these two different ways to process leaves varies strongly with the seasonal cycle.

The geology, general history, general biology, etc. of the site has been discussed elsewhere (e.g., Janzen 1986a,c,b, 1987b–g, 1989b). Below, I focus briefly on site characteristics that are direct concern to caterpillar seasonal biology.

Vegetation Type

The eastern Santa Rosa forests have been described as dry forest (Holdrige et al. 1971), deciduous and semideciduous forest (Gomez 1986), etc. Rather than descend into a vegetation nomenclature of doubtful biological significance, suffice to say that the vegetation of the study site is that which is generally termed tropical dry forest, seasonal forest, deciduous forest, semideciduous forest, wet–dry forest, and/or monsoon forest (Ridpath and Corbett 1985).

The eastern Santa Rosa forests are highly variable from hectare to hectare in stature, appearance, deciduousness, and species composition (e.g., Janzen 1988b). There are three primary climatic sources of this variation, all relating to seasonality. First, small differences in exposure, drainage, and soil depth result in strong differences in availability of water to plants of different sizes and demands. This in turn results in striking differences in plant species composition, within species deciduousness, stature, growth rate, etc. Since the flora is a mix of evergreen and deciduous species, and since many of the deciduous species vary the duration and intensity of deciduousness with the amount of water that they get, the result is a fine-scale patchwork of degree of deciduousness. This in turn creates a fine-scale patchwork of understory moisture levels, shade, temperature levels, etc. Second, the amount and pattern of rain vary strongly from year to year, again directly influencing the pattern of deciduousness through the process just described. Third, when a forest with a given degree of deciduousness is cleared or otherwise severely perturbed, for centuries afterward the subsequent successional stages are much more deciduous than was the "parent" forest. In general, the Santa Rosa dry forests of today are substantially drier than were the same forests just a few centuries earlier, simply because they are successional.

The Santa Rosa dry forests are distinctly different from Costa Rican lowland rain forest on well-drained soil. The dry forests have

- only about 20–40% of the number of plant species per large area (the study site contains only about 600 species of angiosperms, Janzen and Liesner 1980, while an area of Costa Rican lowland rain forest of comparable habitat complexity contains more than 2000 species.
- a lower canopy (5–40 m in height),
- relatively few epiphytes (but are rich in vine biomass and species), and
- several dozen very common species of trees, shrubs, and large woody vines (among which are scattered hundreds of other species).

By way of contrast, Santa Rosa animal faunas are nearly as species-rich as are those of nearby rain forest, except for amphibians and other taxa that require nearly year-round moisture. This dry-wet equivalence appears to be largely because Santa Rosa's vegetation supports both a distinctive dry forest fauna and large numbers of what are thought of as rain forest species (e.g., Janzen 1986a). Likewise, for many major taxa the Santa Rosa biomass of active animals certainly exceeds that of nearby rain forest areas during the first 2–12 months of the rainy season, but active animal biomass during the dry season is substantially lower than that of rain forest throughout the year.

Climate and Weather.

For about 6 months of each year the Santa Rosa dry forest is sunny, hot, and dry. It is extremely windy (from the northeast) during the first 3 months of the dry season (late December through mid-March). There are reliable rainfall and temperature records from the Santa Rosa weather station in the park administration area that extend from July 1979 to the present . From 1980 to 1989, the average

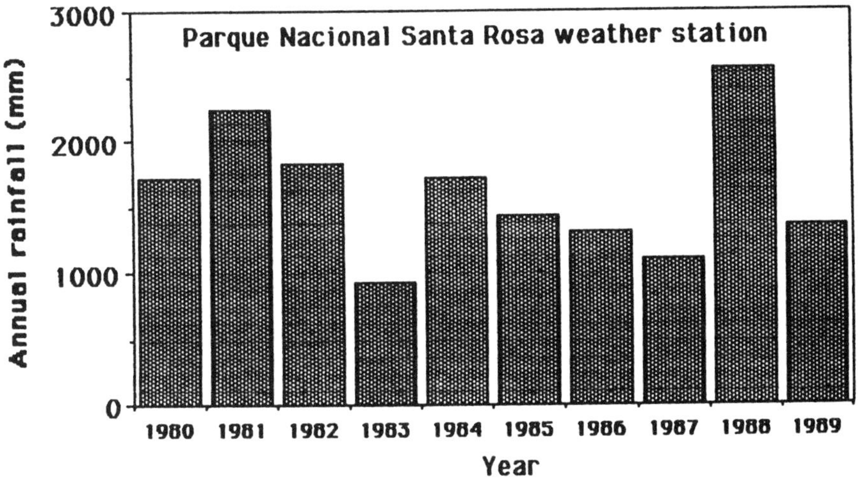

Figure 14.2. The total annual rainfall at the Weather Station in southeastern Parque Nacional Santa Rosa, Area de Conservación Guanacaste, northwestern Costa Rica.

rainfall was 1614 mm per year (Santa Rosa National Park weather station records). The variation in total rainfall ranges from 915 to 2558 mm per year (Fig. 14.2). While such variation in total rainfall is of extreme importance to agriculture and pasture industries, it often seems to be of less importance to the caterpillars and their interactants than is the detailed distribution of the rain within the year. A year that is too dry for an unirrigated dry-land rice crop in Guanacaste Province, for example, may receive quite enough rain for a normal crop of forest caterpillars, especially if that low rainfall is uniformly distributed during the first half of the rainy season and continues to appear after the first rainstorm.

There is an obvious within-year rainfall pattern (Figs. 14.3, 14.4) that is repeated annually. After approximately 6 months with no rainfall, the rains begin sometime between late April and mid-May. They peak during a 1- to 2-month period and then decline during the short dry season ("veranillo"). During September–November the rains intensify again. The long dry season is substantially hotter than is the rainy season (Figs. 14.3, 14.4). During the first half of the long dry season, mild to gale-force winds blow from the northeast; these are strongest during the day and the first half of the night.

There is extreme interyear variation in the stopping and starting dates of the rainy season(s) (e.g., compare Figs. 14.3 and 14.4), the duration of the windy period, the weekly temperatures of the wet or dry season, the continuity of the rainy season once started, the total amount of water to fall during the rainy season, etc. All of these kinds of variation have a conspicuous but complex impact on the caterpillar biomass and species composition in the eastern Santa Rosa dry forest.

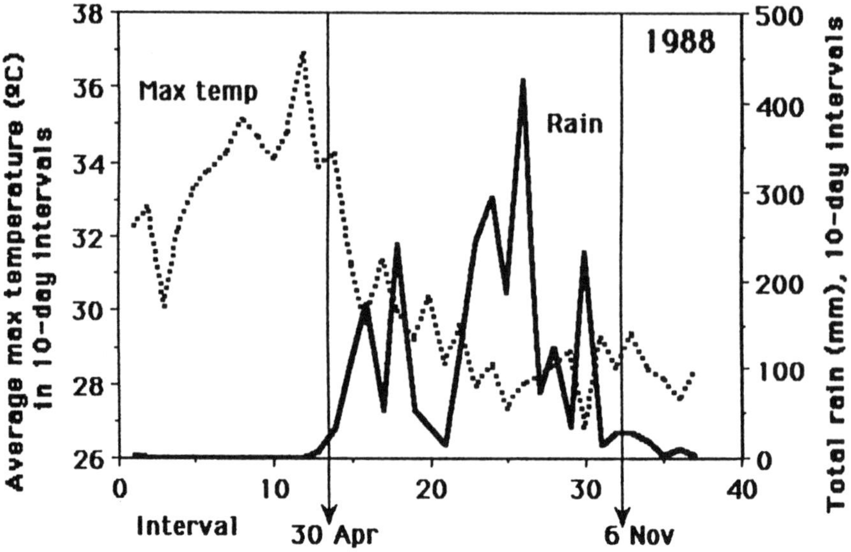

Figure 14.3. Total rainfall during 10-day intervals and the average maximum temperature over 10-day intervals in 1988 at the Weather Station in southeastern Parque Nacional Santa Rosa, Area de Conservación Guanacaste, northwestern Costa Rica.

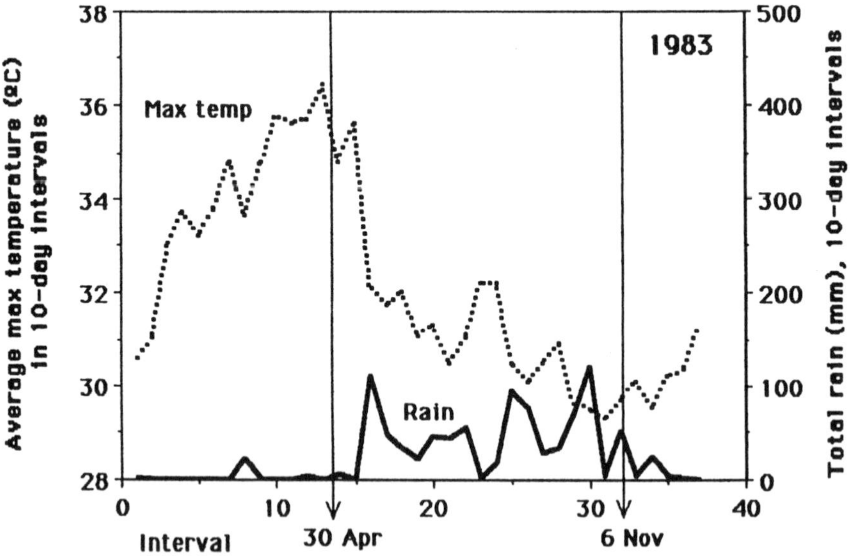

Figure 14.4. Total rainfall during 10-day intervals and the average maximum temperature over 10-day intervals in 1983 at the Weather Station in southeastern Parque Nacional Santa Rosa, Area de Conservación Guanacaste, northwestern Costa Rica.

Caterpillar Fauna.

The 600 species of angiosperms in the study site support about 3140 species of caterpillars (Janzen 1988d). This is twice the number to be found in a similar large area around Ithaca, New York (42° N lat.), and roughly 12 times the number to be found in a similar area around Kevo, northern Finland (70° N lat.) (Janzen 1988d). The Santa Rosa caterpillar fauna is about 37% external plant feeders, with the remainder living in leaf rolls or mines in leaves, stems, seeds, fruits (e.g., Janzen 1983a), flowers, and roots. A very few species live in mammal nests (e.g., Davis et al. 1986), wasp nests, dung accumulations, and other bizarre habitats.

All Santa Rosa Papilionidae, Pieridae, and Nymphalidae have names and are easily identified (DeVries 1987). Probably all the other butterflies also have names, though many species remain to be collected. At least 90% of the Santa Rosa macromoth and Pyraustinae (Crambidae) have been described. The microlepidoptera are less well known taxonomically, but many of them have also been described. Most of the Santa Rosa Lepidoptera names were applied well before 1977 (see Fig. 14.6), the date of the beginning of this study. The healthy taxonomic situation with the Santa Rosa caterpillar fauna is due to most Santa Rosa Lepidoptera (as with other organisms) having ranges that extend across many degrees of latitude and countries. Santa Rosa species were described from the United States, Venezuela, Brazil, Mexico, Guatemala, etc. Lest this situation be thought to be unusual, I should point out that it also applies to macrolepidoptera faunas in most other parts of Costa Rica (and the neotropics). The problem that ecologists have with achieving taxonomic certainty with neotropical butterflies and macromoths is generally not due to a lack of species descriptions per se, but rather a lack of geographic or large-taxon revisions and/or field guides and curated reference collections.

Association of caterpillars with described adults is at a far more primitive state. The effort to rear all Santa Rosa species of caterpillars, and thus associate them with an adult and at least one wild host plant, began in 1977 and has not been repeated elsewhere in the neotropics. By September 1990, 341 species of Santa Rosa macromoths (and 116 butterflies) have been reared and are identified or have their identification in process. All Santa Rosa saturniid larvae and nearly all sphingid larvae are known (e.g., Janzen 1982, 1984a, 1985a).

Perhaps the most conspicuous feature of the Santa Rosa caterpillar fauna is its extreme abundance and omnipresence during the first 3 months of the rainy season (May–July), the large number of quite conspicuous species, the large number of carnivores (species and biomass) that feed on caterpillars during the first 3 months of the rainy season, and the massive amounts of leaf consumption at this time. It is much easier to find caterpillars in the Santa Rosa forest during these 3 months than in any Costa Rican rain forest.

During the second half of the rainy season, the caterpillar fauna is substantially reduced (e.g., Janzen 1980a) but it is still easier to locate caterpillars then than it is in rain forest at any time of year. During the dry season, caterpillars are almost nonexistent except for miners in a variety of substrates and (in a few cases) prepupae in cocoons or underground chambers. The most conspicuous feature of the Santa Rosa caterpillar fauna over successive years is how greatly it changes in biomass from year to year, and that different hosts are heavily fed on in different years (e.g., Janzen 1981). For example, in the past 12 years, no species of plant has been heavily defoliated in more than one year, and in 1977 (Janzen 1980a) there was easily 10 times as much biomass of caterpillars as has been seen in any one of the following 13 years. However, these interyear variations in caterpillars are not the subject of this chapter.

What Really Happens When the Rains Begin?

Cueing

There is a strong temptation to view the seemingly abrupt appearance of moths, butterflies, and caterpillars at the beginning of the Santa Rosa rainy season as "cued by the rain." Those of us from northern latitudes tend to imagine that multitudes of dormant pupae have been wetted by the rains, and then eclosed to produce the egg-laying females. Such cueing does not appear to be the case.

As is clear in Figures 14.3 and 14.4, the first rains occur at the end of the hottest time of the year and are associated with an abrupt drop in the daytime average maximum temperatures. The first days following the arrival of the rains can be as much as 6°C cooler than they were a week before during the dry season. All indications are that it is the drop in temperature that the dormant pupae, inactive adults, and/or incoming migrants are using as their cue.

The eclosion times of an experimental cohort of *Rothschildia lebeau* (Saturniidae) illustrates this. In the end of the 1982 rainy season, 416 sibs of *R. lebeau* (voucher number 83-SRNP-1500) were reared to cocoons (Janzen 1984b). Almost all of the pupae became dormant (probably in response to the warming weather in December). Their cocoons were hung in airtight plastic bags in the ceiling of an outdoor laboratory at Santa Rosa, a laboratory through which there is open air circulation. Each cocoon was in a separate bag. Half of the cocoons were in air-dry bags, and the other half were maintained at 100% relative humidity by soaking a mat of wet toilet tissue in the bottom of the bag every 3–5 days. Figure 14.5 shows the eclosion pattern of the 175 females in relation to the temperature drop at the onset of the rainy season in 1983. They obviously did not eclose in response to the rain per se, since the rain never went near the bags and the dry bag cocoons were dry at the time of eclosion. Being wet or dry did not affect the pupal eclosion times at the scale of resolution displayed in Figure 14.5 (though in fact the moths in wet bags had an average eclosion date 1–2 days later than did those in dry bags).

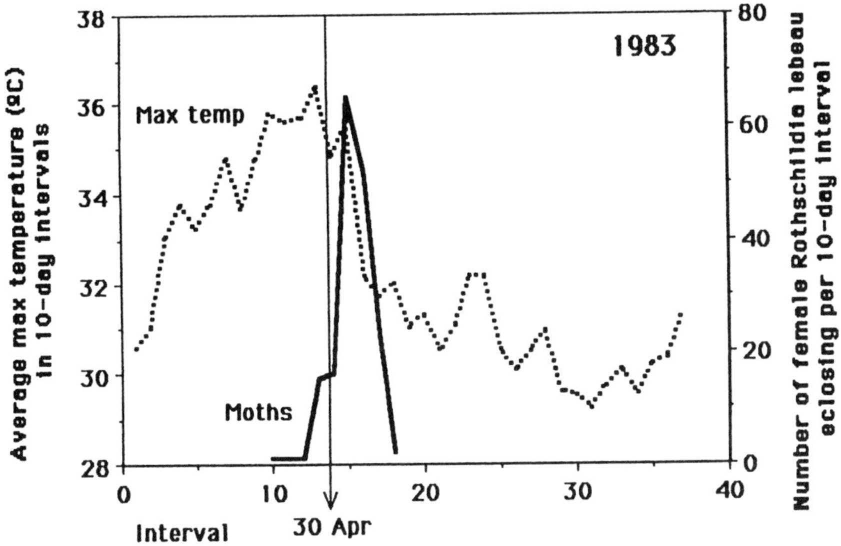

Figure 14.5. The average maximum temperature over 10-day intervals in 1983 and the dates of eclosion of a cohort of 175 sib *Rothschildia lebeau* (Saturniidae) females, summed over 10-day intervals in 1983 (see text).

A *R. lebeau* female mates on the night of eclosion and lays about half of her eggs the following night. The eggs take 6 days to hatch. The pupal eclosion dates (Fig. 14.5) plus 7 days are therefore an exact chronology of the dates of appearance of the first instar larvae in nature (wild-caught females at lights mapped exactly onto the eclosion distribution in Fig. 14.5). However, it is clear from the phenology of this cohort of sibs that the timing with the onset of the rains is not perfect (compare Fig. 14.4 and Fig. 14.5). More than 2 months of variation in the eclosion time from a single cohort (10 April to 24 June in this case) creates substantially overlapping generations during the remainder of the rainy season.

Dependence on a temperature cue is also suggested by univoltine species that normally remain dormant from the end of their annual larval stage (late June to early July) until the beginning of the next rainy season, 10–11 months later. In 1987 the short dry season (July –August) was exceptionally hot and dry. By early July, I had numerous pupae of *Manduca dilucida*, a locally common dry forest univoltine sphingid, hanging in dry plastic bags in the ceiling of the Santa Rosa laboratory "in storage" for use in the following year. However, when the rains began in September, and the temperature simultaneously dropped, about 18% of these moths eclosed within 2 weeks, despite the fact that their bags were full of dry air. Those that did not eclose then remained dormant until the following May (in synchrony with the free-living population). It seems extremely likely that eclosion of *M. dilucida* requires first a long hot period, followed by an abrupt transition to a cooler regime. Normally, this occurs in the annual seasonal cycle, but in 1987 for some of the pupae, the short July–August dry season was sufficiently hot to mimic the long dry season, and therefore the September rains were reacted to as though they were the May beginning of the rainy season. *M. dilucida* caterpillars were ar least theoretically present in the habitat in the second half of the 1987 rainy season, though I was not able to find any.

Cueing to a temperature change is not restricted to pupae. As clouds build up daily to the south of Santa Rosa, with the May rains approaching closer each day, there are often pulses of cooler air that pass through Santa Rosa before the rains actually arrive. Each night following one of these cool air pulses, enormous numbers of species and individual moths appear at the lights. These moths have obviously been present as adults for weeks to months, and are certainly not cued by rainwater, since none has arrived.

The detailed time of appearance of egg-laying females, and therefore the start of the larval period, is also under selection for adult, egg, and pupal survival and health. This timing is not the subject of this chapter, but it is important to recall that the timing of the caterpillar stage is certainly not a fully independent variable in a selection regime. For example, there may well be tons of highly acceptable foliage and a relatively predator-free microhabitat, yet constraints on adult survival prevent the presence of caterpillars at that time.

Time of Caterpillar Appearance Relative to the First Rains

Although most species of adult Lepidoptera make their appearance at Santa Rosa (flying or at lights) in the first 2 weeks after the first soaking rains, caterpillars are not abundant nor is their damage readily visible until 3–6 weeks after the first rains. If the first rains are followed by several weeks of dry weather (see below), the delay can be even longer.

There are two obvious reasons for this delay. First, many species of moths and butterflies do not immediately lay all, or even any, of their eggs on foliage as soon as it is available. Rather, they delay their first egg-laying (or eclosion) and therefore spread their egg-laying over several weeks or more. This behavior is probably very functional in those frequent years when the first rains are followed by several further weeks of dry season (e.g., Fig. 14.4). Second, eggs require 5–15 days to hatch; for example, not only do the hemileucine saturniids (*Hylesia, Automeris, Periphoba, Dirphia, Molippa*) in Figure 14.6 grow slowly, all their eggs require a full 2 weeks to hatch. Third, many of the small species are at very low density (inconspicuous) and do not become visible in samples or censuses until the second generation emerges; this can be many weeks after the rains, even if the species is one that anticipates the rains and oviposits several weeks before them. Fourth, an enormous number of small caterpillars are taken by predators or killed by the weather before they have a chance to consume enough foliage to be conspicuous through their damage.

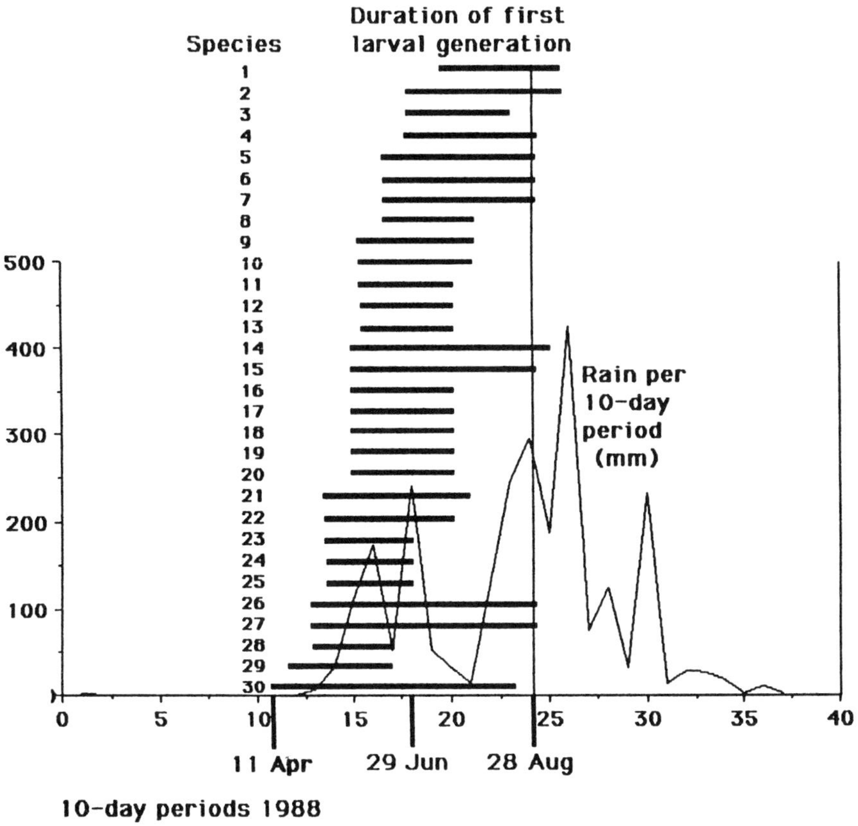

Figure 14.6. The seasonal distribution of the larvae of the first generation of each of the 30 species of Saturniidae that breed in Santa Rosa National Park, ACG, Costa Rica. The date of description of each species is added to emphasize how long the members of this fauna have been known to science (see text). (1) *Automeris io* (1775); (2) *Arsenura armida* (1779); (3) *Syssphinx mexicana* (1872); (4) *Molippa nibasa* (1885); (5) *Hylesia dalina* (1911); (6) *Automeris zugana* (1886); (7) *Automeris zurobara* (1886); (8) *Syssphinx quadrilineata* (1867); (9) *Copiopteryx semiramis* (1775); (10) *Rothschildia erycina* (1796); (11) *Syssphinx colla* (1907); (12) *Titaea tamerlan* (1869); (13) *Citheronia lobesis* (1907); (14) *Automeris metzli* (1853); (15) *Automeris tridens* (1855); (16) *Othorene purpurascens* (1905); (17) *Schausiella santarosensis* (1982); (18) *Citheronia bellavista* (1930); (19) *Caio championi* (1886); (20) *Rothschildia lebeau* (1868); (21) *Eacles imperialis* (1773); (22) *Othorene verana* (1900); (23) *Syssphinx molina* (1780); (24) *Ptiloscola dargei* (1971); (25) *Dysdaemonia boreas* (1775); (26) *Hylesia lineata* (1886); (27) *Periphoba arcaei* (1886); (28) *Adeloneivaia isara* (1905); (29) *Copaxa moinieri* (1974); (30) *Dirphia avia* (1780).

The consequence of this delay is that a very large number of caterpillars and species of caterpillars do not initiate their feeding on very new, still expanding foliage, but rather on fully expanded new leaves that appear to be fully functional at the time they are eaten. While there are species that feed almost exclusively on very new foliage (e.g., *Eutelia, Eulepidotis* and some other noctuids), these species are in the minority. Furthermore, most species that begin their lives feeding on newly expanding leaves in fact can and do eat fully expanded and "mature" leaves as well. It has long been tempting for ecologists to view the caterpillar phase at the beginning of the rainy season as somehow using foliage before it has "hardened up," "matured its defenses," etc. However, in a sample of 80 tree species in Santa Rosa, there was no significant change in the amount of fiber, polyphenolics, or small toxic molecules in the foliage from the time when leaves are generally first fed on (several weeks of age) until very late in the rainy season (6 months of age) (Janzen and Waterman 1984).

Field observations of the overall timing of caterpillar feeding are confounded by feeding by other herbivores. It is commonplace to use the appearance of damaged leaves as a general cue to caterpillar activity. However, Santa Rosa has a rich fauna and high biomass of small melolonthine scarabs that feed on many species of foliage at night. Their feeding damage is characteristic and can be distinguished from that of caterpillars with practice. However, they appear as herbivores from the day of the first rain (or before) and their feeding damage gives the casual observer the impression of much earlier caterpillar activity than is in fact the case.

The dry forest changes from largely brown and leafless to green and fully leafed within a few weeks. Much leaf biomass is therefore aging at somewhat the same rate as the rainy season progresses. However, this does not mean that the larvae present on any given date are consuming leaves of the same age or degree of maturity. First, different species of plants put out their new leaf crop at different times relative to the start of the rains. For example, *Enterolobium cyclocarpum* (Leguminosae) trees flush a major leaf crop 2–6 weeks before the rains, *Crescentia alata* (Bignoniaceae) trees flush a large leaf crop within a week after the first rains, and adult *Hymenaea courbaril* (Leguminosae) makes a new leaf crop in early January, approximately 5 months before the rains.

Second, many species of plants continue to make fewer leaves as the rainy season progresses, both to replace those that have been eaten off and to expand crowns upward and laterally. The ovipositing female has leaves of many different ages among which she can chose, and the caterpillar searching in the crown can also chose among an even greater array of leaf ages during its life. Third, not all conspecifics are highly synchronous in leaf production, even if there is a very obvious seasonal peak. For example, whereas the "evergreen" adult *H. courbaril* trees annually replace their leaf crops in January, small saplings add new leaves throughout the year (with a peak about the time of the first rains in May). Fourth, despite the strong peak of initial oviposition times, many species of moths (or butterflies) oviposit over a period of weeks to months, thereby creating the possibility of adults and caterpillars being able to chose from among almost every possible leaf age.

Within the early rainy season peak of caterpillar abundance, there is substantial variation among the species within a large taxon. For example, there are 30 species of resident breeding Saturniidae in Santa Rosa. The larvae of all (e.g., Lemaire 1987) can be found within a single square kilometer of late successional dry forest in the eastern end of the park. The earliest eggs to hatch are those of *Copaxa moinieri* and *Dirphia avia* in late April, 2–3 weeks before the rains; the last eggs to hatch are those of *Arsenura armida* and *Automeris io* in early July, a full 6 weeks after the rains have come (Fig. 14.6). Such a spread among a species-rich taxon is commonplace and probably should not be thought of as a detailed evolutionary response to the fine details of the Santa Rosa pre- and post-rains microenvironments. Rather, it is a simple outcome of the various species responding to the weather changes according to whatever cue-response system they had when they arrived at Santa Rosa from other sites and habitats.

It is possible to make a general statement as to when most of the members of the first generation of caterpillars will be present in the habitat (e.g., Fig. 14.6). However, there is great interspecific variation in the degree of response to the various variance-inducing interactions among caterpillars and their habitat. For example, the hemileucine saturniid caterpillars mentioned earlier are much more inclined to increase their caterpillar life span in response to low quality food than are the members of the other three subfamilies. Saturniidae as a group seem to have much more tightly defined caterpillar life span lengths than do Sphingidae as a group (and see Janzen 1984a). Saturniidae as a group have a quite tightly defined pupal period (leading to synchrony of eclosion within and between cohorts in the field), while Sphingidae as a whole are more variable in eclosion times. The consequence is that multivoltine species of sphingids that are nonmigratory during the rainy season (e.g., *Pachylia ficus, Protambulyx strigilis, Adhemarius gannascus*) tend to have more overlapping generations of caterpillars than do many multivoltine saturniids. However, some species of multivoltine saturniids eclose gradually over a 2-month period at the beginning of the rains (e.g., *Rothschildia lebeau*) whereas others such as *Caio championi* are highly synchronous; again, variation on this point leads to variation in degree of generational overlap.

Finally, I should mention that even within a single higher taxon, caterpillars of the same body weight can have very different development times (e.g., Fig. 14.6). For example, hemileucine Saturniidae can use nearly twice as long to develop from egg to pupa as do other Saturniidae of the same size; this is presumably because the hemileucines are getting substantially fewer nutrients per food bolus than are the members of the more host-specific other saturniid subfamilies (Janzen 1984a, 1985a; Bernays and Janzen 1988). The even more host-specific sphingids display this phenomenon yet more strongly and in at least one case (*Enyo ocypete* described below) can have two generations in the time that a saturniid of equal body weight has one.

There is a set of moth species whose caterpillars are not tied directly to the weather changes so much as to the flushing of foliage that occurs during the 1–2 months bracketing the first rains. The highly host-specific and stenophagous *Eulepidotis* (Noctuidae: about 30 species in Santa Rosa) is the champion. Two species feed heavily on the very new leaf crops of *Licania arborea* (Chrysobalanaceae), whenever they occur during the last half of the dry season and the first half of the rainy season. The same may be said for the species of *Eulepidotis* that feed on (and sometimes defoliate) the new foliage of *Sloanea terniflora* (Eleocarpaceae), *Sterculia apetala* (Sterculiaceae), *Bombacopsis quinata* (Bombacaceae), *Luehea speciosa* (Tiliaceae), and *Hymenaea courbaril* (Leguminosae) (saplings only). With members of this genus, it is clear that the active adults are present virtually year round in the habitat, but respond to both new foliage and the general time of year to oviposit (there are very few or no *Eulepidotis* caterpillars on new foliage of their hosts during the second half of the rainy season and the first half of the dry season).

The Return of the Killer Dry Season

As is clear in Figure 14.4, the first heavy rains of the rainy season are not always followed by humid, cool, and moist weather. In at least half of the years, the first rains have been followed by 1–2 weeks of hot and dry weather, nearly as severe as is the dry season in March–April. This is because the first rains are often in response to habitat heating (rising heated air sucking moist air in off the Pacific), and subsequently they cool the habitat enough to stop rain-generating air movements for 1–2 weeks until the environment has heated up again.

When the first rains are followed by dry hot weather, it is commonplace to witness massive mortality of first instar larvae from the eggs laid at the time of the first rains. For example, in 1983, I placed hundreds of first instar larvae of the saturniid *Eacles imperialis* and *Rothschildia lebeau* on their wild host plants in the first rainy week in mid-May. The next week turned dry and hot, and at least 98% of these larvae died of desiccation during this week. They quite obviously could not (or would not) eat enough of the newly expanding foliage to maintain their water balance.

It is not surprising in this context that many pupae of nonfeeding Santa Rosa moths such as Saturniidae, Limacodidae, Mimallonidae, and Bombycidae do not eclose until the rainy season has been present long enough to thoroughly soak both the soil and cool down the general environment. The eclosion dates of moths that can feed as adults (which generally have longer adult lives than do the nonfeeders) do not appear to be so strongly affected, but these moths can delay oviposition until the heavy rains start again.

Different species of caterpillars and different habitats are differentially affected by an erratically beginning rainy season. In the *E. imperialis* and *R. lebeau* example above, caterpillars on their *Cochlospermum vitifolium* and *Spondias mombin* hosts plants in highly deciduous young secondary succession were all killed, whereas their sibs on the same species of plants in the semishade of relatively primary forest a few tens of meters away displayed moderate survival.

Migration

One of the most spectacular interactions of caterpillars with seasons in Santa Rosa is the seasonal migration of many species of Lepidoptera back and forth between Santa Rosa and the rainforests to the east of Santa Rosa (Janzen 1984a, 1987a,c, 1988f). About the time of the first rains, more than 100 species of (at least) sphingids, noctuids, nymphalids, and pierids arrive at Santa Rosa (just as they also arrive in other dry forest areas to the south of Santa Rosa). More than 80% of the 64 species of sphingids that regularly breed in Santa Rosa do this. However, in other families, the proportion of migrants is much less. All species of migrants feed as adults and have life spans measured in months.

The migrants have a single generation that occupies 1–3 months of the rainy season (egg–larva–pupa), ecloses, and (mostly) leaves. The adults apparently fly back to the rain forest to the east of Santa Rosa. It is striking that this out-migration occurs in mid-rainy season, at a time of year when the dry forest is humid, cool, and in full leaf (and will continue to be for another 3–4 months). These leaves are clearly edible, as evidenced by occasional larvae of these species that are encountered in the second half of the rainy season. These larvae represent a second generation. They grow at the same rate as do the members of the first generation. These larvae do not show a special predilection for very new foliage, but rather, eat foliage of all ages, just as did their parents' caterpillars in the first half of the rainy season (or, if they specialize on a particular leaf age in the first half of the rainy season, they do likewise in the second half). There is simply no natural history reason to hypothesize that they are leaving Santa Rosa because of inimical weather or a shortage of adult or larval food at the time of eclosion. Ironically, they actually abandon Santa Rosa at its rainiest time of year (e.g., Fig. 14.3, 14.4).

Larval biology interacting with seasonality provides a reasonable suggestion as to why they leave Santa Rosa in August–September. At the time of arrival of the adult moths in late April to mid-May, the carnivore array—spiders, ants, scorpions, bugs, lizards, birds, mammals, fungi, viruses, etc.—is at its lowest point in the yearly cycle, having just endured 6 months of substantial mortality from weather and other carnivores while having no recruitment owing to a general lack of prey. Even the viruses, fungi, and bacteria are at a very low density on the newly produced leaves. The caterpillars—barring a severely irregular start to the rainy season—have a large amount of new and newly mature foliage, high humidity, and moderate temperatures. Finally, the incoming migrants produce their first dry forest generation in synchrony with all the resident dry forest species that are also having their first generation of the year. There are maximum possibilities for inter- and intraspecific satiation of caterpillar predators and parasitoids at this time of year.

However, when the adults eclose in July–August, they are confronted by a rapidly growing (or risen) carnivore array. This set of carnivores has been replenished by feeding on the large biomass of caterpillars present during the first several months of the rainy season and has benefited by the favorable weather as well. Furthermore, that array is now starving and/or desperate for ovipositional opportunities, owing to the downward plunge of caterpillar density as many species pupate and become dormant, eclose, and remain reproductively quiescent, or eclose and migrate away from the dry forest. These starving carnivores will search extra-thoroughly for prey. At this point in the seasonal cycle, a leafy host plant may have quite ample food but is subject to a living sheet of carnivores. Ironically, if a particular species of plant was mostly defoliated, the few leafy individuals are even more likely to represent lethal concentrations of carnivores, since their leafiness is probably an indication of being within the foraging range of some particularly thorough carnivore (such as a large ant colony) (e.g., Janzen 1985a).

When the adult moth (or butterfly) returns to the rain forest, its offspring do encounter a fierce and omnipresent predator/parasitoid community. It is a moot point as to whether this carnivore array is more dangerous for caterpillars than is the dry forest carnivore array in the second half of the dry forest rainy season. It is tempting to suggest that since they migrate out of Santa Rosa in the middle of the rainy season, the rain forest must somehow be safer than is Santa Rosa at this time. However, evaluating the logic of such a (weak) inference would require a discussion of evolutionary biology of substantially greater size than this chapter (and see Janzen 1985b).

The moths may leave the rain forest for the dry forest in April–May both because the rain forest may be more dangerous (for adults and/or immatures) and because there is more food in the dry forest than in the rain forest after the new leaves have been produced. The intriguing question is whether the rain forest is sufficiently friendly and food-rich for caterpillars to maintain populations of the migrant species if these species were deprived of the dry forest part of their annual biology. I think it likely that the rain forest density of a species steadily declines—even if over several generations—and then is replenished by the incoming brood from the dry forest in August–September of each year.

The above scenarios are substantially influenced by whether the dry forest parasitoid array also migrates to the rain forest. There are some hints that they do. W. Haber (personal communication) has found enormous numbers of hymenopterous parasitoids of caterpillars and eggs at the Monteverde Reserve, a site in AC Arenal at 1400–1800 m elevation on the mountains between Costa Rica's dry forest and the Caribbean rain forest. These parasitoids are there during the months when it is the late wet season in the rain forest and dry season in the dry forest remnants on the coastal plain and foothills to the east of Monteverde. These wasps appear to be migrating through Monteverde to the Caribbean rain forests, or waiting there in the cool and moist cloud forest habitat, for later return to the dry forest.

A second example is offered by the *Enyo ocypete* sphingid described below. Its pupae are commonly killed by *Belvosia* nsp. 4 (Tachinidae), a large parasitoid that attacks just the sphingids *E. ocypete, Eumorpha satellita,* and *Unzela pronoe* in Santa Rosa (*Belvosia* taxonomy, N. Woodley, personal communication). All three of these host caterpillars are available only during the first half of the rainy season. This large fly could easily fly back and forth across the mountains separating Santa Rosa's dry forest from the rain forests that are just a few tens of kilometers to the east. *Belvosia* nsp. 4 has now been reared from *E. ocypete* larvae at Estacion Pitilla, a rain forest biological station in the rain forest northeastern end of Guanacaste National Park in ACG (Figure 14.1). The fly probably does migrate, because it ecloses from the *E. ocypete* pupae in Santa Rosa at the same time as the adult moths eclose (there is no dormancy in the fly puparia). Since there are essentially no host larvae in the dry forest for it to parasitize during the following 7 months, the adult flies have to either survive as adults or migrate. I have not been able to locate adults in Santa Rosa after September and before the following June. Furthermore, when large numbers of the flies are eclosing from sphingid pupae in July, the adults are common in the forest understory. However, they abruptly disappear from that habitat in early August.

There are hundreds of species of Santa Rosa dry forest caterpillar/pupae parasitoids that, like *Belvosia* nsp. 4, eclose in large numbers at the end of the first rainy season generation and then disappear from the habitat (as measured by collections with lights and Malaise traps), only to reappear at the beginning of the next rainy season. As a group, these species must be either hiding, relatively quiescent as adults, or migrating to the rain forest side. If they do migrate to the rain forest, the influx of parasitoids may render the rain forest a yet more dangerous place than it is at other times of year. On the other hand, this will be extremely difficult to study today, if for no other reason than the nearly total destruction of Costa Rican dry forest (Janzen 1988b) will have substantially reduced the numbers of parasitoids to migrate annually.

We do not know to what degree the migrant lepidopterans reproduce in the rain forest. However, there are some hints. There definitely are species that have a second (or more) generation in the rain forests. *Perigonia ilus* provides an example. At the beginning of the rainy season, males and females of this small sphingid arrive at Santa Rosa and oviposit on the new foliage of two rubiaceous trees, *Calycophyllum candidissimum* and *Guettarda macrosperma*. About 2 months later, the adults eclose and leave for the eastern rain forests of Costa Rica (where they are caught occasionally at lights). Of more than 200 rearings of wild

P. ilus caterpillars in Santa Rosa, not a single pupa has become dormant (and waited for the next year's rainy season). At the time that they arrived at Santa Rosa (April–May), a sample of 37 adult males caught at the lights had average forewing lengths of 0.327 mm (SD = 0.072 mm), while a sample of 8 newly eclosed wild-caught moths in late June and early July had average forewing lengths of 0.213 mm (SD = 0.057 mm); these differences are significant at the 0.01 level (Janzen 1986a). Although detailed measurements have not been taken, I have observed this difference among *P. ilus* every year (and there are many examples in Costa Rican moths in general where rain forest moths are larger than their dry forest conspecifics). It is clear that the rain forest side of Costa Rica produces bigger *P. ilus* than does the Santa Rosa dry forest. This size differential is of significance here because it demonstrates that there is at least one generation of *P. ilus* on the rain forest side of Costa Rica.

There may be a resident population (in the rain forests) and a migratory population (moving back and forth between the rain forest and the dry forest), or it may be simply that some of the *P. ilus* population "invades" the dry forest annually at the beginning of the rainy season and the offspring return to the rain forest after one generation (a pattern not unlike that of Australian dry forest noctuids, Farrow and McDonald 1987). In either case, there is a background population of *P. ilus* caterpillars present at a very low density in the rain forest throughout the year but a dense population of *P. ilus* caterpillars in the dry forest during June–July. They are a major food item for the trogons to be discussed below.

The very large sphingid *Pseudosphinx tetrio* offers a similar example. This moth appears in Santa Rosa toward the end of the first rainy month. The large aposematic (and/or mimetic) caterpillars (Janzen 1983d) feed only on the leaves and (occasionally) flowers of *Plumeria rubra*. The entire Costa Rican breeding population of *P. tetrio* is therefore restricted to the coasts of both sides of the country, which is the only place that *P. rubra* grows as a wild tree. The last *P. tetrio* larvae disappear from Santa Rosa about the time that *P. rubra* is beginning to drop its leaves in the last month of the rainy season (November). The adults are then (and earlier in the year) encountered at lights throughout Costa Rica's rain forest (from sea level to 2500 m elevation), apparently on their way over to the *P. rubra* trees on the Caribbean coast. Alternatively, these adults may be simply waiting as adults in the inland rain forest until the next May–June. However, there are *P. tetrio* caterpillars on the *P. rubra* trees throughout the year on the Caribbean coast. As with *P. lusca, P. tetrio* in Costa Rica may be two populations (one migratory and the other stationary) or one population, some members of which migrate to the Pacific dry forest coast during the rainy season.

The arrival and departure of migratory species can be more complex than meets the eye. The adults may well arrive weeks before their larvae appear in the habitat or they appear at lights. The small sphingid *Enyo ocypete* is an example. Its dry forest food plant is *Tetracera volubilis* (Dilleniaceae). This dry forest large and woody vine is facultatively evergreen (depending on the amount of shade and soil moisture). However, about a week before the rains begin at Santa Rosa, it begins to produce new leaves in large numbers, and continues through much of the rainy season. About the second week of the rainy season, eggs of *E. ocypete* begin to appear on these new leaves (and adults appear at the lights). A large pulse of *E. ocypete* larvae and pupae pass through development between late May and early August. Almost all *E. ocypete* larvae have disappeared from the *T. volubilis* vines by late July. The pupae from hundreds of wild-caught pupae have never displayed any dormancy. The newly eclosed adults have disappeared from the Santa Rosa habitat by September [they can, however, be collected at flowers in August, apparently gathering food before leaving for the rain forest, where it (and resident conspecifics?) have other generations on the leaves of other dilleniaceous vines].

For 10 years, this appeared to be an accurate description of *E. ocypete* at Santa Rosa. However, in 1990 I found that *E. ocypete* has actually arrived by early April, 6 weeks before the rainy season begins and almost 2 months before the first eggs appear on *T. volubilils* new foliage in the forest. At this time *E. ocypete* lays its eggs on the new foliage of *Curatella americana*, a dilleniaceous tree dotted across the ancient abandoned pastures in Santa Rosa's badly disturbed habitats. The larval generation takes a month. The adults eclose 3–4 weeks later. I cannot know if the new eggs that appear on *T. volubilis* new leaves are from this (first) Santa Rosa generation, or from later incoming migrants, or both. However, since there are two dilleniaceous plants at Santa Rosa that are used by *E. ocypete* during at least part of the year, there are two distinct peaks of caterpillar density—with each peak representing the initial pulse of new leaves by the host. It is particularly striking that both host plants continue to produce new leaves throughout the rainy season, but neither these new leaves nor the older new leaves are used by *E. ocypete*, except for a very few very rare individuals (probably from eggs from the few adults that do not migrate out of Santa Rosa until the second half of the rainy season). During the normal caterpillar season, the caterpillars feed on *T. volubilis* and *C. americana* leaves of all ages (though they begin on new expanding leaves, where the eggs are laid).

There are many ways in which forest clearing or alteration can change the dynamics and demography of dry forest caterpillars of migrants. The *E. ocypete* interaction with its hosts offers an example. In the pre-European Santa Rosa habitat, *C. americana* would have been a very rare and local tree, restricted to rocky outcrops and other areas where the forest canopy does not naturally exceed about 5–10 m height. It could not have been but a quite trivial host for *E. ocypete*, as compared to the common *T. volubilis*. The latter plant is common in Santa Rosa dry forest of all kinds and ages—from deeply shaded understory to full insolation. It is possible that *E. ocypete* has always arrived in April in Santa Rosa and simply waited—while feeding at flowers (?)—for *T. volubilis* to reach the appropriate stage for oviposition in late May–June. However, with forest clearing and the very great expansion of the *C. americana* population in low-grade and frequently burned pastures, this new and earlier food source can probably absorb much of the egg-laying potential of the incoming *E. ocypete* population. It is not possible to know, however, whether this generation on *C. americana* results in a yet higher density of offspring *E. ocypete* caterpillars on *T. volubilis*. It is quite striking, however, that the newly eclosed adults in May oviposit only very rarely on the *C. americana* new leaves being produced at that time.

Univoltine Nonmigrants

The migrant Lepidoptera described above are largely univoltine as far as Santa Rosa is concerned. There is also a large and taxonomically more diverse group of univoltine species that do not migrate (Janzen 1987a). Instead of eclosing and flying out of the habitat, in effect the caterpillars put themselves in tight cocoons, tough pupae, and/or underground chambers and simply wait out the second half of the rainy season and all of the dry season. *Manduca dilucida* is a dry forest sphingid mentioned earlier that behaves in this manner. Interestingly, *M. dilucida* have life spans of only several weeks duration—exceptionally short for feeding sphingids—and carry large numbers of seemingly mature eggs when they arrive at lights—a very saturniid-like trait. *Schausiella santarosensis*, the only saturniid endemic to the Santa Rosa area, has a similar biology.

It is significant in this context that even if the carnivore intensity during the second half of the Santa Rosa rainy season fluctuates from year to year, the univoltine species—migrants and dormant pupae or prepupae—will not be aware of it, and their populations at the beginning of the following rainy season generation will not reflect it (except with respect to predation by extreme generalists such as armadillos and mice that dig up hidden pupae).

An Example of Caterpillar Predation by Vertebrates

The extreme seasonal and annual fluctuation in caterpillar abundance in Santa Rosa dry forest appears to be a major component of the biology of those vertebrates that feed on caterpillars. This is not, however, a simple thing to interpret. I offer the nesting biology of the elegant trogon, *Trogon elegans*, as an example. This medium-sized bird is common in Santa Rosa dry forest and its nestling feeding biology is the subject of an on-going multiyear study (F. Joyce and D. H. Janzen).

T. elegans nests in holes in tree trunks. During the first several weeks of the rainy season, *T. elegans* pairs locate nest holes and incubate eggs. By placing a soft but tight collar around the nestlings' necks, the nestlings are prevented from swallowing the caterpillars (or other food items) brought by the parents. The food is then collected by the observer, the collar removed, the nestling given an alternate and equivalent food item, and the collar replaced. During the past 5 years, data obtained in this manner from 1584 feeding events at 22 *T. elegans* nests in Santa Rosa allow the following general conclusions.

The timing of nestling appearance is such that by the time the nestlings are begging for food, the habitat is rich in the last instar caterpillars of Sphingidae (as well as a variety of other large insects). In a "normal" year, last instar sphingid caterpillars constitute about half of the nestling diet in numbers of individuals brought in, and as much as 70% of the nestling diet in biomass. The remaining diet is pure insects, except for an occasional spider or anolis lizard (*Anolis* or *Norape* spp.). More than 98% of the sphingid caterpillars are in their last instar, irrespective of whether they belong to a small, medium-sized, or large species of sphingid. By about the end of July, the density of last-instar sphingid caterpillars in the forest has declined to a very low level. At the very few remaining (or new) nests at that time, other insects (mostly last-instar saturniid and notodontid caterpillars, and Orthoptera) become the major food items. For the remainder of the rainy season, and all of the dry season, there are no further nesting attempts by *T. elegans*. In the first 3 years of the study (1986–1988), this was the pattern, and it was easy to come to the global conclusion that the presence of a *T. elegans* breeding population in Santa Rosa was dependent on the peak of last instar caterpillars of large moths during a 2-month period that begins about 2 weeks after the beginning of the rainy season.

However, in 1989, a natural experiment occurred. The density of sphingid caterpillars in Santa Rosa declined dramatically. The seasonal peak in larvae disappeared (for quite inexplicable causes). For example, in 1988, sphingid larvae constituted 40.2% of all caterpillar collection records in the Santa Rosa caterpillar inventory (n = 614); in 1989, only 8.3% of the collection records by the same collectors collecting in the same habitat were sphingids (n = 919). The *T. elegans* responded to this decline in sphingids by feeding their nestlings more caterpillars of other species of insects. For example, sphingids made up 49.6, 49.6, 50.7, and 45.6% of the nestling diets in 1986–1988 and 1990, respectively, but only 23.3% in 1989 (n = 133, 232, 493, 338, and 388 in consecutive years). I should add that for the adult trogons to find even this high a percent of sphingids required diligent and directed search for sphingids, since sphingid larvae were far more than 50% reduced in density in Santa Rosa in 1989. In addition, the trogons brought fruit to their nestlings in 1989 (adults trogons eat fruit regularly). For example, 19.5% of the food items in 1989 were fruit, while no fruit were brought in 1986–1988 and 1990.

It is not presently possible to know if permanent removal of the sphingid caterpillars would result in the elimination of *T. elegans* from Santa Rosa or simply produce a change in their nestlings' diet. The fact that they do not breed during the second half of the rainy season (when there are abundant orthopterans, a few other large caterpillars, and some fruit) suggests that the sphingids are a critical resource. However, the fact that they can shift to other food items when sphingid caterpillars are scarce (as in 1989) demonstrates that the sphingids can be interchanged with other items. It does not, however, demonstrate that a breeding population can be maintained without the abundance of food represented by the normally high density of sphingid caterpillars in the first half of the rainy season. The ability to interchange sphingids for other food items appears to be contradictory to the observation that they breed only during the part of the rainy season in which sphingids are present. It is tempting to suggest that the sphingid caterpillars offer the best nestling growth and/or are optimal for adult foraging regimes, but substitutes are acceptable (with perhaps lowered survival or health of the fledged young). In this scenario, it is implied that the second half of the rainy season neither has the sphingid caterpillars nor sufficient alternative foods for nesting, and/or the second half of the rainy season is (also) inimical to nesting per se (e.g., due to an abundance of predators, or due to producing fledglings so close to the dry season that they do not have time for whatever development is necessary for dry season survival.

Unfortunately, there is an awkward alternative hypothesis, based on the concept of "ecological fitting" (Janzen 1985b). This would be that *T. elegans* is genetically locked into breeding only during the first half of the rainy season wherever it is, and feeds its nestlings whatever its genes view as the best food. If sufficient food is present, the population persists; if the food is insufficient, the population is not established. In this scenario, the fact that sphingids make up a large part of the nestlings' diet is a simple reflection of the adults' choice/priority/ability during the time of year when they nest, but does not demonstrate that they are contemporarily basing their nesting phenology on the sphingid caterpillar peak in abundance, or that the Santa Rosa sphingid caterpillar seasonal presence was part of the selective regime that generated or maintains this nesting phenology.

Aseasonality

The most "aseasonal" externally feeding caterpillar in the study site is *Hypercompe icasia* (*Ecpantheria icasia*), a large arctiid whose caterpillar resembles the "wooly bear" of the northern United States. It appears to be rejected by all vertebrate predators. This caterpillar is an extraordinary generalist in its diet (Janzen 1988d). An individual caterpillar feeds for a few minutes to hours to days on a particular plant, and then walks on by ground or foliage to other plants, where it may feed as well. It does not eat all species that it encounters, but it does have a host list of more than 70 species of woody plants to date. Many, but not all, of the plants that it eats can serve as an adequate diet to pupation if the caterpillar is restricted to them in the penultimate or ultimate instar. I have found active and feeding last instar larvae in all months of the year. *H. icasia* is extraordinarily slow growing for two reasons. First, it can easily starve for as long as 1–2 weeks without feeding. Second, even when confined to a seemingly high-quality food (which it eats in copious amounts), it grows more slowly than do the slowest and most generalist hemileucine Saturniidae (e.g., *Hylesia lineata*, Janzen 1984c). It is commonplace for a *H. icasia* caterpillar to use 2–3 months on a mixed or pure diet before pupating. The pupae eclose after about 16–20 days and even during the dry season show little inclination for delayed development.

It appears that the world is not covered with *H. icasia* because when the females lay their batches of hundreds of very small eggs during the dry season, the very small first-instar larvae have very low survivorship at that hot, windy, dry and largely leafless time of year. The very few that survive to become adults at the beginning of the rainy season produce a highly variable number of last-instar larvae by August. For inexplicable reasons, the adults that eclose in August–September do not generate abundant last-instar caterpillars by the end of the rainy season (November). *H. icasia* are often the last large and conspicuous caterpillars to be found in Santa Rosa as the dry season intensifies in December–January.

There are numerous species of evergreen trees, shrubs, and vines in Santa Rosa's deciduous forest. Many of these are host to many species of caterpillars during the rainy season, but none is a host plant to external feeders year-round. In general, even those species of Lepidoptera that have multiple and overlapping generations feeding on these plants during the rainy season do not also feed on them during the dry season. For example, the caterpillars of the large nymphalid butterfly *Archaeoprepona demophoon* are occasional to common (depending on the year) on the understory evergreen treelet *Ocotea veraguensis* (Lauraceae) from about the time that the rains begin until the first month of the dry season. The adult butterflies, however, are present in the habitat throughout the year. They do not oviposit on their sole host in Santa Rosa during the dry months, even though the host plant is fully covered with leaves of all ages.

One-Liners

Space limitation does not permit elaboration on all the patterns that are beginning to emerge in caterpillar phenology associated with the seasons in Santa Rosa dry forest. However, it may be useful to briefly mention some. For example, it is at the end of the rainy season that leaf-mining lepidoptera constitute the largest proportion of the active leaf miners (J. Memmot, personal communication). This may well be due to leaf miners being poor at dormancy (owing to their tiny size) and their populations therefore being severely depressed by the dry, hot, and leafless dry season.

Butterfly caterpillars (Heperoidea, Papilionoidea) as a group show the same phenological patterns as do externally feeding moth caterpillars as a group. Santa Rosa butterflies have highly migratory species with a phenology virtually identical to that of many sphingids (e.g., the nymphalids *Marpesia chiron, M. petreus*, the pierid *Aphrissa statira*), dormant largely univoltine species (e.g., the papilionids *Eurytides philolaus, E. epidaus, Papilio astyalus*), species that are reproductively dormant but active as adults during the dry season (e.g., many Hesperiidae, the nymphalid *Siproeta stelenes*, the pierid *Eurema daira*—and see Odendaal 1990; Jones 1987), etc.

There are many species of small moth caterpillars that live inside of fruits [e.g., *Ectomyelois muricis* (Phycitinae) in *Hymenaea courbaril* fruits, Janzen 1983c], rolled leaves (e.g., tens of species of pyraustine Crambidae), seeds (e.g., the Mexican jumping bean moth, *Cydia deslaisiana*, Olethreutidae; det. J. A. Powell), flowers [e.g., *Margaronia venatalis* (Pyraustinae) in fallen flowers of *Stemmadenia obovata*,] etc. With rare exceptions, none of these larvae displays any kind of dormancy as pupae or prepupae, and the pupae are not desiccation resistant at Santa Rosa temperatures and dry season humidities. As nearly as I can determine, most of these small moths pass the dry season as quiescent and hiding adults (Janzen 1987a) just as do many butterflies (Odendaal 1990; Jones 1987). It is particularly striking that if there is an aseasonal heavy rain in the middle of the dry season, or if cool weather appears for a week or so before the actual rains, adults of many of these species appear at the lights in forest that is leafless, dry, hot, and windy by day.

Although many moths and butterflies display quite uniform periods of duration of the pupal stage during the rainy season, there are a smattering of species that are extremely irregular with respect to this trait. This leads to quite irregular appearance of caterpillars of these species after the first generation of caterpillars following the rains. Various species and genera of Noctuidae, Arctiidae, and Thyrididae are particularly likely to have pupae that eclose after quite unpredictable intervals of weeks to months. Study of this phenomenon is particularly difficult, however, because it is never clear whether laboratory conditions have allowed or generated the eclosion cues used by the pupae.

Not all caterpillar phenology is as one would predict from the weather. Cossidae are the most startling. Adult cossids, which do not feed and therefore lay their eggs within a few days, appear at the lights in largest numbers (species and biomass) in the first month of the dry season (January). This seemingly nonsensical behavior may be related to the extreme windiness in January–February, a windiness that creates frequent wounds in living tree trunks through breaking and falling branches and trees. I suspect that many first instar cossid larvae gain entrance to the tree bark or trunk through these wounds. Additionally, once inside the living tree trunk, cossid larvae are probably the best shielded from the heat and desiccation of the dry season of any group of feeding caterpillars. Other exceptions are caterpillars that are specialists at mining in medium-sized fruits and seeds, many of which are produced only during the dry season at Santa Rosa. For this group of moths, it is clear that the rainy season is the inimical season, just as is the case with bruchid beetles (Janzen 1980b), spiny pocket mice (Janzen

1986b), and other dry forest seed predators (e.g., Janzen 1989b). At this point it should also be emphasized that not even all of the rainy season has physical conditions suitable for some groups of caterpillars. The short dry season in the middle of the rainy season is obviously a time of difficulty for first-instar larvae of external feeders. However, the aquatic caterpillars (e.g., nymphuline Crambidae) at Santa Rosa often find flowing water in the streams only during a few weeks to a month during the entire rainy season.

In Closing

Santa Rosa dry forest caterpillars clearly respond to and/or are affected by the seasonality of a wide variety of abiotic and biotic traits of their forest. My experiences in other dry tropical forests in Mexico, Guatemala, Venezuela, Australia, Kenya, Uganda, and India all suggest that the Santa Rosa situation is generally representative.

I do not see any evidence that there are new major ecological and evolutionary principles to be unveiled in tropical dry forest any more than in tropical rain forest. Of greater importance is coming to understand ecological, behavioral, and physiological interactions in such a manner that the information will be useful in allowing human management, manipulation, conservation, and use of these complex ecosystems without destroying them. For example, seasonal migration of Lepidoptera is hardly a new concept in biology. However, conservation planning can certainly make use of the fact that a conserved tropical dry forest needs to have some kind of sister rain forest area for its migrants to move to during the dry season. A budding butterfly farm industry can certainly make use of pupal dormancy as part of its manufacturing regime. It is difficult to plan tropical grade school exercises with caterpillars as the subjects (e.g., Janzen 1989a) in an area where no caterpillars are readily available for 8 months of the year. The seasonal timing of caterpillars and their parasitoids will definitely influence the planning of dry season irrigation of crops and tree plantations. Research on these considerations and many more like them needs to be done to bring tropical conserved wildlands into productive harmony with the remainder of tropical society.

Acknowledgements

This work was supported by NSF Grants BSR 80-11558, 83-08388, 84-03531, 86-10149, and 87-06155, and by the staff of the Area de Conservación Guanacaste of the Ministerio de Recursos Naturales, Energía y Minas, Costa Rica. M. Johnston, G. Stevens, G. Vega, W. Hallwachs, R. Espinosa, A. Espinosa, C. Moraga, E. Olson, and many others helped with caterpillar location and processing. W. Hallwachs aided greatly with editorial commentary.

Literature Cited

Allen, W. H. 1988. Biocultural restoration of a tropical forest. BioScience 38:156–161.

Bernays, E. A., and Janzen, D. H. 1988. Saturniid and sphingid caterpillars: Two ways to eat leaves. Ecology 69:1153–1160.

Chippendale, G. M., and Mahmalji, M. Z. 1987. Seasonal life history adaptations of a neotropical corn borer *Diatraea grandiosella*. Insect Sci. App. 8:501–506.

Davis, D. R., Clayton, D. H., Janzen, D. H., and Brooke, A. P. 1986. Neotropical Tineidae, II: biological notes and descriptions of two new moths phoretic on spiny pocket mice in Costa Rica (Lepidoptera: Tineoidea). Proc. Entomol. Soc. Washington 88:98–109.

DeVries, P. J. 1987. The Butterflies of Costa Rica and Their Natural History: Papilionidae, Pieridae, Nymphalidae. Princeton Univ. Press, Princeton, NJ.

Gomez, L. D. 1986. Vegetación de Costa Rica. Editorial Universidad Estatal a Distancia, San Jose, Costa Rica.

Farrow, R. A., and McDonald, G. 1987. Migration strategies and outbreaks of noctuid pests in Australia. Insect Sci. Appl. 8:531–542.

Holdridge, L. R., Grenke, W. C., Hatheway, W. H., Liang, T., and Tosi, J. A. 1971. Forest Environments in Tropical Life Zones. Pergamon, London.

Janzen, D. H. 1973. Sweep samples of tropical foliage insects: effects of seasons, vegetation types, elevation, time of day, and insularity. Ecology 54:687–708.

Janzen, D. H. 1980a. Heterogeneity of potential food abundance for tropical small land birds, pp. 545–552. In A. Keast and E. S. Morton (eds.), Migrant Birds in the Neotropics. Smithsonian Institution Press, Washington, D.C.

Janzen, D. H. 1980b. Specificity of seed-attacking beetles in a Costa Rican deciduous forest. J. Ecol. 68:929–952.

Janzen, D. H. 1981. Patterns of herbivory in a tropical deciduous forest. Biotropica 13:271–282.

Janzen, D. H. 1982. Guia para la identificación de mariposas nocturnas de la familia Saturniidae del Parque Nacional Santa Rosa, Guanacaste, Costa Rica. Brenesia 19/20:255–299.

Janzen, D. H. 1983a. Insects, pp. 619–645. In D. H. Janzen (ed.), Costa Rican Natural History. University of Chicago Press, Chicago.

Janzen, D. H. 1983b. Seasonal change in abundance of large nocturnal dung beetles (Scarabaeidae) in a Costa Rican deciduous forest and adjacent horse pasture. Oikos 41:274–283.

Janzen, D. H. 1983c. Larval biology of *Ectomyelois muriscis* (Pyralidae: Phycitinae), a Costa Rican fruit parasite of *Hymenaea courbaril* (Leguminosae: Caesalpinioideae). Brenesia 21:387–393.

Janzen, D. H. 1983d. *Pseudosphinx tetrio*, pp. 764–765. In D. H. Janzen (ed.), Costa Rican Natural History. University of Chicago Press, Chicago.

Janzen, D. H. 1984a. Two ways to be a tropical big moth: Santa Rosa saturniids and sphingids. Oxford Surv. Evol. Biol. 1:85–140.

Janzen, D. H. 1984b. Weather-related color polymorphism of *Rothschildia lebeau* (Saturniidae). Bull. Entomol. Soc. Am. 30:16–20.

Janzen, D. H. 1984c. Natural history of *Hylesia lineata* (Saturniidae: Hemileucinae) in Santa Rosa National Park, Costa Rica. J. Kan. Entomol. Soc. 57:490–514.

Janzen, D. H. 1985a. A host plant is more than its chemistry. Illinois Nat. Hist. Bull. 33:141–174.

Janzen, D. H. 1985b. On ecological fitting. Oikos 45:308–310.

Janzen, D. H. 1986a. Biogeography of an unexceptional place: What determines the saturniid and sphingid moth fauna of Santa Rosa National Park, Costa Rica, and what does it mean to conservation biology? Brenesia 25/26:51–87.

Janzen, D.H. 1986b. Mice, big mammals, and seeds: It matters who defecates what where, pp. 251–271. In A. Estrada and T. H. Fleming (eds.), Frugivores and Seed Dispersal. Dr. W. Junk, Dordrecht.

Janzen, D. H. 1986c. The eternal external threat, pp. 286–303. In M. E. Soule (ed.), Conservation Biology: The Science of Scarcity and Diversity, Sinauer, Sunderland, MA.

Janzen, D. H. 1987a. How moths pass the dry season in a Costa Rican dry forest. Insect Sci. Appl. 8:489–500.

Janzen, D. H. 1987b. Insect diversity of a Costa Rican dry forest: Why keep it, and how? Biol. J. Linn. Soc. 30:343–356.

Janzen, D. H. 1987c. When, and when not to leave. Oikos 49:241–243.

Janzen, D. H. 1988a. Tropical dry forests: the most endangered major tropical ecosystem, pp. 130–137. In E. O. Wilson (ed.), Biodiversity. National Academy Press, Washington, D.C.

Janzen, D. H. 1988b. Guanacaste National Park: Tropical ecological and biocultural restoration, pp. 143–192. In J. Cairns, Jr. (ed.), Rehabilitating Damaged Ecosystems, Vol. II. CRC Press, Boca Raton, FL.

Janzen, D. H. 1988c. Tropical ecological and biocultural restoration. Science 239:243–244.

Janzen, D. H. 1988d. Ecological characterization of a Costa Rican dry forest caterpillar fauna. Biotropica 20:120–135.

Janzen, D. H. 1988e. Management of habitat fragments in a tropical dry forest: Growth. Ann. Missouri Bot. Garden 75:105–116.

Janzen, D. H. 1988f. The migrant moths of Guanacaste. Orion Nature Quart. 7:38–41.

Janzen, D. H. 1988g. Complexity is in the eye of the beholder, pp. 29–51. In F. Almeda and C.M. Pringle (eds.), Tropical Rainforests: Diversity and Conservation. California Academy of Science and AAAS, San Francisco.

Janzen, D. H. 1989a. Affirmative action for insects in tropical national parks, pp. 579–588. In J. H. Bock and Y. B. Linhart (eds.), The Evolutionary Ecology of Plants. Westview Press, Boulder, CO.

Janzen, D. H. 1989b. Natural history of a wind-pollinated Central American dry forest legume tree (*Ateleia herbert-smithii* Pittier). Monog. Syst. Bot. Missouri Bot. Garden 29:293–376.

Janzen, D. H. 1991. How to save tropical biodiversity: The National Biodiversity Institute of Costa Rica. Am. Entomol. 37:159–171.

Janzen, D. H. 1992a. A south-north perspective on science in the management, use, and economic development of biodiversity, pp. 27–52. In O. T. Sandlund, K. Hindar and A. H. D. Brown (eds.) Conservation of biodiversity for sustainable development. Scandinavian Univ. Press, Oslo.

Janzen, D. H., and Hallwachs, W. 1992b. Ethical aspects of the impact of humans on biodiversity. Pont. Acad. Sci. Scripta Varia (in press).

Janzen, D. H., and Liesner, R. 1980. Annotated check-list of plants of lowland Guanacaste Province, Costa Rica, exclusive of grasses and non-vascular cryptogams. Brenesia 18:15–90.

Janzen, D. H., and Martin, P. S. 1982. Neotropical anachronisms: The fruits the gomphotheres ate. Science 215:19–27.

Janzen, D. H., and Waterman, P. G. 1984. A seasonal census of phenolics, fibre and alkaloids in foliage of forest trees in Costa Rica: Some factors influencing their distribution and relation to host selection by Sphingidae and Saturniidae. Biol. J. Linn. Soc. 21:439–454.

Jones, R. E. 1987. Reproductive strategies for the seasonal tropics. Insect Sci. Appl. 8:515–521.

Lemaire, C. 1987. Les Saturniidae Americains. 3. Ceratocampinae. Museo Nacional de Costa Rica, San Jose, Costa Rica.

Murphy, P. G., and Lugo, A. E. 1986. Ecology of tropical dry forest. Annu. Rev. Ecol. Syst. 17:67–88.

Nummelin, M. 1989. Seasonality and effects of forestry practices on forest floor arthropods in the Kibale Forest, Uganda. Fauna Norv. Ser. B 36:17–25.

Odendaal, F. J. 1990. The dry season influences reproductive parameters in female butterflies. Biotropica 22:100–102.

Paarman, W., and Stork, N. E. 1987. Seasonality of ground beetles (Coleoptera: Carabidae) in the rain forests of N. Sulawesi (Indonesia). Insect Sci. Appl. 8:483–487.

Powell, J. A., and Brown, J. W. 1990. Concentrations of lowland sphingid and noctuid moths at high mountain passes in eastern Mexico. Biotropica 22:310–315.

Ridpath, M. G., and Corbett, L. K. 1985. Ecology of the wet-dry tropics. Proc. Ecol. Soc. Aust. 13:1–333.

Tanaka, S., Wolda, H., and Denlinger, D. L. 1987. Seasonality and its physiological regulation in three neotropical insect taxa from Barro Colorado Island, Panama. Insect Sci. Appl. 8:507–514.

Tauber, M. J., Tauber, C. A., and Masaki, S. 1986. Seasonal Adaptations of Insects. Oxford University Press, New York.

Tangley, 1990. Cataloging Costa Rica's diversity. BioScience 40:633–636.

Uhl, C., and Buschbacher, R. 1985. A disturbing synergism between cattle ranch burning practices and selective tree harvesting in the eastern Amazon. Biotropica 17:265–268.

Winston, M. L. 1980. Seasonal patterns of brood rearing and worker longevity in colonies of the Africanized honey bee (Hymenoptera: Apidae) in South America. J. Kan. Entomol. Soc. 53:157–165.

Wolda, H. 1988. Insect seasonality: Why? Annu. Rev. Ecol. Syst. 19:1–18.

15

A Temperate Region View of the Interaction of Temperature, Food Quality, and Predators on Caterpillar Foraging

Nancy E. Stamp

Introduction

For caterpillars in temperate regions, three major factors shape their foraging patterns: temperature, food quality, and organisms that eat them. Because caterpillars are passive thermal bodies, they spend time, often a great deal of it, at temperatures below and sometimes above their thermal optimum (Sherman and Watt 1973; Casey 1976; Knapp and Casey 1986; Casey, Chapter 1). Raising the body temperature (and thus approaching the species' thermal optimum) by basking may speed up physiological processes enabling larvae to consume and digest food faster and, consequently, develop more quickly (Sherman and Watt 1973; Casey 1976; Capinera et al. 1980; Grossmueller and Lederhouse 1985). Leaf quality tends to decline as leaves age, with nitrogen and water concentrations decreasing and fiber and toughness increasing (reviewed by Mattson 1980; Scriber and Slansky 1981; Raupp and Denno 1983). Many insect herbivores have higher survivorship, grow faster, and gain more weight on young leaves than those eating mature leaves (Schweitzer 1979; Cates 1980; Raupp and Denno 1983). So, by both eating high quality food and spending as much time as possible at their thermal optimum, caterpillars can develop faster. Faster development may shorten the period that they are exposed to predators (Evans 1982) and parasitoids (Porter 1983) and maximize intake of high quality food, which may be available for only a limited time in the growing season (Feeny 1970; Stamp and Bowers 1990a). However, basking and acquiring high quality food, such as new leaves that tend to occur on stem tips, may expose caterpillars to predators and parasitoids. Caterpillars are more conspicuous and vulnerable in these locations (Heinrich 1979; Porter 1982, 1983; Grossmueller and Lederhouse 1985) and consequently may be forced into microhabitats that are suboptimal in terms of temperature and food (Damman 1987; Stamp and Bowers 1988).

So we see that the effects of these three factors on caterpillar behavior and growth are likely to interact. It is clear that both temperature and food quality can influence the interactions of caterpillars and predators and parasitoids (Evans 1982; Porter 1983; Campbell and Duffey 1981) and that the presence of predators may determine the food quality and thermal conditions that caterpillars experience (Damman 1987; Stamp and Bowers 1988). Furthermore, it is clear that the effects of food quality can be altered by temperature (Schramm 1972; Stamp 1990; Stamp and Bowers 1990b; Taylor and Shields 1990). But it is as yet unclear what the relative effects of temperature, food quality, and predators are and the degree to which the three factors interact. These questions must be addressed if we are to develop an understanding of how these three factors, which operate simultaneously, constrain foraging of temperate region caterpillars. When, for instance, do the effects of predators dominate the response by caterpillars to temperature and food quality?

My objectives here are to (1) describe some of the consequences for caterpillars of living in temperate regions; (2) address how the effects of temperature, food quality, and predators interact to constrain foraging patterns of caterpillars; and (3) identify areas for future study that are most promising in terms of developing a unified theory of plant–herbivore–predator interactions.

Living in Temperate Regions

Temperate Factors

The major factor differentiating temperate regions from tropical and arctic environments is the degree to which and when temperature fluctuates. For instance, insect herbivores in the deciduous forest region of North America are subjected to (1) mean daily temperatures from 2 to 20°C (National Climatic Center 1979), and average mid-summer temperatures ranging from 21 to 27°C (Greller 1988), (2) a growing season (nonfreezing temperatures) ranging from 5 to 7 months (Greller 1988), which corresponds to the season of their host plants, and (3) daylight during the growing season of 13–16 hours. In contrast, tropical insect herbivores experience daylight of about 12 hours and, regardless of altitude, considerably less annual and monthly fluctuations in temperature (Daubenmire 1978). Typically temperatures in tropical mesic lowland range between 26 and 30°C, with the warmest and coldest months often differing only by 2°C and diurnal variation about 10°C (Daubenmire 1978; Pomeroy and Service 1986). Arctic insects are subject to low annual mean temperature (below 0°C), low summer temperature (e.g., mean mid-summer temperatures of 3–12°C), and a short growing season (1.5 to 4 months) with 17 to 24 hours of daylight (Bliss 1988). These differences among these three environments have important consequences for both the ecology and evolution of insect foraging patterns.

Consequences of Diurnal Fluctuations in Temperature

For temperate lepidopteran species, alternating day:night temperatures can stimulate growth (Cook 1927; Matteson and Decker 1965; Eubank et al. 1973; Foley 1981; Rock 1985; Taylor and Shields 1990). Increased developmental rate can even occur when the minimum temperature in a cycling regime falls below the species' lower thermal threshold (Eubank et al. 1973; Rock 1985). Exceptions to this pattern, such as the flour moth *Anagasta kuehniella*, may reflect adaptations to environments that are relatively constant thermally (Siddiqui and Barlow 1973). An increase in developmental rate due to alternating temperatures may result in earlier maximum fecundity and increase the age for last reproduction (Siddiqui and Barlow 1972; Siddiqui et al. 1973). However, fluctuating temperatures may have such effects only when (1) temperature drops below the threshold for development to occur and (2) the average diurnal temperature is below the thermal maximum tolerated (Campbell et al. 1974). Because development time decreases with increasing amplitude of temperature variation, constant-temperature studies and degree-day models overestimate development time for insects subject to cycling temperatures that do not exceed thresholds (Taylor and Shields 1990).

Although the interactions of lepidopteran prey and invertebrate predators have not been examined under alternating temperatures, aphid predators tend to consume more prey when subjected to varying thermal regimes compared to constant temperatures (Ellingsen 1969) and, correspondingly, developmental rate of predators may be greater at alternating temperatures than at constant temperature (Butler and Ritchie 1970).

Consequences of the Length of the Growing Season

The length of the growing season determines how many generations of an insect herbivore are possible. Thermal conditions, food quality, and predator pressure are the major factors shaping developmental rate of insect herbivores (Scriber and Slansky 1981). The levels of each of these factors change over the growing season. In this section I discuss some of the consequences of the length of the growing season, in terms of these major constraints, on life history of Lepidoptera and foraging patterns of caterpillars.

Effects on Developmental Traits

The developmental rate curve and development accumulation for insects are a function of temperature. The developmental rate curve refers to the pattern of developmental rate with increasing temperature, and development accumulation is the sum at any instant in time of the previous developmental rates, which depends on the varying temperature of the insect (Taylor 1981). Developmental rate increases with increasing temperature to a maximum, achieved at the optimum temperature, above which developmental rate declines. With optimal temperatures ranging from 27.2 to 37.7°C, the maximum development per day for caterpillars ranges from 1.8 to 5.6% (Taylor 1981).

Examining the effect of different developmental rate curves on development accumulation at different locations in spring versus summer indicates one of the consequences of living in a temperate region. For example, in areas such as Oklahoma and Texas, selection may favor either polymorphism for the trait of optimal growing temperature between the generations within the growing season, or species replacement during the growing season, with a species with a low optimal growing temperature favored in the spring and a species with high optimum temperature favored in the summer (Taylor 1981).

Although minimum threshold temperature for development is recognized as playing a critical role in life history patterns and population dynamics of temperate insects, an upper-thermal limit for development may be just as important for temperate organisms. For instance, monarch butterflies (*Danaus plexippus*) have two successive generations between April and early June in northern Florida (where the latitude of the study sites was 29°N), whereas they have three generations between late May and the end of August in Wisconsin (latitude 43°N) (Malcolm et al. 1987). This difference in number of generations reflected their lower (12°C) and upper (30°C) thermal constraints for development. Specifically, by mid-June northern Florida was too hot for monarch larvae to survive, and adults migrated northward. Malcolm et al. (1987) felt that local food quality did not play a role in this pattern, but food quality could potentially modify such a pattern.

Food quality may affect expression of the two developmental traits, optimal temperature for development and the spread of the development rate curve (Taylor 1981). Any effect of food quality on these developmental traits should be greater in the spring, when development accumulation is especially sensitive to changes in these traits, than in the summer (Taylor 1981).

Food quality for insect herbivores is likely to differ between spring and summer (Mattson 1980; Scriber and Slansky 1981; Raupp and Denno 1983). The nitrogen concentration in foliage available to a spring generation of insect herbivores tends to be higher than it is in the summer (Mattson 1980; Scriber and Slansky 1981; Raupp and Denno 1983), in part because more nitrogen is available to plants in the spring (Mattson 1980). The quality of food available for spring versus summer generations may be especially different in years in which spring is early (Campbell 1989). Even if the summer generation feeds preferentially on newly expanded leaves, those leaves may have no higher nitrogen concentration than mature leaves (Stamp and Bowers 1990a). The profile of plant defenses changes over the growing season, so different generations of herbivores may face different arrays and amounts of allelochemicals. For example, in bracken fern, production of cyanogenic glycoside, pterosins, thiaminase, and extrafloral nectaries decline over the growing season, whereas flavonols, condensed tannins, lignin, and silica increase (Jones 1983).

Therefore, the combination of different quality of food in spring versus summer, with such differences often having significant effects on larval performance, and the high sensitivity of development accumulation in the spring to factors (e.g., food quality) affecting developmental traits may together select for different foraging patterns and expression of development traits between spring and summer generations. Potentially these effects could result in use of different plant parts or host plant species. For instance, the spring generation of the geometrid caterpillar *Nemoria arizonaria* feeds on oak catkins (*Quercus* spp.), whereas the summer generation eats mature leaves. At 25°C, the catkin diet promoted faster larval development (by 10 days on average), larger adult size, and greater fecundity than a leaf diet (Greene 1989). What the developmental rates are in the field for these two generations, for which both thermal conditions and food quality vary, is unknown (i.e., spring feeding on catkins vs. summer feeding on leaves), but a combination of different thermal optima and food quality may yield similar growth rates.

An exception to the quality of food resources varying more among generations than within generations in a growing season is found in the response by second and later generations of some species after the first generation defoliates its host plants. Depending on soil and climatic conditions, replacement of leaves may occur and these new leaves may repeat the spring pattern of having initially more water and nitrogen and less fiber than mature leaves (Pullin 1987). Refoliated leaves may (Schultz and Baldwin 1982) or may not (Faeth 1988) have more defensive chemicals than the first set of leaves of the year. If second and later generations of the year prefer regrowth foliage, they may have higher growth rates than they would eating mature leaves (Pullin 1987). Some herb- and tree-feeding insects have been found to prefer regrowth foliage over mature leaves (Carne 1965; Rockwood 1974; Webb and Moran 1978, Pullin 1987). In these cases, food quality may be relatively constant among generations of the year, in which case the developmental patterns should largely reflect thermal conditions.

Degree of Feeding Specialization

In temperate areas, with increasing latitude, especially at the transition zones where number of generations per growing season changes, there are some specific implications of these patterns of developmental rate for host plant exploitation by herbivores. Scriber and Lederhouse (1992) suggested that in areas in which two generations of an insect herbivore are possible on the host plant species that allows the highest growth rate but only one generation is possible on other host plant species, there will be selection for females to specialize by ovipositing on the host plant species providing the high larval growth rate. In contrast, in areas in which, for example, one generation is possible on a variety of host plant species but two generations are not possible on any of them, all of the potential host plant species should be utilized. As a consequence of these temperature-

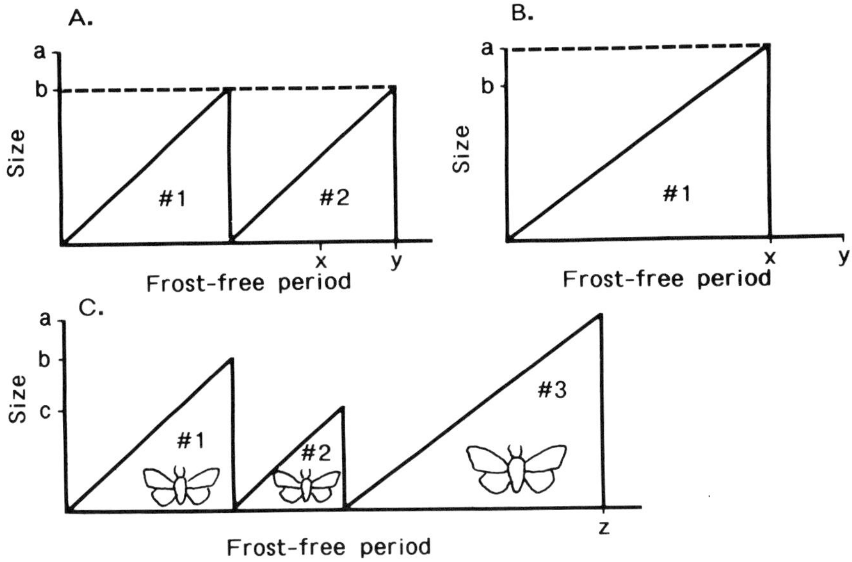

Figure 15.1. Insect size relative to number of generations per year. (A) With a growing season of *y* days, two generations each with mean final (pupal/adult) size of *b* may occur. (B) With a growing season of only *x* days, only one generation is possible, but with the slightly longer developmental period, pupal size is larger (*a*). (C) In South Carolina, fall webworm moths have three generations, which differ in adult size.

imposed and host plant-imposed limitations in temperate regions, alternating zones of specialization and generalization by particular insect herbivores should occur. Furthermore, Scriber and Lederhouse (1992) hypothesized that oviposition in warm microenvironments by females and basking behavior by larvae should be more prevalent in the zones of host plant specialization because it would increase the likelihood of fitting the generation(s) into the growing season. Thus, the combination of thermal and food quality limitations over the growing season may yield different foraging patterns (e.g., dietary specialization, degree of behavioral thermoregulation) regionally for a particular insect herbivore.

Adult Size

Changes in season length along a latitudinal gradient northward or southward may generate simple clines in developmental time for univoltine species or sawtooth" clines for multivoltine species (Masaki 1972; Roff 1980; Fig. 15. 1A). In multivoltine species, as season length decreases, one or more of the generations will decrease in length, with a consequent decrease in adult size. Thus, the fitness value associated with *n* generations per year rather than $n + 1$ will increase as season length decreases; at some point the fitness value associated with n generations will exceed the $n + 1$ generation's value (Roff 1980). Then with only n generations rather than $n + 1$ generations squeezed into a growing season, a shift to longer developmental time per generation occurs and thus larger body size is attained (Fig. 15.1B).

The effects of alterations in number of generations and consequently adult size, reflected in the changing pattern of sawtooth clines in developmental time, should be most apparent in temperate regions. The number of generations per growing season is quite variable there, with 1–4 generations compressed into several months (Slansky 1974). In contrast, in tropical regions most lepidopteran insects are multivoltine (often with 6–12 generations over most, if not all, of a year; Owen 1971; but see Janzen, Chapter 14). In the arctic region, generation time usually exceeds the short growing season, with most lepidopterans requiring 2 years per generation (Scott 1986) but some take even longer (e.g., up to 14 years, Kukal and Kevan 1987; Kukal, Chapter 16).

Generation packing into a growing season in temperate regions may be responsible for the seasonal dimorphism that has been observed in some moths. For example, fall webworm moths (*Hyphantria cunea*) have three generations in South Carolina, with the third generation adults (having pupated in the fall and eclosed in the spring) being the largest, and the second generation, late-summer adults being the smallest (D.C. Ferguson, unpublished data; Fig. 15.1C). The caterpillars of these three generations face different patterns of fluctuating and mean temperatures, different food quality and different sets of predators. Consequently, their exposure to predators and ability to evade predators will vary. Furthermore, the differing body sizes of these three generations of caterpillars are likely to influence their foraging behavior and susceptibility to predators and parasitoids (Feichtinger and Reavey 1989; Reavey, Chapter 8). Presumably the relatively large size of third-generation pupae is advantageous for surviving the winter months. The moderate size of the spring-feeding first generation may reflect the tradeoff between being large and thus more fecund versus fitness increase for individuals when there are three as opposed to two generations per growing season. The relatively small size of the mid-summer (June) second-generation adults may reflect close to the minimum size possible for a viable adult, which yields a relatively short developmental period that ensures that their offspring have time to grow large enough before pupation in the fall. Differing adult sizes between generations was reported for some geometrid species, in these cases with overwintering larvae completing development on new foliage in the spring to give rise to larger adults than those of the summer generation (Schweitzer 1977).

Predator Pressure

The set of enemies, or at least the pressure exerted by individual enemy species, is likely to change over the growing season and therefore be different for each herbivore generation in the season (Rabb and Lawson 1957; Rabb 1960; Morris

1972a,b; Marquis and Passoa 1989). Ants and predatory wasps provide two examples. In temperate regions, many perennials produce extrafloral nectaries for only a few to several weeks in the spring, which draw ants that defend these plants against insect herbivores (Tilman 1978). Thus, ants may be major predators for the spring generation of a caterpillar species, but they may be relatively less important for subsequent summer generations of caterpillars. In contrast, predatory wasps, which are especially important predators of caterpillars, are likely to exert increasing pressure over the growing season as wasp populations increase with the second generation of the summer (Rabb 1960; Gould and Jeanne 1984).

In sum, the length of the growing season in temperate regions may allow 1–4 generations of a species of insect herbivore. When it allows multiple generations, those generations face quite different thermal conditions, food quality, and predator pressures. In particular, the food quality available may vary more among generations of herbivores than within generations. The thermally sensitive developmental factors, coupled with the quality of food available, seem to be the key factors governing whether a species' developmental period (or periods, for multivoltine species) corresponds to all or just part of the climatically favorable period within a year. As a consequence of thermal conditions and food quality, developmental rates and adult size among generations may differ markedly. Larval foraging patterns may vary among generations in significant ways as well. Whether a particular species of insect herbivore is a specialist or generalist locally may depend on the limitations on developmental rate imposed by temperature and food quality.

Developmental Rates of Prey versus Predators

The thermal threshold above which development occurs and the number of degree-days required above that threshold for development to occur may often be higher for invertebrate predators and parasitoids than for their prey/hosts (Campbell et al. 1974; Campbell and Mackauer 1975; Obrycki and Tauber 1982). Consequently, in the spring, the higher thermal threshold of invertebrate enemies could prevent emergence or activity before the appearance of their hosts/prey (Campbell et al. 1974; Campbell and Mackauer 1975). With a lower developmental rate in invertebrate enemies, less of a toll on the host/prey population would occur at any one time and thus the lower developmental rate may increase the likelihood that some hosts/prey would be available.

Developmental rate of predators may also reflect the type of prey used, including its microhabitat. A study of 20 aphidophagous species indicated that for some predator species (syrphids), living in cool microhabitats and laying their eggs in large, ephemeral colonies of prey, the eggs exhibit low thresholds and quick development, whereas for other predators (coccinellids), frequenting warmer microhabitats and depositing eggs at lower prey densities, eggs have a higher threshold and longer developmental period (Honek and Kocourek 1988).

Since the thermal optima for invertebrate enemies of lepidopteran larvae are probably similar to those of invertebrate enemies of aphids, these findings suggest that invertebrate enemies of lepidopteran larvae are likely to exhibit (1) a developmental rate stimulated by fluctuating temperatures, when the temperature drops below the developmental threshold and the average temperature is below the thermal maximum, and (2) a developmental threshold and rate that are either similar to those of host/prey or dissimilar in a way that facilitates timely exploitation of such hosts/prey. Evidence regarding point 1 is lacking. Some evidence for point 2 consists of reports of predator activity lagging behind that of prey in the spring, especially when climatic conditions are somewhat cooler than the norm (Ayre and Hitchon 1986; Evans 1982; Porter 1982, 1983).

It is conceivable that along with global warming (or cooling), there may be other alterations to current climatic patterns. For example, with general warming, there may be an increased probability of extreme heat waves in temperate regions (Schneider 1989). Data representing a fourth of the land area on earth show that, while average annual maximum temperatures have remained unchanged over the last 40 years, minimum temperatures have risen and thus the mean temperature ranges have decreased (Karl et al. 1991). With decreasing amplitude of diurnal temperature variation, mean developmental time of insects increases whereas variation in developmental time decreases (Taylor and Shields 1990). But to what degree such alterations might influence prey and predator population dynamics, given the different developmental thresholds, developmental rates, thermal optima, and thermal maxima of prey and predators, is unclear.

Interactions of Temperature, Food Quality, and Enemies

The previous section indicates some ways in which temperature, food quality, and enemies may act separately and, to some extent, together as constraints on the foraging patterns of insect herbivores. But it does not indicate the relative magnitude of either the effects of each of these or the potential interactive effects. In this section, these three factors are considered in pairs and then together to develop a more realistic picture of the sum of effects of these constraints.

The desire for an assessment of the relative magnitude of the effects of temperature, food quality, and enemies on foraging patterns of insect herbivores is an old one and derived from major questions in ecology about what determines the distribution and abundance of organisms and what shapes community structure. Yet the debate as to which of these factors or what combination of them are primarily responsible for shaping foraging patterns continues (Bernays and Graham 1988). Even for well-studied systems, it is in general unclear (e.g., fall webworms, Morris 1967, 1972a,b; Morris and Fulton 1970).

As for the interactive effects among these factors, the occurrence of such effects seems likely. Numerous reviews and commentaries suggest why that is so and where we should look for interactive effects (Lawton and McNeil 1979; Price et al. 1980; Schultz 1983a; Barbosa and Saunders 1985; Duffey et al. 1986; Faeth 1987; Barbosa 1988a,b: Janzen 1988). But due to the difficulty of addressing multiple factor effects, the potential interactive effects are seldom explicitly examined.

Temperature and Food Quality

Although only a few published studies have examined the effect of food quality as a function of temperature on caterpillar performance, these studies indicate that thermal conditions influence the effect of food quality on caterpillar consumption and growth. Buckmoth caterpillars (*Hemileuca lucina*) are gregarious early-spring feeders on a rosaceous perennial (*Spiraea latifolia*). When reared at 20°C:15°C for 15L:9D, which represents cool overcast spring conditions for this species, the caterpillars grew equally well on new and mature leaves, regardless of whether they were in groups or solitary (Stamp and Bowers 1990b; Fig. 15.2). In contrast, when reared at a daytime temperature of 25°C, solitary larvae grew more slowly than the other diet-group combinations and, when larvae were reared at a daytime temperature of 30°C, those fed new leaves grew faster and gained more weight, regardless of whether they were solitary or in groups, than those given mature leaves. Another example of the effect of food quality as a function of temperature consists of a lepidopteran species having extra stadia when reared at cool temperatures on a particular diet. Fall armyworms (*Spodoptera frugiperda*) have on average 8.6 stadia when reared on cotton at 17°C compared to 7.2 stadia when fed corn; at warmer temperatures, they have similar number of stadia (with an average of 6) on these diets (Ali et al. 1990).

Just as there are few studies of the interaction of temperature and plant tissue on growth of caterpillars, there are few investigations of the interactions of temperature and specific aspects of food quality (e.g., secondary metabolites), yet evidence suggests that these kinds of interactions are important. For instance, lepidopteran larvae (*Laphygma exigua*) and coleopteran larvae (*Phaedon cochleariae*) fed a diet with high cellulose gained less weight at 10°C than larvae on other test diets, but they performed similarly to the other larvae when reared at higher temperatures (Schramm 1972). For the bollworm (*Helicoverpa [Heliothis] zea*) reared at one of four temperatures and seven diets with varying amounts of gossypol, an isoprenoid in cotton, there was a temperature × diet interaction for developmental rate (the reciprocal of time to pupation), but none for developmental time of treatments relative to that of the control (Thomas 1991).

The interactive effects of temperature and the flavonol rutin indicate some ways in which thermal conditions can influence the effects of diet. Rutin is common among terrestrial plants (Harborne 1979). It occurs in tomato plants,

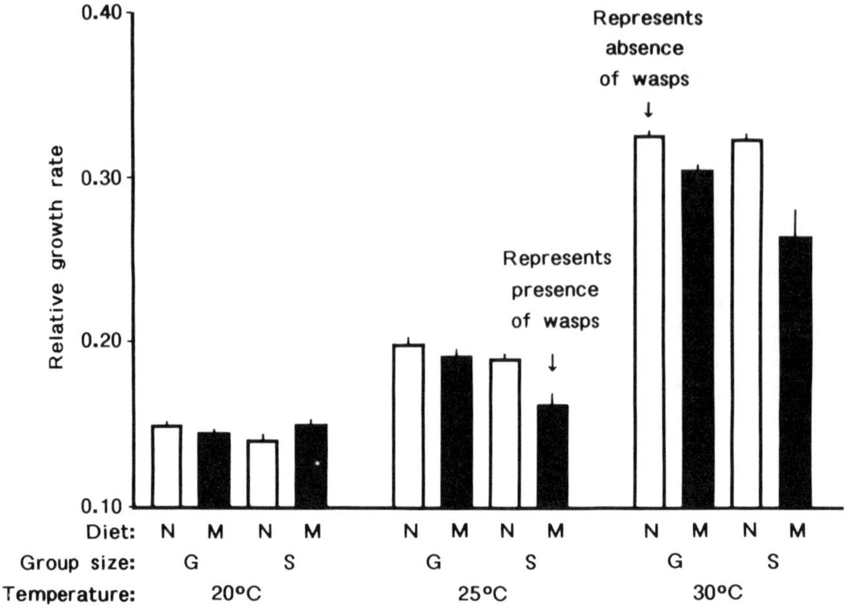

Figure 15.2. Relative growth rate of third instar buckmoth caterpillars when the effects of temperature, diet and larval group size are examined simultaneously. Means ± SE are shown, and $n = 10$. The photoperiod was a 15 hour day:9 hour night, with nighttime temperature of 15°C. Daytime temperatures varied: 20, 25, or 30°C. Larvae were either in groups of 10 (G) or solitary (S). They were fed either new (N) or mature leaves (M). The arrows indicate the most important contrasts: grouped larvae eating new leaves at 30°C represents the absence of predatory wasps, and solitary larvae eating mature leaves at 25°C represents conditions for larvae that seek refuge from the wasps. Data are from Stamp and Bowers (1990b).

with concentrations from 2–20 μmol/g fresh weight (Duffey et al. 1986). With rutin in artificial diet, tobacco hornworms (*Manduca sexta*), which use various solanaceous plants including tomato, experience an increase in stadium duration and a reduction in relative growth rate (mg dry weight of biomass gained/mg larva/day of stadium), and those patterns are a function of temperature (Stamp 1990; Stamp and Horwath 1992).

Because rutin can affect both the feeding and molting phases of the stadium (Stamp 1990; Stamp and Horwath 1992), it is also useful to determine performance indices in terms of the premolt period. Comparison of plots of growth efficiency (relative growth rate, or RGR-PM versus relative consumption rate, or RCR-PM) shows that temperature affects growth efficiency. For caterpillars tested at 20°C:15°C with a photoperiod of 15L:9D, increasing amounts of rutin in the diet up to 3 μmol/g fresh weight resulted in lower growth efficiency due to lower RCR-PM and RGR-PM (with RCR-PM dropping by 22% and RGR-PM by 20%, which were statistically significant by Tukey comparisons, $P<0.05$; Fig. 15.3A). At 6 μmol of rutin, RCR-PM increased by 7%, which probably indicates compensatory feeding. The change in RCR-PM at 6 μmol reflected consumption having increased by 15% while the number of days to premolt increased by only 5% (Fig. 15.4). With yet greater amounts of rutin in the diet, RCR-PM and RGR-PM declined much further. This pattern of initial decline in both performance values, followed by an increase at some point in RCR, which has the effect of holding RGR more or less steady, and then, with increasing concentrations of allelochemicals, followed by yet further decline of RCR and RGR, has been noted for an insect predator, too (Paradise and Stamp 1990). In contrast to this pattern, for caterpillars tested at 30°C:15°C, growth efficiency increased somewhat with 1 μmol of rutin (with RCR-PM increasing by 18% and RGR-PM by 11%), and then growth efficiency steadily decreased with increasing concentrations of rutin in the diet (Fig. 15.3B). The mechanisms generating these patterns are uncertain; that is, the decline in RCR and RGR may reflect deterrence by the allelochemical reducing RCR and thereby RGR, and/or pharmacological activity of rutin reducing RGR and consequently RCR (Slansky 1991). Nevertheless, it is clear that growth efficiency is a function of temperature.

Fluctuating temperatures yield a different growth efficiency plot than constant temperature. For caterpillars tested at 20°C (day and night), growth efficiency increased with 1 μmol of rutin, but with further increases in rutin, the values were clustered with that for the plain diet (Fig. 15.5A). The high value for the diet of 1 μmol of rutin reflects a greater change in growth (up 7% from no rutin in the diet then down 18% with 3 μmol of rutin) than in number of days to premolt (up 0.5% then down 4%), and it may represent feeding stimulation. For caterpillars tested at 23°C:15°C, which provided the same amount of thermal units per day, the RCR-PMs, RGR-PMs, and spread of these indices were greater than those at the constant temperature (Fig. 15.5B). The growth efficiencies for the plain and 1 μmol rutin diets were similar, but the growth efficiencies for increasing amounts of rutin were lower. For the most part, it was consumption and biomass gained rather than number of days to premolt that generated the different patterns of these thermal regimes. The differences in consumption and biomass gained for caterpillars on these thermal regimes are especially apparent for the diets of 6 and 12 μmol rutin (Fig. 15.6).

These growth efficiency plots indicate that response to allelochemicals by insect herbivores is influenced by the thermal regime. In particular, patterns of compensatory feeding and stimulatory feeding may be revealed at some temperatures but not others. If we are to understand the value of these feeding responses for caterpillars under natural conditions, it is necessary to examine these potential temperature–food quality interactions. In general, at constant temperatures below 25–30°C, developmental times for insects tend to be longer than those for fluctuating temperature regimes with the same mean temperatures, whereas for constant temperatures above 25–30°C, developmental times tend to be shorter compared

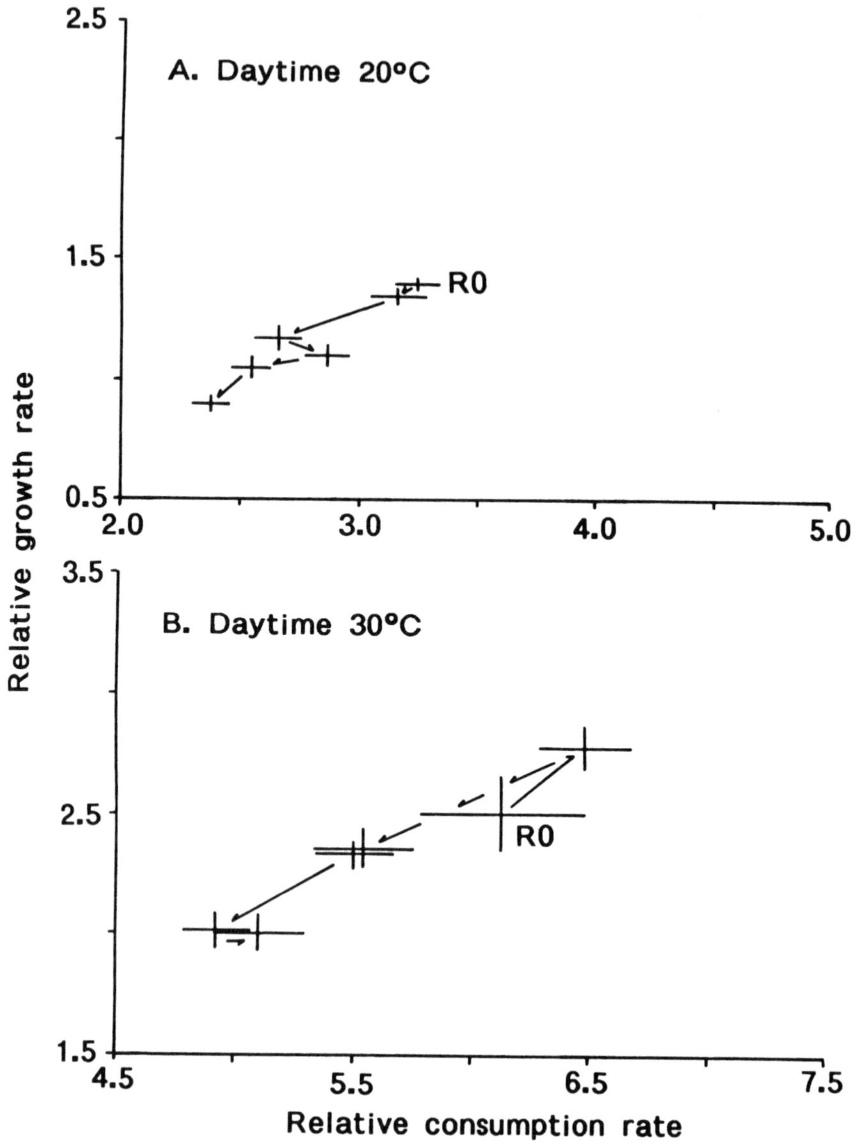

Figure 15.3. Growth efficiency (relative growth rate versus relative consumption rate) for third instar tobacco hornworms (*Manduca sexta*). The larvae were reared at either 20°C:15°C or 30°C:15°C, with a 15 hour day:9 hour night, on one of six diets: 0, 1, 3, 6, 12, and 18 μmol of rutin/g fresh weight of diet. The arrows indicate increasing levels of rutin in the diet, with RO indicating the diet with no rutin. Relative consumption rate (RCR-PM) = mg fresh weight of food consumed/(mg initial fresh weight of larva × days to head-capsule slippage, or premolt). Relative growth rate (RGR-PM) = mg fresh weight biomass gained/(mg initial fresh weight of larva × days to premolt). Means ± SE are shown, with $n = 15$. Both temperature and diet had significant effects on RCR-PM [two-way ANOVA, $F(1,166) = 795.68$, $P<0.001$ and $F(5,166) = 14.17$, $P<0.001$, respectively] and on RGR-PM (two-way ANOVA, $F=757.31$, $P<0.001$ and $F=20.15$, $P<0.001$, respectively).

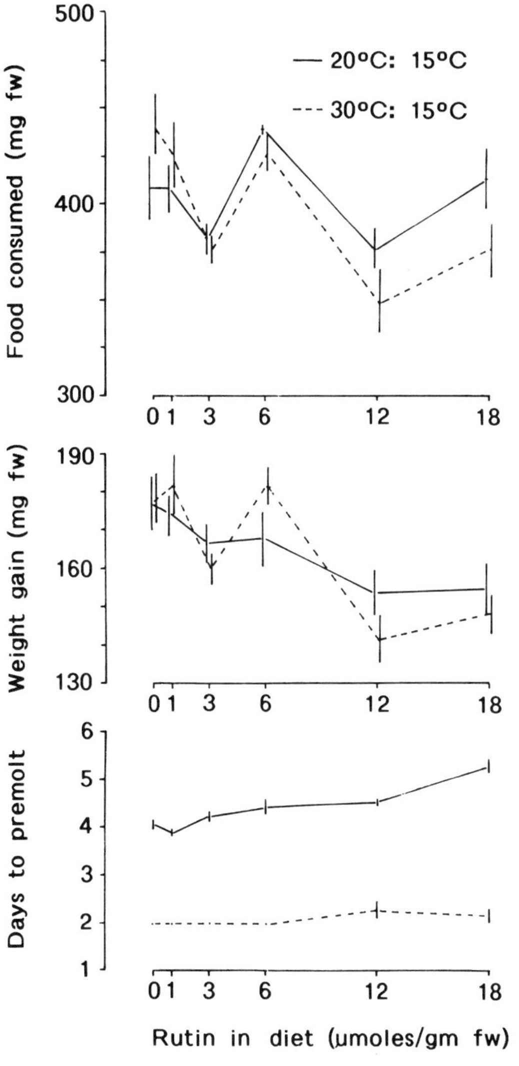

Figure 15.4. Food consumed, weight gained, and days to premolt for third instar tobacco hornworms (*Manduca sexta*). The larvae were reared at either 20°C:15°C or 30°C:15°C, with a 15 hour day:9 hour night, on one of six diets: 0, 1, 3, 6, 12, and 18 μmol of rutin/g fresh weight of diet. Means ± SE are shown, with $n = 15$. Temperature affected the number of days to premolt (two-way Kruskal-Wallis test, $\chi^2=125.25$, $P<0.001$) but not the amount of food consumed (two-way ANOVA, $F=0.46$, $P>0.45$) and weight gained (two-way ANOVA, $F=0.02$, $P>0.85$). Diet influenced all three variables (Kruskal-Wallis for days to premolt, $\chi^2=19.15$, $P<0.01$; two-way ANOVAS for food consumed, $F = 7.96$, $P<0.001$, and weight gained, $F = 10.46$, $P<0.001$).

to those of fluctuating temperatures (Hagstrum and Milliken 1991). Because performance by caterpillars under fluctuating thermal regimes is different than that of caterpillars reared at constant temperature, it is important to examine caterpillar response to food quality under fluctuating temperatures that approximate natural conditions.

Food Quality and Enemies

Avoiding enemies may force caterpillars to use poor quality food even though high quality food is available. Pyralid caterpillars (*Omphalocera munroei*) choose

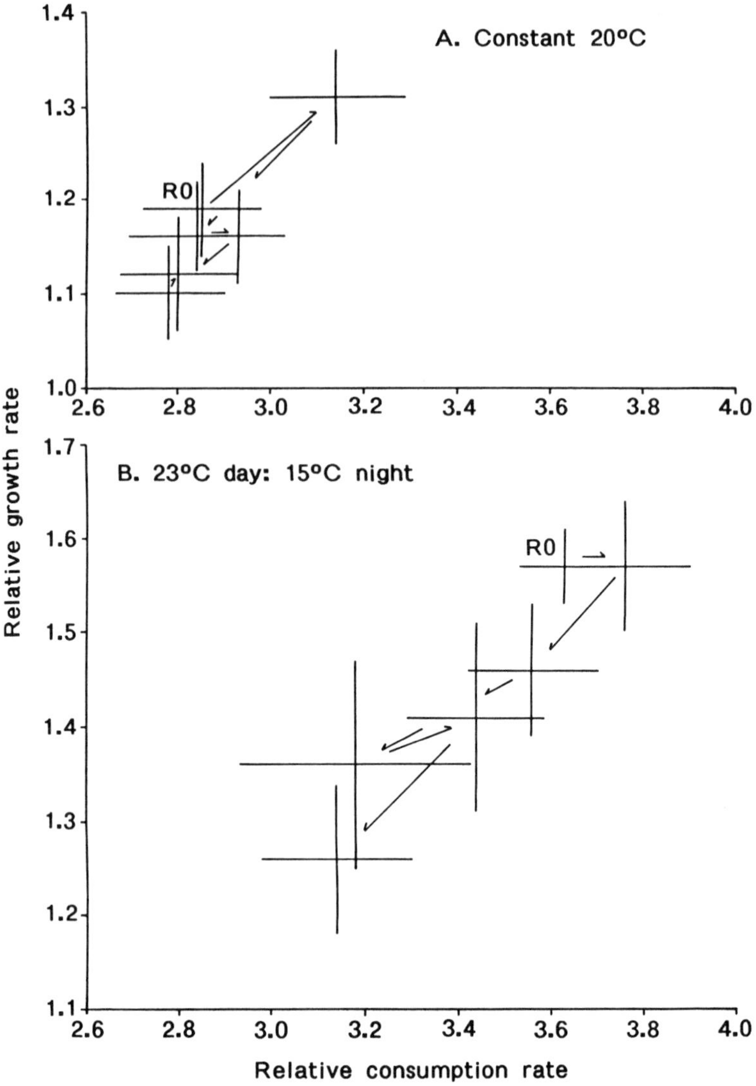

Figure 15.5. Growth efficiency (relative growth rate versus relative consumption rate) for third instar tobacco hornworms (*Manduca sexta*). The larvae were reared at either 20°C:15°C or 23°C:15°C, with a 15 hour day:9 hour night, on one of six diets: 0, 1, 3, 6, 12, and 18 μmol of rutin/g fresh weight of diet. The arrows indicate increasing levels of rutin in the diet, with RO indicating the diet with no rutin. Relative consumption rate (RCR-PM) = mg fresh weight of food consumed/(mg initial fresh weight of larva × days to premolt). Relative growth rate (RGR-PM) = mg fresh weight biomass gained/(mg initial fresh weight of larva × days to premolt). Means ± SE are shown, with $n = 14$. Both temperature and diet had significant effects on RCR-PM [two-way ANOVA, $F(1,151) = 43.59$, $P<0.001$ and $F(5,151) = 2.69$, $P<0.03$, respectively] and on RGR-PM (two-way ANOVA, $F=46.04$, $P<0.001$ and $F=3.76$, $P<0.01$, respectively).

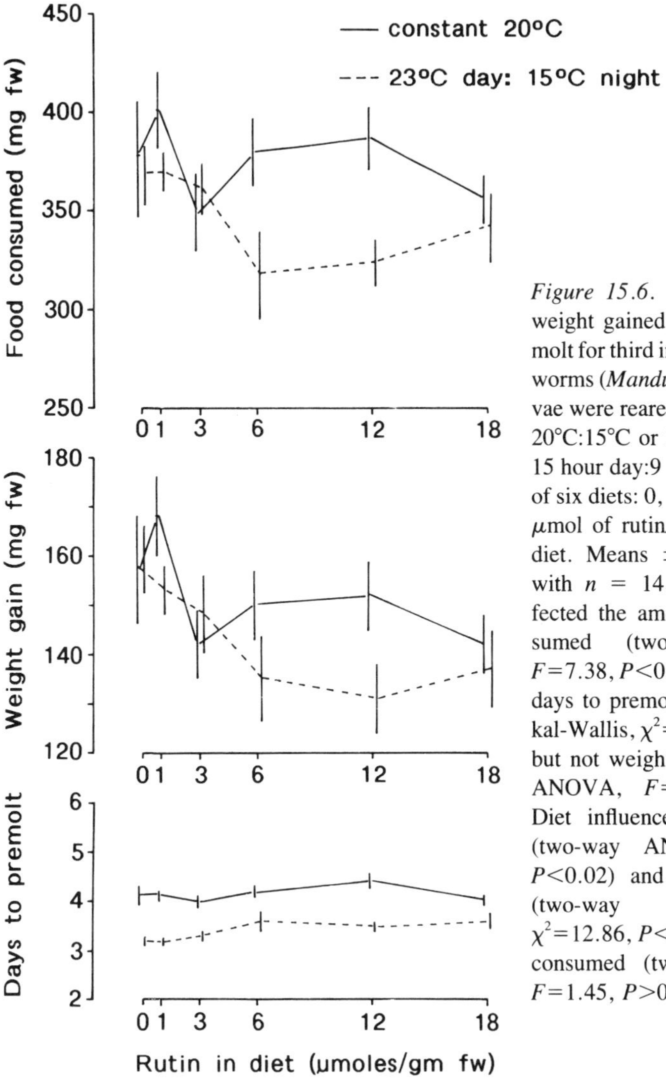

Figure 15.6. Food consumed, weight gained, and days to premolt for third instar tobacco hornworms (*Manduca sexta*). The larvae were reared at either constant 20°C:15°C or 23°C:15°C, with a 15 hour day:9 hour night, on one of six diets: 0, 1, 3, 6, 12, and 18 μmol of rutin/g fresh weight of diet. Means ± SE are shown, with $n = 14$. Temperature affected the amount of food consumed (two-way ANOVA, $F=7.38$, $P<0.01$) and number of days to premolt (two-way Kruskal-Wallis, $\chi^2=91.86$, $P<0.001$) but not weight gained (two-way ANOVA, $F=3.22$, $P>0.05$). Diet influenced weight gained (two-way ANOVA, $F=2.97$, $P<0.02$) and days to premolt (two-way Kruskal-Wallis, $\chi^2=12.86$, $P<0.05$) but not food consumed (two-way ANOVA, $F=1.45$, $P>0.20$).

feeding sites by defensibility rather than nutritional quality of leaves (Damman 1987). These leaf tiers are able to construct better shelters against enemies out of mature leaves than young leaves and consequently feed within the shelters on mature foliage, despite their ability to grow faster on young leaves. In this instance, the use of poorer quality (mature) leaves occurs whether enemies are present or not. It has been suggested for some late-season feeding caterpillars on oaks that avoiding enemies may even cause caterpillars to choose damaged leaves that as food lowers larval survivorship but which may be offset by the ease by which caterpillars can tie such leaves and thus construct a shelter from predators

(Hunter 1987). Rock cresses (*Arabis* spp.), compared to other crucifers, are nutritionally poor food for larvae of *Pieris napi*, but on *Arabis* plants, which usually are concealed by other plants, *P. napi* larvae are seldom parasitized in contrast to those on other crucifers where the caterpillars are more exposed (Ohsaki and Sato 1990). This pattern seems to be due in part to selection of oviposition sites by *P. napi* females (Ohsaki and Sato 1990) but probably also reflects larval behavior.

Although there are a number of studies demonstrating predator avoidance by various insects resulting in the prey moving to microhabitats where feeding and growth rates are lower (Sih 1987, and references therein), few studies report investigations for and evidence of such a phenomenon in caterpillars. Nonetheless, caterpillars may be forced into microhabitats that are suboptimal in terms of food and temperature by the presence of predators (Stamp and Bowers 1988), and that may be especially common for lepidopteran species in which larvae are gregarious and thus often conspicuous.

Temperature and Enemies

The interactive effect between temperature and enemies on caterpillars is likely to be strongest and consequently is best documented in early-spring feeding caterpillars. The thermal conditions during the spring growth period determine the window of time that the caterpillars are available or most vulnerable to enemies. Conditions that favor growth of the caterpillars, yet reduce activity of enemies and consequently slow population increase of enemies, narrow that window of time. In contrast, conditions that promote activity of enemies, even though they facilitate growth of the caterpillars, too, widen that window of time. Two examples illustrate the interaction between thermal conditions and enemies on caterpillars.

Checkerspot caterpillars (*Euphydryas aurinia*) are attacked by braconid parasitoids (*Apanteles bignellii*), with the level of parasitism of the final instar varying from spring to spring (Porter 1983). In springs when air temperature is relatively low and number of sunshine hours high, the pupating parasitoids, which are in cocoons in the shade, develop more slowly than the caterpillars, which bask between feeding bouts and consequently experience elevated body temperatures for several hours a day (Porter 1982, 1983). Consequently, many of the adult parasitoids emerge when their host is no longer at a suitable stage for them. In contrast, when air temperature is relatively high or when sunshine hours are few, the caterpillars and the parasitoid pupae develop at rates that promote synchrony between the adult parasitoids and the vulnerable caterpillar host stage. This link between spring thermal conditions and parasitism probably plays an important role in the dramatic population fluctuations of this caterpillar and others that are subject to similar abiotic and biotic conditions (Porter 1983; Stamp 1984).

In the second case, tent caterpillars (*Malacosoma americanum*), which are also early-spring feeders, are attacked by both ants and predatory stinkbugs, with the level of predation tied to spring thermal conditions. Soon after the caterpillars hatch (e.g., early April in New York), predatory stinkbugs (*Podisus* spp.) appear at the communal webs of the caterpillars (Evans 1982). During cold snaps in the spring (e.g., a 10-day period in May with mean air temperature of 9°C), activity of the stinkbugs is suppressed; but the tent caterpillars continue to feed through such inclement weather and thus grow beyond the size that the predatory stinkbugs can attack successfully (Evans 1982). Large caterpillars can defend themselves effectively and injure their attackers (Sullivan and Green 1950; Morris 1963; Evans 1982). Tent caterpillars are specialists on cherry (*Prunus* species). Black cherry (*Prunus serotina*) produces extrafloral nectaries, which are most active and thus attract ants over the first 3 weeks after budburst (Tilman 1978). The caterpillars are killed by the first ant to encounter them, until the caterpillars become too large for the ants to attack successfully (>2× length of ant), which occurs about 3 weeks after budburst (Ayre and Hitchon 1968; Tilman 1978). Correspondingly, production of extrafloral nectar declines then and ants are less frequent on the trees. Presumably, a cold snap while the extrafloral nectaries are still functional would reduce ant activity on the trees, in the same way that it suppresses activity of predatory stinkbugs, but have little effect on the tent caterpillars.

These two cases indicate that some thermal conditions that allow growth of caterpillars may not be conducive for activity of invertebrate enemies and thus in such situations enemies will have less impact on caterpillar populations. Thermal conditions also affect caterpillar–pathogen interactions. Temperature influences the progression of larval mortality due to pathogens (van Frankenhuyzen 1990). Warm temperatures can promote recovery by infected larvae (Tanada 1967; Inoue and Tanada 1977). In such cases, warm temperature would favor the caterpillars over the pathogens, which is the opposite of the effect of warm temperature on the caterpillar–predator/parasitoid interactions described above.

Temperature, Food Quality, and Enemies

Models of insect–plant interactions, due to the inherent simplification by focusing on only one or two of these factors (temperature, food quality and predators), provide some detailed predictions but also have limitations. For example, Schultz (1983b), in considering the tradeoffs that caterpillars face between obtaining high-quality food and avoiding predators, made some graphic hypotheses for caterpillars with different life-styles, but he had to conclude that a more quantitative approach was not possible because of the lack of sufficient data on the impact of leaf variability on caterpillar growth and behavior. Rhoades (1985) characterized foraging patterns of caterpillars as stealthy (solitary life-style) versus opportunistic (gregarious lifestyle), reflecting the hurdles that plants pose for insect herbivores in obtaining sufficient and appropriate food. Yet another more recent view, dubbed the exploiter model, focuses on the dynamics of leaf characteristics and the differential sensitivity of an insect herbivore to those characteristics, such that population growth is a function of how, when, and where the insect feeds and changes in leaf characteristics that influence the behavior and physiology of the herbivore (Jones and Coleman 1990). Most notably these two views fail to take into account in an adequate way the part that predators and parasitoids may play in constraining feeding by caterpillars. Another model, that considers the multiple factors giving rise to polyphagous herbivorous insects, incorporates the effects of predators and parasites, but it does not specify the effects of thermal conditions (Michaud 1990). The effects of temperature at the host plant and on its phytochemistry could easily be included (e.g., microclimate factor, annual variation factor), but even so this model does not take into account potential interactive effects. The voltinism-suitability hypothesis specifies that for any particular insect species the limitations imposed by thermal conditions and host plants on developmental rate (and consequently number of generations per year) generate latitudinal zones of specialists alternating with zones of generalists (Scriber and Lederhouse 1991). Because of insufficient data, the role of predators and parasitoids in this model was not adequately delineated.

Sufficient data are available for some caterpillar species to draw a composite picture for how temperature, food quality, and enemies individually, and therefore together, may constrain the foraging patterns of these caterpillars (e.g., fall webworms *Hyphantria cunea*, gypsy moth larvae *Lymantria dispar*, tent caterpillars *Malacosoma americanum*). However, few studies have examined two or more of these factors in combination, and there is evidence to suggest that when these factors operate simultaneously they may yield effects that would not necessarily be predicted based on studies examining their separate effects.

Once again, an early-spring feeding caterpillar provides an instructive case. Buckmoth caterpillars (*Hemileuca lucina*) are specialists on a rosaceous shrub (*Spiraea latifolia*) common in meadows in the northeastern United States. By basking at the tops of their host plant, these black, aggregating caterpillars obtain body temperatures as much as 5°C above air temperature (Stamp and Bowers 1990c). As the caterpillars develop, the host plant foliage matures, and the third instar caterpillars are the first stage that encounter both new and mature foliage. These larvae prefer the new leaves, which have higher percentage of nitrogen and water, and in doing so grow faster (Stamp and Bowers 1990a,b). Buckmoth caterpillars are attacked by various insects, including tachinid parasitoids, predatory stinkbugs, and predatory wasps. When vespid wasps (*Polistes dominulus* and *P. fuscatus*) attack these caterpillars, those caterpillars not immediately killed respond by leaving their feeding sites, characterized by new leaves in the sunlight, and moving to the interior of the hostplant, where the caterpillars are shaded and mature leaves predominate (Stamp and Bowers 1988). The caterpillars that escape by moving to the interior microhabitats do not gain as much weight as those that, when protected from the wasps, remain in the sunlight feeding on new leaves.

The effect of the presence of predatory wasps on caterpillar survivorship can be partitioned between the direct effect (immediate killing) and the indirect effect (due to reduced growth) (Stamp and Bowers 1991). For the experiment with buckmoth caterpillars and *Polistes* wasps (Stamp and Bowers 1988), the reduction in caterpillar survivorship due to the indirect effect was one-third of the overall reduction in caterpillar survivorship (Stamp and Bowers 1991). That is , just the presence of predators had a substantial negative effect on caterpillar survivorship, which was a consequence of the reduced growth of the escaping caterpillars, probably because they were forced into a cooler microhabitat with mature leaves.

Another experiment with buckmoth caterpillars corroborates this conclusion that predators can negatively affect caterpillar growth through determining what temperature and food quality caterpillars experience. In a laboratory experiment, the caterpillars were subjected to one of three temperatures, representing the range of spring thermal conditions, fed new or mature leaves, and kept either in groups, as they would be in the absence of predators, or by themselves, as they would be as a result of wasps driving them into the interior of the host plant (Stamp and Bowers 1990b). Caterpillars that were solitary, at the "shade" temperature and fed mature leaves, which represents presence of predators, took 4 more days to complete the test stadium and gained less weight than caterpillars that were grouped, at the "basking" temperature and fed new leaves, which represents absence of predators (Fig. 15.2).

Most gregarious larvae probably contend with relatively more of an indirect effect of predators (i.e., a change in larval behavior in the presence of predators leading to a reduction in larval survivorship due to a change in growth rate) than do solitary larvae. It may be that gregariousness in caterpillars is adaptive largely or in part because it allows them to use, at least temporarily, risky but high quality foraging sites. Because gregarious species tend to forage in exposed locations (where high-quality, new leaves often occur), the larvae frequently are conspicuous to predators and parasitoids. Being in a group lessens the likelihood of being the one attacked when a predator discovers that prey patch (Hamilton 1971; Foster and Treherne 1981) and increases the likelihood of having an opportunity to escape or defend successfully against a predator, in that grouped prey often detect predators sooner than do solitary individuals (Kenward 1978; Treherne and Foster 1980). The advantage of gregariousness increases when predator attacks are more frequent (Caraco et al. 1980), as they may be for prey in exposed locations. Once attacked, such larvae are likely to respond by moving to safer but lower quality foraging sites (Stamp and Bowers 1988). Thus, the indirect effect should be apparent and significant in gregarious species (except perhaps for those that feed within shelters, e.g., Damman 1987).

In contrast to gregarious larvae, a solitary life-style may reduce encounters with enemies but at the cost of limiting feeding periods or locations. Solitary species are usually cryptic in coloration and behavior (Jarvi et al. 1981; Heinrich and Collins 1983). They often feed on the underside of leaves (Greenberg and

Gradwohl 1980). Some spend considerable time moving among feeding sites, never staying long enough to finish a leaf, and some feed mainly at night (Heinrich 1979; Schultz 1983b). Such restricted feeding may slow growth (McGinnis and Kastings 1959; Herrebout et al. 1963; Schultz 1983b; Stamp and Wilkens, Chapter 9).

Species with solitary larvae tend to forage in ways such that the indirect effects of exposure to predators are relatively insignificant. Such larvae are likely to either escape the predator's notice or, when detected, be eaten by the predator. Some solitary species may exhibit behavioral flexibility, so that in the presence of predators, the prey alter their behavior. In such cases, an indirect effect may occur. For example, tiger swallowtail larvae (*Papilio glaucus*) bask, which significantly increases their growth rate; however, the mortality rates are higher at exposed locations (Grossmueller and Lederhouse 1985). The swallowtail larvae should alter their behavior by leaving exposed sites as predator pressure increases. But in other species (e.g., *Pero honestaria* of the forest understory, that move between daytime hideaways and nighttime feeding sites, Schultz 1983b), an indirect effect may not be detectable because the foraging pattern already reflects maximal presence of predators. That is, the antipredator behaviors are genetically fixed for a high level of predation. A fixed avoidance of predators may be a good strategy when prey have a low probability of surviving an encounter with a predator (Sih 1987). Therefore, even without harassment by predators, solitary-type larvae may exhibit behavior and suffer consequences that are similar to those of predator-harassed, gregarious caterpillars.

Thus, on a continuum, solitary summer-feeding caterpillars may be the least affected by potential interactions of the effects of thermal conditions, food quality, and predators, with warm conditions and relatively less thermal fluctuation then, relatively poor but uniform food quality, and relatively high constant numbers of predators. Leaf miners, which tend to complete their development within one leaf, represent the extreme example of this. In contrast, gregarious spring-feeding caterpillars are likely to be the most affected by the interaction of these three factors, due to greater thermal fluctuations, greater differences in food quality available in late spring, and greater variability in presence and numbers of predators.

Future Directions for Developing Plant–Herbivore–Predator Theory

How food quality influences the expression of two developmental traits, optimum temperature for development and spread of the developmental rate curve, is not known (Taylor 1981). Although it is often assumed that because poor food quality prolongs larval development the likelihood of predation and parasitism increases, neither the generality of that nor the part that temperature may play, through the effects of food quality on thermally sensitive developmental traits, has been demonstrated.

Size dimorphism among generations within lepidopteran species within a growing season is fairly common (D.C. Ferguson, unpublished data). Adult size is positively correlated with fecundity and thus affects potential fitness (Slansky 1982, and references therein). The size continuum of a particular species ranges from a minimal size for viability to a maximal size under ideal conditions (i.e., thermally optimal, highest quality tissue on preferred host plant, and absence of enemies). The three major constraints on final lepidopteran size then are (1) thermal conditions, which influence thermally related developmental traits; (2) food quality, which affects growth rate and final size, and with its effects influenced by temperature; and (3) for systems in which predators can exert an indirect effect on growth and survivorship of caterpillars, predation with potentially a negative correlation between adult size and indirect effects of predators. Size dimorphism among generations reflects each generation meeting the combination of these constraints but with a different outcome (i.e., adult size). Careful assessment of the relative magnitude of each of these constraints for each generation would increase our understanding of the ecology and evolution of lepidopteran life history patterns. It would also be interesting to contrast patterns for populations in natural ecosystems versus agroecosystems because in managed systems thermal conditions, food quality, and pressure by enemies are quite different from those in natural systems (Barbosa, Chapter 17).

By necessity, the effect of food quality on caterpillar foraging patterns and performance has been examined for the most part by having larvae feed on leaves classified by age or on artificial diet with the component of interest added to it. But neither approach tells us what synergistic or antagonistic effects result from the dynamic array of phytochemicals ingested by an insect herbivore (Berenbaum and Zangerl 1988). We need more studies that investigate such potential interactions among phytochemicals and, in turn, the likelihood of such interactions influencing the efficacy of the enemies of herbivores. For example, the toxicity of tomatine, an alkaloid in tomato leaves, to an ichneumonid parasitoid (*Hyposoter exiguae*) of the tomato fruitworm (*Heliothis zea*) is a function of the composite levels of tomatine and phytosterols (Campbell and Duffey 1981). Since temperature–allelochemical interactions are likely, such studies conducted under the appropriate thermal range of the herbivores would expand out understanding of "the real world" interactions tremendously. Similarly, in studies of interactive effects of host plant quality and pathogens on larval performance (e.g., Keating et al. 1988) and interactions between parasitoids and pathogens on each other (e.g., Hochberg 1991) and their host caterpillars, it would be quite useful to add the temperature factor, preferrably as a contrast between two representative thermal regimes (e.g., spring vs. summer) with fluctuating day:night temperatures.

The reduction in survivorship due to the indirect effect from the presence of predators is likely to be significant for many caterpillars, but in general, we do not know the degree to which the indirect effect is important in reducing growth rate and survivorship or what conditions magnify or reduce the indirect effect. By comparing prey species, data may be obtained to evaluate the hypothesis that gregariousness in caterpillars is adaptive in that it allows larvae to exploit risky but high-quality foraging sites, in which case individuals risk the direct effect of predators. As long as natural enemies are absent, foraging in sites that are suboptimal, in terms of thermal conditions and food quality, should not be induced in the caterpillars and thus the indirect effect should be zero. In contrast, solitary behavior in caterpillars may be adaptive because it reduces predation risk while accepting the consequences of foraging in microhabitats in which food quality and temperature are low and thus accepting the indirect effect of the presence of predators while reducing the direct effect. If that hypothesis is valid, then in most cases gregarious larvae and asocial larvae are expected to show different patterns in the relative proportions of the direct and indirect effects of predators. Furthermore, it seems likely that many caterpillars, especially the gregarious ones, may assess the gains and risks of foraging and change their behavior to obtain optimal feeding sites when it is advantageous.

As useful as the approach of concentrating on one or two of the factors affecting foraging patterns can be, we do need models that incorporate the effects of temperature, food quality, and predators, because all three can have such an impact on shaping foraging patterns of caterpillars and because the effects of these three factors interact. In addition to standard factorial design, a variety of useful techniques are available to examine the interaction of these factors, such as response surface methods (Box et al. 1978) and net effects diagrams (Jones 1990).

Conclusions

In temperate regions, temperature determines the activity of both herbivores and their enemies, with each having different thermal ranges and optima. Food quality affects larval growth, but that effect may often be a function of temperature. The limitations imposed by thermal conditions and host plants influence developmental rate and consequently the number of generations per year. How the developmental periods of generations fit into a growing season may affect foraging patterns and adult size. Poor quality food, by prolonging the developmental period, may facilitate predation and parasitism. Enemies can force herbivores into microhabitats that are suboptimal in terms of food quality and temperature, which may contribute to a reduction in survivorship of the herbivores. Thus, predators can have both direct and indirect effects on caterpillars; both of these effects are likely to be important in shaping foraging patterns of caterpillars. Adaptations that facilitate avoidance of enemies include occupying microhabitats that are suboptimal in terms of food quality and temperature.

It should be noted that while it seems reasonable to draw these conclusions, each of these statements is based on only a few studies and in some instances the evidence is indirect. Studies that examine these hypotheses directly will contribute markedly to development of the theory of interactions of plants, herbivores, and their predators and parasites.

Acknowledgments

I thank D. Bowers, R. Karban, and R. Wilkens for their helpful comments on this manuscript and D. Ferguson for providing unpublished data. My research has been supported by grants from the Whitehall Foundation and the National Science Foundation (BSR-8906259).

Literature Cited

Ali, A., Luttrell, R.G., and Schneider, J.C. 1990. Effects of temperature and larval diet on development of the fall armyworm (Lepidoptera: Noctuidae). Ann. Entomol. Soc. Am. 83:725–733.

Ayre, G.L., and Hitchon, D.E. 1968. The predation of tent caterpillars, *Malacosoma americana* (Lepidoptera: Lasiocampidae) by ants (Hymenoptera: Formicidae). Can. Entomol. 100:823–826.

Barbosa, P. 1988a. Some thoughts on "The evolution of host range." Ecology 69:912–915.

Barbosa, P. 1988b. Natural enemies and herbivore-plant interactions: Influence of plant allelochemicals and host specificity, pp. 201–229. In P. Barbosa and D. K. Letourneau (eds.), Novel Aspects of Insect-Plant Interactions. Wiley, New York.

Barbosa, P., and Saunders, J.A. 1985. Plant allelochemicals: Linkages between herbivores and their natural enemies. Rec. Adv. Phytochem. 19:107–137.

Berenbaum, M.R., and Zangerl, A.R. 1988. Stalemates in the coevolutionary arms race: Syntheses, synergisms, and sundry other sins, pp. 113–132. In K.C. Spencer (ed.), Chemical Mediation of Coevolution. Academic Press, San Diego.

Bernays, E., and Graham, M. 1988. On the evolution of host specificity in phytophagous arthropods. Ecology 69:886–892.

Bliss, L.C. 1988. Arctic tundra and polar desert biome, pp. 1–32. In M.G. Barbour and W.D. Billings (eds.), North American Terrestrial Vegetation. Cambridge University Press, Cambridge.

Box, G.E.P., Hunter, W.G., and Hunter, J.S. 1978. Statistics for Experimenters. Wiley, New York.

Butler, G.D., Jr., and Ritchie, P.L., Jr. 1970. Development of *Chrysopa carnea* at constant and fluctuating temperatures J. Econ. Entomol. 63:1028–1030.

Campbell, A., and Mackauer, M. 1975. Thermal constants for development of the pea aphid (Homoptera: Aphididae) and some of its parasites. Can. Entomol. 107:419–423.

Campbell, A., Frazer, B.D., Gilbert, N., Gutierrez, A.P., and Mackauer, M. 1974. Temperature requirements of some aphids and their parasites J. Appl. Ecol. 11:431–438.

Campbell, B.C., and Duffey, S.S. 1981. Alleviation of a-tomatine-induced toxicity to the parasitoid, *Hyposoter exiguae*, by phytosterols in the diet of the host, *Heliothis zea*. J. Chem. Ecol. 7:927–946.

Campbell, I.M. 1989. Does climate affect host-plant quality? Annual variation in the quality of balsam fir as food for spruce budworm. Oecologia 81:341–344.

Capinera, J.L., Wiener, L.F., and Anamosa, P.R. 1980. Behavioral thermoregulation by late-instar range caterpillar larvae *Hemileuca oliviae* Cockerell (Lepidoptera: Saturniidae) J. Kan. Entomol. Soc. 53:631–638.

Caraco, T., Martindale, S., and Pulliam, H.R. 1980. Avian flocking in the presence of a predator. Nature (London) 285:400–401.

Carne, P.B. 1965. Distribution of the eucalypt-defoliating sawfly *Perga affinis affinis*. Aust. J. Zool. 13:593–612.

Casey, T.M. 1976. Activity patterns, body temperature and thermal ecology in two desert caterpillars (Lepidoptera: Sphingidae). Ecology 57:485–497.

Cates, R.G. 1980. Feeding patterns of monophagous, oligophagous, and polyphagous insect herbivores: The effect of resource abundance and plant chemistry. Oecologia 46:22–31.

Cook, W.C. 1927. Some effects of alternating temperatures on the growth and metabolism of cutworm larvae. J. Econ. Entomol. 20:769–782.

Damman, H. 1987. Leaf quality and enemy avoidance by the larvae of a pyralid moth. Ecology 68:88–97.

Daubenmire, R. 1978. Plant Geography. Academic Press, New York.

Duffey, S.S., Bloem, K.A., and Campbell, B.C. 1986. Consequences of sequestration of plant natural products in plant-insect-parasitoid interactions, pp. 31–60. In D.J. Boethel and R.D. Eikenbary (eds.), Interactions of Plant Resistance and Parasitoids and Predators of Insects. Wiley, New York.

Ellingsen, I. 1969. Effect of constant and varying temperature on development, feeding, and survival of *Adalia bipunctata* L. (Col., Coccinellidae). Norsk. ent. Tidsskr. 16:121–125.

Eubank, W.P., Atmar, J.W., and Ellington, J.J. 1973. The significance and thermodynamics of fluctuating versus static thermal environments on *Heliothis zea* egg development rates. Environ. Entomol. 2:491–496.

Evans, E.W. 1982. Influence of weather on predator-prey relations: Stinkbugs and tent caterpillars. J. New York Entomol. Soc. 90:241–246.

Faeth, S.H. 1987. Community structure and folivorous insect outbreaks: The roles of vertical and horizontal interactions, pp. 135–171. In P. Barbosa and J.C. Schultz (eds.), Insect Outbreaks. Academic Press, New York.

Faeth, S.H. 1988. Plant-mediated interactions between seasonal herbivores: Enough for evolution or coevolution?, pp. 391–414. In K.C. Spencer (ed.), Chemical Mediation of Coevolution. Academic Press, San Diego.

Feeny, P. 1970. Seasonal changes in oak leaf tannins and nutrients as a cause of spring feeding by winter moth caterpillars. Ecology 51:565–581.

Feichtinger, V.E., and Reavey, D. 1989. Changes in movement, tying and feeding patterns as caterpillars grow: The case of the yellow horned moth. Ecol. Entomol. 14:471–474.

Foley, D.H. 1981. Pupal development rate of *Heliothis armiger* (Hubner) (Lepidoptera: Noctuidae) under constant and alternating temperatures. J. Aust. Entomol. Soc. 20:13–20.

Foster, W.A., and Treherne, J.E. 1981. Evidence for the dilution effect in the selfish herd from fish predation on a marine insect. Nature (London) 293:466–467.

Gould, W.P., and Jeanne, R.L. 1984. *Polistes* wasps (Hymenoptera: Vespidae) as control agents for lepidopteran cabbage pests. Environ. Entomol. 13:150–156.

Greenberg, R., and Gradwohl, J. 1980. Leaf surface specialization of birds and arthropods in a Panamanian forest. Oecologia 46:115–124.

Greene, E. 1989. A diet-induced developmental polymorphism in a caterpillar. Science 243:643–646.

Greller, A.M. 1988. Deciduous forest, pp. 287–316. In M.G. Barbour and W.D. Billings (eds.), North American Terrestrial Vegetation. Cambridge University Press, Cambridge.

Grossmueller, D.W., and Lederhouse, R.C. 1985. Oviposition site selection: An aid to rapid growth and development in the tiger swallowtail butterfly, *Papilio glaucus*. Oecologia 66:68–73.

Hagstrum, D.W., and Milliken, G.A. 1991. Modeling differences in insect developmental times between constant and fluctuating temperatures. Ann. Entomol. Soc. Am. 84:369–379.

Hamilton, W.D. 1971. Geometry for the selfish herd. J. Theor. Biol. 31:295–311.

Harborne, J.B. 1979. Variation in and functional significance of phenolic conjugation in plants. Rec. Adv. Phytochem. 12:457–474.

Heinrich, B. 1979. Foraging strategies of caterpillars. Leaf damage and possible predator avoidance strategies. Oecologia 42:325—337.

Heinrich, B., and Collins, S.L. 1983. Caterpillar leaf damage, and the game of hide-and-seek with birds. Ecology 64:592–602.

Herrebout, W.M., Kuyten, P.J., and de Ruiter, L. 1963. Observations on colour patterns and behaviour of caterpillars feeding on Scots pine. Arch. Neerl. Zool. 15:315–357.

Hochberg, M.E., 1991. Intra-host interactions between a braconid endoparasitoid, *Apanteles glomeratus*, and a baculovirus for larvae of *Pieris brassicae*. J. Anim. Ecol. 60:51–63.

Honek, A., and Kocourek, F. 1988. Thermal requirements for development of aphidophagous Coccinellidae (Coleoptera), Chrysopidae, Hemerobiidae (Neuroptera), and Syrphidae (Diptera): some general trends. Oecologia 76:455–460.

Hunter, M.D. 1987. Opposing effects of spring defoliation on late season oak caterpillars. Ecol. Entomol. 12:373–382.

Inoue, H., and Tanada, Y. 1977. Thermal therapy of the flacherie virus disease in the silkworm, *Bombyx mori*. J. Invert. Pathol. 29:63–68.

Janzen, D.H. 1988. On the broadening of insect-plant research. Ecology 69:905.

Jarvi, T., Sillen-Tullberg, B., and Wiklund, C. 1981. Individual versus kin selection for aposematic coloration: A reply to Harvey and Paxton. Oikos 37:393–395.

Jones, C.G. 1983. Phytochemical variation, colonization, and insect communities: The case of bracken fern, pp. 513–558. In R.F. Denno and M.S. McClure (eds.), Variable Plants and Herbivores in Natural and Managed Systems. Academic Press, New York.

Jones, C.G. 1991. Interactions among insects, plants, and microorganisms—A net effects perspective on insect performance, pp. 7–36. In P. Barbosa, V.A. Krischik, and C.G. Jones (eds.), Microbial Mediation of Plant-Herbivore Interactions. Wiley, New York.

Jones, C.G., and Coleman, J.S. 1991. Plant stress and insect herbivory: Toward an integrated perspective, pp. 249–280. In H.A. Mooney, W.E. Winner, and E.J. Pell (eds.), Response of Plants to Multiple Stresses. Academic Press, San Diego.

Keating, S.T., Yendol, W.G., and Schultz, J.C. 1988. Relationship between susceptibility of gypsy moth larvae (Lepidoptera: Lymantriidae) to a baculovirus and host plant foliage constituents. Environ. Entomol. 17:952–958.

Kenward, R.E. 1978. Hawks and doves: Factors affecting success and selection in goshawk attacks on wood pigeons. J. Anim. Ecol. 47:449–460.

Knapp, R., and Casey, T.M. 1986. Thermal ecology, behavior, and growth of gypsy moth and eastern tent caterpillars. Ecology 67:598–608.

Kukal, O., and Kevan, P.G. 1987. The influence of parasitism on the life history of a high arctic insect, *Gynaephora groenlandica* (Wocke) (Lepidoptera: Lymantridae). Can. J. Zool. 65:156–163.

Lawton, J.H., and McNeil, S. 1979. Between the devil and the deep blue sea: On the problem of being a herbivore. Symp. Br. Ecol. Soc. 20:223–244.

Malcolm, S.B., Cockrell, B.J., and Brower, L.P. 1987. Monarch butterfly voltinism: Effects of temperature constraints at different latitudes. Oikos 49:77–82.

Marquis, R.J., and Passoa, S. 1989. Seasonal diversity and abundance of the herbivore fauna of striped maple *Acer pennsylvanicum* L. (Aceraceae) in western Virginia. Am. Midl. Nat. 122:313–320.

Masaki, S. 1972. Seasonal and latitudinal adaptations in the life cycles of crickets, pp. 72–100. In H. Dingle (ed.), Evolution of Insect Migration and Diapause. Springer-Verlag, Berlin.

Matteson, J.W., and Decker, G.C. 1965. Development of the European corn borer at controlled constant and variable temperatures. J. Econ. Entomol. 58:344–349.

Mattson, W.J., Jr. 1980. Herbivory in relation to plant nitrogen content. Annu. Rev. Ecol. Syst. 11:119–161.

McGinnis, A.J., and Kasting, R. 1959. Nutrition of the pale western cutworm, *Agrotis orthogonia* Morr. (Lepidoptera: Noctuidae). Can. J. Zool. 37:259–266.

Michaud, J.P. 1990. Conditions for the evolution of polyphagy in herbivorous insects. Oikos 57:278–279.

Morris, R.F. 1963. The effect of predator age and prey defense on the functional response of *Podisus maculiventris* Say to the density of *Hyphantria cunea* Drury. Can. Entomol. 95:1009–1020.

Morris, R.F. 1967. Influence of parental food quality on the survival of *Hyphantria cunea*. Can. Entomol. 99:24–33.

Morris, R.F. 1972a. Predation by insects and spiders inhabiting colonial webs of *Hyphantria cunea*. Can. Entomol. 104:1197–1207.

Morris, R.F. 1972b. Predation by wasps, birds, and mammals on *Hyphantria cunea*. Can. Entomol. 104:1581–1591.

Morris, R.F., and Fulton, W.C. 1970. Models for the development and survival of *Hyphantria cunea* in relation to temperature and humidity. Mem. Entomol. Soc. Can. No. 70.

National Climatic Center. 1979. Comparative climatic data for the United States through 1978. Natl. Oceanic Atmospheric Adm., Asheville, NC.

Obrycki, J.J., and Tauber, M.J. 1982. Thermal requirements for development of *Hippodamia convergens* (Coleoptera: Coccinellidae). Ann. Entomol. Soc. Am. 75:678–683.

Ohsaki, N., and Sato, Y. 1990. Avoidance mechanisms of three *Pieris* butterfly species against the parasitoid wasp *Apanteles glomeratus*. Ecol. Entomol. 15:169–176.

Owen, D.F. 1971. Tropical Butterflies. Clarendon Press, Oxford.

Paradise, C.J., and Stamp, N.E. 1990. Variable quantities of toxic diet cause different degrees of compensatory and inhibitory responses by juvenile praying mantids. Entomol. Exp. Appl. 55:213–222.

Pomeroy, D.E., and Service, M.W. 1986. Tropical Ecology. Longman, Essex.

Porter, K. 1982. Basking behaviour in larvae of the butterfly *Euphydryas aurinia*. Oikos 38:308–312.

Porter, K. 1983. Multivoltinism in *Apanteles bignellii* and the influence of weather on synchronisation with its host *Euphydryas aurinia*. Entomol. Exp. Appl. 34:155–162.

Price, P.W., Boulton, C.E., Gross, P., McPheron, B.A., Thompson, J.N., and Weis, A.E. 1980. Interactions among three trophic levels: Influence of plants on interactions between insect herbivores and natural enemies. Annu. Rev. Ecol. Syst. 11:41–65.

Pullin, A.S. 1987. Changes in leaf quality following clipping and regrowth of *Urtica dioica*, and consequences for a specialist insect herbivore, *Aglais urticae*. Oikos 49:39–45.

Rabb, R.L. 1960. Biological studies of *Polistes* in North Carolina (Hymenoptera: Vespidae). Ann. Entomol. Soc. Am. 53:111–121.

Rabb, R.L., and Lawson, F.R. 1957. Some factors influencing the predation of *Polistes* wasps on the tobacco hornworm. J. Econ. Entomol. 50:778–784.

Raupp, M.J., and Denno, R.F. 1983. Leaf age as a predictor of herbivore distribution and abundance, pp. 91–124. In R.F. Denno and M.S. McClure (eds.), Variable Plants and Herbivores in Natural and Managed Systems. Academic Press, New York.

Rhoades, D.F. 1985. Offensive-defensive interactions between herbivores and plants: Their relevance in herbivore population dynamics and ecological theory. Am. Nat. 125:205–238.

Rock, G.C. 1985. Thermal and thermoperiodic effects on larval and pupal development and survival in tufted apple bud moth (Lepidoptera: Tortricidae). Environ. Entomol. 14:637–640.

Rockwood, L.L. 1974. Seasonal changes in the susceptibility of *Crescentia alata* leaves to the flea beetle *Oedionychus* sp. Ecology 55:142–148.

Roff, D. 1980. Optimizing development time in a seasonal environment: The 'ups and downs' of clinal variation. Oecologia 45:202–208.

Schneider, S.H. 1989. Global Warming: Are We Entering the Greenhouse Century? Sierra Club, San Francisco.

Schramm, U. 1972. Temperature-food interaction in herbivorous insects. Oecologia 9:399–402.

Schultz, J.C. 1983a. Impact of variable plant defensive chemistry on susceptibility of insects to natural enemies, pp. 37–54. In P.A. Heden (ed.), Plant Resistance to Insects. Am. Chem. Soc., Washington, D.C.

Schultz, J.C. 1983b. Habitat selection and foraging tactics of caterpillars in heterogeneous trees, pp. 61–90. In R.F. Denno and M.S. McClure (eds.), Variable Plants and Herbivores in Natural and Managed Systems. Academic Press, New York.

Schultz, J.C., and Baldwin, I.T. 1982. Oak leaf quality declines in response to defoliation by gypsy moth larvae. Science 217:149–151.

Schultz, J.C., and Lechowicz, M.J. 1986. Hostplant, larval age, and feeding behavior influence midgut pH in the gypsy moth (*Lymantria dispar*). Oecologia 71:133–137.

Schweitzer, D.F. 1977. Larval hibernation of Geometridae in eastern United States. J. Lepid. Soc. 31:71–72.

Schweitzer, D.F. 1979. Effects of foliage age on body weight and survival in larvae of the tribe Lithophanini (Lepidoptera: Noctuidae). Oikos 32:403–408.

Scott, J.A. 1986. The Butterflies of North America: A Natural History and Field Guide. Stanford University Press, Stanford.

Scriber, J.M., and Lederhouse, R.C. 1992. The thermal environment as a resource dictating geographic patterns of feeding specialization of insect herbivores, pp. 429–466. In M.D. Hunter, T. Ohguishi, and P.W. Price (eds.), Effects of Resource Distribution on Animal-Plant Interactions. Academic Press, San Diego.

Scriber, J.M., and Slansky, F., Jr. 1981. The nutritional ecology of immature insects. Annu. Rev. Entomol. 26:183–211.

Sherman, P.W., and Watt, W.B. 1973. The thermal ecology of some *Colias* butterflies. J. Comp. Physiol. 83:25–40.

Siddiqui, W.H., and Barlow, C.A. 1972. Population growth of *Drosophila melanogaster* (Diptera: Drosophilidae) at constant and alternating temperatures. Ann. Entomol. Soc. Am. 65:993–1001.

Siddiqui, W.H., and Barlow, C.A. 1973. Population growth of *Anagasta kuehniella*

(Lepidoptera: Pyralidae) at constant and alternating temperatures. Ann. Entomol. Soc. Am. 66:579–585.

Siddiqui, W.H., Barlow, C.A., and Randolph, P.A. 1973. Effects of some constant and alternating temperatures on population growth of the pea aphid, *Acyrthosiphon pisum* (Homoptera: Aphididae) Can. Entomol. 105:145–156.

Sih, A. 1987. Predation and prey lifestyles: an evolutionary and ecological overview, pp. 203–224. In W.C. Kerfoot and A. Sih (eds.), Predation: Direct and Indirect Impacts on Aquatic Communities. University Press New England, Hanover.

Slansky, F., Jr. 1974. Relationship of larval food-plants and voltinism patterns in temperate butterflies. Psyche 81:243–253.

Slansky, F., Jr. 1982. Insect nutrition: An adaptationist's perspective. Florida Entomol. 65:45–71.

Slansky, F., Jr. 1991. Allelochemical—nutrient interactions in herbivore nutritional ecology pp. 135–174. In G.A. Rosenthal and M.R. Berenbaum (eds.), Herbivores: Their Interaction with Secondary Plant Metabolites, Vol. II: Evolutionary and Ecological Processes. Academic Press, New York.

Stamp, N.E. 1984. Interactions of parasitoids and checkerspot caterpillars *Euphydryas* spp. (Nymphalidae). J. Res. Lepid. 23:2–18.

Stamp, N.E. 1990. Growth versus molting time of caterpillars as a function of temperature, nutrient concentration and the phenolic rutin. Oecologia 82:107–113.

Stamp, N.E., and Bowers, M.D. 1988. Direct and indirect effects of predatory wasps on gregarious larvae of the buckmoth, *Hemileuca lucina* (Saturniidae). Oecologia 75:619–624.

Stamp, N.E., and Bowers, M.D. 1990a. Phenology of nutritional differences between new and mature leaves and its effect on caterpillar growth. Ecol. Entomol. 15:447–454.

Stamp, N.E., and Bowers, M.D. 1990b. Variation in food quality and temperature constrain foraging of gregarious caterpillars. Ecology 71:1031–1039.

Stamp, N.E., and Bowers, M.D. 1990c. Body temperature, behavior and growth of early-spring caterpillars (*Hemileuca lucina*: Saturniidae). J. Lepid. Soc. 44:143–155.

Stamp, N.E., and Bowers, M.D. 1991. Indirect effect on survivorship of caterpillars due to presence of invertebrate predators. Oecologia 88:325–330.

Stamp, N.E., and Horwath, K.L. 1992. Interactive effects of temperature and concentration of the flavonol rutin on growth, molt and food utilization of *Manduca sexta* caterpillars. Entomol. Exp. Appl. (in press).

Sullivan, C.R., and Green, G.W. 1950. Reactions of larvae of the eastern tent caterpillar and of the spotless fall webworm to pentatomid predators. Can. Entomol. 82:52.

Tanada, Y. 1967. Effect of high temperature on the resistance of insects to infectious disease. J. Sericult. Sci. Jpn. 36:333–339.

Taylor, F. 1981. Ecology and evolution of physiological time in insects. Am. Nat. 117:1–23.

Taylor, P.S., and Shields, E.J. 1990. Development of the armyworm (Lepidoptera:

Noctuidae) under fluctuating daily temperature regimes. Environ. Entomol. 19:1422–1431.

Thomas, W.M. 1991. Modeling the effect of temperature and gossypol concentration on developmental rate of *Helicoverpa zea* (Boddie) (Lepidoptera: Noctuidae). J. Econ. Entomol. 84:466–469.

Tilman, D. 1978. Cherries, ants and tent caterpillars: Timing of nectar production in relation to susceptibility of caterpillars to ant predation. Ecology 59:686–692.

Treherne, J.E., and Foster, W.A. 1980. The effects of group size on predator avoidance in a marine insect. Anim. Behav. 28:1119–1122.

van Frankenhuyzen, K. 1990. Effect of temperature and exposure time on toxicity of *Bacillus thuringiensis* Berliner spray deposits to spruce budworm, *Choristoneura fumiferana* Clemens (Lepidoptera: Tortricidae). Can. Entomol. 122:69–75.

Webb, J.W., and Moran, V.C. 1978. The influence of the host plant on the population dynamics of *Acizzia russellae* (Homoptera: Psyllidae). Ecol. Entomol. 3:313–321.

16

Biotic and Abiotic Constraints on Foraging of Arctic Caterpillars

Olga Kukal

Introduction

Little is known about the foraging of arctic caterpillars. Most of the information available pertains to a single species, the arctic woolly-bear caterpillar (*Gynaephora groenlandica*: Lymantriidae), which will be the main focus of this chapter and will provide a useful contrast to caterpillars foraging in other biomes.

Low temperature, short growing season, and seasonal photoperiod limit the arctic radiation of most insects (Danks 1981). However, I propose that the life cycle and foraging of the arctic woolly-bear caterpillar is biotically, rather than abiotically controlled, although the biotic constraints are closely linked to abiotic constraints, particularly to temperature. I have chosen here to contrast the biotic and abiotic constraints on the foraging of *G. groenlandica* because of the frequent assumption that arctic organisms are primarily under abiotic controls, whereas biotic factors, such as food and natural enemies, play a lesser role.

Gynaephora groenlandica is the most northerly representative of the order Lepidoptera (Curtis 1835; Ferguson 1978) and is the largest terrestrial invertebrate reaching the northern limit of vegetation (Downes 1964, 1965). At high latitudes, *G. groenlandica* is subjected to severe constraints on development including: (1) feeding season curtailed by parasitism, (2) feeding activity regulated by behavioral thermoregulation contingent on sunshine, (3) growth rate dependent on quality and availability of host plant, (4) assimilation efficiency dependent on low temperature threshold for maintenance metabolism, and (5) preparation for overwintering by carbohydrate storage, cryoprotectant synthesis, and tissue changes. These constraints will be examined in the following sections.

Parasitoids and Life Cycle of *Gynaephora groenlandica*

One of the major biotic constraints on *Gynaephora groenlandica*, is parasitism. Two parasitoids attack *G. groenlandica*, an ichneumon wasp, *Hyposoter pectina-*

Table 16.1. Partial Life Table for Gynaephora groenlandica *at Alexandra Fiord, Ellesmere Island, Canada*[a]

Stage	No. surviving to beginning of stage	Cause of mortality	Survival rate	Cumulative mortality (%)
Eggs	1000	Inviability	0.71	28.6
Larvae (I–III)	714	*Hyposoter*	0.81	53.1
Larvae (III–VI)	579	*Exorista*	0.81	42.1
Pupae	469	*Exorista*		
	315	Birds	0.62	70.8
Adults	292		0.57	74.9
Females (× 2)	215[b]			

[a]From Kukal and Kevan (1987).

[b]Adjustment for sex ratio of 43% females.

tus (Thomson) and a tachinid fly, *Exorista* sp. The overall mortality within a developmental cohort caused by parasitism is 56%, most of which is attributable to *Exorista* (Kukal and Kevan 1987). Other causes of mortality include inviability of eggs and predation of pupae by birds (Table 16.1). In contrast to the summer mortality, the calculated winter mortality, presumably caused by low temperatures, is only 13% of the population, which suggests biotic as opposed to abiotic population regulation (Kukal and Kevan 1987). Generally, in severely limiting environments, density-independent controls of an abiotic nature, such as climate, are more important than the biotic density-dependent factors, such as parasitism or availability of food (Andrewartha and Birch 1954; Danks 1986).

The parasitoids also appear to influence the feeding activity of *G. groenlandica*. The caterpillars feed only during June, then spin hibernacula and hide in soil cracks and cushion plants (e.g., *Dryas integrifolia*, *Saxifraga oppositifolia*), before mid-summer in July when the activity of adult parasitoids is at its peak (Kukal and Kevan 1987). This very brief feeding period likely extends the developmental life cycle of *G. groenlandica*, the scheme of which is depicted in Figure 16.1. Pupation, adult emergence, mating, oviposition, hatching, and molting to the second larval instar occur within the month of June of a single summer season. Each of the subsequent instars III–V takes 3 years to complete development and the sixth instar takes 4 years, which in total yields a 14-year life cycle. The length of this life cycle was deduced from the proportion of each larval instar molting within a single summer. The estimate is based on overlapping cohorts and on the observation that larvae molt no more than once per summer (Kukal and Kevan 1987), as has been shown for other arctic insects (Danks and Byers 1972; Sotavalta et al. 1980; Butler 1982). For instance, if 30% of the third instars molt in June of a single summer, then it would take approximately 3 years for the entire cohort to complete development.

The 14-year life cycle and overlapping generations probably increase the

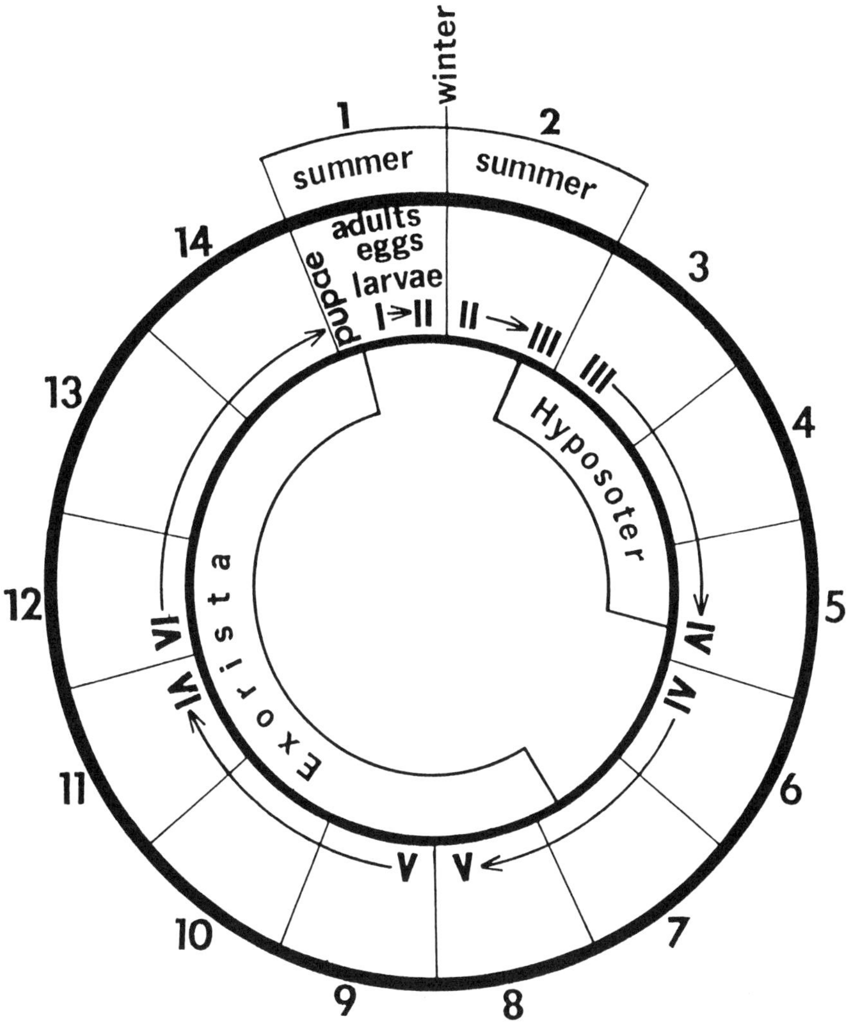

Figure 16.1. The schematic life cycle for *G. groenlandica* and its two parasitoids. The 14 outer circular segments represent 14 summers of development interrupted by winters (crossbars). The life phase of *Gynaephora* is indicated within these sectors (the Roman numerals refer to larval instars) and the number of years required for development of each phase is depicted cumulatively. Two inner circular segments represent the development and emergence of the two parasitoids restricted to a particular life phase of *Gynaephora*. (From Kukal and Kevan 1987.)

reproductive fitness of *G. groenlandica* by enabling a greater developmental flexibility compared to most other species of arctic Lepidoptera. The more commonly encountered uni- or bivoltine populations with synchronized generations can be decimated by inclement summers, even though synchrony, particularly in the bivoltine species, reduces mortality by temporal avoidance of parasitoids (Mikkola 1976). Despite the overlapping cohorts of *G. groenlandica*, parasitoid pressure is alleviated by the larvae becoming inactive and hiding before the peak emergence of parasitoids (Kukal and Kevan 1987). However, hiding from parasitoids abbreviates the feeding season available for growth and development of the larvae.

Food

The principal host plant of *G. groenlandica* is the dwarf arctic willow, *Salix arctica* L. (Salicaceae), a prostrate, clonal shrub capable of reaching hundreds of years in age. The larval foraging pattern is influenced by the distribution of arctic willow plants and by the changing quality of the willow leaves over the summer season. The distribution of *G. groenlandica* is contagious, with greatest densities of larvae found in relatively xeric areas, where the arctic willow is most abundant (Kukal and Kevan 1987). Caterpillars feed only in June when the leaf buds and young leaves of *S. arctica* contain the highest concentrations of macronutrients and nonstructural carbohydrates (Kukal and Dawson 1989). The mid-summer hiatus in larval feeding is coincident with an abrupt decline in the carbohydrate content of leaves and a buildup of plant secondary metabolites (Fig. 16.2A,B). In particular, the decline in carbohydrates stored by the host plant has an impact on the potential energy storage for overwintering caterpillars that require glycogen for cryoprotectant synthesis. These energetic constraints may not be as important in modulating the feeding phenology of caterpillars inhabiting tropical and temperate biomes.

Changes in larval feeding behavior and lower assimilation efficiency in other species of arctic caterpillars have also been correlated with seasonal changes in nutrient, carbohydrate, and caloric content of *S. arctica* tissues (Chapin et al. 1980, 1986; MacLean and Jensen 1986; Ayres and MacLean 1987a,b). It should be noted that decline of carbohydrate content of host plants may not be as crucial to caterpillars foraging in biomes other than the arctic. Increasing levels of secondary metabolites in host plants have been shown to inhibit feeding in numerous insect species (Feeny 1970; Rosenthal and Janzen 1979; Haukioja et al. 1985) and may function similarly in *G. groenlandica*. A combination of decreased nutrient and energy content of *S. arctica* with a buildup of tannins and phenols confines the feeding period to only 2–3 weeks in early summer, with the consequence of retarded development and an extended life cycle of the arctic woolly-bear caterpillars.

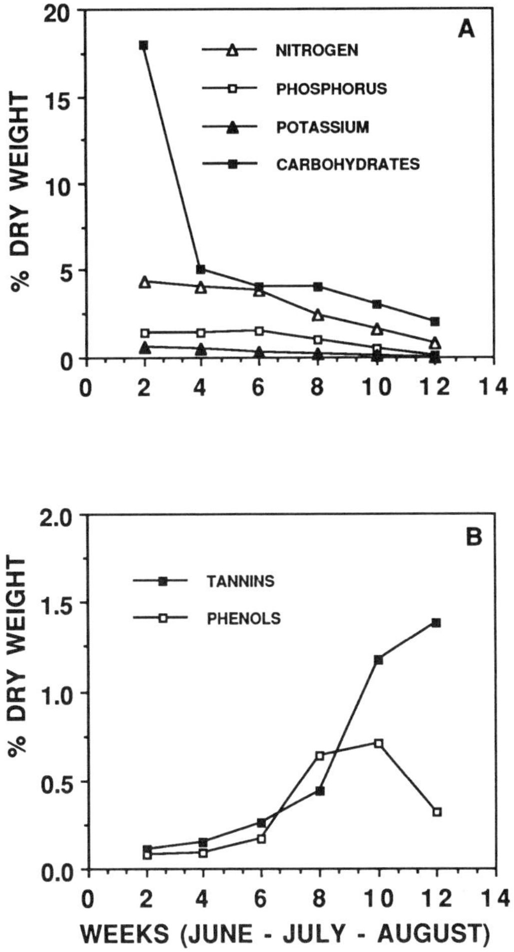

Figure 16.2. Seasonal variation in (A) macronutrients (N = 17; mean values $\pm$ SE $\leq$ 0.1) and nonstructural carbohydrates (SE $\leq$ 1), and (B) secondary metabolites (N = 5; mean values $\pm$ SE $\leq$ 0.1) in the leaves of *Salix arctica*. The larvae of *G. groenlandica* feed only during June of each summer season; they stop feeding at the beginning of July. (From Kukal and Dawson 1989.)

Temperature

The development of *G. groenlandica* larvae is not only a function of food quality, it is also influenced by temperature. In a laboratory study, we assessed the relative growth rate and assimilation efficiency in late instars held at 5, 15, and 30°C in constant light. Larval growth rates were similar at 15 and 30°C and approximately 15-fold greater than at 5°C (Table 16.2). The assimilation efficiency at 15°C was 40% compared to 10% at 30°C, owing to the 7-fold increase in the maintenance metabolism rate at the higher temperature (Kukal and Dawson 1989). Even at the best efficiency of food consumption (i.e., at 15°C), the relative growth rate of *G. groenlandica* was only 4% of the average growth rate for ca. 500 other Lepidoptera (Table 16.3).

Table 16.2. Temperature Influence on Growth Rate and Assimilation Efficiency of G. groenlandica *Larvae (Instars IV–VI) Feeding on* Salix arctica[a]

Incubation temperature	Growth rate[b]	Food ingested[b]	Frass excreted[b]	Metabolic rate	Assimilation efficiency (%)
5°C	0.001±0.0005[c]	0.012±0.002	0.001±0.0004	0.003	7
15°C	0.016±0.007	0.056±0.018	0.017±0.009	0.023	40
30°C	0.017±0.007	0.227±0.006	0.053±0.008	0.157	10

[a]Each incubation temperature contained 3 groups of 10 larvae. Assimilation efficiency = % mass gain per mass ingested; metabolic rate = (food ingested − excreted) − mass gain. Values = mean ±1 SE. Mean dry wt. of larva = 300 ± 5 mg. From Kukal and Dawson (1989).

[b]mg dry wt./mg larva/day.

[c]Corrected for the extended residence time of food in the gut (mean dry wt. of food in the gut = 2.11 ± 0.08 mg/larva/day).

Larvae growth rate and development may be especially inhibited by the effect of low temperatures on the molting process. Molting can be inhibited by near-zero temperatures and comprises 43–60% of developmental time of an instar in at least two other arctic species, *Gaterucella sagittariae* and *Epirrita autumnata* (Geometridae) (Ayres and MacLean 1987b). Molting can also be impeded by presence of phenolic compounds in the host plant (Stamp 1990). Furthermore, studies of gypsy moth larvae (*Lymantria dispar*, Lymantriidae) suggest that molting is energetically expensive, requiring as much as 27% of the nitrogen and caloric content of a caterpillar (Montgomery 1982). These findings offer further support for the idea that the curtailed feeding season of *G. groenlandica* reflects the sharp decline in energy and nitrogen constituents of the host plant.

Table 16.3. Food Utilization by G. groenlandica *(A) Compared to Other Species of Lepidoptera (B)*[a]

	RGR	RCR	% AD	% ECD	% ECI
A. *Gynaephora*					
5°C	0.001	0.012	33.3	9.1	3.0
15°C	0.016	0.056	69.6	41.0	28.5
30°C	0.017	0.227	76.7	9.8	7.5
B. Other Lepidoptera (*N* = 444–629)					
Mean	0.38	2.03	53	40	20
Range	0.03–1.50	0.27–6.90	16–97	2–87	1–78

[a]Data from (A) Kuhl and Dawson (1989) and (B) Slansky and Scriber (1985). RGR, relative growth rate; RCR, relative consumption rate; AD (approximate digestibility), (food ingested-frass)/food ingested; ECD (efficiency of conversion of digested food) = biomass gained/(food ingested−frass); ECI (efficiency of conversion of ingested food) = AD × ECD; RGR and RCR were calculated as mg dry wt./mg larva/day.

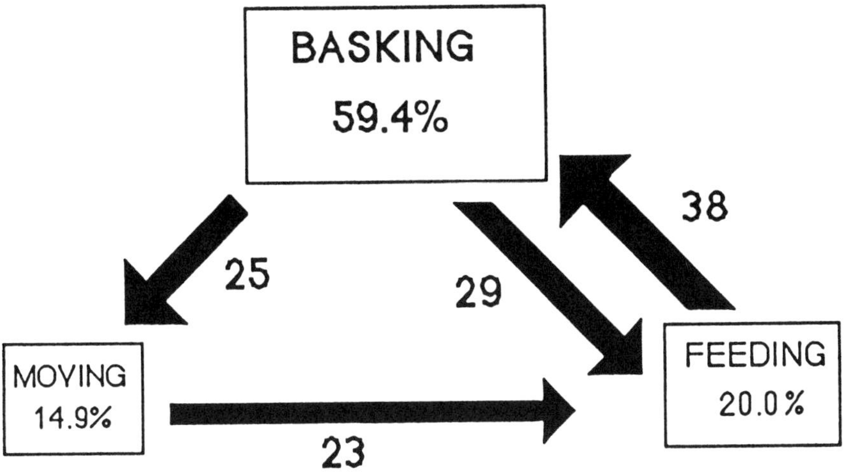

Figure 16.3. Kinematic graph showing the frequency and sequential organization for 3 major behaviors of *G. groenlandica* larvae. Relative sizes of rectangles and arrows indicate the frequency of occurrence by the percentage of time spent in particular behaviors (within rectangles) and also the number of times one behavior followed another (arrows). (From Kukal et al. 1988a.)

Behavioral Thermoregulation

Interaction of biotic constraints (in particular parasitoids and food quality) and abiotic constraints (especially temperature) in modifying the foraging pattern of *Gynaephora* larvae is closely linked to the thermoregulatory behavior of the larvae. The temperature of basking caterpillars is 30.5°C (± 2.2 SD) compared to 26.7°C (± 2.6 SD) when moving and 23.9°C (± 2.2 SD) when feeding (Kukal et al. 1988a).

Unlike the larvae of many Lepidoptera that spend most of the time feeding (Scriber and Slansky 1981), the arctic woolly-bears spend most of their time basking (i.e., orienting their bodies perpendicular to the sun's rays). They show a definite sequential pattern of activity between the three major types of behaviors; basking, feeding and moving (Kukal et al. 1988a; Fig. 16.3). Approximately 60% of their time is spent basking, followed by feeding (20%) and moving (15%). In a behavioral sequence, feeding is most often followed by basking and less often vice versa. The caterpillars appear to move primarily in order to feed since the reverse sequence almost never occurs (Fig. 16.3). Presumably, the larvae can conserve energy by restricting their mobility mainly to the search for host plants. The caterpillars' ability to regulate their body temperature by behavioral change influences the diel pattern of foraging. In June, during the foraging season of *G. groenlandica*, the sun is approximately 33° above the horizon at noon, whereas

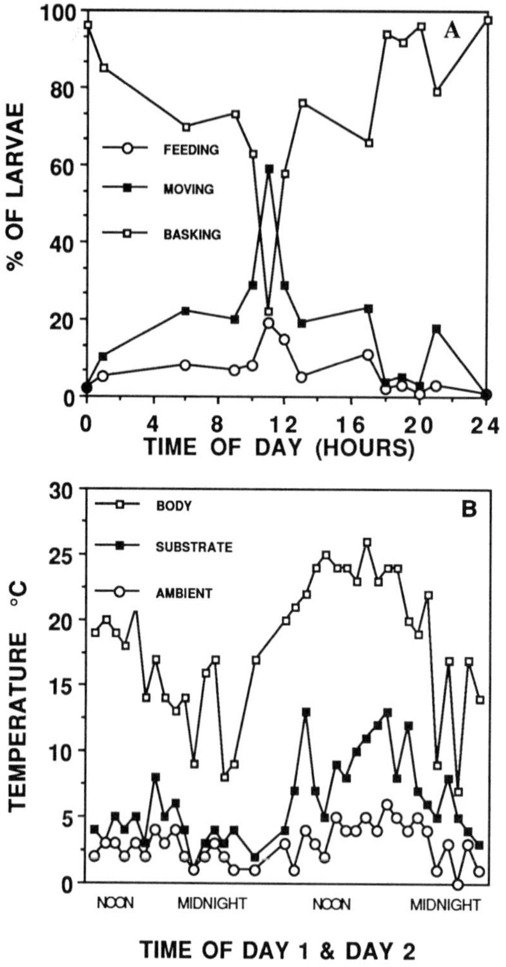

Figure 16.4. (A) Diel changes in the pattern of major behaviors (feeding, basking, moving) in *G. groenlandica* larvae ($N = 500$) on the tundra under sunny, calm conditions in mid-June (T_a at 1 m = 2.0–6.3°C). The sun is at minimum angle of 10° above the horizon at midnight and at a maximum angle of 33° at noon. (From Kukal et al. 1988a.) (B) Diel changes in mean body temperature ($\pm$ SD $\leq$ 5) of 8 *G. groenlandica* larvae relative to substrate and ambient temperatures recorded bihourly over 2 days (an overcast day followed by a sunny one) on the open tundra. (From Kukal et al. 1988a.)

it drops to a 10° angle by midnight. This change in angle of solar incidence results in greater insolation during midday than at midnight and consequently the larvae achieve higher body temperatures during midday, which facilitates their movement and feeding (Fig. 16.4; Kukal et al. 1988a). Consequently, the larvae of *G. groenlandica* feed and bask at different times of the day. At midnight, the caterpillars spend most of their time basking, whereas feeding and searching for food is confined primarily to midday (Fig. 16.4A). In contrast to *G. groenlandica*, temperate lymantriid larvae (*Lymantria dispar*) are not dependent on insolation; they feed mostly in early mornings and evenings and are immobile at midday (Knapp and Casey 1986), an adaptation attributed to the avoidance of predation and parasitism (Campbell 1981). However, unlike the case of *G. groenlandica*,

Table 16.4. Metabolic Rate at 25°C of G. groenlandica *Larvae in Different Metabolic States: Feeding, Digesting, Starved, in Hibernacula, and following Cold Acclimation*[a]

Metabolic state	Oxygen consumption (ml/g/hr)
Resting metabolism	0.06±0.02[e]
Starved larvae	0.06±0.02[e]
Moving larvae	0.11±0.03[f]
Feeding larvae[b]	0.29±0.03[g]
Digesting larvae[c]	0.17±0.02[h]
Larvae in hibernacula	0.07±0.02[e]
Low temperature acclimation[d]	0.07±0.02[e]

[a]$N=10$; values = mean ± SE; means followed by a different letter are significantly different by *t* test, $P < 0.05$. After Kukal and Dawson (1989).

[b]Larvae kept for a week without food but provided with water.

[c]The digestion period estimated from the length of time of elevated metabolism.

[d]Larvae held at 5°C and dark for 2 months.

the ambient temperatures encountered by feeding gypsy moth caterpillars are well above their developmental threshold and hence the larvae are thermoconformers (Knapp and Casey 1986).

The ability of caterpillars to warm up is under the constraint of abiotic factors, such as direct sunlight, cloudiness, time of day, wind, and the presence of snow. For instance, the intensity of incoming solar radiation peaks during the summer solstice, and then gradually declines over the summer season (Bliss 1977; Svoboda and Freedman 1992). Moreover, the presence of snow during most of the month of June creates an albedo effect which enhances the ability of caterpillars to raise their body temperature. The relative capability of larvae to raise their body temperatures under different physical conditions (i.e., overcast vs. direct sunlight, noon vs. midnight) is depicted in Figure 16.4B (Kukal et al. 1988a). Other species of arctic Lepidoptera can raise their body temperatures by basking passively within flowers that serve as parabollic reflectors (Kevan 1975) or actively, like *Gynaephora*, by orienting toward the sun (Kevan and Shorthouse 1970; Kevan et al. 1982).

Metabolism and Feeding Energy Budget

Maintenance metabolism of larvae in different behavioral states is indicative of their energy expenditure. For instance, the metabolic rate of feeding larvae is nearly 5-fold the standard metabolic rate or the rate of larvae deprived of food (Table 16.4). During digestion and movement, the maintenance metabolism is raised with the consequence of increased degradation of stored carbohydrates. It is crucial for the larvae to store enough glycogen and trehalose over the 3- to 4-

week feeding period for the synthesis of cryoprotective glycerol. Although the caterpillars can survive freezing throughout the entire year, their freezing tolerance is greatly increased during the winter months by cryoprotective compounds. The lower lethal temperature (LLT; temperature at which 100% of larvae die after 24-hour exposure) in the summer is −15°C, whereas the LLT in the winter is less than −70°C owing, at least in part, to the presence of glycerol (Kukal et al. 1988b). Unlike most other cold tolerant species of insects (Zachariassen 1985), the woolly-bear caterpillars retain freezing tolerance during the summer season despite the lack of cryoprotective glycerol (Kukal et al. 1988b).

In vivo studies using ^{13}C nuclear magnetic resonance (NMR; Gadian 1982) show the absence of glycerol in the warm acclimated, summer larvae (Kukal et al. 1989). However, an injection of labeled glucose, which is eventually converted into glycogen (Fig. 16.5A,B), indicates their capability to synthesize glycerol. Under anoxic conditions the glycerol is retained, presumably increasing the freezing tolerance of the caterpillars (Fig. 16.5B). This result has important implications for the cessation of larval feeding at the onset of mid-summer. The larvae hide near the permafrost, where temperatures are below freezing and their metabolic rate is greatly depressed (Kukal et al. 1989). Consequently, they are able to conserve their carbohydrate energy storage, which otherwise, at an elevated metabolic rate, would be utilized more rapidly. Furthermore, when the quality of host plant food decreases later in the summer season (i.e., decreased nutrient and carbohydrate content of *Salix*), the larvae may not be able to obtain sufficient energy storage for overwintering. Consequently, during mid-summer, maintenance metabolism may exceed the energy obtained by caterpillars from their food.

While the larvae of *G. groenlandica* remain hidden near the permafrost for the remainder of the arctic summer, the mitochondrial numbers in larval tissues diminish greatly (Table 16.5), with a concurrent decrease in oxidative metabolism (Table 16.6) and glycerol accumulation. The mitochondria are resynthesized within less than one week at the onset of feeding in the spring (Kukal et al. 1989). Mitochondrial degradation and associated glycerol production take place mainly during the period of feeding hiatus when the larvae are hiding near the permafrost. Since glycerol is essential for cryoprotection during overwintering, the abbreviated feeding period may have been subject to selection for winter survival.

Summary

Foraging of arctic woolly-bear caterpillars has evolved in response to the biotic constraints of parasitism and food quality, modified by temperature. The feeding pattern of *Gynaephora groenlandica* involves behavioral and physiological adaptations. The behavioral adaptations include (1) cessation of feeding before mid-summer, which results in feeding restricted to early summer when food quality

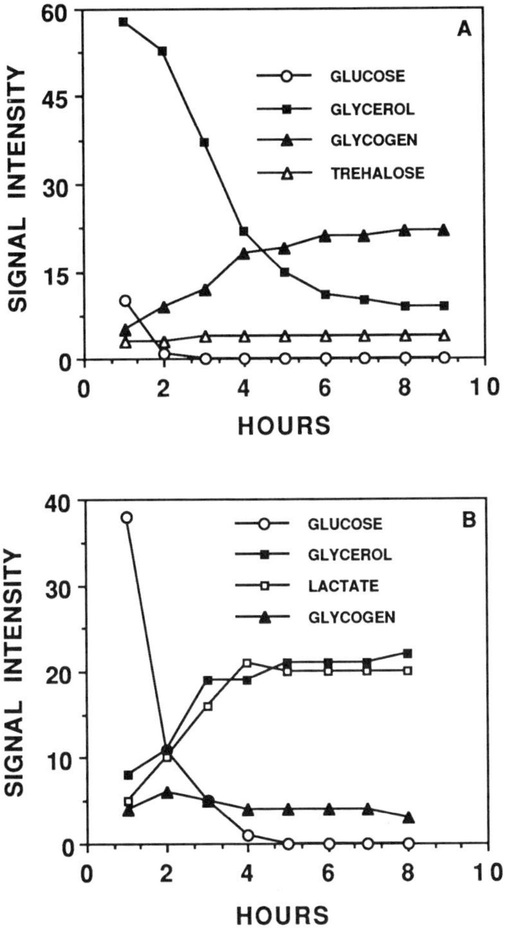

Figure 16.5. Effects of anoxia on glycerol metabolism of *G. groenlandica* at 25°C. Results of a time-lapse ^{13}C-NMR spectroscopy of larvae after an injection of [$1-{}^{13}$C] glucose in the presence of oxygen (A) and under anoxia (B). (From Kukal et al. 1989.)

*Table 16.5. Ratios of Mitochondria per Nucleus (*N = *400–500 Nuclei) in Tissues of 10* G. groenlandica *Larvae following Different Acclimation Regimes for 3 Months*[a]

	Acclimation temperature (°C)		
Tissue	15	5	−15
Brain	100+	1.43	0.04
Fat body	100+	1.18	<0.01

[a]After Kukal et al. (1989).

Table 16.6. Oxygen Consumption by Larvae of G. groenlandica *at 23°C following Acclimation at −15°C for Either 75 Days or 6 Months (Mean ± SD)*[a]

Time spent at 23°C before measurement (hr)	Oxygen uptake (ml/g/hr) following acclimation at −15°C for	
	75 days[b]	6 months[b]
0	0.000±0.005	0.000±0.005
1–3	0.095±0.05[c]	0.084±0.03[e]
24	0.180±0.06[d]	0.038±0.03[f]
48	0.190±0.08[d]	0.038±0.03[f]

[a]From Kukal et al. (1989).

[b]c vs. d, e vs. f, and d vs. f, values significantly different by *t* test ($p<0.05$).

is high, and temporal avoidance of parasitoids, and (2) behavioral thermoregulation by basking, which may be coupled with minimal energy expenditure in other activities. The physiological adaptations involve (1) low-temperature threshold for feeding activity, (2) carbohydrate storage during the feeding period for cryoprotectant production in preparation for winter, and (3) mitochondrial degradation during the feeding hiatus, which reduces the consumption of stored energy.

Future Directions

Gynaephora groenlandica is perhaps the only insect inhabiting the northernmost land masses of the arctic that has been carefully examined in terms of its feeding ecology and ecophysiological adaptations to cold. Even though many questions pertaining to this species still remain unanswered, there is a need for comparative studies, particularly on other species of arctic caterpillars. Information on foraging of arctic insects is lagging behind the investigation of their host plants, to which the plant–herbivore relationships are yet to be identified.

Literature Cited

Andrewartha, H.G., and Birch, L.C. 1954. The Distribution and Abundance of Animals. Chapman, London.

Ayres, M.P., and MacLean, S.F. 1987a. Development of birch leaves and the growth energetics of *Epirrita autumnata* (Geometridae). Ecology 68:558–568.

Ayres, M.P., and MacLean, S.F. 1987b. Molt as a component of insect development: *Galerucella sagittariae* (Geometridae) and *Epirrita autumnata* (Geometridae). Oikos 48:273–279.

Bliss, L.C. (ed) 1977. Truelove Lowland, Devon Island, Canada. A High Arctic Ecosystem. The University of Alberta Press, Edmonton, Alberta.

Butler, M.G. 1982. A 7-year life cycle for two *Chironomus* species in arctic Alaskan tundra ponds (Diptera: Chironomidae). Can. J. Zool. 60:58–70.

Campbell, R.W. 1981. Population dynamics, pp.65–214. In C.C. Downe and M.L. McManus (eds.), The Gypsy Moth: Research Toward Integrated Pest Management. U.S. Dept. of Agriculture, Washington, D.C.

Chapin, F.S., Johnson, D.A., and McKendrick, J.D. 1980. Seasonal movement of nutrients in plants of differing growth form in an Alaskan tundra ecosystem: Implications for herbivory. J. Ecol. 68:189–209.

Chapin, F.S., McKendrick, J.D., and Johnson, J.D. 1986. Seasonal changes in carbon fractions in Alaskan tundra plants of differing growth form: Implications for herbivory. J. Ecol. 74:707–731.

Curtis, J. 1835. Descriptions and c. of the insects brought home by Commander James Clark Ross, R.B., F.R.S., and c, pp. lix–lxxx. In J.C. Ross (ed.), Account of the Objects in the Several Departments of Natural History, Seen and Discovered during the Present Expedition. Appendix to the Narrative of a Second Voyage in Search of a North-West Passage, and of a Residence in the Arctic Regions during the Years 1829, 1830, 1831, 1832, 1833. Webster, London.

Danks, H.V. 1981. Arctic arthropods. A review of systematics and ecology with particular reference to the North American fauna. Entomol. Soc. of Canada, Ottawa, Ont.

Danks, H.V. 1986. Insect plant interactions in arctic regions. Rev. Entomol. Quebec 31:52–75.

Danks, H.V., and Byers, J.V. 1972. Insects and arachnids of Bathurst Island, Canadian Arctic Archipelago. Can. Entomol. 104:81–88.

Downes, J.A. 1964. Arctic insects and their environment. Can. Entomol. 96:279–307.

Downes, J.A. 1965. Adaptations of insects in the arctic. Annu. Rev. Entomol. 10:257–274.

Feeny, P. 1970. Seasonal changes in oak leaf tannins and nutrients as a cause of spring feeding by winter moth caterpillars. Ecology 51:565–581.

Ferguson, D.C. 1978. Noctuoidea, Lymantriidae, pp.17–21. In R.B. Dominick et al. (eds.), The Moths of North America North of Mexico. Fascicle Foundation, London.

Gadian, D.G. 1982. Nuclear Magnetic Resonance and Its Applications to Living Systems. Oxford University Press, London.

Haukioja, E, Hiemala, P., and Siren, S. 1985. Foliage phenols and nitrogen in relation to growth, insect damage, and ability to recover after defoliation in the mountain birch *Betula pubescens* spp. *tortuosa*. Oecologia 65:214–222.

Kevan, P.G. 1975. Sun tracking solar furnaces in high arctic flowers: Significance for pollination and insects. Science 189:723–726.

Kevan, P.G., and Shorthouse, J.D. 1970. Behavioral thermoregulation by high arctic butterflies. Arctic 23:268–279.

Kevan, P.G., Jensen, T.S., and Shorthouse, J.D. 1982. Body temperatures and behavioral thermoregulation of high arctic woolly-bear caterpillars and pupae (*Gynaephora rossii*, Lymantriidae: Lepidoptera) and the importance of sunshine. Arctic Alpine Res. 14:125–136.

Knapp, R., and Casey, T.M. 1986. Thermal ecology, behavior, and growth of gypsy moth and eastern tent caterpillars. Ecology 67:598–608.

Kukal, O., and Dawson T.E. 1989. Temperature and food quality influences feeding behavior, assimilation efficiency and growth rate of arctic woolly-bear caterpillars. Oecologia 79:526–532.

Kukal, O., and Kevan, P.G. 1987. The influence of parasitism on the life history of a high arctic insect, *Gynaephora groenlandica* (Wocke) (Lepidoptera: Lymantriidae). Can. J. Zool. 65:156–163.

Kukal, O., Heinrich, B., and Duman, J.G. 1988a. Behavioral thermoregulation in the freeze-tolerant arctic caterpillar, *Gynaephora groenlandica*. J. Exp. Biol. 138:181–193.

Kukal, O., Serianni, A.S., and Duman, J.G. 1988b. Glycerol metabolism in a freeze-tolerant arctic insect: An *in vivo* ^{13}C NMR study. J. Comp. Physiol. B 158:175–183.

Kukal, O., Duman, J.G., and Serianni, A.S. 1989. Cold-induced mitochondrial degradation and cryoprotectant synthesis in freeze-tolerant arctic caterpillars. J. Comp. Physiol. B 158:661–671.

MacLean, S.F., and Jensen, T.S. 1986. Food plant selection by insect herbivores in Alaskan arctic tundra: The role of plant life form. Oikos 44:211–221.

Mikkola, K. 1976. Alternate-year flight of northern *Xestia* species (Lep., Noctuidae) and its adaptive significance. Ann. Entomol. Fenn. 42:191–199.

Montgomery, M.E. 1982. Life-cycle nitrogen budget for the gypsy moth, *Lymantria dispar*, reared on artificial diet. J. Insect Physiol. 28:437–442.

Rosenthal, G.A., and Janzen, D.H. (eds.) 1979. Herbivores: Their Interaction with Secondary Plant Metabolites. Academic Press, New York.

Ryan, J.K., and Hergert, C.R. 1977. Energy budget for *Gynaephora groenlandica* (Homeyer) and *G. rossii* (Curtis) (Lepidoptera: Lymantriidae) on Truelove Lowland, pp. 395–404. In L.C. Bliss (ed.), Truelove Lowland, Devon Island, Canada. A High Arctic Ecosystem. University of Alberta Press, Edmonton, Alberta.

Scholander, P.F., Flagg, W., Hoch, R.J., and Irving, L. 1953. Climatic adaptation in arctic and tropical poikilotherms. Physiol. Zool. 26:67–92.

Scriber, J.M., and Slansky, F., Jr. 1981. The nutritional ecology of immature insects. Annu. Rev. Entomol. 26:183–211.

Slansky, F., Jr., and Scriber, J.M. 1985. Food consumption and utilization, pp. 87–163. In G.A. Kerkut and L.I. Gilbert (eds.), Comprehensive Insect Physiology, Vol. 4. Pergamon, Oxford.

Sotavalta, O., Karvonen, E., Korpila, S., and Korpila, T. 1980. The early stages and biology of *Acerbia alpina* (Lepidoptera: Arctiidae). Not. Entomol. 60:89–95.

Stamp, N.E. 1990. Growth versus molting time of caterpillars as a function of temperature, nutrient concentration and the phenolic rutin. Oecologia 82:107–113.

Svoboda, J., and Freedman, B. 1992. Ecology of a High Arctic Lowland Oasis, Alexandra Fiord (78°53′N, 76°55′W), Ellesmere Island, N.W.T., Canada. University of Toronto Press, Toronto (in press).

Ont. Zachariassen, K.E. 1985. Physiology of cold tolerance in insects. Physiol. Rev. 65:799–832.

17

Lepidopteran Foraging on Plants in Agroecosystems: Constraints and Consequences

Pedro Barbosa

Introduction

Perspectives on Ecological and Evolutionary Constraints

When reviewing the literature on trophic interactions of herbivores in agroecosystems one is struck by the scarcity of direct and rigorous research that proposes and/or tests hypotheses on ecological and evolutionary constraints on foraging. In addition, a great deal of current agroecology research is designed with the assumption that theories that describe plant–herbivore interactions in unmanaged ecosystems are relevant to agroecosystems. That assumption is rarely tested and often has established irrelevant research priorities. Thus, it may be prudent to await the accumulation of appropriate data from a variety of agroecosystems before making too many broad generalizations. Nevertheless, in this chapter I discuss some of the factors and interactions that may constrain foraging by lepidopteran larvae in an effort to set priorities for much needed research. Further, some preliminary hypotheses and research approaches are suggested to stimulate interest in agroecological research.

My discussion of lepidopteran foraging is limited, for the most part, to the behavior of larvae of species on annual crops rather than that of larvae feeding on tree crops, ornamental plants, and so on. Second, I assume that in many cases host selection, feeding behavior, and subsequent patterns of larval distribution exist as a result of the consistent influence, over many growing seasons, of one or more factors in the herbivore's agroecosystem. For example, utilization of young leaves by larvae may be a response to enhanced foliage quality resulting from long-term, historical use of high levels of fertilizer, or to some agronomic practice that encourages young foliage growth. Further, I assume that the physical and biotic factors in agroecosystems can not only influence but also shape foraging patterns of herbivores in unique ways. For example, specific factors associated with agroecosystems could be responsible for certain feeding behavior of an herbivore that are not observed in its counterpart herbivore in an unmanaged ecosystem.

In this chapter, I use the term constraints to mean circumstances and/or interactions that determine foraging options and level of behavioral performance (i.e., the quantity and quality of foraging) of an organism. Constraints may be (1) functional (and thus determined by biochemical or biomechanical properties of tissues, organs, and/or ontogeny), (2) genetic (and thus determined by previous adaptations), or (3) historical (i.e., phylogenetic influences) (see Gould 1989).

Foraging Patterns of Herbivores in Agroecosystems

The distribution of an herbivore on plants in its habitat is, to a large extent, a consequence of host selection and feeding behaviors, and/or ecological interactions influencing host selection, colonization, and utilization of plant hosts. The foraging patterns of larval Lepidoptera, for example, represent the outcome of behaviors associated with the seeking of food (i.e., of plants) and those behaviors associated with the ingestion of food. Foraging patterns provide indirect insights into the factors and forces that control or influence where herbivores end up (at the end of their quest for appropriate food) and the nature of their choice of diet. For those species whose larvae do not move from plant to plant, patterns of female oviposition, in large part, parallel foraging patterns, except for the slight changes resulting from limited intraplant movements of larvae (Feeny et al. 1983). However, for other species (1) whose larval preferences for certain tissues are specific and different from oviposition sites (Farrar and Bradley 1985), (2) whose larvae are highly mobile (Lance 1983; Terry et al. 1989), or (3) where females oviposit off the plant (Varela and Bernays 1988), egg distribution is of little significance in determining the inter- and intraplant foraging pattern of larvae.

In agroecosystems, the study of foraging patterns also has focused on determining how various traits of crops or cultivars, which confer resistance to herbivory, alter oviposition and feeding by herbivores (Adesiyun 1978; Argandona et al. 1983; Dover 1986). Explanations for observed patterns of distribution can often be sought in the functional relationships between oviposition behavior or mechanisms of feeding (such as those described in Ahmad 1983) and characteristics of host plants (e.g., see Uematsu and Sakanoshita 1989). Thus, the direct causation of patterns of larval distribution, and perhaps abundance, may be related to the way an herbivore bites, bores, rolls, or ties plant tissues (Bernays and Janzen 1988), and the factors that influence these behaviors. Physical and structural differences among cultivars of a crop species (such as *Brassica oleraceae*; see Fig. 17.7) might limit the ability of herbivores (such as, *Pieris rapae*) to find, oviposit, and feed on their hosts, and thereby define the limits of larval distribution.

Similarly, other studies seek to ascertain the influence of agronomic practices such as fertilization (or specifically of nitrogen levels) on patterns of larval distribution and abundance (Chelliah and Subramanian 1972; Hargrove et al. 1984; Heidorn and Joern 1987). Consistent and clear conclusions have been difficult to obtain, and appear to depend on the crop, fertilizer, and herbivore selected for study (Prestidge 1982; Manuwoto and Scriber 1985; Eigenbrode and Pimentel 1988). Still other factors, such as predation (Myers and Campbell 1976; Bernays 1988, 1989; Bernays and Cornelius 1989), parasitism (Ankersmit et al. 1981; Damman 1987; Jaenike 1986; but see Gross and Price 1988), and microclimate (Pinter et al. 1975), also may modify host selection, feeding behavior, and thus patterns of larval distribution. These factors and others will be discussed later in this chapter.

The Feeding Behavior of Leipdopterous Larvae

The time spent feeding by a lepidopteran herbivore may range from as little as 7% to as much as 75% of any given observation period. The time spent feeding does not appear to be correlated, in any consistent fashion, to any obvious factor [including habitat type, plant chemical defense, the periodicity of feeding, amount of food eaten by an herbivore, or its size (see Bowdan 1988; Dethier 1988)]. Obviously, a species may be influenced by one or more of these factors, but the influence varies from herbivore to herbivore. The crop species or cultivar selected by herbivores may cause dramatic changes in feeding behavior. For example, the highly polyphagous beet armyworm *Spodoptera exigua* feeds differently on tomato than on chrysanthemum, both of which are hosts. On chrysanthemum larvae feed predominantly on upper leaves, and prior to the fourth stadium most feeding is on the underside of leaves. Even in the fourth and fifth stadia, only 30 and 49% of the leaves, respectively, are perforated. Only 46 and 37% of the leaves eaten by fourth and fifth instars, respectively, have upper-surface damage. In contrast, on tomato, larvae feed on the lower leaves and although first instars feed on the underside of leaves, feeding by older instars primarily entails perforating leaves (Smits et al. 1987).

The differences in feeding behavior (noted above) among different instars also have been observed in other lepidopteran pest species (Kranz et al. 1977; Theunissen et al. 1985). Feeding habits may change dramatically, switching from phytophagous to carnivorous (e.g., *Heliothis* spp.) (Kranz et al. 1977), external foliage feeding to fruit feeding (e.g., *Heliothis zea*) (Lange and Bronson 1981), foliage feeding to stem cutting (e.g., *Agrotis ipsilon* (Kranz et al. 1977), leaf webbing to free-feeding (e.g., *Spodoptera exigua*) (Pinter et al. 1975), or gregarious to solitary (e.g., *Pieris brassicae*) (Long 1955), as larvae age. There is no reason to believe at present that these behavioral shifts are unique to lepidopterans in agroecosystems. For example, in unmanaged ecosystems, the larvae of *Tyria jacobaeae* (the cinnabar moth) feed on the underside of leaves as a first instar, on leaves and flower buds as intermediate instars, and when they become fifth instars they disperse to the inflorescences of their ragwort host plants (van der Meijden 1976).

Herbivory can induce physiological changes in a host plant, that can alter its nutritional suitability, defensive chemistry, and/or morphology (see Tallamy and Raupp 1991, and references therein). These changes may influence subsequent feeding and foraging patterns of larvae of the species that induced the plant changes and those of other species (Karban 1988). In addition, feeding by early season herbivores may alter crop plants such that foraging patterns of late season herbivores are affected. Agronomic practices, such as the application of insecticides, may add an additional twist to this interaction. If a certain intensity of herbivory is required to induce changes in a plant, the loss of one or more pest species due to the use of selective pesticides (or some other agronomic practice) may eliminate the induction, and prevent certain changes in plant quality. The foraging patterns of subsequent herbivores on the plant might be distinctly different from what they would have been had insecticides not been used. Such a series of events, although feasible, has not been investigated, yet may be responsible for the variability observed in levels of damage and abundance of pest herbivores. Conversely, it may be that the frequent outbreaks that often characterize some agroecosystems may make the frequency of induced chemical changes in plants more likely. The latter speculation awaits experimental testing.

Phylogenetic Constraints and the Diversity of Feeding Habits among Lepidoptera

The feeding behavior of herbivores in selected superfamilies is reviewed by Powell (1980) and Stehr (1987) and is highlighted below to illustrate the diversity of foraging and feeding behaviors.

Noctuoidea

The feeding habits of species in this superfamily are diverse and include leaf feeders, various types of borers, fruit feeders, and cutworms. The diversity of feeding types is most readily observed among species of the Noctuidae, although leaf mining and nest constructing noctuids are uncommon. Larvae of species in the Lymantriidae and many Notodontidae feed primarily on the foliage of woody plants and a few typically cause extensive defoliation. Unlike most Lymantriidae, the Notodontidae are not known to feed on conifers. In many notodontid species, larvae begin as skeletonizers, and then become free-feeders as they age. The Arctiidae feed on a wide variety of plants from grasses to various species of trees and shrubs. Although a few arctiids are common in agroecosystems, most species are not economically important.

Tortricoidea

Within the Tortricidae, the Tortricinae include species with larvae that are externally feeding, leaf tiers and rollers, webbers, and tent makers. The Tortricini feed on the foliage of plant species primarily in the Rosaceae, Salicaceae, Betulaceae, Ericaceae, and Fagaceae. Similarly, the Cnephasiini feed mainly on the Rosaceae, Asteraceae, Betulaceae, and Fagaceae. In contrast, the Archipini feed on more plant species in the Pinaceae than any other tribe in the Tortricinae.

Pyraloidea

Almost all pyralids are concealed feeders, particularly as larvae proceed beyond the first and second stadia. However, feeding behavior varies widely across the many subfamilies of Pyralidae. Species in these subfamilies may feed on a variety of tissues (leaves, stems, roots, seeds, flowers) and in a variety of ways (as free-feeders, leaf miners, borers, leaf tiers, or leaf rollers). Most larvae bore into the host plant or form tents, webs, or cases from which they feed.

Papilionoidea

Among the Papilionoidea, relatively few species feed on economically important crops. Of those that do, few inflict economically significant damage. However, the Pieridae, which are all external feeders, feed on hosts in the Brassicaceae (Cruciferae) and the Fabaceae (Leguminosae) and include several economically important species. Other economically important herbivore species in this group include pests of plant species in the Umbelliferae, Rutaceae, Magnoliaceae, and Lauraceae.

Species in the Lycaenidae exhibit a preference for plant species in the Fabaceae, although they will feed on a wide variety of plant species in several families. This family is worldwide in distribution and mainly a tropical group, but few species within this family are pests. Species within the Nymphalidae are, perhaps, the most diverse family in this group; larvae feed on a wide variety of trees, shrubs, and herbs.

Gelechioidea

The Gelechiidae is the largest family of the Gelechioidea and although most species are only occasional pests some of the more well known pests of agroecosystems are in this family. They include *Pectinophora gossypiella* (the pink bollworm), *Phthorimaea operculella* (the potato tuberworm), as well as tree pests such as *Coleotechnites milleri* (the lodgepole needle miner). The feeding habits of species in this family are tremendously varied and include leaf and needle mining, leaf rolling and tying, free-feeding, stem boring, gall making, flower and seed feeding, and twig and bark feeding.

The Oecophoridae have similarly varied feeding habits. Whereas some species in the Oecophoridae feed only on dead plants and saprophytic microorganisms on plant tissues, others are leaf, flower, or seed feeders, leaf miners, casebearers, and gall makers. Although most oecophorids are of little economic importance, a few species in the Depressarinae and Oecophorinae, such as the parsnip webworm *Depressaria pastinacella*, are important pests. Other subfamilies are restricted to noneconomically important plant families or occur in mesic forests of Central and South America. The Coleophoridae (the casebearers) include a variety of pests of fruit and forest tree crops. Although the first instars of *Coleophora* species are leaf miners, older instars construct cases from pieces of plant material, frass, and silk and feed from the case. Other species are skeletonizers.

Yponomeutoidea

The Plutellidae and the Yponomeutidae are leaf feeders and skeletonizers that may also loosely web leaves together. Some species in both families bore into buds and petioles or mine leaves. Species of economic importance in the Plutellidae feed on plant species primarily in the Brassicaceae (Cruciferae) but also in the Caprifoliaceae, Celastraceae, Fagaceae, Juglandaceae, Oleaceae, Pinaceae, Rutaceae, and Ulmaceae. On the other hand, species in the Yponomeutidae feed on a wide variety of trees, shrubs, and forbs, including species in the Aceraceae, Betulaceae, Celestraceae, Corylaceae, Crassulaceae, Empetraceae, Ericaceae, Fagaceae, Lauraceae, Oleaceae, Rhamnaceae, Rosaceae, Salicaceae, Saxifragaceae, and some others.

Sesiodea and Cossoidea

Families in these superfamilies have similar feeding habits. Most species bore into and consume woody plant parts (ranging in size from small twigs to trunks). Still other species feed under bark, and on the external surface of roots. Although many species have narrow host plant species range, the host range for each family taken as a whole is broad. Several species in these groups are of economic importance.

How Are Agroecosystems Different from Unmanaged Ecosystems?

Introduction: What Is an Agroecosystem?

Agroecology, the study of the interactions between plants and animals in agroecosystems, has gained acceptance in recent years as a distinct subdiscipline of ecology. Perhaps because it is a new focus of interest, agroecological research has been, mostly, confined to research on economically important species and/or the testing of theory using these species. Hill (1987) defined an agroecosystem as the "ecosystem of an area as modified by the practice of agriculture, horticulture or animal rearing" (Table 17.1, Fig. 17.1. Typically, an agroecosystem is described as one requiring a supplemental energy source to enhance productivity, usually in the form of processed fuels or their by-products (Pimentel 1976, 1986). It is an ecosystem in which the dominant fauna and flora are under human (so-called artificial) selection, and where persistence, continuity, and other aspects of ecosystem homeostasis are external and not a function of feedback loops, interdependence and successional processes (Odum 1984; Risser 1986). It is also generally assumed that species diversity in agroecosystems is greatly reduced and, thus, that herbivore populations are inherently unstable and prone to outbreaks (Wilhelm 1976; Edwards 1977; Price and Waldbauer 1982). These and other factors represent environmental elements and interactions that may constrain foraging by herbivores (Table 17.1), along with phylogenetic influences (see Phylogenetic Constraints). These constraints serve as the ecological and evolutionary selective forces that shape the foraging patterns of herbivores in agroecosystems.

Many concepts and hypotheses have been proposed to define and describe agroecosystems and the plant–arthropod interactions therein. These ideas too often have been prematurely elevated to the level of dogma without being rigorously tested (see Table 17.1, Fig. 17.1). These relatively broad generalizations about agroecosystems have gained wide acceptance even though they are unlikely to be applicable to widely diverse types of agroecosystems. Indeed, the great diversity of agroecosystems (ranging, e.g., from rice fields to forest tree plantations) is often paralleled by equally diverse herbivore guilds and foraging strategies.

Generalizations about agroecosystems often tend to create an impression that agroecosystems are unique, which is not wholly deserved. Moreover, the assumptions on which many of these generalizations are based are often debatable (see Van Emden and Williams 1974; Ehrlich et al. 1977; Rabb 1978; Levins and Wilson 1980), if only because ecological theory, such as that linking diversity with stability, is misapplied (see Goodman 1975; Murdoch 1975; Boyce and Cost 1978) or because the concept (i.e., diversity or stability) is defined in so many different ways so as to be less than useful (Boyce and Cost 1978; Risch et al. 1983). Many generally accepted ideas about agroecosystems are highly dependent on the ecosystems being compared. The proposition that herbivore species diversity is lower in agroecosystems than in unmanaged ecosystems is an example of a generalization that has gained wide acceptance, even though examples of both relatively high and low species diversity can be found in both types of ecosystems. Although some unmanaged ecosystems support diverse and species-rich interacting communities, species diversity is often relatively low in other unmanaged ecosystems, such as many marshes and swamps. The latter ecosystems are nearly monospecific and have natural nutrient supplementation (perhaps comparable to supplemental energy sources of agroecosystems) (Mitchell 1984). In addition,

Table 17.1. Selected Structural, Functional, and Biotic Differences between Natural Ecosystems and Agroecosystems[a]

Characteristics	Agroecosystem	Unmanaged ecosystem
General		
Stability (resilience)	Low	High
Entropy	High	Low
Temporal permanence	Short	Long
Habitat heterogeneity	Simple	Complex
Phenology	Synchronized	Seasonal
Maturity	Immature, early successional	Mature, climax
Life history and genetic characteristics of plants and animals		
Life-history strategies	*r*-selected	*K*-selected
Life cycles	Short (annual)	Long (perennial)
Niche breadth	Broad	Narrow
Reproductive allocation	High	Low
Genetic Variability	Low	High
Defensive allocation	Low	High
Mortality	Density independent	Density dependent
Population and community characteristics		
Total organic material	Low	High
Plant architecture	Simple	Complex
Species diversity	Low	High
Symbiotic associations	Few	Many
Chemical diversity	Low	High
Food webs	Simple	Complex
Population dynamics	Fluctuating	Constant
Internal cycling by plants	Low	High
Synchrony of plant–microorganism activity	Low	High

very few, if any, studies are available that provide complete faunistic surveys of agroecosystems; most focus solely on pest species. Thus, an accurate assessment of the species diversity and richness of many agroecosystems must await thorough surveys of all plants and animals.

Pest Species in Agroecosystems

Of all the feeding behaviors in agroecosystems, leaf feeding is the most common. Of crops in two-thirds of the temperate plant families surveyed by Hill (1987), an average of about 36% of all species damaging plants (ranging from about 31 to 44%, depending on the family) were defoliators (Fig. 17.2A). In the remaining one-third of families evaluated, about a quarter of the pests on crops were

Table 17.1. *(continued)*

Characteristics	Agroecosystem	Unmanaged ecosystem
Temporal diversity of organism activity	Low	High
Balance of plant–microorganism activity	<1	1
Structural diversity of plants	Low	High
Genetic diversity	Low	High
Reproductive potential	Low	High
Control	Abiotic, anthropogenic	Biological
Energetics studies		
Net production (yield) per unit biomass	High	Low (or medium)
Standing crop of biomass	Low	High
Biomass supported per unit energy flow	Low	High
Food chains	Simple, short, linear	Complex, long
Nutrient flux studies		
Element cycles	Open (leaky)	Closed (conservative)
Pathways for chemical transformations	Simple	Complex
Exchange rates	Rapid between soil and plant biomass	Slow, efficient
Runoff	High	Low
Erosion	High	Low
Presence of canopy	Low	High
Litter and debris	Low	High
Rocks	Low	High
Soil water loss to transpiration	Low	High
Soil colloids	Low	High
Leaching losses	High	High
Soil temperature	Low	High
Mineral cycles	Open	Closed

[a]Modified from Mitchell (1984), Odum (1984), Risser (1986), Altieri (1987), and Stinner and Stinner (1989): see text of these references for details.

defoliators (Fig. 17.2A,B). Of the 13 plant families analyzed, the only exception was a group of five species of grasses on which damage was relatively evenly distributed over five feeding types and in which defoliators represented only about 17% of the total (Fig. 17.2A,B). Leaf distortion (caused by species with piercing-sucking mouthparts) was also a dominant form of feeding, and represented about 20 to 25% of all the types of damage on crop plants. Damage of other types was highly variable and generally was a very low proportion of all the damage incurred by plant species within any given family (Fig. 17.2A,B). Although the Coleoptera are important defoliators in agroecosystems, the primary defoliators are Lepidoptera (Hill 1983, 1987).

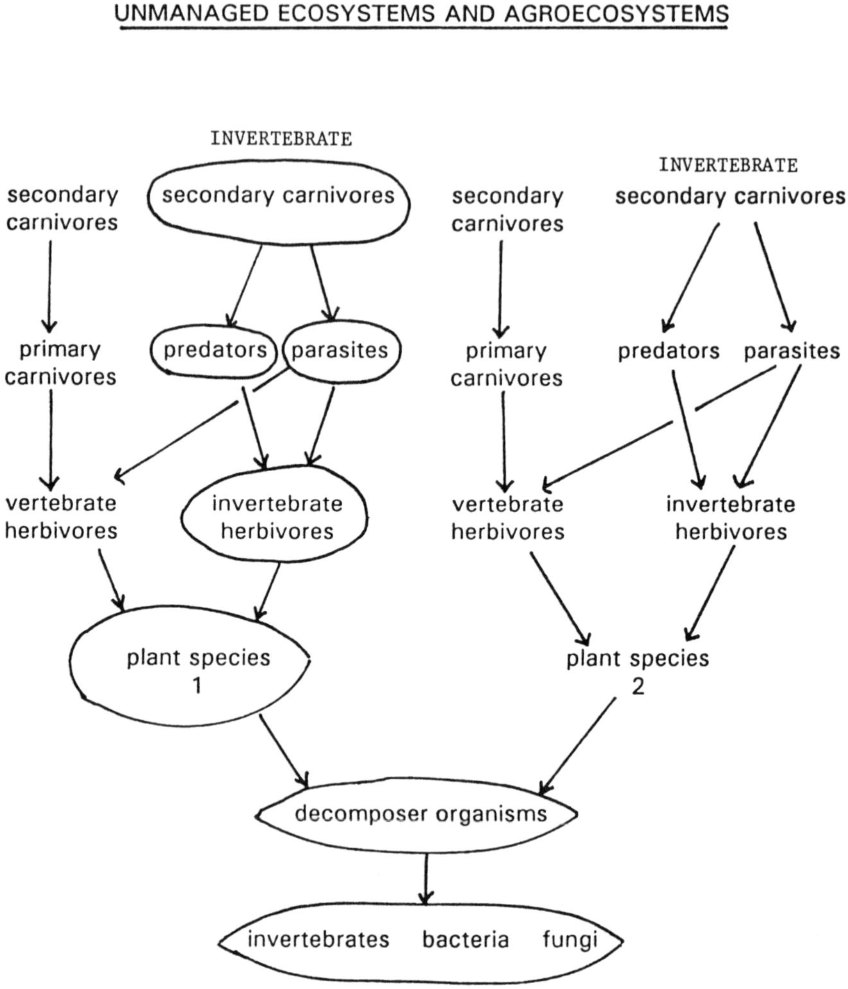

Figure 17.1. Comparison of food webs in agroecosystems and unmanaged habitats. The diagram represents the interactions in unmanaged ecosystems. Those interactions and components that are in circles represent the generalized food web for agroecosystems. (Modification of Stinner and Stinner 1989.)

Among the Lepidoptera in both temperate and tropical agroecosystems, defoliation is by far the dominant feeding behavior (Fig. 17.3). Nearly 40% of all lepidopteran pests are defoliators. Shoot and bud feeders and fruit feeders are a distant second in abundance, representing, in both temperate and tropical areas, about 11 to 15% of pest species (Fig. 17.3). About 75 and 80% of temperate and tropical lepidopterans, respectively, are monophagous (i.e., species that feed on

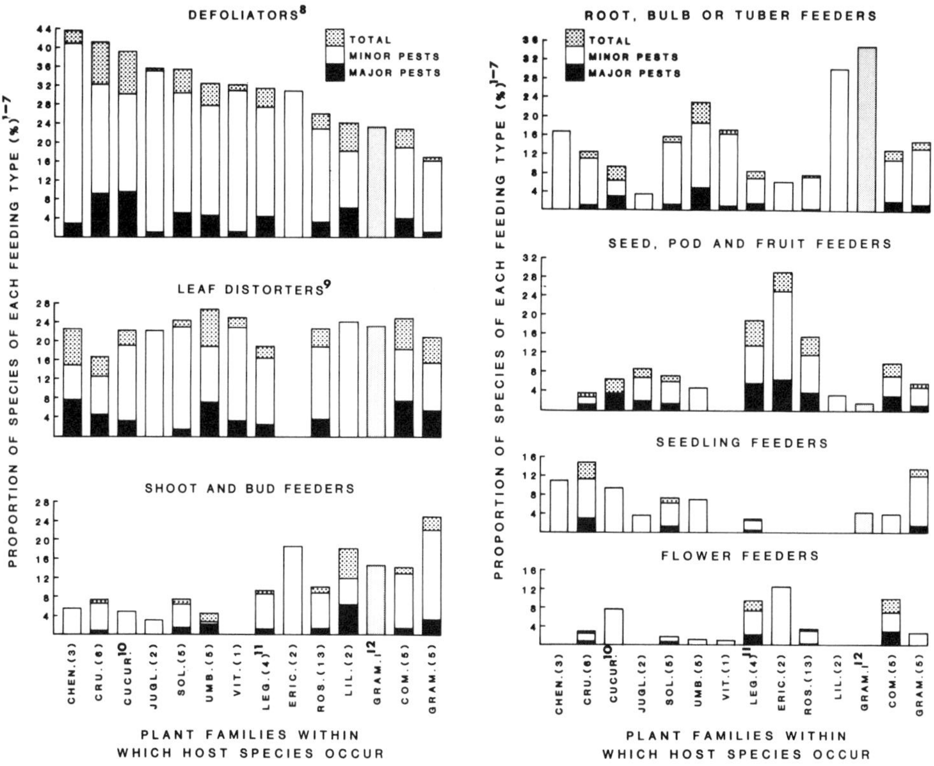

Figure 17.2. Distribution of feeding types among temperate pest species of crops in selected major plant families. (Based on Hill 1987.) (1) Chen., Chenopodiaceae; Com., Compositae; Cru., Cruciferae; Jugl., Juglandaceae; Leg., Leguminosae; Lil., Liliaceae; Sol., Solanaceae; Umb., Umbelliferae; Cucur., Cucurbitaceae; Eric., Ericaceae; Gram., Gramineae; Ros., Rosaceae; Vit., Vitaceae. (2) One pest species may be tallied more than once if host plants are in different families or if adult and larvae cause different types of damage on one host species. (3) Numbers in parentheses represent the number of species per family category considered in the compilation. This number includes particular species such as *Medicago sativa* and generic designations such as *Phaseolus spp*. Note one species, such as, *Brassica* spp. could include a large number of crops such as cauliflower, broccoli, brussels sprouts, kale, turnip, and swede. (4) Stem damage, representing damage to stems, branches, and the trunk of woody crop species, is not shown but is as follows. Major, minor, and total percentage of species for Jugl. is 1.0, 22.1, and 23.1; for Vit. is 0, 16.1, and 16.1; and for Ros. is 1.2, 12.2, and 13.4, respectively. (5) When only the percent of minor species is shown it also represents the total. (6) Gall insects are classified depending or the plant part they affect. (7) Damage may be attributed to a particular pest species or to a pest genera, e.g., to *Epilachna* spp., which, for this figure, is viewed as a single species. (8) Leaf damage includes an assortment of types of injury from removal of pieces of leaves, defoliation, leaf mining, leaf galling, leaf rolling (to create a feeding shelter), leaf scarification, and bronzing. (9) Leaf distortion is restricted to damage by species with piercing-sucking mouthparts that is manifested as stippling, "burn," leaf abnormalities such as curling and rolling (due to injection of toxins), as well as localized or whole plant changes in growth. (10) Cucurbitaceae represents an unspecified number of various cultivated species in nine genera that are termed "biologically similar and have very similar pest spectra" (Hill, 1987). Included are *Cucumis* and *Cucurbita* species. (11) Within the Leguminosae a group classified as "clovers" is not included here but contains crop species in more than 5 genera, that are subjected to damage by 2, 2, 2, and 1 major pest species in the leaf removal, leaf distortion, seed and fruit, and root, bulb, and tuber categories, respectively. Similarly, they are subjected to 14, 7, 2, 8, 7, 1, and 1 minor pest species in the leaf removal, leaf distortion, shoot and bud, seed and fruit, root, seedling, and flower, categories, respectively. (12) Gram. 1, represents an unspecified number of Graminaceae listed as "many species" and not divided into major or minor species.

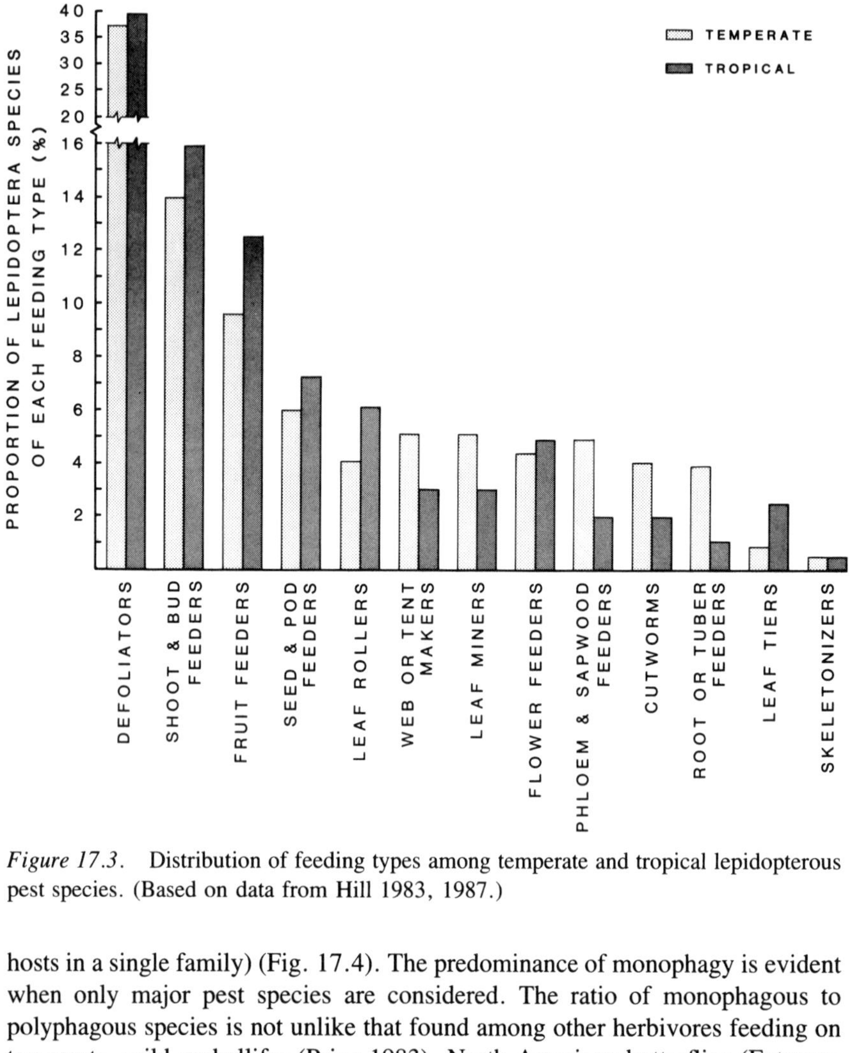

Figure 17.3. Distribution of feeding types among temperate and tropical lepidopterous pest species. (Based on data from Hill 1983, 1987.)

hosts in a single family) (Fig. 17.4). The predominance of monophagy is evident when only major pest species are considered. The ratio of monophagous to polyphagous species is not unlike that found among other herbivores feeding on temperate, wild umbellifer (Price 1983), North American butterflies (Futuyma 1976) and tropical sphingids (Janzen 1981). The abundance of monophagous species in agroecosystems supports Root's (1973) contention that monophagous herbivores are more likely to find and remain in areas where host plants are concentrated, i.e., where they are grown in pure culture. In temperate or tropical agroecosystems, about 75% of all pest species are the only representatives, within their genus, that are economically important (Fig. 17.5). Another 19 to 20% of pest species have one to two congeners that are also pests. About half of all pest species (46 and 51% of temperate and tropical pests species, respectively) belong to the Noctuidae, Tortricidae, and Pyralidae (Fig. 17.6A,B).

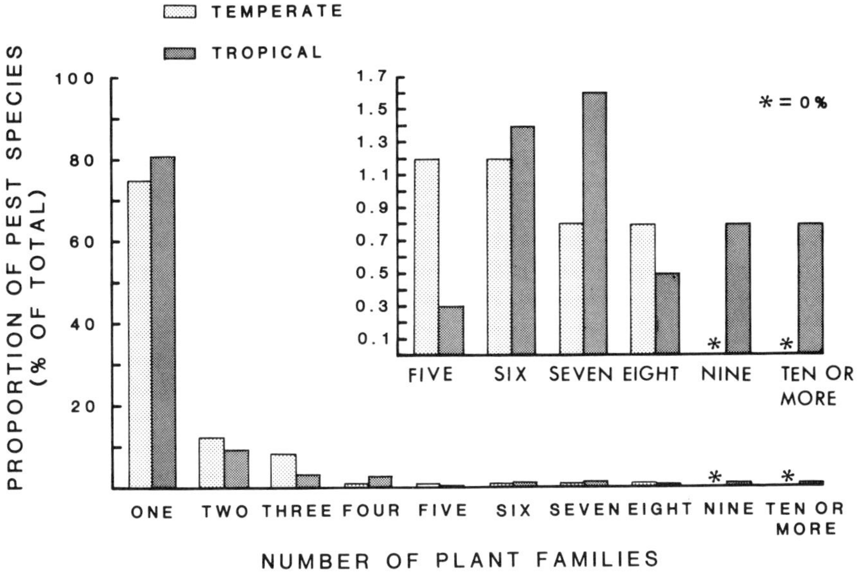

Figure 17.4. Host specificity of Lepidopterous pests in temperate and tropical agroecosystems. (Based on data from Hill 1983, 1987.)

The most specious families are not necessarily the most well represented families in agroecosystems. Although the Noctuidae and Pyralidae are among the largest of Lepidoptera families (21,000 and 20,000 species worldwide, respectively) and are well represented in agroecosystems, other families such as the Geometridae also have about 20,000 species and are relatively poorly represented among agroecosystem pests. In addition, the Tortricidae, which represent about 18 and 13% of temperate and tropical pests, respectively, has only about 5500 species: roughly equivalent to other poorly represented families, such as the Gelechiidae, the Oecophoridae, the Nymphalidae, and the Lycaenidae (Holloway et al. 1987). There appear to be no phylogenetic constraints to the foraging patterns held in common by the Noctuidae, Pyralidae, and Tortricidae, which explain their predominance in agroecosystems since the feeding habits of species in the families are quite diverse.

What Forces Shape Foraging Patterns in Agroecosystems?

Plant Quality

Nutritional

Agricultural plants are the result of breeding and selection for specific traits. Under optimal conditions these traits often include or result in large and water

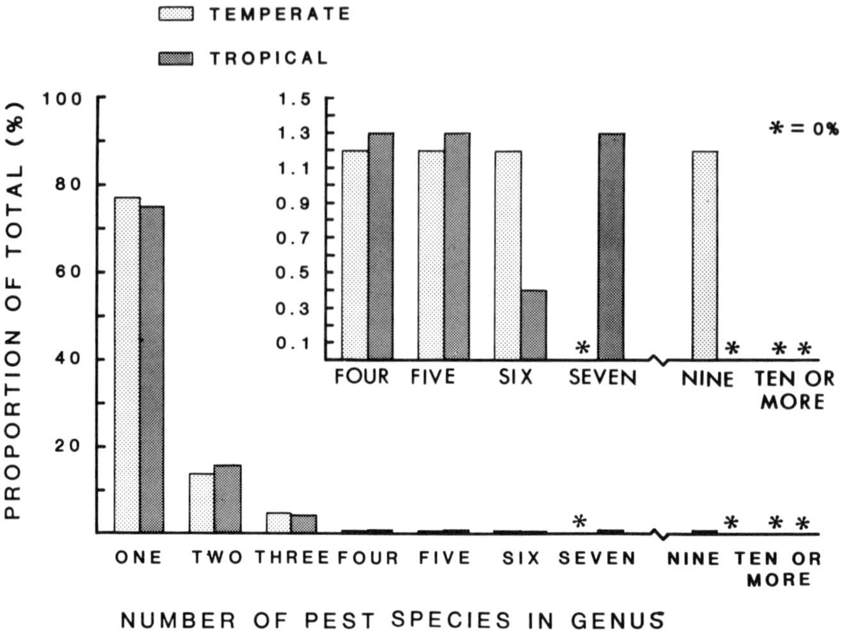

Figure 17.5. Abundance of pestiferous species in genera of pest species of temperate and tropical agroecosystems. (Based on data from Hill 1983, 1987.)

laden fruits, leaves, roots and tubers, and stems. In agroecosystems, nutrient and water supplementation in the form of fertilization and irrigation generally enhances net production and plant tissue quality: providing herbivores with large amounts of high quality food. Water content of plants [determined either by macroclimate, agronomic practices (such as fertilization and irrigation), or crop microclimate] can significantly affect feeding behavior and foraging patterns (Heidorn and Joern 1987, Lewis 1982).

The nutritional physiology of lepidopteran herbivores is affected by food plant water content (Mattson 1980; Slansky, Chapter 2). Growth rates and the efficiency of nitrogen utilization are adversely affected in a variety of larval Lepidoptera, by low water content of leaves (Scriber 1977, 1978, 1979; but see Tabashnik 1982). In addition, to a great extent, preventing water loss in terrestrial organisms such as lepidopteran larvae, which can be as much as 90% water (Brues 1946), depends on the consumption of food with a high water content. Thus, it may be possible that larval distribution may parallel the distribution of plant tissues with preferred moisture content. Similarly, the pattern of distribution of feeding herbivores may be affected by moisture stresses because oviposition preferences may be influenced by leaf water content (Wolfson 1980).

Water stress can have a significant impact on carbohydrate metabolism in

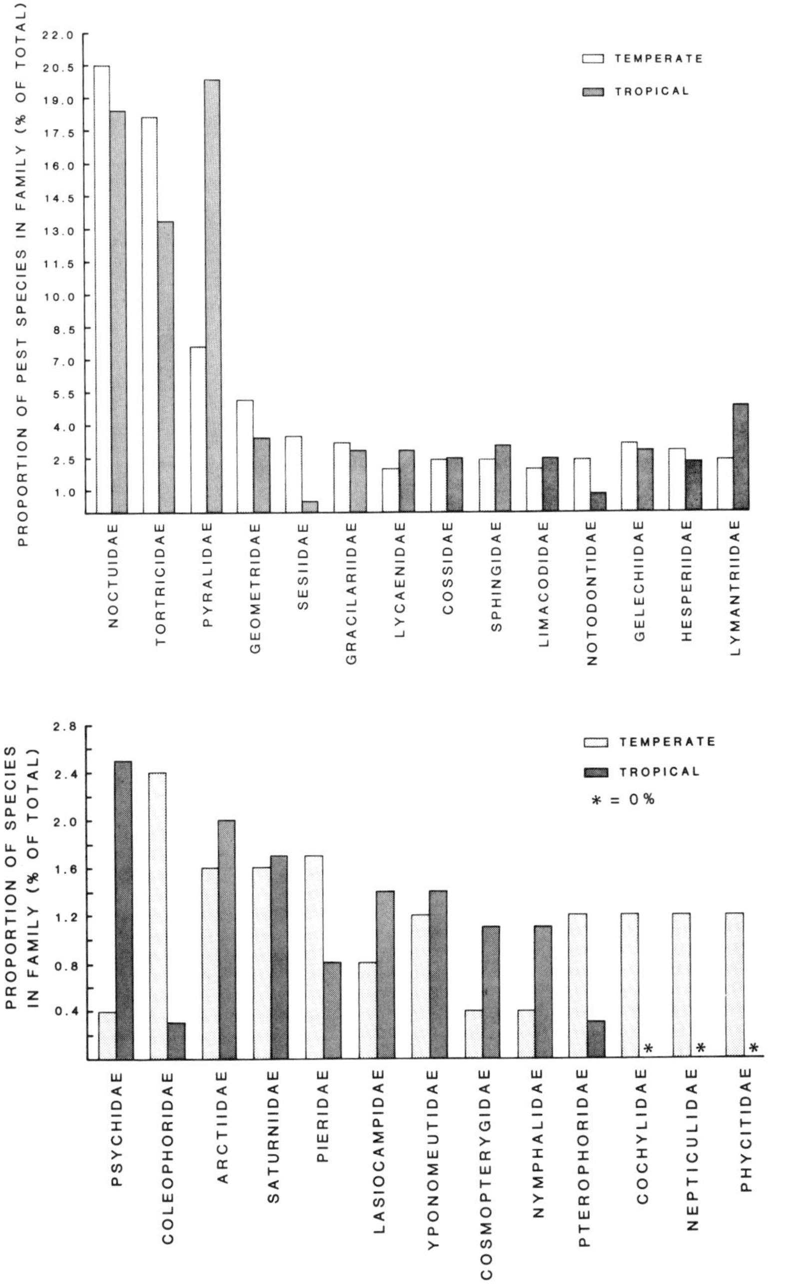

Figure 17.6. Relative abundance of pest species among various families of Lepidoptera feeding on temperate and tropical crops. (Based on data from Hill 1983, 1987.) The remainder of the data set used for this figure is not shown and includes 23 other families whose abundance averages about 0.4% each.

plants (Holtzer et al. 1988). The synthesis, translocation, and partitioning of photosynthate between sugars and starches could significantly influence how phagostimulatory a plant might be or its nutritional value to an herbivore (Holtzer et al. 1988). Because nitrogen content of plants is usually lower than that of insects, nitrogen is a limiting resource for herbivores (Strong et al. 1984) and critical to their survival (Slansky, Chapter 2). The importance of nitrogen and protein levels has been the justification for a great deal of research, much of which is reviewed in McNeill and Southwood (1978), Mattson (1980), and Brodbeck and Strong (1987). Although in many cases water and/or nutrient stresses increase nitrogen levels, in other situations, such stresses may decrease nitrogen content (Gershenzon 1984; Holtzer et al. 1988). Some herbivores, such as *Pieris rapae*, on the crucifer *Brassica oleracea* (Slansky and Feeny 1977) and *P. brassicae* (Courtney 1986), compensate for low leaf nitrogen levels by increased rates of consumption of their host plant. This suggests that for these and similar species, all other things being equal, herbivory levels might be lower in fertilized agroecosystems than in many unmanaged ecosystems that characteristically have plants with low nitrogen (or protein) content. Similarly, one would predict that in any given agroecosystem the intensity of feeding rate by lepidopterous species should increase as seasonal nitrogen uptake and nitrogen levels in plant tissues decline. The need for herbivores to compensate for low nitrogen levels should increase seasonally, particularly in the absence of the reapplication of fertilizer.

Nutrients are not only essential for growth but may act as feeding stimulants (Jones et al. 1972; Slansky and Rodriguez 1987) or may influence the levels of certain behavior-modifying allelochemical cues (Wolfson 1982; Argandona et al. 1983; Scriber et al. 1975). Among the Lepidoptera, glucose, sucrose, and other nutrients act as feeding stimulants (Mattson and Haack 1987). Nutrient (and perhaps allelochemical) cues in many crops bred for rapid growth and biomass accumulation, may provide "supernormal" foraging cues. For example, the preference for fast growing *Brassica chinensis* by the diamond-back moth, *Plutella xylostella*, as well as by other specialists (Finch 1988) may be related, at least in part, to higher levels of nutrient cues or allelochemical signals.

Because many chemical plant defenses are nitrogen based (e.g., alkaloids and cyanogenic glycosides) enhancement of nitrogen levels (by fertilization) may influence the concentration and seasonal patterns of allelochemicals, which induce feeding by monophagous species, and repel polyphages (Berenbaum 1981; Rhoades 1983; but see Eigenbrode and Pimentel 1988; Manuwoto and Scriber 1985). In potatoes grown under poor conditions (cool and cloudy), levels of glycoalkaloids increase with increasing nitrogen availability (Rhoades 1983). Similarly, Berenbaum (1981) noted that the presence or absence of furanocoumarins was linked to total nitrogen content of the soil. She also reported that a similar relationship existed for alkaloids in *Atropa belladonna*, *Nicotiana tabacum*, and *Datura stramonium*. Although these data are highly suggestive, studies are needed in which fertilization (or water stress) is directly shown to alter nutrient levels, water content, and/or allelochemical concentration, and in which these changes, in turn, are shown to alter feeding behavior. In summary, although agronomic practices improve plant growth and make plants more suitable for herbivores, the practice also may make host plants less susceptible to attack (Slansky and Rodriquez 1987, and other chapters in this volume).

Allelochemical

The elimination of chemicals from crops because they impart an undesirable taste (for humans) may be the aim of breeders but a corresponding consequence may be greater susceptibility to herbivores, when the same chemical also serves a defensive role. In contrast, cucurbitacins, a group of bitter compounds, occur at low levels in nonbitter cucurbit varieties, making them considerably less attractive to cucurbit-adapted herbivores. Similarly, cucumbers selected for low cucurbitacin content are unaffected by cucumber-adapted herbivores although they are vigorously fed on by a polyphagous mite (DaCosta and Jones 1971a,b). In general, high levels of cucurbitacins act as arrestants and feeding stimulants for monophagous herbivores (Sinha and Krishna 1970; Howe et al. 1976). Herbivores, such as diabroticite beetles are more readily attracted to wild *Cucurbita* species than to cultivated *Cucurbita* because of the relatively high levels of cucurbitacins in wild species (Carroll and Hoffman 1980; Metcalf et al. 1982). These types of differential responses have also been observed among herbivores of other major crops in families such as the Cruciferae (Finch 1988). In species in this family, compounds such as mustard oil glycosides and even cyanogenic compounds act as feeding stimulants, but primarily for crucifer-adapted herbivores such as the larvae of various lepidopterans (Mattson and Haack 1987). Breeding programs have produced various cultivars that cater to different human "taste" preferences, and growth and/or yield requirements, but that are also differentially defended against herbivory. Variation in defense allelochemistry, in turn, may result in a diversity in patterns of larval (or egg) distribution of herbivores normally associated with the crop, or of any new suite of pests attracted by the new cultivar (Finch 1988).

Many crop plants are effectively defended by a wide variety of chemicals that have significant activity against both monophagous and polyphagous herbivores (Liener 1986; Grainge and Ahmed 1988; Beier 1990), in spite of intense selection for particular traits. However, reduction or elimination of defensive plant allelochemicals through breeding programs could have significant but differential effects on feeding behavior and foraging patterns of monophagous and polyphagous caterpillars; and other lepidopteran–plant interactions in which allelochemicals play key roles. Plant defenses are not automatically reduced or eliminated as a result of selection and breeding, but rather are reduced only in circumstances where the presence of defensive allelochemicals also reduces the quality and/or yield of harvestable parts of the crop. Thus, the leaves of any given cultivar may be as well defended against herbivory as those of its wild congener, if the plant's roots are the harvested portion of the crop (e.g., carrots, or sweet potatoes). This applies, of course, only to those crops in which the presence and/or level of a chemical in different tissues are not correlated. In contrast, level of total glycoalkaloids in the foliage and tubers of wild potato plants are highly correlated (see references in Raman et al. 1979). Human selection for the elimination of glycoalkaloids from tubers is likely to yield low levels of glycoalkaloids in foliage as well.

In still other crops (such as dry beans and cassava) there is little need to change traits that may be associated with defense, even if it involves high levels of toxic allelochemicals. These and other traits may not be selected out of consumable parts of the plants (Ames 1983) because the methods used for processing or cooking the foods deactivate or neutralize toxins or other potentially deleterious traits (Liener 1986). In summary, for many crops species, a more accurate generalization about the consequences of breeding, for defense against herbivory, can be suggested. Whether cultivated plants are more or less defended than their wild counterparts depends on whether the desired crop traits are linked to physical and chemical plant defenses. More importantly, loss of a plant's defensive capabilities is not an automatic consequence of breeding.

The concept that human selection and breeding have resulted in plants with few morphological and chemical defenses appears repeatedly in the scientific literature (Pimentel 1976; Risch 1987; Stinner and Stinner 1989) but may be too broad a generalization to be useful. Support for this contention is minimal, contradictory, and largely anecdotal. Nevertheless, this presumed association between crop breeding and diminished defense against herbivory is often alluded to as the causal mechanism underlying the hypothesis that more outbreaks occur in agroecosystems than in unmanaged ecosystems (Stinner and Stinner 1989). Proponents of this hypothesis suggest that studies reporting a negative correlation between the levels of a specific allelochemical and yield, or between resistance to herbivory and yield, support the hypothesis. Since greater yield is generally the objective of breeding, negative correlations between yield and defensive plant allelochemicals or herbivory by insects would appear to support the hypothesis. Clearly however, these correlations do not constitute evidence that crop plants are less well defended; particularly since similar relationships are observed in both cultivated plants and their wild congeners (see review by Krischik and Denno 1983). In addition, other studies have failed to find any such correlations (Krischik and Denno 1983).

One would predict that if plants in agroecosystems are poorly defended (e.g., contain relatively low levels of defensive allelochemicals), so-called nonadapted, polyphagous herbivores would be strongly represented and relatively abundant, would consume more food, and would be more widely distributed, relative to monophagous species. Conversely, adapted, monophagous species, that often rely on allelochemicals as ovipositional and feeding stimulants or that sequester them for their own protection, should be at a disadvantage and less abundant. This scenario is not supported by my analysis of the information provided by Hill (1983, 1987) on host plant breadth of pest lepidopterans (see Pest Species in Agroecosystems). At least among the Lepidoptera, most species in agroecosystems are monophagous, not polyphagous. This discrepancy may be important in that it suggests that crop allelochemistry may be as varied and complex as that of other plants in unmanaged ecosystems.

Berenbaum (1981) found the greater the chemical complexity of umbellifers the greater the degree of specialization of the insects associated with these plants. She observed that the proportion of monophagous species increased from 0% in plants without furanocoumarins to 43% in plants with both linear and angular furanocoumarins. Conversely, polyphagous species decreased from 64% in plants lacking furanocoumarins to 28.5% in plants with both types of furanocoumarins. If the same phenomenon is true of crop plants than the high proportion of monophagous species may suggest a high degree of chemical complexity among crop plants. A delineation of plant chemicals (of behavioral and physiological importance to lepidopterans) in several crop species and their wild counterparts, and an assessment of the degree of feeding specificity of herbivores using the plants, would provide a direct test of whether chemical complexity and herbivore host specificity are correlated in cultivated and wild plants.

Plant Architecture

Changes in plant and/or leaf color, shape, and structure are not always the objective of selection by breeders but these plant traits may nevertheless be altered by the process. Differences in plant architecture among cultivars of a crop species can be quite distinctive (Figs. 17.7 and 17.8). Although, such changes in shape, color, texture, and size of plants or plant parts may be unintended they nevertheless may have significant direct and indirect consequences on herbivore foraging (Bossenbroek et al. 1977; Hagen and Chabot 1986; Varela and Bernays 1988; Jackai and Oghiakhe 1989). For example, some lepidopterous larvae exhibit directed orientation toward vertical objects, or the larger of two vertical objects, when searching for food (Lance and Barbosa 1982). Similarly, the results of Risch (1980) and Bach (1980, 1984) suggest that the physical structure of the canopy may exert some influence on host plant location and movement within a plant patch, both of which are interrelated components of herbivore foraging. Indeed, the degree of discrimination exercised by foraging herbivores often depends on the opportunities for discrimination that movement provides (Kareiva 1982).

Although changes in plant architecture and associated changes in the plant canopy may also influence herbivore foraging in unmanaged ecosystems, the widespread, abundant, and relatively uniform distribution of plants, more often

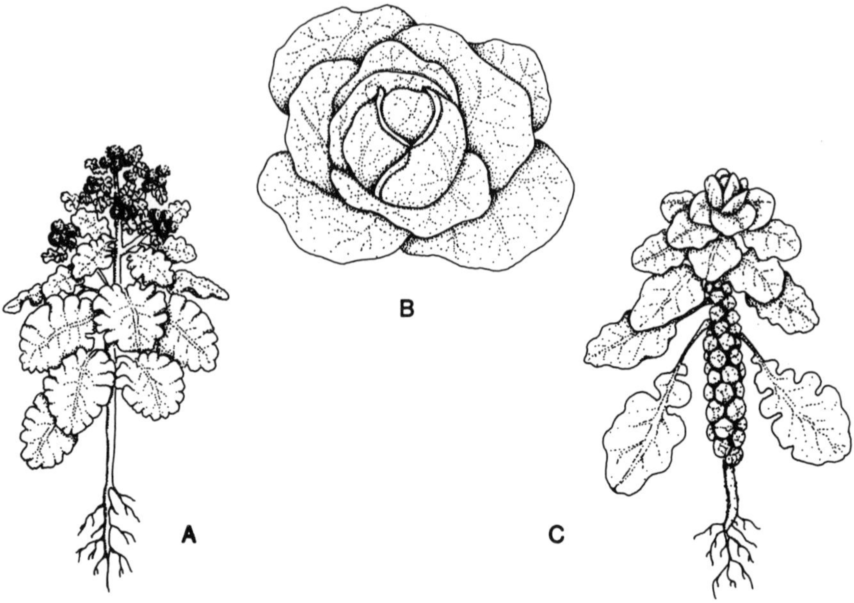

Figure 17.7. Architectural variation among cultivars of *Brassica oleracea*: (A) broccoli, (B) cabbage, and (C) brussels sprouts. (Modified from Hill 1987; illustration by Eileen Hsu.)

associated with agroecosystems, must create distinctive microclimates not found in many unmanaged ecosystems. Changes in plant architecture can have a direct influence (i.e., involving the physical interference of plants or plant parts in food seeking and feeding) or an indirect effect (in which plant shape alters microclimate, for example). Although differences among cultivars may be dramatic (Fig. 17.7) less obvious differences, even between cultivars that appear structurally similar, may produce major changes in microclimate and thus in lepidopteran foraging patterns. For example, air temperature, vapor pressure, and CO_2 levels can be significantly different within the canopies of two soybean canopies (Baldocchi et al. 1983). Differences in canopy microclimate of the relatively similar soybean cultivars, Clark and Harosoy, are due to plant architecture, leaf pubescence, and leaf orientation (Baldocchi et al. 1983). Given the effects of physical factors on feeding one would expect significant qualitative and quantitative differences in foraging behavior.

Climate/Physical Factors

Introduction

The macro- and microclimate to which plants and herbivores are exposed are, in large part, an outcome of the modulation of local and regional weather by

Figure 17.8. Example of the architectural variation among cultivars of bean *Phaseolus vulgaris,* i.e., the bush bean (left) and the pole bean (right) varieties, which may produce distinct microclimatic conditions. (Illustration by Eileen Hsu.)

specific crop characteristics. Agronomic practices may also influence crop microclimate. The use of irrigation, or no-till procedures, the choice of cultivar (and associated differences in plant architecture and canopy; see Fig. 17.7), and planting density and row spacing create or modify the microclimate (Pinter et al. 1975) within which the herbivore lives and feeds. Climatic forces directly and indirectly influence lepidopteran herbivores by altering the quality of the food plant and by influencing the herbivore's nutritional physiology and behavior (Wellington 1949a). In addition, the effects of microclimate on the levels of plant allelochemicals can change the way these chemicals influence feeding behavior of caterpillars. Moisture, light intensity, and associated temperature changes can influence the concentration of many allelochemicals (Loustalot et al. 1947; Winters and Loustalot 1952). For example, the concentrations of cyanogenic compounds, benzoxazolinones, glycoalkaloids, and tannins can depend on whether the plant is shaded or not (Tingey and Singh 1980; Rhoades 1983). Even though insolation may be complete on the upper canopy of crop plants, leaves within the canopy, whether those of soybean or a forest tree, can be shaded and thus affected.

Thermoregulation and Foraging Patterns

Because insects are poikilotherms, searching for suitable food and feeding itself are significantly influenced by changes in physical factors such as temperature and humidity (Knapp and Casey 1986; Casey, Chapter 1). Temperature increases, approaching the optimal temperature for a particular caterpillar's growth, can result in increased consumption rate (and associated reduced digestive efficiency) (Ali 1973; Slansky and Rodriquez 1987). In general, however, herbivores do not commonly respond to single factors but respond and orient to specific ranges of humidity and temperature that maximize fitness. Herbivores may be thermoregulators (Casey and Knapp 1987) and thus avoid regions of relatively low temperature that, e.g., produce detrimental growth rates and perhaps increased susceptibility to biotic mortality agents. In general, thermoregulators adjust (i.e., increase) body temperature by indulging in a variety of behaviors such as communal basking, posture adjustments, and shelter or shade seeking. Resting metabolism of thermal regulators is strongly temperature dependent at normal ambient temperatures.

In contrast, thermal conformers may retreat to sheltered niches and remain inactive but their body temperatures never exceed air temperatures by more than 5°C. They exhibit none of the thermoregulatory behavior noted above (Casey and Knapp 1987; Casey, Chapter 1). Given that thermoconforming or thermoregulating are appropriate adaptations or "a compromise shaped by physical and biotic components of the environment" (Knapp and Casey 1986) both behavioral strategies are observed in lepidopteran larvae of unmanaged ecosystems and agroecosystems. One must assume that thermoregulatory behavior (or the lack there of) has evolved so as to maintain a balance between adjusting to microclimatic constraints and the consequences of that adjustment, rather than by the type of habitat (Knapp and Casey 1986).

Stresses from water loss (Welling 1949a,b; Willmer 1980) may also induce foraging for high water content plant tissues and may result in an herbivore distribution among plants and/or plant parts that enhances the probability of finding such tissues. Thus, the advantages and disadvantages of thermoregulatory behavior may be measured by benefits accrued in the selection of a microclimate, which minimizes water loss but also provides abundant, high quality plant tissues for the herbivore. One might speculate that the evolution of certain thermoregulatory behavior and patterns of foraging may be enhanced when the behaviors also increase overall fitness of the herbivore by, for example, enhancing growth and development (Courtney 1986; Knapp and Casey 1986).

Many of these multipurpose thermoregulatory behaviors also involved feeding. Larvae may bask in large feeding aggregations, web or tie leaves, or make tents. Other multipurpose feeding behaviors may involve lack of behavior, i.e., inactivity and avoidance of solar radiation by feeding from the underside of leaves (Fig. 17.9) or at night (Casey and Knapp 1987). Basking in large aggregations, for example, can further increase survival and development due to earlier and more efficient feeding among aggregated larvae (Matsumoto 1989). Similarly, leaf rolling and leaf tying may modify thermal conditions so as to ensure favorable conditions for growth and survival, but may also protect against phototoxic

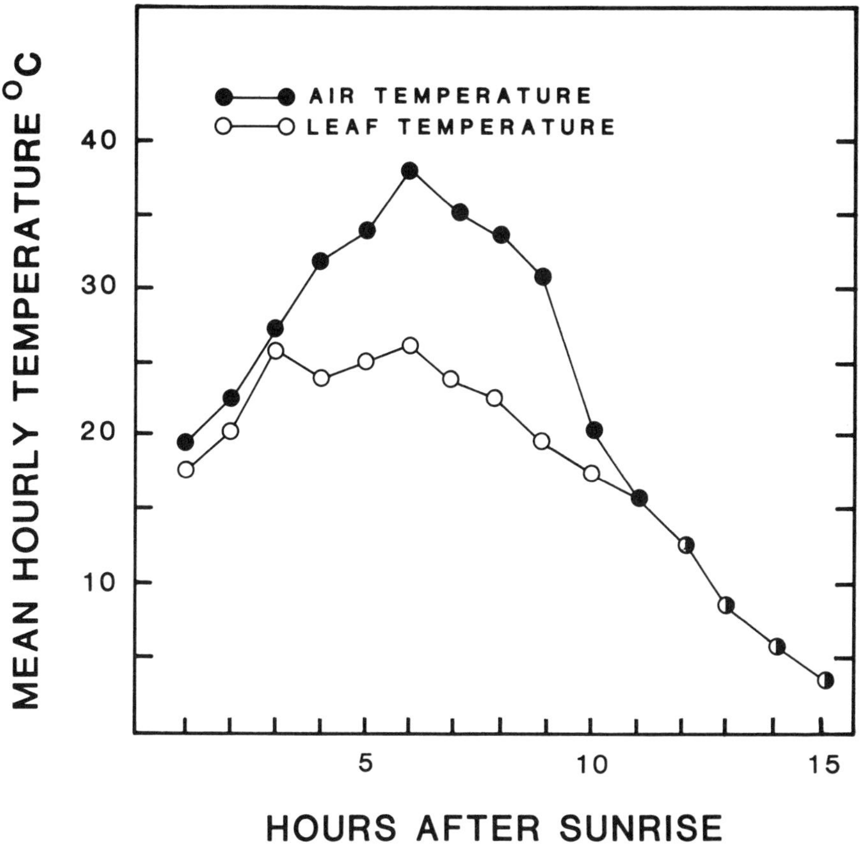

Figure 17.9. Differences between the under leaf surface temperature of apple foliage and air temperature. (From Ferro et al. 1979.)

allelochemicals (Berenbaum 1978; Sandberg and Berenbaum 1989) as other feeding behaviors appear to protect against parasitoids and predators (Heinrich 1979; Heinrich and Collins 1983; Powell 1980).

Light and Herbivore Foraging

Changes in the quality and quantity of light can affect the foraging and feeding behavior of lepidopteran herbivores. Late instars of nocturnally active cutworm larvae are photonegative and respond to very small changes in light intensity. They will cease feeding even in bright moonlight (0.5 lux) (Madge 1964). The length of the photoperiod may affect the amount of food consumed by larvae of species such as the silkworm *Philosamia ricini* (Ali and Salem 1978). Similarly, the feeding rate of the Colorado potato beetle given leaf disks, from young tomato plants grown with short days, was greater than that of beetles given disks from plants grown with long days (Sinden et al. 1978).

Pest species exhibit photopositive and photonegative responses (Archer and Musick 1976) that lead them toward or away from preferred temperature zones, within which feeding and therefore growth is maximized. Light in this circumstance serves as an important token stimulus, i.e., a stimulus that causes a response by an organism, that is an adaptive response to another biotic factor other than the stimulus, but is closely linked to it. Usually the token stimulus itself does not cause any biological change in the responding organism. Thus, e.g., movement away from intense light may be an adaptation to avoid the high temperatures found in sunny areas. It may also be advantageous to avoid such areas because of the inferior quality of the plant tissues that grow under such light conditions, or because of direct detrimental effects of high temperature on the herbivore.

Whether the abundance and distribution of crop plants influence the quantity and quality of light in a fashion unique to agroecosystems has not yet been investigated. However, some conditions that potentially can alter the reflectance and transmittance of light often are more frequent and widespread in agroecosystems. One such factor is plant disease. Although not rare in unmanaged populations, plant disease is often more extensive and severe in agroecosystems (Agrios 1988; but see Barbosa 1991). Typically, diseases, both localized and systemic, cause color changes ranging from localized chlorotic spots to changes affecting the entire plant. Any such color changes can result in significant changes in the response of herbivores to plant parts or entire plants. The vision of caterpillars is usually less developed than that of adults; the focal length is short and the perception of form is rudimentary (Dethier 1988). Nevertheless, lepidopterous larvae can discriminate colors (Beck and Schoonhoven 1980), so it may be possible that color changes caused by plant diseases can be perceived by larvae and affect foraging and plant tissue selection.

Plant Disease

Although plant disease is probably equally common in both managed and unmanaged ecosystems (Burdon 1987), the apparently greater severity and persistence of plant pathogens in agroecosystems are undeniable (Agrios, 1988). Plant diseases often fail to kill crops but instead may cause major, long-term quantitative and qualitative changes in the chemistry, growth, and physical appearance of plants. In general, plant disease may result in a reduction of energy reserves, a reduction in the efficiency of physiological processes, and/or a diversion of resources to defense (Hammond and Hardy 1988; Barbosa 1991). Plant disease and the changes observed as the plant responds to infection often cause changes in the occurrence and concentration of chemicals, which typically influence feeding behavior and thus influence foraging patterns of herbivores. The study of the effects of plant disease on herbivores has focused, primarily on herbivores other than lepidopterans. Although the linkage between changes in the nutritional quality of diseased plants and changes in feeding behavior, development, and survival of insect herbivores has been shown in only a few studies, similar changes in the nutritional suitability of plants (caused by other factors) have been shown to affect the behavior and physiology of herbivores (Slansky and Rodriquez, 1987; Slansky, Chapter 2).

Diseased plants are not only smaller (with correspondingly smaller leaves, stems, buds, etc.) but are biochemically different versions of their healthy counterparts. Changes in both the levels and types of amino acids, as well abnormal carbohydrate accumulation, have been noted in diseased plants (Allison 1953; Cooper and Selman 1974; Fife 1956; Goodman et al. 1965; Jensen 1969; Mainer and Leath 1978; Porter and Weinstein 1960; Selman et al. 1961). Significant changes in the distribution of assimilates can be observed even in newly infected plants. For example, Burdon (1987) noted that not only do diseased plants (such as beans, wheat, radish, and potato) fail to distribute assimilates to various plant tissues but diseased leaves may act as a net energy sink. The result of these types of distortions is reflected in changes in the relative dry weight of leaves, shoots, roots, and reproductive structures and their nutritional value to herbivores. Tobacco hornworm (*Manduca sexta*) growth rate is reduced by 27% when larvae feed on plants with localized tobacco mosaic virus infections. In contrast, growth rate is reduced by only 16% when hornworms consume plants with systemic infections (Hare 1983; but see Wan and Barbosa, 1990). What is most interesting is that the 27% growth reduction was attributed to a reduced consumption rate among hornworms feeding on plants with localized tobacco mosaic virus (Hare 1983). The Colorado potato beetle (*Leptinotarsa decemlineata*) suffers similar detrimental effects when it feeds on diseased plants (Hare and Dodds 1987).

A plant species, when healthy, often produces distinctive volatiles and contains allelochemicals in their tissues that produce characteristic "tastes" and "odors." Many of these compounds affect host plant finding and also significantly influence initiation and maintenance of feeding (Beck and Schoonhoven 1980; Ahmad 1983). In diseased plants, compounds are produced that are either novel or which occur at atypical concentrations. The production of phytoalexins (i.e., compounds that are produced as a result of infection and which inhibit the growth of microorganisms) is an example. Phytoalexins, such as sesquiterpenoids and norsesquiterpenoids, can accumulate in infected solanaceous plants, as do isoflavonoids in many legumes, furanoterpenoids in Convolvulaceae, and isocoumarins in Umbelliferae. Whether a plant is diseased or not may determine what allelochemicals are present in the plant (Kuc 1966; Kuc et al. 1975), their concentrations (Matsumoto 1962), or whether other chemicals are produced that mask chemicals that influence insect behavior. Novel, disease-induced compounds or changes in preexisting chemicals may attract herbivores not normally found on the plant.

When rusty mottle disease affects plants such as cherry trees it induces about three times as much kaempferol and quercitin and two times as much caffeic acid as contained in healthy leaves (Goodman et al. 1967). The induction of compounds such as quercitin may affect the behavior of cherry tree herbivores. Quercitin, for example, induces *Aphis rumicis* to settle and feed on its host plant (Dixon 1985).

Some herbivores prefer to feed on diseased plants over healthy plants. *Melanoplus differentialis* and *Melanoplus mexicanus* (Sauss.) prefer to eat plant tissues infected with *Puccinia helianthii* rather than healthy tissues (Lewis 1979). They consume more sunflower (*Helianthus annuus*) leaves infested with the rust fungus, when given a choice between healthy and diseased leaves (Lewis 1979). A similar preference by *Zonocerus variegatus* for fungus-infected cassava leaves was attributed to the lack of any significant concentration of HCN in diseased leaves (Lewis 1979).

Is it possible that associations of herbivores with diseased plants, even if they are facultative and/or unpredictable, might mediate changes in an herbivore's host range (see Kennedy 1951, Barbosa, 1991)? Studies of the leafhopper *Dalbulus maidis* indicate that infection by a mycoplasma-like organism (MLO) of a plant that is normally unacceptable, makes it acceptable, as well as other plant species that were also formally unacceptable (see Barbosa 1991 for details). Although further research is needed on this phenomenon, the results of work on this leafhopper suggest that, in general, plant disease and the subsequent changes in plants may be important in the feeding behavior and foraging patterns of herbivores.

Changes caused by plant disease may serve as the mechanism facilitating the incorporation of new plant species into the host range of herbivores (particularly polyphagous species) if, e.g., (1) the suitability, for herbivores, of preferred but diseased hosts is reduced (Boiteau and Singh 1982), or the attractiveness of (normally preferred) diseased plants is reduced (Gibbs 1980), and (2) hosts that are normally unfavored (compared to healthy preferred hosts), are nevertheless more suitable than diseased preferred host (Hare and Dodds 1987). Physiological adaptation may not be difficult once the behavioral (preference) barrier has been overcome. No analysis of the relative ease with which physiological barriers are breached after behavioral adaptations are achieved by lepidopterans has yet been undertaken. However, it is conceivable that the process may be facilitated by consistent exposure to plants with low level or short-term infections and may enable slow adaptive physiological adjustments by herbivores. It may be possible that over time, and given the appropriate circumstances, exposure of herbivores to diseased plants may drive the evolution of physiological adaptations necessary for shifts onto new host plants.

Natural Enemies

The feeding behavior of lepidopterans may be affected by natural enemies. The presence of natural enemies is, like a wide variety of others factors, a common feature of agroecosystems and unmanaged ecosystems, and therefore may not play a unique role in agroecosystems. However, it is possible that interactions involving natural enemies and unique to agroecosystems do exist but have yet to be elucidated by appropriate studies. For example, it may be possible in a particular instance that the foraging behavior of a lepidopteran larvae might be affected by the presence and abundance of introduced natural enemies in ways that are distinct from those observed in unmanaged habitats, where introduced natural enemies do not exist. The effects of natural enemies in agroecosystems would be unique if, for example, (1) the impact of introduced natural enemies differs from that of indigenous species, (2) the atypical abundance of natural enemies caused by intentional releases results in predator–prey interactions that are distinct from those between hosts and indigenous parasitoid complexes, or (3) introduced natural enemies exhibit life history traits, behaviors, and a community structure that affects host feeding and foraging behavior and that differs from that of most indigenous natural enemies.

Natural enemies may influence feeding behavior and foraging in various ways. I will discuss a few, to illustrate how natural enemies might affect foraging and feeding. However, the determination of whether these predator–prey interactions are unique to agroecosystems or occur there more frequently must await further research. Feeding behavior and the feeding site of many lepidopterans may be affected significantly by the presence or activity of natural enemies. Heinrich (1979) suggested that free-feeding defoliators tend to consume only partially many different leaves or to cut off partially consumed leaves as an adaptive behavior that keeps larvae away from parasitoids and predators that cue in on damaged leaves (Nealis 1986; Roland 1986; Zanen et al. 1989; Turlings et al. 1990). Indeed, some parasitoids lay their eggs directly along the chewed edge of a leaf (Odell and Godwin 1984).

If natural enemy avoidance is a major ecological force shaping foraging patterns and feeding site selection, one might suggest that in some instances feeding in relatively less exposed locales such as under leaves, within shelters, at night, and so on, may be adaptations for predator–parasitoid avoidance. The larvae of the pyralid moth, *Herpetogramma aeglealis*, construct and inhabit a series of identical shelters on the same plant. Over several weeks many of these accumulate and about 85% may be empty. Ruehlmann et al. (1988) suggested that the high proportion of empty bags would fail to reinforce parasitoid foraging and enhance larval survival. Even in the absence of special adaptations by herbivores, the presence of natural enemies can induce movement and dispersion, and thus alter foraging patterns. The presence of predators may cause herbivores to hyper-disperse, may interrupt feeding, or may induce defensive behaviors with a distinctly different dispersion pattern than that associated solely with feeding (Myers and Campbell 1976; Roitberg et al. 1979; Kidd 1982).

Natural enemies can also have dramatic effects on herbivore feeding behavior. When parasitized, most herbivores feed significantly less or at slower rates than unparasitized immatures. This relationship has been observed among beetle

(Armbrust et al. 1970; Duodu and Davis 1974), aphid (Cloutier and Mackauer 1979, 1980; Couchman and King 1979), and lepidopteran (Guillot and Vinson 1973; Schoenbohm and Turpin 1977; Sajap et al. 1978; Thompson 1982, 1983; Isenhour 1988; Powell 1989; Bentz and Barbosa 1990) herbivores parasitized by hymenopterous parasitoids. Similarly, lepidopterans parasitized by tachinids feed less than unparasitized larvae (Brewer and King 1978; Levine and Clement 1981; Parkman and Shepard 1981; Huebner and Chiang 1982; Mani et al. 1982). Not all parasitoids cause their hosts to feed less. Indeed the same host, e.g., *Diatraea saccharalis* (Brewer and King 1980, 1981) or *Hypera postica* (Armbrust et al. 1970; Morrison and Pass 1974), may feed less when parasitized by one parasitoid species, and more than (or the same as) unparasitized host larvae when parasitized by a second species. The few examples of lepidopterans in which an increased consumption by parasitized larvae was observed are complicated by parallel influences of parasitism, the induction of supernumerary molts, or the extension of the larval period (Rahman 1970; Parker and Pinnell 1973; Hunter and Stoner 1975; Slansky 1978; Jones et al. 1982).

Other natural enemies can similarly alter herbivore feeding. Viruses, bacteria, fungi, and microsporidia have all been shown to reduce host feeding (Subrahmanyam and Ramakrishnan 1981; Sareen et al. 1983; Mohamed et al. 1982; Retnakaran et al. 1983; Oma and Hewitt 1984; Rombach et al. 1989).

Plant Distribution and Abundance

Among the most distinctive aspects of agroecosystems are the abundance (and/or density) of a single (or a few) plant species and their relatively uniform dispersion. These attributes may modify host plant selection behavior of herbivores, and the distribution of larvae within and between plants. A particular pattern of plant density and spacing, or level of plant abundance, can be exploited in very different ways by different herbivore species (Cromartie 1975). Some studies, such as those of Finch and Skinner (1976), Luginbill and McNeal (1958), Adesiyun (1978), and others, suggest that the number of herbivores per plant is lower in densely planted crops than in sparse plantings. But, Crawley (1983) suggests that this relationship may be confounded by the fact that in dense crop plantings, plants are smaller (see Farrell 1976; Mayse 1978) or that, at high density, a fixed number of immigrant pests per unit area is distributed over more plants. In addition, if there is greater interplant movement by foraging herbivores when food plants are close together (Kareiva 1982; Parker 1982) then tenure time on plants of herbivores might be short and the likelihood of finding herbivores on a given plant, at any one time, would be less.

Variation in plant density often results in changes in plant quality and/or microclimate (Kareiva 1983). The direct and indirect effects of planting density and spacing can significantly influence plant quality, and thus herbivore feeding (Mayse 1978). Moisture content of stems, stem diameter, and plant height at peak oviposition time of *Cephus cinctus* (the wheat stem sawfly) were reduced in plots that were densely seeded and that had narrow row spacing (Luginbill and McNeal 1958). The positive correlation between sawfly feeding and plant height, stem diameter, and moisture content led Luginbill and McNeal (1958) to suggest that females prefer to oviposit on larger, lush plants. As the spacing between rows was decreased or rows were densely seeded, feeding by the wheat stem sawfly decreased (Luginbill and McNeal 1958). Similarly, as corn plant density increased and stalk vigor decreased, feeding by the southwestern corn borer *Diatraea grandiosella* decreased (Zepp and Keaster 1977). Changes in microclimate are among the most significant indirect influences on foraging that result from the density and spacing of crop plants. As noted above, microclimate can influence the feeding of herbivores or influence natural enemies (Alston et al. 1991; Sprenkel et al. 1979) that affect mortality or feeding behavior (see Weseloh, Chapter 6).

The abundance of a crop may influence not only the distribution of larvae (as reflected in oviposition patterns) but may determine which herbivores colonize and feed on the crop. Karieva (1983) notes that entirely different herbivores are associated with the same plant species, depending on the spatial dispersion of the plant. The foraging patterns of a given herbivore may also differ dramatically depending on the type of other herbivores that cooccur on its host (Yokoi and Tsuji 1975) or previously fed on the plant (Tallamy and Raupp 1991). Host plant tissue choice in a plant altered by previous herbivory may be quite different from that in an undamaged plant. Similarly, increased movement in response to damaged induced changes may result in more extensive herbivory than that found in undamaged plants (Bergelson et al. 1986)

Insecticides

A large number of other factors and interactions can influence the foraging of herbivores. Among the most important of these is the use of pesticides. The extent of their influence is dependent on circumstance-specific agronomic variables. The direct effect of insecticides on plants and its consequence on pest herbivores, for example, depends on the crop, the insecticide used and the pest species. Nevertheless, the types of changes in the nutritional quality of plant tissues caused by insecticides can significantly affect the growth, survival, and thus presumably the feeding behavior of pest herbivores. Exposure to insecticides can alter plants in many significant ways. The alteration of plant physiology by carbofuran, a carbamate insecticide, is one of the best documented examples of the influence of an insecticide. Carbofuran increases growth and yield in burly tobacco (Pless et al. 1971), and grain yield in corn (Daynard et al. 1975), and inhibits the enzymatic degradation of indole acetic acid (IAA), perhaps allowing this plant hormone to persist in plant tissues and cause increased growth (Lee 1977).

Mellors et al. (1984) found that soybeans treated with carbofuran exhibited enhanced growth, a result similar to that found by Wheeler and Bass (1971).

The use of the insecticide decamethrin has been found to decrease the ratio of carbohydrates to nitrogen and increase the levels of free amino nitrogen in a susceptible rice strain (Buenaflor et al. 1981) and enhance growth of rice plants (Chelliah and Heinrichs 1980). Methyl parathion similarly enhanced growth of rice plants (Chelliah and Heinrichs 1980). Many of these changes in plant chemistry and suitability may influence the feeding behavior and foraging patterns of herbivores. Mineral (or soil) nutrition (Dale 1988) and the use of various other synthetic chemicals such as plant growth regulators (Campbell 1988) may have effects similar to those of insecticides. However, space limitations, and in many cases lack of data, prevent any logical or useful discussion of their influence on foraging patterns of herbivores.

Pesticides may have important direct effects on herbivores in addition to their toxicity. Sublethal levels of pesticides can alter the behavioral responses of herbivores to their host plants. However, some pesticides do not induce behavioral changes, and others alter only specific behaviors. For example, carbaryl and methamidophos possess little or no repellency when presented to diamondback (*Plutella xylostella*) larvae (Kumar and Chapman 1984). Feeding by 20 third instars placed in petri dishes with carbaryl and methamidophos treated leaf disks (at a LC_1 concentration) was not significantly different from that of larvae given untreated leaf disks. However, at a LC_{50} there were significant differences in leaf consumption. In contrast, the pyrethroid insecticides, permethrin and fenvalerate, significantly repelled more larvae, and those larvae consumed less leaf tissue, when disks were treated at LC_1 or LC_{50}, than larvae given untreated leaves (Kumar and Chapman 1984). In addition, the average number of eggs deposited by females on leaf disks treated with pyrethroids at a LC_{50} was significantly lower than that of untreated control females, although no such differences were found at LC_1 or with the other two insecticides. Similar results have been obtained for larvae of *Spodoptera littoralis*, *Pieris brassicae*, and other herbivores (Tan 1981; see references in Kumar and Chapman 1984).

Although these results suggest that sublethal effects of insecticide used in agroecosystems can have a significant impact on foraging patterns and feeding behavior of lepidopterans, no such detailed investigations have been conducted. One can speculate a variety of possibilities. Hyperdispersion after contact with treated foliage, compared to the movements of larvae in untreated patches of host plants, would seem a high probability given the repellency of large portions of the sprayed host plant and assuming larvae will actively search preferred unsprayed tissues. A high degree of aggregation of feeding larvae in certain parts of a plant or in certain areas of a field might be a reasonable expectation for similar reasons. For other crops with a dense canopy, insecticide coverage may not extend beyond the upper canopy foliage and larvae may congregate in lower parts of the crop.

These larvae would then be exposed to a different set of microclimatic conditions and thus, presumably, would exhibit feeding behavior that differs from that in the upper canopy. Finally, whether the repellent effects of insecticides are of any relevance to lepidopteran leaf miners, borers, leaf rollers, or leaf tiers is still an unanswered question.

Lepidoptera Foraging on Plants in Agroecosystems: Epilogue

In this chapter I have discussed some of the aspects of an agroecosystem that potentially influence the foraging behavior of the larvae of lepidopterous pest species. I have suggested a variety of possible trophic interactions involving feeding and foraging behaviors, some with more supporting data than others. It is my hope that the discussion presented will stimulate experimental examination of some of these ideas.

I have argued in this chapter that there are two types of factors that are likely to affect caterpillar feeding behavior and foraging patterns. The first is agronomic practices such as fertilization, irrigation, the degree of tilling, crop spacing, and pesticide application, that directly or indirectly affect herbivore feeding. The second type of factor is biological based factors such as the defensive capabilities of plants against herbivory, the degree of complexity of food webs, the variation in morphology and the architecture of crop varieties, etc. For some of these, and other factors, there is little relevant information upon which to draw conclusions. Nevertheless, I hope I have demonstrated that the actual and potential impact of these factors on feeding behavior is not only intriguing but of importance to the development of theory and to the management of pests of agroecosystems.

Clearly, more detailed research is needed on the influence of factors unique to agroecosystems on foraging behavior and how they act as the driving forces in the trophic interactions among herbivores and between herbivores and the agricultural habitat. However, after reviewing and assessing a great deal of literature on the foraging and feeding behavior of herbivores, I conclude that progress might be best achieved in agroecology by setting, where possible, certain priorities in future research. First, it is imperative to reevaluate, in many different types of agroecosystems, some of the generalizations that have been elevated to the status of dogma by virtue of repetition rather than experimental data. Second, once more applicable theories are generated, research should seek to determine which of the factors associated with agricultural practices act, as selective forces in ecological time, to shape behavioral interactions such as those involved in feeding and foraging behavior. Finally, comparative analyses of agroecosystems and unmanaged ecosystems, at the population and community levels can provide insights into the consequences of feeding and foraging strategies of herbivores such as lepidopterans.

Acknowledgments

I wish to recognize and thank Ms. Leslie Smith for her assistance with data tabulation and Ms. Eileen Hsu for preparing the graphs in this chapter. A special thanks to everyone in the lab, J. Bentz, P. Gross, S. Hight, J. Kemper, K. Kester, and N. Mallampalli for being constant sources of the kind of intellectual stimulation that exercises the mind, shapes ideas, and challenges complacency. Finally, my thanks to the editors for their extremely helpful and constructive comments. In contrast, I would like to acknowledge the obnoxious, occasionally petty, and generally unpleasant comments of an anonymous reviewer.

Literature Cited

Adesiyun, A. A. 1978. Effects of seeding density and spacial distribution of oat plants on colonization and development of *Oscinella frit* (Diptera: Chloropidae). J. Appl. Ecol. 15:797–808.

Agrios, G. N. 1988. Plant Pathology, 3rd ed. Academic Press, New York.

Ahmad, S. 1983. Herbivorous Insects. Host Seeking Behavior and Mechanisms. Academic Press, New York.

Ali, M. 1973. Influence of photoperiod and temperature on the food consumption of the alfalfa beetle, *Subcoccinella vigintiquatuorpunctata* L. (Coleoptera: Coccinellidae). Acta Phytopath. Acad. Sci. Hung. 8:207–215.

Ali, M. A., and Salem, M. S. 1978. Feeding of larvae of the silkworm *Philosamia ricini* Boisd. under the influence of different photoperiods. Acta Phytopathol. Sci. Hung. 13:197–204.

Allison, R. M. 1953. Effect of leaf roll virus infection on the soluble nitrogen composition of potato tubers. Nature (London) 171:573.

Alston, D. G., Bradley, J. R., Jr., Schmitt, D. P., and Coble, H. D. 1991. Relationship of *Heliothis zea* predators, parasitoids and entomopathogens to canopy development in soybeans as affected by *Heterodera glycines* and weeds. Entomol. Exp. Appl. 58:279–288.

Altieri, M. A. 1987. Agroecology. The Scientific Basis of Alternative Agriculture. Westview Press, Boulder CO.

Ames, B. N. 1983. Dietary carcinogens and carcinogens. Science 221:1256–1264.

Ankersmit, G. W., Acreman, T. M., and Dijkman, H. 1981. Parasitism of colour forms in *Sitobion avenae*. Entomol. Exp. Appl. 29:362–363.

Archer, T. L., and Musick, G. J. 1976. Responses of black cutworm larvae to light at several intensities. Ann. Ent. Soc. Am. 69:476–478.

Argandona, V. H., Corcuera, L. J., Niemeyer, H. M., and Campbell, B. C. 1983. Toxicity and feeding deterrency of hydroxamic acids from graminae in synthetic diets against the greenbug, *Schizaphis graminum*. Entomol. Exp. Appl. 34:134–138.

Armbrust, E. J., Roberts, S. J., and White, C. E. 1970. Feeding behavior of alfalfa weevil larvae parasitized by *Bathyplectes curculionis*. J. Econ. Entomol. 63:1689–1690.

Bach, C. E. 1980. Effects of plant density and diversity on the population dynamics of a specialized herbivore, the striped cucumber beetle, *Acalymma vittata* (Fab.). Ecology 61:1515–1530.

Bach, C. E. 1984. Plant spatial pattern and herbivore population dynamics: Plant factors affecting the movement patterns of a tropical cucurbit specialist (*Acalymma innubum*). Ecology 65:175–190.

Baldocchi, D. D., Verma, S. B., and Rosenberg, N. J. 1983. Microclimate in soybean canopy. Agric. Meteor. 28:321–337.

Barbosa, P. 1991. Plant pathogens and non-vector herbivores, pp. 341–382. In P. Barbosa, V. A. Krischik, and C. G. Jones (eds.), Microbial Mediation of Plant-Herbivore Interactions. Academic Press, New York.

Beck, S. D., and Schoonhoven, L. M. 1980. Insect behavior and plant resistance, pp. 115–135. In F. G. Maxwell and P. R. Jennings (eds.), Breeding Plants Resistant to Insects. Wiley, New York.

Beier, R. C. 1990. Natural pesticides and bioactive components in foods, pp. 47–137. In G. W. Ware (ed.), Reviews of Environmental Contamination and Toxicology. Springer-Verlag, New York.

Bentz, J.-A., and Barbosa, P. 1990. Effects of dietary nicotine (0.1%) and parasitism by *Cotesia congregata* on the growth and food consumption and utilization of the tobacco hornworm, *Manduca sexta*. Entomol. Exp. Appl. 57:1–8.

Berenbaum, M. 1978. Toxicity of a furanocoumarin to armyworms: A case of biosynthetic escape from insect herbivores. Science 201:532–533.

Berenbaum, M. 1981. Patterns of furanocoumarin distribution and insect herbivory in the Umbelliferae: Plant chemistry and community structure. Ecology 62:1254–1266.

Bergelson, J., Fowler, S., and Hartley, S. 1986. The effects of foliage damage on casebearing moth larvae, *Coleophora serratella*, feeding on birch. Ecol. Entomol. 11:241–250.

Bernays, E. A. 1988. Host specificity in phytophagous insects: Selection pressure from generalist predators. Entomol. Exp. Appl. 49:131–140.

Bernays, E. A. 1989. Host range in phytophagous insects: the potential role of generalist predators. Evol. Ecol. 3:299–311.

Bernays, E. A., and Cornelius, M. L. 1989. Generalist caterpillar prey are more palatable than specialists for the generalist predator *Iridomyrmex humilis*. Oecologia 79:427–430.

Bernays, E. A., and Janzen, D. H. 1988. Saturniid and sphingid caterpillars: Two ways to eat leaves. Ecology 69:1153–1160.

Boiteau, G., and Singh, R. P. 1982. Effect of potato foliage infected with potato leafroll virus on fecundity and longevity of the Colorado potato beetle, *Leptinotarsa decimlineata* (Coleoptera: Chrysomelidae). Can. Entomol. 114:473–477.

Bossenbroek, Ph., Kessler, A., Liem, A. S. N., and Vlijm, L. 1977. The significance of plant growth-forms as 'shelter' for terrestrial animals. J. Zool. 182:1–6.

Bowdan, E. 1988. Microstructure of feeding by tobacco hornworm caterpillars, *Manduca sexta*. Entomol. Exp. Appl. 47:127–136.

Boyce, S. G., and Cost, N. D. 1978. Forest diversity—New concepts and applications. U.S. Dept. Agric. For. Serv. Res. Pap. SE-194.

Brewer, F. D., and King, E. G. 1978. Effects of parasitism by a tachinid, *Lixophaga diatraeae*, on growth and food consumption of sugarcane borer larvae. Ann. Ent. Soc. Am. 71:18–22.

Brewer, F. D., and King, E. G. 1980. Consumption and utilization by larvae of the tobacco budworm parasitized by the tachinid *Eucelatoria* sp. Entomophaga 25:95–101.

Brewer, F. D., and King, E. G. 1981. Food consumption and utilization by sugarcane borers parasitized by *Apanteles flavipes*. J. Ga. Entomol. Soc. 16:185–192.

Brodbeck, B., and Strong, D. 1987. Amino acid nutrition of herbivorous insects and stress to host plants, pp. 347–364. In P. Barbosa and J. C. Schultz (eds.), Insect Outbreaks. Academic Press, New York.

Brues, C. T. 1946. Insects, Food, and Ecology. Dover, New York.

Burdon, J. J. 1987. Disease and Plant Population Biology. Cambridge University Press, Cambridge.

Campbell, B. C. 1988. The effects of plant growth regulators and herbicides on host plant quality to insects, pp. 205–247. In E. A. Heinrichs (ed.), Plant Stress–Insect Interactions. Wiley, New York.

Carroll, C. R., and Hoffman, C. 1980. Chemical feeding deterrent mobilized in response to insect herbivory and counter adaptation by *Epilachna tredecimnota*. Science 209:414–416.

Casey, T. M., and Knapp, R. 1987. Caterpillar thermal adaptation: Behavioral differences reflect metabolic thermal sensitivities. Comp. Biochem. Physiol. 86A:679–682.

Chelliah, S. and Heinrichs, E.A. 1980. Factors affecting insecticide induced resurgence of the brown planthopper *Nilaparvata lugens* on rice. Environ. Entomol. 9:773–777.

Chelliah, S., and Subramanian, A. 1972. Influence of nitrogen on the infestation by the gall midge, *Pachydiplosis oryzae* in certain rice varieties. Ind. J. Entomol. 34:255–256.

Cloutier, C., and Mackauer, M. 1979. The effect of parasitism by *Aphidius smithi* on the food budget of the pea aphid, *Acyrthosiphon pisum*. Can. J. Zool. 57:1605–1611.

Cloutier, C., and Mackauer, M. 1980. The effect of superparasitism by *Aphidius smithi* on the food budget of the pea aphid, *Acyrthosiphon pisum*. Can. J. Zool. 58:241–244.

Cooper, P., and Selman, I. W. 1974. An analysis of the effects of tobacco mosaic virus on growth and the changes in the free amino compounds in young tomato plants. Ann. Bot. 38:625–638.

Couchman, J. R., and King, P. E. 1979. Effect of the parasitoid *Diaeretiella rapae* on the feeding rate of its host *Brevicoryne brassicae*. Entomol. Exp. Appl. 25:9–15.

Courtney, S. P. 1986. The ecology of pierid butterflies: Dynamics and interactions. Adv. Ecol. Res. 15:51–131.

Crawley, M. J. 1983. Herbivory. The Dynamics of Animal-Plant Interactions. University of California Press, Berkeley, CA.

Cromartie, W. J. 1975. The effect of stand size and vegetation background on the colonization of cruciferous plants by herbivorous insects. J. Appl. Ecol. 12:517–533.

DaCosta, C. P., and Jones, C. M. 1971a. Cucumber beetle resistance and mite susceptibility controlled by the bitter gene in *Cucumis sativus* L. Science 172:1145–1146.

DaCosta, C. P., and Jones, C. M. 1971b. Resistance in cucumber, *Cucumis sativus* L. to three species of cucumber beetles. Hort. Sci. 6:340–342.

Dale, D. 1988. Plant-mediated effects of soil mineral stresses on insects, pp. 35–110. In E. A. Heinrichs (ed.) Plant Stress–Insect Interactions. Wiley, New York.

Damman, H. 1987. Leaf quality and enemy avoidance by larvae of a pyralid moth. Ecology 68:87–97.

Daynard, T. B., Ellis, C. R., Bolwyn, B., and Misener, R. L. 1975. Effects of carbofuran on grain yield of corn. Can. J. Plant Sci. 55:637–639.

Dethier, V.G. 1988. The feeding behavior of a polyphagous caterpillar (*Diacrisia virginica*) in its natural habitat. Can. J. Zool. 66: 1280–1288.

Dixon, A. F. G. 1985. Aphid Ecology. Blackie and Sons, Glasgow.

Dover, J. W. 1986. The effect of labiate herbs and white clover on *Plutella xylostella* oviposition. Entomol. Exp. Appl. 42:243–247.

Duodu, Y. A., and Davis, D. W. 1974. A comparison of growth, food consumption, and food utilization between unparasitized alfalfa weevil larvae and those parasitized by *Bathyplectes curculionis*. Environ. Entomol. 3:705–710.

Edwards, C. A. 1977. Investigations into the influence of agricultural practices on soil invertebrates. Ann. Appl. Biol. 87:515–520.

Ehrlich, P. R., Ehrlich, A. H., and Holdren, J. P. 1977. Ecoscience: Population, Resources, Environment. Freeman, New York.

Eigenbrode, S. D., and Pimentel, D. 1988. Effects of manure and chemical fertilizers on insect pest populations on collards. Agric. Ecosys. Environ. 20:109–125.

Farrar, R. E., Jr., and Bradley, J. R., Jr. 1985. Within-plant distribution of *Heliothis* spp. eggs and larvae on cotton in North Carolina. Environ. Entomol. 14:205–209.

Farrell, J. A. K. 1976. Effects of groundnut crop density on the population dynamics of *Aphis craccivora* in Malawi. Bull. Entomol. Res. 66:317–329.

Feeny, P., Rosenberry, L., and Carter, M. 1983. Chemical aspects of oviposition behavior in butterflies, pp. 27–76. In S. Ahmad (ed.), Herbivorous Insects. Host Seeking Behavior and Mechanisms. Academic Press, New York.

Ferro, D. N., Chapman, R. B., and Penman, D. R. 1979. Observations on insect microclimate and insect pest management. Environ. Entomol. 8:1000–1003.

Fife, J. M. 1956. Changes in concentration of amino acids in leaves of sugar beet plants affected with curly top. J. Am. Soc. Sugar Beet Technol. 9:207.

Finch, S. 1988. Entomology of crucifers and agriculture—Diversification of the agroecosystem in relation to pest damage in cruciferous crops, pp. 39–71. In M. K. Harris and C. E. Rogers (eds.), The Entomology of Indigenous and Naturalized Systems in Agriculture. Westview Press, Boulder, CO.

Finch, S., and Skinner, G. 1976. The effect of plant density on populations of the cabbage root fly (*Erioischia brassicae* (Bch.)) and the cabbage stem weevil (*Ceutorhynchus quadridens* (Panz.)) on cauliflowers. Bull. Entomol. Res. 66:113–123.

Futuyma, D. J. 1976. Food plant specialization and environmental predictability. Am. Natur. 110:285–292.

Gershenzon, J. 1984. Changes in the levels of plant secondary metabolites under water and nutrient stress, pp. 273–320. In B. N. Timmermann, C. Steelink, and F. A. Loewus (eds.), Phytochemical Adaptations to Stress. Recent Advances in Phytochemistry, Vol. 18. Plenum, New York.

Gibbs, A. 1980. A plant that partially protects its wild legume host against herbivores. Intervirol. 13: 42–47.

Goodman, D. 1975. The theory of diversity-stability relationships in ecology. Quart. Rev. Biol. 50:237–266.

Goodman, P. J., Watson, M. A., and Hill, A. R. C. 1965. Sugar and fructosan accumulation in virus-infected plants: rapid testing by circular-paper chromatography. Ann. Appl. Biol. 56:65–72.

Goodman, R. N., Kiraly, Z., and Zaitlin, M. 1967. The Biochemistry and Physiology of Infectious Plant Disease. Van Nostrand, Princeton, NJ.

Gould, S. J. 1989. A developmental constraint in *Cerion*, with comments on the definition and interpretation of constraint in evolution. Evolution 43:516–539.

Grainge, M., and Ahmed, S. 1988. Handbook of Plants with Pest-Control Properties. Wiley, New York.

Gross, P., and Price, P. W. 1988. Plant influences on parasitism of two leafminers: A test of enemy-free space. Ecology 69:1506–1516.

Guillot, F. S., and Vinson, S. B. 1973. Effect of parasitism by *Cardiochiles nigriceps* on food consumption and utilization by *Heliothis virescens*. J. Insect Physiol. 19:2073–2082.

Hagen, R. H., and Chabot, J. F. 1986. Leaf anatomy of maples (*Acer*) and host use by Lepidoptera larvae. Oikos 47:335–345.

Hammond, A. M., and Hardy, T. N. 1988. Quality of diseased plants as hosts for insects, pp. 381–432. In E. A. Heinrichs (ed.) Plant Stress–Insect Interactions. Wiley, New York.

Hare, J. D. 1983. Manipulation of host suitability for herbivore pest management, pp. 655–680. In R. F. Denno and M. S. McClure (eds.), Variable Plants and Herbivores in Natural and Managed Systems. Academic Press, New York.

Hare, J. D., and Dodds, J. A. 1987. Survival of the Colorado potato beetle on virus-infected tomato in relation to plant nitrogen and alkaloid content. Entomol. Exp. Appl. 44:31–35.

Hargrove, W. W., Crossley, D. A., Jr., and Seastedt, T. R. 1984. Shifts in insect herbivory in the canopy of black locust, *Robinia pseudoacacia*, after fertilization. Oikos 43:322–328.

Heidorn, T. J., and Joern, A. 1987. Feeding preference and spatial distribution of grasshoppers in response to nitrogen fertilization of *Calamovilfa longifolia*. Func. Ecol. 1:369–375.

Heinrich, B. 1979. Foraging strategies of caterpillars. Leaf damage and possible predator avoidance strategies. Oecologia 42:325–337.

Heinrich, B., and Collins, S. I. 1983. Caterpillar leaf-damage and the game of hide-and-seek with birds. Ecology 64:592–602.

Hill, D. S. 1983. Agricultural Insect Pests of the Tropics and Their Control, 2nd ed. Cambridge University Press, Cambridge.

Hill, D. S. 1987. Agricultural Insect Pests of Temperate Regions and Their Control. Cambridge University Press, Cambridge.

Holloway, J. D., Bradley, J. D., and Carter, D. J. 1987. CIE Guides to Insects of Importance to Man. 1. Lepidoptera. CAB International Institute of Entomology, Wallinford, U.K.

Holtzer, T. O., Archer, T. L., and Norman, J. M. 1988. Host plant susceptibility in relation to water stress, pp. 111–137. In E. A Heinrichs (ed.), Plant Stress–Insect Interactions. Wiley, New York.

Howe, W. L., Sanburn, J. R., and Rhodes, A. M. 1976. Western corn rootworm adult and spotted cucumber beetle associations with *Cucurbita* and cucurbitacins. Environ. Entomol. 5:1043–1048.

Huebner, L. B., and Chiang, H. C. 1982. Effects of parasitism by *Lixophaga diatraeae* on food consumption and utilization of European corn borer larvae. Environ. Entomol. 11:1053–1057.

Hunter, K. W., Jr, and Stoner, A. 1975. *Copidosoma truncatellum*: Effect of parasitization on food consumption of larval *Trichoplusia ni*. Environ. Entomol. 4:381–382.

Isenhour, D. J. 1988. Interactions between two hymenopterous parasitoids of the fall armyworm. Environ. Entomol. 17:616–620.

Jackai, L. E. N., and Oghiakhe, S. 1989. Pod wall trichomes and resistance of two wild cowpea, *Vigna vexillata*, accessions to *Maruca testulalis* and *Clavigralla tomentosicollis*. Bull. Entomol. Res. 79:595–605.

Jaenike, J. 1986. Parasite pressure and the evolution of amanitin tolerance in *Drosophila*. Evolution 39:1295–1301.

Janzen, D. H. 1981. Patterns of herbivory in a tropical deciduous forest. Biotropica 13:271–282.

Jensen, S. G. 1969. Composition and metabolism of barley leaves infected with barley yellow dwarf virus. Phytopathol. 59:1694–1698.

Jones, D., Jones, G., Van Steenwyk, R. A., and Hammock, B. D. 1982. Effect of the parasite *Copidosoma truncatellum* on development of its host *Trichoplusia ni*. Ann. Entomol. Soc. Am. 75:7–11.

Jones, R. L., McMillian, W. W., and Wiseman, B. R. 1972. Chemicals in kernels of corn that elicit a feeding response from larvae of the corn earworm. Ann. Entomol. Soc. Am. 64:821–824.

Karban, R. 1988. Resistance to beet armyworms (*Spodoptera exiqua*) induced by exposure to spider mites (*Tetranychus turkestani*) in cotton. Am. Midl. Nat. 119:77–82.

Kareiva, P. 1982. Experimental and mathematical analysis of herbivore movement: Quantifying the influence of plant spacing and quality on foraging discrimination. Ecol. Monogr. 52:261–282.

Kareiva, P. 1983. Influence of vegetation texture on herbivore populations: Resource concentration and herbivore movement, pp. 259–289. In R. F. Denno and M. S. McClure (eds.), Variable Plants and Herbivores in Natural and Managed Systems. Academic Press, New York.

Kennedy, J. S. 1951. A biological approach to plant viruses. Nature (London) 168:890–894.

Kidd, N. A. C. 1982. Predator avoidance as a result of aggregation in the grey pine aphid, *Schizolachnus pineti*. J. Anim. Ecol. 51:397–412.

Knapp, R., and Casey, T.M. 1986. Thermal ecology, behavior, and growth of gypsy moth and eastern tent caterpillars. Ecology 67:598–608.

Kranz, Jr., Schmutterer, H., and Koch, W. 1977. Diseases, Pests and Weeds in Tropical Crops. Wiley, Berlin.

Krischik, V. A., and Denno, R. F. 1983. Individual, population, and geographic patterns in plant defense, pp. 463–512. In R. F. Denno and M. S. McClure (eds.), Variable Plants and Herbivores in Natural and Managed Systems. Academic Press, New York.

Kuc, J. 1966. Resistance of plants to infectious agents. Annu. Rev. Microbiol. 20:337–364.

Kuc, J., Shockley, G., and Kearney, K. 1975. Protection of cucumber against *Colletotrichum lagenarium* by *Colletotrichum lagenarium*. Physiol. Plant Pathol. 7:195–199.

Kumar, K., and Chapman, R. B. 1984. Sublethal effects of insecticides on the diamondback moth *Plutella xylostella*. Pestic. Sci. 15:344–352.

Lance, D. R. 1983. Host-seeking behavior of the gypsy moth: The influence of polyphagy and highly apparent host plants, pp. 201–224. In S. Ahmad (ed.), Herbivorous Insects. Host-seeking Behavior and Mechanisms. Academic Press, New York.

Lance, D., and Barbosa, P. 1982. Host tree influences on the dispersal of late instar gypsy moths, *Lymantria dispar*. Oikos 38:1–7.

Lange, W. H., and Bronson, L. 1981. Insect pests of tomatoes. Annu. Rev. Entomol. 26:345–371.

Lee, T. T. 1977. Promotion of plant growth and enzymatic degradation of indole-3-acetic acid by metabolites of carbofuran, a carbamate insecticide. Can. J. Bot. 55:574–579.

Levine, E., and Clement, S. L. 1981. Effect of parasitism by *Bonnetia comata* on larvae of *Agrotis ipsilon*. J. Kans. Entomol. Soc. 54:219–222.

Levins, R., and Wilson, M. 1980. Ecological theory and pest management. Annu. Rev. Entomol. 25:287–308.

Lewis, A. C. 1979. Feeding preference for diseased and wilted sunflower in the grasshopper, *Melanoplus differentialis*. Entomol. Exp. Appl. 26:202–207.

Lewis, A. C. 1982. Leaf wilting alters a plant species ranking by the grasshopper *Melanoplus differentialis*. Ecol. Entomol. 7:391–395.

Liener, I. E. 1986. The nutritional significance of naturally occurring toxins in plant foodstuffs, pp. 72–94. In J. B. Harris (ed.), Natural Toxins. Clarendon Press, Oxford.

Long, D. B. 1955. Observations on sub-social behaviour in two species of lepidopterous larvae, *Pieris brassicae* and *Plusia gamma*. Trans. Roy. Entomol. Soc. London 106:421–437.

Loustalot, A. J., Winters, H. F., and Childers, N. F. 1947. Influence of high, medium, and low soil moisture on growth and alkaloid content of *Cinchona ledgeriana*. Plant Physiol. 22:613–619.

Luginbill, P., and McNeal, F. H. 1958. Influence of seeding density and row spacing on the resistance of spring wheats to the wheat stem sawfly. J. Econ. Entomol. 51:804–808.

Madge, D. S. 1964. The light reactions and feeding activity of larvae of the cutworm *Tryphaena pronuba* L. (Lepidoptera: Noctuidae). Part II. Field investigations. Entomol. Exp. Appl. 7:105–114.

Mainer, A., and Leath, K. T. 1978. Foliar diseases alter carbohydrate and protein levels in leaves of alfalfa and orchardgrass. Phytopathology 68:1252–1255.

Mani, M., Nagarkatti, S., and Narayanan, K. 1982. Influence of parasitism by *Eucelatoria bryani* on the consumption and utilization of chickpea flour diet by *Heliothis armigera*. Entomophaga 27:399–404.

Manuwoto, S., and Scriber, J. M. 1985. Differential effects of nitrogen fertilization of three corn genotypes on biomass and nitrogen utilization by the southern armyworm, *Spodoptera eridania*. Agric. Ecosys. Environ. 14:25–40.

Mattson, W. J. 1980. Herbivory in relation to plant nitrogen content. Annu. Rev. Ecol. Syst. 11:119–161.

Mattson, W. J., and Haack, R. A. 1987. The role of drought stress in provoking outbreaks of phytophagous insects, pp. 365–407. In P. Barbosa and J. C. Schultz (eds.), Insect Outbreaks. Academic Press, New York.

Matsumoto, K. 1989. Effects of aggregation on the survival and development of different host plants in a papilionid butterfly, *Luehdorfia japonica* Leech. Jpn. J. Entomol. 57:853–860.

Matsumoto, Y. 1962. A dual effect of coumarin, olfactory attraction and feeding inhibition, on the vegetable weevil adult, in relation to the uneatability of sweet clover leaves. Jpn. J. Appl. Entomol. Zool. 6:141–149.

Mayse, M. 1978. Effects of spacing between rows on soybean arthropod populations. J. Anim. Ecol. 15:439–450.

McNeill, S., and Southwood, T. R. E. 1978. The role of nitrogen in the development of insect/plant relationships, pp. 77–98. In J. B. Harborne (ed.), Biochemical Aspects of Plant and Animal Coevolution. Academic Press, London.

Mellors, W. K., Allegro, A., and Hsu, A. N 1984. Effects of carbofuran and water stress on growth of soybean plants and twospotted spider mite (Acari: Tetranychidae) populations under greenhouse conditions. Environ. Entomol. 13:561–567.

Metcalf, R. L., Rhodes, A. M., Metcalf, R. A., Ferguson, J., Metcalf, E. R., and Lu, P-Y. 1982. Cucurbitacin contents and diabroticite feeding upon *Cucurbita* spp. Environ. Entomol. 11:931–937.

Mitchell, R. 1984. The ecological basis for comparative primary production, pp. 13–53. In R. Lowrance, B. R. Stinner, and G. J. House (eds.), Agricultural Ecosystems. Unifying Concepts. Wiley, New York.

Mohamed, A. K. A., Brewer, F. W., Bell, J. V., and Hamalle, R. J. 1982. Effect of *Nomuraea rileyi* on consumption and utilization of food by *Heliothis zea* larvae. J. Ga. Entomol. Soc. 17:356–363.

Murdoch, W. W. 1975. Diversity, complexity, stability and pest control. J. Appl. Ecol. 12:795–807.

Myers, J. H., and Campbell, B. J. 1976. Predation by carpenter ants: A deterrent to the spread of cinnabar moth. J. Entomol. Soc. Br. Col. 73:7–9.

Nealis, V. G. 1986. Responses to host kaironomes and foraging behavior of the insect parasite *Cotesia rubecula*. Can. J. Zool. 64:2393–2398.

Odell, T. M., and Godwin, P. A. 1984. Host selection by *Blepharipa pratensis*, a tachinid parasite of the gypsy moth, *Lymantria dispar*. J. Chem. Ecol. 10:311–320.

Odum, E. P. 1984. Properties of agroecosystems, pp. 5–11. In R. Lowrance, B. R. Stinner, and G. J. House (eds.), Agricultural Ecosystems. Unifying Concepts. Wiley, New York.

Oma, E. A., and Hewitt, G. B. 1984. Effect of *Nosema locustae* on food consumption in the differential grasshopper. J. Econ. Entomol. 77:500–501.

Parker, F. D., and Pinnell, R. E., 1973. Effect on food consumption of the imported cabbageworm when parasitized by two species of *Apanteles*. Environ. Entomol. 2:216–219.

Parker, M. 1982. Herbivore foraging movements and plant population dynamics: The impact of a specialist grasshopper on two arid grassland composites. Ph.D. Dissertation, Cornell University, Ithaca, N.Y.

Parkman, P., and Shepard, M. 1981. Foliage consumption by yellowstriped armyworm larvae after parasitization by *Euplectrus plathypenae*. Flor. Entomol. 64:192–194.

Pimentel, D. 1976. World food crisis: Energy and pests. Bull. Entomol. Soc. Am. 22:20–26.

Pimentel, D. 1986. Agroecology and economics, pp. 299–319. In M. Kogan (ed.), Ecological Theory and Integrated Pest Management Practice. Wiley, New York.

Pinter, P. J., Jr., Hadley, N. F., and Lindsay, J. H. 1975. Alfalfa crop micrometeorology and its relation to insect pest biology and control. Environ. Entomol. 4:153–162.

Pless, C. D., Cherry, E. T., and Morgan, H., Jr. 1971. Growth and yield of burly tobacco as affected by two systemic insecticides. J. Econ. Entomol. 64:172–175.

Porter C. A., and Weinstein, L. H. 1960. Altered biochemical patterns induced in tobacco by cucumber mosaic virus infection, by thiouracil, and by their interaction. Contr. Boyce Thompson Inst. 20:307.

Powell, J. A. 1980. Evolution of larval food preferences in Microlepidoptera. Annu. Rev. Entomol. 25:133–159.

Powell, J. E. 1989. Food consumption by tobacco budworm larvae reduced after parasitization by *Microplitis demolitor* or *M. croceipes*. J. Econ. Entomol. 82:408–411.

Prestidge, R. A. 1982. The influence of nitrogenous fertilizer on the grassland auchenorrhyncha. J. Appl. Ecol. 19:735–749.

Price, P. W. 1983. Hypotheses on organization and evolution in herbivorous insect communities, pp. 559–596. In R. F. Denno and M. S. McClure (eds.), Variable Plants and Herbivores in Natural and Managed Systems. Academic Press, New York.

Price, P. W., and Waldbauer, G. P. 1982. Ecological aspects of pest management, pp. 33–68. In R. L. Metcalf and W. H. Luckmann (eds.), Introduction to Insect Pest Management. Wiley, New York.

Rabb, R. L. 1978. A sharp focus on insect populations and pest management from a wide-area view. Bull. Ent. Soc. Am. 24:55–61.

Rahman, M. 1970. Effect of parasitism on food consumption of *Pieris rapae* larvae. J. Econ. Entomol. 63:820–821.

Raman, K. V., Tingey, W. M., and Gregory, P. 1979. Potato glycoalkaloids: Effect on survival and feeding behavior of the potato leafhopper. J. Econ. Entomol. 72:337–341.

Retnakaran, A., Lauzon, H., and Fast, P. 1983. *Bacillus thuringiensis* induced anorexia in the spruce budworm, *Choristoneura fumiferana*. Entomol. Exp. Appl. 34:233–239.

Rhoades, D. F. 1983. Herbivore population dynamics and plant chemistry, pp. 155–220. In R. F. Denno and M. S. McClure (eds.), Variable Plants and Herbivores in Natural and Managed Systems. Academic Press, New York.

Risch, S. J. 1980. The population dynamics of several herbivorous beetles in a tropical agroecosystem: The effect of intercropping corn, beans, and squash in Costa Rica. J. Appl. Ecol. 17:593–612.

Risch, S. J. 1987. Agricultural ecology and insect outbreaks, pp.217–238. In P. Barbosa and J. C. Schultz (eds.), Insect Outbreaks. Academic Press, New York.

Risch, S. J., Andow, D., and Altieri, M. A. 1983. Agroecosystem diversity and pest control: Data, tentative conclusions and new research directions. Environ. Entomol. 12:625–629.

Risser, P. G. 1986. Agroecosystems—Structure, analysis, and modelling, pp. 321–343. In M. Kogan (ed.), Ecological Theory and Integrated Pest Management Practice. Wiley, New York.

Roitberg, B. D., Myers, J. H., and Frazer, B. D. 1979. The influence of predators on the movement of apterous pea aphids between plants. J. Anim. Ecol. 48:111–122.

Roland, J. 1986. Parasitism of winter moth in British Columbia during build-up of its parasitoid *Cyzenis albicans*: Attack rate on oak v. apple. J Anim. Ecol. 55:215–234.

Rombach, M. C., Aguda, R. M., Picard, L., and Roberts, D. W. 1989. Arrested feeding of the Asiatic rice borer by *Bacillus thuringiensis*. J. Econ. Entomol. 82:416–419.

Root, R. B. 1973. Organization of a plant-arthropod association in simple and diverse habitats: The fauna of collards (*Brassica oleraceae*). Ecol. Monogr. 43:95–124.

Ruehlmann, T.E., Matthews, R.W., and Matthews, J.R. 1988. Roles for structural and temporal shelter-changing by fern-feeding lepidopteran larvae. Oecologia 75:228–232.

Sajap, A. S. B., Beegle, C. C., and Lewis, L. C. 1978. Effect of parasitism by *Microplitis kewleyi* on the cutting ability of its host, *Agrotis ipsilon*. Environ. Entomol. 7:343–344.

Sandberg, S.L., and Berenbaum, M.R. 1989. Leaf-tying by tortricid larvae as an adaptation for feeding on phototoxic *Hypericum perforatum*. J. Chem. Ecol. 15:875–885.

Sareen, V., Rathore, Y. S., and Bhattacharya, A. K. 1983. Influence of *Bacillus thuringiensis* var. *thuringiensis* on the food utilization of *Spodoptera litura*. Z. Ang. Entomol. 95:253–258.

Schoenbohm, R. B., and Turpin, F. T. 1977. Effect of parasitism by *Meteorus leviventris* on corn foliage consumption and corn seedling cutting by the black cutworm. J. Econ. Entomol. 70:457–459.

Scriber, J. M. 1977. Limiting effects of low leaf-water content on the nitrogen utilization, energy budget, and larval growth of *Hyalophora cecropia* (Lepidoptera: Saturniidae). Oecologia 28:269–287.

Scriber, J. M. 1978. The effects of larva feeding specialization and plant growth form on the consumption and utilization of plant biomass and nitrogen: An ecological consideration. Entomol. Exp. Appl. 24:494–510.

Scriber, J. M. 1979. Effects of leaf water supplementation upon post-ingestive nutritional indices of forb-shrub- and tree-feeding Lepidoptera. Entomol. Exp. Appl. 25:240–252.

Scriber, J. M., Tingey, W. M., Gracen, V. E., and Sullivan, S. L. 1975. Leaf feeding resistance to the European corn borer in genotypes of tropical (low-DIMBOA) and U.S. inbred (high-DIMBOA) maize. J. Econ. Entomol. 68:823–826.

Selman, I. W., Brierley, M. R., Pegg, G. F., and Hill, T. A. 1961. Changes in the free amino acids and amides in tomato plants inoculated with tomato spotted wilt virus. Ann. Appl. Biol. 49:601–615.

Sinden, S. L., Schalk, J. M., and Stoner, A. K. 1978. Effects of daylength and maturity of tomato plants on tomatine content and resistance to the Colorado potato beetle. J. Am. Soc. Hort. Sci. 103:596–600.

Sinha, A. K., and Krishna, S. S. 1970. Further studies on the feeding behavior of *Aulacophora foveicollis* on cucurbitacin. J. Econ. Entomol. 63:333–334.

Slansky, F., Jr. 1978. Utilization of energy and nitrogen by larvae of the imported cabbageworm, *Pieris rapae*, as affected by parasitism by *Apanteles glomeratus*. Environ. Entomol. 7:179–185.

Slansky, F. J., and Rodriguez, J. G. 1987. Nutritional Ecology of Insects, Mites, Spiders, and Related Invertebrates. Wiley, New York.

Slansky, F., Jr., and Rodriquez, J. G. 1987. Nutritional Ecology of Insects, Mites, Spiders, and Related Invertebrates. Wiley, New York.

Smits, P. H., van Velden, M. C., van de Vrie, M., and Vlak, J. M. 1987. Feeding and dispersion of *Spodoptera exigua* larvae and its relevance for control with a nuclear polyhedrosis virus. Entomol. Exp. Appl. 43:67–72.

Sprenkel, R. K., Brooks, W. M., Van Duyn, J. W., and Deitz, L. L. 1979. The effects of three cultural variables on the incidence of *Nomuraea rileyi*, phytophagous Lepidoptera, and their predators on soybeans. Environ. Entomol. 8:334–339.

Stehr, F. W. 1987. Immature Insects. Kendall/Hunt, Dubuque, IA.

Stinner, B. R., and Stinner, D. H. 1989. Plant–animal interactions in agricultural ecosystems, pp. 355–393. In W. G. Abrahamson (ed.), Plant–Animal Interactions. McGraw Hill, New York.

Strong, D. R., Lawton, J. H., and Southwood, T. R. E. 1984. Insects on Plants. Harvard University Press, Cambridge, MA.

Subrahmanyam, B., and Ramakrishnan, N. 1981. Influence of a baculovirus infection on molting and food consumption by *Spodoptera litura*. J. Invert. Pathol. 38:161–168.

Tabashnik, B. E. 1982. Responses of pest and non-pest *Colias* butterfly larvae to interspecific variation in leaf nitrogen and water content. Oecologia 55:389–394.

Tallamy, D. W., and Raupp, M. J. 1991. Phytochemical Induction by Herbivores. Wiley, New York.

Tan, K-H. 1981. Antifeeding effect of cypermethrin and permethrin at sub-lethal levels against *Pieris brassicae* larvae. Pestic. Sci. 12:619–626.

Terry, I., Bradley, J. R., Jr., and Van Duyn, J. W. 1989. Establishment of early instar *Heliothis zea* on soybeans. Entomol. Exp. Appl. 51:233–240.

Theunissen, J., den Ouden, H., and Wit, A. K. H. 1985. Feeding capacity of caterpillars on cabbage, a factor in crop loss assessment. Entomol. Exp. Appl. 39:255–260.

Thompson, S. N. 1982. Immediate effects of parasitization by the insect parasite, *Hyposoter exiguae* on the nutritional physiology of its host, *Trichoplusia ni*. J. Parasitol. 68:936–941.

Thompson, S. N. 1983. The nutritional physiology of *Trichoplusia ni* parasitized by the insect parasite, *Hyposoter exiguae*, and the effects of parallel-feeding. Parasitology 87:15–28.

Tingey, W. M., and Singh, S. R. 1980. Environmental factors influencing the magnitude and expression of resistance, pp. 87–113. In F. G. Maxwell and P. R. Jennings (eds.), Breeding Plants Resistant to Insects. Wiley, New York.

Turlings, T. C. J., Tumlinson, J. H., and Lewis, W. J. 1990. Exploitation of herbivore-induced plant odors by host-seeking parasitic wasps. Science 250:1251–1253.

Uematsu, H., and Sakanoshita, A. 1989. Possible role of leaf wax bloom in suppressing diamondback moth *Plutella xylostella* oviposition. Appl. Entomol. Zool. 3:253–257.

Varela, L. G., and Bernays, E. A. 1988. Behavior of newly hatched potato tuber moth larvae, *Phthorimaea operculella* in relation to their host plants. J. Insect Behav. 1:261–275.

van der Meijden, E. 1976. Changes in the distribution patterns of *Tyria jacobaeae* during the larval period. Neth. J. Zool. 26:136–161.

Van Emden, H. F., and Williams, G. C. 1974. Insect stability and diversity in agroecosystems. Annu. Rev. Entomol. 19:455–475.

Wan, X., and Barbosa, P. 1990. Growth, development, feeding preference, and food consumption and utilization by tobacco hornworm on tobacco mosaic virus-infected and non-infected tobacco leaves. Experientia 46:521–524.

Wellington, W. G. 1949a. The effects of temperature and moisture upon the behavior of the spruce budworm, *Choristoneura fumiferana* Clemens (Lepidoptera: Tortricidae) I. The relative importance of graded temperatures and rates of evaporation in producing aggregation of larvae. Sci. Agric. 29:201–215.

Wellington, W. G. 1949b. The effects of temperature and moisture upon the behavior of the spruce budworm, *Choristoneura fumiferana* Clemens (Lepidoptera: Tortricidae) II. The responses of larvae to gradients of evaporation. Sci. Agric. 29:216–229.

Wilhelm, S. 1976. The agroecosystem: A simplified plant community, pp. 59–70. In J. L. Apple and R. F. Smith (eds.), Integrated Pest Management. Plenum, New York.

Willmer, P. G. 1980. The effects of a fluctuating environment on the water relations of larval Lepidoptera. Ecol. Entomol. 5:271–292.

Winters, H. F., and Loustalot, A. J. 1952. The effect of light and nitrogen levels on growth and alkaloid content of young plants of *Cinchona ledgeriana*. Plant Physiol. 27:575–582.

Wolfson, J. L. 1980. Oviposition response of *Pieris rapae* to environmentally induced variation in *Brassica nigra*. Entomol. Exp. Appl. 27:223–232.

Wolfson, J. L. 1982. Developmental responses of *Pieris rapae* and *Spodoptera eridania* to environmentally induced variation in *Brassica nigra*. Environ. Entomol. 11:207–213.

Yokoi, S., and Tsuji, H. 1975. Experimental studies on the movement of the final instar noctuid larvae of the cabbage armyworm and the tobacco cutworm. Jpn. J. Appl. Entomol. Zool. 19:157–161 [in Japanese: English Abstract].

Zanen, P. O., Lewis, W. J., Cardé, R. T., and Mullinix, B. G. 1989. Beneficial arthropod behavior mediated by airborne semiochemicals VI. Flight responses of female *Microplitis croceipes*, a braconid endoparasitoid of *Heliothis* spp., to varying olfactory stimulus conditions created with a turbulent jet. J. Chem. Ecol. 15:141–168.

Zepp, D. B., and Keaster, A. J. 1977. Effects of corn plant densities on the girdling behavior of the southeastern corn borer. J. Econ. Entomol. 70:678–680.

Taxonomic Index

Subject Index